A History of Life-Extensionism
In The Twentieth Century

ILIA STAMBLER, PHD

Copyright © 2014 Ilia Stambler

All rights reserved.

ISBN-10: 1500818577
ISBN-13: 978-1500818579

DEDICATION

I thank and dedicate this work to my parents – Samuel and Victoria-Chaya Stambler – may they live long, and grandparents – Israel and Rosa Beigel-Dachis and Zalman and Sarah Stambler-Prosmushkin – I wish they have lived longer.

CONTENTS

Abstract 1
Summary - Introduction 2
   1. Aims 3
   2. Argument 8
Chapter 1. France: The origins of the life-extensionist movement. Maintaining the Equilibrium and dissolution into the Whole 10
   1. Chapter summary 11
   2. The French tradition. The legacy of the Enlightenment. 12
   3. "Father" Metchnikoff (1845-1916) 15
   4. Attractions of Paris and the "Nationality" of life-extension 17
   5. Competing factions 19
   6. Optimistic biology 21
   7. "To live long, you must live poor". 23
   8. Methods of rejuvenation 25
   9. Adding to the balance: Rejuvenation by supplements 26
   10. Masculine Forever – Charles-Édouard Brown-Séquard (1817-1894) 28
   11. Rejuvenation by Grafting – Serge Voronoff (1866-1951) 31
   12. Rejuvenation World Wide. France and Austria – the seedbeds for the rejuvenation movement. Eugen Steinach (1861-1944) 34
   13. The reductionist roots of French life-extensionism 38
   14. Life-extensionism in France in 1930-1950. The dissolution into the Whole 40
   15. Rethinking rejuvenation – Auguste Lumière (1862-1954) 41
   16. The perpetuation of the whole – Alexis Carrel (1873-1944) 48
   17. The "Neo-Hippocratic" holistic movement - Official Medicine and Heretical Medicines 56
   18. Longevity and holism. The theory of Sergey Metalnikov (1870-1946) and the practice of Alexandre Guéniot (1832-1935) 65
   19. Chapter conclusion. Between reductionism and holism 70
Chapter 2. Germany: The preservation of the "National Body" and "Connection to Nature." Allies and neutrals – Austria, Romania, Switzerland 71
   1. Chapter summary 72
   2. The German tradition. "And Nature, teaching, will expand the power of your soul" 73
   3. Rethinking rejuvenation – Benno Romeis (1888-1971) 75
   4. Going the "Nature's way" 79
   5. Life-extension imposed from above – Ludwig Roemheld (1871-1938) 81
   6. Life-extension driven from below – Gerhard Venzmer (1893-1986) 85
   7. Basic research – Hans Driesch (1867-1941), Emil Abderhalden (1877-1950) 93
   8. Institutionalization of gerontology - Max Bürger (1885-1966) 97
   9. Allies – The Kingdom of Great Romania. Dimu Kotsovsky (1896-1965?) 100
   10. The Romanian People's Republic – Constantin Ion Parhon (1874-1969) and Ana Aslan (1897-1988) 106
   11. Neutrals – Switzerland. Paul Niehans (1882-1971) 108
   12. Respectable gerontology – Fritz Verzár (1886-1979) 112
   13. Chapter conclusion. Between "artificial" and "natural" life-extension 114
Chapter 3. The USSR. The perpetuation of Socialism and triumph of Materialism 115
   1. Chapter summary 116
   2. The Russian tradition. Life-extensionism integrated into the Russian Monarchy 117
   3. The emergence of the Soviet state and the creation of a new long-lived man 119
   4. Life-extensionism integrated into the Stalinist order. General Characteristics 121
   5. "Russia is the birthplace of elephants" – Ivan Mikhailovich Sarkizov-Serazini (1887-1964) 123
   6. Socialism as a condition for progress and renewal – Alexander Vasilievich Nagorny (1887-1953) 126
   7. Opposition 130
   8. The triumph of materialism 134
   9. The downfall of idealism – Porfiry Korneevich Ivanov (1898-1983) 135

10. Scientific life-extensionism 138
11. The Soviet "physiological system" – Alexander Alexandrovich Bogomolets (1881-1946) 141
12. The world's first conference on aging and longevity – Kiev, 1938. Controversies 148
13. The quest for reductionist rejuvenation continues 152
14. The rule of the collective – Bogomolets' conference (1938), Lysenko's conference (1948), Bykov's ("Pavlov's") conference (1950) 156
15. Khrushchev's 'spring'. 160
16. Brezhnev's rule and the last years of Soviet orthodoxy 164
17. New Russia 171
18. Chapter conclusion. The perpetuation of the current social order 174

Chapter 4. The United States: In the name of capitalism, religion and eugenics – commercial enterprise, basic research and the struggle for funding. The United Kingdom: philosophical and scientific discussions on life-extension, for and against **175**
1. Chapter summary 176
2. The American tradition. The field of unlimited possibilities. 177
3. The invasion of non-invasive methods 180
4. Capitalism, Religion and Eugenics 182
5. The rejuvenators 186
6. The triumph of the medical establishment 189
7. The life-extensionist as elite scientist – Charles Asbury Stephens (1844-1931) 191
8. Basic longevity research. Immortal Soma – Jacques Loeb (1859-1924), Leo Loeb (1869-1959), Alexis Carrel (1873-1944), Raymond Pearl (1879-1941) 193
9. Theories of Aging 196
10. Rectifying "Discord" and conserving "Vital Capital". 198
11. Institutionalization of longevity research and advocacy – The Life Extension Institute 202
12. The Great Depression, the New Deal, WWII – Projections of socio-biological stability. Walter Cannon (1871-1945) 205
13. "Further research is needed" – Clifford Cook Furnas (1900-1969) 208
14. The movement toward consolidation. Cowdry's Problems of Ageing (1939, 1952) 210
15. Consolidation continues 213
16. The 1950s-1960s. The evolution of rejuvenation methods: From organotherapy to replacement medicine. The cycle of hopefulness 215
17. Theories of Aging. A new longevity research paradigm based on contemporary scientific advances: the advent of molecular biology and cybernetics 218
18. British Allies – Literary and philosophical life-extensionism: The optimistic vs. the pessimistic view. The reductionist vs. the holistic approach 222
19. British longevity research: practice, theory and programs 229
20. Man and His Future 236
21. The present time: Life extension programs in the US 239
22. Chapter conclusion. The US leads the world of longevity research, practice and advocacy 242

General conclusion. Life-extensionism as a pursuit of constancy **244**
Supplemental Materials **251**
Figures 1&2. Publications on Rejuvenation (1) and Theories of aging (2) 252
Table 1. Distribution of publications on Rejuvenation & Theories of Aging according to periods 254
Main thematic classification 255
Name Index 256
References and Notes **259**

ACKNOWLEDGMENTS

I would like to express my gratitude to the distinguished scholars for their kindly encouraging and constructive comments and help at different stages of this work:

Dr. Oren Harman, Dr. Jeffrey Perl, Dr. Noah Efron, Dr. Raz Chen-Morris, Dr. Aubrey de Grey, Dr. James Hughes, Dr. Stanley Shostak, Dr. Alexey Olovnikov, Dr. Alexander Khokhlov, Dr. Suresh Rattan, Dr. Vera Gorbunova, Dr. Vadim Fraifeld, Dr. Natasha Vita-More, Dr. Alfred Tauber, Dr. Michael Gates, Dr. Edna Rosenthal, Dr. Reuven Tirosh, Dr. Etienne Lepicard, Dr. Elena Afrimzon, Dr. Aden Bar-Tura, and Dr. David Blokh.

I thank my comrades from life-extensionist organizations around the world for their personal support and their support of the cause of healthy longevity for all.

NOTES

Please quote this work as:

Ilia Stambler, *A History of Life-Extensionism in the Twentieth Century*, PhD Dissertation, Bar-Ilan University, Ramat Gan, Israel, 2014.

Or

Ilia Stambler, *A History of Life-Extensionism in the Twentieth Century*, Longevity History, Rison Lezion, Israel, 2014.

For more information, please visit www.longevityhistory.com

Title Image: Albrecht Dürer (1471-1528), "St. George killing the Dragon".
Source: Wikimedia Commons

ABSTRACT

This work explores the history of life-extensionism in the 20th century. The term life-extensionism is meant to describe an ideological system professing that radical life extension (far beyond the present life expectancy) is desirable on ethical grounds and is possible to achieve through conscious scientific efforts. This work examines major lines of life-extensionist thought, in chronological order, over the course of the 20th century, while focusing on central seminal works representative of each trend and period, by such authors as Elie Metchnikoff, Bernard Shaw, Alexis Carrel, Alexander Bogomolets and others. Their works are considered in their social and intellectual context, as parts of a larger contemporary social and ideological discourse, associated with major political upheavals and social and economic patterns. The following national contexts are considered: France (Chapter One), Germany, Austria, Romania and Switzerland (Chapter Two), Russia (Chapter Three), the US and UK (Chapter Four).

This work pursues three major aims. The first is to attempt to identify and trace throughout the century several generic biomedical methods whose development or applications were associated with radical hopes for life-extension. Beyond mere hopefulness, this work argues, the desire to radically prolong human life often constituted a formidable, though hardly ever acknowledged, motivation for biomedical research and discovery. It will be shown that novel fields of biomedical science often had their origin in far-reaching pursuits of radical life extension. The dynamic dichotomy between reductionist and holistic methods will be emphasized.

The second goal is to investigate the ideological and socio-economic backgrounds of the proponents of radical life extension, in order to determine how ideology and economic conditions motivated the life-extensionists and how it affected the science they pursued. For that purpose, the biographies and key writings of several prominent longevity advocates are studied. Their specific ideological premises (attitudes toward religion and progress, pessimism or optimism regarding human perfectibility, and ethical imperatives) as well as their socioeconomic conditions (the ability to conduct and disseminate research in a specific social or economic milieu) are examined in an attempt to find out what conditions have encouraged or discouraged life-extensionist thought. This research argues for the inherent adjustability of life-extensionism, as a particular form of scientific enterprise, to particular prevalent state ideologies.

The third, more general, aim is to collect a broad register of life-extensionist works, and, based on that register, to establish common traits and goals definitive of life-extensionism, such as valuation of life and constancy, despite all the diversity of methods and ideologies professed. This work will contribute to the understanding of extreme expectations associated with biomedical progress that have been scarcely investigated by biomedical history.

Summary – Introduction

1. Aims

The current study explores the history of life-extensionism in the 20th century, with a primary focus on the first half of the century. The term life-extensionism is meant to describe an ideological system professing that radical life prolongation (far beyond the present life expectancy) is desirable on ethical grounds and is possible to achieve through conscious scientific efforts. Champions of the life-extensionist intellectual movement extrapolated on contemporary scientific and technological achievements and perceived human progress to be unlimited, capable of radically extending the human life-span. Seminal biological developments – such as Jacques and Leo Loeb's and Alexis Carrel's concepts of potential cell immortality, Elie Metchnikoff's theory of phagocytosis or Charles-Édouard Brown-Séquard and Eugen Steinach's hormone replacement therapies – inspired far reaching popular expectations of radical longevity, even of a salvation by biomedical science. In fact, the desire to radically prolong human life often constituted a formidable, though hardly ever acknowledged, motivation for biomedical research and discovery. It will be shown that novel fields of biomedical science often had their origin in far-reaching pursuits of radical life extension. Thus, the development of endocrinology owed much to Eugen Steinach's "endocrine rejuvenation" operations (c. 1910s-1920s). Probiotic diets originated in Elie Metchnikoff's conception of radically prolonged "orthobiosis" (c. 1900). The world's first institute for blood transfusion was established by Alexander Bogdanov to find rejuvenating means (1926). Systemic immunotherapy derived from Alexander Bogomolets' "life-extending anti-reticular cytotoxic serum" (1930s). And cell therapy (and particularly human embryonic cell therapy) was conducted by Paul Niehans for the purposes of rejuvenation as early as the 1930s. Thus the pursuit of life-extension has constituted an inseparable and crucial element in the history of biomedicine.

And yet despite this broad significance for the history of biomedicine, the subject has until now been marginalized and there have been relatively few attempts to research the history of life-extensionism and its underlying ideological and social motives.[1] The present research aims to redress this historiographic gap and to examine the major lines of life-extensionist scientific and philosophical thought, in chronological order, over the course of the 20th century. Seminal works in the field considered in their social and intellectual context, as parts of a larger contemporary social discourse, associated with political upheavals and social and economic events, with state ideologies and cultural fashions.

This work pursues three major aims. The first is to attempt to identify and trace throughout the century several generic biomedical methods whose development or application, were associated with hopes for life-extension. In other words, this work inquires what kinds of biomedical interventions (actual or potential) raised the expectations of radical life-extension enthusiasts over the years. There exists an extensive number of sources containing suggestions for possible methods of life prolongation, written by leading scientists and science popularizers, which when studied carefully reveal a taxonomy: idealistic/holistic/hygienic approaches emphasized the importance of psychological environment and hygienic regulation of behavior, whereas, on the other hand, materialistic/reductionist/therapeutic methodologies sought ways to eliminate damaging agents, to introduce biological replacements, to maintain homeostasis, and to bring about man-machine synergy. The apparent relative weight of each method in public discourse (in terms of notoriety and prestige, funding, amount and dissemination of relevant publications) will be shown to change with time, reflecting the initial hopes, disappointments and reactions to those disappointments in a variety of scientific programs.

The second goal is to investigate the ideological and socio-economic backgrounds of the proponents of radical life extension, in order to determine how ideology and social conditions motivated the life-extensionists and how this affected the science they pursued. Their specific ideological premises (attitudes toward religion and progress, pessimism or optimism regarding human perfectibility, ethical imperatives, adjustments to prevalent state ideologies) as well as their socioeconomic conditions (the ability to conduct and disseminate research in a specific social or economic milieu) will be examined in an attempt to find out what conditions have encouraged or discouraged life-extensionist thought.

The third goal is to attempt to find common defining characteristics of life-extensionism in spite of all the diversity of the professed methods and socio-ideological backgrounds. This will be done partly through examining shared traits among the varied forms of life-extensionism and partly through examining the relation of life-extensionism to general biomedical research and practice. Despite the wide variety, what seems to unify the diverse life-extension advocates is an assertion of the unconditional value of human life, unmitigated optimism and belief in progress, perceived as reaching a long-lasting social and biological equilibrium, and a striving toward the absolute goal of maximal life prolongation for as many people as possible, or at least for the proponents themselves or for the groups to which they felt belonging. This desire was neither trivial nor self-explanatory, and was commonly frustrated. Its very expression required a certain daring on the part of life-extensionist writers. In the present work, the open expression of this desire is what defines adherence to "life-extensionism" or the "life-extensionist movement."

The precise definition of "life-extensionism" or "the life-extensionist movement" is difficult to articulate. The term "life-extensionism" is relatively recent (its precise origins are uncertain).[2] But of course the terms "life-extension" (or "prolongation of life" in early texts) or "rejuvenation" are very old.[3] The prolongation of life and rejuvenation were pursued by alchemy[4] and gerocomia[5] in the Middle Ages, and by experimental gerontology[6] and anti-aging medicine[7] in our time. Obviously they are not the same. In the second half of the 20th century, the advocates of life-extension were alternatively called prolongevitists,[8] life-extensionists, immortalists[9] or transhumanists,[10] and did not seem to have an agreed title before that. How then can "life-extensionism" or the "life-extensionist movement" be defined in a way which is not "Whig-historical" – imposing contemporary terms on earlier phenomena? The current study proposes to use the term "life-extensionist" generally to designate "proponents of life-extension" and then to seek various and often conflicting ways in which this common and definitive aspiration was expressed in particular historical contexts.

Well into the late 1930s there appeared to be no common organizational affiliation of the seekers of life-extension whatsoever. In that early period, a wide variety of thinkers joined the quest: materialists and idealists, scientists and men of letters, socialists and conservatives. The research was multi-focal and multi-lingual. The proponents of life extension constituted a congeries rather than a synthetic entity. And even now, the advocates of life-extension are very loosely and disparately affiliated, if they are affiliated at all. Still, I would suggest that they constituted a movement, that is to say an *intellectual* movement defined by a common aspiration. The model is that of other intellectual movements, such as the "Romantic Movement," the "Enlightenment Movement," or the "Feminist movement," having no clear organizational affiliations, but expressing similar aims. As in the latter cases, the writings of the proponents of the life-extensionist *intellectual* movement show an intricate dialogue and inter-textual influence, cross-fertilization and mutual encouragement. Moreover, the authors expressed an almost universal yearning for a broader cooperation and massive public support. No such broad cooperation and support seem to have occurred until the late 1930s in any of the countries under consideration, yet the striving for their establishment constituted a common ground which may have eventually led to the actual institutional cooperation.

Still, essentially, in the first half of the century, the model of this intellectual movement was "top down": After the publication of works by elite scientists and philosophers, some of their suggestions became adopted or propagandized by the public, or left at the top level – a subject of a learned and restricted discourse. A laboratory here, a club there, the movement could hardly claim any massive affiliated membership. The proponents had very clear and similar objectives and moral imperatives, yet they enjoined an almost endless variety of methods: theories of aging counting by the dozens and rejuvenating nostrums by the hundreds. As will be exemplified, the proponents also diverged dramatically in their political philosophies, each in line with the native dominant socio-political paradigm in which the proponents were most closely integrated. Several central figures and general trends will be here delineated. Yet, despite all this diversity, the expressed common aspiration allows one to see its proponents as belonging to a "movement."

The very notion of a "pursuit of life-extension" also requires qualifications. It might be safe to assume that few people in the world would oppose healthy life extension *per se* (though some authors have expressed such opposing statements). The universal drive of medicine to prolong human life for some periods of time is also generally implied. How then can "life-extensionism" be distinguished as an intellectual movement apart from the aspirations of the whole of medicine or the whole of humanity? The difference may just consist in the extent of hopes, the openness with which such hopes were expressed, and the amount of effort directed toward their fulfillment. When the hopes are high and openly expressed, and the effort toward their implementation is great, the protagonist may be described as a "life-extensionist." In this sense, aspirations for life-extension for very limited periods of time, as for example in terminal cases, would not be subsumed under the heading of "life-extensionism" (though these are highly compatible with life-extensionist goals, representing, so to say, a "first step"). As a rule of thumb, when the earnestly expressed aspirations amount to 100-120 years, the life-span attained by humanity's longest-lived,[11] the person who expresses them might be characterized as a "moderate life-extensionist" or simply a "life-extensionist." Beyond that period, the aspirant may be termed a "radical life-extensionist." And those who envision virtually no potential limit to the human life-span may be categorized as "immortalists." Yet, even without mentioning any specific time periods, life-extensionists can be identified by such expressions as "defeating/reversing aging," "fighting/overcoming death" or even by a prevalent emphasis on the "prolongation of life" or "longevity" generally. These emphases are prominent in the writings of life-extensionists, but almost conspicuously absent in those who might not be categorized as such. The desire to prolong human life may be generally implied in medicine, but it is not always *expressed*. And certainly, speaking of "radical life-extension" is often considered bad taste among physicians and biomedical researchers.[12] Furthermore, biology textbooks often do not include aging and dying, not to mention longevity, among the processes of life.[13] For the life-extensionists, these topics are central.

A further distinction may be expected to be found in the specific methods proposed. Yet, it appears that the distinction may consist not so much in the specifics of the methods, but rather in their specific purposes. As will be exemplified, several biomedical methodologies were developed for the explicit purpose of life-extension and rejuvenation, rather than for the treatment of particular diseases, even though these methods also drew on and were applied to other fields of medicine. The non-identity of the pursuit of health/eradication of diseases and the prolongation of life was first expressed by Christoph Wilhelm Hufeland (1762-1836), the renowned German hygienist, physician to the King of Prussia, Friedrich Wilhelm III, and to Goethe and Schiller. Hufeland coined his own term for life-extension – "macrobiotics." (The word "macrobiotes," designating the extremely long-lived, appears as early as Pliny the Elder (23-79 CE) who also notices the interest in this subject in Herodotus, c. 484-425 BCE, and Hesiod, c. 750-650 BCE.[14]) The term "macrobiotics" has survived to the present. In *Macrobiotics or the Art of Prolonging Human Life* (1796), Hufeland thus distinguished the art of life-extension from the general medical art:[15]

> This art [of prolonging life], however, must not be confounded with the common art of medicine or medical regimen; its object, means, and boundaries, are different. The object of medical art is health; that of the macrobiotic, long life. The means employed in the medical art are regulated according to the present state of the body and its variations; those of the macrobiotic, by general principles. In the first it is sufficient if one is able to restore that health which has been lost; but no person thinks of inquiring whether, by the means used for that purpose, life, upon the whole, will be lengthened or shortened; and the latter is often the case in many methods employed in medicine. The medical art must consider every disease as an evil, which cannot be too soon expelled; the macrobiotic, on the other hand, shows that many diseases may be the means of prolonging life. The medical art endeavors, by corroborative and other remedies, to elevate mankind to the highest degree of strength and physical perfection; while the macrobiotic proves that here even there is a maximum, and that strengthening, carried too far, may tend

to accelerate life, and consequently, to shorten its duration. The practical part of medicine, therefore, in regard to the macrobiotic art, is to be considered only as an auxiliary science which teaches us how to know diseases, the enemies of life, and how to prevent and expel them; but which, however, must itself be subordinate to the highest laws of the latter.

Indeed, the emphasis on the treatment of particular diseases may distinguish general medical practice from life-extensionism. The cure of a disease might be more readily and immediately perceived, while the ascertainment of human life-extension may be a more lengthy and confounded process. Moreover, the possibility of "radical life-extension" is not yet subject to empirical confirmation. These might be some of the reasons why "life-extension" or even "longevity" is not often mentioned in biological or medical discourse.

Yet, it would be a great mistake to think that life-extensionism is somehow watertight and separated from general biology or medicine and that it can be defined by some particular and exclusive "method." Life-extensionism is not a method; it is an aspiration and a motivation. Or more precisely, it is a reason to develop and apply a method, primarily for the purpose of life-extension, which however may also involve treatment of particular diseases. Hence, life-extensionists can be distinguished by their goals, rather than by their methods. Proponents of the very same methods can be perceived as life-extensionists or not, based on the expressed motivations. Moreover, most authors under consideration were not exclusively involved in life-extension research, but also in other fields of biomedicine, often achieving high prominence in these fields. The research of aging, or "gerontology," was a primary field of study, but in no way the only one. Prominent and often world famous scholars and scientists from different fields who sympathized with the life-extensionist goals are the focus of the current study.[16] The interest of these authors in radical life extension has received little attention in biomedical history. The current work addresses these omissions. The scientific contributions of the protagonists were extensive, as was their public appeal, and their life-extensionist views and motivations need to be considered to create a more rich and balanced biomedical history. Insofar as the authors extrapolated on and created contemporary biological and medical advances, the history of life-extensionism represents an integral, though until now under-appreciated, part of the general history of biology and medicine.

When referring to the general difficulty of defining "holism" in twentieth-century medicine, the American medical historian Charles Ernest Rosenberg pointed out that "twentieth-century medical holism has to be understood primarily in terms of what it was not."[17] In a similar fashion, life extensionism might be understood by "what it was not." That is to say, life-extensionist programs may be better appreciated by analyzing the reactions and criticisms raised against them and by their counter-reactions. I would like to suggest three general types of reactions to life-extensionist programs. The first (and very common) type is simply ignoring the topic of life-extension, the kind of reaction which omits aging and longevity from processes of biological development. (Such a reaction might be due to the simple reluctance to think about dying or about struggle with the apparently inevitable end.) In this sense, life-extensionism stands out simply by emphasizing the topics which other authors do not. The second type is the principal opposition to the task of life-extension, seeing life-extension far beyond the present life-expectancy as ethically undesirable and theoretically impossible. The American historian Gerald Joseph Gruman, the author of the best available history of early life-extensionism (or "prolongevitism" to use the term coined by Gruman) – *A History of Ideas about the Prolongation of Life. The Evolution of Prolongevity Hypotheses to 1800* (1966)[18] – termed such principal opposition "apologism," an attitude rationalizing and even apologizing for our mortality. Earlier, the British philosopher Herbert Spencer defined the opposition as the "pessimistic" as contrasted to the "optimistic view" of increased longevity (1879). "Legislation conducive to increased longevity," Spencer wrote, "would, on the pessimistic view, remain blameable; while it would be praiseworthy on the optimistic view."[19] Several principal ethical and political objections are commonly raised, such as "overpopulation," "boredom," "injustice" and a few others. Such objections were reviewed and countered by the bioethicists

Robert Veatch, John Harris and others.[20] The historical tradition of these ethical objections questioning the very desirability of life extension will be considered in greater detail, in Chapter 4, in the section "British Allies – Literary and philosophical life-extensionism: The optimistic vs. the pessimistic view. The reductionist vs. the holistic approach" and mentioned *passim* throughout the text.

Another branch of the principal opposition asserts the theoretical impossibility of radical life-extension. The most common argument has been that there is a "limit" to the human life-span which cannot be overcome. The various perceptions of this limit will be focused on and referenced in Chapter 4, in the sections "Theories of Aging" and "Rectifying 'Discord' and conserving 'Vital Capital'" and *passim* throughout. Yet, it should be noted from the outset, that even when proposing a "limit" to the life-span, it was often realized by the proponents that this "limit" is quite flexible and theoretically not very limiting. As stated by the Nobel Prize winning physicist Richard Phillips Feynman, "there is nothing in biology yet found that indicates the inevitability of death."[21] Yet, *practical* limits, the constraints in our ability to greatly increase the human life-span with the current technological means, have been realized even by the most ardent life-extensionists. Their only distinctive feature appears to be the desire and the striving to overcome those limits.

The third type of reaction was the specific response (critical or accepting) to particular theories or methods proposed for life-extension, or to particular research programs and therapeutic modalities directed toward this purpose. The discussions of this type will occupy the bulk of the present history. It appears that the main responders to the specific life-extension programs were life-extensionists themselves (for whom this was indeed a major topic of concern). Their responses to particular programs were often severely critical and fiercely controversial. Among the enthusiasts of life-extension, the disagreements have been wide. A battle has been waged throughout the century between "reductionist" and "holistic" approaches, with alternate success. "Spiritualists" scorned what they perceived as the ineptitude of modern medicine and science; while "materialists" despised what they saw as unscientific quackery. Even within the "materialistic" branch, there has been much controversy: The more academic life-extensionists argued between themselves on theories, "limits" and funds, and all together attacked unproved remedies which were in turn defended by their providers. Yet, despite all these controversies, life-extensionism may still be seen as a significant intellectual movement, defined by the common goal. Beside the shared goals, other common denominators of life-extensionism will be sought in this study, such as a pursuit of constancy, stability and adaptation.

In summary, the three aims of this survey are, first, to fill an important lacuna in biomedical history by looking at the understudied history of life-extensionist research and its diverse methodologies; second, to examine the motivating factors for life-extensionist research by looking at the people behind the work, their biographies, psychologies, philosophies, and social and political contexts; and the third, overarching aim, is to collect a broad register of life-extensionist authors, and, based on that register, to find common and definitive characteristics of the life-extensionist intellectual movement and its role in science and society.

2. Argument

With specific reference to the scientific projects initiated by life-extensionists (the first aim), the current study examines an interrelation between holistic/hygienic approaches and reductionist/therapeutic approaches. The primary focus of the current research is on the formative first half of the twentieth century, with an extension to a later period. I argue that the failure of reductionist "endocrine rejuvenation" attempts, that began at *fin-de-siècle* and culminated in the 1920s, impacted profoundly on the development of longevity research. The reactions to the failures of earlier reductionist rejuvenation were characteristic. A large number of researchers made a transition to a more holistic approach (such a reaction was particularly pronounced among life-extensionists in France and Germany of the 1930s-1940s). Others conceived of the failures as building blocks and signposts for a continued pursuit on the same path, viewing the human body as a machine in need of repair, and searching for new reductionist methods for its prolonged maintenance by surgery or pharmacological supplements (this type of reaction was prominent among many life-extensionist researchers in the US and the USSR of the 1930s-1940s). Thus, the disillusionment with reductionist endocrine rejuvenation exemplifies varied responses to a scientific failure and modifying research approaches in response to that failure. The conflict between reductionist and holistic approaches will be shown to continue throughout the 20th century. Though, it should be noted that several researchers succeeded in combining reductionist and holistic methods. Moreover, several lines of biomedical research will be shown to owe their beginnings and changing forms to particular life-extensionist enterprises.

With reference to the ideological and social determinants (the second aim), I argue that the hopes for life-extension have been coupled to a wide variety of nationalities and ideologies. Insofar as the conditions at home had the most effect, the data are organized according to national contexts. The following contexts are considered: France (Chapter One), Germany, Austria, Romania and Switzerland (Chapter Two), Russia (Chapter Three), the US and UK (Chapter Four). (Unless otherwise specified, all the excerpts are in my translation.) No ideological system or nation seems to have had a monopoly, however strongly it asserted that it constituted the rock-solid foundation for the pursuit of longevity. It may even be that, rather than providing such a foundation, political ideologies enlisted the hope for life-extension to increase their appeal. It therefore appears that radical life extension is a cross-cultural value, with a common humanistic appeal above and beyond any particular ideology. Nevertheless, life-extensionism was a strongly ideologically and socially constructed enterprise: In different national contexts, different, and often conflicting, ideological schemes – secular humanism or religion, socialism or capitalism, materialism or idealism, elitism or egalitarianism – yielded different justifications for the necessity of life prolongation and longevity research and impacted profoundly on the way such goals were conceived and pursued. This work investigates such ideological, socio-economic and national backgrounds, and exemplifies the integral adjustment of the specific scientific pursuits to prevalent state ideologies.

In attempting to establish common defining characteristics of life extensionism (the third aim), I will argue that the persistent striving for adjustment indeed constituted such a defining trait. The term "adjustment" is used here as a general heading in its common dictionary sense. Thus, for example, the Merriam-Webster dictionary defines "adjustment" (synonymous with "adaptation") as achieving "balance" within a given environment. And "balance" (synonymous with "equilibrium") entails "stability," "steadiness" and "constancy."[22] All these terms have been key in life-extensionist writings, and they have been often used interchangeably. The emphasis on these notions could be expected. If adaptation is a defining feature of life, and if a harmonious, balanced state of equilibrium entails durability and constancy, it is hardly surprising that the proponents of the extension of life were determined to adapt and maintain equilibrium and stability without limits, for their own bodies, for their research projects, and for the societies in which they lived. Yet the task of defining specific local adaptations and equilibria is daunting, either for the body or the society. This work will attempt to examine some of the more general forms of adaptation characteristic of particular national contexts and will instantiate these general forms by more nuanced examples of adaptation of particular life-extensionist projects.

In relation to the society, I will examine the adaptation of life-extensionist programs to what might be termed "dominant" or "hegemonic" socio-ideological orders. Such "dominants" can be most clearly perceived in the 1930s (a major focus of this work): Socialism in Russia, National Socialism in Germany, Capitalism in the US. In France, the "dominant" was more difficult to see, yet the strong rise of political "traditionalism" in the 1930s-early 1940s may be significant. I would like to suggest three common types of adjustment of the life-extensionist thought, which will be exemplified in particular contexts. The first is the rhetorical support of the ruling socio-ideological order (if only to ensure the continuation of the research). This is a kind of mimicry, "when in Rome, doing as the Romans." Sometimes it was difficult to distinguish whether such support was purely opportunistic or honestly believed in. (Often this distinction did not seem to have any implications for the authors' words or deeds; when a person was compelled to believe in something to survive, he would seem to believe in it with all his heart, if only to avoid cognitive dissonance.) The second form of adjustment was the positing of metaphorical socio-biological parallels between the workings of the body and of the society in which the authors lived. The society provided the frame of reference for scientific metaphor ("as above, so below") and when the perceptions of the "above" changed, so did those of the "below." And thirdly, and perhaps most importantly, specific research projects were often favored as compatible with the ruling socio-ideological order. As will be exemplified, under particular national ruling regimes, certain lines of research were simply not allowed to flourish or were discouraged. In all these senses, life-extensionism was "adjusted" to the ruling national orders or to more local orders. Yet, the striving for adaptation, that is, maintaining stability within a particular environment, appears to be universal for the life-extensionists. Moreover, the support of the existing ruling regime, whatever it may be, may derive from the nature of life-extensionism that seeks stability and perpetuation. In this regard, a question may be raised regarding the forms of society that would indeed merit such a perpetuation.

In relation to the body, the striving for stabilization and equilibration has been equally persistent. Insofar as the stability and equilibrium of the body have been perceived to be under continuous threat of disruption and destruction, means for preserving the constancy of the body (or the constancy of the "internal environment") have been relentlessly sought through a variety of methods. And if some particular methods – "holistic" or "reductionist," "hygienic" or "therapeutic" – failed to maintain this constancy, new methods for stabilization would be sought, either by embracing a scientific or therapeutic paradigm opposed to the one that failed, or by continuing in the earlier paradigm hoping for its gradual perfection. Still, the desire for constancy appears to be universal and only sought by varied and often novel means. Thus, the underlying conservative (or conservationist) proclivity of prominent life-extensionist scientists in many countries, seeking stability and perpetuation, may have been a source for their diverse and often unorthodox scientific and medical developments.

Chapter 1.

France: The origins of the life-extensionist movement. Maintaining the Equilibrium and dissolution into the Whole

1. Chapter summary

France was a fertile, perhaps even a primary, ground for the life-extensionist movement, since the Enlightenment and even earlier. In the late 19th – early 20th century, the pursuit of life-extension was encouraged by peaceful and prosperous social conditions, and by the philosophical traditions of positivism, liberalism and progressivism. In this period, France produced some of the world's ground-breaking advancements in the theory of aging, methods of rejuvenation and life-extensionist philosophy, epitomized by the works of Elie Metchnikoff, Charles-Édouard Brown-Séquard, Serge Voronoff and Jean Finot. In the early 20th century, the practice of reductionist endocrine rejuvenation, essentially viewing the body as a machine in need of repair and refueling, formed the main stem of the French life-extensionist movement. Yet, with the growing realization that the effectiveness of reductionist rejuvenation methods falls short of their initial promise, a shift of emphasis occurred in the 1930s toward more "holistic" (or "Neo-Hippocratic") approaches to life extension, emphasizing the organic unity of the human being in integration with one's physical environment, social and psychological milieu, and with a shift of focus from interventionist therapy to life style improvements, physical and psychological hygiene. The "holistic shift" is epitomized by the works of Auguste Lumière, Alexis Carrel and Alexander Guéniot. The shift of emphasis towards "holistic" approaches corresponded with the contemporary resurgence in France of traditionalism, Christian revival and political conservatism, particularly during the Vichy regime. With a withdrawal from reductionist rejuvenation endeavors, France appears to have lost its leading position in the life-extensionist movement.

2. The French tradition. The legacy of the Enlightenment

The radical prolongation of life has been a desire of humanity and a mainstay motif in the history of scientific pursuits and ideas throughout history, from the Sumerian Epic of Gilgamesh and the Egyptian Smith medical papyrus, all the way through the Taoists, Ayurveda practitioners, alchemists, hygienists such as Luigi Cornaro, Johann Cohausen and Christoph Wilhelm Hufeland, and philosophers such as Francis Bacon, René Descartes, Benjamin Franklin and Nicolas Condorcet.[23] However, the beginning of the modern period in this endeavor can be traced to the end of the 19th – beginning of the 20th century, to the so called *"fin-de-siècle"* period, denoted as an "end of an epoch" and characterized by the rise of scientific optimism and therapeutic activism, with life-extensionism representing their most radical form.[24]

The *fin-de-siècle* was a time of peace, yet with a widely felt apprehension of stagnation, of a crisis, even of an imminent extinction of humanity.[25] At the same time, contemporary scholars delighted in the astonishing scientific, technological and industrial achievements of the period: the advances in transportation, energy supply, manufacturing, agriculture and general medical care.[26] It appeared to them evident that science, perhaps for the first time in history, did have the genuine ability to ameliorate social plights, cure diseases and extend human life. Advocates of radical life extension extrapolated on the technological advances and were motivated by them. The scientific and technological achievements, they believed, would be limitless, and the core values of the Enlightenment, of reason and progress, would eventually triumph, making the human beings masters of their destiny, even of their own mortality.

The contemporary hopes were summarized by Jean Finot (1856-1922, born Jean Finkelstein in Warsaw, a prominent French-Jewish journalist and social scholar, and an activist of the anti-racial movement). In *The Philosophy of Long Life* (1900)[27] – positively reviewed by Gustave Kahn, Paul Margueritte, Otto Horth, and by Finot's close friend Max Nordau – Finot projected scientific and social progress *ad infinitum*:

> The progress of hygiene; the increased comforts of the working classes; the results obtained by serum therapy, which has revolutionized medical science by giving it the means of fighting infectious diseases, that most important factor in human longevity, all these are so many elements which may perhaps allow us to draw near to the beautiful dream fondly imagined by the authors of *Genesis*. Methusaleh, ancestor of Noah, was, according to the latest Bible criticism, only a myth, but who knows whether, thanks to the progress shown above, this myth may not some day become a reality? When liquid air shall have destroyed the evil effects of the unhealthiness of big towns, and synthetic chemistry have delivered us from the poisons contained in adulterated food; when electricity facilitates life by reducing its labor; when universal peace rids us of mortality on the battle-field; when humanity at last, thus freed from misery and its warlike instincts, as well as the debilitating principle of hate, shall have found its end in the life-giving domain of love and universal fraternity, then we may see longevity again drawing near to its natural limits.

The observed improvement in the quality of life, the decline in mortality, the steady increase in the average lifespan, as well as the existence and the reported growing number of centenarians, reassured Finot in the future success of prolonging the human life to 150 years and beyond.[28] Many of Finot's contemporaries were similarly convinced: the Russian/French/Jewish biologist Elie Metchnikoff, the Russian/French/Jewish surgeon Serge Voronoff, the Austrian/Jewish physiologist Eugene Steinach, the Russian physician and politician Alexander Bogdanov, the British playwright Bernard Shaw, the American physician and writer Charles Stephens, to name just a few proponents of super-longevity that will be studied here. Mere "hoping" was not sufficient – an active search for life prolonging means became for them an imperative. Representatives of many countries joined the search. And yet despite this multi-focality, France was very notably at the forefront of the global progressivist movement and *fin-de-siècle* France may be well considered a birthplace of modern life-extensionism.

France was a fertile, perhaps even a primary, ground for the life-extensionist movement, since the Enlightenment and even earlier. French Aristocracy had been traditionally quite fond of the idea of a significant prolongation of their lives, gladly acquiring the means made readily available. Thus, Andre du Laurens (1558-1609), physician to the French king Henry IV, supplied hopeful health regimens for the retardation of old age, much in the Galenic tradition.[29] In 1612, there was published *Livre des figures hiéroglyphiques* (The Exposition of Hieroglyphic Figures) by the Parisian alchemist Nicolas Flamel (1330-1417), seeking to "prolong life to the moment ordained by God."[30] In 1667, Jean-Baptiste Denis, physician to King Louis XIV, transfused blood from sheep to men for the patients' "placation" as well as for their rejuvenation.[31] René Descartes too might be considered among life-extensionists. In the *Discourses* (1645) he pledged to advance medical science, and asserted that "we could free ourselves from an infinity of maladies of body as well as of mind, and perhaps also *even from the debility of age*" (emphasis added).[32] Latter on, Louis XV (1710-1774) and Madame de Pompadour sponsored the great "master of rejuvenation" and claimant to immortality, Count St. Germain (1710-1784), a fact mentioned in Giacomo Casanova's *Memoirs* (1797). Following suit, Louis XVI (1754-1793) and Marie Antoinette pampered Giuseppe Balsamo (Count Allesandro Cagliostro, 1743-1795), a man of equal notoriety.[33] Stories of these men's super-longevity and miraculous rejuvenating nostrums filled Versailles.

After the subversion of the monarchy, the rationales for life-extension were adjusted to the changing political situation. Life-extension was no longer perceived primarily as a task of preserving royalty and high aristocracy, but rather as a more common, progressive humanistic task. By the end of the 18th century, Nicolas Condorcet (1743-1794), the great philosopher of the Enlightenment and ideologist of the French Revolution, formulated a consistent "progressive" philosophy and eschatology in which the striving for life extension played an integral part. According to Condorcet, "melioration in the human species [is] susceptible of an indefinite advancement." Hence, human life "will itself have no assignable limit" and "the mean duration of human life will for ever increase."[34] By the mid-19th century, the positivist philosophy of Auguste Comte reaffirmed the ideals of progress, and included increasing longevity as one of the elements of progress, even though this element was not overwhelmingly emphasized. In *The Catechism of Positive Religion* (1852), Comte referred to a long life as a blessing, and claimed that an "ideal existence [is] the simple continuation of our real life," that "death is not in itself the necessary consequent of life," and that the inheritance of beneficial characteristics may improve the health and longevity of future generations.[35] Comte's general notion of positive progress, "the improvement of man's moral and physical nature," chimed with the aspirations of French life-extensionists from the *fin-de-siècle* onward, from Elie Metchnikoff through Alexis Carrel.

Several late 18th-early 19th century French scientists, such as François-Joseph Broussais (1772-1838) and Marie-François-Xavier Bichat (1771-1802) developed models of human longevity in response to external stimuli. For Broussais, physiological inhibition was the path to conserving the vital energy and hence increasing longevity. In contrast, for Bichat, internal stimulation was the key to resistance against the constant threat of death from the environment.[36] As a middle ground, Jean-Baptiste Lamarck (1744-1829) believed that the duration of life is determined by the balance between "disintegration and destruction followed by recuperation and renewal" and this balance "prolongs the life of the individual so long as the equilibrium between these two opposed elements is not too rudely disturbed."[37]

Other French biologists attempted to determine the natural limit of human life, which, in their calculations, should be far longer than the life-span commonly attained. Thus, Georges-Louis Buffon (1707-1788), posited the 7:1 ratio between the maximal longevity and the period of growth, hence reaching the figure of 140 years for a normal human life-span. Later, Marie-Jean-Pierre Flourens (1794-1867), employing a similar principle, arrived at a somewhat lower figure (100 years). According to his calculations, the life-span of mammals equals 5 times the period of their growth (~20x5 for man). The century life-span, according to Flourens, however, should be a norm of "ordinary life." In "extraordinary" cases, human life could be prolonged to two centuries or a century and a half. "A first century of ordinary life," Flourens wrote, "and almost a second century, half a century (at least) of extraordinary life, is then the prospect science holds out to man."[38]

By the end of the 19th century and in the first quarter of the 20th century, France remained the epicenter of the life-extensionist movement. In the 19th century, major contributions to the study of aging were made by French pathologists.[39] Working in the Parisian hospices for the elderly – the Salpêtrière for women and the Bicêtre for men – Charles-Louis Durand-Fardel (1815-1899) and Jean-Martin Charcot (1825-1893) – had at their disposal almost unlimited clinical and autopsy materials. Charcot, Durand-Fardel and their followers carefully investigated the pathological tissue changes in old age: changes of the lungs, the brain, and the blood vessels.[40] This work was summarized in treatises, such as Durand-Fardel's *Traité clinique et pratique des maladies des vieillards* (Clinical and practical treatise on the diseases of the aged, 1854) and Charcot's *Leçons sur les maladies des vieillards et les maladies chroniques* (Lectures on Senile and Chronic Diseases, 1868). Some hygienic measures were suggested to alleviate the suffering of the aged (such as Charcot's baths). However the mid-19th century *médecine de vieillards* (medicine of the aged) was quite limited in combating senescence, hence any explicit hopes for radical life-extension are absent from these writings. Nonetheless, the work of the French physicians laid the foundations for the understanding of the pathology and physiology of senescence and outlined measures of intervention.

Perhaps an even more substantial impetus was given by the Therapeutic Activist approach, advanced by Louis Pasteur (1822-1895) and Claude Bernard (1813-1878). Pasteur, having confirmed the germ theory of disease, having developed measures for preventing food spoilage, and having produced vaccines (using attenuated bacterial and viral strains) against a panoply of diseases – chicken cholera, rabies, anthrax, etc. – vindicated the ability of experimental medical science to eradicate disorder and prolong human life. And Claude Bernard, in the pioneering works on "internal secretion" and "internal milieu," established that "the fixity of the internal environment is the condition for free, independent life."[41] Such a fixated stability (later termed "homeostasis") could, in theory, be maintained for prolonged periods, through conscious interventions into the "internal milieu." Even though any specific aspirations for radical life prolongation were not to be found in either Pasteur or Bernard, the Therapeutic Activist approach that they promoted, greatly contributed to the advent of modern life-extensionism in *fin-de-siècle* France. Thus, the foremost researcher of life-extension at the turn of the century, the Nobel Prize winning biologist Elie Metchnikoff (1845-1916), was Pasteur's protégé and député at Institut Pasteur. Another crucial contemporary proponent of life-extension, Charles-Édouard Brown-Séquard (1817-1894), the president of the French Biological Society, one of the founders of modern endocrinology and the inventor of rejuvenative hormone replacement therapy, was Claude Bernard's pupil and successor at Collège de France.

3. "Father" Metchnikoff (1845-1916)

The indisputable leader of the life-extensionist movement, its most respected spokesman, was Elie (Ilya Ilyich) Metchnikoff, whose teachings formed the crux of the great part of the contemporary discussions on longevity. The renowned immunologist and microbiologist, a vice director of the Pasteur Institute in Paris, and the Nobel Laureate in Physiology/Medicine of 1908 for the discovery of phagocytosis (a major contribution to the cellular theory of immunity), he was also credited as "the father" of gerontology (the disciplinary term he coined).[42] To the present day, his scientific reputation has remained high, and was almost incomparable in the USSR and the 'Eastern' bloc.

After battling for two decades for the acceptance of his theory of phagocytosis (the German leader of microbiologists, Robert Koch, was its mighty opponent),[43] since the 1890s Metchnikoff concentrated on mechanisms and counter-measures of aging.[44] This was a direct continuation of his immunological research, as he suggested the "devouring phagocytic cells" to be the major culprits of senescence. This in fact became a first (if not the first) scientific, empirically based theory of aging and longevity, rooted in observations of dynamic cellular behavior. From the scientific theory of aging, there followed practical propositions for increasing longevity. Hence Metchnikoff can be considered the "father" of the modern life-extensionist movement.[45] He summarized his work in *The Etudes on the Nature of Man* (first published in 1903)[46] and *The Etudes of Optimism* (1907)[47] – representing contemporary state-of-the-art scientific "bio-materialistic" methodologies for life-extension. The works were grounded in uncompromising materialism and positivism, Darwinism, liberalism and pacifism, and a profound optimistic belief in human progress and perfectibility.

As Metchnikoff recurrently admitted, he was not always optimistic.[48] After the study abroad, in 1870 he became a full professor of zoology (at the age of 25) at Odessa University, Ukraine. He stayed there until 1882, when he was forced to retire due to student riots and the reactionary policy of the university authorities, following the assassination of Tsar Alexander II in March 1881. Due to the lack of funds, he almost accepted a position of a district entomologist in Poltava, Ukraine. But thanks to a small inheritance of his second wife Olga, he was able to conduct independent research on comparative embryology in Messina, Italy, where he stumbled upon his most important discovery of phagocytosis. As a result of the sojourn in Messina, he "transformed from a zoologist into a pathologist and bacteriologist."[49] After coming back to Ukraine, he struggled for two years as an independent researcher, in a laboratory he set up in his flat. That enterprise having failed, he established a bacteriological station in Odessa, to produce Pasteur's vaccines against Rabies, Anthrax, etc. But the unending bureaucratic red tape, revisions and persecutions by the district authorities, and the incessant ridicule by the press and by the local medical associations, made his work unbearable. In 1887, he sought the patronage of Louis Pasteur, who gladly placed at Metchnikoff's disposal a division at the newly established Pasteur Institute, which became Metchnikoff's home to the end of his days.

Metchnikoff's personal life was also quite tempestuous (he was married twice, though never had children). After his first wife, Ludmila Fedorovitz, died from tuberculosis in April 1873, he attempted his first suicide, taking a large dose of opium. After his second wife, Olga Belokopytova, contracted typhoid, he made another suicide attempt, in April 1881, in a more scientific fashion, by injecting himself with relapsing typhoid bacteria. Surprisingly, following this attempt, his generally failing health, especially the eyesight, radically improved. (To the end of his research career, Metchnikoff remained quite fond of treating himself and his colleagues to a hefty dose of potentially harmful sera.) Having recovered his wife's and his own health, having left behind the unbearable and pointless bureaucratic and political struggles in Odessa, having assured his scientific reputation, having found financial security and a productive, cooperative scientific environment at Pasteur's Institute in Paris – Metchnikoff entered the self-termed "Optimistic" or "Life Asserting Phase of Life" which directly entailed his life-extensionist philosophy. This pattern will recur in many other instances, where life-extensionism arose from the desire to have 'more of a good life,' under conditions of economic and social security, individual and national. Such prosperous conditions produced a wish to perpetuate the present agreeable state.

4. Attractions of Paris and the "Nationality" of life-extension

Fin-de-siècle France provided a fertile ground for life-extensionism generally, and for Metchnikoff in particular. Metchnikoff loved Russia, he remained in close contact with leading Russian intellectuals (including Tolstoy), and advocated the development of Russian science. Yet, in "The Story of How and Why I Emigrated" (1909), he explained his reasons for leaving the motherland. At the epidemiological station, his work was thwarted by bureaucratic hurdles, by unjustified, excessive control of government auditors, by the misunderstanding and opposition of the medical establishment and the press. He could not continue in the university either, partly because of the political "fermentation" among students, disrupting the normal course of study, partly because of political repressions on the part of the university authorities (for example, disqualifying socialist applicants). He could not stand the deceptions and betrayals within the academia. Another reason was Russian Anti-Semitism (not necessarily aimed at Metchnikoff himself, since, even though his mother was a converted Jewess, his father was a Russian nobleman).[50] Metchnikoff explained:[51]

Dr. [Eugène] Wollman became my excellent assistant [at the Pasteur Institute]. I owe this to the fact that he is a Russian Jew; therefore a scientific career is closed for him in Russia. If he had stayed in Russia, he – like my other colleague, Dr. [Alexandre] Besredka, who is already highly reputed in science – would have led the abject existence of a general medical practitioner. His attraction to science would have remained fruitless, and would have faded, leaving in the soul the bitter sense of an undeserved insult. This is how they get rid of talent in Russia.

The situation in Paris was much more auspicious:[52]

In Paris, thus, I could achieve the goal of scientific work, unencumbered by any political or other social activities. In Russia, on the other hand, the obstacles from above, from below and from aside, made the accomplishment of this dream impossible. It might be thought that the time has not yet come for science to be useful in Russia. I disagree. I believe, scientific work is absolutely essential for Russia, and I wish, whole-heartedly, that its conditions improve in the future.

In Paris, Metchnikoff remained involved in political and ideological debates, but there, unlike in Russia, these did not encumber his research. Notably, the criticisms of the Russian scientific establishment were expressed by Metchnikoff in France, not in Russia. In Russia, he could not afford to openly oppose the regime. While living in France, Metchnikoff became a Russian dissenter and a staunch French patriot.

Despite the changing national allegiances (or rather thanks to them), Metchnikoff may be considered a foremost representative of both the Russian and French life-extensionist traditions (see Chapter 3 of the present investigation on Russian life-extensionism). The American historian Peter Wiles, in "On Physical Immortality" (1965),[53] argued for the inherent or predominant "Russianness" of the life-extensionist movement. He contrasted the Russian attitude with the attitude of the "highly rationalistic" French who presumably neglect this pursuit. There seems to be little evidence for such assigning of a Russian 'national trait.' In fact, a much stronger case can be made for the predominant "Jewishness" of life-extensionism, listing luminaries and adducing quotes from traditional texts on the valuation of life in Judaism. (Wiles never identified Metchnikoff, Voronoff, Steinach, or any other life-extensionist as Jewish.) Yet such a claim might be equally fallible. Indeed, there were many Russians and Russian (or Jewish) immigrants in the movement at the beginning of the 20th century. But there were also many British, German, Austrian, American, Romanian, Japanese, Spanish and Italian proponents, *et cetera*. And certainly, historically the "highly rationalistic French" were far from being averse to the idea of radical life extension.

As we shall see, the struggle for life-extension has been coupled, in specific ways, to a widest variety of cultural identities and ideologies.

5. Competing factions

The long tradition of rejuvenation attempts and hygienic regimens for the retardation of old age, the ideology of progress and enlightenment, the in-depth pathological and physiological examinations of senescence, the rise of Therapeutic Activism, and the peaceful and optimistic atmosphere – rendered *fin-de-siècle* France a welcoming soil for the growth of life-extensionism. However, the advancement of biomedical science and super-longevity were not the only aspirations in France in that period. In the introduction to the second edition of *The Forty Years in Search of a Rational Worldview* (February 19, 1914),[54] Metchnikoff outlined the contemporary intensifying competing ideological and social trends. He felt the "ubiquitously growing antagonism to the scientific world view. In Russia, this antagonism manifests in the attraction to mysticism. … Even in much more positive France, in recent time, antagonists of the positivist philosophy (the philosophy that had developed on the French soil) have become increasingly vocal." Metchnikoff perceived the futility, even danger of infatuation with militant socialism. He further lamented the strengthening of anti-intellectualism, anti-liberalism and war-mongering. Such tendencies, according to him, gradually progressed:

Soon after the Frankfurt Treaty [1870, following the defeat of the French in the Franco-Prussian war], with a complete inability to take revenge, many progressive Frenchmen turned to peaceful activities. Hence their admiration of intellectual progress, the cult of science and art…. But gradually there developed a reaction to this trend… The possibilities of alliance [against the Germans] were encouraging…. The advances in automobilism and especially aeronautics, gave a new impetus to the warlike temperament of the French.

The "cult of sport" was replacing the "cult of science and art," and the "youth of the brain" was pushed aside by the "youth of the brawn" for whom "war seemed to inspire the most noble human virtues: energy, self-control, self-sacrifice for the higher goal."

Even more emphatically, Metchnikoff opposed the waxing spiritualism, epitomized, according to him, by the teachings of the philosopher Henri Bergson (1859-1941): "Bergson preaches the limitations of knowledge and valorizes intuition, struggling to convince us of the existence of a soul independent of the brain function and of the existence of a free will."[55] According to Metchnikoff, this "religion of action, of energy, of human will" suited the current aspirations of the French youth, who "spurned positivist professors and rushed to hear Bergson's lectures." Metchnikoff recognized the sources of Bergson's popularity: "Science, having promised to alleviate human suffering, has proved, time and again, its impotence …. Having lost faith in knowledge, no wonder many turned to metaphysics with its hopes for the immortality of the soul and happiness in an afterlife." For Metchnikoff, Bergson's proofs of an immortal, independent soul carried no scientific evidence, and were mere comforting "lullabies." Metchnikoff offered an alternative – the scientific pursuit of life-extension:

The second of Bergson's questions "What are we doing in this world?" should be formulated differently: "What *should* we do in this world?" Our answer to this, presented in this work and elsewhere, can be stated as follows: "We should, by all means, strive that people, ourselves included, live their full life cycle in harmony of feeling and of mind, until reaching, in the ripest old age, a sense of saturation with life. The main misfortune on earth is that people do not live to that limit and die prematurely." This statement is the basis of all moral actions… It is difficult to imagine that, in some more or less distant future, science will not accomplish this goal and will not solve the problem of the prolongation of human life to a desired limit, as well as rectify other disharmonies of the human nature.

Still, in contemporary France, the receptivity to the ideals of scientific and technological progress was high, and Metchnikoff believed that, in such a milieu, scientific rationalism and life-extensionism would be victorious, while anti-intellectualism and aggression would become things of the past.

Five months after the writing, with the outburst of The First World War on July 28, 1914, Metchnikoff's hopeful forecasts were shattered, while his apprehensions were vindicated beyond any measure. Millions were killed in battle, with no small help from science and technology. Millions more died in the concomitant pandemics of typhus and influenza, against which science and technology were largely powerless (the epidemics apparently subsided by themselves, after the war was over and living conditions improved).[56] Yet Metchnikoff's program of life prolongation remained valid for a "more or less distant future," for more prosperous times. But at the time of war, as Metchnikoff recurrently admitted, his research became "almost impossible."

6. Optimistic biology

The Etudes on the Nature of Man (1903), were written earlier, at the height of what Metchnikoff termed a "life-asserting phase," and epitomized his optimistic philosophy. The work envisioned that extreme longevity can be achieved through the progress of medical science, requiring a massive collective effort. Metchnikoff believed that it is our duty as conscious human beings to fight death, the main disharmony and evil of nature. He strongly emphasized that each death has an identifiable, combatable cause and in this sense every death is "violent" and not "natural." The fact that everyone must succumb to it does not make it right or even acceptable.[57] From the assertion of the desirability of increasing the life-span, Metchnikoff proceeded toward formulating a theory of aging, and, based on the theory, toward suggesting actual or potential means of anti-aging intervention.[58] Metchnikoff's practical suggestions for life prolongation through material/biological mediators, included microbiological and toxicological methods (which can be tentatively termed "elimination of damaging agents"), and he also alluded to hormone therapy and gland transplantations (which could be generally termed "biological replacements"). The goal was to achieve a long-lasting substance and energy balance.

In summary, according to Metchnikoff's generalized systemic theory of aging, in the body there occurs a constant struggle between "noble" (differentiated, functional) "elements" (tissues) and "primitive" (non-functional, undifferentiated, "harmful") cells and tissues damaging the harmonious function of the former. That general description fitted various contemporary conceptions of aging and death: loss of protein elasticity, desiccation, autoimmunity and cancer. Accordingly, the general path to life-extension was to "strengthen the noble elements," while "attenuating" or destroying the primitive harmful ones; in other words, working by "subtraction" and "addition" toward "balance."[59] Essentially, a conflict was posited between the degenerative and regenerative processes.

Metchnikoff specified germ-related diseases as a major cause of premature ("violent") death and maintained that these diseases are tractable. Several anti-microbial and anti-viral agents were listed, among which the most notable were inoculations and antiseptics. These medical contrivances have saved millions of lives and showed beyond any doubt that medicine does have the real ability to cure illness and extend life. Metchnikoff mentioned the success in fighting such scourges as smallpox and syphilis. At the turn of the twentieth century, arsphenamine ("Salvarsan") was used against syphilis by Paul Ehrlich (who received the Nobel Prize in 1908, in the same year as Metchnikoff). This proved to be the first effective specific chemotherapy. As a vice-director of the Pasteur Institute, Metchnikoff also actively tested arsenicals and worked on a vaccine against syphilis. He suggested an infectious etiology of cancer and expected a cure, perhaps a vaccine. Since then, the existence of cancer-inducing viruses has become established, and anti-cancer vaccines begin to appear.[60] Among the anti-microbial agents mentioned by Metchnikoff, some seem to anticipate antibiotics, in accordance with Louis Pasteur's principle "*La vie empeche la vie*" (life against life). Alexander Fleming's penicillin, Felix d'Herelle's anti-bacterial phage therapy and Gerhard Domagk's "sulfa-drugs" appeared later, in the 1920s-1930s.[61] These almost synchronous discoveries may seem serendipitous, yet Metchnikoff's earlier study attests to the massive effort on the part of the contemporary scientific community searching for anti-microbial agents, that eventually led to the discoveries and a substantial increase in human life-expectancy.[62]

Besides exogenous infection, endogenous intoxication and infection were said to be most potent though controllable causes of degeneration and death, acting through blocking or interfering with vitally important structures and processes. Over-proliferating phagocytes were specified as the agents "devouring" the noble elements, while the build-up of "non-functional" connective tissue was said to replace the noble (parenchyma) tissues (e.g. the tissues of the muscle, kidney, lung and brain) and to contribute to the development of sclerosis. Among the toxic products of body metabolism, special attention was given by Metchnikoff to those produced by intestinal micro-flora. These toxins were said to contribute to functional incapacitation and general degeneration of the body, and specifically to the development of tissue hardening. The toxins (such as indole) produced by intestinal bacteria (e.g. *Clostridia*) were believed to, on the one hand, stimulate the "devouring" macrophages' activity, and, on the other, weaken the noble parenchyma tissues and make them easy prey for the macrophages.

Parallels between Metchnikoff's visions of a harmonious body and a harmonious society can be clearly observed: in both cases, stability and longevity were to be achieved by the protection of the "noble" functional elements from the disruptive or oppressive elements from "above, below and aside."

7. "To live long, you must live poor"

According to Metchnikoff, the most practical and readily available means to fight the intoxication is through an appropriate diet. In Metchnikoff's theory, fermented/acidified dairy products (such as yogurt), containing lactic acid bacteria, are able to suppress putrefactive microflora and detoxify the body. Metchnikoff himself followed this dietary practice and attributed to it his relatively high longevity (though coming from a very short-lived family). Generally, Metchnikoff advocated a simple and restricted diet. He believed luxurious cuisines favor toxigenicity, and therefore are detrimental to health and longevity, to the "correct course of life" or "orthobiosis." This line of thought has continued in the multitude of modern dietary practices aimed at life-extension and life-enhancement, including bio-active, vitamin-rich, pro-biotic, anti-toxic, anti-inflammatory, differential and calorie-restricted diets.[63]

The importance of simplicity in diet was further expanded by Metchnikoff and his followers to include the simplicity and moderation of the entire life style. Indeed, Metchnikoff might in the main agree with the English philosopher Herbert Spencer's dictum that "the highest conduct is that which conduces to the greatest length, breadth, and completeness of life" (*The Principles of Ethics*, 1897).[64] But Metchnikoff's vision of progress differed from that of Spencer who defined it as increasing complexity, or that of another English philosopher, Winwood Reade (*The Martyrdom of Man*, 1872) who saw it as a gradual improvement of human capabilities. For Metchnikoff, progress only meant life as long and healthy as possible for as many people as possible, and the way toward it was not through increased complexity, but rather through greater simplicity. The life-prolonging properties of fermented milk products were suggested to Metchnikoff by the robust and long-lived Bulgarian peasants for whom this was a traditional diet. After Metchnikoff's endorsement, this nutritional regimen spread from Europe to the US and Japan, giving birth to "probiotic" diets.[65] The success was greatly due to the fact that such products were cheap and could be consumed massively by almost all strata of society, including the very poor.

Other contemporary life-extensionists took moderation to extremes, and claimed that in order to live long, the person must be poor. This was the professed conviction of Dr. Arnold Lorand (1865-1943?), a highly successful French/Austrian physician, working in the 1910s-1920s at Carlsbad (Karlovy Vary Spring Resort, Czechia, then part of the Austro-Hungarian Empire) and in the 1930s in the resorts of Nice, France. In *Old Age Deferred: The Causes of Old Age and its Postponement by Hygienic and Therapeutic Measures* (1910), Lorand preached:[66]

> If we were asked for the best means of living to be 100 years old we would say: become a peasant or a pauper and be received into an English work-house… They have no anxieties about getting their daily bread, and oftentimes are fed better than they would have been in their homes, although only the minimum amount of hygienic food is given… Workhouse inmates lead a very regular and frugal life, rising in the small hours of the morning and retiring to bed early in the evening. Thus, in winter time, they can never contract pneumonia by coming home late from the overheated theatre, concert, or club-house. They also need not worry about their fortunes, for they have none.

This comes from a wealthy, elite physician, proud of his friendship with members of the European aristocracy, who prescribed to his clientele, for dinner (against constipation) "Roast or boiled meat, two sorts of green vegetables (by preference spinach), French beans, carrots, boiled lettuce, one course of stewed compote of fruit, and finish with dessert of grapes, figs (dried or green), or preserved plums (California or Bordeaux)."[67] "A peasant or a pauper," having to subsist on carbohydrate-rich "poverty foods" – bread, potatoes and cereals – might have benefited little from this advice.

Jean Finot too concurred with the theory that "to live long you must live poor."[68] But unlike Lorand, Finot admitted to the fact that "statisticians seem as a rule hostile to this theory." According to statistics, the rich live longer. Finot sought a middle course: "The truth is always found in the middle. If riches spare us certain privations which decimate the poorer classes, they deaden, on the other hand, our powers of resistance." This argument was later developed by uncompromising Soviet gerontologists. Thus, Prof. Vladimir Nikitin of the Kharkov Institute of Biology, Ukraine, claimed in 1962 that "the pathological shortening of human life is mainly the result of class antagonism; therefore the first requirement for orthobiosis [prolonged healthy life] is the replacement of capitalism by socialism."[69] That is to say, to help the rich not to die of gluttony and the poor of starvation, classes should be abolished, and a healthy "mean" achieved – an idea inherent in the Soviet ideological paradigm. Fin-de-siècle life-extensionists went to no such lengths as to suggest a de-classification of society; by no means did they intend to upset the existing social order. They did, however, preach moderation and poverty, a check on hedonistic aspirations, a message of contentment to the poor and of warning to the rich liable to abuse their wealth. The preaching of moderation may thus be viewed as an effective device for the conservation of the existing social order. And the emphasis on moderation appears to be as old as life-extensionism itself.[70]

8. Methods of rejuvenation

Acidified milk (yogurt) consumption was primarily advocated by Metchnikoff and later by a host of rejuvenators as an easily practicable, inexpensive and noninvasive way both for providing necessary nutrients and for "combating the agents of damage" – the putrefactive micro-flora. Other methods of combat were much more cavalier. In Metchnikoff's theory, the large intestine is the main seat of putrefactive bacteria, and is therefore a major source of intoxication. Accordingly, a possible method of eliminating such a source of damage was to extirpate the large intestine by surgery, once and for all. The Scottish surgeon William Arbuthnot Lane (1856-1943) undauntedly went through with this method, in the period when triumphant surgeons routinely removed the appendices, tonsils and uteri, just in case, as a prophylactic measure.[71] Metchnikoff understood the logic behind such interventions, but cautioned (1910):[72]

> Dr. Lane, the uncommonly skillful and courageous English surgeon, dared to resort to operation, instead of a prolonged and ineffectual internal treatment… Naturally, such an operation is very dangerous and presently yields very many lethal cases. Dr. Lane did over 50 such operations, and at the beginning of the last year described in detail 39 cases. He lost 9 patients, or about 23%, which is a high mortality rate. However, this intervention proved highly beneficent for the other 30 patients. Such patients, who had been suffering tremendously for many years, after the surgery became revitalized and productive. Dr. Lane's system was, of course, met with strong opposition. His opponents condemn the entire method, citing the high mortality rates. Nonetheless, it should be noted that the operation technique can be improved and thus the mortality rates reduced.

Elsewhere Metchnikoff further valorized minimally invasive methods. "In the present state of surgery," he wrote, "we cannot expect a success from a direct extirpation of these intestines, but we can rely on an artificial change of intestinal flora, that is, the replacement of harmful bacteria by beneficial bacteria."[73]

Yet another "moderately" invasive method, suggested by Metchnikoff as a means to eliminate the agents of damage, was by serum therapy (immunotherapy). At first, he attempted to develop a serum to destroy the "devouring" phagocytes. These attempts having failed, Metchnikoff started developing mildly cytotoxic sera to elicit an immune response of "noble" parenchymal tissues and induce their stimulation. He did not succeed in these attempts either. None the less, this line of research was later carried on by the leading Soviet gerontologist and life-extensionist, Academician Alexander Bogomolets (1881-1946), who developed the Anti-Reticular Cytotoxic Serum (ACS) precisely for the purpose of immunizing and stimulating connective tissues, for the prolongation of life.[74]

Yet, the primary technique for rejuvenation practiced at the beginning of the century was "endocrine rejuvenation." The seekers of rejuvenation invested in "far-reaching" biomedical technologies with uncertain prospective results. These were the first developers of "replacement therapy." They were not averse to radical surgical interventions, but instead of "subtraction," "destroying the agents of damage," they placed their hopes in "addition," in "strengthening the noble tissue," through replacing or regenerating organs and tissues. Some "rejuvenators" promised immediate and radical results, and for that were relegated by their critics to the realm of quackery.[75] Most of them, however, presented their methods as the first experimental steps in a research program for the future, based on the first emerging theories of aging which, they hoped, would be developed further on.

9. Adding to the balance: Rejuvenation by supplements

Since the seminal work of August Weismann *Über die Dauer des Lebens* (The Duration of Life, 1882),[76] through the 1920s and later, senescence and death were widely explained as due to tissue differentiation.[77] According to these theories, differentiated, functional tissues lose the ability to regenerate (in contrast to the immortal "germinal plasma"), hence general "degeneration" or "involution" ensues. Among the differentiated tissues, some were assigned a primary status in causing aging. For some, the heart and the blood vessels played the primary role, in accordance to the dictum of the French physician Henri Cazalis (c. 1870) *"on a l'âge de ses artères"* (the man is as old as his arteries).[78] According to others, the degeneration of the brain and the nervous system were the chief culprits in aging.[79] For some, it was specifically the middle brain (mesencephalon).[80] For great many others, the primary root of senescence lay in the degeneration or involution of various endocrine glands. For the Austrian physiologist Wilhelm Raab, the pituitary gland played the key part, for the French/Austrian physician Arnold Lorand it was the thyroid, for the Austrian surgeon Eugen Steinach (1861-1944) and the French surgeon Serge Abramovich Voronoff (1866-1951) it were the sex glands.

The assignment of a primary role in aging to a single endocrine gland, led to suggesting the enhancement (supplementation) of that gland deficiency by corresponding glandular extracts, transplants or stimulations.[81] The regimens of supplementation were designed to "supply a demand" or "compensate for a deficit" in the body "economy." Thus, Arnold Lorand observed the similarities between Myxoedema (thyroid deficiency) and senescence. Hence he designated a causative role in senility to the thyroid gland impairment, seeing aging as a peculiar form of disease that could be treated by supplying thyroid extracts from animals, their only available source. Generally, Lorand prescribed rather standard hygienic regimens: much fresh air and sunshine, breathing deeply, exercising, consuming meat once a day and lots of vegetables (Metchnikoff's "acidulated milk" was also recommended), masticating properly, bathing, ensuring regular elimination, sound sleep, avoiding overstrain by work (reserving a day in a week for complete rest), avoiding mental stress, benefiting from marriage and neither overdoing nor completely avoiding sex, temperance with alcohol, tobacco and other stimulants. But the crowning point in Lorand's life-extending approach was the suggestion "to replace or reinforce the functions of the organs which may have become changed by age or disease, by means of the extracts from the corresponding organs of healthy animals; *but only to do this under the strict supervision of medical men* who are thoroughly familiar with the functions of the ductless glands"[82] (Lorand's emphasis). Even though the greatest weight in Lorand's work was placed on the thyroid, supplementations of other glands were also important: the extracts of the kidneys, pancreatic extracts, ovarian and testicular preparations. A good understanding of complex endocrine functions, of the dangers of their hypo-activity or hyper-activity, and a personal approach, were said to be paramount. According to Lorand, only a member of the physicians' guild is qualified to recognize hormonal deficits and supply the demands, though the responsibility for implementing the rest of the hygienic suggestions is placed in the hands of the individual enthusiast.

There may seem to be nothing radical about Lorand's advice, it was all standard hygiene (with the possible exception of glandular extracts). Lorand strongly negated the possibility that "the aged can be transformed into sprightly adolescents." At the same time, he claimed that "We need no longer grow old at forty or fifty; we may live to the age of ninety or one hundred years, instead of dying at sixty or seventy. All this can be brought about by the observance of certain hygienic measures, and by improving the functions of a certain few of the glandular structures in our body."[83] This was indeed a radical statement, at the time when the average life expectancy was 40-50 years, that we can *all* and *now* double it. This hope sharply contrasted with the assertions of the contemporary American physician, the editor of the *Journal of the American Medical Association*, Morris Fishbein (1889-1976), who basically advised the same hygienic measures as Lorand, including glandular supplementation, yet claimed that we "may now confidently look forward under all ordinary circumstances to reaching the age of fifty to fifty-five years" and that "there has been, however, but little average prolongation of life beyond the age of seventy, and there is not the slightest scientific reason to believe that there ever will be."[84]

Lorand's claim may have been even more drastic than Metchnikoff's. Metchnikoff believed that human beings may reach the 150 years mark, but that will occur in "a more or less distant future." In the mean time, the life span can be increased but moderately, and much further research will be needed to prolong it radically.[85] These might have been the principal contrasts between the early century attitudes to life extension: those of the "Skeptics" like Fishbein who made a case for "therapeutic nihilism" and against a "Medical Utopia" versus the "Theorists" urging for extensive research and its massive public support, versus "Practitioners" offering readily available means here and now, according to the laws of the market. Lorand's approach is representative of a host of rejuvenators practicing at the beginning of the 20th century, singling out a particular failing component (gland) in the human machine and attempting to restore its function by replenishing or replacing this component.

10. Masculine Forever – Charles-Édouard Brown-Séquard (1817-1894)

Among the methods of glandular supplementation practiced in the first quarter of the century, the sex glands were by far the most important targets. The energy expended on sex played a central part in the bio-economic models of aging. Thus, Freud's *Beyond the Pleasure Principle* (1920) made an equation between the life drive and libidinal/sexual energy or "Eros" defined as an instinct "exercising pressure towards a prolongation of life."[86] The replenishment of the sexual "active principle" has been deeply rooted in the humoralist tradition. The idea of "gerocomia" (literally "care of the aged," involving the prolongation of life by proximity to young healthy individuals, or sexual stimulation) goes back at least two thousand years. It was prescribed by Galen and practiced by King David (1 Kings 1:2).[87] The strengthening and conservation of the vital/sexual energy, the *chi* (often associated with semen), was the basis of Taoist techniques for life extension.[88] Animal tissue preparations containing a "revitalizing substance" (e.g. from the horns of rutting bucks or animal testicles) were widely used by the Chinese "external alchemists." The German Physician Johann Heinrich Cohausen (1665-1750), distinguished by his meticulous collection of prescriptions for longevity and rejuvenation, *Tentaminum physico-medicorum curiosa* (1699),[89] in the treatise *Hermippus Redivivus* (Hermippus Revived, 1742) speculated on the rejuvenating influences of exhalations from young maidens. He traced this type of gerocomia to ancient Rome and discussed its uses up to his time (in the Middle Ages and the early modern period).[90] Sexual stimulation and conservation of sexual energy were also discussed by the German life-extensionist physician of a later date, Christoph Wilhelm Hufeland (1762-1836), in *Macrobiotics or the Art of Prolonging Human Life* (1796).[91]

Though there has been a long tradition associating sexual vigor with longevity, sex gland supplementation rose to unprecedented prominence in the late 19th century thanks to the work of the French scientist, Charles-Édouard Brown-Séquard. In Brown-Séquard's work, the ancient "revitalizing principle" assumed a very tangible chemical and physiological substance. Serving as the President of the French Biological Society and Chair of Medicine at the Collège de France in Paris (form 1878 to 1894, the post formerly occupied by his teacher, Claude Bernard), Brown-Séquard was a foremost authority in endocrinology, having proven specific effects of internal secretions, particularly those of the sex glands. In the widely publicized presentation to the French Biological Society of June 1, 1889, entitled the "Effects in man of subcutaneous injections of freshly prepared liquid from guinea pig and dog testes,"[92] Brown-Séquard announced his first attempts at hormone replacement therapy for rejuvenation, introducing longevity and rejuvenation research as an integral part of scientific discourse, and in fact establishing the field of therapeutic endocrinology.

In that seminal address of 1889,[93] Brown-Séquard proceeded from the observation that "true eunuchs are remarkable in their feebleness and their deficit in physical and intellectual activity" and the conviction that "analogous defects are observed in men who abuse coitus or masturbate" to the assumption that "these along with numerous other facts, clearly show that the testicles furnish to the blood, … principles which give energy to the nervous system and probably also to the muscles." It followed that the supplementation of these deficits by animal sex gland extracts can retard senility. Their almost miraculous reinvigorating effects on his own person were described by Brown-Séquard, ending with a call for further research. (For Brown-Séquard the reinvigoration was not lasting, he died 5 years later, at the age of 77.)

Brown-Séquard's line of reasoning will reappear in the writings of his most loyal followers – Eugen Steinach and Serge Voronoff. Contemporaries met Brown-Séquard's suggestions with mixed admiration and distrust. Some critics attacked the author personally. Thus, *Deutsche medizinische Wochenschrift* wrote "[Brown-Séquard's] fantastic experiments with testicular extracts must be regarded almost as senile aberrations." And the *Wiener medizinische Wochenschrift* echoed, "The lecture must be regarded as further proof for the necessity of retiring professors who have attained their threescore years and ten."[94] Other criticisms were more scientifically grounded. Thus, Dr. Amédée Dumontpallier (1826-1899) noted in 1889 that the results of

Brown-Séquard's experiment may be due to a non-specific stimulation: "M. Brown-Séquard better than anyone, knows that common peripheral irritations, more or less repeated, non-inflammatory irritations in many physiological and therapeutic experiences often show dynamic effects manifesting themselves by some degree of return of major (physiological) functions."[95] In other words, no matter what you inject, the body will respond by general stimulation. Another, the most common, objection against Brown-Séquard's experiments, was that the results were possibly due to auto-suggestion.

In the initial address, Brown-Séquard anticipated these concerns:[96]

I hope that other older physiologists will repeat these experiments and demonstrate whether or not these effects in me depend only on my idiosyncrasy… The interesting work of Dr. Hack Tuke is full of facts showing that the majority of the changes which I have observed in myself, after the injections, can be brought about solely by suggestion in the human organism. I wish not to deny that a part at least have occurred in that manner, but since they follow the introduction into the organism of a substance capable of producing them, it must be granted that the injections have at least contributed to their origin.

He later defended his method with greater confidence. He ruled out non-specific stimulation, since the effects of the testicular extracts were constant and lasting. Further, he quoted similar case studies by other investigators to negate the possibility of auto-suggestion. For example, in a controlled experiment, Dr. Gaston Variot injected "blood tinted water" (with no effect) followed by injection of animal testicular fluid (with a positive reinvigorating effect) – "without the patient's knowledge [!]". Brown-Séquard concluded that "the works I have cited and numerous others no longer permit me to suppose that the effects I observed in myself depended partly or entirely on one special idiosyncrasy or an auto-suggestion."[97]

He further refuted priority challenges from other users of glandular extract therapy (or as it was then commonly termed "organotherapy" or "opotherapy," from the Greek *"opos"* – juice). The opotherapists' methods of tissue preparation, he maintained, were unsound and destructive, their applications were narrow and bordered on "charlatanism."[98] They "exploit the deep desire of a large number of individuals" and expose them to "greatest risks."[99] It remained the heavy task of Brown-Séquard's followers, in particular Voronoff and Steinach, to rule out auto-suggestion, exclude the treatment's non-specificity, and validate the experimental methods.

Despite the criticisms, Brown-Séquard's work was recognized as trail-blazing, by contemporaries and successors. Even his opponent, Dumontpallier, admitted that Brown-Séquard's work was "extremely interesting" and potentially "of tremendous significance." In that work, Brown-Séquard first explicitly introduced the concept of Replacement Therapy, the "supplementation of the deficit from inadequate spermatic secretion." Without venturing any figures for a maximally attainable longevity, he formulated the fundamentally pro-active ideology of life-extensionism:[100]

They show great ignorance who maintain that it is impossible in old men to reverse their organic state so that they resemble that of an earlier age, especially since the organic changes resulting from better nutrition are possible at all ages… Critics of my ideas have said that it is well known that senile degeneration and wasting present insurmountable obstacles, especially return of neural center function both in the sensory and the motor apparatus. A study of the excellent work of Charcot on Aging (*Studies of Diseases of Old Men*, Paris, 1868) and a number of other works show that nothing about senility is constant nor absolutely characterized. … If the degenerations, if the senile alterations are diseases, a day will come when it will be possible to cure them.

At the fin-de-siècle, Brown-Séquard stood at an epicenter of ideological and scientific debates on life-extension. A decided materialist, he loathed to embrace "ethereal" mind-over-body influences, and struggled to entirely rule out auto-suggestion. He fought both against powerful opponents in the scientific establishment and against popular opotherapy practitioners unaffiliated to the academia, abjured by him as "charlatans." The sage President of the French Biological Society strove to establish a movement within the academy, rather than to incite a revolt against it.

Another point of debate was the importance assigned by Brown-Séquard and his successors, Steinach and Voronoff, to sexual function. In many earlier works on longevity – such as that of the Italian preacher of moderation Luigi Cornaro (1467-1566, *Discorso sulla vita sobria* - Discourse on a sober life, 1566) or the Flemish Jesuit priest Leonardus Lessius (1554-1623, *A Treatise of Health and Long Life - Hygiasticon*, 1613) – sexuality was brushed under the carpet. In the later *Macrobiotics* by Hufeland (1796), only one of its 19 chapters was directly dedicated to sex life, and basically amounted to an advice on sexual moderation.[101] In Brown-Séquard, and later Steinach and Voronoff, sex function assumed the absolute centrality.[102] The "rejuvenators" had both to assert the "sex-centric" perspective and pay tribute to conservative mores, by affirming the value of moderation and by recurrently emphasizing that sexual reactivation is mainly intended for the general physical and intellectual reinvigoration, and not primarily for a pursuit of sexual pleasures.

11. Rejuvenation by Grafting – Serge Voronoff (1866-1951)

Perhaps the most ardent follower of Brown-Séquard was Serge (Samuel) Abramovich Voronoff. Born in 1866 near Voronezh, Russia, a son of a wealthy Jewish manufacturer, he immigrated to Paris in 1884 at the age of 18 and became a naturalized French citizen in 1895. (The drive and ability to adapt in the new country must have been strong, as Voronoff was often said to have become "more French than the French."[103]) In *Rejuvenation by Grafting* (1925),[104] Voronoff related the history of his method. While serving as a personal physician of the Egyptian viceroy Abbas II, in 1898 he observed the enfeeblement of eunuchs, which led him to believe in the invigorating and rejuvenating power of the sex glands. In 1913, he began experimenting with tissue grafting, first perfecting the technique at College de France in Paris (the workplace of Claude Bernard and Brown-Séquard) and at the Rockefeller Institute for Medical Research in New York. In 1917, he moved on to animal experiments with sex gland transplantations (ovaries, but predominantly testes). The technique basically involved grafting pieces of testicular tissue into the scrotum (not replacing the entire organ as might be imagined). After testing the technique in animal models, in 1920 he started male sex gland transplantations in humans for the explicit purpose of rejuvenation and life prolongation. Between 1920 and 1923, 52 such testicular grafting operations were performed: in one case the graft was taken from man (a homograft), in all the other cases from apes (xenografts). In 1930, Voronoff reported 475 cases of testis transplantations from apes.[105] Refinements on Voronoff's technique were offered by his collaborators: Édouard Retterer, Placide Mauclaire, Lois Dartigues, Raoul Baudet and others. The sex gland tissue grafts were intended to renew the supply of "revitalizing" sex hormones, to provide a renewed "source of life."

Voronoff thus became a pioneer of live tissue transplantation in humans, a fact which is seldom acknowledged. Even though his name has remained widely known (especially in Russia, proud of the fame of its native son[106]), it is carefully omitted from Western medical histories. In the rare cases it is mentioned, Voronoff's technique is either vilified or marginalized. Voronoff's side of the story remains veiled.[107] A sympathetic scholar of prolongevity, Gerald Gruman (1966)[108] suggested the possible reasons for a marginalization of longevity research. One is the long tradition of reconciliation with mortality. Another, perhaps even more powerful explanation, is the failure of many rejuvenation and life-prolongation attempts (such as Voronoff's), earning the field a stigma of charlatanism:

> Another reason is the fact that there are few subjects which have been more misleading to the uncritical and more profitable to the unscrupulous; the exploitation of this topic by the sensational press and by medical quacks and charlatans is well known. Furthermore, the past fifty years have seen the failure of at least three highly publicized remedies for aging: at the turn of the century, there was the fermented-milk fad [Metchnikoff's proposition]; in the 'twenties, there were transplants of sex glands [Voronoff's method]; and, in the 'forties, there was the cytotoxic serum advocated by Bogomoletz.

Yet, elsewhere Gruman acknowledged Metchnikoff's crucial role in establishing gerontology as a scientific field. Bogomolets's role in developing immunotherapy and in the institutionalization of gerontological research, and Voronoff's role in developing hormone replacement therapy and transplantation, as well as their contribution to the philosophy of life-extensionism, should also be acknowledged.

In the works *Rejuvenation by Grafting* (first published in English in 1925), *The Conquest of Life* (1928), and *The Sources of Life* (1943), Voronoff justified the pursuit of longevity. According to Voronoff, Death cannot be reconciled with, and the goal of human progress is to preserve life for as long as possible, by all means afforded by human ingenuity:

Would it be possible to instill increased vitality into the weakened organs and exhausted tissues of a body worn out by age? Up to quite recent years scientists did not believe that this question could be answered in the affirmative, and they always advised poor humanity, thirsting with a desire to live, that it should accept old age without hope and die without revolt. This is the law of Nature, we were told, and we must submit to it. ... Our minds cannot accept this verdict any longer. All human progress is due to the Triumph of Man over Nature...To subdue Nature is to ensure the progress of Humanity. ...We now have greater ambitions; we aspire to bring life itself under the domination of our will.[109] The ideal towards which all our efforts are tending is to preserve life with the plentitude of its physical and intellectual manifestations, to abridge the duration of old age, to put off Death to the last limit. LIVE YOUNG![110] (Voronoff's emphasis)

In Voronoff, the immediate aims set for human longevity range from 120 to 140-150 years. A fairly standard arsenal of arguments is deployed to prove that radical life extension is possible and that human progress leads in the direction of life prolongation: medicine has advanced rapidly and brought a multitude of diseases under control; senescence is not 'written in stone' but is due to identifiable and potentially controllable material causes; death is not an indispensable part of life, since some (unicellular) life forms are immortal, moreover the primordial ability to regenerate is retained in some degree in the human organism, and can be improved. Voronoff was apparently not religious, according to him, "Religion itself brings us only trifling consolation."[111] Life-extensionism was, however, not exclusively associated with atheism or agnosticism. (Arnold Lorand, for example, was deeply religious and saw Faith as a life-prolonging measure.) Rather than plunging into religious disputes, Voronoff asserted that life-extension is a universal value: "Believers and unbelievers alike call on God or on Science to prolong their existence on earth, and spare them the pitiful infirmities of old age."

The key word in Voronoff's writings, as in those of great many other life-extensionists (Finot, Metchnikoff, Lorand, Brown-Séquard, et al.) is *"Equilibrium"* – both bodily and social. It is the Balance, the Stability, that Voronoff strove to preserve. The social metaphor for the body is unambiguous and is almost identical to Metchnikoff's and Metchnikoff's pupil Sergey Metalnikov (another Russian emigrant working in Paris at the period).[112] In Voronoff, as in Metchnikoff, the "specialized, productive" elements contribute to the prosperity of "the whole society," but sacrifice their ability to regenerate, their "own means of resistance," and their harmonious, regulated functioning is threatened by "primitive elements":[113]

> Thus, as in human society, there is established in the human organism a selection, a hierarchy between the various elements which constitute it, from the humble intestinal cell which, so to speak, prepares our daily bread, up to the delicate and highly perfected cells of the cerebral substance which co-ordinate the labor of all the artisans of our organism, stimulating some, checking others and forming a kind of Roman senate, which governs our cellular republic....By the side, however, of all these more or less perfected and specialized cells, by the side of these industrious citizens, each following a special craft, are beings that are incapable of accomplishing any function requiring a professional education.... Sturdier than any of the other cells, they continually encroach upon the places occupied by the noble cells, which, sooner or later, wear themselves out... The conjunctive cell ["the plebeian"] brings into an organized society a kind of anarchy which causes its death.

In agreement with Metchnikoff's model of "Subtraction and Addition," Voronoff believed that a prolonged organic balance can be achieved by "attenuating" the "primitive elements" and/or "strengthening the noble elements." Voronoff ultimately depreciated the "Subtraction" and valorized the "Addition," claiming that "it is the connective tissue, in fact, which forms the supporting framework for the other tissues and serves them as an intermediary for the passage of liquid nutriments." It follows that "we must not, therefore, find a means to destroy the conjunctive cells, as we have to do in regard to harmful agents which come from the outside." In contrast, all efforts should concentrate on "reinforcing the noble cells, increasing their vitality and their resistance. … to stimulate some, to replace others which have grown old and become worn out, is to render ourselves masters, so to speak, of our lives."[114]

Voronoff further emphasized that the future of medicine and life-prolongation lies with "addition," with supplementation or replacement therapies, and perceived his own technique as such a preliminary method.[115] He made every exertion to present his sex gland grafting technique (quite invasive as it was) as minimally invasive, as just restoring the natural state of balance of the body, by no means upsetting the status quo. Both Voronoff's socio-biological metaphors and his insistence on maintaining a "natural state" of the body, demonstrate that for Voronoff (as well as for Metchnikoff and Brown-Séquard, and, as will be exemplified further on, for the vast majority of life-extensionist authors) the conservation of stability and order, both biological and social, was a primary aspiration.

Voronoff was a successful capitalist. His family wealth allowed his education abroad. The position of Abbas II's personal doctor was well paid. When starting the gland grafting experiments in the 1910s, he was already a prominent, wealthy Parisian physician. The human graft operations were also quite lucrative, the case studies including mostly middle to upper class clientele. The inheritances from his first two wives added more millions to his estate, more than enough to buy a castle in the Italian Riviera (Château Grimaldi).[116] Life-extensionism thrived in prosperity. And the social order Voronoff strove to maintain was the capitalist order. Perceiving the body in terms of capitalist economy, Voronoff recurrently spoke of the body being "in need of new capital with which to defray the expenditures inseparable from life. The idea of bestowing fresh capital upon these bankrupts is of relatively very recent origin."[117] Yet, Voronoff expressed a poignant concern that replacement therapies should not be restricted to a narrow elite. He sought 'expanding the market,' making the therapy more accessible. In the 1920s, life-extensionism was often accused of elitism, of being a pursuit for the rich. It was commonly feared that replacement therapies would never be widely shared.[118] With the advent of synthetic hormones, however, the accessibility of hormone treatments expanded.[119] It was Steinach who envisioned that replacement therapies are to become more accessible through the development of cheap synthetic hormones.[120] Voronoff, rather, hoped salvation would come from the use of animal organs. He saw organ transplants from humans as hardly attainable, most of all, due to the unwillingness of potential donors. The use of organs from accident victims appeared to him ethically and legally unacceptable: "this method is perhaps inevitably forbidden by law." Obtaining glands from people on the death bed in hospitals or "from people under sentence of death" too seemed to him "so difficult of realization that they could not be counted upon to afford a practical solution." Hence, organs from animals seemed a most feasible means "which would assure the benefits of the graft *to all who would profit by it*" (emphasis added).[121] Thus, Voronoff sought to avoid the threat to the social stability that might arise due to an unequal access to the rejuvenation procedures, but he sought this within the framework of the current capitalist paradigm which he never wished to upset. The propositions for a wider access were based on a 'market expansion model' and on the present set of legal statutes that he did not think could ever change.

12. Rejuvenation World Wide. France and Austria – the seedbeds for the rejuvenation movement. Eugen Steinach (1861-1944)

Sex gland stimulation, supplementation and replacement formed a mainstay of the life-extensionist movement of the 1900s-1920s. Such interventions indicated that the physiological state of the elderly could be manipulated, and some form of "rejuvenation" or restoring the functional capacity and productivity, could be, at least temporarily, effected. Beside Voronoff, analogous methods were exploited by a great number of physicians. In 1919-1920, Dr. Leo Leonidas Stanley performed sex gland transplants among prisoners of San Quentin penitentiary, California. Since the 1910s, human testes were also grafted by the Chicago doctors Victor Darwin Lespinasse, George Frank Lydston and Max Thorek.[122] At the same time (c. 1918-1920), John Romulus Brinkley of Kansas transplanted goat testicles into humans (according to Voronoff, this was "going much farther than I have gone" and ineffective, since the grafted testicles from goats and rams "become absorbed" in humans). Voronoff did not seem bothered by priority disputes and asserted that science is a cumulative, collective enterprise:[123]

The notion of grafting in general, and of testicular grafting in particular, is not entirely new. So many ideas have emanated from the human brain that it is always difficult to award to any one person the palm for absolute originality. What is new in the idea is its practical application, and in this the last hundred years have shown a marvelous development.

A host of prominent physicians worked on sex gland transplants:[124]

As early as 1767, Hunter grafted the testicles of a cock into a hen. The experiment was repeated by Berthold in 1849, Philippeaux in 1858 and Montagazza in 1864. In our day, Steinach of Vienna holds the first place; he performed testicular grafts on rats... Pezard has grafted testicles into hens and castrated cocks.... Experiments on cocks, guinea-pigs and white rats have been carried out by Zavadovski, Knud Sand, Lipschutz, Nusbaum, Maissenheimer, Cuthrie, Mushan, and Payer. Mauclaire, the author of a remarkable work on grafting, Lichtenstern, Thorek, Jahnu, Hammond, Enderlen, Sutton, Lespinasse, Morris, Lydston, McKennen, Stanley, Keller, Lissmann, Gregory, Mariotti, and Falcone have made grafts with human testicles which they were able to obtain from the human subject, under special circumstances similar to my own experience with a cryptorchid testicle.

In the 1920s, "endocrine rejuvenation" became a world wide movement, spreading from the US to Russia. Yet, France and Austria emerged as the primary and competing rejuvenation superpowers. Beside Voronoff, a crucial figure in the movement was the Viennese physician Eugen Steinach. Life-extensionism in German-speaking countries will be discussed in greater detail in the subsequent chapter. Yet, the centrality of Austrian rejuvenators, particularly their patriarch Eugen Steinach, needs to be emphasized from the outset, as the "Steinach operation" was a most widely publicized method of endocrine/sexual rejuvenation in the 1920s beside Voronoff's sex gland transplantations.

First performed in a human patient on November 1, 1918, the Steinach operation involved the ligation of seminal ducts ("vasoligation" or vasectomy) to suppress the sperm-producing activity and thereby to stimulate the hormone-producing activity of the "interstitial tissue" of the testis. Such enhanced sex hormone production was assumed to effect "rejuvenation," "revitalization" or "reinvigoration." The general rejuvenating effects were ascribed by Steinach to the whole-body increase in the blood flow (hyperemia) produced by the sex hormones (though Steinach recognized other methods of blood flow increase, such as diathermy, massage, exercise and baths). To avoid infertility, the operation was commonly performed only on one of the two spermatic ducts. This method enjoyed an incredible vogue. William

Butler Yeats (1865-1939) and Sigmund Freud (1856-1939) were among its beneficiaries.[125] Steinach specified the methods in great detail in *Verjüngung durch experimentelle Neubelebung der alternden Pubertätsdrüse* (Rejuvenation by Experimental Revitalization of the Ageing Puberty Gland, 1920). The results were fully summarized in his *Sex and Life. Forty Years of Biological and Medical Experiments* (1940).[126] The latter book was written in Zurich, where Steinach, having Jewish ancestry, had to remain until his death after Austria's annexation in 1938.

The technique emerged from Steinach's animal experiments, first reported in 1910, that determined the role of sex gland secretions for the appearance of secondary male and female sex characteristics. Steinach further proceeded to attempt sex gland surgery "to heal the effects of castration and homosexuality"[127] – making a fundamental contribution to general endocrinology and, among other implications, establishing the foundation for later "sex change" operations.[128] Yet, insofar as Steinach was also interested in the duration of the sex hormone action, his experiments expanded into attempts to prolong this action by surgery. These gave birth to the techniques of "Rejuvenation" (Verjüngung) or "Combating Aging" (Altersbekämpfung) either by "autoplastic" surgery (e.g. vasoligation, on the patient's own sex glands to stimulate hormone secretion) or "homoplastic" surgery (e.g. implantation of sex gland tissue from a donor). In men, the vasoligation of *vas deferens* was the primary "auto-plastic" procedure; whereas women were "rejuvenated" by fallopian tube ligation, transplantation of ovarian tissue or irradiation of the ovaries. Steinach's conclusions after the initial ten years of research (in 1920), even though rather cautious, were nonetheless highly optimistic. "Senescence, within certain limits, can be influenced," he asserted. In men "premature deterioration can be fought against," and in women "rejuvenating effects" can be produced. Following these experiments, Steinach was nominated for a Nobel Prize in 1922 for the "work on transplantation of reproductive glands and particularly on rejuvenation." In fact, Steinach was nominated for the Nobel Prize 11 times, from 1921 to 1938, but only in 1922 he was nominated with specific reference to "rejuvenation."[129]

Steinach embarked upon the rejuvenation research program in 1912, which since then became his main subject and expanded drastically.[130] After the first vasoligation on the "prematurely aged" 44 year old "A.W." in November 1918, thousands of operations were performed by Steinach and by the growing number of his followers worldwide, in the first place in Steinach's native Austria.[131] Steinach vigorously championed the priority of Austria in all matters of sexual rejuvenation, renowned by the works of Erwin Last, August Bier, Karl Doppler, Emerich Ullmann, Paul Kammerer, Robert Lichtenstern, Otto Kauders, Gottlieb Haberlandt and others, and in no small measure his priority. (Patriotism and valorizing one's own cohort may have been important elements in Steinach's social adjustment.) And so Steinach did vigorously valorize the "autoplastic" vasoligation that, according to him, was much more "natural" and less subject to "rejection" by the host than the "homoplastic" sex gland transplants that were widely practiced by the French Voronoff. In other parts of the Austro-Hungarian Empire, there were rejuvenators and life-extensionists too: Zoltán von Nemes-Nagy and Istvan Szabo in Hungary, Arnold Lorand and Vladislav Ruzicka in Czechoslovakia. But Austria, particularly "liberal Vienna," occupied the center stage, in the decades before and after the breakdown of the Empire in 1919-1920.

For Steinach, the sense of vindication was long delayed. Apart from the fierce disputes and competition with other rejuvenators, Steinach had to withstand massive opposition on the part of the medical establishment: "It appeared [in the period 1910-1920]," he recalled in 1940, "as if the whole theory and problem of reactivation and the fight against old age were to be outlawed by official science." Steinach, the Director of the Biological Institute of the Viennese Academy of Sciences, could not afford to be implicated in quackery or violation of the scientific method. Like Brown-Séquard and Metchnikoff, Steinach worked to conquer the academy from within rather than subvert it. "Rather than waste energy on endless literary controversies," he related, "I preferred to widen the circle of my friends and followers by establishing one compelling proof after another, thereby disarming criticism and prejudice."[132] While Voronoff claimed that a primary motivation for his research was to find a means of rejuvenation, Steinach asserted that his work started as purely basic research which gradually revealed a possibility of "reinvigorating" the aging body.

Like Brown-Séquard and Voronoff, Steinach had to contend with the accusation that the results in patients were due to auto-suggestion. Steinach confronted the accusation by a wide range of animal experiments and by devising an "objective" test of activation (according to creatine levels). Like Brown-Séquard and Variot (long before the Helsinki guidelines for human experimentation of 1964), he often performed the surgery without the patients' knowledge (during a concomitant procedure) and then observed the "objective" rejuvenating effects.

Another very common objection against Steinach's methods, in fact against the majority of rejuvenation attempts by sex gland stimulation, was that the observed results were not those of "true rejuvenation" but only of "erotization." The semantics had practical implications: by substituting the terms, even highly positive therapeutic results could be dismissed, and the entire program of rejuvenation research undermined. Steinach argued against the identification of "general revitalization" with "improvement of sexual function":[133]

> The real significance of the hormonic hyperemia which is induced by vasoligature is proof of the falsity, and even malice, of the claim that the effects of sex hormones are confined to the field of sexuality, and that the methods of reactivation are, in fact, no more than a revitalization of sexuality. One cannot say that the contrary is actually the case; but the improvement in the sexual functions is only a small part of a general revitalization which embraces the whole of the organism. For example, the weight increases, the skin becomes smooth and supple, there is improved growth of hair, and the senses are remarkably sharpened.

According to Steinach, such a restoration of the physiological and functional state of the aged *is* rejuvenation and *does* enhance longevity.

Multiple forms of rejuvenating sexual stimulation were investigated in the 1920s worldwide – by Jürgen Harms (Germany), Harry Benjamin (US), Norman Haire (UK), Boris Zavadovsky (Russia), Knud Sand (Denmark), León Cardenal (Spain), Vincenzo Pettinari (Italy), Dimu Kotsovsky (Romania), Paul Niehans (Switzerland), Ottmar Wilhelm (Chile), Yasusaburo Sakaki (Japan) – to name a few among dozens. Diathermy or tissue heating was widely employed for invigorating the blood flow (hyperemia induction) in the body generally and in the sex organs particularly. The diathermic heating and hyperemia were mainly induced by electric currents to the tissue (as in the practice of Erwin Last of Vienna), but also by a variety of other means: elastic bandages, suction cups, hot fomentations and blistering agents, or hot air (August Bier, Vienna). In the late 1920s, Dr. Karl Doppler of Vienna practiced chemical "Sympathectomy" by irritating the spermatic cord with toxic phenol. In this way, the sympathetic nerves of the testis' blood vessels were damaged, and an increased blood flow into the testis produced. This procedure became extremely popular, especially in Austria, Germany, France, Italy, Argentina and Spain.[134] In Italy, rejuvenation by injection of testicular blood serum was attempted by Francesco Cavazzi.[135] Naum Lebedinsky (Latvia) explored smashing a part of an animal testicle to stimulate the rest of it.[136] Rejuvenation was a "male-dominated" endeavor: the studies mainly focused on male sex glands. Experiments on female sex gland transplantation, stimulation and supplementation were performed as well (by Steinach and others) though with less reported successes and in a much lesser scope.

Though "endocrine rejuvenation" was practiced world wide, and Austrian researchers were very prominent, France was still one of the most dominant centers of the movement, where rejuvenation was practiced among the widest and earliest. Yet, as it later became evident, the actual effectiveness of the rejuvenation techniques fell far short of their initial promise. As will be demonstrated below, as the "endocrine rejuvenation" started among the earliest in France, so its popularity ended there among the earliest. This loss of popularity may be explained not just as due to specific technical shortcomings of the rejuvenation methods, but perhaps also due to a perceived failure of their underlying reductionist theoretical basis.

13. The reductionist roots of French life-extensionism

Since the mid-19th through the early 20th century, reductionism and materialism, essentially viewing the human body as a machine in need of repair, underscored most of the ground-laying works of French life-extensionists and rejuvenators. Thus, the primary methodology of the French founders of *Médecine de Vieillards* (medicine of the aged) and anti-aging interventions in the mid-19th century – Charles-Louis Durand-Fardel, Jean-Martin Charcot and others – was dissection. Through the autopsy dissection, they established the age-related pathological changes in separate organs and tissues, in fact, reducing the entire process of aging to these specific organ changes. Though Durand-Fardel spoke of the life-span being determined by a "vital principle of limited duration"[137] – finding tissue-specific degeneration was for him by far more determinative. One of the most prominent French researchers of longevity of the mid-19th century, Marie-Jean-Pierre Flourens (1794-1867), asserted that "Just as the duration of growth, multiplied a certain number of times, say five times, gives the ordinary duration of life, so does this ordinary duration, multiplied a certain number of times, say twice, give the extreme duration. A first century of ordinary life, and almost a second century, half a century (at least) of extraordinary life, is then the prospect science holds out to man" (1854).[138] He thus determined that human longevity is preset by a mechanical body buildup, as if by winding of a clock-work. And the main line of Flourens' research on the localization of brain functions (the field he in fact founded) was an epitome of reductionism. Furthermore, one of the first modern scientific theories of aging, proposed by Édouard Robin of the French Academy of Sciences in 1858, posited that aging is due to body "mineralization" or accumulation of "alkaline residues," "calcification" or "ossification." Robin's theory considered lactic acid and "vegetable acids" as possible means to dissolve the "mineral matters" and thus prolong life. Thus, the body was essentially viewed as a rusting and clogging machine subject to a cleanup.[139]

The teachings of both Brown-Séquard and Metchnikoff were profoundly materialistic and reductionist. Brown-Séquard suggested that through the supplementation of deficient hormones, bodily equilibrium can be restored, youth returned, and life prolonged. Brown-Séquard's chief concern was to rule out any psychosomatic influences and to reduce medical intervention to a subtraction or addition of matter. Metchnikoff, in turn, divided the body into "noble" or "functional" elements and "primitive" or "harmful" elements: the former needed to be strengthened or replenished, the latter destroyed or attenuated. According to Metchnikoff, the direct effects of the mind on the body were limited to "some nervous disorders."[140]

Materialism and reductionism underwrote the work of yet another prominent *fin-de-siècle* French life-extensionist, the social scholar Jean Finot. In *The Philosophy of Long Life* (1900), Finot did speak of "Will as a means of prolonging life," yet for him reductionism held the key for understanding, manipulating and extending life. In Finot's philosophy, biology is reducible to chemistry and physics, and the complexity of a living organism is reducible to an interrelation of its components. Such a reduction, according to Finot, opens the possibility for engineering life, and eventually for life's indefinite maintenance:[141]

What is the life of a man? The result of the lives of millions of plastides. For each plastide lives its own life, and there are even cases in which the man dies whilst the plastides composing him continue to live. Now, biology proves to us that among the phenomena observable at a given moment in a living plastide there is none which has no affinity to physics and to the chemistry of inert bodies. Nothing in them permits us to separate them from the body of elements already studied and *possible of reproduction*. (Emphasis in the original.)

Through "fabrication of living matter," Finot believed, sentient, immortal beings can be created.[142] (Finot was apparently among the first to discuss this possibility in terms of modern biology and organic chemistry, far in advance of the emergence of the Transhumanist intellectual movement.[143]) Even if the appearance of the "homunculi" may be too remote, the progress of reductionist biology will surely enable life-enhancement and life-extension:

> The possibilities of nature are infinite, as [Thomas] Huxley has so justly said. Nothing then authorizes us to doubt that the intensity of life will be some day rendered more powerful by science. It may not perhaps succeed in creating new life. No matter, so long as it can preserve and greatly strengthen existing life. And that will be enough.

Equally in favor of materialism and reductionism were Metchnikoff's French followers, Albert Dastre and Sergey Metalnikov. According to Albert Dastre (1844-1917), Claude Bernard's pupil and Chair of the Department of General Physiology at the Sorbonne, reductionism opens the possibility to profoundly manipulate life's components, since biology "is a particular chemistry, but chemistry none the less. ... The vital action is not distinct in basis from physicochemical action, but only in form."[144] Subjected to physicochemical manipulation, human beings may "remain forever in full health and guarded from disease."[145] And according to Metchnikoff's pupil at Institut Pasteur, Sergey Metalnikov (1870-1946), senescence and death arise from a disharmony of differentiated body components. (This fundamental tenet was shared by the majority of *fin-de-siècle* theorists of aging, from Weismann to Metchnikoff.) The basic unit of life, the cell, however was seen as potentially immortal. Therefore, according to Metalnikov, the body, composed of such potentially immortal units, can be made potentially immortal, if only learning to bring the components into harmony.[146]

Essentially, reductionism formed the theoretical basis for rejuvenation attempts. The "father" of rejuvenative replacement therapy, Brown-Séquard, vigorously defended the specificity of sex hormone injections and countered critics (such as Dumontpallier) who argued that their effects are due to auto-suggestion or non-specific stimulation (i.e. that any injection would produce the same stimulating effect on the body). Brown-Séquard's follower, Serge Voronoff, continued in his master's footsteps. Voronoff recapitulated the basic theoretical premise that the deterioration of aging is due to an imbalance of the components comprising the body machinery, and that the balance can be restored through supplementing or replacing failing components. Voronoff's methodology of sex gland "grafting" was reductionist almost by definition: by substituting a single crucial element in the body mechanism, its 'run-time' could be increased. The grafting did affect the whole organism, but the effect was believed to be analogous to replacing an energy carrier (a principal component or a "battery") to sustain the operation of the entire machine.

The grafting appeared to be a novel, unorthodox intervention, but Voronoff's (as well as Steinach's) central claim was that their operations restored the natural equilibrium of the body. The ultimate aim of rejuvenation techniques was, in accordance to Claude Bernard's dictum, to achieve "the fixity of the internal environment" which is "the condition for free life."[147] And the means to attain this fixity were through supplementing or replacing those components whose deteriorative change would otherwise threaten the overall body stability. Reductionism might thus be pivotal to the rejuvenation enterprises of the early 20th century: the supplementation or replacement of an isolated component appeared to the rejuvenators a feasible task, perhaps more feasible than attempting to comprehend and/or manipulate the whole of the human reaction to the whole of the external environment. In later assessments, however, reductionist rejuvenation techniques did not appear to live up to their promise.[148] With regard to Voronoff's method, the problem of graft rejection by the host appeared almost insurmountable. Replacing or supplementing a single gland did not appear to durably forestall the deterioration of the entire organism, and no conclusive evidence for extending the life-span by such means was offered. Consequently, as we shall see, a recoil from immediate rejuvenation attempts occurred.

14. Life-extensionism in France in 1930-1950. The dissolution into the Whole

As we shall see, in the 1930s-1940s, an increasing number of French life-extensionists began to espouse holistic perceptions of the unity of the mind and body, and the subordination of an individual to society, gradually replacing the earlier prevalent notions of reductionism, physicalism and individualism. The holistic tone in French life-extensionism of the 1930s-1940s was set by the leading French longevity researcher, Alexis Carrel (1873-1944). Echoing Auguste Comte, who asserted that "No sound treatment of either body or mind is possible, now that the physician and the priest make an exclusive study, the one of the physical, the other of the moral nature of man,"[149] Carrel urged:[150]

> Man is much more than a sum of analytical components. One has to embrace at the same time both the parts and the unity of man, because he reacts like a unit, and not like a multiplicity, to the cosmic, economic and psychological milieu. The solution of grand problems of civilization depends on the knowledge not only of different aspects of humanity, but of the human being as a whole: as an individual within a group, a nation and a race. This is the true science of man…. The conquest of health is not sufficient. It is the progress of a human person that is sought, because the quality of life is more important than life in and of itself.

Such a holistic vision moved away from the earlier materialistic and reductionist proclivities of French life-extensionists.

As we shall see in the following sections, the recoil from reductionist rejuvenation took many forms. First and foremost, since immediate rejuvenation appeared at the time untenable, the scholarly focus seems to have shifted to basic research of aging, predicated on the assumption that only after a comprehensive, lengthy and costly investigation, the complexity of the aging processes can be gradually unraveled, and consequently, in some distant future, actual life-extending interventions may be found. The concepts of bodily "equilibrium" of the early rejuvenators were rather qualitative and vague, and it was necessary to establish precisely what components and quantities constitute "steady states" or deviations from them. To enable such an extensive research, scientific collaboration and public support were deemed necessary, and, since the 1930s, the process of institutionalization of aging research, first national and later international, took place. The first institutions for aging research, in turn, became parts of the concurrent social "equilibria" – adapting to various state regulations and ideologies. The practical aims of aging research became more modest: rather than attempting to effect an immediate and thorough rejuvenation, it became more presentable to seek a thorough understanding of the aging process and perhaps some mitigation of age-related diseases. Such an emphasis on basic research and caution in goals have been expressed by many gerontologists since the beginning of the institutionalization process, during the establishment of gerontology as an international discipline in the late 1930s through the early 1950s.

Another form of withdrawal was a rejection of "surgical" means of rejuvenation, in favor of more "natural" improvements in the life style. Yet another was an abandonment of actual rejuvenation research and practice in favor of more literary, science-fictional or philosophical treatments of the subject of extreme longevity – a futuristic discussion of its potential ethical and social impacts.

Yet perhaps one of the central forms of withdrawal from rejuvenation attempts appears to have been a movement away from their underlying reductionism, and toward a more holistic perspective, emphasizing the mind-body connection. This trend seems to have been salient in the "post-Voronoff" French longevity research community, and was epitomized by the work of Auguste Lumière, the crusader for the revival of "humoralist" medicine[151] and Alexis Carrel, the protector of "wholeness."[152]

15. Rethinking rejuvenation – Auguste Lumière (1862-1954)

One of the leading actors in the "holistic turn" in French life-extensionism of the 1930s-1940s was Auguste Lumière. Auguste and Louis Lumière (1864-1948) are renowned for the creation of cinematography in 1895. Perhaps less known is their deep involvement in biomedical research, especially that of Auguste. A fundamental contribution of the brothers Lumière to medicine was their pioneering work on medical photography.[153] In the late 1890s-early 1900s they were the first to produce high quality color medical photographs. Their invention of auto-chrome plates for medical photography proved indispensable for medical instruction, anatomic, microscopic and histological examinations, embryological and bacteriological research. This work paved the way for all future biomedical imaging, a contribution on a par with Wilhelm Roentgen's discovery of X-ray radiography in 1895. The photo-stereo-synthesis plates, developed by Louis Lumière, created an impression of image depth and thus contributed to the origin of tomography. Cinematography itself was enlisted by the Lumières to the service of medical research, the first medical film having been produced by Auguste Lumière in 1895. The contribution of the brothers Lumière to biomedicine was not exhausted by imaging: during WWI Louis constructed articulated arm prostheses for amputees, while Auguste developed a non-adherent anti-septic bandage (*tulle gras*) that dramatically reduced the time of wound healing, introduced oral anti-typhoid vaccination, anti-tetanus serum booster (sodium persulfate), and more. Auguste's engagement in biomedicine was especially pronounced. While Louis concentrated more on research and development of photographic technology, Auguste fully dedicated his efforts to biomedical research proper, with a notable emphasis on aging, rejuvenation and life-extension.

For Auguste Lumière, the pursuit of biomedical research was made possible thanks to his excellent financial standing. The Lumière family business (started in Lyon by Louis and Auguste's father, the painter and photographer Antoine Lumière, 1840-1911) was highly successful. By the 1900s it was setting up cinemas around the world and selling millions of photographic plates each year. Being well established financially, Lumière was not only able to endow medical institutions and charities (such as the Hôtel-Dieu Hospital in Lyon), but also to freely pursue his own clinical and basic studies. In 1896 he set up the Lumière Laboratory (among others sponsoring Alexis Carrel's first experiments on blood vessels suture). The laboratory became the Lumière Clinic in 1910 and further expanded into the interdisciplinary Lumière Institute in 1930. The Institute comprised cabinets for medical specialists (from cardiology to dentistry), patient facilities (including a massage cabinet and operating room), laboratories for radiological, photographic, electro-diagnostic, bacteriological, histological and biochemical analysis, and therapeutic facilities, including radiotherapy (irradiation with short, ultra-violet and infra-red waves), electrotherapy (negative ionization or ozone therapy), immunotherapy (artificial fever induction), etc.[154] This was apparently one of the first such interdisciplinary clinic and research institutes in the world, driven by Lumière's vision and financed largely out of his own pocket.[155]

In *Sénilité et Rajeunissement* (Aging and Rejuvenation, 1932), Auguste Lumière somewhat regretfully noted that before the age of 40, due to the preoccupation with the "industrial, technical, administrative and commercial" affairs of the Lumière Company, he had little time for instruction in biomedicine. Only by his 50s he was beginning to catch up on learning, and only in his 60s and 70s he was able to start making a contribution to the field.[156] And what a prolific contribution it was: over 30 books and hundreds of scientific papers! As he recurrently emphasized, formal medical accreditation was not necessary for him to think and experiment. Moreover, formal instruction, according to him, might even stifle the "native faculties of reason, initiative and curiosity" and only foster "memorization."[157] Though he became a member of the Paris Academy of Medicine in 1928, he was not particularly impressed with medical associations, congresses or centralized research centers that, in his view, promoted "dogmatism." According to him, questions in research and therapy ought to be decided on the basis of clinical and experimental evidence, and not according to rank, titles or affiliations. Being self-funded, he did not depend on established institutions – he owned an institution.

As Auguste Lumière was well established financially, so was he well established politically. For decades he was a pillar of the Lyon community, sitting on the Board of Directors of the Hospitals of Lyon and at the Lyon City Council. His loyalty to his country was absolute, whatever the ruling regime may be. He fist served in the army in 1880 (at the age of 18), and after the onset of WWI, at the age of 52, he wished to be mobilized again. During the war, he dedicated an enormous effort both to treating soldiers personally at the Hôtel-Dieu Hospital in Lyon, as well as making massive donations to the army, ranging from vaccinations to bandages to radiography exams. After the Nazi occupation of France and the establishment of the collaborationist Vichy regime under Marshal Philippe Pétain (from July 1940 to August 1944), Lumière's loyalty did not seem to change in the very least. In a letter published in *Le Petit Comtois*, on November 15, 1940, Louis Lumière quoted (and fully endorsed) Auguste's exalting remarks on the "incomparable prestige, the indomitable courage, the youthful ardor of Marshal Pétain, and his sense of reality that must save our homeland."[158] Furthermore, Auguste was quoted as saying in that letter that "In order to achieve that desirable era of European harmony, it is obviously necessary that the conditions imposed by the victor do not produce the ferment of irrevocable hostility against him. And no one could better achieve this goal than our wonderful Head of State, assisted by Pierre Laval, who has already given us evidence of his vision, his ability and his dedication to the real interests of the country." In July 1941, Auguste became a member of the patronage committee of the pro-collaborationist, "anti-Bolshevik" Legion of French Volunteers (Légion des Volontaires Français) affiliated with the French Popular Party (Parti Populaire Français) led by Jacques Doriot. He was noted by the party officials as "our friend at the Municipal Council of Lyon."[159] However, it should be noted that the main bulk of Lumière's scientific work was produced in the 1930s, before the time of Vichy. In the period 1940-1944, Lumière, then in his late 70s-early 80s, continued to research and publish, though on a greatly diminished scale, essentially going "underground." The verbal support of the ruling Vichy regime may have been an act necessary to sustain his long-established social status and ensure the continuation of his research. After the liberation, his social prestige did not diminish either. Thus, the firm socio-economic foundations, the social prestige and prosperity, the advocacy of stability in all spheres, the loyalty to any social regime currently in place (what may be termed 'adaptive conservatism') were inseparable concomitants of Auguste Lumière's life-extensionist enterprise.

To match Auguste Lumière's political conservatism, his scientific theory may be seen as conservative as well. Lumière often described himself as a medical "innovator" struggling against the "ostracism" of the "dogmatic" medical establishment. He saw himself as an heir to the rebellious Austro-Hungarian physician Ignaz Semmelweis (1818-1865), whose sound hygienic advice on anti-sepsis met with fierce opposition and persecutions by the medical guild.[160] But, in fact, Lumière's medical theory and practice were based on the most ancient and widespread medical tradition of all – the "Humoralism." For thousands of years, the balance and stability of different body liquids (or "humors") was seen as the necessary condition for health and longevity. The balance of the three "doshas" (humors) was sought in Ayurveda. The balance of the bodily fluids, of the five elements, or of the soft and flowing "yin" and the rigid and abrupt "yang" was sought in traditional Chinese medicine. The "Four humors" corresponding to the "Four elements" needed to be in equilibrium according to the teachings of Empedocles, Hippocrates and Galen. The balance of the four humors was just as much desired by medieval physicians worldwide, from Avicenna in Persia to Arnold of Villanova in Italy. And the alchemists searched for the "philosopher's stone" whose function was to restore the balance and stability of bodily elements and fluids.[161] Auguste Lumière's most ambitious and pervasive project was "The Renaissance of the Humoral Medicine."[162]

In *La Renaissance de la Médecine Humorale* (1935), Lumière reviewed the earlier humoralist tradition: from Hippocrates and Galen through Jan Baptist van Helmont (1580-1664) and Antoine Lavoisier (1743-1794) to Sigismond Jaccoud (1830-1913). According to Lumière, the classical "four humors" – the blood, the phlegm (pituita or mucus), the yellow bile, and the black bile – are nothing more than "views of the spirit," that is, theoretical or operational constructs that do not correspond to physiological reality, to the actual content and properties of body fluids. Nonetheless, Lumière did emphasize the enormous importance of liquids in physiology. This importance, according to him, was almost entirely overlooked by the triumphant contemporary "solidist" paradigm in biomedicine. Contemporary cytology, histology, microbiology, pathology and anatomy, Lumière claimed, considered almost exclusively rigid structures without due regard to changes in the liquid medium.

It was Lumière's task to revive the humoralist tradition, to create a new "scientific humoralism" that would focus not only on blood, but on liquids throughout the body, especially intracellular and interstitial liquids. Unlike the classical "hypothetical" humoralism, the new "scientific humoralism" was to be based on empirical observations, on measurements of the liquids' density, pH, viscosity, surface tension, electric conduction, light refraction, John Tyndall's and John Rayleigh's particle light scattering, the biochemical content analysis of the salts, proteins, lipoids, urea, pigments, glucose, serological reactions, etc. Despite the modern methods, Lumière's theory can be easily traced back to the ancient concepts of the "congestion" of humors, of the prevalence of "earth" over "water" or "rigidity" over "fluidity" in the humors' composition, and most importantly to the concept of their stable "equilibrium." Lumière admitted that progress can only build and improve on ancient notions (and therefore the study of medical history has a practical utility).

As for the majority of life-extensionist authors discussed so far, for Lumière, the key concepts were "balance," "equilibrium," "stability" and "fixity." He just introduced other constituents of "equilibrium," different from the equilibrium of "essential elements" in Roger Bacon, Paracelsus and other alchemists, or the equilibrium of "cell types" in Metchnikoff, Steinach, Voronoff and great many other life-extensionist physiologists in the first quarter of the century. Instead, Lumière's theory considered the equilibrium and stability of cell colloids – dispersed liquid suspensions of macro-molecules (particularly proteins), or micelloids (colloid droplets). And this is the gist of his theory: When the colloids are stable and balanced (maintained dispersed in suspension), this state is characteristic of health and vitality; but when the colloids become unstable or imbalanced, they precipitate and flocculate – and this is the state of pathology and aging.[163] For Lumière, at the infancy of molecular biology, the colloids were essentially blobs of matter that "congest" or "dissolve," with molecular composition and mechanisms unknown.[164] Nonetheless, he was able to detect the colloids' "stability" or "perturbations" and incorporated that knowledge into a vast explanatory apparatus and therapeutic methodology, including anti-aging strategies.

In the 1920s-1930s, the entropic transformation and condensation (hysteresis) of cell colloids (chiefly represented by proteins), forming flocculates and precipitates that clog the cell machinery, was a dominant theory of aging. It was first proposed in 1913 by the Romanian gerontologist Georges Marinesco (Metchnikoff's long-time scientific opponent) and developed by the Czech physiologist Vladislav Růžička (Ruzicka) who published a series of articles on the subject in the 1920s.[165] However, Marinesco and Ruzicka largely considered the process of colloid condensation, and hence aging and dying, to be quite inexorable, even though they did attempt to retard the process, Ruzicka by using lecithin as a "protective colloid" and Marinesco by using hormonal supplements. Lumière took a more proactive approach, both with regard to acute and chronic diseases, and the aging process. According to Lumière, during the process of normal aging, the colloids become unstable and precipitate (the internal cause of deterioration), but the process can be greatly hastened by diseases (external causes). Therefore, the task of life prolongation, according to Lumière, must consist in combating diseases, macrobiotic health regimens, and rejuvenating interventions proper, all pursuing the same objective – the "stabilization" of cell colloids.[166] Lumière's medical theories and practices, especially in the field of anti-aging, now seem to be almost entirely forgotten. Even in his own time, only one of his books, *Tuberculosis, Infection, Heredity* (1933), was translated into English, and none other since. Hence, a brief exposition is in order.

One of Lumière's favorite "stabilizing substances" (essentially emulsifiers) was Magnesium Hyposulfite ($Mg-S_2O_3$), an anti-shock substance, capable of "reestablishing humoral equilibrium" and "inhibiting the disorganization of colloids, with all the cortege of disorders this entails." It was also supposed to exert a general "desensitizing" (immunizing or tempering) effect on the entire body.[167] A wide variety of other "desensitizing" and "stabilizing" agents were tested and clinically applied by Lumière: "Antibacterial desensitization" (the regular immunization by specific antigens, e.g. Koch bacillus extracts), "Auto-hemotherapy" (injection of the patient's own blood, presumably exerting "desensitizing" effects), as well as a wide assortment of metal compounds: magnesium benzoate, copper glycocholate, sodium undecylate, etc. etc. A special place was reserved for compounds of gold, partly realizing the ancient dream of alchemists. Chrysotherapy, using salts of gold, such as Allochrisine, was employed to stabilize innumerable "humoral imbalances" (and gold particles are still used today in the treatment of rheumatoid arthritis and other diseases, even in "nano-medicine"[168]). "Granulotherapy" or "Anthrotherapy" employed small carbon or other particles, not just to absorb toxins, but to produce a "mild mechanical irritation" in order to achieve general desensitization and stimulate phagocytosis. Changing the blood volume was yet another mechanical means to influence the colloidal state. Changing fluid pressure would change the vasomotor sensitivity to the pressure of flocculates, and consequently affect the sensitivity or immunity to shock. Accordingly, the most ancient methods of "balancing the humors" – the blood-letting or water intake – remained in the arsenal and were given a new rationale: "When the quantity [of blood] is augmented or diminished," Lumière wrote, "the phenomena of shock are no longer produced or greatly attenuated, which confirms the capital importance of vasomotor activity." Yet another therapy dear to Lumière's heart was "negative ionization," using the "aero-ionization lamp" introduced by the Russian biophysicist Alexander Chizhevsky in 1919. In Lumière's view, negative ions presumably stabilize the colloids' electrical charge. Endocrine extracts too were employed for colloids' stabilization, for maintaining the "equilibrium of humors."

By the time Lumière was writing *La Renaissance de la Médecine Humorale* (1935, with the second edition appearing in 1937) and *Les Horizons de la Médecine* (1937), sulfamides (sulfa-drugs, such as "Prontosil") were just discovered by Gerhard Domagk in Germany (c. 1935) and successfully put to use against streptococcal infections. Lumière immediately delved into the testing of the new substance.[169] He emphasized that the sulfamides do not exert their anti-bacterial action *in vitro* (as regular antiseptics do), but only in the entire organism. Somehow the organism's internal environment becomes refractive to the spread of the infection (presumably due to preventing streptococci encapsulation in the body and facilitating their capturing and destruction by phagocytosis). Therefore, Lumière's central thesis was that just attempting to destroy bacteria, the "microbe-hunting" that had prevailed in the "solidist" medical paradigm, is not sufficient to combat infectious diseases. Rather, a very fruitful method would be the "modification" of the internal environment, the organism's "terrain," so as to make it resistant to the spread of pathogens. Substances like sulfamides, as well as the thioderivatives of gold, "granulotherapy," "auto-hemotherapy," in fact all the interventions he employed, were designed to make the internal terrain refractive to change, to achieve "humoral equilibrium" and "fixity."

As discussed so far, Lumière's main efforts were dedicated to combating acute and chronic diseases by chemical means, by the injection of "humor stabilizing" or "desensitizing" substances for the elimination of the "external causes" that hasten our untimely end. But health regimens for the prolongation of life and rejuvenating interventions proper, too played a considerable part in his pursuits. In *Sénilité et Rajeunissement*, Lumière gave practical advice on "how to prolong the existence." In fact, he briefly recapitulated and endorsed the earlier recommendations given by the chair of the Faculty of Forensic Medicine in Lyon, one of the founders of the field of forensic medicine, Alexandre Lacassagne (1843-1924) in *La Verte Vieillesse* (The Green Old Age, 1924), the suggestions made by the nutrition scientist Jean Frumusan in *La cure de rajeunissement. Le devoir, la possibilité et les moyens* (Rejuvenation: The Duty, The Possibility and the Means of Regaining Youth, 1923), and by the centenarian president of the French Academy of Medicine Alexandre Guéniot (1832-1935) in *Pour Vivre Cent Ans. L'Art de Prolonger ses Jours* (To Live a Century: The Art of Prolonging the Days, 1931). The recommendations have not changed much since the days of the early life-extensionist hygienists: Hufeland, Cornaro, or even Hippocrates. Lumière emphasized the importance of sufficient sleep, skin care and baths, moderate work and exercise, moderation in food and drink, sexual moderation (in accordance to the old French proverbs that "Each sacrifice to Venus is a spade of earth over the old man's grave" and "A good cook and a young female are two enemies of the aged"), the "resistance to passions" generally. Moderation – the watchword of all conservatives – was seen as the key to longevity. Lumière was only too eager to affirm Lacassagne's dictum that "sobriety is the only recipe for a long life" or Hippocrates' prescript "Work, food, drink, sleep and love – all in small measure."[170] Perhaps the only novel element, introduced by Lumière into the discussion of longevity regimens, was his insistence that these regimens fully agree with his colloidal theory, with his reinterpretation of the "stabilization of humors." In Lumière's theory, moderate nutrition minimizes colloidal "disturbances"; exercise acts to "desensitize" the body to stress; and rest allows the colloids to "stabilize."[171] Mental equilibrium was related to physiological equilibrium.[172]

In addition to exploring therapies for age-related diseases and longevity regimens, Lumière also carefully investigated the rejuvenation methods proposed up-to-date (*Sénilité et Rajeunissement*, 1932). According to Lumière, nothing in theory precludes the possibility of reversing the detrimental changes of aging or prevents artificial life-extension. To illustrate the possibility of rejuvenation, Lumière drew an analogy between aging and disease. Both derive from an "imbalance" of colloids, and as a person can be cured of disease and reverted to "balance" and normal functioning, the reversion from a state of aging may not be dissimilar:[173]

> Consider a diseased individual suffering from a serious infectious disease… His tissues are dehydrated, skin wrinkles, he loses his strength, his sexual appetites, he assumes the appearance of an old person, but the moment the infection is suppressed, the person begins to gain weight, his wrinkles disappear, his strength returns, he regains all former faculties, in summary, he is rejuvenated.

Even though, Lumière admitted, the current state of knowledge does not yet allow us to practically accomplish the goal of rejuvenation or even point a definite path toward it, after an extensive research the hope may become a reality. Lumière recognized current limitations, but there appeared to be no intractable barriers to overcoming them. The path to rejuvenation, he believed, would lie in the "stabilization" of cell colloids, "inhibiting the destruction of the colloidal structure that occurs through precipitation of proteins, on the one hand, and arresting the evolution of micelloids toward flocculation, on the other," or else adding "in place of used up materials, new colloids and thus conferring on colloid-generating cells a new power of multiplication."[174]

The (endocrine) rejuvenation methods developed thus far, according to Lumière, even though not very effective, offered promising directions.[175] Lumière reviewed recent advances in rejuvenative therapy: the injection of sex gland extracts by Brown-Séquard, vasoligation by Steinach, sex gland transplantations by Voronoff, chemical sympathectomy by Karl Doppler (a development of René Leriche's sympathectomy), the transfusion of young blood by Paul Busquet and Ottmar Wilhelm, and the injection of testicular blood serum by Francesco Cavazzi. Lumière was quite critical of these attempts: their results were said to be, in most cases, of a short duration and rather inconsistent among different individuals. The precise calibration of these interventions remained elusive. However, according to Lumière, these techniques provided a proof of principle: by impregnating the organism with sex hormones, it appeared possible to increase cellular growth and proliferation, the only processes which, according to Lumière, could bring about the renovation of cell colloids. It is this continuous renovation of colloids during cell growth and proliferation that explains, in Lumière's view, the immortality of protozoa and germ cells. The "uneven distribution" of colloids during cell division in multicellular organisms explains cell differentiation, whereas in immortal cells the colloids are distributed "evenly" and are capable of eternal renovation, yet essentially maintaining the "fixed" form of the organism.[176]

Despite the general optimism regarding the future of rejuvenation techniques, Lumière broke away from the "rejuvenators" of the 1920s, such as Voronoff, Steinach and Doppler. The dissent manifested in the criticism and skepticism of the actual effectiveness of their methods, and in the fact that hormone replacements played only a very minor part in Lumière's clinical practice. But perhaps the strongest point of departure was Lumière's withdrawal from the mechanistic reductionism that underscored the majority of rejuvenation techniques. Thus, the grandmaster of French rejuvenators, Serge Voronoff, spoke of the "essential mechanism of our body" where "each organ performs its part," the thyroid gland provides "the brain-motor's ignition spark," all the endocrine glands are "wonderful little factories" that "regulate the action of each organ" and, when some of these controls fail, the body is "put out of gear" and disintegrates. The sex glands, the main object of rejuvenating interventions, are akin to a battery, supplying the body with "vital energy." The removal of particular parts brings about disarray and death, while their replacement provides new "sources of energy," reestablishes the "controls" and restores the body's "equilibrium of functions." Hence, a major task of rejuvenative medicine is to establish "a stock of spare parts for the human machine." A central place in Voronoff's writings was reserved for anatomical and histological examinations, for detailing the surgical technique, which conformed to what Lumière called the "solidist" approach, rearranging the solid parts of the mechanism.[177] Indeed, Voronoff attributed great importance to the "psychic" effects of the grafting operation, but only to demonstrate how a material, "mechanistic" interference positively affects the mental sphere: improving intelligence, productivity, interest in life. But the opposite influence, from the mind onto the body, was thoroughly depreciated. Such mind-over-body effects would obscure the results of treatment and had to be ruled out.

Lumière's approach was quite the contrary: not only did he emphasize the importance of purely psychological motivation for longevity, but the body itself was seen as much more than a combination of its parts. In *Sénilité et Rajeunissement*, Lumière wrote:[178]

> The most important [characteristic of living beings] consists in the prodigious faculty of synthesis that only the living cells possess and that does not appear in any measure and in any degree in other molecular arrangements… Experience and observation, unaffected by all the reasoning of logicians, demonstrate that the properties of a substance essentially depend on the arrangement, the assemblage, the aggregation of atoms and molecules that compose it, and that these assemblages give birth, out of all the pieces, to novel properties that are present in no way and in no degree in the constituent parts.

Lumière's model of the body just could not be easily broken down into parts, because it was mainly composed of "balanced fluids." The therapeutic implications of this "holistic" theory followed.

In 1936, the Hungarian/Canadian endocrinologist Hans Selye, the founder of stress physiology and the great authority among all subsequent proponents of the "holistic" approach, published his concept of the "General Adaptation Syndrome," describing a non-specific reaction of the organism as a whole to a variety of perturbations. (The direct implications of this concept for longevity research, as envisioned by Selye, will be discussed later on.) Yet, the notions of non-specificity were no news to Lumière by 1935 (in fact they were no news to Dumontpallier in 1889). For Lumière, the organism's capability for a non-specific response – that can be manifested in either "shock" or increased "resistance," depending on the dose of the "nocuous agent" and the degree of the organism's "sensitivity" – was the basis for his medical theory and practice.[179] In Lumière's theory, a local change in colloid density (flocculation) could affect the entire "fluidic" system of the body. Moreover, all the influences of colloid "precipitates" and "flocculates" were linked to and responded by the cardiovascular and central nervous systems throughout the body, as the precipitates exerted their "mechanical irritation on the vascular sympathetic nerve terminals." Much in the same way, the "desensitizing" or "stabilizing" pharmaceuticals that Lumière employed, were supposed to affect the entire body. In addition, Lumière insisted, all therapy should be multi-faceted or "polyvalent":[180]

> One grand principle must dominate the methods of treatment of chronic functional afflictions: in

pathological states, the disequilibrium and the instability of humors depend on accidents that have multiple causes; with rare exceptions, the stabilizing and curative therapy directed against them cannot achieve its goal except when it addresses simultaneously all the causes, that is to say, it must be "polyvalent" in order to be completely efficient.... [It is necessary] to remedy, in the same time, all the dysfunctions and eliminate all the factors that are involved in their production.

Both Voronoff and Lumière sought "equilibrium" and "fixity." According to Lumière, "the great principle of life appears to be fixity" that must be maintained in all life forms, from individual cells to organisms, to species to societies.[181] The task of therapy, for Lumière as well as for Voronoff, consisted in maintaining the fixity, safeguarding it against catastrophe or degenerative change. But, paradoxically, in the quest for "fixity," in the struggle against formidable change, both scientists became great medical innovators, introducing new "stabilizing" treatments that may appear unorthodox even by contemporary standards.

Yet, an important difference may be pointed out between Lumière and Voronoff. In Lumière, the "vital equilibrium" appears to be much more complex than Voronoff's "human machine." In Lumière's vision, the equilibrium must involve the responses of the body as a whole, including diverse environmental and psychological factors. Compared to Voronoff, Lumière was much more willing to admit to his almost complete ignorance of these intricacies:[182]

If, generally, we perceive the existence of a relation between matter and intelligence, we are completely ignorant of the mechanisms that govern this relation, and the cause for psychic equilibrium completely evades us. Not only in this order of phenomena has the vital equilibrium remained an enigma: the processes of the regulation of all organic functions, the thermal, the respiratory, the cardiac, etc.... remain entirely obscure. We are ignorant!

It was perhaps this realization of the immense complexity of the "vital equilibrium," of integrating elements that could not be readily "removed" or "supplemented," that contributed to Lumière's withdrawal from the current rejuvenation methods, and caused him to seek solace in anticipating the results of "future research."

Intriguingly, despite the humility before the grandeur of the "organism as a whole," Lumière remained a "therapeutic activist" and "scientific optimist" to the end of his nonagenarian life. A similar coexistence of therapeutic and scientific optimism with the great "holistic" awe before the complexity and wholeness of the human being, in his/her infinite connectedness to the society and the universe, is salient in the work of another French contemporary longevity researcher and seeker of "equilibrium," a chief authority among both reductionist surgeons and holistic philosophers – Alexis Carrel.

16. The perpetuation of the whole – Alexis Carrel (1873-1944)

The works of Alexis Carrel, the Nobel Laureate in Physiology and Medicine of 1912, were an inspiration to the life-extensionists in the first half of the twentieth century. The Nobel Prize was given "in recognition of his work on vascular suture and the transplantation of blood vessels and organs."[183] Carrel pioneered the technique of blood vessels suture (anastomosis) while working at Auguste Lumière's laboratory in Lyon in the early 1900s. In 1904 he immigrated to the US, and continued this research at the University of Chicago, before moving on in 1906 to the Rockefeller Institute for Medical Research in New York where he remained for most of his career. During WWI, he stayed in France and served as a major in the French Army Medical Corps, deploying a novel antiseptic solution and irrigation apparatus to treat battle wounds (developed in collaboration with Henry Dakin), and studying wound-healing and regeneration together with Lecomte du Noüy. At the war's end, in 1918, he returned to the Rockefeller Institute and continued to work there until he was forced to resign in 1939 due to the mandatory retirement age (set by the Institute at 65). Even while at the Rockefeller Institute, he never severed ties with France, and spent all his summers in study and contemplation at his private island of Saint Gildas, in Brittany, at the shores of Northern France. In 1939, he again returned to France, where on November 17, 1941, he established and presided over the "Fondation Française pour l'Étude de Problèmes Humains" (the French Foundation for the Study of Human Problems) – the world's first multi-disciplinary research foundation dedicated to the study and amelioration of the entirety of human problematics: from experimental biology to psychology to sociology to "prevention of precocious aging." This establishment, that was initially decreed and sponsored by the Vichy government and Marshal Pétain personally, was since 1945 transformed into the Institut National d'Études Démographiques (INED, The National Institute for Demographic Studies). Carrel died on November 5, 1944, after the liberation of France by the allied forces, awaiting trial for collaboration. Controversy continues as to the degree of Carrel's "collaboration." Some researchers suggest it was purely nominal, aimed to ensure the operation of the Foundation.[184] At any rate, he was never a political dissident. Called "the Frenchman" in the US and "the American" in France, often highly judgmental of both the French and American societies, he remained a patriot and a prominent, highly respected and well-established figure in both countries, closely tied to both cultures, and a crucial contributor to both the American and French life-extensionist thought.

Carrel's major contribution to life-extensionism was his experimental demonstration that body parts could be replaced as they wear out, just as the worn-out parts of a machine could be replaced for its sustained maintenance. Carrel's initial success in blood vessels anastomosis that earned him the Nobel Prize – including the development of aseptic conditions (as infection was a major cause of thrombosis and operation failure), the performance of suture on all layers of the vessels, and the development of the "triangulation method" of suture – were crucial for successful organ transplantations, paving the way for all subsequent grafting operations in humans, from Serge Voronoff's "rejuvenation by grafting" onward. Using the perfected anastomosis techniques, since the early 1900s, Carrel pioneered the transplantation of kidneys, limbs, thyroid, ovaries and heart in animals. The preservation of organs for transplantation was vital, and Carrel was among the first to achieve this by cold storage of blood vessels and entire organs.[185]

In Carrel's vision, tissues and organs for transplantation could be artificially grown. For that purpose, around 1910, he started a series of experiments on growing tissues, attempting to find the culture medium conditions and stimulants for their speedy regeneration and indefinite maintenance and growth, pioneering the field of tissue engineering. His famous experiment on culturing the chicken embryonic heart tissue was started in 1912.[186] That tissue culture (in fact composed of fibroblast connective tissue, rather than muscle tissue) was maintained in continuous growth by Carrel's assistant Albert Ebeling until 1946. Thus the longevity of the tissue exceeded many times the life-span of the organism from which it was taken. The tissue culture appeared to be virtually immortal. The finding of the somatic tissue's "potential immortality" was a tremendous inspiration for the life-extensionist movement. Between the 1910s and 1950s, there appeared to be not a single author writing about longevity who did not mention that outcome from Carrel's lab (Metchnikoff, Voronoff, Steinach, Kammerer, Lumière, Metalnikov, Bogomolets, etc. etc.). As somatic cells appeared to be potentially immortal, the organism composed of them, it was believed, could become potentially immortal as well (the tenet alternately disputed and reaffirmed later on).[187]

Even though growing entire organs was not feasible at the time, Carrel succeeded in maintaining existing organs outside of the body. He began this research in 1912 with sustaining a "visceral organism" – a system of animal heart, lungs, liver, stomach, intestines and kidneys – in an artificial medium. But the highest success was achieved in the 1930s, thanks to the development of the "Perfusion Pump" in collaboration with Charles Lindbergh (1902-1974). Lindbergh, the aviator celebrated for his first trans-Atlantic flight from New York to Paris in 1927, offered Carrel his engineering expertise in 1930 to build a perfusion pump, capable of providing a flow of nutrients and oxygen to excised organs. By 1935, the perfusion apparatus (the "Lindbergh Pump") was perfected, applying a more "natural" pulsating pressure to circulate the perfusion fluid, and having effectively solved the problem of infection by using a sterile, filtered system of air and nutrient medium supply to the organs. A wide variety of organs were thus maintained: heart, kidney, spleen, pancreas, fallopian tubes, thyroid, and more.[188] This system was an essential step in a series of developments that eventually led to the creation of "heart-lung" and "dialysis" machines, making open-heart surgery possible, and generally enabling the maintenance of organs for transplantation. Bur Carrel's vision extended further.

In 1938, Carrel and Lindbergh coauthored *The Culture of Organs*, where they described the technique, Lindbergh specifying the perfusion apparatus, its assembly and operation, Carrel elaborating on perfusing media, organ preparations, their physiological behavior outside of the body, and the practical purposes of these studies. *The Culture of Organs* gives the following descriptions of "The Ultimate Goal" of this research. Using out-of-body organs, natural, non-rejectable bio-pharmaceuticals can be manufactured. Furthermore, diseased organs can be removed from the body, treated and then re-implanted:[189]

> Human organs would manufacture *in vitro* the substances [hormones and anti-bodies] supplied today to patients by horses or rabbits. ... We can perhaps dream of removing diseased organs from the body and placing them in the Lindbergh pump as patients are placed in a hospital. Then they could be treated far more energetically than within the organism, and if cured replanted in the patient. ... The replantation would offer no difficulty, as surgical techniques for the suture of blood vessels and the transplantation of organs and limbs were developed long ago [by Carrel].

And perhaps crucially, by examining extracorporeally cultured organs, the optimal nutrition requirements can be established for each organ, and these needs can be provided to achieve the organs' optimal growth, regeneration and organic equilibrium within the body.

Contemporary press further exaggerated the potential applications of the "Culture of Organs," writing about the possible artificial maintenance of a brain, of an entire human embryo or even of an adult organism, about Carrel and Lindbergh introducing "immortality in a physical sense" and revealing "the secret of life itself."[190] But Carrel's own vision for the possibilities of life-extension, described in his own words, was grand enough. Not only could worn-out parts of the body be removed, treated and replaced, but the aging process of each and every organ could be understood and countered through appropriate nutrition. Carrel announced:[191]

> A new era has opened. Now anatomy is capable of apprehending bodily structures in the fullness of their reality, of understanding how the organs form the organism, and how the organism grows, ages, heals its wounds, resists disease, and adapts itself with marvelous ease to changing environment. The ultimate goal of the culture of organs is to obtain this new knowledge and to pursue it through the

complexity of its unpredictable consequences.

Carrel's longevity research was not exhausted by tissue cultures or organ transplantations. He built a giant "Mousery" where, by 1933, over 50,000 mice were kept. Their longevity was studied and manipulated by nutrition, lodging conditions, exercise and rest, heredity, chemicals, etc. (After 1933 the project funding was terminated, presumably because the data base became too extensive to manage.[192]) Carrel also conducted experiments with long-term drying and reviving of small animals, mostly rotifers. (Similar experiments were conducted at the Rockefeller Institute by Jacques Loeb.) Speculations abounded, fostered by Carrel himself, about a possible future induction of a similar "resting state" in humans, perhaps to preserve them for centuries.[193] In 1925, Carrel attempted to revivify a mummy.[194] He was keenly interested in the possibility of lowering the overall metabolic rate as a means for life-prolongation. Lindbergh was equally inspired by the idea, having heard stories of the Yogis capable of lowering their body temperature, breathing and heart rate, entering a state of "suspended animation."[195] Yet Lindbergh's and Carrel's attempts to induce such a state by artificial changes of temperature, pressure, oxygen and CO_2 levels, failed. (Nowadays, reversible drug-induced or hypothermic coma, used for recuperation, is a well-established medical practice.[196])

Carrel's interest in life-extension was encompassing, and was fully shared by Lindbergh (who shared many of Carrel's philosophical views). Nonetheless, Carrel seems to hardly merit the title of an "Immortalist," given him in David Friedman's biography.[197] In the first half of the 20th century, very few real "immortalists," the true believers in the possibility of physical immortality for humans, could be found, except perhaps for the Russian Nikolay Fedorov and the American Charles Asbury Stephens. The majority of life-extensionist authors of the period did have some limits to their aspirations (usually ranging from a century life-span as a short-term goal to 150-200 years as a longer-term project). In 1935, Carrel published *L'Homme, Cet Inconnu – Man, the Unknown* – "an intelligible synthesis" of "information on human beings," where he made his views on life-extension explicit.[198]

Carrel had no illusions about the current abilities of medical science to significantly extend the human life-span or restore youth: "we have not succeeded in increasing the duration of our existence. A man of forty-five has no more chance of dying at the age of eighty years now than in the last century."[199] Carrel was quite critical about the earlier attempts of rejuvenation by Steinach and Voronoff (that were in fact made possible by Carrel's success in developing the blood vessel anastomosis technique). He acknowledged that the rejuvenation operations might have been "followed by an improvement in the general condition and the sexual functions of the patients." But he considered these results as "doubtful" and "not lasting." He did not believe Voronoff's sex gland grafts could take (due to rejection) and explained the apparent reinvigoration by a degeneration of the grafted gland in the body that "may set free certain secretory products" for a short period.[200] He never attributed to the degeneration of sex glands a causative role in aging, but saw it as a consequent or concomitant of general deterioration. According to Carrel, modern human beings may have become more skillful at keeping youthful appearances, but looks are not tantamount to a true restoration of youth or longevity.[201]

Carrel summed up all preceding attempts at life-prolongation and rejuvenation as a failure: "The greatest desire of men is for eternal youth. From Merlin down to Cagliostro, Brown-Séquard, and Voronoff, charlatans and scientists have pursued the same dream and suffered the same defeat."[202] Moreover, according to Carrel, there are principal restrictions to radical life-prolongation or immortality for humans. He saw senescence as unavoidably originating from the very organization of the human body, from the inherent disaccord between its differentiated parts. He, as well as the majority of contemporary longevity researchers, perceived the fixity of the internal environment and the harmonious interrelation of the parts of the organism as the necessary conditions for longevity. But he realized how difficult, almost impossible, it is to maintain such a harmony. Therefore, Carrel concluded: "Man will never tire of seeking immortality. He will not attain it, because he is bound by certain laws of his organic constitution. … Never will he vanquish death. Death is the price he has to pay for his brain and his personality."[203]

All these difficulties, however, did not mean that some extents of life-prolongation or rejuvenation would be impossible and that longevity research should be abandoned:[204]

> It is probable that neither Steinach nor Voronoff has ever observed true rejuvenation. But their failure does not by any means signify that rejuvenation is for ever impossible to obtain. We can believe that a partial reversal of physiological time will become realizable. ... If an old man were given the glands of a stillborn infant and the blood of a young man, he would possibly be rejuvenated. Many technical difficulties remain to be overcome before such an operation can be undertaken. We have no way of selecting organs suitable to a given individual. There is no procedure for rendering tissues capable of adapting themselves to the body of their host in a definitive manner. But the progress of science is swift. With the aid of the methods already existing, and of those which will be discovered, we must pursue the search for the great secret.

Far from being daunted by the impotence of contemporary longevity science, Carrel perceived its shortcomings as a hurdle to be overcome, a powerful stimulus for further, more extensive, research. "A better knowledge of the mechanisms of physiological duration," obtained through empirical studies of heredity and environment, nutrition and life-style, "could bring a solution of the problem of longevity."[205]

According to Carrel, such research could only be conducted in a cooperative manner (though he believed a single person was needed to synthesize the data), in a well-equipped, well funded, truly interdisciplinary scientific institution, capable of continuous studies for a very long period:[206]

> Our life is too short. Many experiments should be conducted for a century at the least. Institutions should be established in such a way that observations and experiments commenced by one scientist would not be interrupted by his death. Such organizations are still unknown in the realm of science. ... Institutions, in some measure immortal, like religious orders, which would allow the uninterrupted continuation of an experiment as long as might be necessary, should compensate for the too short duration of the existence of individual observers.

Carrel attempted to establish such an institution in the US, while working at the Rockefeller Institute (where his experiments would be discontinued and he himself forced to retire in 1939 due to age limitations). But the idea came to partial fruition in France with the establishment of the "Fondation Française pour l'Étude de Problèmes Humains." As will be elaborated in the following sections, the increasing institutional banding together of longevity researchers, from Eastern Europe to the US and France, was a definitive mark of the 1930s.[207]

Institutions were a means to provide the necessary permanence, stability and continuity for aging and longevity research. As Carrel recurrently noted, life-prolonging interventions in humans are impossible to ascertain except over a period of several decades. And generally, permanence, stability and continuity appeared to be the driving concepts in Carrel's reasoning about the longevity of both biological and social organisms.[208] According to Carrel, "the law of constancy" is fundamental for any sustainable system: "when a system is in equilibrium, and a factor tends to modify the equilibrium, there occurs a reaction that opposes this factor." In biological systems, "a steady state, and not an equilibrium, persists with the help of physiological processes." In summary, "steadiness of the inner medium is, without any doubt, indispensable for the survival of the organism."[209]

At the first glance, it may appear that Carrel valorized progress, and the effort necessary to achieve progress, over constancy: "The law of effort is still more important than the law of the constancy of the organic states." Furthermore, "the physiological and mental progress of the individual depends on his functional activity and on his efforts." To survive, the organism needs to adapt to a changing environment, and this adaptive ability is exercised by effort and exposure to hardships: "the exercise of adaptive functions is as necessary to the development of body and consciousness as physical effort to that of the muscles." And yet, ultimately, progressive effort becomes subservient to maintaining constancy. Carrel pointed out that "Science has supplied us with means for keeping our intraorganic equilibrium, which are more agreeable and less laborious than the natural processes. …the physical conditions of our daily life are prevented from varying." Yet, while providing this artificial stability, the natural exercise of the adaptive function, which essentially amounts to an ability to maintain constancy against all odds and changes, should not be discarded: "By doing away with muscular effort in daily life, we have suppressed, without being aware of it, the ceaseless exercise required from our organic systems in order that the constancy of the inner medium be maintained." The ultimate end of adaptation is to achieve a sustainable equilibrium and constancy under new conditions, to arrive at "permanent modifications of body and consciousness." Essentially, "Adaptation, considered in its various manifestations and its oneness, appears as an agent of stabilization and organic repair, as the cause of the moulding of organs by function, as the link that integrates tissues and humours in a whole enduring in spite of the attacks of the outer world."[210]

As a living organism, according to Carrel, needs to be stable, the society too needs to be preserved in a steady state. Carrel devoted extensive passages to advocacy of "progress," "change," "reform," "the remaking of man," even "revolution." "We must modify our mode of life and our environment," he wrote, "even at the cost of a destructive revolution. After all, the purpose of civilization is not the progress of science and machines, but the progress of man."[211] But how much change did Carrel really want? Born into a wealthy capitalist family, married into aristocracy, and accepted into the social and scientific elite, Carrel never wished those statuses to disappear. The rhetoric of progress might conceal a deep-seated conservatism, the striving to perpetuate the existing social order. Or more precisely, it was the upper rank of the existing social order, of which Carrel undoubtedly and justly saw himself to be a part, which was in need of perpetuation.

All the "modifications" Carrel spoke of, were intended to prevent the collapse of "civilization," or rather of its elite stratum, which, according to Carrel, was under a threat of destruction by external or internal nocuous influences.[212] First of all, "civilization" meant "Western civilization," synonymous with "civilized races" or "white races," that needed protection against faster-breading non-whites, as well as against "the newcomers, peasants and proletarians from primitive European countries." "Their offspring are far from having the value of those who come from the first settlers of North America" and their encroachment on the existing demographics needed to stop. Secondly, the existing class system had to be preserved indefinitely. The classes that have already formed, emerged due to biological differences, and now it was necessary to make the divide permanent:[213]

> In democratic countries, such as the United States and France, for example, any man had the possibility during the last century of rising to the position his capacities enabled him to hold. Today, most of the members of the proletarian class owe their situation to the hereditary weakness of their organs and their mind. Likewise, the peasants have remained attached to the soil since the Middle Ages, because they possess the courage, judgment, physical resistance, and lack of imagination and daring which render them apt for this type of life. These unknown farmers … were, despite their great qualities, of a weaker organic and psychological constitution than the medieval barons who conquered the land and defended it victoriously against all invaders. Originally, the serfs and the chiefs were really born serfs and chiefs. Today, the weak should not be artificially maintained in wealth and power. It is imperative that social classes should be synonymous with biological classes.

Thirdly, the society had to be forever male-dominated (to recall, in the US, women were given voting rights in 1920). "Women should receive a higher education, not in order to become doctors, lawyers, or professors, but to rear their offspring to be valuable human beings." "The differences existing between man and woman" are of "fundamental nature." Women, therefore, "should not abandon their specific functions." Fourthly, private property and nuclear family should forever remain the bedrocks of society:[214]

> Modern society must, therefore, allow to all a certain stability of life, a home, a garden, some friends. ... It is imperative to stop the transformation of the farmer, the artisan, the artist, the professor, and the man of science into manual or intellectual proletarians, possessing nothing but their hands or their brains....All forms of the proletariat must be suppressed. Each individual should have the security and the stability required for the foundation of a family.

Finally, and perhaps crucially, all the existing hereditary strengths, or traits perceived at the time as strengths, had to be accentuated, and thus carried forward indefinitely: "Modern nations will save themselves by developing the strong. Not by protecting the weak. Eugenics is indispensable for the perpetuation of the strong. A great race must propagate its best elements."[215] (Notably, Carrel mainly advocated "voluntary" rather than coercive eugenics. According to him, only the propagation of the "insane" or "feeble-minded" should be forcefully prevented.) Thus, perpetuation was in order for the existing elite, for the group at large, as well as for its individual members, for those who, like Carrel, would merit membership. Carrel wrote:[216]

> We must not yield to the temptation to use blindly for this purpose [of life-prolongation] the means placed at our disposal by medicine. Longevity is only desirable if it increases the duration of youth, and not that of old age. The lengthening of the senescent period would be a calamity. The aging individual, when not capable of providing for himself, is an encumbrance to his family and to the community. If all men lived to be one hundred years old, the younger members of the population could not support such a heavy burden. Before attempting to prolong life, we must discover methods for conserving organic and mental activities to the eve of death. It is imperative that the number of the diseased, the paralysed, the weak, and the insane should not be augmented. Besides, it would not be wise to give everybody a long existence. The danger of increasing the quantity of human beings without regard to their quality is well known. Why should more years be added to the life of persons who are unhappy, selfish, stupid, and useless? The number of centenarians must not be augmented until we can prevent intellectual and moral decay, and also the lingering diseases of old age.

This passage is ambiguous. It can be understood as simply saying that the efforts for life-prolongation should be widely coupled with efforts to improve the quality of life, to make the society as a whole more durable. It can also imply that life-extension *per se* could cripple western society, and therefore should rather not be attempted. But such an interpretation would contradict Carrel's life-long persistence at longevity research, his conviction that "we must pursue the search for the great secret." Or rather, it could be understood, the prolongation of life is to be reserved for a defined upper class, for those who are not "stupid and useless." In either case, the ultimate goal is the perpetuation of the existing social order, just as the ultimate goal of biomedical intervention is to maintain the constancy of an individual organism.

A researcher at the cutting age of biomedical science, Carrel was very fond of traditional, "natural" ways of life, constantly lamenting the grievances brought in by modernity. For all the discussions of social "progress," the "perfect society" envisioned by Carrel was a mere emphasis of the circumferential culture, as his vision of medical progress amounted to perfecting the means for sustaining a steady state of the body. For Carrel, the social, biological and psychological realms were not separated. If there could be a general definition of Carrel's aspirations, it would be the 'preservation of the whole' – body, mind and society included. In the 1930s, there was perhaps no stronger proponent of "holism" than Carrel. Even though the term "holism" was first introduced by the South African philosopher and statesman Jan Smuts in 1926, in the book *Holism and Evolution*, Carrel did not use this specific term. Nonetheless, *Man the*

Unknown was imbued with the thought that the human being is a whole that cannot be reduced to an interrelation of components, that human beings react as a whole to the external environment, that they form a whole of a higher order with the environment, with the "cosmic, economic and psychological milieu,"[217] and that they should be studied and influenced in such an integrity.

The mind becomes an indispensable factor of wholeness.[218] "In disease as in health, body and consciousness, although distinct, are inseparable," Carrel argued. "The whole consisting of body and consciousness is modifiable by organic as well as by mental factors. Mind and organism commune in man like form and marble in a statue." Carrel well recognized that the mind is profoundly affected by physiological activities, and those are not restricted to the brain: "In fact, the entire body appears to be the substratum of mental and spiritual energies. Thought is the offspring of the endocrine glands as well as of the cerebral cortex. The integrity of the organism is indispensable to the manifestations of consciousness. Man thinks, invents, loves, suffers, admires, and prays with his brain and all his organs." Yet, an even stronger emphasis was placed on the effects of mental states on the body: "Man integrates himself by meditation, just as by action."[219] Physiological equilibrium, and consequently rejuvenation and increased longevity, can be achieved by such a direct, integrating and balancing control of the mind: "A sort of rejuvenation may be brought about by a happy event, or a better equilibrium of the physiological and psychological functions. Possibly, certain states of mental and bodily well-being are accompanied by modifications of the humours characteristic of a true rejuvenation."[220]

The fascination with the power of the mind led Carrel to the uncharted territories of mysticism and faith-healing, the phenomena of "ecstasies, thought transmission, visions of events happening at a distance, and even of levitations."[221] For Carrel, the interest in the paranormal began in the early 1900s. In 1903, he visited Lourdes, a holy site in Southern France, where Virgin Mary was said to have appeared to Saint Bernadette in 1858, and which since then became a place reputed for numerous miraculous cures. Carrel set out on the voyage a skeptic, intent to "examine the facts objectively, just as a patient is examined at a hospital or an experiment conducted in a laboratory." He returned a believer, as he observed seemingly instantaneous cures of organic disorders – "peritoneal tuberculosis, cold abscesses, osteitis, suppurating wounds, lupus, cancer, etc."[222]

Such an interest could damage Carrel's fledging career in France, and this was one of the considerations that led to his emigration. Until the mid-1930s, he was little disposed to publicly discuss mysticism, except perhaps among a close circle of friends, members of the "Philosophers Club" in New York, such as Frederic Coudert, Cornelius Clifford and Boris Bakhmeteff, who encouraged him to write the philosophical treatise and to whom *Man, the Unknown* was dedicated. In 1935, Carrel felt confident enough to widely announce his mystic beliefs, encouraged by the increasing interest in the psychic, occult and miraculous, even among the medical profession.[223] As per Carrel's confession, he "does not hesitate to mention mysticity in this book, because he has observed its manifestations."[224] According to Carrel, the control of the mind over the body, or rather the integrated relation of the mind to the universe achieved through spiritual or mystic activity, especially through prayer, could produce immediate and prolonged physiological healing, the restoration and maintenance of the whole:[225]

> The miracle is chiefly characterized by an extreme acceleration of the processes of organic repair. There is no doubt that the rate of cicatrization of the anatomical defects is much greater than the normal one. The only condition indispensable to the occurrence of the phenomenon is prayer. But there is no need for the patient himself to pray, or even to have any religious faith. It is sufficient that some one around him be in a state of prayer. Such facts are of profound significance. They show the reality of certain relations, of still unknown nature, between psychological and organic processes. They prove the objective importance of the spiritual activities, which hygienists, physicians, educators, and sociologists have almost always neglected to study. They open to man a new world.

How could this mystical and holistic belief in an almost unlimited direct influence of the mind over the body, coexist with Carrel's scientific methodology, the very reductionist, almost Frankensteinian, assemblage of body parts into a 'living machine'? A 'functional personality split' might be suspected: a reductionist at the lab, Carrel could well be a "holist" among friends and family, plunging into mystic contemplation on holidays and vacations. There could also be a temporal split. Between the early 1900s – mid-1930s, Carrel might have deferred mysticism, being too involved with tissue engineering, and only in the mid-1930s, toward the end of his scientific career, he let the mysticism surface. However, the writing of *Man, the Unknown* temporally coincided with the construction and testing of the "Perfusion Pump" and the "Culture of Organs," the very paragons of reductionist biology. In *Man, the Unknown*, the research of psychic activities and tissue engineering were discussed side by side. Definitely, Carrel perceived no antagonism between the diverse fields of study. No Cartesian mind-body dualism existed for him. Rather, he recognized the existence of many "unities" at different "levels of organization" – from individual molecules, to cells, to tissues and organs, to individual organisms, to mind-body unification in an individual human being, in relation to larger social and, ultimately, cosmic unities.[226] He did recognize the necessity for specialists to investigate different levels of organization, some "studying the larger structures resulting from the aggregation and organization of molecules, the cells of the tissues and of the blood – that is, living matter itself," others focusing on "cells, their ways of association, and the laws governing their relations with their surroundings; the whole made up of the organs and humours; the influence of the cosmic environment on this whole; and the effects of chemical substances on tissues and consciousness." Carrel did not discard the need for analysis in scientific investigation. What he proposed, however, was that reductionist analysis should not be the only method of study. A larger synthesis is necessary: "We now possess such a large amount of information on human beings that its very immensity prevents us from using it properly. In order to be of service, our knowledge must be synthetic and concise."[227] Such a synthesis cannot be restricted to a study of body parts; it must also include higher forms of organization: the organism as a whole, mental activities, society and environment.

This thought really comes to the foreground between the mid-1930s-mid-1940s, in Carrel as well as other prominent contemporary French physicians and life-extension advocates. However, the fact is that with holistic synthesis coming to the foreground, reductionist analysis is displaced to the background. The immense scope of *Man, the Unknown*, of necessity, makes limited room for any particular field of study. And thus, in *Man, the Unknown*, the subject of rejuvenation occupies a very restricted corner in the grand edifice of human knowledge. Indeed the preservation of "the whole," the equilibrium and constancy at different levels of organization – individual and social – is a permeating thought in the book; and mental and social influences are shown as powerful factors for increasing longevity. But the subject of rejuvenation by replacing or rearranging body parts, so prevalent in the former French life-extensionist thought, becomes overshadowed by the recognition of the immensity of other factors, mental, social or environmental, which go beyond the tinkering with organs and tissues. Of such factors of wholeness we have but little knowledge and little control. Thus, the quest for mechanistic rejuvenation 'dissolves in the whole.'

17. The "Neo-Hippocratic" holistic movement - Official Medicine and Heretical Medicines

On his return to France in 1939, Carrel assumed the leading position in the French holistic medical movement that was increasing in strength in the 1930s and reached its height of influence under the Vichy regime. With forty million francs allocated by the Vichy government to Carrel's interdisciplinary "French Foundation for the Study of Human Problems" (almost as much as to the National Center for Scientific Research – Centre National de la Recherche Scientifique – CNRS, with about F50M budget), Carrel made many allies. The joint publication *Médecine Officielle et Médecines Hérétiques*[228] (Official Medicine and Heretical Medicines) provides a remarkable testimony to the great reinforcement of the holistic medical movement in Vichy France. The book was published in 1945, after the liberation of France and Carrel's death, but the articles were written in the last years of the Vichy regime. The authors were prominent physicians, some of them members of the French Academy of Sciences and the Medical Academy, heads of medical departments. They were united against the "analytical," "materialistic," "mechanistic" and "dehumanizing" approaches in medicine, and unanimous in their advocacy of "synthesis," treating the human being as "a whole," with due consideration of his mental or "psychic" activities, "imponderable" factors in healing, the recognition of the patients' individuality, the integration of an individual with the grander social and physical environment. These emphases are now firmly associated with a "holistic" medical paradigm. The authors of *Heretical Medicines* did not use the term "holism," but rather saw themselves as "Neo-Hippocratists." From the teachings of Hippocrates, they derived the basis for considering the human being as a "whole" in an unbreakable rapport with the environment, with season and place. Following Hippocrates, they sought to maintain a physiological and mental equilibrium, never considering a disease as an entity, but as a temporal imbalance of the equilibrium. The equilibrium, according to them, could be restored not so much by chemicals and operations, but rather by a more "natural" and "moderate" way of life. Depreciating heroic interventions, they strove to assist the "healing power of nature" through adjusting the life-style, and most of all, by cultivating the healing power of the mind. These principles were, according to them, neglected by the "official" or "materialistic" medicine and must be restored. The heretics were on a move to conquer the orthodoxy.

Even though perceiving themselves as dissenters, the 'heretics' could not emphasize strongly enough how traditional their views are, their goal being the restoration and development of the old, more "natural" and perennial ways of treatment, in contrast to more modish "materialistic" interventions. As one of the authors, the Marseille surgeon Jean Poucel claimed in the article "La Médecine Naturiste" (The Naturist Medicine):[229]

> It has not been many years, one remembers, since the rise to grandeur and fall to decadence of many treatments that had raised such great hopes: supplemental nutrition, tuberculines, vaccines, salts of calcium, salts of lime, salts of rare earth elements, salts of gold, etc… But there remain aeration galleries, rest coupled with moderate exercise, varied and fresh diet.

If this was a revolution, it was a conservative one. As in Carrel, increased longevity was mentioned as a goal by most authors, yet the subject was somewhat dissolved in the rather metaphysical discussions of "wholeness." No specific therapy or element of treatment was given a predominant weight. The more pervasive concern was the general restoration and maintenance of the "wholeness," of physiological and mental equilibrium, via restoration and maintenance of medical tradition. The reductionist rejuvenation by organotherapy, pioneered by French physicians at the *fin-de-siècle*, now became thoroughly discarded as an unsubstantiated novelty, as the French physicians made a daring move into "holistic" health-care, sanctified by tradition and return to "natural ways."

The book opened with an article by the leader of the movement, Alexis Carrel, "Le rôle futur de la

Médecine" (The future role of medicine),[230] where he briefly reiterated the main tenets of *Man, the Unknown*: the necessity to treat the human being as a whole, without neglect of the mental sphere, and including the social and physical environment, in order to save the individual and the race. Auguste Lumière was there beside him, summarizing his life-long work on the resurrection of Humoralist Medicine and its results ("La Médecine Humorale et ses Résultats").[231] Several other authors reinstated the Hippocratic tradition to its proper glory.[232] Dr. Pierre Galimard of Lyon spoke of the Hippocratic tradition and the medicine of correspondences ("La tradition hippocratique et la médecine des correspondances"). According to Galimard, Hippocratism, that "monument of extraordinary duration" teaches us to "unite apparently distant things into a universal harmony," to seek universal "correspondences or analogies" of the "Microcosm" and the "Macrocosm," the same "symbolic laws, and the same cycles of formation" that govern all forms of being, and which are united by the "universal spirit."[233] Through seeking symbolic analogies between different forms of creation (minerals, plants and animals), we can hope to extract from them "pure curative powers" that can "renew homologous powers that degenerate in a human being." From this journey into the medical scholastics of the Middle Ages, Galimard arrived determined to "reconstitute" or "resurrect" the ancient knowledge.

The same desire was expressed by Dr. Marcel Martiny (chief of the Leopold Bellan Hospital in Paris) in "Nouvel Hippocratisme" (New Hippocratism). Martiny was a dedicated champion of Neo-Hippocratism, of that "passionate and controversial medical movement assuming great importance in Europe, and particularly in France." The Neo-Hippocratic medicine is "at the same time humanistic and scientific," and is permeated by the "spirit of synthesis." It is "general," "imaginative," "unitary," tending toward "finalism and vitalism." According to Martiny, the unity of an individual human being can be understood in terms of biological "types," correlated to ancient "constitutions." Together with other contemporary French Neo-Hippocratists – Claude Sigaud, Léon Mac Auliffe, Alfred Thooris, Paul Carton, Louis Corman, René Allendy, René Biot, and Pierre Winter – Martiny constructed a four-pronged physiological typology (the square, of course, being the symbol of stability and equilibrium). The "prevalence" of one of the embryological layers (the endoderm, ectoderm, mesoderm or their balanced state termed the blastoderm), determined the four morphological types, corresponding to four classical or Hippocratic-Galenic humoral constitutions, as well as four "reactive modalities" and ultimately four stages of "racial evolution."[234] Only when considering the biological types, Martiny proposed, can a human being be treated effectively as an individual and as a whole. Martiny valorized "weak" (homeopathic) therapeutic modalities, acting both on the entire organism and local biological unities, over "strong" and "intermediate" (allopathic) modalities attempting to destroy the agents of damage.[235] In the "weak" modality, the remedy is a catalyst that allows for a full utilization of the ambient natural energy, working through "syntonization" (fine tuning) or "resonance" with existing organic components. With such a utilization of the organism's own "ambient energy," Hippocrates' hope of stimulating *Vis Medicatrix Naturae* (The healing power of nature) may become realized. And only by utilizing "weak modalities," including mental affirmations, the individual may remain "exactly the same" and the treatment may become personal.

The Parisian professor Léon Vannier (1880-1963), the founder of the French Society for Homeopathy (1927) and of the Homeopathy Center of France (Paris, 1932), in "La Tradition Scientifique de l'Homœopathie" provided an even stronger advocacy for homeopathy or "homeotherapy." According to Vannier, the "healing power of nature" can be stimulated in two ways: either "passively" (e.g. by rest) or "actively" (e.g. by exercise or proper nutrition). And bodily equilibrium can be established in two ways of therapy: by agents either contrary or similar to those causing the disease or disequilibrium. Hence, one of the principles of treatment is that of the Opposites, "*Contraria Contrariis Curantur*" (opposite cures opposite) using allopathic remedies directly antagonistic to agents of the disease, "expunging that which is in excess and supplementing that which is in deficit." The second principle is that of the Similitude, "*Similia Similibus Curantur*" (like cures like) employing homeopathic remedies whose action is similar to that of the disease, yet whose application suppresses the disease, effecting immunization.[236] Building on a long historical tradition, from Hippocrates to Samuel Hahnemann, Vannier considered homeopathy as an integral part of Neo-Hippocratism. Neo-Hippocratism was seen as an eclectic and holistic system, where both allopathic (opposite) and homeopathic (similar) treatments have their proper indications, where the body, spirit and soul are perceived as a unity, and where individual "types," "constitutions" and "temperaments" are taken into account. The cause of disease was sought not only in the "apparent microbial action, but also in the transmutation of the subject, in whom the normal rhythm is troubled through multiple conditions depending on the cosmic, social and professional environment, life-style and nutrition, and finally heredity."

According to Martiny, Vannier, and most other authors of *Heretical Medicines*, the Neo-Hippocratic medicine, first and foremost, seeks health in a communion with nature. Thus, Martiny emphasized the importance of "the contact of the body with the air, water and light, moderate and varied nutrition, accompanied by fasting, a harmonious alteration between the activities of the body and of the spirit." The importance of "Natural Ways" was further emphasized in "Naturist Medicine" by Jean Poucel (interestingly, a specialist in surgery, a field not commonly associated with "holism").[237] According to Poucel, "Naturism" (not to be confused with philosophical or literary Naturalism, neither with nudism or vegetarianism) implies "obedience to physiological laws and application of natural agents," the emphasis derived, of course, from Hippocrates. By observing animals that supposedly do not show signs of decrepitude in old age, and healthy, long-lived humans living in "natural conditions" unadulterated by the evils of modern civilization, rules for a healthful and long life can be deduced. Poucel credited Carrel for propagating this idea, though, he emphasized, it had been expressed much earlier in Hufeland's *Art of Prolonging Human Life* (1796), among other sources. According to Poucel, naturist medicine must fight "intoxication" (alcohol, tobacco, other narcotics), as well as social poisons that intoxicate the spirit: unhealthy literature and spectacles, passion for money, play and pleasures. The human being must live in harmony with the external milieu. He must become exposed to fresh air, to sunlight, to water. Nutrition is of primary importance, though Poucel recognized the immense controversy on what constitutes a proper or "natural" nutrition. According to him, "naturism" must not at all be equated with vegetarianism, or any other particular diet, but must rather emphasize the "quality and freshness" of the products, if possible, to be consumed in their naturally occurring form. "Natural exercises" need to be done in close communion with nature, as in "scoutism," rather than using ingenious gymnastic devices. According to Poucel, Naturism is not to be reduced to mere "animal" or "vegetative" functions, but needs to develop psychological and moral strength, emphasizing the notion of the "unity of the individual" and the "immorality of disease." The use of naturist hygiene is, according to him, "the only means to create a virile race."

Martiny, Vannier and Poucel were no marginal, isolated voices. In the 1930s-early 1940s, France was at the forefront of the "Neo-Hippocratic" and "Naturist" movement, upheld by such prominent physicians as Claude Sigaud, Paul Carton, René Biot, Louis Corman and Pierre Winter. In 1933, the world's first journal dedicated to natural and integrative medicine – *Hippocrate* – was inaugurated by Prof. Maxime Laignel-Lavastine of Paris. In the same year, on the initiative of Dr. Jérome Casabianca, there was organized in Marseille "La Société de Médecine Naturiste de Marseille – Médecine Préventive et Néo-hipppocratique" (the Marseille society for naturist, preventive and neo-Hippocratic medicine). The first International Congress on Neo-Hippocratism was held in July 1937 in Paris, under Laignel-Lavastine's presidency. Then, the first grand, national conference in the field took place in Marseille in November 1938, presided over by Prof. Lucien Cornil, dean of the Marseille medical faculty. In 1939, there was founded the "Union pour la defense de l'espèce" (The union for the defense of the species") dedicated to developing agriculture "conforming to our physiological needs." At the time, these were pioneering institutions on the world scale.[238]

At about the same time, the holistic movement also became consolidated in Germany, sometimes referred to as the "New German Science of Healing (or Medicine)" – *Neue Deutsche Heilkunde*. The "Verein Deutsche Volksheilkunde" (the Association of German Folk Medicine) and the "Paracelsus Institute" in Nuremberg were formed in 1935. The Congress of the New German Science of Healing was held in

Wiesbaden in 1936. In March 1940, Dr. Leonardo Conti, the Health minister of the Reich, charged Dr. Ernst Günther Schenk of Munich with establishing a scientific society dedicated to research and promotion of the "methods of life and medical care conforming to nature, in order to increase the health of the people." And in 1941, the German "Committee of Associations for Natural Methods of Living and Healing," was reorganized into the "German People's Health Association."[239] Yet, in several important instances (such as the world's first specialized scientific journal and the first international congress on Neo-Hippocratism), the French preceded the Germans. The simple enumeration of the methods enjoined by the French "Neo-Hippocratic" medical movement, reveals an uncanny similarity with what in some 20-30 years later came to be celebrated in the English-speaking world as the "holistic" or "New age" approaches to health-care.[240] In summary, French physicians of the 1930s-1940s led the world in promoting "natural" and holistic health-care and life-extension, as they formerly led the world in developing reductionist rejuvenative medicine.[241]

In addition to the Hippocratic tradition, several French authors of the 1930s-1940s, imported conceptions of "wholeness" from Oriental traditions, the Chinese in particular. Here too, French intellectuals appear to be at the forefront. At the *fin-de-siècle*, many European, particularly British researchers, exhibited great contempt for Chinese medicine and its Taoist religious foundations.[242] French researchers of the 1930s were among the first to speak of the Chinese medical tradition with respect, including the pursuit of physical immortality so pervasive in Taoism. The French sinologist Henri Maspero (1882-1945) provided a pioneering sympathetic account of Taoist immortalism, of its philosophical rationales, as well as of the Taoist "external" alchemy, seeking life-prolongation by external remedies (such as cinnabar, gold or opotherapy), and "internal" alchemy employing respiratory, dietary, gymnastic, sexual and spiritual techniques.[243] Maspero's work long preceded the rise of interest in Taoist science in the English-speaking world, popularly associated with the works of Joseph Needham (*Science and Civilization in China*, 1954) and Fritjof Capra (*The Tao of Physics*, 1976).

Other French researchers of the 1930s, such as George Soulié de Morant (Paris, 1878-1955) and Paul Ferreyrolles (La Bourboule, 1880-1955), were more concerned with immediate practical applications of Chinese medicine, such as acupuncture. Soulié de Morant was a leader in the field, proud of having introduced acupuncture to Europe in the early 1930s, with the publication of his seminal *Précis d'Acuponcture* (summary of acupuncture) in 1934. (Indeed, reports about acupuncture appeared in Europe since the 17th century, but never before was it consistently and widely used in European medical practice. In the US, acupuncture was not widely known before the 1960s, and the interest peaked in the early 1970s with the improvement of American-Chinese relations.[244]) In *Heretical Medicines*, Soulié de Morant treated of "Acupuncture, vital energy and cosmic electricity" ("Acuponcture, énergie vitale et électricité cosmique"), in the wider context of Taoist philosophy.[245] However, the Taoist message of extreme longevity was watered down. Rather, the work emphasized the general concepts of the vital energy *chi*, of the balance of "yin" and "yang" (explained as relative poles on a continuum), the wholeness of human beings and their integral relationship with the environment. A great effort was made to relate the principles of the "yang" or activity (the high, hot, hard, dry, vigorous, male, right) and the "yin" or nourishment (the low, cold, soft, moist, passive, female, left) with the notions of electromagnetism, electrophysiology or "radioesthesia" – the study of the effects of electromagnetic fields on the organism, initially including the study of dowsing (wand divination) – a forerunner of radiobiology. The "yin" was related to electrical "negativity" (as the nurturing earth has a net negative charge) and the "yang" was equated with "positivity" (as the sky has a net positive charge). Together with other contemporary French electro-physiologists and radioesthesists – doctors Elio and Hugo Biancani, Jacqueline Chantereine and Camille Savoire, Albert Leprince, and others – de Morant sought the most electromagnetically propitious environments for human health, resurrecting and rationalizing the ancient art of Feng-Shui.[246]

The existence of bioelectricity, of electromagnetic field emanations by the human body, and their relation with surrounding fields, were the bases for de Morant's theory and practice. The body was perceived as an integrated energy system. Imbalances of the "yin" and "yang" energies, that is, the local overabundance or deficit of the two kinds of energy, were seen as the major cause of disease. Therefore, "the purpose of acupuncture is to displace [redirect] energy between yin and yang." The metallic needles (sometimes bimetallic, made of gold and silver, acting as galvanic elements) were supposed to act either as conductors for the discharge of excessive energy (dispersion) or supply deficient energy (tonification). According to de Morant, acupuncture "does not create energy, but adduces energy in infinitesimal quantities from metallic needles or atmospheric electricity." De Morant further noted, apparently for the first time, the existence of anomalous electric resistance at acupuncture points. These ideas preceded those of American electrophysiologists.[247]

In de Morant, the body was seen as an integrated energy system, affected by cosmic energies, and responding to "alterations and variations of the environment, the atmosphere and the universe."[248] Hence, an extensive discussion was dedicated to the correlation of human activities with fluctuations of the weather, planetary cycles, seasons of the year, phases of the moon, times of the day and places of the earth. By recognizing and adapting to the environmental factors, based on the traditional Chinese precepts, the human being was treated as a whole and as a part of the cosmos. Thus, de Morant joined the avant-garde of French holists, champions of the conservative revolution. Indeed, the rejuvenators of the 1920s, notably Voronoff and Steinach, also credited the Chinese as the initiators of opotherapy, but they made no reference to Taoist concepts of unity. De Morant too presented a rather abridged version of Taoist alchemy. Its more metaphysical and obscure notions, such as the partaking in the properties of the eternally unchanging but flowing Tao by the immortal adept (*hsien*) or the principle of "effortless action" (*wu wei*) were left out. Yet the introduction of the basic concepts of organic unity and communion with nature derived from Taoism was an essential component in the "holistic turn" of French physicians and life-extensionists of the 1930s-early 1940s.

Others built their constructs of wholeness on the foundations of Indian tradition. The Parisian internist Pierre Winter, in "What must be the traditional medicine?" (Que devrait être une médecine traditionelle?"),[249] promoted the Ayurvedic medical tradition, founded on the principles of humoral balance, that "remained alive in India without change for millennia," as well as the Yogic tradition leading to bodily and mental purification and endurance. The principle of unity of all being was emphasized, and the paths toward its recognition and spiritual liberation involved the physical preparation, purification and disciplining of the body for higher forms of spiritual meditation. The influence of the mind over the body was seen as almost unlimited, leading to sizable health benefits, enhanced physical abilities and longevity. Yet, the improved regulation of physiological functions, health improvement or extraordinary physical feats performed by yogis and fakirs, were said to be only inferior and subservient to spiritual enlightenment. Winter's central thesis was that traditional medicine must be fundamentally spiritual and "sacerdotal" – the medical profession needed to assume the function of priesthood. In India, Winter noted, the practice of medicine was performed by the elite Brahmin caste of priests and scholars. Winter provided further examples of the fusion of the pursuit of health and longevity with spiritual pursuits, and of the union of medicine with priesthood, based on sacerdotal-medical traditions from the Egyptian to Roman Catholic priests.[250] It was Dr. Winter's goal to restore the traditional "priestly" character of medicine.

Dr. Pierre Merle of the Montpellier Academy of Sciences and Letters, in "Rationally inexplicable cures" (Guérisons rationnellement inexplicables") further linked healing with the Christian faith. The particular emphasis was on faith in the Catholic doctrine, as the influence of Catholicism increased in France in the 1930s and during the Vichy regime.[251] Merle methodically recounted the miraculous cures produced by the power of prayer and the laying on of hands: those produced by Christ,[252] by the Church fathers, and by modern Christian devotees, as confirmed by Vatican investigations for canonization of particular church figures.[253] A special emphasis was given to miraculous cures reported at the sanctuary of Lourdes, where a large Bureau of Medical Records has been operating since 1884 to verify the miracles. Meticulous medical anamneses were provided, describing how the blind begin to see, the lame walk, cancers disappear, bones consolidate, and so forth, often extending the life of patients for decades.[254] Merle raged against earlier "materialist" and "determinist" researchers, such as Jean-Martin Charcot and Julien Mareuse, who denied the very possibility of miraculous cures that "contradict all the laws of biology and pathology," who disputed the authenticity of reported miracles, and, in cases of apparent authenticity, sought their explanation in some yet unknown physiological or neurological functions. According to Merle, Carrel too, though he acknowledged the phenomena of miraculous cures, was insufficiently faithful. Since, in *Man, the Unknown*, Carrel said that miraculous cures imply "the reality of certain relations, of still unknown nature, between psychological and organic processes," Merle concluded that "Carrel did not lose the hope to explain, based on determinism, on a correlation between organic and mental processes, the extraordinary facts [of the cures at Lourdes] that he has been incessantly pondering since 1902." Instead, Carrel should have acknowledged the fundamentally "imponderable" direct intervention of the Holy Spirit, invoked by prayer. Mechanistic explanations being entirely out of the question, not even the power of mental suggestion could be accepted as an explanation. According to Merle, "psychological suggestion cannot explain the instantaneous disappearance of certain organic lesions, fractures, Pott's disease, pulmonary tuberculosis, or malignant tumors, which Charcot considered as pseudo-diseases originating in hysteria." With reductionism thrown out of consideration, space was vacated for full-fledged spiritualism.

The unity of the spirit and the body was also the central theme in "Medicine and imponderable agents" (La médecine et les agents impondérables) by one of the leading Parisian homeopathic doctors and psychoanalysts, René Allendy. The spirit-body connection was emphasized, even though this work contained little reference to religion.[255] According to Allendy, the primitive conceptions of dualism – matter vs. energy, body vs. soul – need to be rejected in favor of a synthetic, unified vision. Appropriating terminology from contemporary radiation and quantum physics, Allendy claimed that both matter and energy, mind and body, are forms of energy that cannot be dichotomized, but rather represent a "gradual continuity." It is this continuity that can explain the profound action of thought on biological tissues. It may explain the reported "miraculous therapeutic actions" of "magnetizers, healers, occultists" as well as faith cures. The entire world consists of "electric charges and radiation," through which all beings are interrelated, and the therapeutic effects are achieved via "wave resonance." Each individual organism emits structured radiation, comprising the "luminous ethereal body" or "ethereal double" spoken of in mystical traditions throughout the world, from the ancient Egyptians to contemporary Spiritists. The individual "ethereal body" is connected to those of others, to the environmental and cosmic energies at large, and is considered as the primary target of intervention. Within the "ethereal body," the thought can enact a positive physiological change "not only on the tissues of the thinking individual himself, but also on other individuals toward whom it is directed," thus rationalizing reports of telepathy, distant therapy and divinatory radioesthesia. Changes in the "ethereal body," in its "energy levels" or "quantum states," induced by thought and mediated by emotion, were said to be capable of producing immediate and all-embracing healing, i.e. the restoration of "wholeness," the "instantaneous *restitutio ad integrum*." Such notions have remained the staple of holistic, "mind-over-body" life-extensionism to the present. [256]

For Allendy, as well as for other contemporary French holists, the notions borrowed from electromagnetism – "radiation," "energy field," "electric charge," "vortices," "vibrations," "waves," "quantum states," "resonance" – formed the theoretical basis of discussion. The studies of cosmic and environmental electromagnetism and of electrotherapy – by Robert Tournaire, Étienne Pech, Auguste Lumière, Gaston Sardou, Maurice Faure, and many others – sought to trace and manipulate those "imponderable," intangible agents. The International Institute of Cosmobiology of Nice (Institut International de Cosmobiologie de Nice), apparently the world's first institute in the field, was in the 1930s actively researching the effects of environmental radiation on health, as well as attempting radiation therapy and electrotherapy.[257] Allendy conducted laboratory experiments on the "physico-electric displacement of vital energy" in cases of metastatic transfer, attempting to cure one lesion by inducing another. He noted that the current knowledge of bioenergetics was rudimentary, but believed it opened a field of unlimited possibilities. Yet, Allendy did not go into technical details on radiation and quantum physics, such as Maxwell's equations or wave functions, and was very likely incapable of doing so. Rather, the electrophysical allusions seem to have provided a terminological framework for Allendy's main subject of interest: the manipulation of the mind. And in that area Allendy was an acknowledged expert, being one of the founders of the Psychoanalytic Society of Paris (Société psychanalytique de Paris, established in 1926). "Mental representations," "images," "suggestions" were the true means to program the individual for health and longevity, rather than electromagnetic coils.

The holistic connection between mental affirmation, health and longevity, seems to have been well grounded in contemporary psychoanalysis. To recall, Freud too considered the organism and the mind as an entirety, discussed in terms of energy balance, undergoing energy "tensions" and "discharges," where the "Life Drive" or "Eros" were defined as instincts "exercising pressure towards a prolongation of life" and "integrating the living substance."[258] In *Heretical Medicines*, the power of the mind was further emphasized in "Psychological Medicine" (Médecine Psychologique) by the psychiatrist Marc Guillerey of Lausanne, Switzerland, one of the founders of the "guided imagination" technique (rêverie dirigée).[259] Guillerey concluded that psychological therapy "must aid medicine in resuming contact with its object, the real human being, his personality, spirit and body, with profound realities, incessant changes and influences of the social milieu, the universe, and the world of values."

And that was indeed the unifying message of all the 'heretics' of the 1930s - early 1940s: emphasizing the need to treat the human being as an individuality and as an ensemble, including mental, social and environmental factors, involving a combined action of all therapeutic means available, physical and psychological. As succinctly stated by the founder of the "Lyon Group for Medical, Philosophical and Biological Studies" (1924), René Biot (1889-1966), in the article "Toward the unity of medicine" (Vers l'unité de la médecine):[260] "[Therapy] must aid in the restoration of equilibrium of the entire human being, resorting to all the means by which one can hope to attain this goal, by remedies and regimens, by operations, serums, irradiations, by those infinitely complex and delicate affairs one so nicely calls 'care' – from the aspect of the room and a flower on the table, to smoothing a wrinkle on the pillow." Moral values, even the faith of the patient, as well as those of the therapist and researcher, are indispensable components of biomedical research and practice. As Biot pointed out, physicians and researchers "hardly have any illusions about the imperfection of their knowledge." Nonetheless, "the most sceptical and disillusioned ones respond to the call of the suffering and return to their laboratory to patiently resume experimentation." According to Biot, "The belief in Providence, far from impeding medical effort, encourages scientific research."

The message of reconciliation and unification of various forms of therapy recurred in the article of Dr. Paul Jottras of Elbeuf, "Scientific analysis and humane medicine" (Analyse scientifique et médecine humaine).[261] The majority of authors in *Heretical Medicines* valorized the power of the mind over the body, highlighting "synthesis," and taking up arms against mechanistic, materialistic, deterministic or reductionist views. Jottras begged not to discard the reductionist, "analytical" approach altogether. He pointed out that "pre-scientific," vitalist and anti-materialist views often masked lacunas in the understanding of physiological mechanisms, and sometimes amounted to superstition and charlatanry. He reminded the readers that thanks to the conquests of "materialist," "quantitative" and "analytical" medicine, such as antiseptics and anesthesia, "since the beginning of the scientific era, the average duration of human life approximately doubled." Yet, according to Jottras, "*exclusive* analysis" is "sterile" and "insufficient." Rather, analysis must be complemented by synthesis. The classical antitheses – patient and disease, contagion and heredity, humanism and physical chemistry, intelligence and *élan vital* – are not mutually exclusive. Rather, these are "complementary" laws and disciplines which limit each other, but do not exclude each other. According to Jottras, medicine must expand to become a "method for happiness, recognizing esthetic and moral values," for "what is the purpose of the prolongation of human life, if one does not know how to use it?" Having embarked upon the way of unification and complementarity, human beings can overcome their limitations and perhaps become "superhuman."

The message of reconciliation between the so-called "official" and "heretical" medicines was most vehemently expressed by Rémy Collin, professor at the Medical Faculty of Nancy and a corresponding member of the French Academy of Medicine, in "Is there an official doctrine?" (Existe-t-il une doctrine officielle?).[262] Collin loathed the opposition between "innovators" and "classics." Very often, the so-called "innovators" – independent practitioners, researchers and healers – were represented by their detractors as charlatans and ignoramuses. On the other hand, the members of the so-called establishment, the proponents of the "classical" or "official" doctrine, were depicted by opposition as petrified snags on the road to progress and arrogant, self-serving bigots. According to Collin, such a dichotomy is far from representing the actual state of affairs and at any rate cannot be normative. Indeed, Collin attributed a privileged position to the "establishment." According to him, it is the obligation and the right of the State to ensure public health, and for that purpose to regulate medical instruction and practice, ensuring compliance with a certain set of standards via established institutions. Such institutions are the origin of the "traditional" or "official" doctrine. However, they are also the origin of the largest part of innovation ever produced. Moreover, unaffiliated "independent" researchers are defined by their relation to the existing institutions: they either need a certain level of accreditation from the "establishment" to give credence to their propositions, or, at the very least, be aware of the existing standards to challenge them. Conversely, the official doctrine derives from the independents "the nourishment to increase its growth," though retaining its "preeminence." According to Collin, there is no conflict of goals between "classics" and "innovators" whatsoever:

We must trust the official doctrine and education with providing the practitioners with classical means of action. It is clearly desirable that therapy accelerates in pace, that social evils are defeated, that human beings avoid suffering, and that human life is extended to extreme biological limits. This generous desire supports the most ardent of innovators in their efforts. But is it not also shared by the classics?

The desire to maintain equilibrium and constancy, I argue, has been the underlying drive for the majority of life-extensionists, even though they sometimes envisioned quite unorthodox and varied methods to sustain the constancy. Collin expressed this desire for conservation in most unambiguous terms. According to him, biomedical research and practice, including the pursuit of longevity, are born within established institutions and can never exist apart from them. They evolve toward strengthening of institutions and toward unification and stability of the doctrine. The unified doctrine, solidified and perpetuated in established institutions, "provides us, here and now, with applicable means in the fight for health, it also gives us an idea of permanence, of necessity, which is a factual contributor to science." Rather than excluding or annihilating unorthodox schools and doctrines, Collin proposed that they should be tested and subsumed into the edifice of the "official" medical establishment. Collin, together with professors Pierre Delore and Marcel Martiny, fully supported the (unrealized) proposal for "The Institute of Control" (Institut de Contrôle) intended to "methodically explore para-medical treatments, subject them to scientific criticism and severe control, to distinguish between the true, the false and the doubtful, between the useful, unuseful and harmful, to establish a balance, to appropriate what works and reject the rest." Collin fully agreed with the principle "*Ars una, species mille*" – Art is one, its species are thousand. And, of course, psychological and social factors are never to be excluded from the unity of the "official doctrine," as it must incorporate not only "original cognitive content" but also "moral and social values" in the "unanimous effort toward procuring well-being, alleviation of suffering, preservation of health – the conditions for the infinitely more precious achievement, the realization of human personality and its ultimate purpose." Thus, the existing medical establishment, replenished rather than torn apart by diverse holistic streams and traditions, can itself be made more durable and produce lasting effects.

In summary, a great, pioneering effort was made by the authors of *Heretical Medicines* to incorporate "holistic" approaches into the foundations of the French medical establishment, drawing on the "perennial" Hippocratic, Oriental and Catholic traditions, and rejecting more "modish" materialistic and reductionist interventions. It may not be accidental that the "holistic shift" occurred concomitantly with the strengthening of the so-called French political "traditionalism," reinforcing in the 1930s and culminating during the Vichy regime. The term "traditionalism" (*traditionalisme*) generally refers to an attempt to return to pre-Enlightenment values, for example, replacing the values of "Liberty, Equality, Fraternity" with "Labour, Family, Fatherland."[263] Yet, its extension into the field of longevity research is conceivable. It is difficult to speculate whether the attempts of rejuvenators to dissect and make far-reaching "progressive" intrusions into the "human nature" became increasingly suspicious within the political "traditionalist" paradigm; or whether the failure of the "rejuvenative" intrusions somehow contributed to the general rise of "traditionalism." In any case, the concomitance of the "holistic shift" with the rise of "traditionalism" may be significant.

18. Longevity and holism. The theory of Sergey Metalnikov (1870-1946) and the practice of Alexandre Guéniot (1832-1935)

The majority of authors of *Heretical Medicines* (most notably Carrel, Lumière, Collin, Jottras, Martiny, Allendy, and Poucel) expressed some hope for increasing longevity and combating aging. They spoke of the desirability of a dramatic increase in the "mean duration of human life," "prolongation of human life to extreme biological limits," reaching extreme old age "free of decrepitude," preventing "precocious senescence." Yet, for them, extending longevity was not the central theme. Rather, the ubiquitous subject was the preservation of "wholeness" and "equilibrium" of human existence that must involve not only organic, but also psychological, social and environmental components, emphasizing hygiene and life-style improvements. Temporal limits for maintaining such an equilibrium were not posited, and it was implied that it could be sustained for prolonged periods. The analysis of *Heretical Medicines* indicates that the pursuit of longevity in France in the 1930s-early 1940s existed in the context of a substantial strengthening of these "holistic," or as it was then more commonly termed "Neo-Hippocratic," approaches to prevention and therapy.[264] As it can be clearly seen from the works of Carrel and Lumière, the "holistic" precepts emerged in part as a counter-reaction to "reductionist," "materialistic" and "mechanistic" approaches to biomedicine that had informed the "rejuvenation attempts" of the 1920s.

The shift toward a "holistic" perspective can be further exemplified by the works of other contemporary French authors for whom longevity was indeed the central theme. The writings of Sergey Ivanovich Metalnikov provide a striking example. Metalnikov was Metchnikoff's pupil at the Pasteur Institute, and then professor of zoology at St. Petersburg University (1907-1917). Descending from a wealthy Russian aristocratic family, he immigrated to France after the revolution of 1917, and to the end of his career continued to work at the Pasteur Institute. The "holistic turn" in his work can be shown by comparing two editions of Metalnikov's book on the subjects of rejuvenation and longevity, of 1924 and 1937. The Russian title of 1924 was *Problema Bessmertia i Omolozhenia v Sovremennoy Biologii* (*The Problem of Immortality and Rejuvenation in Modern Biology*), and the French edition of 1937 was entitled *La Lutte Contre La Mort* (*The Struggle against Death*).[265] Metchnikoff's loyal disciple, Metalnikov fully shared his teacher's dream of radically extending human longevity (to the age of 150 at least), as he shared many of Metchnikoff's theoretical precepts: the supposition of potential immortality of living matter and the central role of intestinal intoxication in aging. A major portion of Metalnikov's research concerned combating intestinal putrefactive micro-flora to the point of sterility.

The edition of 1937 reproduces in bulk the text of 1924, with identical overarching conclusions. According to Metalnikov, potential immortality, and not death, is the most characteristic feature of living organisms:[266]

> All the stated facts convince us that immortality is the fundamental property of living organisms, and that the multiplication, growth and succession of generations are only possible thanks to immortal sexual and somatic cells that incessantly nourish themselves, accumulating new reserves of living matter.
> … The living cell, under present conditions of terrestrial existence, can divide and multiply to infinity. Aging and death are not necessary stages of terrestrial existence.

And if fundamental life components are potentially immortal, their ensemble can be made potentially immortal, and thus the efforts toward immortalization are "practically feasible and scientifically grounded." Even though Metalnikov acknowledged that the present knowledge of the factors regulating cell growth and immortality is rudimentary, and their practical manipulation is presently unavailable, the problem is well within the scope of experimental science and biomedical technology:

The human organism is composed of living cells that have the potential for infinite growth and multiplication. If we could restore to the human being that property that we observe in so many primitive organisms (hydras, worms, etc.) – the ability to restore lost parts, then the problem of aging and death would be solved. If we found the factors that induce the division of cells in a living organism and could regulate their multiplication, the problem would be solved.

These are fairly reductionist statements, carried forward from 1924 to 1937, underscoring Metalnikov's fundamentally materialist worldview. Yet, a substantial revision was made in the edition of 1937. In that latter edition, Metalnikov added the brain to the body.

Already in 1924, Metalnikov briefly indicated that "we insufficiently consider the role of neural centers and psychic activity in the life of the organism and even its individual cells and tissues…. Through the education of the will, through certain training, through auto-suggestion or external suggestion, we could obtain important results."[267] In the edition of 1937, he expanded that thought into several chapters. He proposed a central role of the nervous system in immunity, to the point of suggesting that immunity could be a conditioned reflex. Similarly to Lumière, Metalnikov believed that phenomena of immune hypersensitivity and desensitization are regulated by the nervous system, and can be subject to conscious control. (Lumière, however, was critical of Metalnikov's expositions. He thought Metalnikov's title "The problem of immortality and rejuvenation in modern biology" should be replaced by "The problem of immortality and rejuvenation in 'Ancient biology'" because Metalnikov did not make account of Lumière's cutting edge "humoral-colloidal" theory.[268]) More generally, in 1937, Metalnikov attributed great importance to conditioned reflexes as principal factors in diseases, nervous and chronic, viral and toxic. According to Metalnikov, "a simple change in the conditions of life, that is to say, the suppression of some conditioned reflexes, is enough to produce a salutary effect on the patient." The action of speech, or "verbal excitation," was considered as a conditioned reflex, as a means "to invoke in the organism certain defensive reactions." The examples of profound organic changes, either destructive or healing, effected by verbal suggestion "demonstrate the possibility of the utilization of speech as a very important therapeutic means that intervenes in the function of the entire organism."

The final chapter of the 1937 edition was dedicated to Yoga, in place of the chapter on Steinach's and Voronoff's operations that concluded the book in 1924. In that chapter, Metalnikov catalogued feats of physical endurance and health benefits attained by yogis and fakirs. He admired Yogis' conscious mastery of the body, attained in their ascetic effort to unify with the Universe, the very term "yoga" meaning "unification." Exercises in "Hatha Yoga" (literally the "sun and moon yoga") or the physical preparation for spiritual ascension, impart to the "ensemble of the body the flexibility, power and harmony, the factors of a perfect equilibrium." Yogic practices demonstrate that "vegetative life" is inseparable from and controllable by "consciousness." They show that "the will" and "mental concentration" can have "infinite power for regulation and order." Even though the yogis are ignorant of the structure of their organs, they are incontestable masters of their functions. Most remarkably, they masterfully control the respiratory function and hence the overall body energy. According to Metalnikov, yogic mental, physical and alimentary practices should be extensively studied and appropriated into a new branch of longevity science. "Surely," Metalnikov concluded, "this science, involving many secrets and mysteries, can open for us new ways in the struggle against diseases and death."

In addition to giving a greater weight to mental factors, the 1937 edition also placed a stronger emphasis on social and environmental factors for longevity. Indeed, Metalnikov remained quite fond of reductionist rejuvenation:[269]

The works of Brown-Séquard, who for the first time showed that hormones of seminal glands act as a stimulant on the aging organism; the works of Metchnikoff who studied the phenomena of aging in different animals and showed the role of intestinal flora in diverse infirmities and in premature senescence, and who outlined new ways for the struggle against aging; finally the works of Steinach, Voronoff and many others, who showed that the transplantations of living seminal and sexual glands into the aging organism act in a rejuvenating manner – all these are such acquisitions of science that will never be forgotten.

Yet, he judiciously removed from the 1937 edition the chapter on human applications of Steinach's and Voronoff's procedures, and only left the chapter on animal experiments. Apparently the "rejuvenation experiments in humans and higher animals" of the 1920s, did not produce sufficiently conclusive results that Metalnikov could still flaunt in 1937.

In 1924, the chapter entitled "Is the struggle against aging and death possible?" ended with an account of experimental regeneration, rejuvenation and life-prolongation in hydras and flatworms (among other factors, by starvation, cold, wounds, chemical exposure, and other kinds of stressful milieu) and with the hope that further extensive research will discover the factors regulating the multiplication of cells within the human organism, and thus it will become possible to control cell life and effect unlimited damage repair. In 1937, the chapter ended with a consideration of environmental and social factors in human longevity:[270]

> Each year, we make new discoveries in the study and treatment of diseases. It will not be long before all diseases will be vanquished. However, exterior conditions also play a very important role: profession, housing, resources, medical and sanitary organization, and finally education. The spreading of well-being, of hygiene, the progress of medicine, considerably reduce general mortality.

Metalnikov supported the notion that heredity and social factors are of about equal importance for longevity. His ultimate conviction was that "human age limit can significantly exceed 150 years, if we could live in perfect hygienic conditions." To the last, Metalnikov believed in the unlimited potentialities of experimental, reductionist science in fighting senescence. To accelerate progress in this area, he advocated the establishment of "an institute for experimental studies of aging and death."[271] The radical prolongation of life, he believed, is "not a fantasy, but a real scientific problem" that could be resolved by "collective work of many scientists." Yet, the increasing emphasis on the mental, social and environmental determinants in his work indicates that, despite his entrenched materialism, he was no exception from the "holistic turn" in French life-extensionism of the 1930s.

Another crucial figure in French life-extensionism of the period, the centenarian president of the French Academy of Medicine, Alexandre Guéniot of Paris (1832-1935), placed even less stock in reductionist rejuvenation. His book *Pour Vivre Cent Ans. L'Art de Prolonger ses Jours* (To Live a Century. The Art of Prolonging the Days, 1931)[272] valorized hygiene and the direct power of the mind over experimental, reductionist biology. According to Guéniot, the mechanistic views of former French longevity researchers were too narrow and overlooked a large area of potential means for life-prolongation. Thus, according to Guéniot, nothing can be farther from truth than the claims of Georges-Louis Buffon (1707-1788) who stated that "the duration of human life does not depend either on habits, or morals, or the quality of nutrition, nothing can change the laws of mechanics that determine the numbers of our years" and those of Marie-Jean-Pierre Flourens (1794-1867) who posited that "the duration of life does not depend either on climate, or nourishment, or race, does not depend on anything exterior; it depends only on the intimate constitution of our organs." In contrast, it was precisely the adjustment of the life-style and environmental factors that Guéniot advocated.

Indeed, Guéniot attributed great significance to the "hereditary vital force" that, according to him, determines the natural duration of human life at no less than a hundred years. Yet, he emphasized that very few attain this limit, and the task of practical life-extension must consist in hygienic measures to help everybody attain it:

The human life has a *natural* duration of about hundred years. There can be no proper discussion of life-prolongation, except for attaining this term. But when we consider that the *ordinary* duration does not exceed 75 to 85 years, the expression is exact, because this is a veritable prolongation of life for 15 to 25 years. (Emphasis in the original.)

Though recognizing the power of heredity, Guéniot was far from advocating eugenics. Even though he acknowledged that there might be a rationale for encouraging healthy parentage, eugenics is "not limited to healthy practices," and becomes fraught with "culpable pretensions." It can rapidly expand into racism, interfering with interracial marriages, sterilizing "bad producers" and "undesirables," putting to death severely ill and handicapped. Hygiene provides an alternative: from the moment a new being is conceived, during and after the birth, "both the mother and infant must become objects of enlightened care that would insure for them suitable existence in their particular conditions." In this way, "without recurring to practices that are as much criminal as they are radical … the application of special hygiene, applicable for each particular case, will prevent the eventuality of an undesirable or harmful generation."

Guéniot was not particularly inspired by reductionist rejuvenation either. Indeed, the vitalizing effects of sex hormones were acknowledged, and their supplementation deemed reasonable. However, according to Guéniot, continuous injections of hormones were not equivalent to a continuous functioning of sex glands within the human organism. The transplantation of sex glands, as practiced by Mauclaire, Voronoff, Dartigues and others, in turn, raised almost insurmountable difficulties. Transplants from domestic animals were impossible due to rejection, and those from apes or humans were so difficult to obtain, rare and costly, that it made their use generally impracticable and only possible for a very "few privileged" subjects. At any rate, insofar as such transplants had been only practiced for a dozen years, their long-term effects and potentialities were difficult to assess. Hence, presently, endocrine grafts "cannot be held for a solution. Let us wait for a solution in the future." According to Guéniot, the vitalizing power of sex hormones can only be practically utilized within the framework of personal hygienic regimens. "No excess in the expenditure of this matter during our maturity!" Guéniot urged.

Indeed, hygienic regimens occupied the central place in Guéniot's exposition. According to him, these were perhaps the only means currently available for life-extension, and may remain the only means for some indeterminate time. The traditional assortment of measures was discussed: physical exercise, massages, baths, rest and sleep, pure air, moderate and balanced, vitamin-rich nutrition. The simultaneous "holistic" application of diverse and concerted means was preferred over any reductionist tweaking of a particular target.[273]

According to Guéniot, within the ensemble of hygienic measures, psychological and moral factors play a crucial role. Guéniot noted the "astonishing conservation of mental faculties in ultra-centenarians" and hypothesized that the "exceptional vitality of ultra-centenarians enacts in an equal measure all the parts composing their organism," achieving a balanced exercise of the body and mind. Following a moral code was believed to be highly conducive to longevity. The entire book was in fact a set of codified rules of conduct, and moral rules were not excluded. In the best tradition of the early life-extensionist hygienists – Cornaro, Lessius, Cohausen and Hufeland – Guéniot's macrobiotic moral code involved a strong Christian sentiment: "To become free of defects, faithfully follow the precepts of the Gospel. 'The code of happiness, a wise man said, is all in these precepts.' Knowing them, you will advance in the right path; you will guard the peace of your soul at the same time as your natural happiness." And furthermore, "It is from the point of view of the divine, that human life really acquires all its value and its splendid beauty." The code emphasized the values of frugality and industry, and addressed the middle-to-upper class: "scholars," "magistrates," "officers," "honorable merchants." In summary, Guéniot's work was conservative in morals, as it was conservative in methods. The morals stemmed from and were addressed to his immediate social surroundings. The methods built on centuries-long hygienic traditions, emphasizing "natural" regimens, not much different from anything we do every day, in counter-reaction to more farfetched methods of intervention. Yet, Guéniot proposed an overarching moral imperative for life-extension that would apply to all social settings and that would be served by all interventions: the absolute value of human life and its preservation. "Yes, for man, life is a Good. It is the primary of all benefits, without which all other benefits are as much as inexistent."

19. Chapter conclusion. Between reductionism and holism

Guéniot's work is a prime example of the general trend found in the attitudes to life-extension in France in the 1930s-early 1940s. The methods proposed by Guéniot are essentially the same as for the majority of the proponents of Neo-Hippocratism or Naturism. Given the dates of publication, age and status, Guéniot may well be considered the patriarch of the movement. Guéniot, as well as other Neo-Hippocratists, moved away from the reductionism underlying the rejuvenation attempts of the 1920s, toward a more "traditional," "natural" and "holistic" perspective, aiming to integrate psychological, societal and environmental factors into a comprehensive therapy for the 'perpetuation of the whole.' The Neo-Hippocratists also appear to have embraced a transition from French progressivism and positivism toward more traditionalist and religious views (the transition that may have been generally characteristic of the French society of the 1930s and the Vichy regime.[274])

After the "holistic turn" France seems to have lost its leading position in the life-extensionist movement. The diminishing impact of French life-extensionists after the departure from reductionist rejuvenation endeavors is understandable, as it signified a departure from daring interventions toward a more protective and conservative attitude. Since the 1930s through the 1950s, the (predominantly critical and cautious) discussions of rejuvenation continued in France, notably by the founders of the French Society of Gerontology (formed in 1939) – Leon Binet (1891-1971) and Francois Bourlière (1913-1993). Leon Binet, who became president of the French Academy of Sciences in 1957, focused on clinical geriatrics and experimented with oxygen therapy and embryonic extracts. Francois Bourlière, who became the founding director of the Claude Bernard Center of Gerontology in Paris in 1956, conducted comparative studies of average longevity from various animal species and supported the use of hormones (mainly sex hormones) in geriatric therapy.[275] In the late 1940s, Michel Bardach of the Pasteur Institute in Paris advanced a "rejuvenative"/"orthobiotic"/"stimulating" cytotoxic serum.[276] In the 1950s and 1960s, the biologist and philosopher of science Jean Rostand (1894-1977) was an outspoken supporter of radical life-extension and pioneer of cryo-preservation.[277] But none of these researchers seems to have had the notoriety and ambition of the French "founding fathers" of life-extensionism – Brown-Séquard, Metchnikoff and Voronoff.[278]

Presently, the discussion of radical longevity in France appears to have become little noticeable. And yet, despite the apparent relative decrease of emphasis, longevity research does continue in France. There are well established gerontological research centers and associations.[279] Anti-aging medicine is widely spread in France, with applications ranging from cosmetics to stem cells.[280] There are vibrant (though apparently not very numerous) transhumanist and life-extensionist communities.[281] And of course, several famous longevity-related issues have been discussed: the French super-centenarian woman Jean Calment (1875-1997), holding the certified world record of human longevity at 122 years and 164 days[282]; the relatively high average life-expectancy in France (rising from about 81 years in 2010 to about 82 in 2013)[283]; the "French Paradox" (the alleged low incidence of heart disease, which is commonly explained by a relative moderation in diet and by a moderate consumption of red wine),[284] etc. Despite all these, few groundbreaking longevity research sensations seem to have lately emerged out of France, compared to the trail-blazing times of the early 20th century. Was the pioneering spirit lost in the "holistic turn" of the 1930s? Are there now signs of catching up?[285]

As will be shown in the next chapter, a very similar trend occurred in Germany in the 1930s, though it may have been more burdened by what Guéniot called "culpable pretensions." As both French and German attitudes would exemplify, the discourse on the prolongation of life implied radical conservationism, the drive for the preservation of current perceptions of personality and the perpetuation of the current social order.

Chapter 2.

Germany: The preservation of the "National Body" and "Connection to Nature."
Allies and neutrals – Austria, Romania, Switzerland

1. Chapter summary

Similarly to France, the discussion of life-extension in Germany in the 1930s showed a notable shift of emphasis from "materialistic-reductionist" or "artificial" rejuvenation toward more "natural" and "holistic" macrobiotic hygiene. The enhancement and prolongation of individual life were often seen as necessary conditions for the strengthening and maintenance of the National Body (*Volkskörper*) and eventually for the perpetuation of the ruling order. The "natural" hygienic life style improvements were emphasized, as the German society was supposed to be reverting to a "natural," "healthy" state of the nation. Besides the valorization of the "natural" hygienic approach, the second central characteristic of contemporary German life-extensionism appears to be its "national" emphasis. Within the "national" discussion framework, several elements are conspicuous: the almost exclusive focus on German achievement, as well as the authoritarian imposition of the power of the medical-political establishment from above, and the valorization of the collective will power of the entire nation from below. Rather than antagonistic, the massive enthusiasm from below and the authoritarian rule from above appear to be mutually reinforcing in the attempt to create a long-enduring "National Body." Similarly to France, after the "holistic/natural" turn, life-extensionism is Germany and Austria appears to have become largely deemphasized, in contrast to Romania and Switzerland where the tradition of reductionist rejuvenation remained strong.

2. The German tradition. "And Nature, teaching, will expand the power of your soul"[286]

Similar to the case of France, in Germany of the 1930s, "materialistic-reductionist" or "artificial" rejuvenation attempts were increasingly displaced by more "natural" and "holistic" hygienic approaches to life-extension. The dichotomy was not new for German thought. The ground-laying works in German life-extensionism – Johann Heinrich Cohausen's *Hermippus Redivivus or the Sage's Triumph over Old Age and the Grave* (1742),[287] Johann Bernhard Fischer's *On Old Age, its Degrees and Diseases* (1754),[288] and most influentially Christoph Wilhelm Hufeland's *Macrobiotics or The Art of Prolonging Human Life* (1796)[289] – emphasized "natural" (*natürliche / natürgemässe*) hygienic life-style improvements and scorned the alchemists' "artificial" (*künstliche*) rejuvenating nostrums. The same distinction was made by one of Hufeland's patients – Johann Wolfgang Goethe (1749-1832) in *Faust* (1808-1832). Before Faust embarks on his epic adventure, he considers the means that would lend him the youthful strength necessary for the effort: "Has not a noble mind found long ago Some balsam to restore a youth that's vanished?"[290] Mephistopheles knows "a natural way to make thee young":

Well! Here's one that needs no pay,
No help of physic, nor enchanting.
Out to the fields without delay,
And take to hacking, digging, planting;
Run the same round from day to day,
A treadmill-life, contented, leading,
With simple fare both mind and body feeding,
Live with the beast as beast, nor count it robbery
Shouldst thou manure, thyself, the field thou reapest;
Follow this course and, trust to me,
For eighty years thy youth thou keepest!

Faust finds the "natural" method too inconvenient and laborious. Therefore he needs to resort to a rejuvenating potion to be prepared by the witches. Though drinking the potion is easy, its preparation is devilishly difficult. "To tell each thing that forms a part would sound to thee like wildest fable" and the preparation "costs a still spirit years of occupation." Mephistopheles himself would not undertake the task. The witch produces the potion, with a great show of humbuggery (the "hocus-pocus" or "witches' arithmetic"). Though initially scared by the flames the brew exudes, Faust eventually drinks the potion and is infused with vital heat, eroticizing him and inducing his passion for Margaret: "Soon will, when once this drink shall heat thee, In every girl a Helen meet thee!" Ultimately, the potion's effect proves to be short lasting, and at the end of the play Faust dies in complete decrepitude. This short passage foreshadows many of the controversies in the pursuit of longevity in Germany.

As the works of the early hygienists exemplify, "natural" and "holistic" macrobiotics, "the art of prolonging human life" based on personal hygiene, physical work, diet and mental suggestion, rather than pharmacology or surgery, had been well established in the German-speaking world since the 18th century, and even earlier. Yet, from the beginning of the twentieth century and through the 1920s, "artificial" reductionist rejuvenation, using organotherapeutic (endocrine gland) supplements, gland transplantations or gland stimulation (predominantly those of the sex glands), surged in Germany and Austria, on a scale comparable to that of France and in a constant competition with French rejuvenators. Against the French pioneers of rejuvenation – C.E. Brown-Séquard, Elie Metchnikoff, Serge Voronoff, Placide Mauclaire, Lois Dartigues, Raymond Petit, Léopold Lévi, Henri de Rothschild, and others – the German-speaking world pitted its own champions. Austria led the way with the pioneering works of Eugen Steinach on sexual rejuvenation, most famously the "Steinach procedure" involving spermatic cord vasoligation, but also various forms of sex gland transplants and supplements. Steinach was followed by a host of other Austrian rejuvenators – Karl Doppler, Robert Lichtenstern, Paul Kammerer, Erwin Last, August Bier, Emerich Ullmann, Otto Kauders, and others – performing the "Steinach Procedure" by the hundreds and thousands, as well as other modified versions of sexual stimulation. Sigmund Freud (1856-1939) was enthusiastic enough about Steinach's rejuvenating operation to have it performed on himself in 1923 by the surgeon Victor Blum. In Germany too, in the first quarter of the century, the rejuvenation movement boomed, though perhaps to a lesser extent than in Austria. The German physiologist Jürgen Harms (1885-1956) performed the first testis transplantations in guinea pigs at about the same time as Steinach, in 1911, while Voronoff conducted the first testis transplantations in animals (he-goats and rams) only in 1917. In the 1920s, Harms was one of the most active proponents of sexual rejuvenation in humans, though Voronoff's work stirred a much greater sensation. One of the most energetic practitioners of Steinach's procedure in Germany was Peter Schmidt, having performed hundreds of operations in his clinic in Berlin. Other Berlin practitioners, such as Richard Mühsam and Ludwing Levy-Lenz, followed suit. Indeed, in the 1920s, Steinach's procedure was widely applied all across the world.[291] In Germany, however, the application of Steinach's operation was among the widest.

    Beside operative interventions, by the late 1920s, hormonal preparations for rejuvenation became an object of keen interest in Germany. Very soon after Brown-Séquard's seminal announcement of 1889 on the self-injection with testicular extracts for rejuvenation, in 1893 the Danish/German physician Frederick Vermehren already actively employed thyroid extracts for the same purpose.[292] In the early 1900s, adrenal, pituitary and pancreatic extracts were widely administered. Sex gland extracts, however, enjoyed by far the greatest vogue in the rejuvenation business, becoming the modern equivalents of Faust's reinvigorating and eroticizing brew. By the 1930s, sex gland extracts and stimulants became widely available for men and women: Testiglandol from Grenzach and Testifortan from Promonta, Novotestal from Merck and Plazentaopton from Kalle, Pituglandol and Neosex, Testogan and Yohimbin, Progynon and Menformon, Follikulin and Unden, dried ovaries, testis and erectile tissue powder.[293]

    The majority of German founders of endocrinology (many of whom incidentally were Jewish) actively researched, developed and advertised the rejuvenating and eroticizing effects of such supplements. The illustrious Jewish proponents included Magnus Hirschfeld (the founder of the *Institut für Sexualwissenschaft* – the Institute for Sex Research in Berlin and one of the first gay rights activists), Bernhard Schapiro (Hirschfeld's co-worker and co-developer of Testifortan), Max Hirsch (editor of the monumental *Handbuch der Inneren Sekretion* - Handbook of Inner Secretion, 1926-1933), Bernhard Zondek and Selmar Ascheim (developers of the first reliable pregnancy test), and Hermann Zondek (Bernhard's older brother, also a prominent endocrinologist). Their "rejuvenation research" in Germany terminated with the ascension of National Socialism. After the German-Austrian "Anschluss" in 1938, the work of many Jewish rejuvenators in Austria, such as Eugen Steinach and Otto Kauders (1893-1949), also became impossible.[294] The fact that so many German rejuvenators of the 1920s were Jewish (not to mention that the principal founders of the rejuvenation movement – Metchnikoff, Steinach and Voronoff – had Jewish origins) will be shortly brought to bear on the discussions of life-extension in Nazi Germany. At this point, it is important to note that in the 1920s the rejuvenation through "artificial" operative or pharmacological/organotherapeutic interventions flourished in Germany and Austria, and from this intertwined research into sex and rejuvenation a large part of modern endocrinology was born.[295]

3. Rethinking rejuvenation – Benno Romeis (1888-1971)

Despite the initial advances, by the early 1930s it was, however, becoming increasingly clear that the "rejuvenation" methods did not quite live up to their promise. A striking testimony to the disenchantment with "rejuvenation" can be found in *Altern und Verjüngung. Eine Kritische Darstellung der Endokrinen "Verjüngungsmethoden," Ihrer Theoretischen Grundlagen und der Bisher Erzielten Erfolge* (Aging and Rejuvenation: A Critical Presentation of Endocrine "Rejuvenation Methods," Their Theoretical Foundations and Up-to-Date Successes), by Benno Romeis (1931).[296] Prof. Dr. Benno Romeis, at the time director of the Department of Experimental Biology of the Institute of Anatomy at the University of Munich, was an active member of the rejuvenation movement, having conducted in the 1920s extensive original studies on the effects of rejuvenating interventions (particularly Steinach's operations and hormonal supplements). *Altern und Verjüngung* (initially published as a part of Max Hirsch's *Handbuch der Inneren Sekretion*) was one of the most authoritative and thorough accounts of the worldwide rejuvenation research up-to-the-date, a watershed monograph, summarizing the rejuvenators' successes and failures and pointing out possible directions for future study.

For most instances, Romeis's criticisms were devastating. First of all, the extensive anatomical and physiological data on age-related changes of various endocrine organs (thyroid, parathyroid, pituitary, adrenals, testes and ovaries) did not permit to ascribe to any of these organs the predominant or primary causative role in senescence, as many reductionist rejuvenators would have liked to believe.[297] Furthermore, the most widely used and journalistically acclaimed rejuvenation methods came under fire, Steinach's procedure of the ligation of spermatic ducts ("vasoligation") in the first place.[298] Equally unfavorable were the conclusions regarding the methodology and clinical success of other rejuvenating operations: Eugen Steinach's and Victor Lakatos' "albugineatomy" (removal of the tunica albuginea layer of the testicles); Emerich Ullmann's "decortication of testicles"; Naum Lebedinsky's smashing and tearing of animal testicles;[299] Ivan O. Michalowsky's method of testicle ligation; Karl Doppler's "phenolization" (sympathico-diaphtheresis) destroying sympathetic vasoconstrictive nerves of the testicles to improve local blood flow; diathermy (electric heating) and irradiation of sex glands.[300] A special section in Romeis's critique was dedicated to the most sensational "rejuvenating" method of all – the transplantation of sex glands (male and female), performed by Voronoff and others.[301] As Romeis demonstrated, in all these operations, the results were either dubious or short lasting, inconclusive, contradictory and confounded, the sample size was too small and the observation time too short, autosuggestion difficult to rule out and the presentation faulty.

Next to the operative interventions, Romeis's work testified to and warned against the enormous spread of various "rejuvenating" hormone preparations. Similarly to the rejuvenating surgery, from their very inception, hormone potions and injections fueled journalistic enthusiasm and empowered charlatans. Even Brown-Séquard, the inventor of rejuvenative hormone therapy, felt obliged to caution against "many clinicians but more often quacks and charlatans [who] have exploited the deep desire of a large number of individuals and have exposed them to the greatest risks" (1889).[302] Soon after Brown-Séquard's report of 1889, testicular extracts were used not only to combat aging, but were also said to work wonders against "general weakness, tuberculosis, ataxia, neural degeneration (tabes), carcinoma," etc.[303] Since 1889 until 1893, at least 10,000 flasks of the "testicular extract" were produced and at least 200,000 injections administered. The successes of such preparations were "seeming successes, whose weak foundations could not be defended even by the greatest of optimists." The end of Brown-Séquard's "rejuvenation therapy" was a "deep disappointment." Fin-de-siècle studies on the anti-aging effects of thyroid hormones (by Frederick Vermehren) or adrenalin (by Anastasios Aravandinos) were too scant to provide any further grounds for hopefulness. In Romeis's account, by the 1920s, the interest in hormones subsided, but rekindled in the 1920s in the wake of successful operative interventions. As the obstacles to the general application of rejuvenating operations were immense (even for those that Romeis considered to be most effective, such as transplantation), the rejuvenators naturally sought cheaper and more widely applicable oral or subcutaneous preparations that would achieve similar results. Yet, after the surge of interest in "organotherapeutic" hormone preparations in the 1920s, the state of the art was that "until now the number of ascertained successes is inversely proportional to the number of advertised preparations."[304]

The general conclusion regarding all the rejuvenation methods that flourished in the 1920s was sobering:[305]

> The exaggerated hopes, raised [by the endocrine rejuvenation methods] in wide circles, especially by Steinach's vasoligation procedure and Voronoff's transplantation of apes' testicles, were not fulfilled. There can naturally be no talk about the 'Defeat of Aging,' the catchphrase touted by uncritical journalism. 'Rejuvenation,' in the true meaning of this word, is today impossible for vertebrates.

Yet, Romeis retained a mighty measure of optimism. "It would be mistaken," he thought, "to deny the endocrine reactivation methods any general objective value and dismiss their successes as pure workings of mental suggestion." It was, perhaps, the excessive initial hopefulness that led to the excessive disappointment, to a general dismissal of all that was valuable in these methods. Despite all the deficiencies in observation and presentation, the accumulated materials nonetheless demonstrated beneficial effects in a considerable proportion of cases, both in animals and humans. "Undoubtedly," Romeis concluded, "endocrine organs and their secretions became a valuable aid in the struggle against age-related phenomena, especially when these phenomena depend on endocrine function."[306] Such means were seen as particularly successful in cases of premature aging. Still, at the present, the risks far outweighed the benefits.

The mechanisms of action of these interventions became clearer by the time of Romeis's writing, helping to specify the relative advantages of the interventions, the rates of their success and failure, the indications and forecasts for their use. Thus, the rejuvenation methods were divided into two groups, according to their mechanism of action. The first ("autoplastic") group involved the destruction of extant sex gland tissues and subsequent release of stimulating "breakdown products." The group included Steinach's vasoligation and albugineatomy, Ullmann's decortication, Michalowsky's testicle ligation, Lebedinsky's testicle smashing, X-ray irradiation, and intensive diathermy. With these methods, the effects would only be temporary, as there was only a limited amount of tissue to be destroyed. The "temporary refreshing is bought at the price of a dangerous lasting deterioration of testicular tissue" and in cases of highly atrophied or wasted tissue, i.e. in advanced senescence, no effects could be expected. The second ("homoplastic/heteroplastic") group involved true supplementation, that is, the introduction into the host of additional reserve materials via transplantation of sex glands (as done by Harms, Steinach and Voronoff) or injection of sex gland extracts. Such methods had the advantage that they left the host's own sex gland tissue intact and could be used even in cases of advanced atrophy and performed repeatedly. Yet, with these methods, there was a constant danger of rejection by the host and uncontrollable "breakdown" of the transplants and supplements. Doppler's chemical sympathectomy, with a subsequent improvement of sex gland blood supply and secretion, was in a category of its own, though it was closer to the first ("autoplastic") category. For this method as well, clinical evidence was scant, and there arose the possibility of deregulating the entire organism by preferentially stimulating one of its subsystems. In all these methods, the optimal effect could be achieved when the host's own sex gland secretory function could be stimulated. The possibility of such a "revitalization" of the host's own glands was suggested by animal experiments for Steinach's vasoligation and for the transplantation of ovaries and testicles. Unfortunately for humans, neither for men nor for women was there any objective evidence for such a "revitalization" or "regeneration." The possibility also arose that the pituitary, regulating the activity of the sex glands, could be a primary target for anti-aging interventions, yet clinical evidence was lacking and the mechanism remained uncertain.[307]

Romeis further emphasized the temporary nature of all the present rejuvenation methods. In fact, even such an enthusiastic rejuvenator as Harms posed some temporal limits on the effectiveness of rejuvenation, due to the constancy of nerve and sensory cells in the body and their irreversible degeneration. The possibility of immortality for human beings, similar to the immortality of unicellular organisms, was unequivocally rejected. Nonetheless, the frequent assumption that the rejuvenation methods lead to an exhaustion of the organism's "last reserves" due to over-stimulation was said to be in no way supported by evidence (no such "exhaustion" had been shown in animal studies, rather the opposite). The fact that the aging processes are subject to endocrine manipulation, of course, did not mean that the involution (degeneration) of certain endocrine glands, e.g. sex glands or the pituitary, is the primary cause of aging. No morphological or functional evidence had yet been given for such a "primitive conception of the aging process." Nonetheless, Romeis concluded, the works of Harms and Steinach, as well as other proponents of endocrine rejuvenation, "founded a fruitful field of work in the study of aging processes. And this is evidenced by the great number of problems posed, most of which however remain unresolved."[308]

Romeis's work illustrates the beginning and the end of the rejuvenation movement of the 1920s. It showcases the prevailing explanations for precisely why this was considered a "failure." The rejuvenation methods were found thoroughly wanting. They did not fulfill the high popular expectations associated with them. Indeed, these methods demonstrated "some effects," such as increased appetite, improved libido and work power, in a large percentage of cases. Yet, these effects were mostly temporary and unpredictable, confounded by an immensity of other contributing factors and relativized by (mostly unknown) individual idiosyncrasies of each and every patient. The evidence presented was often doctored or even falsified, and in any case contradictory and inconclusive. Surgical rejuvenation could not in principle be widely practicable, as the operations were complex, costly, time and skilled-labor-consuming (by the 1930s, perhaps only a few thousand such operations were performed worldwide). As for the cheaper and more widely available hormonal supplements, their effects were so little known that the consumers took them at their own peril. At any rate, neither operative nor supplementary rejuvenating methods provided any real evidence for a significant life-extension. The dominant theories of aging of the 1920s, emphasizing the primary role of particular (mainly endocrine) organs in the aging process and in the reestablishment of organic equilibrium upon rejuvenation, were also found to be too rudimentary and restricted, requiring a profound elaboration. These might be some of the reasons why, in the 1930s, the very term "rejuvenation" was rapidly disappearing from the scientific literature. According to Nathan Shock's *Classified Bibliography of Gerontology and Geriatrics* (1951, 1957, 1963), the number of publications on "rejuvenation" dropped from a maximum of 27 in 1928 to a minimum of 2 in 1940.[309]

The disappointment of biomedical researchers with the existing surgical or organotherapeutic rejuvenation methods was overwhelming. With the original research programs failing, where was the life-extension movement to go from there? Several options emerged and were pursued: Giving up entirely on the dream of the prolongation of life; lowering the expectations and aspirations to almost a zero; or else continuing to peddle unproved remedies. Another option was to embark on extensive "further research" that could take indefinite time (perhaps centuries) until producing any real means to combat aging. A converse option was to disengage from any "reductionist" and "artificial" surgical and pharmacological interventions, but rather embrace a "holistic" and "natural" macrobiotic life-style. In its extreme manifestation, the latter option would involve discarding scientific research altogether and instead cultivating the meditative power of the mind in order to wish longevity into existence. In a more practical and common manifestation, the abandonment of "reductionist" methods would lead to an emphasis on the practice of such health and longevity regimens that have not changed much since the time of Hippocrates, and with similar, by no means negligible, but nonetheless limited effects. Another way out was, instead of actual research and practice, to immerse in entertaining literary fantasies, spawning utopian and dystopian visions, trying to prove to the readers and to oneself whether and why we "should" or "should not" extend our existence. The life-extensionists of the 1930s reacted in all these different ways, though combinations of the various responses were not uncommon.

Setting any definitive temporal or local "trends" in the life-extensionist intellectual movement is extremely difficult and necessarily involves a large measure of impressionism, as many "tendencies" existed in one and the same time, and the continuity and interrelatedness of different approaches were often unbreakable. However, it appears that in particular countries, some reactions were more popular than others. Thus, as we shall see further on, in the UK from the 1920s to the 1930s, from Bernard Shaw to Aldous Huxley, literary representations, pondering the social and ethical implications of possible increases in longevity, seem to have been more prominent than actual research and practice. In the USSR and the US, in the 1930s, the slogans demanding "Further Research" were conspicuous. In contemporary France and Germany, a shift of emphasis seems to have occurred from the "reductionist" and "artificial" rejuvenation to "holistic" and "natural" ways of life-preservation. The shifts of emphases in the discourse on longevity were not exclusively determined by internal scientific arguments (deriving from the scientific failure of endocrine rejuvenation), but were to a large extent influenced by the social and ideological atmosphere of the period in each particular context.

4. Going the "Nature's way"

"Natural" approaches to life-extension indeed rose to preeminence in Germany in the 1930s. The opposition between "natural" and "artificial" methods for life-prolongation in early twentieth century Germany has been noted earlier. The German historian Heiko Stoff in *Ewige Jugend. Konzepte der Verjüngung vom späten 19. Jahrhundert bis ins Dritte Reich* (Eternal Youth. Concepts of Rejuvenation from the late 19th century until the Third Reich, 2004)[310] distinguished between "natural rejuvenation" involving "strengthening of the body mainly by naturist health measures, physical culture and public health practices" and "artificial rejuvenation" based on "surgical and hormone-therapeutic optimization."[311] Furthermore, he distinguished between "productivist" and "consumerist" perceptions of a rejuvenated body.[312] The "productivist" approach implied that the restored youth and healthy longevity were mainly seen as a means toward social productivity and were associated with "natural" rejuvenation (that is, with work on the individual and public body). The "consumerist" approach, on the other hand, perceived youth-restoring and life-extending means as something to be bought and enjoyed and was linked with "artificial" rejuvenation (operations and supplements being the commodity). According to Stoff, "the end of the rejuvenation hype at the beginning of the 1930s is due to the scientific-sociological fact that rejuvenation, through so many substitutions, could not achieve any stability."[313] "The peak of consumerism," Stoff noted, "was over by the end of the 1920s in all transatlantic societies." From that period onward, "rejuvenation meant work, separation of sexes, purity, sobriety. It was the 'German Rejuvenation' which translated productivist rejuvenation into selective treatments, the separation of the dead from the living, the young from the old, the Aryan from the Jew."[314]

However, as mentioned, the distinction between "natural" and "artificial" means of rejuvenation was made in Germany as early as Goethe, and even earlier. It is extremely difficult to set in stone any definitive temporal and local "trends" and separate "periods" in the pursuit of life-extension. It should be noted, for example, that even at the peak of the "rejuvenation hype" in the 1920s, endocrine rejuvenation was practiced by only a negligible proportion of the population (perhaps thousands operated and tens of thousands taking organotherapeutic supplements worldwide), leaving the rest of the world unaffected or even unfamiliar with these methods. In contrast, some forms of physical culture and "healthy diet" could always be practiced by almost anybody. Thus "natural" means could have always predominated over the "artificial" ones in popular practice. In the literature, however, the trends might be different. Indeed, in the scientific and popular-scientific literature, the "artificial methods" may "leap to the eye" in the 1920s and disappear from view later on. On the other hand, with all the valorization of the "natural means," hormonal and other pharmacological supplements for rejuvenation and life-extension did not vanish from the German market, in fact their consumption only increased (as will be discussed). The attitudes to life-extension in Nazi Germany may perhaps be better described by what the historians Paul Weindling and Robert Proctor termed a corporate "polycentric" and "polycratic"[315] system of views, when they referred to Nazi health care and Nazi society generally, rather than by any definitive "trend." Heiko Stoff's book virtually did not discuss the attitudes to rejuvenation, even less the general attitudes to life-extension in Nazi Germany. Nonetheless, the life-extension movement did exist in Nazi Germany and many of its central proponents indeed weighed heavily (though not exclusively) in favor of "natural" macrobiotic health regimens, despite many antagonistic undercurrents.

The war-mongering atmosphere of the Nazi period was not especially conducive to research and advocacy of radical life-extension. The survival of the group (the society) was often valorized over individual survival. Moreover, the public discourse dominant at that period praised a "noble" and "heroic" death, as can be seen in the contemporary propaganda in public media and literary fiction. The leitmotif of individual sacrifice was ubiquitous: from propaganda posters through leaders' speeches to biology textbooks.[316] The old battle cry of Friedrich the Great: "*Hunde, wollt ihr ewig leben?*" (Dogs, do you want to live forever?),[317] could well describe such a sacrificial attitude. Notably, not only was there the demand to literally lay down one's life during the war, but also to give up convenience, rest and leisure in favor of production and war effort. Honors and ideals of various kinds, including such fetishes as the glory of the leader or the land, were considered more valuable than life. The Nazi theorists appropriated a very selective view of the value of an individual life. The life of a less physically capable person was considered worth less than that of a young healthy individual (hence the extensive Nazi use of euthanasia, sterilization, and concentration camps for the physically and mentally disabled). Most notoriously, the tenets of racial hygiene assigned less or no value to the lives of persons considered racially inferior.[318] The elimination of "parasites" and conquest of "living space" from others were advocated.

However, the lives of the members of the "superior" race were considered valuable. Massive hygienic, epidemiological and prophylactic measures were undertaken in Germany to improve public health-care and health-span. Such hygienic measures are comprehensively described in Robert Proctor's *The Nazi War on Cancer* (1999).[319] As Proctor pointed out, "the Nazi campaign against tobacco and the 'whole-bread-operation' are, in some sense, as fascist as the yellow stars and the death camps." It were perhaps first and foremost the hygienic measures that made Nazism attractive: "Appreciating these complexities may open our eyes to new kinds of continuities binding the past to the present; it may also allow us better to see how fascism triumphed in the first place."[320] I would like to note, however, that the aspiration to strengthening the body (e.g. making people fit for battle) was by no means equivalent to aspirations to increasing longevity. As Hufeland first suggested, health and vigor are not tantamount to a long life, as immoderately vigorous activity may burn out the reserves of the life force and cause an early demise.[321] The powerful Nietzschean *Übermensch* is not necessarily long-lived; on the contrary, he is more likely to be short-lived, as he burns in the ecstasy of life's struggle (which follows from Nietzsche's own teachings).[322]

Nonetheless, despite the diverse ideological undercurrents antagonistic to the very idea of life prolongation, advice on individual life extension formed a substantial part of the Nazi "polycentric" health movement. In these counsels, "natural" hygienic life style improvements were emphasized, as the German society was supposed to be reverting to a "natural" and therefore "healthy" state of the nation.[323] The strengthening and prolongation of individual life were often seen as necessary conditions for the strengthening and maintenance of the National Body (*Volkskörper*), mainly understood as the national society at large, integrating the bodies of every member of the nation. The ultimate hope was the perpetuation of the ruling order, the "Thousand Year Reich." Beside the valorization of the "natural" hygienic approach, the "national" emphasis appears to be defining, the "natural" and "national" becoming closely linked. The "national" discourse focused almost exclusively on German accomplishments, and praised at the same time the authoritarian power of the medical-political leadership and the collective will power of the entire nation. Rather than antagonistic, the massive enthusiasm from below and the authoritarian rule from above were posed as mutually reinforcing in the attempt to create a long-enduring "National Body."

5. Life-extension imposed from above – Ludwig Roemheld (1871-1938)

A number of works on life-extension were published in Nazi Germany. A prime example is the book by the prominent German physician Ludwig Roemheld (1871-1938), *Wie verlängere ich mein Leben?* (How Do I Prolong My Life?).[324] Dr. Ludwig Roemheld of Gundelsheim in Baden-Württemberg, the discoverer of the "Roemheld Syndrome" (or gastro-cardiac syndrome, where bowel flatulence produces symptoms similar to a heart attack), was a highly respected figure in the German medical establishment. His longevity manual *How Do I Prolong My Life?* was first presented on March 17, 1933, at the initiative of the executive committee of the Württemberg regional Red Cross Association (shortly after National Socialism officially came to power on January 30, 1933). It was a call to action for German health professionals and laymen to improve the health and prolong the life-span of the nation. The book enjoyed extreme popularity and underwent several reprints. During the war, it was still used as an inspirational manifesto. The last (third) edition appeared at the height of Nazi expansion, in May 1941, after the author's death, and was accompanied by an introduction by his son, L. Roemheld, a military field physician. Roemheld Jr. recommended the book as a means to improve national strength and endurance, and basically implied that the German people have to be fit and vigorous to demolish/outlive their enemies:[325]

> Indeed, in today's time of war, the pieces of advice of this concise guide to a healthy and life-prolonging life-course have a particular significance. The present time imposes on every German person special challenges. The impending danger of increased and untimely consumption of irreplaceable health assets, for the individual and for the whole nation, becomes a crucial problem. Indeed, in these times requiring from the entire nation the highest daily performance, the physicians have a heightened duty to act as attentive and knowledgeable guardians for the good of health of every single fellow national. But also, every single individual has, in such times, the undeniable duty to practice a healthy and expedient life-style in order to maintain and increase his productivity. Thus, in this time of war, the new edition of my father's work will be a commendable contribution to our struggle.

According to Ludwig Roemheld, a radical prolongation of life could be achieved in the future through eugenics, through the "promotion of marriage of young people stemming from particularly healthy and long-lived families."[326] (To recall, the "Law for the Prevention of Hereditary Diseased Offspring" – *Gesetz zur Verhütung erbkranken Nachwuchses* – was enacted in Germany on July 14, 1933.) But before that happy, eugenically enhanced future comes, the presently-available means toward longevity was a healthy, "natural" hygienic life-style. The short-term goals of life-prolongation were relatively modest. Roemheld quoted the rising life-expectancy in European nations as proof of a continuous trend toward increasing longevity, and as an indication that the life-span of 70-80 years is "achievable by almost everyone."[327] Hygiene was stated to play a crucial role in increasing longevity. Sanatorium health care was mentioned in the first place. Roemheld, himself a sanatorium physician, did not wish his work to be seen exclusively as "propaganda of sanatorium care."[328] Yet he did claim that, "undoubtedly there is a causal relation between sanatorium care and life-extension."[329] He admired the practice of American insurance companies who obliged the ensured to undergo sanatorium care, and reported the plans to enact such practices in Germany. And indeed, from 1933 to 1938, sanatorium and tourism programs, such as *"Kraft durch Freude"* (Strength through Joy) of the German Labor Front (*Deutsche Arbeitsfront*), involved millions of German citizens. Roemheld's more general aim was to show "how, on the basis of today's state of medical science, our life can be prolonged and indeed is being prolonged."

The progress of medical science and improving general social conditions were attributed the leading role in increasing longevity: "The medical science," Roemheld asserted, "in fact succeeded in considerably prolonging the life of people in cultured countries, as the overall material progress of wide masses was advanced due to the increasing social consciousness in these countries." Among the social conditions and regimentations conducive to longevity, Roemheld named the avoidance of "disorderly life" (e.g. alcoholism), entering the life-prolonging bond of marriage, practice of certain professions characterized by higher longevity, especially the agricultural work. "Beside the general improvement of social conditions," Roemheld wrote, "it is first and foremost the grand achievements of hygienic science, particularly the successes of bacteriology in combating infant mortality and epidemics that markedly reduced general mortality."[330] Apart from inoculations and antiseptics, the struggle with infectious diseases was furthered by "measures of prophylaxis and isolation." The contribution of German scientists to this struggle was said to be decisive: "Indeed, we Germans can be proud of what we contributed to the world in the medical field in terms of life extension." Also in combating life-shortening chronic diseases thanks to internal medicine, 'Germany comes first.'[331]

Nationalist undertones apart, Roemheld's major thesis was that the prolongation of life, "this main wish and main purpose of humanity," is an integral, perhaps even a definitive product of general medical progress. "Medical science," he emphasized, "in the matter of fact, very considerably prolongs people's lives." Therefore, he called to stop all those who "decry our so-called 'scholarly medicine' and drivel about a 'crisis of medicine'… Science will continue to achieve further successes in this direction."[332] He believed that "the first condition for life-prolongation" is that "every single person undergoes, from time to time, medical examination and consultation, even when he feels healthy." Similarly to regular maintenance checkups of an automobile, "the most perfect machine" – the human body – also needs recurrent checkups and maintenance. "Only continuous medical control and surveillance" can achieve "rational prevention of diseases" and, as a consequence, can lead to a successful "postponement of death" and "prolongation of the life-span." It was thus Roemheld's central premise that in order to achieve life-extension, each and every individual needs to submit to medical authority. He expressed this imperative repeatedly, and in most unambiguous terms. The means for such a constant control may be massive state-sponsored and insurance-company-sponsored screenings, as well as the surveillance, from "cradle to grave," by the "old fashioned" family physician. "Only when the entire life-course follows modern medical knowledge, when effort and relaxation, work and rest are supervised by the physician – it will be possible to balance out the dangers of our civilization and defend ourselves in advance against their health-damaging effects."[333] Despite this praise of centralized medical control, the individual was assigned an essential responsibility for prolonging his or her own life. The physician, however, would provide the guiding directions. The physician and the patient must work in close rapport, to "approach the goal of life-extension for the individual and for the entire nation." The use of a set of simple hygienic rules by individuals who "adhere to life, however bad their conditions" may help achieve this goal.

Roemheld's practical recommendations, self-admittedly, stemmed from Hufeland's *Macrobiotics* (which was at the time of Roemheld's writing massively reprinted), with some minor modifications derived from recent research. According to Roemheld, Hufeland's basic rule for life extension remained valid: "we need to avoid everything that shortens life, and do everything that serves to prolong life."[334] The "unnatural" life-shortening factors that "diminish our life-force" or "damage our organs" fully complied with the traditional perspective.[335] These factors included:

Hereditary burden with predisposition to diseases and short life span, weakly upbringing, softening, overexcitement of the nervous system, depravities of any kind, excessive exertions, especially the hurry and unrest so often present in our modern life, insufficient balancing by sports, holidays and mental relaxation, overburdening of the body, especially of the aged and untrained individuals, unhygienic life-style, bad housing conditions, undernourishment on the one hand and lack of moderation in food and drink on the other, life-shortening moods and passions, poisons such as nicotine, concentrated alcohol, morphine, cocaine, diseases of every kind.

The life-prolonging means, on the other hand, were proposed to begin at birth, or even before birth. The choice of healthy parentage, though without any direct benefits to the parents themselves, would ensure "healthy heredity" for the children and would improve their chances for a long life. And the survival ability of the offspring would be facilitated by their "strict and energetic physical and mental upbringing," by the "hardening and strengthening of the body and soul" from an early age.[336] For all ages, "the struggle against aging rests on these three pillars: [1] sufficient and all-round movement in fresh air, if possible in the sun; [2] correct, healthy and moderate nutrition; and [3] efficient and complete rest periods with sufficient sleep." Several particular measures were recommended: first and foremost exercise, "hardening" (*Abhärtung*) by cool air and water baths, moderate and balanced nutrition rich in fiber and vitamins.[337] Roemheld's favorite recommendation for life-extension was "Diaphragmatic breathing" or "Abdominal breathing" whereby, he suggested, "gymnastic exercise of the circulation system" can be achieved. Roemheld self-confessedly borrowed the idea from Taoist and Yogic breathing techniques, yet thanks to his popularization efforts, "diaphragmatic breathing" became a massive practice in Europe and carried Roemheld's name.[338]

The most universal suggestions for life-prolongation were the regularity and constancy of the life-style, moderation and an optimistic worldview: "Equanimity, inner contentedness and stability, stand in the first line of defense against early wearing out."[339] This suggested a general mental stance for a society desiring a prolonged existence: the stance of contentedness, even indifference (*Gleichmut*), the stance of "inner and outer calmness, regularity and constancy of life." The society was required to promote this stance, to establish a balanced, constant and regular life-course for its citizens. "Work is good for people, idleness is bad" Roemheld preached, "nonetheless, overwork and rash, exhausting, irregular life under constant stress and anxiety shorten the life-span." German social legislation was said to have achieved much in regulating work hours, giving each individual the opportunity for restorative rest and exercise, for the "hygienic enjoyment of all that nature and art have to offer." The massive activities of the German Labor Front to provide "every individual with necessary means of recreation" were highly praised. The "efforts of the Reich's health leadership against excessive consumption of stimulants" ("enjoyment poisons" - *genussgifte*) were also hailed, first of all the fight against the poison of nicotine, but also the crackdown on narcotics: morphine and cocaine. (Moderate consumption of alcohol and coffee was tolerated.) Through the combined efforts of the state as the health-legislative authority, the physicians as the advisory, supervising and executive authority, and each and every individual as a compliant and responsible agent, the goals of "People's health and prolongation of life for each individual" were hoped to be accomplished. Thus life-extensionism was adjusted to the ruling political regime.

The promotion of "natural" hygienic regimens was said to be central to this joint effort. The emphasis on "natural" hygienic regimens for life-extension may have partly arisen from the failure of earlier "artificial" and reductionist rejuvenation attempts. Roemheld was no therapeutic nihilist and perceived clinical medicine as a crucial force in the struggle for longevity. Nonetheless, he was acutely suspicious of "magic means to defeat aging and death" actively promulgated in contemporary popular and medical press.[340] Indeed, according to Roemheld, rejuvenating power may reside in sex glands, as the immortality of the human race is insured by the immortal germ cells. However, "the experiments of Steinach and Voronoff only succeeded to occasionally reawaken the reproductive ability and prolong the life of single individuals for a few years." The rejuvenation operations were only given to a few, and successful for even fewer "particularly suitable" patients. "For the general prolongation of life," Roemheld contended, "these operations that are still in experimental stages, give nothing." Indeed, certain chemicals and operations might prove effective in animal experiments, and could sometimes be used in humans under proper medical control. "Yet all these treatments only work temporarily and only in connection with a rational life-style." Thus pharmacological and operative interventions were removed to the background, and the hygienic life-style became central.

Furthermore, it was implied, the "natural" hygienic life-style must necessarily be holistic. No single hygienic measure alone was said to be sufficient for life-extension, but only their combination. No single measure could be a "cure-all." The true path to longevity, Roemheld affirmed, lies in "eclecticism" – the avoidance of "one-sidedness" and "overexertion" during the entire life course.[341] Another crucial factor in the pursuit of life-extension was said to be the consideration of each person as an individual and as a whole. "There can be no two people with the same daily regimen," Roemheld observed. For some, a certain amount of physical exertion or a certain diet can be beneficial; for others this can be deadly. It was the proclaimed duty of the physician to provide general guidelines and render judgment for each individual case. Thus, a clear shift of emphasis occurred from "reductionist" and "artificial" life-extensionism, emphasizing particular pharmacological or operative means, to a more "holistic" and "natural" hygienic approach seeking to deploy a combination of life-style improvements personally suited for each individual.

A question may arise: How could this admission of individuality coexist with the praises of central authority, with assigning to the state the leading role in regulating people's lives for the purposes of stability, health and longevity, for the nation and every one of its members? An answer may derive from the fact that Roemheld's discussion of individuality was primarily addressed to physicians. The medical apparatus and its practitioners needed to be knowledgeable and powerful enough to understand and carry out a verdict for each individual case, while the individual patient was seen as largely incompetent. Another possibility is that the mention of individuality was really negligible: the set of general "guidelines" of the medical/hygienic science would be sufficiently valid to apply to very wide segments of the population, and there would be no place for any outstanding idiosyncrasies. A compromising answer could be that, while the central (medical) authority was seen as the most knowledgeable and powerful, lay individuals were still granted a (minor) part of the responsibility for their own longevity and that of the nation, especially in areas not requiring particular expertise. But their actions should better conform to the generally prescribed guidelines.

More general questions arise: Would people buy their longevity at the cost of a complete submission to the authority and the unswerving obedience to prescribed regimens and guidelines of conduct? (Or would they even buy it at the cost of devoting too much effort to their study and adoption?) And if the central guidelines proved to be wrong or disastrous, who would carry the responsibility and consequences: the guideline givers or those who implement them? These are some of the questions that arise from the consideration of Roemheld's teachings. Whatever the answers, Roemheld's work represented a consistent form of life-extensionist philosophy under National Socialism, where macrobiotic hygiene was perceived as a practicable joint effort of the state, medical authority, and each and every individual, united for the purposes of personal and social stability and endurance and for the perpetuation of the existing social order.

6. Life-extension driven from below – Gerhard Venzmer (1893-1986)

Another life-extension manifesto – Gerhard Venzmer's *Lang leben und jung bleiben!* (Live long and stay young!)[342] – was written along very similar lines. However, Venzmer's book assigned a greater role to each and every individual for extending their longevity, than to the physicians' guild. Venzmer's work was perhaps one of the most widely read life-extension manuals in Nazi Germany, having undergone several reprints from 1937 to 1941. The Doctor of Philosophy and of Medicine Gerhard Venzmer of Stuttgart was one of the best known popularizers of life sciences in Germany, since the 1920s well into the 1950s, having written many books on diverse aspects of biomedicine, with a peak of acclaim during the Nazi period, with hundreds of thousands of copies in circulation.

According to Venzmer, the pursuit of life-extension is the fundamental desire of human beings and derives from the primordial, unalienable "will to life."[343]

This will to life ... is the only unchangeable, unconditional property of every living creature, the core of its existence. ... Its force puts to shame all the opinions that teach otherwise. Philosophers and poets may try to convince us that 'Life is not the highest value' and that the 'Gods let die those whom they love.' If mortals had not the will to live, and the primordial drive for self-preservation had not run in their veins, there would have long been not a single living creature on our planet.

The will to life-extension is a natural extension of the will to life: "No wonder then, that the desire for the prolongation of life beyond its natural limits and for the attainment of eternal youth has been the oldest desire of humanity." Indeed, Venzmer believed that "immortality does not avail human beings"; folklore throughout history attests to the popular aversion against "emulating gods," and attempts to achieve "eternal youth" are doomed to failure. Nonetheless, a radical prolongation of life is both desirable and possible: "presently, for the first time in history, we have succeeded in winning considerable ground in the struggle against the inexorable death." Public hygiene and medical science were said to be the most significant contributors to the increase in longevity, with a long road of achievements ahead.[344] Venzmer admitted that the estimates of the "natural" human life span vary greatly among different researchers. According to some, we are far from reaching our full potential for life given us by nature, hence a long road ahead. According to others, we have already partly overcome our "natural" limitations, demonstrating the great human capacity for progress. In both cases, the prolongation of life is both desirable and possible: "we really still stand in the midst of a rise in the average term of life; and who shall tell us that this development is at its end?"[345]

Venzmer's writings again point to two hallmark concepts in German life-extensionism of the period: the praise of the existing social order and the valorization of a "natural" hygienic life-style. There is an implied adoration of the existing social conditions. "Never before," Venzmer claimed, "was life so inexhaustibly rich in events and forms as now, never before were we so full of hopes, having so short a time for their fulfillment; with the progress of our technologies and our transportation, never before had man so many opportunities, if only he had the time." Such an admiration of present-day achievements can be found in many life-extensionist authors across the world and throughout the century. In Venzmer, however, the "richness of events and forms" was considered to be inherent to contemporary Germany. Healthy life-extension was a tool to conserve the present happy state. It was, above all, an instrument for the strengthening and perpetuation of the "national body." It was a means to harmonize and stabilize the society by integrating the elderly. Their continuous productivity and contentedness, through healthy life extension, were considered to be some of the principal aims for the general social stabilization:[346]

What [the national community] needs is that the aged classes of society remain fresh and vigorous, ready for deployment and capable to fight for existence by their own strength. ... Good health and long life are gifts that cannot be valued highly enough, and must be carefully protected and cultivated. Whoever does not take this into account in his life-course, but rather behaves carelessly with this good granted him, transgresses not only against himself, but no less against the national community to which he belongs.

As a stabilizing social mechanism, the pursuit of life-extension first and foremost focused on the German national society. It seems that no ostensible "Nordic supremacism" can be found in Venzmer's writings. Rather, the will to life and life-extension were considered to be universal throughout the world and throughout history: "In ancient Greece, Persia and India, all wisdom began when scholars recognized that a healthy, long life is the first human duty and the first condition of human dignity."[347] Moreover, the way of life of "cultured nations" is not necessarily superior, perhaps even inferior, to that of "natural nations" and non-Westerners generally, particularly the "vast populations of the Asian part of the world." Their life-style was presumably healthier, better agreeing with natural laws, and more conducive to a long life. Westerners, and Germans in particular, it was asserted, could learn much from the life-style of these peoples (including the Chinese, Indians, Bulgarians, Eskimos, etc.) to increase the durability of the German people. Researchers from many countries were cited, including the French Louis Pasteur and C.E. Brown-Séquard, and the English Joseph Lister and Karl Pearson. American eugenicists and life-extension researchers were admired the most: Alexis Carrel, Raymond Pearl, Eugene Fisk, Alexander Bell, James Slonaker and Thomas Card. One can but remember the famous Nazi propaganda poster *Wir stehen nicht allein*: "We do not stand alone," of March 1936, listing the countries where, as in Germany, eugenic laws were passed (the US, Denmark, Norway, Sweden and Finland) or considered (Hungary, the UK, Switzerland, Poland, Japan, Latvia, and Lithuania) – with the American flag in the upper position.[348] Also, at about the same time, the Nazi sports movement culminated with the World Olympic Games in Berlin in August 1936. Venzmer's work (written in 1936) too testifies to the desire to see the German nation as a powerful and proud member of the multi-national community.

The displayed attitude of the Germans to other nations was that of a mixed desire for recognition, pride and competitiveness. Thus, when discussing the relative longevity advantages of different professions, Venzmer cited British and French data about the unequal life-expectancy for different social strata. "Such profound differences in the average life-expectancy," Venzmer claimed, "can of course only be possible in countries with strong social oppositions."[349] In the presumably more egalitarian National Socialist state, such inequalities are leveling out. The high mortality rates for certain professions can only occur in societies where "protections against accidents and professional diseases are not nearly as highly developed as in our country now." The current German social laws were said to be vastly superior. The contemporary German society provided to its citizens facilities for recreation, accessible to "people of lesser means." On the other hand, as early retirement might pose grave dangers to individual health and induce high mortality risks, the state encouraged a continuous professional occupation of the elderly.[350]

There appeared to be an on-going competition between the Germans and other peoples in terms of health and longevity. For example, Venzmer noted the low blood pressure of the Americans, the high of the French, the Germans occupying a middle position.[351] Venzmer did admire the American eugenic science and the successes of American life insurance companies who obliged the insured to undergo preventive medical examinations, prophylactic measures and health education programs, gaining immense dividends from the resulting healthy life-extension. (Such practices by the American Metropolitan Life Insurance Company and the health education campaigns by the American Life Extension Institute were hailed by Roemheld as well.) Nonetheless, some sense of superiority over the Americans is noticeable:[352]

The author must recall New York fast food restaurants where the hectic stock-market people, standing in front of a sort of a bar, swallow up their dinner, constantly looking at the watch and holding a stock-market bulletin in the free hand. No wonder the incidence of stomach diseases is the highest in the United States, exacerbated by the Americans' love for cold water and cold drinks – all the things that people should not do!

This repelling picture was countered by the portrayal of sturdy, long-lived German peasants. Besides the peasants, Prussian soldiers (such as the Prussian soldier Peter Mittelstedt, allegedly having lived for 112 years, from 1681 to 1792, out of which serving in the army for 67 years and participating in 17 major battles) were particularly noted for their longevity.[353] Venzmer allowed no equivocation that the primary goal of the German life-extension movement was to uphold the German national body and the National Socialist order.

The nationalist attitude was further exemplified by the fact that, in Venzmer, German researchers were assigned the primary role in the development of biomedicine generally and longevity research in particular. To the German achievements in hygiene and microbiology, it was claimed, we largely owe the actual increases in our life-expectancy, thanks to the studies of Robert Koch, Emil Adolf von Behring, Max Joseph von Pettenkofer and Carl von Voit.[354] In the research of aging and longevity, German scientists too were said to occupy the central place, the hall of fame spanning from Christoph Wilhelm Hufeland (1762-1836) to Emil Abderhalden (1877-1950).[355] In contemporary times as well, German researchers were declared to stand in the frontline of the struggle for life-prolongation, doing ground-breaking work in the research and application of epidemiology, improvements in life style and nutrition, particularly the administration of vitamins.[356]

Among the many German life-extending projects, the anti-tobacco campaign was waged by Fritz Lickint. The life-prolonging breathing exercises or "gymnastics of the cardiovascular system" were recommended by Lothar Tirala and Ludwig Roemheld. The importance of physical exercise (particularly walking) against age-related cardiovascular diseases was emphasized by Julius Hermann Greeff. (Greeff was the author of apparently one of the world's first dedicated studies of centenarians, published in 1933.[357] Some of his findings may appear surprising even now.[358]) Relaxation techniques against cardiovascular and nervous diseases were suggested by Karl Fahrenkamp.[359] All these means were recommended by Venzmer as essential props to strengthen the vitality of the individual, and to maintain the entire "National Body." The emphasis on "natural" preventive measures was overwhelming and was well in line with the "natural" and "holistic" teachings of the "father of Nazi medicine" Erwin Liek.[360] Not only German scientists, but also German men of letters were hailed as heroes of the life-extension movement. The foremost among them was Goethe (1749-1832) whose long life, marked by productivity, mental acuity and sexual potency to the very end, was the example to follow. Goethe's own reflections on a long, healthy and creative life were quoted throughout.[361] Thus, German intellectuals appeared to be the central agents in the multi-national research community. Foreign intellectuals and non-German folk customs were also given some space, though only a subordinate one.

It seems the only exception to this multi-national conglomerate were the Jews. Jewish scientists seem to be completely written out of Venzmer's account, whether they are from Germany or elsewhere. Thus, for example, when speaking of suspended animation, the groundbreaking contributions of Jacques Loeb (a German-born American Jew) were not acknowledged. When discussing rejuvenative transplants in animals (or parabiosis), the work of Paul Kammerer (an Austrian Jew) was not referenced. When promoting vitamins, Tadeus Reichstein (a Polish/Swiss Jew) was forgotten. When speaking of the "struggle with premature aging by active substances," Venzmer did not refer to the original works of Jewish scientists on endocrine rejuvenation, first and foremost, the French/Russian/Jewish Voronoff and the Austrian/Jewish Steinach, but also many German Jews, such as Magnus Hirschfeld, Bernhard Schapiro and Max Hirsch. When discussing the proliferation of connective tissue during aging or the benefits of fermented milk products, the Russian/French/Jewish Metchnikoff was left out. In Venzmer's account, it appeared that the main contribution of Metchnikoff (referred to as a "Russian zoologist and bacteriologist") to longevity research was the finding of living spermatozoa in centenarians, which were also found by the German biologist Waldeyer.[362] (Heinrich Wilhelm von Waldeyer, 1836-1921, was indeed a very prominent researcher, having consolidated the neuron theory of the nervous system and having coined the term "chromosome."[363] However, Metchnikoff's contribution to the theory of aging evidently was somewhat greater in scope and impact.) The very concept of rejuvenation (or "mechanistic/artificial/reductionist" rejuvenation), in Venzmer's account, underwent a thorough process of reformulation, even euphemization. It may not be accidental that the writing out of "rejuvenation" coincided with the writing out of Jewish researchers who were uncommonly active in that area. (This coincidence would be hardly surprising, since, as shown by the American historian of science Anne Harrington, in Nazi Germany, anything reductionist and "mechanistic" was commonly associated with the Jews, and it was believed that "the Jewish mind was fundamentally analytic, dissolutive and materialistic."[364]) The omission of the Jews may be yet another example of the first hallmark feature of German life-extensionism of the period – its nationalism.[365]

Venzmer's attitude to "rejuvenation" may also be a prime example for the second pervasive feature of the contemporary German struggle for longevity – the valorization of "natural" holistic hygiene over "artificial" and reductionist rejuvenative surgery and pharmacology. The very term "rejuvenation" was vilified: "The word 'rejuvenation', Venzmer wrote, "was earlier severely abused. Overreaching reports, much more 'sensational' than factual, awakened false hopes, which were necessarily followed by bitter disappointments. All the endeavors to fight premature aging were thereby made 'unsavory.'" In place of the "unsavory" (*anrüchig*) rejuvenation procedures, there comes the "natural" and holistic self-help:[366]

> The desire to "rejuvenate" a person who ages harmoniously *as an entire organism* is as absurd as it is superfluous. However, the situation is different when, due to whatever influences, the entire organism or its individual parts age *before their time*. Then, one can attempt to counter the premature aging processes through diverse *natural self-help* means described here. For now, these means are still insufficiently effective and medical science constantly seeks new special biological ways, known only to nature itself, to refresh and revitalize the organism which aged prematurely. The natural means described here, have already returned to many prematurely aged people their zest for life, freshness and work capacity. These means, however, have not the least to do with the unsavory concept of 'rejuvenation' of the past which only deserves ridicule. (Emphasis in the original.)

Thus, with a few rhetorical moves – the verbal substitution of "revitalization" for "rejuvenation" and the "combat against premature aging" for "reversing the aging process" – Venzmer steered clear of earlier failed practices and paved the way toward the emphasis on natural hygiene. Of course, the distinction of "premature aging" from "natural aging" would be extremely difficult, and Venzmer did not provide this distinction. But to combat a "premature" (and thus unnatural) process sounded much more respectable than trying to reverse or postpone "natural aging" which would mean going against nature itself.

These renunciations of "artificial rejuvenation," however, did not mean that Venzmer entirely rejected hormone supplements.[367] He recognized the immense role of hormones (particularly sex hormones) in human vitality and was one of the strongest supporters of "the science of human types" (*Typenlehre*) largely based on different hormone levels in different human types and races.[368] In *Live Long and Stay Young*, the hormonal, vitamin and mineral supplements were said to be "by no means unnatural interventions," no "foreign chemical poisons, but natural substances produced by the organism itself."[369] The administration of these "natural" supplements could balance out "discords of civilization" – "haste, competition, ever increasing demands for productivity." Such means would eliminate "disorderliness," "depression," "fatigue," "sense of self-insufficiency," "dissatisfaction." Instead, they would enhance "posture, decisiveness, concentration, zest for work, responsibility, general satisfaction with life, improve sleep and blood pressure, in short 'refresh' the entire personality."[370] Thus, the supplements would induce a person into an adaptive and compliant state. "The purpose [of supplements]," Venzmer affirmed, "is to maintain and improve the vital activity of the *entire organism* for as long as possible, and so preserve or renew the *harmony of the whole* that can be endangered by premature aging" (emphasis in the original). In short, the supplements could be a harmonizing, stabilizing and equilibrating force for the individual and for the German society as a whole.

Venzmer's 'Ode to Nature' and to social stability may appear paradoxical. On the one hand, its conservative proclivities were unmistakable: modern society was said to bring in its wake a host of "unnatural" life-shortening conditions that needed to be corrected or adapted to. Yet, at the same time, modern science (particularly medical science) and current social health programs were hailed as the primary contributors to increasing longevity. The second apparent contradiction was that although a "natural" hygienic life-style was valorized over "artificial" surgery and pharmacology (including hormonal supplements), the latter were nonetheless included in the macrobiotic arsenal. These contradictions are not too difficult to resolve. It is easy to acknowledge that "science" and "civilization" may have both beneficial and adverse effects on health and longevity. The benefits (e.g. subduing infectious diseases or establishing "facilities for recreation accessible to people of lesser means") are to be enjoyed and promoted. The detriments (such as "haste, competition, and ever increasing demands for productivity") are to be adapted to and endured. Notably, Venzmer did not advocate the elimination of "competition" or "demands for productivity" but rather suggested the use of hygienic regimens and supplements so that the individual may become better adapted and compliant with these conditions. The individual body was seen as a "cell state," and the "national body" (or the state) was composed of individuals who must "feel a part of the whole" and work compliantly with that whole. Thus, again, life-extension can be seen as an instrument for the perpetuation of the existing social order in its entirety.

As for extolling "nature" over artificiality, this appears to be a largely semantic distinction. Any medical intervention (including pharmacology and surgery) could be claimed to be based on "natural laws" or assisting the "healing power of nature." The supplements, in particular, could always be declared "natural" substances. Modern science could always be purported to "confirm the ancient wisdom" of "peoples living naturally"[371] or help humanity to live in accord with natural laws, or to master them. In any case, whatever the degree of the supplements' 'naturalness,' they only represented a very minor and highly contentious element among the suggested "natural" hygienic means for life-extension.

One such "natural" means was eugenics, or "choosing parents" for the better health and longevity of the offspring. (The idea could be traced back to Hufeland, who posited "good physical descent" as the first among the "means which prolong life."[372]) Venzmer acknowledged a strong hereditary component in longevity, based on studies of twins and on the relatively high longevity of descendants of long-lived families. He was also the author of a popular sympathetic commentary on the recently passed "Law for the Prevention of Hereditary Diseased Offspring" (*Gesetz zur Verhütung erbkranken Nachwuchses*, 1933). Under that law about 400,000 German citizens were to be sterilized.[373] The commentary was entitled *Erbmasse und Krankheit. Erbliche Leiden und ihre Bekämpfung* (Heredity and Disease. Hereditary Suffering and its Elimination, first published in 1933, republished in 1940), and was endorsed by Dr. Karl Ludwig Lechler, head of the Racial-Political Office of Württemberg-Hohenzollern. Of course, eugenics could also be branded "natural," as selection for vigor has been occurring in nature for eons, and eugenics was seen just as its logical continuation. As Venzmer stated bluntly in *Heredity and Disease* (1940), "in Nature, the eradication of the weak, the sick, the incapable of life, is generally self-understood." Moreover, he sympathized with the exterminations of the weak among the "natural peoples" – the warring and hunting-gathering tribes of Africa, Australia and the Far North, battling for survival. He very much resented the intervention of "colonial governments" into such "natural" and "time-honored" practices. Though, Venzmer made sure to add, "of course what applies to a primitive natural people cannot be readily applied in a civilized national community. And so the eradiation of hereditary diseased individuals, when they already exist, obviously cannot be conducted in a similar, sometimes cruel and inhumane manner." Only the sterilization of the "hereditary diseased" was permissible and desirable, both to prevent the "passing on of their suffering to their children" and to decrease their "burden" on the "healthy population."³⁷⁴

Yet, in *Live Long and Stay Young*, interventions into heredity were assigned even less importance than rejuvenative supplements:³⁷⁵

> Is the duration of our life only determined by heredity? Does the time during which we wander this earth depend exclusively on the predispositions given us by the reproductive cells of our parents? Is the day ordained when the Reaper will beckon us with his bony finger? If it was so, we should, like Muslims believing in "Kismet" [Fate], resignedly wait for our last hour. How we live our life would mean nothing for life's duration, whether we are restful or hectic, calm or constantly irritated, whether we spend our days rationally and naturally or untimely ruin our body by unnatural influences and poisons. It would all be the same, … our end would be determined already in the cradle, in the womb, or even at the moment of conception. … Fortunately, such reasoning is wrong. Personal life-style plays no lesser role than heredity in determining our life-span. The gods, according to the Greek proverb, not only reward sweat with success, but also reward a rational and natural life-style with great longevity and good health until very late years.

The weightiest components in the "rational and natural life-style" were the good old Hippocratic/Hufelandian macrobiotic rules. The rules focused on diet, weighing the relative merits of vegetarian, unprocessed (*Rohkost*), mixed (*Mischkost*) and alternate (*Wechselkost*) foods.³⁷⁶ They advised that the "means of enjoyment" should not become "poisons."³⁷⁷ They further recommended rest and exercise, cleanliness, moderation and peace of mind – a good-natured piece of advice for good-natured German citizens.³⁷⁸ Venzmer literally begged his readers to dedicate a few minutes a day to exercise and consideration of a rational life-course, to extend both the personal and national life-span (apparently not all German citizens were at the moment marching and saluting, but quietly reading books about health and longevity). Many of these bits of advice may be not entirely out of place even now.

Generally, mental self-affirmation and "spiritual hygiene" were said to be the most important factors for life-prolongation.³⁷⁹ "Anger and dissatisfaction," on the other hand, were the primary factors shortening life. The first recommendation "when annoyed" was "to ask oneself, whether the upsetting situation is worth destroying our brain cells, and furthermore, whether the situation will be improved by our anger." The prolonged maintenance of both the individual and national body required that the citizens should constantly maintain a stance of contentedness. "A cheerful, happy state of mind," "enjoyment of life," constant smile, courage, and, most importantly, trust in the society and social involvement, are life-prolonging; while "mistrust," "wariness," "secretiveness," "withdrawal," "loneliness," "sulkiness," "sense of self-insufficiency and failure" damage the individual and the nation. The person must "affirm and not deny, reconcile and not disgruntle, be cheerful and not moody, enlighten and not just burn." In short, to live long, the person needs to be a cheerful and accommodating part of the current social order. As for virtually all life-extension advocates under consideration, for Venzmer too, the sustained bodily and social equilibrium was the ultimate goal. Here, specifically, the equilibrium of the national body was sought for. Individual mental balance, the "inner equilibrium and order," involving a calm and jolly attitude, were the first conditions for the attainment of the general bodily and social equilibrium.

The ability and readiness to work to extreme old age were further highly prized conditions for the individual and national stability: "The awareness that anyone of us can carry on his duty, within the limits of one's capacity, that anyone can strive for perfection, ensures a balanced state of the soul, the true inner harmony which is the necessary condition for becoming old and yet remaining young."³⁸⁰ This referred not only to the spiritual sphere, but also to manufacturing: "The leaders of economy determined that the workers' productivity, between ages 30 to 70, does not substantially change, if they behave healthily." The life-extensionist envisioned an idyllic, harmonious social state where the aged would be perfectly integrated into the society, forming a "union with the young":³⁸¹

> Only the cooperation of the parts can lead to a harmony of the whole. ... As necessary is the youth, with its zest and fire, for the development of the nation, so indispensable is the old age with its experience and wisdom of life... A nation without the youth is like a body without a heart, but a nation without the old is like a body without a head.

Yet, Venzmer asserted that the integration and utilization of the aged should not be "overreaching," but must recognize their physical limitations. Furthermore, the productive activities of the aged must be based on the principle of "constancy," according to which "everything which falls beyond the boundaries of the usual or upsets the established life-habits... endangers the aged organism." The fulfillment of all these conditions was hoped to lead to the ultimate goal of the life-extensionist program in Germany: the conservation of its social order.

All of Venzmer's recommendations may be summarized by two slogans: "Feel a part of the National Body" and "Connect to Nature." The nationalist emphasis and the praise of a "natural" macrobiotic life-style appeared to be the major themes in German life-extensionist thought of the period, recurring in several other contemporary authors. Both themes were underscored by the desire to indefinitely maintain the social and bodily equilibrium. Thus, Dr. Sigismund Thaddea from the Charité University Medical Center in Berlin spoke in the Monthly of the Main Office of the NSDAP and the National Socialist Medical Association about the importance of healthy longevity and continuous employment of people of higher age for sustaining the national economy. He claimed that "the deficit of work power in the work market necessitates the increased employment of people of higher age" (1940). The goal was to attain "natural aging" through the "hygienic ordering of daily life."[382]

Dr. Johannes Steudel of Leipzig, in "Zur Geschichte der Lehre von den Greisenkrankheiten" (On the history of the study of the diseases of old age, 1942),[383] also emphasized national survival. Like Roemheld and Venzmer, he assigned to German scientists (such as Johann Bernhard Fischer, Burkhard Wilhelm Seiler, Carl Canstatt, Carl Mettenheimer, Christoph Wilhelm Hufeland and others) the crucial role in the history of aging and longevity research, overshadowing the works of other nationals, particularly the French. The emphasis was on "hygienic-dietetic regulations" and "morphological and functional changes" and there was no mention of rejuvenative interventions at all. According to Steudel, the purpose of the entire field of aging and longevity research is to facilitate the employment of the aged in national economy:

> As the age statistics show, the average life-span of people in all cultured nations grows. Even without considering the impact of the war, we have to deal with the increase of the elderly in the population. The proportion of their demand for help must be at least as large as their proportion in the population. Therefore the efforts of the research of aging and clinical study of age-related diseases make sense. The necessity to make useful the work power of the aged in the economy, imparts on these studies a great practical value.

Fundamental Health Improvement: Full Power, Success, Rejuvenation (Gründliche Gesundung. Vollkraft, Erfolg, Verjüngung, 1937) by Dr. Wladislaw Klimaszewski of Munich, represented the same tendencies. In the section on "Rejuvenation," the "connection to nature" was strong: virtually all the recommendations concerned natural hygienic, dietary and psychological regimens, with an overwhelming emphasis on the revitalizing power of the mind. The only thing said about Steinach's rejuvenative operations was that "even though interesting, they are not yet ripe for practical human application." The nationalist elements were also strong: "The profound will to life, to durable full power, rises in millions of hearts" pointing "the way to longevity for our race." And, as in all the above contemporary German authors, increasing healthy longevity was perceived to be absolutely vital for the German national economy: "through the correct way of life ... everyone will remain capable of work."[384]

The necessity to integrate the aged into the German national economy also justified the studies of Emil Abderhalden and Max Bürger who, in 1938, established a journal dedicated specifically to aging and longevity – *Zeitschrift für Altersforschung* (Journal for Aging Research, Dresden-Leipzig) – the third such specialized journal in the world.[385] They too were highly suspicious of any "unnatural"/"rejuvenative" intrusions. The first two specialized gerontological journals emerged in other fascist-allied countries: *Acta Gerontologica Japonica* grounded in Japan by the Yokufukai society in 1930,[386] and *Altersprobleme: Zeitschrift für Internationale Altersforschung und Altersbekämpfung* - Problems of Aging: Journal for the International Study and Combat of Aging, founded in 1937 by Dimu Anatoli Kotsovsky, in Kishinev, Moldova, then part of the Kingdom of Great Romania.[387]

7. Basic research – Hans Driesch (1867-1941), Emil Abderhalden (1877-1950)

Abderhalden and Bürger were among the central figures in German longevity research in the 1930s-1940s. The commendation of a "natural" hygienic life-course was eminent in their conceptions of aging. In the introduction to the first issue of the *Zeitschrift für Altersforschung*,[388] they spoke about the "natural" or "normal thread of life" leading to inexorable death. Yet they suggested that this "natural" life-span is greatly curtailed by "unnatural" influences and that it is the task of the researchers of aging to identify and counter these influences. The practical aim of this research was to suggest a "natural" life style that would lead to a significant extension of health and productivity. Hence, applied and basic gerontological research occupied the central place in the journal. Not only Abderhalden and Bürger, but Venzmer too spoke about the importance of basic research and theories of aging for discovering life-extending regimens. Yet, in fact, the actual input from the current theories of aging (such as the colloidal condensation theory) into actual macrobiotic regimens was minimal. Venzmer's manual included virtually no practical hygienic recommendations that would have been unheard of by Hufeland or even by Hippocrates. Similarly, Max Bürger's definitive work *Altern und Krankheit*[389] (Aging and Disease, first published in 1947, republished in 1954, with most of the materials collected up to 1945) was a massive and authoritative monograph, containing a vast amount of basic information on the morphology and physiology of age-related processes and diseases. Yet, its practical recommendations virtually consisted of a few pages reaffirming Hufeland's "natural" macrobiotic principles of exercise, rest, moderation in diet and optimism.[390]

Nonetheless, basic longevity research did continue in Nazi Germany on a large scale, as a part of its massive basic research in life sciences.[391] Some of the leading German basic longevity researchers of the 1920s faded into oblivion, but others continued to flourish well until the end of the Nazi regime, and even later. Benno Romeis (1888-1971), in the 1920s an enthusiast and in the 1930s a critic of rejuvenation, was promoted in 1944 to become a full professor and director of the Institute for Anatomy, Histology and Experimental Biology at the University of Munich. The surgeon Ernst Ferdinand Sauerbruch (1875-1951) continued to experiment with rejuvenating sex gland operations, at least as late as 1939, while directing the Surgical Department at the Charité University Medical Center in Berlin.[392] One of the foremost researchers of sex hormone therapy, Adolf Butenandt (1903-1995), joined the NSDAP in 1936, and in the same year became the director of the Kaiser Wilhelm Institute (later the Max Planck Institute) for Biochemistry in Berlin. After the war, Butenandt remained highly influential, serving as the president of the Max Planck Society for the Advancement of Science, from 1960 to 1972. Another keen researcher of hormone therapy and rejuvenation was the Nobel-prize winning Austrian neurologist, president of the Austrian League for National Regeneration and Heredity, Julius Wagner-Jauregg (1857-1940). He continued to work on the problems of aging and longevity to his last days.[393]

Several basic researchers of aging were prominent in the 1920s. Eugen Korschelt (1858-1946) focused on comparative longevity and regeneration. Samson Hirsch elaborated on "changes in the cooperation of activities, rhythm and regulation" during the aging process. Jürgen Wilhelm Harms (1858-1956) excelled not only in rejuvenative operations on humans, but also in basic life-extension experiments on lower animals. Max Hartmann (1876-1962) proved the possibility of *individual* immortality. By preventing the spatial expansion or reducing the size of simple multi-cellular organisms, such as flatworms, he was able to maintain them virtually indefinitely. Hans Driesch (1867-1941) posited the immaterial life-force or "entelechy" as the primary determinant of the material life-span, and argued for the ability of the life-force to regenerate the body, or even perpetuate it indefinitely, as in the case of immortal protozoa. Driesch further speculated about the conservation of the life-force or the soul after death.[394] In 1926, these authors joined forces to produce a highly popular compendium, *Leben, Altern, Tod* (Life, Aging, Death), altogether confirming the theoretical possibility of "potential immortality of living matter," "rejuvenation" and "life-prolongation."[395]

The authors' paths diverged in the 1930s. Hirsch immigrated to Belgium and continued his research there in relative obscurity.[396] The fate of Korschelt and Harms is unclear, except that they continued to publish well to the end of their days, though in a diminishing extent.[397] Hartmann and Driesch, on the other hand, rose in prominence. In 1934, Max Hartmann was appointed an honorary professor at the University of Berlin, and in 1939 he became a co-editor of the journal *Der Biologe* (The Biologist), the monthly of the Reich Association for Biology, Division of Life- and Racial Sciences of the National Socialist Teachers League (*Nationalsozialistischer Lehrerbund, NSLB*) and the Research Society of the German Ancestral Heritage (*Forschungsgemeinschaft Deutsches Ahnenerbe*) affiliated with the *Schutzstaffel* (SS – Defense Corps). (Hartmann apparently failed to advise his affiliates on the life-shortening effects of spatial expansion.)

Hans Driesch, by the early 1930s, seems to have lost interest in the problems of aging and longevity, and firmly established himself in the realm of the occult and paranormal. The importance of extending this earthly life was overshadowed by the promise of spiritual immortality and by the fascinating world of extra-sensory perception, telepathy, clairvoyance, telekinesis, and contacts with the "field of souls" beyond, which he purported to study empirically. As, according to Driesch, the physical development and regeneration are determined by the workings of the immaterial "entelechy," so, he argued, "materialization is just paranormal embryology."[398] Driesch's excursions into the paranormal were frowned upon by many Nazi officials during the first years of the regime. Yet, it is also known that many of the Nazi leadership and party base were far from being averse to the occult and paranormal.[399] And Driesch remained a foremost authority in that area of study well through and after the end of the Nazi period.

Extensive research of the paranormal was carried out by Driesch's devoted follower, the parapsychologist Hans Bender (1907-1991), at the "Paracelsus-Institut" in Strasbourg, around 1943-1944, in Germany-annexed Alsace. Besides parapsychology, the Paracelsus Institute in Strasbourg was also involved in the research of various types of healing: by magnets, geological and cosmic influences, natural herbal remedies and the revitalizing power of the mind.[400] (Do not confuse Bender's Institute with the "Paracelsus Institute" established in 1935 in Nuremberg, Bavaria,[401] under the direction of the oncologist Dr. Wilhelm von Brehmer (1883-1958),[402] or the "Paracelsus Institute" founded in 1951 in Bad Hall, Austria,[403] or several other "Paracelsus Institutes" across the world[404] – all having a strong relation with "natural," "spiritual," "vitalistic" or "holistic" healing, rejuvenation and life-extension, in accord with the strong interest in these subjects by the German-Swiss rejuvenator Paracelsus himself, 1493-1541.) Despite Driesch's shift of emphasis from this-worldly to other-worldly life, the influence of his vitalistic teachings on aging and longevity research continued throughout and beyond the Nazi period, particularly through the work of his most dedicated proponent, the Leipzig internist and gerontologist Max Bürger.

The founders of the *Zeitschrift für Altersforschung*, Emil Abderhalden and Max Bürger, indeed played a crucial role in the German longevity research and life-extension movement of the 1930s-1940s. Emil Abderhalden (1877-1950) was higher in rank. Though Swiss by birth, he served as the chair of physiology at the University of Halle-Wittenberg, Germany, for most of his career, and was the president of the German Academy of Natural Scientists "Leopoldina" from 1931 officially until 1950 (he was de facto replaced in the post by Otto Schlüter in 1945).

The historian Paul Weindling singled out Abderhalden's research as particularly supportive of the Nazi ideology:[405]

> The biochemist, Abderhalden, regarded proteins as racial characters. He argued that proteins took a key role in heredity and attempted to explain genetic mutations by an interaction of proteins in the reproductive glands. He undertook experiments on different species of rabbit, and suggested research on twins with comparison of each twin's blood protein and injections to modify the protein structure.... Between April 1943 and March 1944 Abderhalden co-operated with the Kaiser Wilhelm Institute for Anthropology for research on the racial specificity of proteins. ... Abderhalden... provides a classic example of the transition of science from being a means of social emancipation to one of racial

persecution with his commitments to temperance, family welfare and medical ethics, since he ultimately provided the scientific rationale for the human experiments by [Josef] Mengele and [Otmar] Verschuer.

Recently, several historians accused Abderhalden of a lack of scientific rigor or outright scientific fraud, particularly with regard to his theory of *Abwehrfermente* ("defense enzymes," namely, protein-degrading proteases) as primary agents of the immune defense. The detection of the "defense enzymes" via detection of tissue protein breakdown products ("the Abderhalden reaction"), Abderhalden claimed, could be used for diagnosis (from blood tests for pregnancy to infectious diseases and cancer, to the degree of senescence) and for immunotherapy. These claims have been rejected as unfounded.[406] Earlier accounts, however, treated Abderhalden with much greater respect, both with regard to his ideological and scientific legacy.[407] Abderhalden's scientific studies – on the protein degradation into polypeptides and amino-acids during digestion, the activity of various proteases, the synthesis of amino-acids into peptides, the discovery of essential amino-acids, the determination of the role of B vitamins for carbohydrate metabolism, the research of the thyroid hormone – were recognized as fundamental.[408]

These studies concerned the human organism generally, and the aging organism particularly (the latter field is usually omitted in the accounts of Abderhalden's research). Yet, Abderhalden's investigations into the "colloidal theory of aging," his studies of changes in the water content of aging cells and tissues, changes in protein metabolism, the utilization of hormones and vitamins by the aged organism, age-related changes in morphology and pathology, the use of the Abderhalden reaction to detect senile degradation – all studied for the purposes of age-specific therapy and productive life extension – were among his central scientific pursuits.[409] But perhaps Abderhalden's greatest contribution to aging and longevity research was political. By lending his authority to this poorly established field, he became instrumental for its institutionalization in Germany and worldwide.

The fact remains that Abderhalden, occupying one of the highest positions in the Nazi scientific establishment, coexisted perfectly with the regime and only reinforced its prestige. After the end of the war, in 1945, Abderhalden was forced to leave Germany and return to Switzerland, where he served as a teacher of physiological chemistry at the University of Zurich until the end of his days. Nonetheless, the implication of Abderhalden as an intellectual mentor of Mengele appears to be an exaggeration. Abderhalden's work may have influenced Mengele in no greater measure than it influenced scores of other researchers of enzymology and protein biology around the world and throughout the century (by the 1950s, studies testing the Abderhalden reaction counted by the hundreds). In a more charitable account, Abderhalden's research of "proteins as racial characters" might be seen as pioneering for modern human molecular genetics, where the consideration of ethnic differences is believed to be crucial both for the fundamental understanding of heredity and for specifying and personalizing therapy for particular ethnic groups.[410]

Paul Weindling described the "purpose" of Josef Mengele's "research" in Auschwitz as follows:[411]

Mengele's scientific aims drew on Verschuer's method of twin research as the basis for hereditary pathology. [Miklós] Nyiszli related how experiments on living twins were complemented by pathological comparisons of healthy and diseased organs in twins, whom Mengele had killed for the purpose of simultaneous evaluation. Growth defects, reproductive biology resulting in twin births, variations like eye-color in twins, and characteristics of racial degeneration such as endocrine and anatomical anomalies were of interest.

According to the protocols of the Nuremberg trials of Nazi War Criminals, particularly the "Medical Case" (1946-1947), the horrendous human experiments in the Nazi concentration camps also included "Freezing experiments." According to the indictments, "From about August 1942 to about May 1943 experiments were conducted at the Dachau Concentration Camp primarily for the benefit of the German Air Force to investigate the most effective means of treating persons who had been severely chilled or frozen. In one series of experiments the subjects were forced to remain in a tank of ice water for periods up to three hours...." Another set of "experiments" were "Bone, Muscle, and Nerve Regeneration and Bone Transplantation Experiments." "From about September 1942 to about December 1943 experiments were conducted at the Ravensbruck Concentration Camp for the benefit of the German Armed Forces to study bone, muscle, and nerve regeneration, and bone transplantation from one person to another. Sections of bones, muscles, and nerves were removed from the subjects. As a result of these operations, many victims suffered intense agony, mutilation, and permanent disability."[412]

As the protocols make clear, the experiments in the concentration camps were conducted "for the benefit of the German Armed Forces," to enhance the survival of soldiers on the battlefield. No reference to longevity could be found in these protocols. Yet, Venzmer's popular scientific work attests to the wide general interest in prewar Germany in twin research as a means to determine environmental and hereditary components of longevity. (Venzmer made a special reference to the investigation of age-related deterioration in the eyes of twins, originally conducted at the University of Zurich.) Venzmer also discussed at length transplantation experiments in animals (salamanders), involving transplants from young to aged animals with general or local rejuvenating effects, as well as experiments inducing regeneration in these animals.[413] Suspended animation, particularly by freezing, was also noted as a possible future means to radically extend the individual life-span. It might be an historical injustice to draw a direct continuum between comparative twin longevity studies, general investigations on regeneration, transplantation, hypothermia or suspended animation by freezing – such investigations began at least as early as the 19th century and are now conducted all over the world and believed by many to hold a key to improving health and longevity[414] – and the horrendous concentration camp experiments. Yet, the apparent topical similarities make one wonder whether, during the slaughter of the innocents in the concentration camp experiments, some thought has not been given to a general "rejuvenation" or "life-prolongation" for the "master race," which would thus represent the most abhorrent form of "life-extensionism" one could imagine.[415] However, the limited data do not allow me to further speculate on this issue.[416]

8. Institutionalization of gerontology - Max Bürger (1885-1966)

The legacy of the second cofounder of the *Zeitschrift für Altersforschung*, Max Bürger, is much less controversial (possibly due to an almost complete lack of interest in his person by recent historical studies). Max Bürger was one of the leading German internists and gerontologists. He began his professorial career at the University of Kiel (from 1920 to 1928).[417] There he developed "Osmotherapy" where by administering hypertonic solutions he was able to drain fluid-swollen tissues, particularly lung edema. (He began developing this therapy as a military field physician during WWI, to treat German soldiers poisoned by Phosgene gas in the chemical warfare.) In Kiel, he also worked on metabolic diseases (such as diabetes and protein deficiency) and "metabolic regeneration" or nutritional therapy (including the use of nutritional yeast as a supplement in cases of emaciation). He was also the doctoral advisor of Gerhard Domagk (1895-1964), the future inventor of anti-bacterial "sulfa drugs" and Nobel Laureate (the Nobel Prize was awarded in 1939 and accepted in 1947). From 1929 to 1931, Bürger worked at the Osnabrück hospital; and from 1931 to 1937, he served as a professor on the medical faculty of the University of Bonn. In Bonn, his research on aging became full-fledged. In 1937 he joined the NSDAP and was thereafter selected for the post of the director of the Leipzig University Medical Center. He remained in the post until his retirement in 1957. He led the Leipzig Medical Center with a firm hand through its initial modernization, the partial destruction by bombing raids, and reconstruction after the war. It became one of the major outposts in the struggle against outbreaks of epidemics (such as the outbreak of hepatitis in 1938 in Thalheim), and a major center of medical instruction, particularly for military physicians during the war.

Apparently, Bürger was no great Nazi ideologue or anti-Semite. His closest collaborator in Kiel in the 1920s was Georg Schlomka (the name indicates a Jewish origin[418]) with whom Bürger cooperated extensively in the research of the aging processes, particularly on the morphological and chemical-physiological deterioration of "braditrophic" (slowly nourished) tissues (such as cartilage) and on the use of embryonic extracts as a rejuvenating medium.[419] Schlomka was also one of the contributors to the first issue of the *Zeitschrift für Altersforschung* in 1938. In that first issue, Bürger additionally cited several Jewish researchers, such as Eugen Steinach and Moisey Mühlmann, though omitting their nationality and criticizing their reductionist approach that reduced aging to an "involution" of a particular organ system (the sex glands or the brain, respectively). Bürger, moreover, took under his protection several colleagues threatened by Nazi persecutions, such as Karl Matthes who was married to a half-Jewish woman (this action would help to save Bürger's career after the war). Nevertheless, Bürger's adjustment to and support for the ruling regime were impeccable.

After the war, he rapidly adjusted to the new Socialist regime. After the occupation of Leipzig by the American troops on April 18, 1945, Bürger was slotted for transfer to the West (together with about 50 other Leipzig scientists), but refused to leave his hometown. On July 1, 1945, Leipzig was given over to the Red Army, to ensure the territorial continuity of the occupation zones. Bürger had to blend in. As a former NSDAP member, he was suspended from his post, but continued to work at his medical center as part of *notdienstverpflichtung* (necessary service duty). On October 1, 1947, he was officially reinstated, thanks to his dedication and professional capacities, but perhaps in no small measure thanks to his former acts of kindness, as in the case of Karl Matthes.

As Bürger had done nothing to destabilize the Nazi order, he thereafter did nothing to destabilize the new Socialist order, but strove to coexist with it in perfect harmony. On the occasion of the renewed publication of the *Deutsche Zeitschrift für Verdauungs- und Stoffwechselkrankheiten* (German Journal for Digestive and Metabolic Diseases, established in 1938 and renewed in 1949, where Bürger was one of the chief editors and contributors), Bürger expressed his great gratitude to the Soviet authorities: "We are thankful for this opportunity to the considerate officials of the Soviet Military Administration."[420] At the congress of internists in Leipzig in 1955, Bürger nodded agreeably to the assertions of the minister of healthcare of the

German Democratic Republic, Luitpold Steidle, who spoke about the problems of aging and longevity: [421]

> These are the questions that concern equally Materialists and Christians. The Christian sees this world as subordinate, but the Marxist works to fulfill the following demands: Redesign the world anew and better. Instead of making theories about the possible demise of humanity, actively master and reform the Earth. Engage all the means against the negative forces of destruction.

Bürger was later highly admired by a leading Soviet longevity researcher and official, president of the USSR Gerontological Society, Dmitry Chebotarev (1908-2005). Thus, Bürger's biography provides yet another example of the high adaptability of life-extensionists to the current ideological and social milieu, as well as their role as advocates and agents for the stabilization and maintenance of the current social order.

Bürger's basic experimental research of the aging processes, as well as his expertise in pharmacology and surgery, were extensive. Yet, his views on aging and longevity, as in fact the views of all the life-extensionists from the Nazi period discussed so far, weighed heavily toward "holistic" and "natural" hygiene, rather than "reductionist" pharmacology or surgery.[422] A crucial role in Bürger's theory was played by the "entelechy," the good old Aristotelian/Hufelandian/Drieschian "whole-making, wholeness-ensuring vital factor" or "vital force" driving the transition from the potentiality to the actuality of life, its initial impulse and ultimate exhaustion.[423] In his adherence to the theory of "entelechy," Bürger was perhaps the last follower of Driesch's vitalism in the field of gerontology. (Interestingly, Bürger's "vitalistic" and "holistic" views were tolerated by the East German and Soviet scientific establishments.) In Bürger's definitive concept of "biomorphosis" or "biorhesis" – considering the entire development of the organism, instead of just "aging" late in life or "differentiation" early in life, and advocating the adjustment of treatments to different stages of the life-course – it is the inherited force of "entelechy" that determines the natural (and limited) duration of the human life-span.[424] Nonetheless, Bürger suggested that under the current harmful habits, we are yet far from achieving our "natural life-span" and the current life-expectancy can be significantly prolonged. These were the central theses of Bürger crowning work in the field of gerontology, *Altern und Krankheit* (Aging and Disease, 1954, first published in 1947), yet he expressed similar views already in 1938.[425]

Bürger was definitely not someone who could be called a "radical life-extensionist" or "immortalist." In fact, no prominent "radical life-extensionists" or "immortalists" could be found in Germany, before, during or after National Socialism (unlike the US, UK, Russia and France). During the National Socialist period, the boldest aspiration that German longevity advocates expressed was a 200 year "natural" life-span, sanctified by Hufeland's authority, but usually the desire of life-prolongation was limited to a century. Bürger too ventured no far-reaching forecasts, though he admitted the general theoretical possibility of life-extension above the current value. He was generally skeptical about "finding an elixir of life" or the possibility that an individual can "significantly exceed the life span allocated to the human species." A far-reaching life-prolongation by "hormonal-chemical, organo-therapeutic or pharmacological means" was, according to him, not to be expected.[426] Nevertheless, in *Altern und Krankheit*, in the chapter on "Macrobiotics," Bürger did offer several suggestions toward a "moderate" life prolongation. *Altern und Krankheit* was a compendium of physiology. Yet, its practical advice, almost identical to that of the other contemporary German life-extensionists discussed so far, essentially amounted to Hufeland's traditional rules of macrobiotic hygiene: moderation and keeping out of harm's ways, exercise in fresh air, baths, avoidance of stress, good breeding, no culinary excesses, no poisons, and a good sleep. The ultimate suggestion was to have the "Wisdom of Life and a Happy Mind." Thus, Bürger followed the general "holistic" and "naturalistic" hygienic trend characteristic of the contemporary German life-extensionism.

Yet, Max Bürger's major contributions to aging and longevity research appear to be political. Besides the establishment of the journal *Zeitschrift für Altersforschung*, he also founded in 1938 in Leipzig the "Deutsche Gesellschaft für Altersforschung," shortly after renamed "Deutsche Gesellschaft für

Alternsforschung" – the "German Society for Aging Research" – apparently the first such specialized society in the world.[427] After the war, the society branched into the "Society for Aging Research of the German Democratic Republic" in East Germany (assuming this name in 1964 and initially presided over by Bürger's student and historian, Werner Ries) and the "German Society for Gerontology" in West Germany (founded in 1966, initially chaired by Bürger's colleague, René Schubert). After the reunification of Germany in 1990, the societies merged into the German Society of Gerontology and Geriatrics (DGGG).

    To the present, gerontology remains strongly established in Germany, though radical life-extensionist aspirations appear to be hardly perceptible. Still, despite the relative absence of "radical" aspirations, longevity research does continue in Germany on a large scale: in experimental and clinical gerontology,[428] cryopreservation,[429] demographics and genetics of longevity,[430] bioinformatics of aging[431] and bioengineering,[432] with the involvement of anti-aging facilities[433] and life-extension advocacy communities.[434] Nonetheless, radical life-extensionist hopes appear to have never been strongly pronounced in Germany, as compared to the other countries under consideration. At the present time as well, Germany's impact in the global life-extensionist movement appears to be comparatively limited, and even smaller, relative to other Western countries, than it was early in the 20th century. The apparent trend toward a diminishing relative impact may not be dissimilar to that observed earlier for the case of France after the "holistic turn" and WWII.

9. Allies – The Kingdom of Great Romania. Dimu Kotsovsky (1896-1965?)

Bürger's establishment in 1938 of the German Society for Aging Research predated that of the "British Club for Research on Aging" founded by Vladimir Korenchevsky in 1939 in the UK (renamed "The British Society for Research on Aging" in 1946),[435] as well as the American Geriatrics Society (1942) and the Gerontological Society of America (1945) led by Edmund Vincent Cowdry. The British and American associations would later become the basis for the establishment of the International Association of Gerontology (IAG) in 1950, which heralded a new period in the internationalization and institutionalization of gerontological research. Yet in 1938, Bürger's society was the world's first, and his journal was the world's third and the West's second, after the Japanese *Acta Gerontologica Japonica* (1930) and the Romanian *Zeitschrift für Internationale Altersforschung und Altersbekämpfung* (1937). Bürger's society and journal became major venues for the cooperation of longevity scientists, namely, of those ready to cooperate with the National Socialist scientific establishment. WWII broke out shortly after the journal's inauguration, in 1939, and, during the period of conflict, the vast majority of authors in the *Zeitschrift für Altersforschung* were German, yet also included authors from Japan,[436] Italy,[437] Spain,[438] Hungary,[439] France,[440] Sweden, Denmark,[441] Romania, etc. Indeed, aging and longevity research spanned all the Axis powers and sympathetic regimes.

But perhaps the firmest stronghold of aging and longevity research was in Romania.[442] The Kingdom of Great Romania officially joined the Axis in 1940, with the establishment of the National Legionary State. But even before that, Romania maintained strong ties with Germany, including the scientific establishments. Aging and longevity research in Romania was traditionally among the strongest in Europe. One of the world-leading Romanian gerontologists from the *fin-de-siècle* well into the 1930s was Georges (Gheorghe) Marinesco (1863-1938) who predominantly studied the age-related deterioration of the brain. Marinesco was Metchnikoff's primary contemporary scientific opponent, disputing Metchnikoff's findings about the destructive activity of macrophages in the aging brain.[443] Instead, Marinesco focused on the balance between nerve cells and glial cells, the formation of granules in nerve cells as initial signs of aging, chemical synthesis and breakdown in the nerve cells. He also attempted to promote the chemical synthesis of the nerve cells, and thus retard their aging, by serums of young animals, extracts from young organs, and "mitogenic radiation" (the latter concept originated with the Russian biologist Alexander Gurwitsch in 1923). In 1913, Marinesco pioneered the colloidal condensation theory of aging,[444] which became the dominant theory of aging (upheld by such authors as the Czech Vladislav Ruzicka and the French Auguste Lumière) well until the emergence of molecular-biological theories in the 1950s. Another prominent Romanian longevity scientist was Grigore Benetato (1905-1972) whose research started in the late 1930s and focused on age-related changes of cell colloids and their reversal by hormonal substances. Marinesco's and Benetato's laboratories, however, did not develop into institutes.

The world's first Institute for The Study and Combat of Aging (*Institutul Pentru Studierea si Combaterea Batranetii - Institut für Altersforschung und Altersbekämpfung*) was established in Kishinev (Moldova, then part of Romania) in 1933 by a single man – Dimu Anatoli Kotsovsky. The institute was initially sustained by Kotsovsky's own means, and was subsequently recognized by the Romanian government.[445] Kotsovsky also founded in 1936 the first European (and the first Western) journal dedicated to the subjects of aging and longevity: the Institute's *Monatsberichte* (monthly reports), renamed in 1937 *Altersprobleme: Zeitschrift für Internationale Altersforschung und Altersbekämpfung* (*Problems of Aging: Journal for the International Study and Combat of Aging*). Thus, Kotsovsky's journal emerged before the inauguration of the German *Zeitschrift für Altersforschung* (1938) and much longer before the American *Journal of Gerontology* (1946) or the *Journal of the American Geriatrics Society* (1953). Kotsovsky's *Monatsberichte* and *Altersprobleme* published predominantly in German and in a lesser extent in French and English. More than 100 scientific institutions contacted and exchanged publications through the journal, including the US Rockefeller Institute for Medical Research, the New York Neurological Institute, the London Royal Society of Medicine, the German Gesellschaft Deutscher Naturforscher und Ärzte, the French Académie des Sciences, the Academy of Sciences of the USSR, the Kiev Physiological Institute, etc. Like the journal, Kotsovsky's Institute also became a center of international cooperation and knowledge exchange. The Institute's honorary members included the renowned biochemists Emil Abderhalden and Casimir Funk, the physicians Max Bürger and Eugen Steinach, the philosopher Oswald Spengler, the Nobel Laureate in Physiology Hans Spemann, the Nobel Laureate in Chemistry Theodor Svedberg, and some 80 other prominent scientists, from the US and South America, throughout Europe, to Russia and Japan.[446] According to Nathan Shock's *Classified Bibliography of Gerontology and Geriatrics* (1951, 1957, 1963),[447] by the early 1960s, Kotsovsky was by far the most prolific author in the field of aging and longevity research (many of his publications appeared in his own journal). Despite Kotsovsky's contributions, despite his priority in establishing the world's first gerontological institute and the West's first gerontological journal, his vast, painstaking work is now as much as entirely forgotten, even by the gerontological community.

Dimu Anatoli Kotsovsky was a confirmed life-extensionist who believed in the possibility of future life-prolongation to 200 years and beyond. He became active in the field of longevity research and advocacy since the early 1920s, with his first work on "The Origin of Senility" (*Genezis Starosti*) written in 1923.[448] A man of encyclopedic learning who wrote in German, French, English, Romanian, Italian and Russian, he published scores of articles on all aspects of rejuvenation and life-prolongation, including organotherapy and theories of senescence. Kotsovsky may be well considered as one of the pioneers of a "cataloguing" approach in aging research. In one review, he presented and synthesized over a hundred contemporary theories of aging.[449] He realized that the aging process can be countered by no "magic bullet," by no single "rejuvenative operation," and that massive collection of dispersed data from various fields of biomedicine is necessary before even beginning to understand, let alone intervene into such an immensely complex process as aging. In contrast to other contemporary Romanian gerontologists (Marinesco, Benetato, Parhon), who concentrated on particular aging processes and organ systems, Kotsovsky described his own research and that of the Institute for the Study and Combat of Aging as the "study of age from biological, medical, psychological, and sociological standpoints."[450] Thus, he may be considered one of the pioneers of the interdisciplinary approach in the study of aging and longevity.

In accord with his interdisciplinary orientation, Kotsovsky's research interests varied greatly. He wrote on medical history and futurism, neo-Vitalism and neo-Darwinism, centenarians and abortions, thanatology and juvenology, heredity and environment. He wished to determine both the causes for the "acceleration" and "inhibition" of aging in men.[451] He investigated both the "fundamental origin of senility" (disputing the inevitability of aging in multi-cellular organisms) and the aging of particular organ systems. He studied age-related damage in the brain and the nervous system (applying such diverse methods as surgical neuropathology, Pavlov's reflexology, and biochemistry of brain lipoids), activation and inhibition of inner secretion (particularly considering the reactivating properties of sex hormones and insulin), age-related deterioration of the heart and the circulation system, blood biochemistry and gall formation. He was interested both in the course of specific diseases in the aged (e.g. coronary insufficiency and syphilis) and the "general symptoms" and "general pathology" of aging. He examined both the aging of living beings (comparative biology of aging) and non-living systems (the aging and reactivation *in vitro* of vitamins, hormones and enzymes). He attempted to discover general biochemical markers of aging, or the dynamic "aging reaction" (*Altersreaktion*) that would not only enable the evaluation of physiological age, but also aid in testing rejuvenative treatments. Kotsovsky's "aging reaction" was largely determined by "acidity as an index of the aging reaction" and by "precipitation of proteins."[452]

Kotsovsky's interests included basic experiment and theory, as well as practice. His practical recommendations for life-extension included both pharmacology and hygiene. He sought "substances for integral reactivation," experimenting with specific nutritional supplements, such as blood and heart tissue from young animals. But hygiene too played a vital role in Kotsovsky's considerations. He was a strong proponent of "Racial hygiene" as a practical means toward life-extension. In fact, his major programmatic article in *Monatsberichte* was on the relation between "racial hygiene" and "combating aging," though the emphasis was on hygiene and the role of the central nervous system for life-extension, rather than the race.[453] Both in Kotsovsky's basic studies and hygienic recommendations, sleep was given a privileged position. From the late 1920s onward, Kotsovsky suggested rest and sleep to be major practical means for rejuvenation and longevity,[454] noting the direct relation between advanced aging and diminished sleep requirements, and terming rest and sleep as the principal "life defense function" in the struggle against disintegration and death.[455] Besides biological treatments, Kotsovsky also strongly emphasized the importance of psychological hygiene and social stability for healthy life-extension.[456] It appears that the only unifying characteristic of all of Kotsovsky's diverse areas of study was the unrelenting search for life-prolonging means.

During WWII, Kotsovsky's research continued to flourish and his cooperation with German scientists was strengthened. In fact, German-speaking journals were then the almost exclusive venue for his publications. Thus, among others, he contributed to the German journal *Der Biologe* (The Biologist), affiliated with the SS-Ahnenerbe and the Division of Life- and Racial Sciences of the National Socialist Teachers League. (The study was rather innocuous, concerning "Hormone effect on the aging of plants,"[457] published in the journal alongside other harmless articles about birds' flight and vision, pond and sea ecology, kinetic cell measurements, etc.). In that period, he also propagandized life-extension in Austrian journals (such as the *Wiener Klinische Wochenschrift*).[458]

Perhaps due to this cooperation with German scientists or due to his support of "Racial hygiene" or, most likely, due to his outspoken anti-communist views, with the advance of Soviet troops in 1944, Kotsovsky had to flee Kishinev. (Even during the first Soviet occupation of Moldova, in 1940, Kotsovsky's institute was closed and reopened again only with the temporary withdrawal of the Soviet government.) After 1944, Kotsovsky continued his academic pursuits "in exile" in Munich. In the 1950s, his studies proceeded along earlier lines, and had as great a scope as before. He continued to discuss the general theoretical questions about endogenous and exogenous factors in aging, its retardation and hastening, and the possibilities of rejuvenation and life-extension. Yet, the weight of his studies seems to have shifted toward practical problems of geriatrics. He wrote summaries of various therapies of the renal excretion, respiration and circulation systems in the aged, the use of sedatives and analgesics, cardiac drugs such as Angifin and Miroton, psychological adjustment in advanced age, and the importance of work for longevity.[459] Some of his works advocated the social acceptance of the aged and the "elimination of antagonism between age classes."[460] His last books (self published) were on more general questions of history, philosophy and sociology: *The Tragedy of Genius – Genius, Aging, Death* (1959); *The Problem of Aging in History – Attempt of a Biosocial Synthesis* (1960); *Dostoevsky, Tolstoy and Bolshevism – On the Historical Responsibility of Writers* (1960).[461] But history, philosophy and sociology too were seen through the prism of longevity.

In *The Tragedy of Genius – Genius, Aging, Death* (1959), Kotsovsky analyzed the lives of scores of artists and scholars and established correlations between their psychological profiles and life-span, noting the longer life-span of the "classics" and the shorter of the "romantics."[462] In his attempt to understand the nature of creative genius, Kotsovsky came to the conclusion that all creative spirits were driven by the fear of death (*Todesangst* or *Thanatokomplex*) and by the desire for perpetuation, in accordance with Cicero's dictum "Tota vita philosophorum mortis commentatio est" – "the whole life of a philosopher is the meditation on his death." Life-extensionism is just a ramification of this perennial aspiration, attempting to treat the problem of death at its biological root: "Based on this desire for immortality (eternity), we can understand the striving of medicine to preserve the health and strength of youth against destructive natural forces and against the 'entelechy of death' of modern technology."[463]

In *The Problem of Aging in History – Attempt of a Biosocial Synthesis* (1960), Kotsovsky drew parallels between aging bodies and aging societies. As the aging and death of an organism result from the "progressive, irreversible accumulation of toxic products of the organism's own life activity, especially in highly differentiated tissues, and the impossibility of their excretion as a function of the organism's development," so the aging and demise of societies follow from the "progressive, irreversible accumulation of products of economic and cultural creative activity in highly differentiated social classes, and the impossibility of their excretion and exchange as a function of development." In Kotsovsky's holistic theory, "minimal excitations" stimulate the functional activity (either of the organism, the psyche, the nation, the state or the culture), "intermediate excitations" inhibit it, and "maximal excitations" destroy it. According to Kotsovsky, the aging of organisms and societies is, in most cases, inexorable. Nonetheless, there is a theoretical possibility for their rejuvenation and prolonged durability. The rejuvenation of societies can be achieved by "crossing of different nations" (parallel to biological conjugation), or by resettlement in a new territory (parallel to a "change of nutritional medium"). Kotsovsky believed the "crossing" or "resettlement" are difficult to achieve, but even without such measures, a significant extension of durability is possible through rational social conduct: "Insofar as the processes of aging in the individual can be slowed down by a rational life-style, and, in the modern times, even by artificial interventions, so it must be possible, if the correct diagnosis is made, to postpone the senile decay of nations. And conversely, erroneous leadership can bring about a premature destruction of the nation."[464]

Two elements stand out in Kotsovsky's treatments of individual and social longevity, and they are very similar to those of his German colleagues of the late 1930s-early 1940s. One is the valorization of a "holistic" and "natural" approach to life-extension. Kotsovsky was a strong believer in the progress of science, in its ability to significantly prolong our lives. At the same time, his criticisms of the negative impacts on health and longevity by "civilization," by science and technology, were recurrent. "Civilization" shortens life by intensifying the "life struggle," by rapidly improving the means of mass destruction, by strengthening the feelings of insecurity and fear, by introducing harmful nutrition, by poisoning the body and the environment. According to Kotsovsky, the situation in the past was much more auspicious: "Certainly, our ancestors were biologically stronger and stouter than we are." "Natural," preventive hygiene was attributed by Kotsovsky the primary role in combating the life-shortening effects of "civilization," as expressed in the following, somewhat paradoxical statements: [465]

> The civilization reduces the vitality of people and therefore shortens their life-span. Civilization accelerates 'aging,' even though, thanks to the progress of preventive medicine, it indirectly, in the course of the past hundred years, almost doubled the human life-span. ... Man, that created culture and civilization, turns day by day into a slave of scientific and technological discoveries, leading to negative selection. This tragedy of modern humanity, created by the progress of science and technology, can be only eliminated by science.

Thus, according to Kotsovsky, science and technology solve the problems that they create; civilization shortens life while indirectly doubling it; and civilized men need to struggle both against "destructive natural forces" and the "entelechy of death [dead end] of technology." Furthermore, the current technological "civilization" is rotten to the core, but needs to be vigorously protected. How can this paradoxical thinking be reconciled? One obvious solution may be the recognition by Kotsovsky of both positive (life-prolonging) and negative (life-shortening) effects of science and technology; the former need to be fostered, the latter opposed. Another solution (to which Kotsovsky seems to tend) may be the ultimate subordination of science and technology to a "natural," hygienic way of life. Science may help to discover a life-style that would accord with "human nature," but it cannot design or modify this nature at will. Kotsovsky seems to acknowledge both the hygienic "rational life-style" and "artificial [pharmacological/surgical] interventions" as possible means for life prolongation, yet he ultimately put more trust in "natural" macrobiotic hygiene (sleep, exercise, nutrition, cleanliness and optimism). And so weighed the preferences of Romeis, Roemheld, Venzmer and Bürger. After having vigorously pursued basic laboratory research, after having experimented with diverse "rejuvenative" interventions, Kotsovsky may have arrived at a partial disappointment and his focus shifted toward metaphysics and medical naturalism.

The second telling feature of Kotsovsky's life-extensionist philosophy (as also the philosophy of virtually all the life-extensionists discussed in the present work) is the obsession with the "perpetuation of ideas" and the preservation of the social order, equilibrium and stability, or more precisely, the perpetuation of a particular social order to which the author felt belonging. Kotsovsky fully agreed with the poet Emanuel Geibel's saying that "what the epoch possesses, hundreds of talents advertise."[466] In Kotsovsky's last books, some of the views appeared as if still addressing a National Socialist audience, extending a nostalgic apology for an extinct ideological and social paradigm. (Kotsovsky admitted that his *Tragedy of Genius* and *The Problem of Aging in History* recapitulated his ideas expressed in the 1930s.) Thus, Kotsovsky spoke of the "civilization arriving from the forests of Germany that destroyed the Roman civilization. Will they now master the entire world or must all civilizations run through a certain course of development, marching through the times of flourishing, aging and death?"[467] "The society, as well as the culture," Kotsovsky believed, "is a function of an interaction between two categories of factors: the racial characters of the nation or nations on the one hand, and their bio-social environment on the other." The most urgent task is to fight the decay and demise of the white, Western nations:[468]

> Comparative statistics and demography show that the majority of great powers populated by white nations already passed their zenith, and, in fact, in Western and Central Europe, the problem of the aging population is very urgent. When we also take into account certain colored races (the Japanese, the Negroes), then the problem of aging is actually no longer an abstract problem, but a mighty threat. Therefore, it is the duty of leading statesmen and scientists to do everything to postpone the senile decay of their nations.

Such statements could well have been made at a meeting of the "National-Socialist Teachers' League." Yet, they were written in 1960, in free, denazified West Germany. Some ideological adjustments had to be made. After all, according to Kotsovsky's own definition, longevity is determined by the "ability to adjust to one's environment, to persist steadfastly and maintain equilibrium in the face of destructive influences."[469] For example, the praise of freedom may have been highly likeable. And so, according to Kotsovsky, "the loss of freedom in all its forms depresses the entire life tonus and leads to premature vegetative fatigue, decrease of vitality, premature aging and finally shortening of life." Kotsovsky, however, made sure to add that "there is no absolute freedom [independence of environment]."[470]

Another notion that might have chimed well with many in West Germany was the unrestrained denunciation of Communism. According to Kotsovsky, Communism is either outright destructive or at the very least life-shortening. (As will be shown shortly, according to the majority of Soviet gerontologists of that period, there was not a single thing in the world that was more life-affirming and life-prolonging than Socialism. True Communism would be virtually synonymous with immortality.) As a true physiologist, Kotsovsky sought destabilizing and destructive agents in the individual and social organism. And he found such subversive agents in the entire ideology of Communism, and even more specifically in the persons of Tolstoy, Dostoevsky and Berdiaev who, according to Kotsovsky's (quite original) theory, were largely responsible for the rise of Communism (*Dostoevsky, Tolstoy and Bolshevism. On the historical responsibility of writers*, 1960, written in Russian).[471] He also denounced Fascism (though this denouncement seems to somewhat jar with his former concern about the expansion of "certain colored races"). Communism, however, has the greatest destructive potential, it leads to "dehumanization, the denial of value of a person, narrowing of consciousness" and bloodshed. The underlying conservative, anti-revolutionary tone of Kotsovsky's philosophy is unmistakable.

Kotsovsky's works provide yet another example for the adjustment of life-extensionist philosophy to the ruling socio-ideological regime. This philosophy was both a product and advertisement of a particular social and ideological paradigm. Kotsovsky was perfectly adjusted to live and work in the Romanian

Kingdom and the Romanian National Legionary State allied to the Axis. Yet, he was unable to survive in the Moldavian Soviet Socialist Republic. Neither would he have a place in the Romanian People's Republic. Emigration was the solution. In Western Germany, he found a society where he could fit in perfectly, and whose dominant anti-communist ideology was compatible with his own.

10. The Romanian People's Republic – Constantin Ion Parhon (1874-1969) and Ana Aslan (1897-1988)

While anti-communists were purged from Romania, communist life-extensionists rose there to prominence. One of Romania's and the world's leading gerontologists and life-extension advocates was the endocrinologist Constantin Ion Parhon, one of the initiators of modern endocrinology, having coauthored with Moise Goldstein a definitive monograph on "Inner Secretions" in 1909.[472] Parhon was a veteran Marxist (self-allegedly from his teens), he was one of the founders of the Romanian Laborers Party (*Partidul Muncitor*) in 1919, and from 1921 onward was a staunch supporter of the Communist Party. Parhon's pronounced communist predilections did not seem to greatly interfere with his career. From 1912 to 1933, he was Chair of the Department of Neurology at the University of Iasi, and from 1933 onward Chair of the Department of Endocrinology at the University of Bucharest. Even during the National Legionary regime and WWII, he retained his professorial post and membership in the Romanian Academy of Sciences.

Yet, after the war, thanks to his academic acclaim combined with his loyalty to the Communist Party, he was propelled to unprecedented distinction. After the abdication of King Michael I, in 1947 Parhon became President of the Presidium of the Romanian People's Republic, and from 1948 to 1952 Chairman of the Great National Assembly (head of state), not to mention such posts as Honorary President of the Romanian Academy of Sciences and founding member of the Romanian societies for Biology, Anatomy, Neurology and Endocrinology. Not surprisingly, in Communist Romania, Parhon's research programs had the force of law, both scientific and political. And life-extension and rejuvenation were the definitive elements in his scientific program. For example, he demanded that the biology of aging and longevity be made a compulsory subject in schools. In 1951, Parhon founded the world's second dedicated Institute of Gerontology and Geriatrics in Bucharest (or the world's first, as believed by some gerontologists who are unaware of Kotsovsky's existence[473]). The institute exists to the present.[474] The next in line were Switzerland's Institute for Experimental Gerontology, founded in 1956 by Fritz Verzár (1886-1979), under the inspiration of Parhon's success, and the Institute of Gerontology of the USSR Academy of Medical Sciences, established in 1958 in Kiev, after studying the experience of the Romanian comrades.[475] The research at Parhon's institute was explicitly aimed at finding rejuvenating substances. Parhon's reaction to the failure of earlier rejuvenation methods was straightforward: Search for new ones!

Parhon recognized the tremendous importance of preventive hygiene for life extension. In *The Biology of Lifespan* (1955, *Biologia Virstelor*), he asserted that "the first prerequisite for the prophylaxis of premature old age is the creation of favorable living conditions. Social and economic conditions must be created in which, on the one hand, premature aging cannot take place as result of general exhaustion and, on the other, the aging organism can be stimulated as a whole."[476] And yet, he believed that "true" artificial rejuvenation is possible and has been observed. In Parhon's view, in order to accomplish radical rejuvenation and life-extension, pharmacological interventions will be indispensable. The rejuvenative substances he experimented with included vitamin E, folliculin and thyroid hormones, even alkaline soda baths (of the kind proposed by the Soviet biologist, one of Stalin's favorites, Olga Lepeshinskaya, 1871-1963).

Yet, perhaps the major product of Parhon's research, and the most famous outcome of the Institute's activity, was "Gerovital" or low concentration Novocain solutions ($C_{13}H_{20}N_2O_2$). Most researchers credit Dr. Ana Aslan (1897-1988), Parhon's protégée and Director of the Bucharest Institute of Gerontology and Geriatrics since 1952, for the discovery of the rejuvenative properties of Novocain (around 1949).[477] Kotsovsky, however, directly credited Parhon for this discovery, made when trying Novocain in animals as a potential therapy for arthritis, and Aslan was only credited for its development. Undoubtedly, no test-tube would have been moved in the Institute without Parhon's blessing. Undoubtedly also, Aslan was chiefly responsible for perfecting the technique and even more for its popularization worldwide.

The major effects of Procaine (a.k.a. Gerovital / Aslavital / Vitamin H3 / low-concentration Novocain), were claimed to be rejuvenating trophic and stimulatory effects. Clinical results of injections into veins or muscles included tightening of the skin, growing and renewed pigmentation of the hair, improving memory, sight and hearing, regaining the power of movement, increasing blood circulation, etc. Clinical observations were accompanied by a massive battery of physiological and biochemical tests, including changes in muscle and nerve tone, blood vessels elasticity and dilation, ATP production, protein and lipoprotein balance, etc.[478] The effects of Novocain were chiefly attributed not to its anesthetic (nerve-numbing) properties, but to a vitamin-like "eutrophic" (nourishment-improving) action, presumably via a transformation of Novocain into para-aminobenzoic acid and folic acid. Hence the substance was branded "Vitamin H3." Many of these studies were published by Aslan in the rather obscure East-German journal *Die Therapiewoche* (Therapy Weekly) in the 1950s.[479] As with earlier rejuvenation methods, observations showed that "Gerovital" produces notable "reinvigorating effects." But so does Cocaine, Novocain's derivative. "Gerovital" (Novocain) was not found to be addictive or fatal. Yet, as with earlier rejuvenating methods, no substantial life-prolonging properties were evidenced either.[480]

Nonetheless, following the initial reports of "rejuvenating" phenomena, Gerovital turned into a world sensation. It became world-famous around 1957, after Aslan's presentation that year at the Fourth World Congress of Gerontology in Merano, Italy. Publicity became overwhelming after the treatment of Archbishop of Canterbury Geoffrey Francis Fisher (1887–1972) at Aslan's institute in 1958. Other dignitaries treated by Aslan's method in the late 1950s through the 1960s included the West German Chancellor Konrad Adenauer (1876-1967), president of the French Republic Charles de Gaulle (1890-1970), president of the Socialist Federal Republic of Yugoslavia Josip Broz Tito (1892-1980), and many more. The Soviet leader Nikita Khrushchev (1894-1971) underwent the treatment in 1959. The General Secretary of the Romanian Communist Party and the leader of the Socialist Republic of Romania (from 1965 to 1989) Nicolae Ceaușescu (1918-1989) too was one of Aslan's enthusiastic patients and a great patron of the Bucharest Institute of Gerontology. (Ceaușescu was generally quite a "pro-life" ideologue, banning abortions in 1966.) With the continuous support from the Communist Party, Aslan's institute massively capitalized on the treatment. The socialist institution became a successful capitalist unit, providing paid treatments at the institute itself, in-place manufacturing and distribution or licensing (in the West, licenses were given in West Germany, Belgium, Switzerland and Mexico). In the 1960s, Gerovital was available in all countries of the Eastern bloc, but was in deficit. In the West it was much easier to come by (and it is still widely available today, mainly in Europe). Aslan received a great many honors, including the Romanian title of a Hero of Socialist Labor (1971) and the Italian "Dama di Collare Del Santo Graal" (Lady of the Collar of the Holy Grail, 1978).[481] Thus, under the Communist blessing, Romania became a new stronghold of "Rejuvenation."

11. Neutrals – Switzerland. Paul Niehans (1882-1971)

Yet another stronghold of rejuvenation was in Switzerland, where rejuvenators stood fast throughout the 1930s-1940s, through the general debunking of endocrine rejuvenation methods and through the disruptive period of world confrontation. Switzerland, historically, was one of the oldest and most fertile grounds for rejuvenation research in Europe. This research tradition was there long sanctified by the elder of Swiss rejuvenators and the founder of medical chemistry, Paracelsus (Phillip von Hohenheim, 1493-1541) and by one of the founders of physiology, the author of the concept of "irritability" Albrecht von Haller (1708-1777). Haller, surely, inherited his interest in longevity and rejuvenation from his teacher, one of the most renowned physicians and physiologists of his time, the Dutch Professor Herman Boerhaave (1668-1738). Both Paracelsus and Haller affirmed the theoretical possibility of a radical prolongation of life (well extending into centuries) and produced quite a few rejuvenating "elixirs" for the purpose.[482] Thus, the tradition of life-extensionism and rejuvenation research was long legitimized in Switzerland by their formidable authority.

Besides the long entrenched tradition, there might have been additional social causes for the persistence of radical life-extensionism and rejuvenation research in Switzerland in the 1930s-1940s. During the period of conflict, neutral Switzerland had the greatest political stability and economic prosperity in Europe.[483] Consequently, Switzerland was spared the disruptions that plagued rejuvenation and longevity research in the belligerent countries (in Germany and the USSR, for example, publications on rejuvenation and longevity were reduced to a minimum in 1943-1944[484]). Another possible reason may have been that, in Switzerland in that period, the life-expectancy at birth was among the highest in Europe and in the world.[485] The relatively high number of the long-lived among Alpine mountaineers was traditionally a matter of national pride.[486] The old saying "the more one has, the more one wants" might have been the motto.[487] Whatever the explanation might be, from the 1930s through the 1950s, Switzerland became a safe haven for many rejuvenators. For Serge Voronoff (1866-1951) Switzerland became the last stop in his wanderings during and after the war (through Portugal, United States, France and Italy); he died in Lausanne. In Switzerland, Eugen Steinach (1861-1944) found refuge from Nazi persecutions. From there he fired the last volley in defense of his endocrine rejuvenation methods: *Sex and Life; Forty Years of Biological and Medical Experiments* (1940). Steinach's work concluded with this statement: "Here is the foundation upon which the proud structure of present-day hormone research is reared. ... On the tree of knowledge, as with every other tree, fruit can ripen only if the roots are strong and well secured."[488]

Several Swiss researchers cultivated the tree of rejuvenation planted by Steinach. In 1916, Siegfried Stocker of Lucerne conducted sex gland transplantations in humans.[489] And in 1922, Karl Kolb of Zurich conducted such transplantations in goats.[490] (To recall, the first sex gland transplantations in animals were performed by Steinach in 1911.) But perhaps the most prominent Swiss rejuvenator was Paul Niehans of Vevey (1882-1971). He became interested in endocrine rejuvenation methods in the late 1920s, and in the early 1930s developed his own method: the Cell Therapy (*Zellular-therapie*).[491] In *20 Jahre Zellular-Therapie* (Twenty Years of Cell Therapy, 1952), Niehans described the gist of his method:[492]

> The new healing method, that today may still seem revolutionary, employs specific embryonic or youthful fresh cells, cell cultures, conserved cells or cell-rich liquids, to treat the sufferings against which contemporary medical art had been powerless.

For diagnosis, Niehans extensively used the Abderhalden reaction, determining the levels of Abderhalden's "protective enzymes" and protein degradation in particular organs and tissues (to recall, the Swiss-born Abderhalden, 1877-1950, too found refuge in his homeland after the war). After the organ diagnosis, Niehans injected the patients with cell suspensions from corresponding organs. Thus, cell therapy was applied to regenerate the nervous system: the brain cortex, the middle brain, thalamus and hypothalamus, etc. (Niehans noted that contrary to the dominant conviction that nerve cells do not regenerate, his practice showed that they can indeed undergo some degree of regeneration). Success was reported in the treatment of blindness, deafness and idiocy. Dysfunctions of endocrine glands – pituitary, thyroid, parathyroid, thymus, pancreas, adrenals, and sex glands – were treated by cells from corresponding organs. The same was done for the heart, liver, intestinal mucosa, the reticulo-endothelial system, bones (using injections of osteoblasts), bone-marrow, kidneys, etc. Injections of placental tissue and "cell-rich" blood, particularly enriched fractions of leukocytes, were among the favorite treatments.

Niehans believed that cell therapy can be applied against virtually all chronic organic diseases, including cancer. But its major purpose was rejuvenation proper, the "mitigation of the deterioration of aging." According to Niehans, "premature senescence is a pathological problem, which can be in a large degree solved through the revitalization of damaged organs, particularly the sex glands." In Niehans's view, "youthful cells serve the tired, the ill, the old, and the weak. Here science bestows on us a new capital of life. Many valuable people thus avoid a disabled existence and enjoy life afresh, overcoming their depressions."[493] It is of course necessary to note that the efficacy of Niehans's treatments was far from being generally proven or acknowledged, to say the least. Even Niehans's own descriptions of the results (such as "suffering disappears") were rather vague.

As Niehans's reports make us understand, the cells for treatments were taken from domestic animals, either animal fetuses (whose cells were believed to be the least rejectable or the least toxic), or young animals (in case particular organs were not sufficiently developed in the fetus). Sheep were the donors of choice (as they were believed to be the most resistant to diseases). Yet, Niehans let it slip that in some cases human embryonic tissues, most likely from abortions, were used. Thus "cell cultures of the pancreas of a human fetus" were injected to Herr F.A., born 1894, suffering from Diabetes Mellitus.[494] (For Niehans, the use of such tissues did not seem to raise any ethical concerns whatsoever.) Thus, Niehans may be well considered as one of the primary instigators of "Regenerative Cell Therapy" in general and "Embryonic Stem Cell Therapy" in particular. Today these therapies are among the most hotly debated fields of biomedical research, both in terms of their potential efficacy and ethical implications. Some 70-80 years after Niehans's first attempts, the promise of these therapies is yet to be validated, much "further research" being needed.[495]

Yet another ramification from Niehans's cell therapy may have been "Gene Therapy." Niehans preferred to use fresh cells for his treatments, as he believed that these were the most biologically potent and needed to be administered almost immediately after the animal was slaughtered. Niehans's clinic in Vevey was the chief provider of fresh cell therapy (and it still operates today[496]). However, as the immediate delivery of fresh cells from slaughter houses to hospitals was somewhat cumbersome and expensive, factories were set up to prepare lyophilized cells (dried by freezing in vacuum) for shipment all over the world in ampoules. Niehans's factory "Siccacell," set up in Heidelberg, Germany, became the world's largest producer of such preserved cells (which are still available today) and these were widely applied by Niehans's followers in Mexico, Argentina, France, Belgium, Italy, Holland, West Germany and Britain (in the US their use was banned in 1984). Some of Niehans's disciples in the 1960s, such as the German Wolfgang Goetze-Claren and the British Peter Stephan, in place of using entire cells, sought to extract their active, rejuvenating ingredients. DNA and RNA seemed to be likely candidates, so they were extracted and administered, giving rise to "Gene/Genetic Therapy."[497] Of course, those methods of "Gene Therapy" were rather crude, using direct ingestion or injection of DNA/RNA extracts, without employing any sophisticated viral transmission vectors that are tried today. Yet, the title of "Gene Therapy" did they carry. As with "regenerative cell therapy," "gene therapy" is still in research and development.[498]

The ideological underpinnings of Niehans's work may be characterized as a mixture of aristocratism and capitalism. In *The Youth Doctors* (1968),[499] the American journalist and health care activist Patrick Michael McGrady[500] related that Niehans was a scion to the Hohenzollern's royal family, a grandson of the

German Emperor and King of Prussia Friedrich III (1831-1888) and Countess (Elisabeth?) von Fürstenberg (the latter allegedly settled in Switzerland around 1853, perhaps to cover up for the birth of her and Friedrich's illegitimate daughter Anna, Paul Niehans's mother).[501] During WWI, Niehans served as an honorary aid-de-camp to his uncle, German Emperor and King of Prussia Wilhelm II (1859-1941). As McGrady testified, both Niehans's manners and the luxurious trappings of his mansion in Vevey were imbued with the spirit of aristocratism (flaunting, among other regalia of noble birth, his grandfather's medallion at his desk). Aristocratism was also displayed in the treatment of Niehans's clients. When presented with a choice of patients, Niehans preferred those higher in standing. Celebrities were red-carpeted, not only because they were generally more able to pay (the treatment courses were in the range of thousands of dollars), but perhaps also because Niehans craved prestige and the association with celebrities would have great image-boosting and advertising power. Perhaps it were, first and foremost, the celebrities to whom Niehans referred as "valuable people" who should "avoid a disabled existence" thanks to his therapy.

Niehans indeed could boast an illustrious list of beneficiaries. Among the subjects treated by Niehans's cell therapy in the 1950s-1960s, McGrady attested, there were heads of state: the Saudi Arabian king Ibn Saud, Winston Churchill, Charles de Gaulle and Konrad Adenauer. De Gaulle and Adenauer apparently received both Aslan's gerovital and Niehans's cell therapy. Several world-famous intellectuals and artists were enthusiastic recipients of the therapy, such as the writers Somerset Maugham and Thomas Mann, the actors Gloria Swanson and Charlie Chaplin, the French fashion designer Christian Dior. The American financial mogul Bernard Baruch (1870-1965) too was among the beneficiaries.

But perhaps the most glory came from the treatment of Pope Pius XII (Eugenio Maria Giuseppe Giovanni Pacelli, 1876-1958), which propelled Niehans to world fame. Pius XII was first administered cell therapy by Niehans in 1954, and since then resorted to Niehans's services on several occasions. Photographs of Pius XII together with Niehans graced Niehans's dossiers (and they can still be seen in Clinique Paul Niehans' advertisements[502]). In 1955, Pius XII made Niehans a fellow of the Pontifical Academy, in place of the late Alexander Fleming (1881-1955), the discoverer of penicillin. The Pope was also reported by McGrady to try other rejuvenating therapies, such as royal jelly (or "bee's milk") and dishes from chick embryos.[503]

The treatment of the Pope seems to be significant on several levels. First, this seems to be a striking example of how public relations can be leveraged in passing judgment in matters of science. Of course, the likely event that the Pope (or any other celebrity) took an aspirin in their days, would not bear any witness as to the drug's efficacy. However, the fact that the Pope received the relatively less known and less tried "rejuvenating therapy," was a great reassurance, worth any number of clinical trials, both for the public who demanded an immediate access to the therapy and to medical practitioners who were too happy to oblige. The Pope's implicit blessing propelled the specific therapy from relative obscurity to a spot-light of investigation and application. The Pope's treatment perhaps tells even more about the relation of religion and science. It has been a common conviction among atheist life-extensionists that religion generally, and particular branches of Christianity, are somehow intrinsically averse to far-reaching biomedical interventions or even to the idea of life-extension, placing a greater emphasis on faith-healing and life in the world to come.[504] But the Pope's case demonstrates that when it comes to prolonging one's life by yet another day and for as long as possible, by whatever means necessary, the religious can be quite zealous. And the same attitude can be often found in the long history of the Pontifical office.[505] (Niehans himself, in his youth, went to a Protestant divinity school and received a doctorate in theology; though in his later life he was not particularly observant.)

The well-known attitude of the Catholic Church to abortions, or the use of human embryonic tissues for medical purposes, is complete rejection.[506] However, it is difficult to believe that the Pope, or his retinue, or electors of the Pontifical Academy, did not know about Niehans's use of human fetal tissues for therapy. This did not seem to hinder employing Niehans as the Pope's treating physician. The type of cells the Pope was administered remains a secret. The Pope's treatment demonstrates, above all, that life-extensionism can be adjusted to religion, when the interested parties have religious backgrounds.

12. Respectable gerontology – Fritz Verzár (1886-1979)

The efficacy of Niehans's cell therapy was widely disputed. One of its major detractors was the Swiss (Hungarian born) professor of physiology, Fritz Verzár, himself a powerful figure in the life-extension movement.[507] In 1952, at the age of 66, he switched from bacteriology and nutritional medicine to gerontology (apparently under the persuasion of the Russian/British gerontologist and life-extension advocate Vladimir Korenchevsky). Verzár remained active and influential in the field of aging and longevity research to the end of his days at the age of 93. In 1956, he founded the Institute for Experimental Gerontology in Basel, one of the world's first.[508] In 1957, Verzár founded the specialized journal *Gerontologie*, also one of the world's earliest. Verzár was chiefly responsible for developing the cross-linking theory of aging. The formation of cross-links was initially observed between molecules of elastin and collagen. This explained the loss of tissue elasticity, particularly the hardening of blood vessels, and suggested the breaking of cross-links as a potential method to combat aging.[509] However, as Verzár's background was in nutritional medicine, one of his main research emphases was on the use of proper nutrition for extending longevity, including the use of vitamins and the study of nutritional habits of long-lived Swiss mountaineers (as well as climatic factors for their longevity, such as oxygen consumption at high altitudes). Verzár's Institute for Experimental Gerontology became a training ground for many aspiring longevity researchers and advocates.

One of Verzár's pupils, the Belgian physician Herman Le Compte (1929-2008) took the idea of proper nutrition to an extreme. According to Le Compte, aging generally derives from nutritional deficiencies. Therefore, by supplying or balancing these deficiencies, people may be able to live virtually indefinitely: "The causal treatment of aging could thus be brought about in two ways: firstly by restoring the deficiencies, secondly by removing the degradation substances." Hence, "Youth is a possession you can keep if you choose."[510] Verzár, however, made no such far-reaching assumptions. If anything, he represented the 'respectable face' of gerontology: as the founder of an academic institution and a journal, the key note speaker at scientific conferences, the venerable mentor of many scientists. Appropriately, he was at war with Niehans, with rejuvenators generally, with anyone who would sell us untested snake oil.[511]

Verzár represented the approach to life-extension that would first conduct extensive research, and the more extensive the better, in order to understand everything there is to be understood about the aging process, and then maybe, just maybe, suggest strategies for intervention. In the meantime, to increase our longevity (if only slightly), the proponents of this approach would recommend following simple rules of hygiene and maintaining an active and optimistic attitude (the suggestions that do not seem to require a very extensive basic research). Indeed, Verzár's nonagenarian life furnished to many a telling example of how activity and optimism can prolong one's life, regardless of Verzár's research findings. Thus, Le Compte emphasized: "What a mercy for [Verzár] that Gerontology exists, i.e. that he was able to begin a second career at 70 after his retirement. For this work keeps him young!"[512] The president of the French Gerontological Society and Verzár's biographer, Francois Bourlière, made the point even more explicit: "When some participants were gravely discussing the meaning of science, Verzár unexpectedly remarked 'For me, science means mainly fun for the scientist!' I have the feeling that this never ending delight in unveiling the unknown is the very essence of his Fountain of Youth."[513]

The distinction between reductionist "rejuvenators" (who would devise a specific tweaking for the human engine and give us a rejuvenating pill here and now), holistic "hygienists" (who would recommend working, sleeping and eating in moderation and being happy as prescribed by Hippocrates, looking forward to reaching a "natural" life-span that we do not normally reach) and "basic researchers" (like rejuvenators valorizing reductionism, but who would first conduct extensive study and build a theory and then, perhaps, arrive at conclusive recommendations) seems to have been persistent in the history of the life-extension movement. However, the distinction has been often blurred, and combinations of approaches have not been uncommon. Thus, both Niehans and Verzár were great enthusiasts of hygiene and physical activity, both conducted extensive laboratory tests, and both prescribed some forms of life-extending products. The distinctions may only be observable in different degrees of emphasis.

13. Chapter conclusion. Between "artificial" and "natural" life-extension

The responses to the failure of endocrine rejuvenation of the 1920s were manifest in different shifts of emphasis toward different potential life-extending means. Moreover, in different countries, the accents appeared to differ. In France and Germany, "holistic" and "natural" macrobiotic hygiene was emphasized. In Romania and Switzerland, attempts at artificial rejuvenation (with new methods) and basic research were conspicuous. As will be shown in the next chapters, in the Soviet Union and the United States, the approach was similar to Romania and Switzerland. The importance of physical and mental hygiene was recognized in the SU and US. But there was a great readiness to try new rejuvenative pills and injections, and voices proliferated demanding ever greater basic research in order first to find such remedies and then to test them.

The ideological justifications for longevity research in the Soviet Union and the United States will provide further examples (perhaps even more salient than France and Germany) of how ideology shapes research programs. It will be again shown that longevity researchers and life-extension proponents acted as true advocates for the perpetuation of the social and ideological order of which they considered themselves to be a part. In attempting to relate the "internal" evolution of longevity research with the "external" social determinants, it may not be an accident that the strengthening of "holistic" and "natural" approaches in France and Germany were concomitant with the strengthening of traditionalist and nationalist attitudes in these countries. Perhaps a simultaneous attempt was made to revert to some sense of "native," "inborn" and "primordial" community and body, unencumbered by any "foreign" and "modish" intrusions. Presently, some western proponents of "natural" or "holistic" approaches would not wish to have anything to do with "traditionalism" and especially with "nationalism," yet in the early period under consideration, in France and Germany, these notions were strongly correlated. In the Soviet Union and the United States, on the other hand, socialist aspirations and capitalist concerns appear to be respectively the chief determinants for pursuing diverse basic and applied life-extension research programs.

Chapter 3.

The USSR: The perpetuation of Socialism and triumph of Materialism

1. Chapter summary

In this chapter, the works and ideas of several prominent Russian life-extensionists will be presented, with a special emphasis on the period of Stalin's rule. First, with regard to the social milieu of science, Russia, I will argue, constitutes perhaps the most salient example of a link between the ruling political, ideological and social scheme and longevity research and advocacy. The role of life-extensionists as advocates for the perpetuation of the existing ideological and social order will be shown. Secondly, with regard to specific scientific projects initiated by Soviet life-extensionists, their emphasis on reductionist, materialistic methodology will be demonstrated. The materialistic approach will be shown to be one of several central elements compatible with the prevailing communist ideology. The Russian reaction to the failures of the reductionist rejuvenation attempts of the 1920s was characteristic: rather than perceiving these failures as a stimulus to change the course of scientific work (such as making a transition to a more holistic approach which was pronounced among life-extensionists in France and Germany in the 1930s), Russian life-extensionists conceived of the failures as building blocks and signposts for a continued pursuit on the same path. This attitude may be considered as yet another element of the influence of the dominant ideology on science, a manifestation of the frequent Soviet disregard of current hardships in view of grander future prospects.

2. The Russian tradition. Life-extensionism integrated into the Russian Monarchy

Traditionally, Russian life-extensionists had very strong ties with the ruling regimes. Before the Socialist Revolution of 1917, the connections of longevity researchers to the monarchy were conspicuous. Thus, Empress Anna Ioanovna[514] employed the German-Latvian proponent of life-extension, Johann Bernhard Fischer (1685-1772), as her personal and court physician and appointed him the "Archiatrus" (head of the ministry of medicine) of the Russian Empire. Fischer's investigations on aging and longevity culminated during the rule of Empress Elisaveta Petrovna,[515] with the publication of his book *On Old Age, its Degrees and Diseases* (*De Senior Eiusque Gradibus et Morbis*, 1754, republished in 1760). In Nazi Germany, the "German-Baltic" Fischer was honored as a forefather of German life-extensionism. Thus the German medical historian Johannes Steudel claimed in 1942 that to Fischer "belongs the merit of opposing the [earlier] therapeutic skepticism" regarding the combat of aging. Steudel emphasized Fischer's priority in describing the morphology and pathology of aging (before the works of the Italian pathologo-anatomist Giovanni Battista Morgagni, 1682-1771) as well as in studying the physiology of aging (before the Swiss physiologist Albrecht von Haller, 1708-1777). Fischer was also credited by Steudel for specifying hygienic rules for healthy aging (before Christoph Wilhelm Hufeland, 1762-1836).[516] Soviet Russian life-extensionists, however, claimed Fischer as one of their own (Sergey Tomilin, 1939),[517] without any reference to Fischer's being a German or a court physician, and only mentioning his title as the head of the Russian medical service. Fischer died during the rule of the Russian Empress Catherine II the Great.[518] During Catherine's rule, in 1778, the Russian court entertained the Italian "master of rejuvenation" Count Alessandro di Cagliostro (Giuseppe Balsamo, 1743-1795). Catherine's successors, the royal couple Paul I and Maria Feodorovna[519] too had a strong relation to life-extensionism. After the assassination of Paul I in 1801, Maria Feodorovna had her husband's body preserved and built a mausoleum in his memory, presumably with some hopes for future resurrection.[520] Under Alexander II and Alexander III,[521] the Russian longevity researcher and pioneer of geriatrics, Sergey Petrovich Botkin (1832-1889) served as a court physician and occupied high posts in the political establishment (in the Parliament and in the Ministry of Internal Affairs).

During the subsequent rule of Nikolay I,[522] a world authority on rejuvenative organotherapy was Prof. Alexander Vasilievich Poehl of St. Petersburg (1850-1908), whose "rejuvenating" preparation "spermin" (an extract from animal testicles) was supplied to the Royal Court. Another renowned rejuvenator, Ivan Romanovich Tarkhanov (Georgian prince Tarkhan-Mouravov, 1846-1908), who believed that it should be a "shame to die before the age of one hundred," belonged to the highest aristocracy. Tarkhanov and Poehl co-authored a definitive monograph on *Organotherapy* in 1905.[523] Another prominent Russian life-extensionist during the rule of Nikolay I was Porfiriy Ivanovich Bakhmetiev (1860-1913) who was apparently the first to achieve suspended animation (anabiosis) in animals (bats) by freezing (he conducted these experiments since 1897, when working at the University of Sophia, in the Russia-allied Bulgarian Kingdom).[524] A great adept of artificial "resurrection" (*ozhivlenie*) was Alexey Alexandrovich Kuliabko (1866-1930) who was the first to revive the human heart after death (around 1902). In 1913, Fyodor Andreevich Andreev (1879-1952) revived a dog using the first life-support system.[525] Though Bakhmetiev, Kuliabko and Andreev did not seem to have any ostensible ties to high aristocracy, they were no revolutionaries either. Neither were the physiologists Ivan Mikhailovich Sechenov (1829-1905) and Ivan Petrovich Pavlov (1949-1936), whose views were strongly compatible with life-extensionism.[526] To recall, Metchnikoff, both in Russia and in emigration in France, denoted himself as a "Russian nobleman."[527]

One of Russia's greatest life-extensionist philosophers during the rule of Nikolay I was Nikolay Fedorovich Fedorov (1829-1903), the Russian Pravoslav religious philosopher and founder of "Russian Cosmism," respected by Lev Tolstoy and Fyodor Dostoevsky, among many great Russian thinkers.[528] According to Fedorov's *Philosophy of the Common Task* (most of his works appeared posthumously in 1906 and 1913 under this title), the Christian doctrine of salvation dictated a practical program toward individual and social immortality, even resurrection of past generations, which, he believed, would be achieved by collective, scientific effort.[529] In setting these goals, Fedorov presented himself as a devoted Russian monarchist. In Fedorov's works, nationalist and totalitarian undertones are unmistakable:[530]

> Russia and the Russian people can (and must) call all peoples of the world to an alliance against this common enemy [death]. Absolute monarchy will play the highest role in this struggle, and Pravoslav Christianity, that will sanctify this union, will become the common religion.

And furthermore, the "common task" of fighting death requires universal conscription, and must be directed by a "Psychocracy" grounded in absolute monarchy:[531]

> Regulation is not restricted to the physiological aspect of the conscripted, but expands onto the internal, psychic aspect, and the latter becomes the foundation of society (Psychocracy). Psychocracy cannot coexist with judicial forms of government, with aristocratic or democratic republics, not even with constitutional monarchies, but only with absolute, patriarchal monarchy, with a King, standing in place of the Fathers, as a sovereign of the two kinds of regulation, the internal and the external.

Notably, absolute monarchy was presented by Fedorov as a symbol of power for the unification of equals, rather than as the rule of high aristocracy. Nonetheless, Fedorov did build on and advocate the conservation of his native social and ideological institutions: Pravoslav Christianity, universal conscription and absolute monarchy. Not only would the present social institutions remain in the future, but they would remain indefinitely.

3. The emergence of the Soviet state and the creation of a new long-lived man

After the Great October Socialist Revolution of 1917, all associations with Christianity and Monarchy were immediately wiped out from the Soviet Russian life-extensionist ideology. Instead, the purported ideological foundations of the emerging socialist state – central planning, creation of a "new man,"[532] equality, collectivism and support of the weakest members of the social organism – were heralded as the definitive elements of a new life-affirming social order that was hoped to continue far into the future.[533] These, for example, were the beliefs of the Marxist politician, economist and physician Alexander Alexandrovich Bogdanov (1873-1928, b. Alexander Malinovsky), the creator of the science of "tectology" – the "universal science of organization" (1913-1928), and a leader of the "Scientific Organization of Labor" movement.[534] Since 1918, he was a professor of political economy at Moscow University, a member of the presidium of the Communist Academy, and member of the Central Committee of the Proletcult - the Proletarian Culture movement. Since 1926, Bogdanov was also the founding director of the State Institute for Blood Transfusion in Moscow, apparently the first such dedicated institute in the world, established for the explicit purpose of using blood transfusions for rejuvenation.[535]

In *The Struggle for Viability* (1927), Bogdanov summarized his general views on life-extension, asserting that vitality deteriorates due to an impairment of cells' "organizational relations" and "internal milieu." According to him, social imperfections and inequalities largely contribute to bodily dissonance and life-shortening. The "organizational relations," both social and biological, are adjustable in a new socialist society, their equilibrium can be enforced, whereby "our life should last 120-140 years" at the least.[536] In such a society, a "new human type" would emerge that would develop harmoniously and in communion with others, and hence avoid the attrition inherent in capitalist specialization and seclusion.[537] (Similar ideas on the harmonizing, equilibrating and sustaining power of socialism were expressed by Leon Trotsky.[538])

In the paper "On Physiological Collectivism" (1922), Bogdanov asserted that by supporting the "weakest elements" of the biological and social system, by sharing resources with them, a prolonged existence of the entire system can be accomplished. Such a support of the "weakest elements" can be "only systematically achieved by transcending the limits of physiological individuality, as foreign as this thought may seem to the individualist worldview of our epoch." According to Bogdanov, such collectivism "is now only seldom present. But it is present, nonetheless, and it is augmenting with the progress of culture." The rejuvenative blood transfusion was considered as one of the highest manifestations of collectivism and its revitalizing action (even if blood was transfused from an old to a young person) was explained by the simple gift of sharing resources (1922).[539] Notably, Bogdanov expressed similar views before the revolution, while in political exile in Europe.[540] Yet, it was in new Socialist Russia that he was able to profess radical collectivism widely and authoritatively.

In the first years of the socialist state, similar collectivist rationales for rejuvenation and life-extension were posited by one of Russia's foremost geneticists, president of the "Russian Eugenics Society" and founder of the Moscow Institute of Experimental Biology, Nikolay Konstantinovich Koltsov (1872-1940). Koltsov believed rejuvenation is a "problem of high importance," though he maintained that for humans, despite the present "positive results," "caution in the application of a new method … would be necessary" and the problem requires a "detailed and prolonged research." However, for animal husbandry, Koltsov recommended the use of rejuvenative operations (Steinach's and Voronoff's) immediately and on the widest possible scale, in order to prolong the reproductive periods of prized breeders. Koltsov feared the use of rejuvenation and eugenics for class segregation and exploitation, while at the same time recommending them as a potential means to strengthen the socialist society generally and the communist party base particularly (1923, 1924).[541] Despite Koltsov's advocacy, however, no eugenic laws were ever passed in Russia.

The life-extensionists who would not comply with the dominant communist ideology, were expelled from the country in 1919-1922. This was the fate of Vladimir Georgievich Korenchevsky (1880-1959), a former counterrevolutionary White Army medical officer, who fled to England in 1919 and would later become the founder of institutional gerontological research in Great Britain. Abram Solomonovich Zalmanov (1876-1964), the proponent of rejuvenative and stimulating "capillary therapy" (for example, by turpentine baths) and the personal physician of Vladimir Ilyich Ulyanov-Lenin (1870-1924), fell out of favor and defected to Paris in 1921.[542] Naum Efimovich Ischlondsky, one of the prominent investigators of rejuvenative operations, found refuge in Berlin in 1922. Another prominent emigrant was Alexander Alexandrovich Maximov (1874-1928) who introduced the concept of "stem cells" as early as 1909, while working on blood formation (hematopoiesis) at St. Petersburg Military Medical Academy.[543] He immigrated to the US in 1922 and his last studies at the University of Chicago (1922-1928) concerned embryonic extracts as stimulants of rejuvenation.

4. Life-extensionism integrated into the Stalinist order. General Characteristics

A new stage in the development of the Soviet state occurred after Lenin's death and Stalin's coming to power in 1924, and particularly after the suppression of the New Economic Policy in 1928 (NEP, enacted between 1921-1928, which involved some economic liberalization and encouragement of business initiative). With the enactment of the first "Five Year Plan" in October 1928, the era of "collectivization" of agriculture and general "industrialization" began, with the strengthening of the authoritarian government and Stalin's personality cult. Since that time, the glorification of the socialist state became the all-pervasive motif in Soviet life-extensionist ideology, which continued in various forms well until the end of the socialist system.

In the "industrializing" and "collectivizing" Soviet Union, the support of scientific progress and the egalitarian care for the well being of each member of the socialist collective, were parts of the professed ideology. However, despite the widely announced support for science, the fear of technological advancement was widely spread in Stalin's times, during the massively supported persecutions of educated professionals, when an engineer was often seen as a saboteur, a geneticist a polluter of nature, a cyberneticist a madman and enemy of humanity, and a doctor a poisoner.[544] Despite the rhetoric of personal care, similarly to Nazi Germany, the rhetoric of individual sacrifice for the state was not uncommon.[545] Nevertheless, extensive social and biological experimentations were under way, in line with the official "where there is a wish, there is a way" ideology, comprising the wish and the way toward a radical extension of longevity. Atheism and the absence of belief in an after-life contributed to the popularity of life-extensionism, as a part of a new, 'this-worldly' progressivist eschatology, forming a new 'secular religion.'

On the popular level, various hygienic, physiotherapeutic, exercise and recreational techniques became widely practiced for life-extension. The adoption of such techniques and the universal, free provision of medical services, as well as the construction and deployment of prophylactic and recreational facilities, were propagandized as signs of the growing prosperity of the Soviet people. The fact that, in the first decades of the Soviet state, the general life-expectancy increased dramatically was taken as a proof positive of the advantages of the socialist system.[546]

The importance of hygienic measures for life-extension was emphasized, for example, in the works of Ivan Mikhailovich Sarkizov-Serazini (1887-1964), one of the founders of Soviet therapeutic physical culture and sports medicine, serving from 1944 to 1964 as a professor at the Moscow Institute of Physical Culture. As will be discussed shortly after, according to Sarkizov-Serazini, life-extensionist hygiene could truly flourish only under socialism, that is to say, in a society that cares for each and every one of its citizens and where life-shortening inequalities are rectified.[547] Longevity through "social hygiene" was also propagandized by Academician Zakhary Grigorievich Frenkel (1869-1970), one of the founders and leaders of the Leningrad Sanitary Hygienic Medical Institute, and others.[548] Yet, the importance of "following nature" or hygienic "natural regimens" was not nearly as strongly emphasized as in contemporary Germany. The new Soviet man was no slave to nature, but a master of nature, bending and shaping nature to his will. And Soviet science was called on to provide the tools for such mastery, including the tools to manipulate and retard the aging process. Nevertheless, some 'naturalism in disguise' often crept in when the Soviet gerontologists spoke of combating "premature aging" and achieving "physiological aging." The state promotion of gerontological and general life-extension research was then seen as yet another hallmark of the socialist society, a society willing to invest in far-reaching studies that may bear fruit in the very long term, without any immediate promises of profit.

The attitude of the socialist state under Stalin's regime to science was quite complex and selective. Certain lines of research were favored (especially in physics and engineering), while others were forcefully suppressed (formal genetics, cybernetics).[549] Longevity research, however, was generally favored by the authorities. Stalin's personal support was not negligible. Iosif Vissarionovich Jughashvili-Stalin (1878-1953) seems to have been quite fond of the idea of extreme longevity. In his letters to his mother, he wished her a life span ranging between a thousand and ten thousand years: "May you have a thousand years of life, vigor and health" (June 25, 1925), "Live ten thousand years." (October 9, 1936), "Our breed is, evidently, a strong breed. Wishing you health, live many years, my mother" (March 10, 1937).[550] (Stalin's mother, Ekaterina Georgievna Jughashvili, died in June 1937 at the age of 77.) Further attesting to Stalin's interest in life-extension, the dissident Russian gerontologist Zhores Medvedev (in exile in the UK since 1974) related:[551]

> In December 1936, [Ivan] Valedinsky [Stalin's personal physician] was summoned to the dacha at Kuntsevo where Stalin lay ill with tonsillitis and a high temperature. The patient was also examined by Professor Vladimir Vinogradov, a cardiologist, and by Professor Boris Preobrazhensky, a specialist in throat infections. Stalin greeted Valedinsky as an old friend and asked him about the work of the recently established All-Union Institute of Experimental Medicine. In Stalin's view, the scientists of the Institute 'spend a lot of time on theory but come up with very little in practice and are not working on the problem of life-extension.' Soon after this remark, which of course was relayed to those in charge of the Institute, life-extension became a central subject of Soviet medical research.

Throughout Stalin's rule (1924-1953), cases of fabulous longevity were celebrated, especially those from the Caucasus area, Stalin's birthplace.[552] In the autumn of 1937, the Kiev Institute of Clinical Physiology, under the direction of the most influential Soviet life-extensionist – Academician Alexander Bogomolets, dispatched an expedition, led by Prof. Ivan Basilevich, to Abkhazia (Georgia), to study persons of very advanced age and super-centenarians, assumed to be particularly abundant in that area, and to discover the secrets of their longevity.[553] This was apparently the first such concerted expedition in the world. One may conjecture whether the death of Stalin's mother had anything to do with his increasing support for longevity research after 1937 (or whether the psychological trauma sustained was somehow connected to the massive repressions of 1938). In any case, Stalin's patronage of Soviet longevity science was paramount, and panegyrics to Stalin were all pervasive in the Soviet life-extensionist literature of the period. The acclaim of Stalin formed an essential part of the general acclaim of the socialist system. References to Stalin were dropped after his death and the demolition of his "personality cult," but the praises of the socialist government remained well until its end.

5. "Russia is the birthplace of elephants" – Ivan Mikhailovich Sarkizov-Serazini (1887-1964)

An old Soviet joke told that in honor of the "International Year of the Elephant" the USSR Academy of Sciences produced four volumes: "Elephants and classics of Marxism-Leninism," "Elephants in the light of the decisions of the latest communist party conference," "Elephants and increasing socialist productivity" and finally "The USSR as the birthplace of elephants." Replace "elephants" with "longevity," and one will have a summary of the ideological underpinnings of life-extension research and advocacy in the USSR, since the foundation of the Soviet Union until its collapse.

A striking example of the professed support of the socialist system by Soviet life-extensionists can be found in the works of the sports physician Prof. Ivan Mikhailovich Sarkizov-Serazini.[554] In Czarist Russia, he struggled for bare survival: a son of the poor, he worked as a post-man, sailor, fisher, apothecary apprentice. For the involvement in the revolutionary newspaper, *The Crimean Riviera*, he was exiled from his native Yalta, Crimea Ukraine, to Ekaterinodar at the Northern Caucasus. As Sarkizov-Serazini recalled, "the years 1910-1913 were the years of wanderings in the struggle for a piece of bread, across cities in the south of Russia."[555] The revolution made him a respectable man. In 1917, he was able to receive a high school diploma (at the age of 30), and finished medical school by 1922. In 1923 he became the scientific secretary of the High Council for Physical Culture, and felt sufficiently confident to issue an appeal in the journal *Fizicheskaya Kultura* (Physical Culture) "to all scientific workers," calling all scientists and physicians in the Soviet Republic to promote physical culture, to "nip in the bud the degeneration of the youth in the Republic" that may result from the neglect of physical culture, and admonishing physicians for not participating in this effort.[556] Subsequently, posts and honors of various kinds were lavished on Sarkizov-Serazini. From 1923 until his death in 1964, he was a leading figure in the Institute of Physical Culture in Moscow, which he helped to establish and where, from 1944 to 1964, he served as the dean of the faculty of therapeutic physical culture. He received two Orders of Lenin, the title of a "Distinguished Worker of Science," and various other medals and titles. His acceptance by the Soviet establishment was wide and firm. A list of his friends in the Soviet political, intellectual, scientific and artistic elite would be extensive and illustrious indeed.[557] He also became fabulously rich by Soviet standards, able to amass an immense collection of paintings by renowned Russian artists (hundreds of paintings, some purchased, some given as personal gifts) and a 10,000 volume library, both of which he donated to the state later in life.

In the 1930s, Sarkizov-Serazini became a top authority in all the matters of "practical life-extension" or as it was then commonly termed "active longevity," in therapeutic physical culture, sports massage, sanatorium care, physiotherapy, climate therapy or environmental therapy, and various systems of "tempering" by exercise and exposure to sun, air and water. Largely due to his emphasis on the healing power of sun rays or "light therapy" (the idea that originated with the Danish physician, the Nobel Laureate of 1903, Niels Finsen, 1860-1904), Sarkizov-Serazini received the common appellation of "Sun Doctor." In 1939, during the Soviet-Finnish war, he devised for the Soviet soldiers an ointment against frost-bites (for which he received a high commendation from Clement Voroshilov, then the USSR minister of defense, who will later play a decisive role in the institutionalization of Soviet longevity research). During WWII, massage and exercise techniques developed by Sarkizov-Serazini were widely used in the physiotherapy of wounded Soviet soldiers. Sarkizov-Serazini became a cultural icon, a celebrity, and often a butt of friendly satire. Thus, the foremost Soviet caricaturists Mikhail Kuprianov, Porfiry Krylov and Nikolay Sokolov (collectively nicknamed the "Kukrinix"), who were all personal friends of Sarkizov-Serazini, depicted him in 1939 as a bureaucrat buried in medical tomes, with a crowd of petty-bourgeois-looking patients rushing toward him exclaiming "Save us! Resurrect us! Heal us! Rejuvenate us!" In the 1939 movie "Chirurgia" (Surgery, based on Anton Chekhov' stories "Surgery" and "Rural Asclepiuses"), the protagonist (played by Igor Ilyinsky) bears a photographic resemblance to Sarkizov-Serazini, and is great at reassuring his patients and suggesting life-style improvements, but is unable to pull out the correct tooth. (Sarkizov-Serazini was also a friend and physician of Chekhov's family, Chekhov's widow, Olga Knipper-Chekhov, and sister, Maria Pavlovna Chekhov.)

Under the socialist system, Sarkizov-Serazini rose to prominence, and was only too eager to repay the socialist system by glorifying it. In the article "Our Motherland – the country of strong, robust people" (1951), his enthusiasm for Soviet Russia abounded. The Soviet state, according to him, both provides the conditions and makes demands for increased longevity:[558]

> We live in a wonderful epoch. Led by the wise Communist party, the Soviet people, greatly inspired, are involved in vast, creative labor. ... The Communist party and the Soviet government daily care for the human being, for his health, for creating the conditions favorable for the prolongation of life of every Soviet citizen. There is no country in the world where these questions receive as much attention, care and material resources, as in our country. ... In the Soviet state, where the exploitation of a human being by a human being has been liquidated, where broad social measures have been implemented, radically improving the conditions of labor and of life, where the level of culture has been highly raised, and where the newest achievements of medical science are fully utilized – every person should do everything in order to prolong his life, maintain his productivity, guard against diseases.

Several characteristic arguments were adduced by Sarkizov-Serazini to demonstrate why the socialist system was particularly conducive to longevity. Socialism was supposed to level out life-shortening inequalities (i.e. scarcity and famine of the poor and excesses and obesity of the rich) and to rectify inequalities in health care. It was said to improve people's material conditions (providing comfortable, but not luxurious accommodation, sufficient but simple nutrition) and to ameliorate the conditions of labor (including the prevention of overwork and work-related accidents assumed to be due to exploitation). Moreover, the socialist state was able to conduct massive, centrally directed hygienic, prophylactic and therapeutic measures, and as a result produce a continuous increase in mean life-expectancy, pointing to a long road ahead. And finally, the socialist state established and supported research programs and "institutes dedicated to solving the problems of longevity and developing means for life-prolongation."[559] It is beyond the capacity of the present work to evaluate how these arguments/theories/slogans were implemented in real life. But these arguments are absolutely characteristic of Soviet life-extensionism and exemplify the strong adaptation of life-extensionist philosophy to the currently ruling social and ideological regime and the supportive role of this philosophy within the regime.

When reading Sarkizov-Serazini's articles, such as "Our Motherland is the country of strong, robust people" or "On the long-lived" (written in 1951) and comparing them with life-extensionist works from Nazi Germany (such as Roemheld's and Venzmer's), one may wonder whose socialism is more "national" – the Russian or the German. First of all, there is a sense in Sarkizov-Serazini's writings of a high superiority over foreigners, and at the same time a deep desire to be recognized and respected by these very same people who have the detriment of living outside of Russia. The physical superiority of the Russians, according to Sarkizov-Serazini, has been a long established historical fact: "The historical legends of ancient Slavs, our ancestors, tell of the great power, endurance, productivity, braveness, perseverance and beauty of this people." The struggle for survival developed in the Russians the "qualities admired and envied by other nations." "Strong, freedom-loving, hardened, fearless of cold and heat, easily enduring hardships of work and combat, unspoiled by excesses, luxury and depravity – such were our ancestors even according to the descriptions of their enemies."[560]

Furthermore, according to Sarkizov-Serazini, the contributions of Russian scientists to life-extension research were most weighty and were made far in advance of other countries. Thus, eighteenth century Russian scientists – Mikhail Lomonosov, Alexander Protasov, Semen Zybelin, I. Fischer, etc. – proposed the main macrobiotic rules long before Hufeland, to "the amazement of prudish Germans who then comprised the [Russian] Academy of Sciences" (cf. Roemheld's and Venzmer's glorification of the role of German scientists in the pursuit of life-extension). The dictum "Russia is the birthplace of elephants" really comes to mind. Whatever ills befell the Russian people (despite their innate physical superiority) were the fault of the Czarist aristocratic regime which "ignored the demands of the best sons of the Russian people."[561]

Under the Soviet rule, it was implied, the innate superiority was only intensified. All the necessary conditions for physical development were provided by the Soviet state, and the social plagues curtailing the longevity in capitalist countries were eliminated. In the article "Aging and causes of premature aging," Sarkizov-Serazini stated:[562]

> It is unnecessary to cite the numerous statistical data, demonstrating the increase of mortality, incidence of disease, poverty and forced prostitution, hunger or half-hungry existence, the catastrophically increasing number of mental illnesses and suicides in capitalist countries and their dependants, the colonial and half-colonial countries. The expression 'capitalism is the killer of peoples' finds its support in the cruel exploitation, in the wars waged by capitalists for profit, in their dreams of world domination.

Sarkizov-Serazini's own experiences illustrated the (unquoted) statistics: "I have seen these people, sadly standing near the London and Hull docks, having lost any hope to find work… I have seen the extreme poverty of Greece and Turkey, witnessed how the carabineers kicked out poorly dressed workers from the piazza in Venice … I cannot forget the slums of London, Istanbul, Naples, Genoa, Alexandria. … I still hear the songs of beggars on the streets of fancy resorts in Italy…" Thanks to such descriptions, the Soviet citizens (whose vast majority did not have the slightest chance to travel abroad) should have felt truly fortunate to have been spared the horrors of capitalism. Sarkizov-Serazini's main point is that living in Soviet Russia is the most essential contributor for increasing longevity, as Russian Socialism is the most life-affirming and life-prolonging social order on the planet, or in Sarkizov-Serazini's own words: "The entire Soviet Union is the land of longevity."[563]

6. Socialism as a condition for progress and renewal – Alexander Vasilievich Nagorny (1887-1953)

The propagandistic recipes concocted in the 1920s and perfected in the 1930s served well for decades to come. The current chapter will present examples, consecutively, from the periods of Stalin's, Khrushchev's and Brezhnev's rule. One of the telling cases from Stalin's period (1924-1953) is that of Alexander Vasilievich Nagorny (1887-1953), a leading Soviet life-extensionist, since 1929 head of the Department of Physiology at Kharkov University, Ukraine. Nagorny's ideological drive was explicit. In his book *Starenie I Prodlenie Zhizni* (Aging and Life-extension, 1951), he claimed that "the capitalist system – bringing about overwork, unemployment, undernourishment, diseases – creates all the conditions for the degeneration and dying out of millions of working people."[564] Furthermore, according to Nagorny, the assertions that life-extension will lead to overpopulation, to the overburdening of the society by decrepit old people, and that therefore life-extension should not be pursued – are definitive marks of capitalism and fascism, both "German fascism" as well as "American fascism." Nagorny referred to such assertions as "delusional ideas" and "cannibalistic assumptions":[565]

> Every honest human being will understand the theoretical falsehood, the anti-scientific and anti-humanist nature of such reasoning. Which honest person, based on everything known about the capitalist society, shall not agree that the struggle for longevity is, first and foremost, the struggle against class exploitative society, and for a classless society, for communism?

In contrast to capitalist societies, in the Soviet Union, it was claimed, the struggle for life-extension was a natural consequence of socialism. In the socialist formation, great achievements in the quality of life and health care have already been made, manifested in the increased birth rates, alongside with the reduction of diseases and mortality, and the resulting prolongation of life. And these indices of health and longevity will be further improved with the further strengthening of the socialist state:[566]

> The face of the new land is becoming beautiful, having been washed by the tears and blood of millions and purified by the holy fire of the Great Socialist revolution. And a beautiful new man is being created, the all powerful master of innumerable secrets of live and inanimate nature… Yes, people of the past were entirely powerless in this [struggle against aging and death], yet their distant descendants, armed by knowledge, will become stronger than their ancestors: they will eradicate diseases and senile decrepitude, and will make human life so long and so full of joy of existence, that the problem of individual immortality will no longer trouble human beings… We, people of the Soviet Union, already see the dawn of this wonderful world. … The Soviet science takes on this problem, and it will solve it, driven by the desire for human happiness and by the faith in the omnipotence of free human genius.

Other Soviet life-extensionists glorified the Socialist State with an even greater pathos. Thus Georgy Vladimirovich Folbort (1885-1960, between 1926 and 1946 chair of the department of physiology at Kharkov Medical Institute) enthusiastically announced (1938):[567]

> The scientific work in the USSR is not abstract. In our country, life itself poses questions that need a scientific solution. … At the time when, in fascist countries, the lack of rights, unemployment and hunger lead people to despair, to contempt for their own life, to suicide – in our socialist country we daily and hourly feel the care of our government and our party, our life is becoming ever more bright and joyful. Nobody wishes to reconcile with the thought that, in old age, he may become a deficient member of our Soviet society; nobody wants to think that he should part with this beautiful, bright,

satisfying life. With the high authority granted in our Union to science and its workers, we necessarily feel that the Soviet people place this demand [to solve the problems of aging and longevity] on its Soviet science.

In yet another ode, Zakhary Grigorievich Frenkel (1869-1970) – one of the leading Soviet hygienists, epidemiologists and gerontologists, founder and chair of the department of Social Hygiene at the Leningrad Sanitary Hygienic Medical Institute (1919-1951) – went even as far as to suggest that life-extension is predominantly not a biological, but a social problem, socialism, of course, providing the solution (1949):[568]

In the study of the dynamics of longevity, we should not concentrate on the human organism, on human nature, on its hereditary endogenous properties, but on the structure and organization of human society. … The problem of longevity and of life-extension in the human society is not an individual question, not a biological question, but a socio-historical question, a question of social hygiene.

Nagorny believed Frenkel went too far in deemphasizing the role of "biological factors" in life-extension. Nonetheless, the crucial role of "social factors" (that is to say, of socialism) for longevity was accepted by virtually all Soviet life-extensionists, without exception.

Beside the blatant glorification of the socialist system, more subtle ideological rationales were presented to justify the search for life-extending means in the Soviet Union. These included the tropes of regulation and social equilibrium, the ideals of human progress, of personal development and individual contribution to society, secular humanism, mastery over nature and environment. All these ideas were presented as necessary, fundamental components of the communist ideology.

First, the trope of regulation was pervasive in the Soviet life-extensionist writings, with a strong parallelism between biological and social regulation. Individual longevity entailed a thorough regulation and equilibration of all physiological systems, just as the durability of the socialist state depended on firm, central regulation and planning. Folbort expressed this parallelism in most unambiguous terms, when speaking of the regulation of the daily regimen of the aged, of the equilibrium between material and energy expenditures and their recovery: "The crucial moment determining the productivity of an aged organism is the sufficient time for recovery processes, the sufficient rest. For us, this conclusion is realistic, since our work has been conducted in the Soviet Union, where the right to rest is one of the main rights of every citizen – it is included in our constitution, which regulates our whole life."[569]

One of the dominant concepts in Soviet life-extensionism (in Bogomolets, Nagorny and many others) was that of "physiological aging" where all systems of the organism age "harmoniously" and whereby the productivity and enjoyment of life is maintained by the aged to the very latest. According to Nagorny's and Bogomolets' reckonings, such a "physiological" life-span should last somewhere around 150-180 years, and its attainment was posited as the primary goal of Soviet longevity science. These suppositions regarding the regulation of individual longevity were directly analogous to the ideal regulation of a socialist society, where the resources should be distributed "harmoniously" and "equitably" through optimal planning, thereby achieving maximal productivity and sustained development of the society.

The second pervasive ideal, aligned with the life-extensionist pursuit, was that of individual and social development. According to Nagorny, traditional schools of philosophy preach deadening concepts and represent futile "attempts to overcome the desire of life." Religion inculcates "a false teaching of the afterworld," displacing human hopes for this-worldly life. Pessimistic philosophy "sees a solution in self-destruction." Hedonists are preoccupied in momentary enjoyments and suppress the thought of impending mortality. Stoics plunge into indifference and calmly reconcile with death. In contrast, communism offers a life-affirming alternative, a philosophy of striving for personal and social development. An extended life is a prerequisite for such an ongoing and lengthy progress. This argument largely recapitulated Metchnikoff's philosophy as expressed in *On the Nature of Man* (1903).[570] Yet, in Nagorny's exposition, the progressivist ideology was presented as a necessary component of the communist world-view. Nagorny fully agreed with the writer Maxim Gorky's dictum: "Everything is in the Human Being. Everything is for the Human Being!"[571] And if "everything is for the Human Being," there can be no nobler task than extending human health and human life. The potential for life-extension, in the same way as the general potential for human development, was seen as limitless: "The possibilities of human struggle with diseases, with senile deterioration, with the shortness of human life, are truly boundless ... They open before human beings unprecedented, wonderful perspectives and refute any disbelief in the power of free science and in the all-overpowering force of human genius."[572]

Thirdly, extended longevity was not seen exclusively as a goal for an individual, but rather as a means of social development, as a driving force toward increasing the contribution of each and every individual to the society. Personal development indeed required an extended life-span. But healthy life-extension was also (and perhaps even mainly) desired to maximize the productivity of the aged, to enable a greater accumulation of knowledge and skill in order to more fully contribute to the socialist economy and culture. Thus, half a year before the end of his 87 year life, the great Russian physiologist Ivan Pavlov (1849-1936) expressed his undying desire for life-extension, the desire to continue both his personal development and his contribution to the scientific community:[573]

> The normal duration of life inherent in the human organism is at least 100 years. By our own lack of restraint, disorderliness, atrocious treatment of our own organism, we reduce our normal term of life to a significantly smaller value. ... I will endeavor to live to 100 years. I will fight for this.[574] ...
>
> I very, very much want to live long... at least to 100 years... and even longer. Why do I want so much to live very long? First of all, for my dear, only treasure – my science. I definitely want to complete my works on conditioned reflexes, to strengthen that bridge from physiology to clinics, to psychology, which can already be considered roughly established. I definitely want to go to the congress of neuro-surgeons in England, even before the international physiological congress in Leningrad, and in the forthcoming year I hope to present at the conference of psychologists in Madrid.[575]

Indeed, in actuality, the Soviet citizens were often required to sacrifice individual benefits for the good of the society as a whole. Yet, in the communist life-extensionist philosophy, the good of the individual and the good of the community were seen as mutually reinforcing, in agreement with Marx and Engels's fundamental dictum that "the free development of each is the condition for the free development of all."[576]

Finally, strongly linked to the idea of personal and social development was the idea of mastery over nature, over the external environment and over one's own body. In fact, the broad concept of personal "development" mainly implied a continuous learning, productivity and contribution, but also the increasing mastery over the internal and external environment. Once again, the possibilities of human mastery over the environment were perceived to be limitless. The external environment was believed to be infinitely modifiable by human agency; and man-made changes in the environment were believed to bear direct influence on the very nature of organisms. In this regard, Nagorny fully agreed with Trofim Lysenko's (1898-1976) general assertion that "external conditions, being incorporated, assimilated by a living body, cease to be external conditions, they become internal conditions, that is, they become parts of the living body. ... In this way, it is possible to govern heredity, substituting the necessary organic and inorganic conditions for the assimilatory activity of the organism."[577] Indeed, Lysenko's statement referred to living organisms generally, not to human beings specifically. Furthermore, Nagorny, in fact the vast majority of Soviet life-extensionists of the 1930s-1940s, had very little interest in heredity, but mainly focused on physiology, on tissue nourishment and regulation, on the processes of stimulation, exhaustion and recovery. Yet, for Nagorny, there was "not the slightest doubt that the external environment created in the interests of health and longevity will exercise a mighty influence on all the properties of the human organism, will change its metabolism, and thereby will change a number of its present natural properties." And so the Soviet "beautiful new man is being created, the all-powerful master of innumerable secrets of live and inanimate nature," who "reshapes nature, society and himself."[578] The prolongation of the human life-span was seen as one of the primary parameters of such a "reshaping" and as a powerful aid for the mastery in other areas of endeavor.

7. Opposition

During the first years of the Soviet state, there was also a wide apprehension that the experiments in "regulation," "mastery," "reshaping human nature" and "life-extension" may go wrong. The writer Maxim Gorky (1868-1936, the author of the progressivist slogan "everything is for the Human Being") allegedly said that scientists could not even draw a human being, let alone improve human nature. And the physicist Peter Kapitsa (1894-1984) anecdotally argued that despite centuries of intervention, women are still born virgins, and Jews uncircumcised. Such apprehensions were particularly strongly felt in the Soviet science-fictional literature of the 1920s. In the famous anti-utopian novel by Evgeny Ivanovich Zamyatin (1884-1937) *We* (1920), which will later strongly influence Western anti-utopias such as George Orwell's *1984* (1948) and Aldous Huxley's *Brave New World* (1932), the future "One State" is governed by the life-extensionist principles of maximizing human health and life-span. Every citizen ("number") has the self-evident "obligation to be healthy" and all life-shortening habits are outlawed. The protagonist "Number D-503" thus describes the imperative social regulation for life-extension:[579]

> I have read and heard many incredible things about those times when people still lived in a free, that is, unorganized, savage state. But this always seemed to me most incredible: how could the state power of those times, even though rudimentary, allow that people should live without any resemblance of our Tablets, without compulsory walks, without the precise regulation of meal times, rising and going to bed whenever they pleased; some historians even say that in those times streets were lighted all night, and people walked and traveled all night. I cannot grasp this. However limited their reason might have been, they should have understood that such a life was a real mass murder – a slow, daily murder. The state (humanism) forbade the complete murder of an individual, but did not forbid the half murder of millions. To kill an individual, that is to reduce the sum of human lives by 50 years, was a crime, but to reduce the sum total of human lives by 50 million years – that was not a crime. Isn't it ridiculous? ... Like animals, blindly, they bore their children... Our Maternal and Paternal Norms did not occur to them...

In the "One State," the "regulation" necessary to achieve life-extension and control of population numbers, reaches such a mathematical precision that all manifestations of individuality, chance, freedom, emotion and change are abolished.

Furthermore, Soviet fiction writers also doubted whether the "beautiful, new Soviet man" subject to life-extension, will really be that "beautiful." In Vladimir Mayakovsky's (1893-1930) play *Klop* (*The Bedbug*, 1929),[580] the protagonist is "a former worker, former party member" Ivan Prisypkin, who becomes frozen in 1929 and revived in 1979 (together with his bedbug) by the "Institute of Human Resurrections," in a world where communism has won worldwide and the "World Federation" has been established. By the Federation "decree of November 7, 1965, human life is sacred" and "every life of a worker should be used to the last second." Accordingly, since the subject appeared to belong to the "working class," the community votes for his revival for the purposes of "comparative study of working habits and life-style." Yet, the resurrected subject is in no way improved by the extension of his life. He is still a depraved, vain and shallow person, imbued by kitsch culture, burdened by unhealthy and disgusting habits, such as swearing, drinking and smoking, and is an anachronism in the new society. It is announced that the "thawed mammal" has been mistakenly assumed to be a "homo sapiens" and a member of its "highest species – the working class." Rather, he is a "horrific humanoid simulant and a most striking parasite. Both he and the bedbug are of different sizes but of the same essence." The entire difference between them is that while the one "*bedbugus normalis*, having fattened itself on the body of a single person, falls under the bed," the other "*philistinus vulgaris*, having fattened himself on the body of all humanity, falls on the bed."

The new society where Prisypkin is revived is not particularly attractive either. In this new world, life is strictly regulated and individual life "belongs to the collective." The human emotions that the subject displays are entirely foreign to this society. "Infatuation" is considered to be an "ancient disease, where human sexual energy, rationally distributed through the entire life-span, suddenly condenses in a week into a single inflammatory process, leading to irrational and extravagant deeds." The subject almost causes epidemics of "romances," "servility" and "drinking," but these are fortunately suppressed in time. Eventually, the efforts of the society to "elevate [the subject] to a human condition" fail, and he is put in a cage for common derision, and is lucky not to have been shot.

Further examples of the fear that life-extension attempts may misfire, can be found in the works of Mikhail Bulgakov (1891-1940). In his novel *Sobachie Serdze* (*The Heart of a Dog*, written in 1925),[581] the protagonist Prof. Preobrazhensky is a rejuvenator by trade. In the manner of the actual contemporary procedures of Serge Voronoff, he routinely reinvigorates his wealthy aging clients and high-standing Soviet officials by sex gland transplants. Preobrazhensky's subjects are ridiculed by Bulgakov as vain, lecherous, silly old people, whose rejuvenation does not appear to be of great use to anybody. But this does not seem to be Bulgakov's main critique of endocrine rejuvenation. As a side line of his trade, Prof. Preobrazhensky transplants human pituitary and testes into a dog, taking the donor organs from Klim Chugunkin, a thief, alcoholic and hooligan, killed in a brawl. As a result, the dog is transformed into a semblance of a human being, named Polygraph Sharikov. Yet, this "new man" retains all the appalling traits (and memories) of the dead donor: incorrigible swearing, drinking, stealing, debauchery, ignorance and cruelty. At the end Sharikov threatens to destroy his creator, he is ceased by Preobrazhensky and forcefully "de-evolved" back into a dog. In this scenario, Bulgakov seems to suggest that, rather than improving human nature, endocrine manipulations may only serve to perpetuate ancient vices in the best case, or mutilate human character in the worst. In both cases, the endeavor is not worth attempting. The rejuvenators, Prof. Preobrazhensky and his assistant Dr. Bormenthal, despite their better manners and higher intelligence, also do not evoke much sympathy. First they restore the existence of a human being (however badly mannered), and then they destroy him when dissatisfied with his behavior. From any angle, Bulgakov seems to present rejuvenation as a futile, immoral and dangerous enterprise.

In another of Bulgakov's novels, *Rokovye Yaytza* (*The Fatal Eggs*, published in 1925, also known as *Luch Zhizni - The Ray of Life*),[582] the fear of the attempts to "enhance" or "reinforce" life is even deeper. In the novel, the "ray of life" discovered by Prof. Persikov, is capable of tremendously stimulating all life processes. (And indeed, the "stimulating," "rejuvenating" and "life-extending" effects of irradiation at different wavelengths were actively studied in Soviet Russia in the 1920s by Leonid Vasiliev, Alexander Lubishev, Alexander Gurwitsch, Alexander Chizhevsky, and many others.[583]) Yet, instead of the intended beneficial use for raising vigorous live-stock, the ray accidentally gives rise to deadly, voracious monsters. The Russian title of the book, "Rokovye Yaytza," is commonly translated into English as "The Fatal Eggs" – "Rok" meaning "fate" and "yaytza" – "eggs." "Rokk" is also the name of the collective farm manager in the novel who, due to ignorance, mistakenly applies the ray to raise giant reptiles, instead of poultry. Yet, it seems to me that Bulgakov may have also alluded in the title to the giant, ferocious (and in some accounts immortal) bird of pray Roc (or Rukh) from Arabian and Persian mythology. Exactly such a monster, Bulgakov seems to suggest, is what we shall likely obtain when attempting to "improve nature" and "enhance life." Bulgakov's works clearly imported the anti-utopian and apocalyptic visions of Herbert George Wells' *The Island of Doctor Moreau* (1896) and *The Food of the Gods* (1904) to the Soviet soil.

In *Abolishing Death: A Salvation Myth of Russian Twentieth-Century Literature* (1992), the American literary scholar Irene Masing-Delic argued:[584]

> Official Marxism-Leninism ... denies any form of personal immortality in either transcendental or earthly regions... After the Bolshevik victory of 1917, any expectations of physical or spiritual personal immortality have lost credibility. This was not the case, particularly within the creative and artistic

sectors of the intelligentsia. These circles saw old death not only as out of place in a new world but also as incongruous in a world where material justice had allegedly triumphed.

The above examples, however, indicate, on the contrary, that it was within the Soviet "artistic intelligentsia," particularly men of letters of the 1920s, that the opposition to life-extensionism mainly originated. Curiously, Masing-Delic did not mention Zamyatin's and Bulgakov's critical works at all, and referred to Mayakovsky only parenthetically. Many examples that she gave for the expressions of physical immortalism in Russian/Soviet literature – such as Maxim Gorky's *Chelovek* (The Human Being, 1898) extolling the fight of human thought against death and coining the slogan "Everything is for the Human Being" as well as Gorky's *Task* (1930) praising constructive, life-affirming, collective labor; Boris Pasternak's Doctor Zhivago (1955) shooting at a dead tree; or the resurrection of Jesus in the Socialist Revolution in Alexander Blok's *The Twelve* (1918) – are rather allegorical and bear no direct relation to scientific biomedical programs. Also, Masing-Delic virtually did not refer to Soviet biomedical scientists, gerontologists in particular, who constituted the core of the Soviet life-extensionist movement.

Contrary to Masing-Delic's thesis, the widely disseminated writings of these highly authoritative Soviet scientists, as well as the massive support of longevity science from the highest to the lowest ranks of the state, demonstrate that radical life-extensionism was flesh of the flesh of the "official Marxism-Leninism." The life-extensionists glorified the ruling ideology and found in it the justifications for their research, while the ruling ideologues fostered this research to increase their appeal. Indeed, a direct mention of physical immortality could hardly be found in either scientific, popular scientific or fictional Soviet literature of the 1920s-1940s. Yet the directives of "fighting/defeating death," "solving the problem of aging", or, as suggested by Nagorny, extending human life to "two or three times the value of the mean life-span" were virtually equivalent to a pursuit of immortality. Even though the primary task was to achieve a limited extension of life, this was truly tantamount to approaching immortality, as any such expected postponement of the death sentence would give a temporary relief and inspire hopes for further extensions. It may be also significant that the literary opposition to life-extensionism appears to have flourished in the 1920s, before the solidifying of the totalitarian state. Even then, Zamyatin's *We* and Bulgakov's *Heart of a Dog* were not allowed for publication by Soviet censorship (and they were not widely known to the Soviet readership until 1987, the time of "Perestroika").[585] In the 1930s, with the strengthening of totalitarianism, no such conspicuous literary criticisms of life-extensionism could anymore be found. Life extensionism became an inseparable part of the Soviet ideological and political system.

Masing-Delic credited the American historian Peter Wiles as one of the first Western scholars to note "'the inordinate interest in physical immortality on earth' in the USSR (1965)."[586] "Although intrigued by this quest for a world without death," Masing-Delic maintained, "he does not consider it of great importance, since its repercussions are largely limited to 'curious phrases' in some literary works and oddities in medical research policies." Masing-Delic agreed that "That evaluation of the search for immortality is justified in the case of the social historian, but for the literary scholar, 'curious phrases' in literary works offer interpretative challenges." In that article "On Physical Immortality" (1965) Wiles further stated: "Finally, does this subject matter? I cannot think it does. Nothing serious is going to happen because many Russians want to make physical immortality practicable." And further, Wiles argued for the inherent or predominant "Russianness" of the life-extensionist movement. From the other side of the "Iron Curtain" he raged against the "Russo-Marxists" and their excessive dreams of changing the world and human nature and achieving super-longevity. In Wiles' view, these dreams threaten Western sensibilities. According to Wiles, the search for radical life-extension or physical immortality is "specifically Russian" and "is simply not an issue in other cultures."

It seems that both Wiles and Masing-Delic undervalued the influence of life-extensionism in the Soviet scientific, ideological and political system. I would argue that the incorporation of life-extensionist aspirations went far beyond "curious phrases" and literary "interpretative challenges," but constituted a definitive element in Soviet biomedical research and practice. As will be exemplified below, the scope of research motivated by life-extensionist aspirations was massive. This definitive ideological element has been commonly omitted by many historical works on Soviet biomedicine.[587] Also, life-extensionism formed not a cursory, but an indispensable constitutive element in the communist teleology, as well as in the actual social regulation, as evidenced both by the writings of authoritative Soviet scientists and high-standing officials, including Stalin and Khrushchev (see below). Nonetheless, Wiles also seems to have exaggerated when describing radical life-extensionism as a "specifically Russian" phenomenon. As the present work argues, life-extensionist aspirations were present all across the world, yet in particular countries they were shaped by particular native ideological and social environments; and furthermore, life-extensionists acted as champions for the conservation of their particular native environments. In the USSR, unlike elsewhere, life-extensionism was shaped by communism and served as its prop and mouthpiece.

8. The triumph of materialism

One of the crucial characteristics of Soviet life-extensionism was its professed materialism. Several conceptual paradoxes emerged with reference to the "materialist" vs. "idealist" dichotomy. Indeed, hard-core idealistic life-extensionism, professing a direct influence of the mind over the body, without the mediation of labor, science or art, would be unacceptable by Soviet materialism. At the same time, pronounced behaviorist, holistic, neo-Lamarckist and vitalist tendencies, emphasizing the supremacy of the mind over matter, can be observed in the Soviet life-extensionist works of the 1930s-1940s, further strengthening in the early 1950s, though perhaps in disguise (as in Sarkizov-Serazini, Olga Lepeshinskaya, and Alexander Nagorny[588]).

The 'Soviet Man' was an avowed materialist. Yet the call to 'follow nature' that was so pervasive in contemporary German life-extensionism, received no great emphasis among the Soviets, as the communist ideology considered man to be the master of nature. Material nature needed to be understood in order to master it, but the current natural limitations (such as the processes of aging) were not seen as something insurmountable or entirely binding. Nonetheless, despite such an obvious valorization of the human spirit over nature, the direct influence of the mind over the material body was never explicitly emphasized. Instead, such a power was acknowledged under various materialistic 'guises': the "mind" was replaced by the "highest nervous activity." The great Russian luminaries of physiology – Ivan Sechenov, Vladimir Bekhterev, and foremost Ivan Pavlov – were commonly invoked as the pillars of the scientific view of this activity of the "highest form of organized matter." In practice, the power of "aspiration" (*stremlenie*), "enthusiasm," "mental setting" (*ustanovka*), or more scientifically sounding "neuro-humoral regulation" or "influence of the brain" (another euphemism for the mind, very common in Sarkizov-Serazini's writings) were seen as paramount for achieving health and longevity, without any real need for understanding the mechanisms of reflexive nervous activity. Still, any suggestion of a direct influence of the mind over matter would have been completely out of the central materialistic line, and might even lead to dire consequences for those who expressed it. Such a suggestion could lead to being branded as an enemy of "Pavlov's materialistic teachings" and consequent sanctions (the inquisitorial "Pavlov's session" of 1950 and its impact on Soviet life-extensionism will be discussed in greater detail later on).

There seemed to have been no direct reference to "holism" either. Nonetheless, the importance of the psychological and social environment was recognized by virtually all Soviet life-extensionists. Vitalism too was officially reviled. But what was, for example, Olga Lepeshinskaya's "vital substance" (*zhivoe veshestvo*), whose longevity she hoped to increase by alkaline immersions,[589] but yet another expression of vitalism? And what was Alexander Nagorny's "power of self-renewal" but the good old Hufelandian/Drieschian "restituting" vital force? Interestingly, Nagorny openly attacked "mechanistic views" and expressed his agreement with Hufeland's vitalistic statement that "the energy of life is inversely proportional to its duration."[590] Nagorny, however, had a better Marxist authority for his concept of the "power of self-renewal," namely Friedrich Engels' statement in *Anti-Dühring* (1878) that "*Life is the mode of existence of albuminous bodies, and this mode of existence essentially consists in the constant self-renewal of the chemical constituents of these bodies.*"[591] And indeed, this statement by Engels was one of the most widely cited mottos by Soviet life-extensionists, since Nagorny in the 1940s to Vladimir Frolkis in the 1980s.

Whatever vitalistic or idealistic undercurrents there might have been, materialism needed to be avowed, and avowed convincingly. Those who failed to assert their affiliation with materialism were likely to be repressed. This is exemplified by the fate of Porfiry Korneevich Ivanov.

9. The downfall of idealism – Porfiry Korneevich Ivanov (1898-1983)

Porfiry Ivanov's teachings advocated a rigorous ascetic regimen, unity with nature, the power of the spirit over the body and the striving toward unlimited longevity. He began his personal experiment on healthy life-extension in 1933. In 1934, he spent half a year wandering barefoot across Ukraine and Southern Russia, exposing himself to the rigors of cold, fasting and meditation, in the tradition of saintly Russian ascetics or Yogis. In the following fifty years he developed and propagandized his system of "tempering-training" (*zakalka-trenirovka*). The system emphasized, first of all, the communion with nature, living in accordance with natural laws. It urged human beings to overcome their life-shortening "dependence" on technology and artificial comforts, minimize their demands for food, clothing and shelter. Rather, human beings must derive strength directly from nature: "The restoration of human health and strength consists not only in the restoration of the health and strength of our own organism, but also in the restoration of living connections and unity with the external environment and nature" (1951).[592] From such a direct communion with nature, according to Ivanov, radical life-extension, even physical immortality will follow: "We all have the task to prolong life, to maintain health into a very advanced age. And I found such a means, I found it in Nature, through my quest, through my work" (1958).[593] "I write of nature, of its properties leading to health. Nature wishes that man should never die" (1978).[594] "[I] bow to you deeply and ask you to undertake this task, to act in nature without bounds, then we shall not die" (1983).[595] In Ivanov's somewhat paradoxical views, through the initial subjugation to and communion with nature, man can eventually become its master, overcome its destructive influences, become a "god of the earth."

The ideas of the human communion with nature, and the life-extensionist, even immortalist aspirations were also major themes in the philosophy of "Russian Cosmism," since Nikolay Fedorov through later major proponents: Vladimir Vernadsky (1863-1945) – the author of the concept of the "Noosphere"; Konstantin Tsiolkovsky (1857-1935) – the visionary of rocket science and space exploration; and Alexander Chizhevsky (1897-1964) – the founder of Cosmobiology. Thus, in *The Scientific Thought as a Planetary Phenomenon* (1938), Vernadsky maintained that one of the humanity's crucial "tasks" is the "prolongation of life, and diminishment of diseases for all humanity."[596] And according to Tsiolkovsky, "Life has no certain boundaries and can be prolonged to thousands of years. … Science will sooner or later achieve the indefinite prolongation of life" (1928).[597] Chizhevsky, in turn, was convinced by Tsiolkovsky of the theoretical possibility that "having transformed into a ray form of a high level, humanity will become immortal in time and infinite in space" (1920).[598] Ivanov may be well considered as a part of this tradition.

It is unclear, however, to what extent Ivanov was actually influenced by "Russian cosmists." In his writings, Ivanov referred to the works of Marx, Engels and Lenin, the Old and New Testament, the works of the Soviet life-extensionists Alexander Bogomolets and Ivan Sarkizov-Serazini, but not to the "Russian cosmists." Whatever the possible influence, both for Ivanov and the "cosmists" the idea of a direct interrelationship between human beings and the universe or nature was central. But unlike the major "Russian cosmists," Ivanov's views on the human relationship with nature were decidedly a-scientific (Ivanov's formal education was four years of elementary school), and clearly bordered on pantheistic nature-worship. Thus, Ivanov requested that the follower of his regimen "must, in his mind, ask Nature to give him life and health."[599] Having propitiated Nature, and having established amicable relations with her, man can command her, may become her master, a "god of the earth." Furthermore, Ivanov clearly considered himself to be such a "god": "I am a god of the earth, having come to earth to save human life."[600]

Ivanov's method of "tempering-training" showed further, even more conspicuous signs of mysticism, magical and paranormal thinking:[601]

The main thing in my method is not the usual gradual accustoming of the organism to adverse external conditions, and not a system of external physical exercises and influences on the organism, which are lengthy and not very effective, but psycho-technical methods of awakening, the development and conscious mastery of internal forces and abilities of the nervous system, especially its reserve forces and mechanisms of heating and immunity. ... In transferring the power of my will at a distance, as if by unseen radio waves, I launch my thought first high, to the depths of the Universe, and then low, to the depths of water and earth, and then to all hurting living creatures, and then to man. I penetrate his organism by my internal vision and sensation...as if by electric current, I awaken and enact hidden forces and defensive capabilities.

During a course of treatment on a person, Ivanov would conduct "the same regimen on myself, thinking about that person." The belief in the direct power of the mind over the body was absolute in Ivanov's philosophy, as the ultimate purpose of his regimen was the "development of the will and conscious control of our organism in life." As Ivanov progressed in years, the mystical (and immortalist) elements in his teachings seem to have intensified. By the end of his life (the early 1980s), his following assumed all appearances of a sect, centered at the village of Verkhny Kondruchy, near Lugansk, Ukraine. The most devout followers often told wonders of the teacher's deeds and referred to him as "Master of the Universe."

Naturally, such appearances of idealism, pantheism, mysticism, beliefs in the paranormal, "priestliness" (*popovshina*), a-scientific naturism, and rejection of industry and technology, were unacceptable by the Soviet establishment. Ivanov was imprisoned many times, put under house arrest, evicted and hospitalized in psychiatric hospitals (and reasons for a diagnosis were not difficult to find). Altogether, in 50 years since 1933 to 1983, he spent 12 years either arrested for "anti-Soviet propaganda" or institutionalized. (Through most of this time, Ivanov lived in Krasny Sulin, near Rostov, in the Donbass Region of Southern Russia.) Though Ivanov's teachings would not seem anything unusual in a "New Age" context,[602] within the Soviet establishment Ivanov's views were a clear case of encroaching "idealism," "quackery" and/or "madman's medicine," and were suppressed.

Indeed, Ivanov attempted to blend with the system, ever since Stalin's times. Thus, he asserted:[603]

In general, there is a center, with which all parts are connected and on which their life depends. As in the human organism the center is the consciousness and the brain, so the consciousness, brain and center of our society is our leader, comrade Stalin and our party, and our government. In the love for them and in the connection to them – there is the connection to life, the detachment from them is the detachment from life.

These lines were written on November 2, 1951, after Ivanov had been incarcerated for 10 months, "living without freedom, under a compulsory regimen, doing physical work, developing my body, not minding difficulties, breaking iron walls." He further glorified "the simple, poor people" as the most likely ones to enter into the communion with nature and to achieve great longevity. He advocated his system as a means to strengthen the health of the Soviet people. He urged the elimination from the socialist society of "egotism, dishonesty, religiosity, depravity, etc." He requested that his method of "tempering" should be examined scientifically. (And indeed, Ivanov's abilities to withstand cold were tested both by the Soviet physicians Nikolay Kurchakov and Alexander Kogan at the South Russian Rostov Medical Institute, and by the Germans during their occupation of the Donbass Region.) But generally, his appeals and petitions were dismissed out of hand, and in most cases led to further arrests and hospitalizations. The expressions of loyalty did not help him; both he and his teachings were suppressed as incompatible with the dominant ideological system.

A change of attitude came in 1982, with the publication of Sergey Vlasov's article about Ivanov, entitled "An experiment lasting half a century" in one of the Soviet Union's most popular magazines, *Ogonek* (The Sparkle), with a circulation of over a million copies.[604] The article gave a sympathetic account of Ivanov's health system, carefully avoiding anything that might clash with Soviet ideological sensibilities, but rather describing Ivanov's and his followers' feats of endurance, vigorous appearance, low incidence of diseases. Thanks to the article, Ivanov's house arrest was lifted, and his health system (briefly summarized in Ivanov's set of 12 rules – "Detka"/ "The Child") became known to millions of Soviet people.[605] Many began practicing some of its elements (particularly cold baths and fasting). Ivanov's health system almost entered the main stream, but never entirely entered it, as great many of his tenets were clearly incompatible either with "dialectic materialism" or with established medical or communal practices. Indeed, throughout his life, Ivanov was an advocate of the existing social order, an ardent supporter of socialism, that he hoped would continue to flourish indefinitely. Unfortunately for Ivanov, the Soviet system did not need his advocacy. The perceived discrepancies between his views and the established ideology were sufficient to brand him as a charlatan or a mad-man, and to repress him.

10. Scientific life-extensionism

Avoiding Ivanov's mistake, the majority of Soviet life-extensionists preferred to steer clear of anything that would make them suspect of "anti-materialistic," "anti-scientific" or "anti-Soviet" views. Hence they steered clear of the mind-matter problem, of naturism, of holism, and rather focused on the body proper. They relied on the sheer power of materialistic reductionism to comprehend the workings of the human machine. Based on that knowledge, they hoped to find ways to engineer or regulate this machine for a prolonged service. Such life-extension experiments were conducted on a wide front.

Some of the interventions studied in the 1920s-1940s could appear quite unorthodox even by today's standards.[606] The rise and fall of endocrine rejuvenation techniques of the 1920s in Europe served not as a deterrent, but rather as a powerful stimulus to search for new, advanced methods of "reinvigoration" and "rejuvenation." The "stimulation of physiological function" by small doses of nicotine, cocaine, chloroform, ether, mercury chloride, adrenalin, cupper, platinum, silver, gold, and other chemicals, were tested by Nikolay Pavlovich Kravkov in the early 1920s. Kravkov (1865-1924), the head of the Pharmacology Department of the Leningrad Military Medical Academy, was honored as the "founder of Soviet pharmacology," and was in 1926 posthumously awarded the Lenin prize "for the revival of dead tissues."[607] Since 1930, "stimulating salts" such as $MnSO_4$, $MgSO_4$, $MgCl_2$, $CuSO_4$, K_2SO, and so forth, were investigated by Ivan Nikolayevich Bulankin (1901-1960, Kharkov University). In 1935, the USSR Vitamins Institute was established in Leningrad, led by Academician Alexander Schmidt and Prof. Edgar Lederer (student of the German Nobel Laureate Richard Kuhn), that coordinated vitamins production in the Soviet Union and studied their effects, including their anti-aging effects.[608] The preservative action of alkaline drinks and baths, such as those of mineral water or baking soda, were extensively studied by Olga Borisovna Lepeshinskaya (1871-1963) in the late 1940s at the Moscow Institute of Experimental Biology.[609]

"Self-renewal of the protoplasm" was sought by Alexander Vasilievich Nagorny (1887-1953) of Kharkov University, who chiefly employed for this purpose alternating regimens of rest and activity, as well as dietary adjustments, particularly the adjustment of protein intake. The search for "the stimulation of protoplasm synthesis" was continued in the 1950s by Nagorny's pupil, Vladimir Nikolayevich Nikitin (1907-1993), focusing on dietary caloric restrictions. Other stimulants of "protoplasm synthesis" in Nikitin's arsenal included royal jelly (bee's milk) and the use of various high energy compounds ("macroergica") to reenergize the body (ATP, various intermediaries of the Krebs cycle, etc.).[610]

Psycho-regulatory and psycho-stimulating effects on health and longevity were widely discussed. Thus, Maria Kapitonovna Petrova (1874-1948), Ivan Pavlov's pupil and collaborator at the Institute of Physiology in Leningrad, conducted animal experiments on life-prolongation by "training," "sparing conditions of brain activity" and neurotropic substances, such as salts of bromine (between 1938-1946).[611]

The rejuvenating effects of small "stimulating" doses of radiation, light in the infrared, visible and ultraviolet spectrum areas, negative ionization of the air, the use of electromagnetic fields for tissue regeneration (or reinforcing the "mitogenic radiation" in the ultraviolet area) were proposed by Alexander Gurwitsch, Alexander Chizhevsky, Leonid Vasiliev, Alexander Verigo, Alexey Voynar, Alexander Lubishev, and others.[612]

Since the early 1920s organotherapeutic/opotherapeutic/hormonal preparations for rejuvenation were developed by Vasily Yakovlevich Danilevsky of Kharkov (1852-1938). At the same time, a large number of Soviet surgeons began experimenting with endocrine rejuvenation operations, particularly Steinach's vasoligation and Voronoff's sex gland transplantation. The rejuvenators included B.M. Zavadovsky, V.G. Shipachev, A.V. Nemilov, G.M. Kogan, A. Damsky, D.K. Kustria, A.V. Martynov, and Lenin's surgeon, Vladimir Nikolaevich Rosanov, 1872-1934, since 1929 serving as the chief physician of the Kremlin Clinic.[613] Transplantations of testes were also performed by Yuri Yurievich Voronoy (1895-1961) of Kherson, Ukraine, alongside the transplantations of limbs in animals and one of the first human kidney transplantations conducted in 1933.[614]

Since the early 1930s, regeneration was induced by mechanical, chemical or electrical irritation by Boris Petrovich Tokin, Lev Vladimirovich Polezhaev, G.P. Gorbunova, P.P. Kanaev and others. These studies were continued in the 1950s by A.A. Zavarzin, M.A. Voronzova, L.D. Liozner, A.N. Studitsky and others. Different pathways toward regeneration were examined by these researchers: de-differentiation and subsequent re-differentiation (the idea first proposed by the St. Petersburg physiologist Evgeny Alexandrovich Schultz in 1904); renewed formation of tissues from non-differentiated cells existing in the adult organism; regeneration from homogenized tissues; enhanced growth of differentiated tissues, etc.[615]

Since the late 1920s through the 1940s, embryonic and placental tissues were tested for their ability to promote regeneration and rejuvenation – by Vladimir Filatov, Nikolay Krauze, Semen Khalatov, and others. Vladimir Petrovich Filatov (1875-1956), one of Russia's foremost ophthalmologists, since 1936 director of the Center for the Research of Eye Diseases and Tissue Therapy in Odessa, Ukraine, was a pioneer of reconstructive and replacement surgery, having developed new methods of skin grafts to heal injuries. He was the first to perform cornea transplants from corpses to return sight to the blind. He was also the author of a peculiar system of rejuvenation: the "tissue therapy" that he developed since 1933. In Filatov's theory, under extreme conditions (such as cold or darkness), conserved or living tissues produce substances (which he termed "biogenic stimulators") that counteract the stress, and therefore can be employed for the preservation and stimulation of the human organism. Filatov had very little idea of what these "biogenic stimulators" might be or what was their mechanism of action. Yet, his injections with tissues conserved under "extreme conditions" or transplants from such tissues appeared to produce some reinvigorating effects.[616] This was of course yet another ramification of opotherapy or the therapeutic use of animal parts, that was suggested as early as the ancient Chinese Mawangdui medical manuscripts (c. 200 BCE),[617] the Ayurvedic *Sushruta Samhita* (c. 300 BCE)[618] and the Talmud (c. 200-500 CE),[619] and that flourished at the end of the 19th century (with a high manifestation in Brown-Séquard's rejuvenative sex gland supplements). Yet, Filatov's theory and the method of tissue preparation "under stress," as well as the use of isolated human (live or dead) tissues, were novel. Preparations from plants under "extreme conditions" of cold and darkness were employed by Filatov as well.

Yet another theory of "tissue therapy" (derivative from "opotherapy") was advanced between 1926-1935 by Academician Mikhail Pavlovich Tushnov (1875-1935), who served as the chair of the faculty of microbiology at the Moscow Zoological-Veterinary Institute. As in Filatov's method, Tushnov's "tissue therapy," also known as "protein therapy" or "lysotherapy," involved the parenteral (intra-venous or intra-muscular) injection of tissue protein preparations. According to Tushnov's theory, the subsequent reinvigoration of the human organism was due to the physiological activity of the products of tissue degradation (histolysates).[620] The idea was based on Claude Bernard's (or rather alchemical) notion that "The two operations of destruction and renewal, the one the opposite of the other, are absolutely connected and inseparable in this sense that destruction is the necessary condition for renewal; the acts of destruction are the precursors and the instigators of those by which the parts are reestablished and reborn, that is to say those of organic renovation."[621] In Tushnov's theory of protein therapy, the products of tissue protein degradation either stimulate native protein synthesis via negative feedback (in the so-called "auto-catalysis" process) or provide building blocks for such a synthesis. Thus the "waste products" of protein degradation were seen as powerful, even indispensable, agents for tissue maintenance and renewal.[622] Since 1940, the regeneration researcher Boris Petrovich Tokin (1900-1984), working at Tomsk and Leningrad universities, used "protein therapy" to treat cancer, as he believed that regeneration is antagonistic to cancer, and that by stimulating regeneration by protein therapy, cancer may be suppressed.[623]

The "revival of the dead" (*ozhivlenie*) was yet another powerful direction in Soviet life-extensionism of the period. Some of the first experimental resuscitation devices were developed in the 1920s-1930s by Soviet Russian scientists. In the early 1920s, experiments on extracorporeal revival of organs were conducted by Nikolay Pavlovich Kravkov (Leningrad). A little later, in 1928, Sergey Sergeevich

Bryukhonenko and Sergey Ionovich Chechulin of Moscow constructed and applied in dogs the "autojector" – the first effective whole-body artificial circulation device, capable of keeping alive a severed dog's head.[624] In 1933, Nikolay Nikolaevich Puchkov, working at Bryukhonenko's laboratory, first performed extracorporeal blood cooling using an "artificial blood circulation apparatus" and a heat exchanger. Another Russian scientist, Vladimir Petrovich Demikhov (1916-1998, Moscow) constructed and implanted the first artificial heart in a dog in 1937. Demikhov's experiments went as far as head transplantations in dogs in 1954 and joining halves of dogs' bodies in 1956.[625]

In the USSR, the art of mummification was developed to high perfection, since Lenin's mummification in 1924 by Profs. A.I. Abrikosov, V.P. Vorobiev and B.I. Zbarsky. (The mummification was conducted under the directive of Felix Edmundovich Dzerzhinsky (1877-1926), head of the Extreme Committee for the Struggle against Counterrevolution and Sabotage and Commissar of Internal Affairs. A popular poem of 1924 by Vladimir Mayakovsky (1893-1930) claimed that "Lenin is even now more alive than anyone of the living.")[626] At the Institute for the Study of the Brain and Psychic Activity in Moscow, founded in 1918 and initially led by the neurophysiologist and psychiatrist Vladimir Mikhailovich Bekhterev (1859-1927), the collection, preservation and analysis of the brains of outstanding personalities (Lenin's in the first place, but also Mayakovsky's and others) was a major research project, apparently conducted with the hope of future revival or at least with the hope of future reconstitution of the material basis of their genius. Even more daringly, Bekhterev, in his work on the *Immortality of Human Personality as a Scientific Problem* (1918) based on the law of energy conservation, suggested the theoretical possibility of immortalization of human "nervous-psychic activity."[627] Furthermore, the physicist Lev Sergeevich Termen (better known in the West as Leon Theremin, 1896-1993, the inventor of the first electrical musical instrument "Termenvox" and of the spying/tapping system "Buran"), in 1926, while working in Abram Ioffe's Physical-Technical Institute in Leningrad, investigated the possibility of preserving the human body by freezing.[628]

As these examples demonstrate, in the Soviet Union of the 1920s-1940s, life-extensionist aspirations, involving various degrees of hopefulness and various methodologies, represented a central theme in Soviet biomedical science.

11. The Soviet "physiological system" – Alexander Alexandrovich Bogomolets (1881-1946)

Perhaps the best known and most influential Soviet gerontologist and life-extensionist of the 1930s-1940s was Academician Alexander Bogomolets, since 1930 head of the Institute of Experimental Biology and Pathology and of the Institute of Clinical Physiology in Kiev, from 1930 to 1946 president of the Ukrainian Academy of Sciences, and from 1942 to 1945 vice president of the Academy of Sciences of the Soviet Union.[629] Life-extension was the defining motive in his scientific pursuits, as he believed that "even one hundred and fifty years is not the limit of man's life," while "one hundred years, far from being the limit of human life, cannot even be considered as the limit of his full physiological activity and ability to work."[630] Under Bogomolets' leadership, the promotion of life-extension rose to a peak in 1938, with the publication of his book *Prodlenie Zhizni* (*The Prolongation of Life*), and the convening of the conference on the "Problem of the Genesis of Aging and Prophylaxis of Untimely Aging," in Kiev, on December 17-19, 1938, the first publicized gerontological conference in the world, organized on Bogomolets' initiative and conducted under his presidency.[631] The conference established the grounds for the institutionalization of biogerontological research in the USSR, and the aims of life-extension were explicitly posited by the majority of participants (as described in greater detail in the next sections).[632] According to the conference presenters, and Bogomolets in the first place, the promotion of longevity research was a measure of the care of the Soviet State for the well being of its citizens and a sign of the growing prosperity of the socialist society. And reciprocally, longevity research was seen as a means to strengthen the socialist state and to ensure its continuous flourishing.[633].

In his studies, Bogomolets synthesized most of the 'materialistic' methodologies for life-extension available at that time. Among the means for the "prevention of untimely aging" recommended in *The Prolongation of Life*, "prophylactic" hygienic means played a significant role: moderate diet, work, exercise and massage, sex function "not overtaxed," rest and sleep. Yet, the bulk of the book concerned not "prophylaxis" but "treatment of old age." The "treatment" included immunological and toxicological means designed to eliminate damaging cells and toxins. It also included biological replacements, such as hormone replacements and blood serum transfusions. By eliminating harmful agents, and supplementing beneficial agents, Bogomolets hoped to achieve a robust and balanced "regulation" of all bodily functions. Bogomolets' signature technique for the "treatment of old age" was the "anti-reticular cytotoxic serum" (ACS) designed to stimulate the "reticulo-endothelial system" or the "physiological system of connective tissue," that came to be known as "Bogomolets' serum."[634]

The anti-reticular cytotoxic serum method combined several influences, most predominantly (and self-admittedly) Metchnikoff's. To recall, in Metchnikoff's generalized systemic theory of aging, the "noble" (differentiated, functional) "elements" (cells and tissues) are opposed and disrupted by "primitive" (non-functional, undifferentiated, "harmful") cells and tissues. Over-proliferating phagocytes act as the agents "devouring" the noble elements, while the proliferation of non-functional connective tissue was said to replace the "noble" (parenchymal) tissues (e.g. muscle, kidney, lung and brain), leading to their sclerosis, and generally upsetting the functional harmony.

Bogomolets essentially turned Metchnikoff's theory on its head. He disputed Metchnikoff's tenet about the non-functionality and detrimental role of connective tissue build-up, but rather saw connective tissue as the sustaining "root of the organism." "Metchnikoff's basic error in presenting the picture of senescence," Bogomolets contended, "consisted not so much in his describing the phagocytes as some sort of cut-throats as his failure to evaluate the importance for the organism of a healthy reticulo-endothelial system." In Bogomolets' view, the trophic and immune functions of the "reticulo-endothelial system" or "physiological system of connective tissue" – comprised of "fixed histiocytes [mononuclear phagocytes] residing in or on (1) connective tissue; (2) reticulum of spleen, lymph nodes and bone marrow; (3)

endothelial linings of the sinuses of liver, spleen, bone marrow, adrenal and pituitary glands; and (4) microglia; plus the wandering histiocytes in tissue spaces and the blood" – were essential for health and longevity.[635] Thus, even though Bogomolets accepted Metchnikoff's general idea about the necessity of stimulating or inhibiting particular organ systems in order to reach a sustained harmonious organic equilibrium, he differed from Metchnikoff in the small detail of precisely what organs and systems needed to be inhibited or stimulated.

Thus, Bogomolets asserted, "my point of view regarding the importance of the activity of reticulo-endothelial system for longevity is directly opposite to that of Metchnikoff."[636] Yet, the use of "cytotoxic sera" as a principal means to inhibit or stimulate particular organ systems (in Bogomolets' experiments, the "reticulo-endothelial" system) was directly derived from Metchnikoff, or rather from Metchnikoff's assistant at Institut Pasteur (from 1894 to 1900), the Nobel Laureate in medicine of 1919, Jules Bordet (1870-1961). Metchnikoff believed that cytotoxic sera (whose discovery he credited to Bordet) can be used as a double-edged sword in the fight against aging: in large doses the sera can be used to eliminate or inhibit undesirable tissues, but in small doses they can be applied to stimulate these very same tissues if their proliferation is desired. Metchnikoff thus described the principle of the preparation and action of cytotoxic sera:[637]

> The principle of preparing these serums is as follows: specific cells, red blood corpuscles or spermatozoa or kidney or liver cells are injected into an animal of a different species. After several injections the serum of the animal becomes active in relation to the cells that have been injected into its body. Bordet discovered these serums. The serums are called cytotoxic, i.e. toxic for various cells. … It has been proven that small doses of cytotoxic serums, instead of killing or dissolving specific elements of tissues, strengthen them. This is similar to a reaction observed in connection with many poisons, namely, large doses kill, whereas small doses cure and improve the condition of certain elements of the organism.

This was the birth of systemic immunotherapy. However, as the precise active dosages of the serum could not be established either by Metchnikoff or by his pupil and collaborator at Institut Pasteur, Alexander Mikhailovich Besredka (1870-1940), this line of research was abandoned. Bogomolets prided himself on solving the problem of establishing the exact doses of the serum (by modifying the Jules Bordet - Octave Gengou "complement-fixation" method of serum titration, formerly used in diagnosing infectious diseases), and thus making the application of "cytotoxic sera" practicable (around 1924).

Bogomolets also changed the primary target of "cytotoxic sera" from Metchnikoff's "parenchymal tissues" (nervous or muscle tissues, kidney, liver, etc), to elements of the "connective/reticulo-endothelial" tissue. Thus, for clinical applications, Bogomolets' anti-reticular cytotoxic serum was most commonly produced from the blood of horses injected with cells of the human spleen or bone marrow (defined as elements of the reticulo-endothelial system and predominately obtained from human cadavers), and the resulting serum was injected into patients in small doses calculated to stimulate the activity of the corresponding elements of their "reticulo-endothelial system."

Bogomolets' use of "cytotoxic sera" was, of course, a ramification of the more general and ancient idea of using blood for rejuvenation. Bogomolets' studies in this field were a direct continuation of the research on rejuvenation by blood transfusion started by Alexander Bogdanov at the Moscow Institute for Blood Transfusion. Bogomolets worked at that institute since its inception in 1926 (developing, among other methods, techniques of blood banking for transfusion) and, after Bogdanov's death, directed it (from 1928 to 1930). The theory of the activity of the anti-reticular cytotoxic serum was also greatly influenced by the concept of colloidal destabilization as a primary cause of aging and disease (professed in the 1920s-1930s by Vladislav Ruzicka and Auguste Lumière). It was also largely based on the contemporary theory of "auto-catalysis," proposed in the early 1920s by the Austrian researcher Gottlieb Haberlandt (1854-1945) and the Japanese Yoneji Miyagawa (1885-1959) and later developed by the Soviet researchers Mikhail

Tushnov and Vladimir Filatov.

The general principle of rejuvenative and regenerative autocatalysis was defined by Bogomolets as follows: "When cells disintegrate, substances of a rather complex structure are formed. ... They act on analogous cells, stimulating their functions – nutrition, growth and multiplication, i.e. regeneration."[638] These substances are the "autocatalysts," defined as "enzymes that digest or otherwise modify cells similar to the cells that produced them." When discussing the earlier endocrine rejuvenation methods and the "disillusionment which replaced the enthusiasm for rejuvenation by means of transplanted sex glands," Bogomolets did not dwell too long on the failure, but undauntedly proceeded toward exploring new means of rejuvenation based on the principle of "autocatalysis" and explaining whatever effects might have been observed in the earlier rejuvenation methods by the same principle:[639]

> The search for methods of preserving the ability of the cells to regenerate is one of the primary problems in the fight against untimely aging. In this connection autocatalysis requires particular attention. When sex glands are transplanted for the purpose of rejuvenation it is autocatalysis that stimulates the dwindling function of the analogous glands in the recipient. The products of disintegration of the transplant in some cases succeed in sustaining the regenerative and functional properties of the various organs which have shown signs of decline in activity. This question deserves the most careful experimental study.

According to Bogomolets, regenerative autocatalytic products can be generally produced in the organism thanks to blood transfusion. As suggested by Alexander Bogomolets and his collaborators/assistants at the Kiev Institute of Clinical Physiology – Nina Medvedeva, Natalia Yudina, and Oleg Alexandrovich Bogomolets (Alexander Bogomolets' son and successor as the Institute director) – during blood transfusion, the autocatalytic products are formed in the process of "colloidoclasia," that is "damaging the colloids, in this case the particles of albumin that enter into the composition of the blood and cytoplasm."[640] The immediate result of the "colloidoclastic shock" due to transfusion is the formation of easily dissolvable and removable protein/colloid aggregates, whose dissolution produces the stimulating "autocatalysts."

Not only bulk blood transfusion, but also, perhaps even preferably, the administration of the anti-reticular cytotoxic serum in minute, "stimulating doses" was supposed to disintegrate the target tissue in such a measure as to produce sufficient amounts of stimulating "autocatalysts." These rudimentary (or rather pioneering) concepts of biochemical allosteric feedback regulation were perceived by Bogomolets as a potential basis for a wide range of future regenerative and life-extending therapies:[641]

> A new science is now in process of formation, a science of biochemical, autocatalytic self-regulation of functions. It opens wide perspectives for discovering methods of stimulation of cellular functions, which would intensify their ability for regeneration without causing untimely exhaustion. Some of these methods may be blood transfusion, cytotoxins, etc.

The primary purpose of the anti-reticular cytotoxic serum therapy was to combat "untimely aging," strengthening the immune (defense), trophic (nutritive), formative (plastic) and functional state of the aged organism. Yet the serum was also employed in the treatment of a host of severe diseases. The ACS was used to increase the resistance of the organism to infectious diseases, such as scarlet fever, typhus, tularemia and brucellosis. It was also used as an adjunct treatment of cancer. "This study will require a long time," Bogomolets maintained. "Nevertheless, we can already claim that, with the help of stimulating doses of anti-reticular cytotoxic serum, in the majority of cases of cancer patients, we are able to restore the property of the blood to cause dissolution of the cancer cells."[642] Insofar as Bogomolets believed that the reticulo-endothelial system plays a crucial role in the nourishment of all tissues, including the nervous tissue, the

ACS was employed in the treatment of neurological diseases.

But perhaps the widest application of the ACS, and the most urgent during WWII, was for the healing of wounds and fractures, where the ACS was utilized as a stimulant of the "plastic function of the connective tissue." Such application of the ACS began before the war, but during the war it became central.[643] Indeed, the aspirations to radical life-extension were hushed down under the massive deployment of economic and human resources to the war effort. As Bogomolets once regretfully said, "I dreamed to prolong human life. The war interrupted this work."[644] During the evacuation of the Ukrainian Academy of Sciences to Ufa, in the Southern Ural region, Bogomolets' chief occupation was to coordinate the activity of blood banks and provide blood transfusions to wounded soldiers, which saved the lives of many. Yet, the production of the ACS continued during the war, but instead of "combating untimely aging," its proclaimed purpose was the healing of battle wounds and fractures, in order to expedite the return of soldiers to the front: "Thanks to the application of ACS during the war," Bogomolets proudly attested, "many wounded with serious fractures have been saved from becoming invalids and have been returned to the front."[645]

The wide variety of therapeutic applications, as well as the general life-prolonging capacity of the anti-reticular cytotoxic serum derived, according to Bogomolets, from the underlying importance of the connective tissue system and from the singular ability of the ACS to either suppress or stimulate this system, depending on the dosage. Yet, according to Bogomolets, the development of this therapy was far from being complete, and a long road lay ahead of it, seeking new therapeutic targets and calibrating proper dosages.[646] And thus Bogomolets expressed what might be seen as a definitive element in the contemporary Soviet life-extension science: the unending search for new material agents for life prolongation, unimpeded by former failures. The value of this search is never so much in the present as in the future: [647]

If it should prove that the small doses of the anti-reticular cytotoxic serum, on repeated introduction into the body are able to prevent untimely sclerosis of the reticulo-endothelial system, this serum will become a very valuable aid in the struggle for prolongation of life. We have begun to study this problem. Here a great deal of circumspection and long experimentation first on animals and later on human beings is needed.

In other words, it may not become such a life-prolonging aid, but if the research does not continue, we shall never know and a potential for life-extension will be lost.

Bogomolets' work became a central hub in the contemporary international cooperation and in the historical continuation of longevity research. In the 1940s, the ideas of Bogomolets became widely popular in the West, especially in the US. His book, *The Prolongation of Life* (published in English translation in 1946) was a best-seller. During the Soviet-American anti-fascist alliance, Bogomolets' serum was massively supplied to the American allied forces. Generally, during this period, the extent of amity between Soviet and American life-extensionists seemed almost incredible by Cold War standards. As Bogomolets wrote in October 1944 to Prof. Irving Fisher, president of the American Life Extension Institute, "I would like the anti-reticular cytotoxic serum, which proved so profitable to us, especially in the struggle with the after-effects of shock, to prove equally useful to our Allies." And reciprocally, there was a deeply felt enthusiasm of American researchers with regard to Bogomolets' work. Thus, George P. Robinson, president of the Robinson Foundation that sponsored the translation of Bogomolets' *The Prolongation of Life*, hoped "this book may be the first of many steps it will take to promote understanding of the problems of aging." And the translator, Dr. Peter V. Karpovich affirmed that "with the recent, tremendous development in means for the destruction of human life it is only natural that our attention should turn again to means for the prolongation of human life. ... In this connection, the work of the Russian academician, Alexander A. Bogomolets, is of special interest." Bogomolets' work, Karpovich believed, "will undoubtedly provoke discussion and may stimulate other investigators to undertake further research in problems related to the conservation and prolongation of human life."[648] And indeed, among others, Bogomolets' work inspired

Denham Harman (the American originator in the 1950s of the free radical theory of aging, which became the dominant theory of aging for several decades) to embark on the field of aging and longevity research in the first place.[649] In the 1940s, Bogomolets' research and the use of the ACS were promoted by the American National Council of American-Soviet Friendship, and by the American-Soviet Medical Society in New York.

Bogomolets' experiments were reproduced by several American investigators (Reuben Straus, Joseph Skapier, 1946, 1947, and others).[650] As the journalist Patrick Michael McGrady testified, as late as the mid-1960s, Bogomolets' serum was broadly available in the US and on the Continent, as well as all across the Eastern Bloc.[651] Yet, the majority of investigators of the 1950s (such as Max Bürger, Chauncey Leake and Thomas Gardner) failed to observe any far-reaching therapeutic effects.[652] And certainly, no significant extension of human life by the use of the anti-reticular cytotoxic serum was evidenced. Gradually, the serum disappeared from laboratories and pharmacy shelves. This disappearance may be partly attributed to the general superseding of the somewhat cumbersome and not particularly specific anti-sera production, involving horses and cadaver materials, as in Bogomolets' methods, by the more convenient use of monoclonal anti-bodies for specific immunotherapy, involving the immunization of mice and subsequent culture of human-mouse hybrid cell lines, as proposed by Georges Köhler, Cesar Milstein and Niels Kaj Jerne in 1975.[653] Yet Bogomolets' work might be considered an important stepping stone in the development of specific immunotherapy.

Generally, Bogomolets' work may be viewed as an essential link in the development of reductionist rejuvenation techniques. As this work has argued, the failure of reductionist endocrine rejuvenation methods of the 1920s greatly impacted on the progression of life-extensionist thought. Many life-extensionists (particularly in France and Germany) turned away from reductionist rejuvenation by pharmacological and surgical means toward more holistic and hygienic approaches. This, however, did not occur to Bogomolets, and in fact to the majority of Soviet life-extensionists. Bogomolets' work was a direct continuation of earlier rejuvenation methods: Metchnikoff's and Bordet's cytotoxic sera, endocrine stimulation (Bogomolets' first use of cytotoxic anti-sera in 1915 was designed to stimulate the function of the adrenal glands), blood transfusion, and opotherapy (using "auto-catalytic" products of tissue degradation). Bogomolets did not perceive the failures of these methods as prohibitive debacles, but rather as directive signposts for further investigation and improvement. And so, Bogomolets, at the head of Soviet life-extensionists, steadfastly proceeded in the quest for reductionist rejuvenation – an approach indeed definitive of contemporary Soviet life-extensionism. The American biologist Bernard Strehler once likened life-extensionist gerontologists to "Captain Ahab" from Herman Melville's *Moby Dick* (1851).[654] They would continue to hunt the white whale of aging, even if they die trying. Bogomolets was "Captain Ahab" of Soviet gerontologists, persistent and resourceful and undeterred by disappointments, whose work became yet another in the succession of failures or, in an 'Ahabian' view, not a failure but a directive signpost for the continuation of the search.

With the abandonment of the ACS by medical practitioners and basic researchers, Bogomolets' name disappeared from the headlines and there have been no recent attempts in the West to research his activities. In the Soviet Union too, the production of Bogomolets' serum was being gradually reduced until complete disappearance in the 1960s and his work was steadily turning into a vaguely remembered historical curiosity, yet another failed attempt to find a cure for aging. The discontinuation in the Soviet Union of the ACS research might have been due not so much to scientific counter-indications, but to the simple fact that in 1946 Bogomolets died at the age of 65 from tuberculosis that he had battled since his youth. Consequently, his influence and authority in the Soviet scientific establishment ceased. Anecdotally, when Bogomolets died, Stalin felt grossly deceived.[655] Since 1950, "The scientific committee on the problems of physiological teachings of I.P. Pavlov" headed by Academician Konstantin Mikhailovich Bykov (1886-1959), then director of I.P. Pavlov's Institute of Physiology in Leningrad, undertook a crusade to defend the purity of Pavlov's legacy and eliminate heresy from Soviet physiology. The main targets of the committee were a founder of evolutionary physiology Academician Leon Abgarovich Orbeli (1882-1958), as well as the longevity researchers Alexey Speransky and Peter Anokhin. In 1952, among others, the committee conducted an investigation into the works of Bogomolets and Ukrainian physiologists. Bogomolets was posthumously accused of "inhibiting in Ukraine the development of Pavlov's teachings, support of [the genetic theory of] Morganism-Weismannism, idealistic worldview, inculcation of the anti-scientific theory of the physiological system of connective tissue." (Interestingly, Bogomolets was now subsumed into the same "anti-Pavlov's" camp as his former adversary Speransky, see below.) Bogomolets' Institute of Experimental Biology and Pathology (then headed by Alexander Bogomolets' son, Oleg) was threatened with closure. Nevertheless, thanks to the supplications of Oleg Bogomolets and his collaborator, Valentin Tkachuk, to the first secretary of the Central Committee of the Ukrainian Communist Party, Leonid Melnikov, the resolution of "Pavlov's committee" was softened and the work at the institute continued.[656]

Yet, in the late 1930s-early 1940s Bogomolets ruled supreme. His background was ideologically impeccable. He was a son of Russian revolutionaries, members of the "South-Russian Workers' Union," and was born in prison during his mother Sofia's arrest. In 1917, he immediately accepted the Socialist Revolution, and during the Civil War served as the chief epidemiologist for the South-Eastern front of the Red Army. Since his appointment as president of the Ukrainian Academy of Sciences and until his death, he had Stalin's full support. In 1941, he received the Stalin Prize (of the first degree, personally approved by Stalin) and in 1944 received from Stalin the highest distinctions of the Hero of Socialist Labor and the Order of Lenin. In May 1943, Bogomolets had a friendly meeting with Stalin (though, according to Oleg Bogomolets' testimony, their discussion mainly concerned the contribution of Ukrainian scientists to the struggle against fascism, rather than life extension). Bogomolets was also a close friend of Nikita Khrushchev, who was in the period 1938-1947 the First Secretary of the Central Committee of the Ukrainian Communist Party (coinciding with the peak of Bogomolets' prestige). Khrushchev had the "greatest respect" for Bogomolets and referred to Bogomolets' death "as a disaster that struck us all, a disaster for Ukraine and for science."[657] Thus, Bogomolets might be considered as yet another, perhaps perfect example for the adjustment of life-extensionists to the ruling social and ideological establishment.

Bogomolets often used his firm standing in the establishment, not only to promote his research programs, but also to defend many prominent members of the Ukrainian Academy against purges: the demographer Mikhail Ptukha, orientalist Agatangel Krimsky, economist Konstantin Vobly, physicist Alexander Leipunsky, mathematician Nikolay Krylov, and many others.[658] Nonetheless, there was perhaps no greater acolyte of the Soviet state, no person more eager to glorify the ruling regime that has done so well with him, no greater advocate for its continuation, than Bogomolets. Life-extension research, according to Bogomolets, was equally a beneficiary of the present felicitous social order and a major contributor to its further thriving.

In his opening statement at the Conference on the Problem of the Genesis of Aging and Prophylaxis of Untimely Aging, Bogomolets emphatically asserted the superiority of the Soviet system in all matters pertaining longevity:[659]

> The gloomy view has been firmly established abroad that the organism is given by nature a certain amount of restorative energy for a preordained number of years. This view excludes the very possibility of fighting the premature wear-and-tear of the organism. We, Soviet scientists, think differently. We believe that senescence, in the majority of cases, comes much before the organism exhausts its capabilities for constant renewal. We also know that the causes of untimely aging are, first of all, social causes. In countries not satisfied with usual forms of capitalist exploitation, in countries where the robbery and murder of defenseless populations, women and children, are becoming habitual, in countries where unbridled cruelty and terror, overwork and poverty reign – there are no, and can never be, basic conditions for the attainment of normal longevity by human beings. It is, of course, not by chance that the idea to unite the scientific medical thought and direct it to the combat of untimely aging

originated under the life-giving rays of Stalin's Constitution, in our Soviet land, where the human being, his happiness, is the highest value.

Similar pledges of allegiance were expressed by the majority of the conference participants, who saw it as their solemn duty to glorify the socialist state, the communist party, Stalin, and Bogomolets and his anti-reticular cytotoxic serum – in this order or variations thereof (and they had better do this). Socialism was seen as a necessary condition for successful longevity research, and successful longevity research was a necessary condition for the development and strengthening of socialism.

12. The world's first conference on aging and longevity – Kiev, 1938. Controversies

The Conference on the Problem of the Genesis of Aging and Prophylaxis of Untimely Aging, held on December 17-19, 1938, in Kiev under Bogomolets' presidency, was a pivotal event in the history of aging and longevity research. It was the world's first publicized scientific conference dedicated to these subjects. And generally, it was a major (though presently scarcely remembered) milestone in the history of Soviet biomedicine. As Dmitry Chebotarev, the long-time president of the USSR Gerontological Society, pointed out as late as 2001, the conference proceedings and Bogomolets' book *The Prolongation of Life* were "the main sources of information in the country on the questions of gerontology for all subsequent years."[660]

The conference purpose was summarized by Nagorny:[661]

The Soviet Science in fact has long declared war on everything that shortens human life. In our conference, the war is declared openly. We, the representatives of Soviet medicine and biology, throw down the glove to such seemingly invincible enemies as aging and death. The way to victory over untimely, morbid senescence and shortness of life requires an immense collective work of representatives of different specialties, a work directed by a unified plan and with a single leading idea. It must be believed that our conference will provide an impetus to such purposeful work.

The major aim of the conference was the achievement of "physiological aging" which is characterized by "atrophic changes in the organism, gradually and evenly developing in all physiological systems and leading to a harmonious reduction of functional and reactive capabilities, with their adaptation to the reduced capabilities of the aging organism"[662] and prevention of "untimely aging." Yet radical and specific goals of life-extension were posited throughout. As Nagorny summarized, "It is known with certainty that the maximal life span can exceed the mean life span 2 or 3 times and even more." Thanks to the concerted effort of Soviet researchers, longevity would be extended to "no less than its maximal limits, presently known to us, namely 150-180 years."[663]

It was the duty of Soviet medical scientists to dedicate an immense research effort to make radical life-extension a reality. And the effort was made, though the results were far from approaching anywhere close to the realization of the tasks that the scientists were setting. The need for "further study" was the general conclusion of the conference. Instead of clearing the path toward the prolongation of life, the conference presentations seemed to have complicated the matters almost intractably. As will be exemplified below, the studies presented at the conference were far from providing any conclusive recommendations, and were widely controversial: with reference to themselves, other contemporary studies, past studies and, as will turn out, future studies.

Perhaps the only decisive recommendation was given by Moisey Samuilovich Mühlmann (1867-1940), then professor at the University of Baku, Azerbaijan. According to Mühlmann's theory (first published in 1900 in Wiesbaden, Germany[664]), aging is due to a diminishing supply of oxygen to the tissues. The closer the tissues are to the "nourishing surface," that is to the body periphery, the more is their oxygen supply; while "central tissues" receive less nourishment and oxygen, hence their degeneration. Mühlmann decisively concluded:[665]

The only factor, distinguishing 130 year old persons from common people of our age, is that they always lived free, in fresh air, in the mountains, where there is much oxygen, especially ozone. Oxygen is the decisive factor, which is so lacking to us, breathing dust and car gasoline, people of the city with so little vegetation. This is what shortens life in the first place, and we end with what we started: the prolongation of life is assured by pure air. ... Thus, let the motto of our conference be: much fresh air for a long life.

As decisive were Mühlmann's recommendations, so controversial were his premises, even in his own time. It was known to the German chemist Justus Liebig already in 1843 that in the mountains, the seat of the long lived, the oxygen consumption is actually lower: "At the level of the sea," Liebig wrote, "a cubic foot of air contains more oxygen than the same volume of air does on high mountains."[666] And the possibility of oxygen toxicity was first suggested by the French zoologist Paul Bert in 1878. In the 1930s, Mühlmann' theory was forcefully attacked by the leader of German longevity researchers, Max Bürger. Bürger's criticism of Mühlmann was a central subject in the very first introductory article of the *Zeitschrift für Altersforschung* of 1938. And in 1954, regarding Mühlmann's premise that the brain is most remote from the body periphery, and hence should be least supplied by oxygen and therefore most susceptible to degeneration, Bürger maintained that "no organ is better protected against oxygen deficit and sugar deficit than the brain"[667] Hence, Bürger admitted, he "could hardly follow the explanations and justifications for Mühlmann's 'physical growth theory.'" Later, the decisively beneficial effects of enhanced oxygen consumption, professed by Mühlmann, were further questioned by the oxidative damage theory of aging, where oxygen and reactive oxygen species were considered as the main causes of age-related damage.[668] Furthermore, in 1952, Konstantin Pavlovich Buteyko (1923-2003, since 1958 working at the Institute of Experimental Biology and Medicine in Novosibirsk), suggested that increasing oxygen consumption via deep breathing can upset the O_2/CO_2 balance and lead to various cardio-respiratory pathologies. He subsequently developed a method of breathing designed to increase CO_2 content in the lungs and decrease oxygen consumption.[669] Furthermore still, the exact threshold at which oxygen becomes toxic for the human body remains widely debated.[670] Thus, even such a matter of fact proposition as breathing "much fresh air" was controversial within the context of the conference.

The situation was no clearer with regard to physical activity and diet. Nagorny related in complete bewilderment to the findings of the American researcher James Rollin Slonaker (1866-1954) considering the relation of exercise to longevity in rats. In Slonaker's experiment, the animals were subjected to different levels of physical exertion. It was found that, in the exercising group, the more exercise the animals had, the longer they lived. However, the animals that did not exercise at all, but had "usual mobility," lived longer than those who exercised.[671] The question then arose whether exercise *per se* is beneficial, or rather which amounts of exercise are beneficial. Nagorny utterly rejected Slonaker's conclusion that work may shorten life (this conclusion, according to him, was also wholly inconsistent with the goals of socialist productivity and communal involvement). Nagorny did admit that excessive work (as imposed on the workers in the capitalist world of exploitation) can be life-shortening. Yet, Nagorny was entirely unable to provide an exact threshold at which physical work becomes exhaustive, and at which the beneficial effects of stimulation and training are superseded by fatigue and wear and tear.

A series of studies presented at the conference considered at length the questions of the relation of "exhaustion" vs. "recovery," and associated problems of material and energy "deficits" vs. "excesses," "fatigue" vs. "training," "wear and tear" vs. "restoration," "dissimilation" vs. "assimilation," "entropy" vs. "extropy." The regulated maintenance of material and energy equilibrium was the primary objective.[672] The authors emphasized the tremendous importance of prolonged rest for the recovery and conservation of the functional reactive activity of aged organisms, yet at the same time stressed the importance of exercise for its improvement. However, the exact prescriptions for the amounts of rest vs. exercise to optimize the functional activity were lacking.

Several researchers considered various aspects of biochemical/metabolic recovery (N.B. Medvedeva, G.V. Derviz).[673] Alexander Nagorny and his pupil Vladimir Nikitin focused on age-related changes in the synthesis (assimilation) and degradation (dissimilation) of proteins.[674] In the general view of Nagorny's school, the processes of material renewal and assimilation, particularly of proteins, are the main factors for longevity, and "an organism that could entirely replace anything that has lost the ability for self renewal by elements capable of such a renewal, would reveal no aging and would know no death."[675] Accordingly, a vast panel of experiments on material renewal was conducted, at the functional, chemical, physico-chemical and morphological levels, in the blood, various tissues, and in entire organisms, studying a vast array of biological compounds and measures of metabolism available at the time. Yet, the complexity of renewal processes appeared to be too immense for any practical conclusions. As Nagorny admitted, "the factors increasing dissimilation can at the same time be the factors increasing assimilation." The exact distinctions between factors favoring "assimilation" or "dissimilation," "recovery" or "exhaustion" were impossible to determine, hence the impossibility of precise recommendations. According to Nagorny, the "paths for the efforts for life prolongation are clear… these are the paths of enhanced self-renewal."[676] Unfortunately, neither Nagorny nor Nikitin quantitatively specified any such paths of biochemical renewal. Neither did the physiologists studying the central nervous system specify any concrete, quantitative paths toward its functional renewal.

A central focus at the conference was on the equilibrium and function of lipids, particularly cholesterol, in aging. This was perhaps one of the first concerted discussions of this issue. Yet the controversies regarding age-related changes of lipid content, their direction, function and origin, raged almost beyond reconciliation. Thus, Prof. Semen Sergeevich Khalatov (1884-1951, in 1929-1947 director of the Department of Pathophysiology of the First Moscow Medical Institute) suggested that excessive cholesterol in the blood of the aged is the result of "leakage of cholesterol from the brain."[677] Khalatov's theory was however strongly opposed by V.I. Solntzev of the Kiev Institute of Clinical Medicine led by Nikolay Strazhesko. Solntzev found it "difficult to agree with Khalatov's opinion that such a cholesterinemia originates exclusively in the brain." According to Solntzev, "it can in a large measure depend on an impairment of endocrine glands, … it can be related to decreased oxidative processes, … it can be conditioned also by a diminished function of the liver … fighting various intoxications of exogenous and endogenous origin."[678] The complexity of the lipid economy in aging was further emphasized by M.D. Gatzanyuk of the Kiev Institute of Experimental Biology and Pathology directed by Alexander Bogomolets. In his experiments on animals, Gatzanyuk found that, in the course of aging, in some organs and tissues the amounts of fatty acids increase, and in some decrease. Cholesterol was found generally to increase with age, yet with very irregular rates in different organs (ranging from 200% to 10%), and with a wide dependence on diet. Hence an "all-through" manipulation of cholesterol, or of fatty acids, as a means for life-extension appeared difficult. Accordingly, no conclusive recommendations were made, except for asserting the "significance of cholesterol and fatty acids" and the "necessity to study changes in their amount and relations."[679]

In contrast to Gatzanyuk's findings of a general increase of cholesterol with age in animals, I.M. Turovets and L.I. Pravdina of Strazhesko's Institute of Clinical Medicine, cited studies either affirming or denying the increase of cholesterol in humans. They presented their own findings that until 80-90 years of age the level of blood cholesterol is relatively high, but after 90 years it decreases. In the long-lived, they found, the amount of cholesterol is normal.[680] Further complicating the role of cholesterol in aging, V.I. Solntzev disputed the view that rising cholesterol levels can be an exclusively adverse process or symptom of aging, and in particular Khalatov's view that increased blood cholesterol is a sign of brain impairment (no distinctions in "kinds" of cholesterol were as yet made).[681] According to Solntzev, increasing cholesterol can actually be a protective mechanism in the aging organism, signifying the organism's defensive capability: "We have succeeded in showing that hyper-cholesterinemia can be a reflection of the tissue defense reaction against a toxic insult of various substances; it can signify the degree of immunity." Solntzev further emphasized the complexities involved in relating cholesterol to the rest of physiological parameters: "The most important issue is not the value of cholesterol of itself – cholesterinemia needs to be related to other clinical manifestations, and only from the entire combined clinical picture one may make this or other conclusion."[682] It was beyond Solntzev's capacity to create such a "combined" multi-parametric clinical picture, let alone make any decisive conclusions or recommendations.

It appears that the only decisive recommendations on diet were given by Prof. V.A. Elberg (also from Strazhesko's Institute of Clinical Medicine in Kiev). These included dietetic suggestions to combat obesity and emaciation, as well as considerations of endocrine therapy and mineral supplements (such as iodine and calcium, but no mention of vitamins). Elberg was well aware of the controversies about fats, and particularly cholesterol. In fact, his own findings showed no hyper-cholesterinemia in the aged. Nonetheless, his 'halachic' rule was in favor of anti-cholesterol diet:[683]

> Even though neither we personally, nor Turovets, nor several other investigators succeeded in demonstrating hyper-cholesterinemia in the aged, but rather received values below the norm (Turovets), nonetheless we can but accept as absolutely correct the point of view of Khalatov and Anichkov that cholesterol is an inseparable companion of aging, accumulating in the organism and precipitating in various sites of the organism, mainly in the arterial wall. [Semen Khalatov and Nikolay Nikolaevich Anichkov, 1885-1964, were the first to suggest the cholesterol etiology of arteriosclerosis, c. 1913.[684]] ... Therefore we consider it entirely irrational to give the aged in their diet an excessive amount of egg yolk and butter.

However, shortly afterwards, Elberg claimed that "the aged well tolerate fats, especially in the form of butter. ... In most cases, administration of easily melting fats, in the form of sandwiches with butter, goose-grease, palmin, cream with milk, etc, is recommended." In the daily dietary regimen for the aged (to consist of 7 small courses), Elberg included milk with an addition of lactose sugar, porridge, coffee, tea, chocolate, wine, bakery (cookies or white bread toast with butter), honey, fruit mousse. Many of these recommendations might raise a few brows in some more recent "cholesterol-conscious," "carbohydrate-conscious" or "animal-protein conscious" life-extensionist circles. Even at Elberg's time, some of these recommendations contradicted those of others. Furthermore, as the above examples show, the recommendations were not entirely consistent within themselves, and with experimental findings. Elbert did strive to reach a dietary "equilibrium," yet quantitative thresholds between "excesses" and "deficits" were far from being generally established, even less so for individual cases. Elberg admitted that "our dietary recommendations for the aged are limited." Thus, despite the tone of decisiveness, the recommendations were far from being decisive. More research was required.

13. The quest for reductionist rejuvenation continues

The jury was also out on the questions of endocrine treatments of aging. The endocrine rejuvenation methods that had boomed in the 1920s were given a thorough critical evaluation at the conference. Almost intractable complexities were suggested with reference to these hormonal methods of "revitalization" and "life prolongation." The difficulties were expressed by Rostislav Evgenievich Kavetsky (1899-1978), at the time working in Bogomolets' Institute of Clinical Physiology, and later founding and directing the Kiev Institute of Experimental and Clinical Oncology (since 1960). In "Aging and Cancer," Kavetsky seems to have been entirely baffled by the role of hormones in aging generally and cancer particularly. In the wake of the "endocrine rejuvenation" boom, Kavetsky accepted the possibility that "sex hormones can be categorized as 'growth hormones' in the sense that, acting as carriers of tissue differentiation stimuli and thus facilitating the exhaustion of cells' idioplastic energy, they must exert a certain inhibiting effect on the processes of [cancer] growth." At the same time, the possibility of a carcinogenic effect of sex hormone supplements was also suggested:[685]

> On the other hand, by prolonged injections of large doses of folliculin, it was possible to induce breast cancer, and recently also to induce tumors at the site of female sex hormone injection. It can be assumed that, in these cases, the carcinogenic effect is produced not by folliculin itself, but by some products formed from it under given conditions. Analogous products are perhaps formed in the organism under disrupted hormonal function of the sex glands.

Despite Kavetsky's present perplexity, he nonetheless envisioned quite optimistic prospects for further research of the role of hormones in anti-aging and anti-cancer treatment:[686]

> It should be thought that the struggle with factors leading to an untimely aging of the organism, causing a disturbance of the equilibrium between its systems and tissues, the struggle for a healthy connective tissue, will lead to an elimination of age-related dysfunction between the epithelium and the mesenchyme, and will create preconditions for the prophylaxis of cancer.

The findings of Prof. Nikolay Adolfovich Shereshevsky of Moscow, and of Profs. V.H. Vasilenko and R.M. Mayzlish of Kiev, further questioned the wide applicability of sex hormones for anti-aging. At the same time, the authors demanded an expansion of research on the transplantation of pieces of healthy endocrine organs, and on the therapeutic use of plant hormones.[687]

The reaction of Soviet life-extensionists to earlier reductionist "endocrine rejuvenation" methods is perhaps best exemplified by the work of Prof. Vasily Gerasimovich Shipachev (1884-1957) of Irkutsk Medical Institute in Siberia, summarized in his presentation "On the question of the struggle with the organism's aging" (*K voprosu o borbe so stareniem organisma*). Shipachev fully shared Bogomolets' slogan "To live to 150 years" and made quite an effort toward that purpose. Shipachev derived much optimism from the "achievements of the past two decades," such as "the discovery of [anti-microbial] bacteriophages by D'Herelle[688] and the discovery of mitogenic rays by Gurwitsch.[689]" "During this short time, science became enriched with new data on the physiology and pathology of the endocrine system, which enabled us to find new ways in developing the study of various diseases and to outline ways toward their treatment."[690] At the same time, Shipachev passionately criticized earlier methods suggested for life prolongation and rejuvenation. He rejected Metchnikoff's view of intestinal bacteria as predominantly harmful for the human organism, and accordingly undervalued Metchnikoff's methods for suppressing such bacteria (by colectomy or "more conservative methods" such as acidic milk products). Rather, Shipachev sided with the views of the 19th century biologists, Louis Pasteur and Eduard Strasburger, who emphasized the positive role of a large part of intestinal flora for human health.

As regards "endocrine rejuvenation" – grounded on the works of Brown-Séquard, Steinach and Voronoff – Shipachev was even more critical. In the 1920s, Shipachev was one of the first and most enthusiastic partisans of endocrine rejuvenation, particularly sex gland transplants (first suggested by Steinach), yet became disillusioned later on:[691]

> Long before the works of Voronoff (in 1921 and 1922), I together with Prof. I.N. Perevodchikov, replicated Steinach's experiments on animals. Having become convinced in the reinvigorating effect of sex gland transplants on the aging organism, we moved this operation to the clinic and performed it in approximately 100 cases, mainly in aged persons, with the purpose of "rejuvenation," almost exclusively in men. ... It became entirely obvious that the sex gland transplants produce a therapeutic effect, but in no way restore youth... consequently we began to use this operation with therapeutic purposes against cardiovascular diseases... However, the organism's "rejuvenation" in the sense of restoring youth, that so many had expected, was never achieved in any case.

The sex gland transplants failed to produce any far reaching effects apart from a mild and temporary functional and sexual reinvigoration, failed to restore the youth and prolong the life of the patients. No conspicuous signs of senility were arrested, and within 15 years after the transplantations all the old transplantees were dead. These facts induced Shipachev to search for new ways of rejuvenation: "The results of sex gland transplantations to the aged with the purpose of rejuvenation and to the middle-aged patients with therapeutic purposes, forced me to change at the root the direction of my work."

The change of direction, however, does not seem to have been radical at all: the subsequent rejuvenation attempts directly developed the earlier rejuvenation methods of Steinach and Metchnikoff. At any rate, they were grounded on the same fundamental reductionist principle, namely, that the human body is a machine and its maintenance can be prolonged by a replacement of its parts. First, Shipachev attempted "pluri-glandular transplantations," i.e. the transplantation of many young endocrine glands at once, instead of a single sex gland transplant, but soon abandoned these attempts as well (around 1925). Notably, Shipachev never entirely discarded the "high therapeutic properties" of sex gland transplants and sex gland extracts, and their possible use as "additional means in the struggle with aging, but of course not in the way recommended by Steinach, Voronoff and others." "Much work," he believed, "needs to be done on pluri-glandular transplants, studying quantitative relations of transplants (or extracts) from different glands, determining the times of use of such preparations in different age groups, etc." He himself, however, was unable to carry such research through.

After the failed endocrine rejuvenation attempts, Shipachev "gave up the idea of 'rejuvenation' of a decrepit old organism and moved the center of gravity to the struggle against the aging of the organism, beginning to search for the means that would extend the boundaries of youth and help to maintain the organism in a flourishing, vigorous state to a very advanced age." (Here the "change of direction" seems to be purely semantic and hardly perceptible.) Shipachev's subsequent efforts in "the struggle with aging" were a direct continuation of Metchnikoff's ideas of combating putrefactive intestinal bacteria (despite Shipachev's criticisms of Metchnikoff). The only difference was that, instead of suppressing such putrefactive bacteria by diet or their removal by colectomy, Shipachev suggested their replacement in the intestine by young and active bacterial cultures. In Shipachev's method of "active bacteriotherapy," "the old micro-flora is washed out of the large intestine by purgatives and enemas, and in its place a new, young culture of intestinal micro-flora is injected." Shipachev admitted that his clinical observations were very limited, yet reported promising, revitalizing effects for the aged ("the sense of vigor, increased physical strength, ease of joint movements, calm and refreshing sleep, improved mental productivity"). He even envisioned that all across the Soviet Union, "all therapeutic procedures for people above 40 should begin with such active intestinal bacteriotherapy." (Nowadays, similar methods of intestinal cleansing with an addition of probiotic bacteria are practiced in anti-aging and complementary medicine clinics all across the world.[692])

Shipachev's quest for the prolonged maintenance of the human machine did not end there. During the war, he developed a method of plastic reconstruction of the fingers of wounded soldiers from transplanted bones. In all of Shipachev's endeavors, as perhaps in those of the majority of Soviet longevity researchers, the lack of efficiency of earlier rejuvenation methods was seen not as a deterring factor, but as a stimulus for further work. The underlying premise was entirely reductionist: the maintenance of the human machine must, can and will be extended. If earlier attempts were unsuccessful, that was only due to an imperfect knowledge of the mechanism and an imperfect skill of the mechanic. In Shipachev's view, and in the view of the majority of Soviet life-extensionists, it was the task of Soviet biomedical science to perfect its skills in order to help engineer a new durable man in a new durable Soviet society.

The search for the mechanisms of aging and longevity reached out much further. Thus Academician Alexander Vasilievich Leontovich (1869-1943, head of the histology department at Bogomolets' Institute of Clinical Physiology) measured the weight of the brain in human cadavers of different ages and found that the weight of the brain decreases with age. Yet, beside the "simple conclusion" that the aging brain diminishes in weight, Leontovich also proposed that "it is not impossible that people with an especially weighty brain are simply less durable."[693] Even more provocative points were raised, as in the work of Natalia Dmitrievna Yudina of the department of pathological chemistry (directed by Nina Borisovna Medvedeva) in the same Institute. One of the findings of Yudina's study of the "morphological content of the blood during physiological aging" was that the longest-lived of Abkhazia (allegedly aged 90-135) had the highest percentage of type "O" blood. According to other researchers that she cited, the long lived (aged 70-95) were characterized by a prevalence of type "A" blood. Yudina acknowledged at the outset that all the "data on changes of blood in the aged are quite controversial."[694] Even though no practical recommendations followed from such findings, the resolve of the conference participants to continue the search for all imaginable life-prolonging and life-shortening factors was absolute.

The conference's formal resolution expressed precisely such a determination to continue the effort toward the life-extension of the Soviet people:[695] "In the country of victorious Socialism," the resolution stated, "there have been created all the political and socio-economical preconditions for reaching a normal duration of life by the majority of the population. The Conference sets before the Soviet science the task of a planned work for the effective struggle with premature aging." The immediate practical tasks were first of all the "surveillance" and "patronage" of the long-lived, and then, research, research and more research into the mechanisms of aging, including "the analysis of cytological and physico-chemical changes in the state of tissues in aging," the "study of the role of exogenous and endogenous factors accelerating and inhibiting the development of senile changes," and the "special study of the role of the physiological system of connective tissue, nervous tissue and endocrine organs in the process of aging." The conference indeed exposed the immense complexity of the aging processes and the tremendous difficulties in finding effective life-extending means. But these difficulties were by no means perceived as reasons to abandon longevity research, but rather as a stimulus for further investigations. This perception of present difficulties as temporary and relatively insignificant in view of the path toward future grandeur, might be considered as yet another characteristically Soviet ideological or psychological trait, as the Soviet ideology very often blocked out of the mind the present hardships, scarcities and persecutions, in the grander vision of the forthcoming perfect communist society.

The same ideological elements – the glorification of the Soviet state and the call to continue on the same path toward an ever more profound reductionist understanding and engineering of the human body, despite the present almost insurmountable complications – were pronounced in the concluding statement of Academician Nikolay Dmitrievich Strazhesko (1876-1952, Kiev):[696]

I believe that, in our country, not a single scientific worker, not a single physician or even a common citizen, should stand apart from the study of the genesis of aging and prophylaxis of untimely senescence, and that the entire problem can be solved by the joining of forces of the workers of science with all the workers of our great Union.

According to Strazhesko, the conference's primary purpose was "to present [for the first time] the state of the problem of the genesis of aging and longevity, … correct the errors made by science in this regard, and set new ways for the subsequent study of the problem." But the present achievements (or scarcity thereof) were dwarfed in comparison to the achievements of the future. What was required for those grander achievements was the steadfast perseverance on the set course of study, with more scientific collectives involved and more conferences convened. And eventually the continuation of this research, it was implied by Strazhesko, would contribute to the continuation of the socialist state, and vice versa:

I believe that we shall more than once meet at conferences dedicated to the problem of longevity. Our works and results will of course be of extremely vivid interest to the party and to the government, and especially to our leader, comrade Stalin, who dedicates so much attention to the care for the human being. This fact should inspire all who are concerned with the problem of longevity. I believe that by tackling this work with enthusiasm, we shall soon achieve significant results that will be used in the interests of the workers of our great Union.

Thus, with the promise of achievements coming up soon, the convention was adjourned, firmly resolved to convene again. However, no such additional large scale conferences, in fact no major organizational efforts in the field of aging and longevity research, took place in the Soviet Union until 1958, the time of the establishment of the Kiev Institute of Gerontology. And in the world, no such large-scale gerontological conferences took place until the first conference of the International Association of Gerontology in Liege, Belgium, on July 10-12, 1950. Nonetheless, the impact of the Kiev conference for the subsequent development of aging and longevity research, in the USSR and the world, appears to be significant, despite of (or thanks to?) its ideological drive. Not only was it the first concerted discussion of the state of gerontological science, paving the way for all future gerontological conferences and associations, but it outlined problems that would continue to be discussed for the next 70 years.

14. The rule of the collective – Bogomolets' conference (1938), Lysenko's conference (1948), Bykov's ("Pavlov's") conference (1950)

The Kiev longevity conference also involved a small-scale Lysenko-type witch hunt – a miniature pilot model for future scientific "purges." The major target of attack was the neurophysiologist Alexey Dmitrievich Speransky (1887-1961), at the time head of the department of pathophysiology at the All-Union Institute of Experimental Medicine in Moscow (*Vsesoyuzny Institut Experimentalnoy Medizini*, VIEM). It seems the entire Bogomolets' camp was displeased with Speransky, and Bogomolets was discontented in the first place. Speransky declined the invitation to participate in the conference, and his school was only represented by his pupil, Prof. Nikolay Vasilievich Puchkov. Bogomolets was clearly offended: "I would like first of all to ask Prof. Puchkov to pass it on to Prof. Speransky that we invited him to take part in the conference and are truly sorry that he did not come."[697] The disagreement between Bogomolets' and Speransky's camps obviously originated earlier. While Speransky emphasized the predominant role of the nervous system in aging, Bogomolets emphasized the physiological system of connective tissue. According to Speransky's school, championed by Puchkov, nervous stimulation is of paramount importance for life-prolongation, as it explains the "rejuvenative actions" of a host of agents: transplantations of sex glands, pieces of liver and cadaver skin, transfusion of blood and hemolytic serum, even regeneration induced by tar irritation as in the experiments of Prof. Alexander Vasilievich Vishnevsky (1874-1948, VIEM, Moscow).

In the conference debates, Bogomolets took an opposing stand: "Unfortunately [neither Speransky] nor Prof. Puchkov made a presentation of how Speransky's school perceives the genesis of aging, so as to particularly emphasize the role of the nervous system." Bogomolets' views, in contrast, were better grounded, both scientifically and ideologically:

> Here in the Soviet Union, we have a beautiful custom, which I have never witnessed abroad – in our scientific sessions we decorate the podiums with flowers. What is the most essential part in these plants? You would say – the flower of course. But try to cut its root, and the flower will soon wither. We have spoken of the connective tissue as the root of the organism ... Observations show that the signs of aging in many systems, and first of all the physiological system of connective tissue, are manifested incomparably earlier than signs of aging of the nervous system.

As Bogomolets asserted, the emphasis on connective tissue was also more promising therapeutically, as its stimulation by the anti-reticular cytotoxic serum produced marked anti-aging effects, including improved nervous function. (It seems Bogomolets' anti-reticular cytotoxic serum was the only means for life-prolongation undisputed at the conference; the opposition to it would come later, after Bogomolets' death, see above).

The Bogomolets-Speransky debate then rapidly descended onto a predominantly political-ideological plane. The attack on Speransky was spearheaded by Prof. Semen Khalatov. Khalatov's own scientific theories, particularly regarding the brain as the main source of blood cholesterol increase in the aged, were broadly countered at the conference, most strongly by Profs. V.I. Solntzev and A.G. Dinaburg, and Academician N.D. Strazhesko. Yet, in a political-ideological debate, Khalatov was unmatched. A party member since 1921, he knew well how to play on ideological sentiments. According to Khalatov, the views promulgated by Speransky's school were entirely inconsistent with the aspiring, progressive ideology of the socialist society:[698]

> In our time, when the life in our Socialist country is flowing in full speed, when every one, enjoying life, wishes to prolong it for as long as possible, the work of our Conference should attract a deep interest of our entire community. Indeed, how inspiringly optimistic is the question of the natural prolongation of

life and achieving physiological aging, advanced by Academician Bogomolets, based on a deep understanding of science! What wide paths it does open to us in the struggle of science for longevity, for which we have all the possibilities and conditions!

This is especially necessary to notice here because such an understanding of these questions is foreign not only to scientists of the capitalist countries, but also to some, quite authoritative circles of our Soviet scientists. What, for example, does Prof. Puchkov's speech signify, who could only tell us here that our conference does not sufficiently consider the role of the nervous system, while according to A.D. Speransky's theory the nervous system plays a great role in the development of aging in the organism because even the severance of a nerve can cause aging (?!)

We know well the value of this theory of aging and the premises regarding it in Speransky's theory, in whose direction most works at the All-Union Institute of Experimental Medicine develop. It will be sufficient to quote what Speransky says regarding these issues in his well known book, whose directives are compulsory for the contemporary ruling elite of the All-Union Institute of Experimental Medicine and obviously for Prof. Puchkov as well.

Khalatov quoted Speransky's *Elements in the Construction of a Medical Theory* (1937):[699]

"…From this point of view, there is no difference between the 'immortality' of ameba and the 'mortality' of higher animals. *In principle, both are 'immortal'*. But ascending the scale of perfection, higher animals lose the ability to apply their right [to immortality], since in the complex conditions of life, there is no environment that would guarantee against random, uncommon irritation. *It is enough that it should occur once*. The system of established relations will do the rest. Eventually, every scratch, every injection, can become a stimulus for aging; especially if a chemical irritation is added to the scratch, particularly by a substance of protein nature that has the ability to cause specific forms of irritation." [Speransky's emphasis.]

From this view of aging and deterioration as possibly caused by the smallest irritation, Speransky derived the need to be extremely cautious in all medical procedures:

"One of the notable contemporary hygienists [Max Rubner[700]] said that the art of prolonging life consists in learning not to shorten it. The more complex the cultural life becomes, the more factors emerge whose biological benefit or damage cannot be determined immediately. It will be sad if medicine itself, devoted to preserving the life of individuals and of the entire collective, will begin to produce such damage. Therefore, clinical practice, especially clinical practice for children, should reevaluate the actual need for skin probes and all sorts of inoculations, achieve a clear understanding regarding their actual harmlessness, otherwise the so called *achievements of science can be easily transformed into one of the ways of crippling humanity*." [Speransky's emphasis.]

In Khalatov's view, nothing could be more incongruous with the progressive Soviet therapeutic activism:

These confused directives, detached from life and science, breathe incredibly dark pessimism, altogether denying the importance of all most significant therapeutic-prophylactic achievements of scientific medicine. Do we need to include such a dissonance into the works of our Conference that are based on understanding of science and permeated by a deep faith in the unlimited open possibilities to utilize the achievements of science in the struggle for a happy, joyful life of our motherland?

Thus, once again, the ostensible support of the ruling political-ideological paradigm was an essential factor for operating the Soviet life-extensionist enterprise.

Speransky, however, seems to have been little affected by the attack of Bogomolets-Khalatov's faction. He too knew well how to work within the Soviet political system and how to use ideology to his advantage. In 1939 he became a member of the USSR Academy of Sciences, and since 1945 directed the Moscow Institute of General and Experimental Pathology of the USSR Academy of Medical Sciences. In 1943, he joined the Communist Party and shortly afterwards received the Stalin Prize. He immediately donated the prize (100,000 rubles, perhaps 10 to 20 average yearly salaries) to the Defense Fund of the Chief Command (*Fond Oborony Glavnogo Komandovania*). In an accompanying letter to Stalin, he used this opportunity to emphasize his loyalty to the state: "Having been born and educated in the old times, I really only began to think and act in the Soviet period... I am glad that, together with others, I have the opportunity to materially support the strengthening of our defense." To which Stalin replied: "Accept my greetings and thanks of the Red Army, Alexey Dmitrievich, for your care about the defense forces of the Soviet Union."[701]

Speransky later received another blow at the notorious "Pavlov's session" of the USSR Academy of Sciences and Academy of Medical Sciences (June 28 - July 4, 1950), which was devoted to defending the purity of truly materialistic Pavlov's teachings.[702] Speransky, together with Academician Leon Abgarovich Orbeli, and yet another life-extensionist, Academician Peter Kuzmich Anokhin (the future author of the concept of hereditary immortalization of human personality)[703] were the major targets of attack. Yet Speransky survived that attack as well and continued at his post, after some apologies for not sufficiently referring to the classics and assertions that there was no stronger supporter of Pavlov's teachings particularly and of the materialistic foundations of the Soviet Science generally than himself. Thus, virtually all Soviet life-extensionists of the period, including Speransky and his opponents such as Bogomolets and Khalatov, were avowed supporters of the present social-ideological order, advocating nothing but its strengthening and perpetuation. Perhaps the only competition in this regard consisted in the degree of cogency and conviction with which they would express their loyalty. Interestingly, both Speransky's and Bogomolets' theories of aging were also metaphorically compatible with the ruling socialist system. While Speransky emphasized the central regulation of the organism by the central nervous system (analogous to the central regulation by the government), Bogomolets focused on the supportive infrastructural basis of the connective tissue (analogous to a harmonious, sufficient supply system of the society).

The Soviet life-extensionists indeed had to tread very carefully in order not to be suspected of stepping out of the central political and ideological line. In the famous Session of the All-Union Academy of Agricultural Sciences of 1948 (the "Lysenko session," July 31 - August 7, 1948), several veterans of the Russian-Soviet life-extensionism came under fire. Specifically targeted was Academician Ivan Ivanovich Schmalhausen (1884-1963), formerly director of the Kiev Institute of Zoology and Biology (1930-1941), and at the time of the session chair of the department of Darwinism at Moscow University, a pioneer in the research of the evolutionary biology of aging and longevity and the author of *The Problem of Death and Immortality* (Problema Smerti I Bessmertia, 1926).[704] Another target was Academician Boris Mikhailovich Zavadovsky (1895-1951) of the Institute of Neuro-Humoral Physiology in Moscow, a patriarch of Soviet rejuvenators and the author of *The Problem of Aging and Rejuvenation in the Light of Recent Works of Steinach, Voronoff and other Authors* (1921).[705] There was also no mercy for Academician Nikolay Konstantinovich Koltsov (1872-1940) of the Moscow Institute of Experimental Biology, the foremost Soviet geneticist and author of *Death, Aging, Rejuvenation* (1923).[706] In fact, Schmalhausen, Zavadovsky and Koltsov were the main targets of attack in Lysenko's presentation, and in the presentations of virtually all of Lysenko's allies.[707]

The assault was opened by Trofim Denisovich Lysenko (1898-1976, Moscow), in the periods 1938-1956 and 1961-1962 president of the All-Union Academy of Agricultural Sciences. Lysenko's central thesis was that "*The immortal hereditary substance, independent of the qualitative characteristics of the development of the living body, that governs the mortal body, but is not generated by it* – this is the openly idealistic, essentially mystical concept of Weismann, promulgated by him under the screen of 'neo-Darwinism'" (Lysenko's emphasis). It then appears natural that anybody who would speak in terms similar to August Weismann's about the potential immortality of living matter, as that of germinal plasma or protozoa, and about the mortal body (soma) whose aging and death are determined by tissue differentiation, as did Schmalhausen, Zavadovsky and Koltsov, would be classed with "Weismannists." And "Weismannists" or "Weismannists-Morganists," in Lysenko's jargon, were synonymous with enemies of Darwinism, of "Michurin's teachings" in agriculture, and more sweepingly synonymous with enemies of dialectic materialism, of Marxism-Leninism, of "people's economy" and of the Soviet people generally, deserving to be expurgated by all the force necessary.

Indeed, the "Lysenko session" had little to do with longevity research *per se*, even less with human longevity. The main issue was "the possibility of inheriting the characteristics and properties acquired by plant and animal organisms in the course of their life," of which possibility Lysenko was a strong believer. The session protocols seem to have included only two references to aging and longevity. One was made by the Moscow zootechnologist Prof. Dmitry Andreevich Kislovsky (1894-1957), who jeeringly remarked that genetic variations in blood pressure and longevity would, according to the Weismannists-Morganists, class people into different species according to these parameters. And the other was made in Schmalhausen's self-defense when he noticed that the rate of mutations accelerates with aging. Despite the little emphasis on the problems of longevity, the major significance of the "Lysenko session" for the Soviet life-extensionist movement seems to have been that in that session the old guard of Soviet life-extensionists (Schmalhausen, Zavadovsky and Koltsov) were completely obliterated. Following the session, Zavadovsky was stripped of all posts and his subsequent publications were prohibited. Shortly afterwards he suffered a stroke and died in 1951. Schmalhausen, even though he lived for 15 more years after the session, was relieved of his professorial position at Moscow University in 1948 and lost much, if not all, of his influence. Koltsov was not alive at the time of the conference, and the hostile impact was absorbed by his student, Iosif Abramovich Rapoport (1912-1990), who was desperately trying to defend the material existence of genes and called Lysenko's followers "obscurants." But Koltsov's name was nonetheless thoroughly tarnished (needless to add, Rapoport was purged from the Communist Party and relieved of all academic duties). Immediately after the session, for a very brief period, Lysenko's theory entered into the considerations of Soviet longevity researchers, as evidenced by Nagorny's *Aging and Life-extension* (1950) where Nagorny upheld Lysenko's tenet about the "assimilation of external conditions."[708]

But perhaps the most general significance of the Lysenko session and the following "purifying" campaign was that all scientists, including longevity researchers, were reminded in no ambiguous way that any deviation from the dominant political-ideological paradigm is fraught with tangible dangers of being accused of Anti-Sovietism and consequent sanctions. Indeed very notably, not the slightest opposition to Marxism-Leninism and the Socialist State was voiced at the "Lysenko session," but rather all the participants, both Lysenko's proponents and opponents, presented themselves as the strongest supporters of the dominant regime. This is particularly noticeable in Zavadovsky's self-defense, who argued with Lysenko in the capacity of a "biologist-Bolshevik" and out of the "duty of a party member," furthering "the benefit of the state" and the "needs of Soviet science" and "people's economy." (Zavadovsky's ideological position appears to be very sincere.) He expressed a strong opposition against "idealistic-mechanistic concepts" and was perhaps even more adept than Lysenko at calling to his aid the authorities of Russian-Soviet biology, such as Ivan Michurin and Clement Timiriazev, and the classics of Marxism-Leninism – Marx, Engels, Lenin and Stalin. According to Zavadovsky, "under the conditions of victorious socialism, there is only one general line of our party, the line of Marxism-Leninism." This general central line requires "the struggle on two fronts": "with mechanistic vulgarization of Marxism on the one hand, and Menshevik idealism, formalism and metaphysics on the other." Lysenko's camp, however, was more successful in its show of loyalty. Crucially, the loyalty pledge, the glorification and support of the present ideological and social order, were indispensable for Soviet scientists, and longevity researchers expressed their loyalty in one of the most hyperbolic forms.

15. Khrushchev's 'spring'

During Nikita Sergeevich Khrushchev's rule (1953-1964), the expressions of support for the present form of socialism as the basis for moderate life-extension, and future communism as the condition for radical life-extension, continued. The only change in the basic rhetoric, compared to Stalin's epoch, seems to have been the dismissal of Stalin as the glorious patron overseeing and encouraging the struggle of the Soviet people for better health and longevity. This role was ascribed to the more anonymous "Communist Party" and "Soviet Government."

The ideological constructs of Soviet life-extensionism during Khrushchev's time are exemplified by the commentaries on Metchnikoff's works by Vladimir Nikolayevich Nikitin (Nagorny's pupil and continuer of Nagorny's school of onto-physiology at the University of Kharkov, Ukraine, Metchnikoff's *alma mater*) and Rafael Isaacovich Belkin (a leading Moscow physiologist, since 1929 director of the Moscow Institute of Experimental Morphogenesis, a major center of Soviet regeneration research). During Khrushchev's period, Metchnikoff's complete works were published in two magnificent editions, in 1954 and 1962, by the USSR Academy of Medical Sciences, his works on life-extension occupying special volumes. In fact, most Soviet gerontologists saw themselves as executives of the grand task of life-extension bequeathed by Metchnikoff. Yet many disputed Metchnikoff's main tenet of the centrality of gastro-intestinal intoxication as a cause of aging, and perceived some of his suggested methods for life-extension, such as pro-biotic drinks, stimulating cytotoxic sera, and colectomy, as of only limited effectiveness. There was even less agreement over Metchnikoff's "biologism," his oversight of the crucial role of "social factors" (that is to say, of "socialism") for extending longevity.

Indeed, Metchnikoff was no Marxist. In the introduction to the second edition of his *Forty Years in Search of a Rational Worldview* (1914), Metchnikoff asserted that "No sooner the adepts of most radical socialism approach its practical implementation, than they approximate the reigning capitalist regime." Furthermore, "The experience of the recent revolutionary attempts in Russia [the Revolution of 1905] proved that, instead of creating a Paradise on earth, they bring about brutal, unabashed reaction." Yet, in an earlier introduction to the same book (1912), certain socialist precepts were posited by Metchnikoff as preconditions for healthy life-extension or "orthobiosis." "Orthobiosis" was defined as "correct" hygienic living, enabling the person to live a full natural cycle of life (which should be far longer than the present life-span, somewhere around 150 years), until reaching a stage of complete satiation with life and the emergence of a "death instinct" making further life-prolongation undesirable. Metchnikoff's socialist suggestions for orthobiosis were thus summarized: "Orthobiosis requires a work-loving, healthy, moderate life, free of any luxury and excess. Therefore, we need to change the existing customs of life and eliminate the extremes of wealth and poverty that now cause so much suffering."[709]

In his commentary of 1962, Nikitin overlooked such 'socially conscious' statements by Metchnikoff and claimed that "Metchnikoff was attempting to solve the problem of aging and death as just and only a biological problem."[710] Instead, Nikitin maintained, Metchnikoff should have realized that "longevity has its social preconditions … under capitalism, these preconditions produce an artificial shortening of life." In the world of capitalism, "life for the major part of the population is full of burdens, scarcity, inadequate nutrition, joyless, overexerting labor for others, socially conditioned diseases (especially tuberculosis and syphilis, that are so often mentioned by Metchnikoff) and frequent psychological traumas in the conditions of class-antagonistic society." In contrast, in the socialist state, such life-shortening social plagues and antagonisms are being eliminated. Under the improved social conditions, Soviet medical science will soon be able to realize Metchnikoff's dream and accomplish "the task of a significant prolongation of full human life. In this sense, it will make a new enormous contribution to the successful building of communism."[711] Belkin echoed: [712]

If the Soviet state succeeded in a relatively short historical period to so radically reduce early child mortality, it is doubtless that now the next target can be the problem of longevity, which can be successfully solved on the basis of the elimination of diseases characteristic of old age and therefore shortening the lives of people who reached the age of 60-70. ... There can be not the slightest doubt that, with the further successes in creating the material basis of communism in our country, the number of the long-lived will increase further and further. ... Longevity, and not just as a rare exception, but mass longevity characteristic of the majority of the population, can be achieved only in countries where socialism has won, where the exploitation of a human being by a human being has been eliminated, and where conditions have been created and continue to be created for a maximal optimization of labor and life-style. The good and noble dreams of I.I. Metchnikoff can be realized only in the USSR and in countries of the socialist camp.

It is difficult to distinguish whether such precepts were honestly believed in or only adduced as a necessity to survive within the particular ideological milieu. In most cases, the authors' expressions appear sincere. Whether voluntary or coerced, the tributes to the socialist state were pervasive, showing that life-extensionism was an inseparable part of the existing social order and represented perhaps one of the most convincing forms of advocacy for its continuation.

Essentially, such communism-compatible life-extensionist ideals were shared not only by many biomedical researchers, but even at the highest levels of Soviet government, demonstrating that life-extensionism was not a sectarian but a prevalent element in Soviet culture. Thus, at the XXI congress of the Communist Party of the Soviet Union (January 27 - February 5, 1959), Nikita Khrushchev glaringly declared:[713]

As a result of the growing prosperity of the people and the improvement of health care in our country, the life-span of the people increased. In the USSR in the past years, the mortality has been the lowest in the world, and the increase in population was greater than in the vast majority of world countries. The general mortality in the USSR decreased, compared to the pre-Revolutionary period, by a factor of four, and child mortality by a factor of six.

Khrushchev was also known to have tried on himself and approbated (in 1959) the rejuvenating product "gerovital" developed by the Bucharest Institute of Gerontology. Even before his coming to power as the First Secretary of the Communist Party of the Soviet Union in 1953, while serving as the First Secretary of the Central Committee of the Ukrainian Communist Party (1938-1947), he was a personal friend and great patron of the chief Soviet life-extensionist, Academician Alexander Bogomolets. Khrushchev's government in Ukraine coincided with the peak of Bogomolets' prestige (see above).

Life-extensionism fitted well with Khrushchev's progressivist ambitions epitomized in his reforms of agriculture and the space program. Under Khrushchev, the task of life-extension was written into the Program of the Communist Party. This third (and last) program of the Communist Party, approved by the XXII Party Congress (October 17-31, 1961), set the general task to "complete the building of the material-technological basis of communism" within the forthcoming 20 years (by 1980). An indispensable section of that program was "Care for health and prolongation of life," which stated:[714]

The socialist state is the only state that takes on itself the care for the preservation and continuous improvement of health of the entire population. This is provided by a system of social-economic and medical measures. A broad program is being implemented, directed toward the prevention and radical reduction of diseases, liquidation of mass infectious diseases, and further increase in the duration of life.

Naturally, the concurrent statements of Soviet longevity researchers, such as the above statements of Nikitin and Belkin of 1962, were immediately adjusted to the recent party directives for the accelerated "building of the material-technological basis of communism."

Yet another high-standing sponsor of Soviet life-extensionism (though of course less significant than Khrushchev) was Clement Efremovich Voroshilov (1881-1969), a renowned general during the Russian Civil War (1917-1922), from 1925 to 1940 People's Commissar for Military and Navy Affairs and Commissar for Defense, commanding marshal during WWII, and in the period 1953-1960 President of the Supreme Soviet (the official head of state). Already in 1939, he was outraged by "people who consciously decide to become old by forty five and enforce their own senility."[715] In 1958, on the urging of Clement Voroshilov and the Minister of Health, Maria Dmitrievna Kovrigina (1910-1995), the Institute of Gerontology of the USSR Academy of Medical Sciences was established in Kiev, giving rise to the USSR Academy of Medical Sciences' Scientific Committee on Gerontology.

The Kiev Institute of Gerontology was established under the leadership of Professors Nikolay Nikolayevich Gorev (1900-1992, the institute's first director, during 1958-1961) and Dmitry Fedorovich Chebotarev (1908-2005, the second director, during 1961-1988).[716] The establishment of that first Institute in Ukraine was not accidental. Voroshilov was born in Lisichansk, Ukraine. Ukraine was the former seat of power of Khrushchev. But above all that, Ukraine had perhaps the strongest tradition of life-extensionism in the entire Soviet Union.[717] Elie Metchnikoff (1845-1916), the spiritual father of Russian life-extensionists, was born and educated in Kharkov, and served as a professor at the University of Odessa. Alexey Nikolayevich Severzov (1866-1936), the founder of both the Ukrainian (Kiev) and Russian (Moscow) schools of evolutionary morphology, was keenly interested in the problems of longevity. While working in Kiev (1902-1910) he began researching the evolution of longevity, and later, while in Moscow University (in 1917), published a work on "The Factors Determining the Longevity of Multicellular Animals," where he proposed a relation between longevity and reproduction.[718] Academician Ivan Ivanovich Schmalhausen (1884-1963), Alexey Severzov's pupil, was born and educated in Kiev, and from 1930 to 1941 was the director of the Kiev Institute of Zoology and Biology. He was recognized as a foremost researcher in evolutionary morphology and evolutionary embryology, coining the concept of "holomorphosis" – the development of the organism as a whole. Schmalhausen was also one of the pioneers in the research of the evolutionary biology of aging and longevity, and wrote in 1926 a book, *The Problem of Death and Immortality*, in which he correlated between limitations of growth and maximal life-span.[719] Later still, in the 1930s, Ukraine was the home of two major schools of Soviet aging and longevity research: the Kiev physiological clinical school led by Academician Alexander Alexandrovich Bogomolets (1881-1946, director of the Institute of Experimental Biology and Pathology and the Institute of Clinical Physiology) and Academician Nikolay Dmitrievich Strazhesko (1876-1952, since 1936 director of the Ukrainian Institute of Clinical Medicine); and the Kharkov school of onto-physiology led by Prof. Alexander Vasilievich Nagorny (1887-1953, head of the Institute of Biology and Department of Physiology at the University of Kharkov) and Prof. Georgy Vladimirovich Folbort (1885-1960, between 1926-1946 chair of the Department of Physiology at Kharkov Medical Institute). Despite what was then commonly termed "the socialist competition" between Kiev and Kharkov, the two streams merged with the establishment of the Kiev Institute of Gerontology in 1958, that included Bogomolets' pupils (Nikolay Gorev), Strazhesko's pupils (Dmitry Chebotarev), as well as Nagorny's pupils (Vladimir Nikitin) and Folbort's pupils (Vladimir Frolkis).[720]

During Khrushchev's rule and throughout Brezhnev's rule, the Kiev Institute of Gerontology was the major base of Soviet life-extensionism. If anything, the institute represented "moderate" life-extensionism. The leading Soviet gerontologists took great pains to assert that their work had nothing to do with the wild dreams of immortalists. Yet they did work and did advocate for a very significant extension of life, which may presently seem just a little less ambitious than the age of Methuselah. Chebotarev summarized the aspirations of Soviet gerontologists: "We believe that human beings not only can, but must live more than 100-120 years on the average. However, factors of premature aging lead to a decrease of this physiological norm to 70 years. … I want to believe that the immense work done by my colleagues and myself will be used for the further development of gerontology and geriatrics in Ukraine."[721]

16. Brezhnev's rule and the last years of Soviet orthodoxy

During Leonid Ilyich Brezhnev's rule (1964-1982), life-extension research continued in the USSR on a large scale, and the mutual support of the life-extensionists and the socialist state continued. The aspirations of some authors amounted to outright immortalism. Thus the renowned botanist and microbiologist Vasily Feofilovich Kuprevich (1897-1969), president of the Byelorussian Academy of Sciences since 1952, issued a manifesto in 1966 in the journal *Technika – Molodezhi* (Technology for the Youth), circulated in 1.5 million copies, entitled "An Invitation to Immortality" (Priglashenie k Bessmertiu).[722] In that manifesto, Kuprevich suggested:

> By learning to decode DNA, by learning to build these molecules according to a predetermined plan, we can introduce into the organism such DNA molecules that will repair the errors constantly occurring in the organism. It has even been suggested to create a 'virus of immortality.' It is known that, when entering a cell, virus DNA disrupts the cell's genetic program, forcing the cell to build ever new viruses instead of materials necessary for the cell. In exactly such a way, the 'virus of immortality' can enter each cell of our organism, and will immortalize or rejuvenate it.

Kuprevich further asserted:

> Man has overcome the power of natural selection. He no longer adjusts to the conditions of external environment, but creates around him an artificial, beneficent environment, remaking nature. He does not need death as a factor accelerating the improvement of humanity from generation to generation. ... There are no theoretical prohibitions to raising the principal possibility of immortality. I am deeply convinced that, sooner or later, the era of longevity will arrive. ... As in any task, enthusiasts are needed for this, unfortunately these are very few; we are hindered by the deep rooted conviction that death is inevitable and that the struggle with it is futile. This is a sort of psychological barrier that must be overcome.

With the 1.5 million copies in circulation, Kuprevich's manifesto inspired many a Soviet youth to dream of radical life-extension. And the "virus of immortality" was not the only means studied for the purpose.[723]

Yet another renowned immortalist during Brezhnev's period was Lev Vladimirovich Komarov (1918-1985) – head of the Laboratory of the Problems of Functional Regulation in Human and Animal Organisms of the Moscow Institute of Highest Nervous Activity and Neurophysiology of the USSR Academy of Sciences. Komarov was considered one of the chief architects of Soviet radical life-extensionism. Since the late 1950s he conducted a broad series of experiments on the artificial extension of the "species-specific" life-span in tissue cultures and model animals, using pharmacological means (NaBr, vitamins B12 and E, dibazol, diosponine, ATP, fructose, sucrose, magnolia vine, ginseng, theobromine, chlorophyll, petroleum growth substance/"mumijo" and more) and physical means (hypothermia, electromagnetic fields), setting out to prove that the species-specific life-span can be experimentally prolonged two or three times.[724] Since 1962 he served as the scientific secretary of the "Biology of Aging" section of the USSR Academy of Sciences, and under his initiative the research of the "problem of the artificial extension of the species-specific life-span" was included into five-year plans of the USSR Academy of Sciences.

Insofar as Komarov considered intensive international cooperation to be crucial for success, in 1973 Komarov and the Belgian immortalist gerontologist Herman Le Compte (1929-2008) organized an Association and a journal "Rejuvenation" dedicated to the problem of radical life-extension.[725] Komarov's immortalist aspirations were outspoken. In an article entitled "People can and must live not by decades but by centuries" (1972) he claimed:[726]

> The end has come to the idea, that the Man [sic] is not able to change the natural span of his life. Thus, now in sociology and biology there comes forward the goal of practical realization of the artificial postponing of the onset of aging and death beyond the present natural (specific) limits.

According to Komarov, the traditional (hygienic) approach to life extension, that is, by improving general health care and social conditions, has practically exhausted its potential. Even the defeat of cancer, he believed, would not prolong the average life-expectancy by more than a couple of years. Therefore, he asserted, the research effort toward radical life prolongation using reductionist biochemical and biophysical means should assume the first priority in gerontology.

Another eminent proponent of radical life-extension was Academician Nikolay Mikhailovich Amosov (1913-2002) of Kiev, reputed since the 1960s as a foremost polymath authority in cardiology, gerontology, bioenergetics, bioinformatic modeling and social planning. (He worked since 1960 as the director of the Biocybernetics Department of the Kiev Institute of Cybernetics and since 1983 served as the director of the Kiev Institute of Cardiovascular Surgery.) Amosov let his immortalist aspirations be known in his science-fictional novel *Notes from the Future* (1965).[727] The protagonist Prof. Prokhoroff, clearly modeled after Amosov himself, develops a technique for "anabiosis" (suspended animation by freezing) and has himself frozen in 1970 and revived in 1991, in the ever more powerful and prosperous Soviet Union. (Such research has indeed been carried on in the Soviet Union since the late 1920s, also by Amosov himself.[728]) Immediately after the revival, Prof. Prokhoroff sets out to develop his "pre-freezing" idea of achieving radical life extension:[729]

> Of course, theoretically it is possible to design a machine to watch over the body, more complicated and durable than any yet designed. It would check on various functions, safeguard organs from accumulation of harmful residues, eliminate most of them. In some areas this would work, and the average span of life could be greatly increased.

The first necessary step would be to establish an Institute dedicated to achieving physical immortality, euphemistically termed "the study of increased longevity" (or rather "immortality" is a trope for "increased longevity"):

> I saw our Director. Mentioned to him my "immortality" idea; of course, I did not use this word. Simply, "the study of increased longevity." This is what it really is. "Immortality" is just a poetic allusion, and I use it for its shock value. ... I told [the Director], scaling down my requirements to the minimum. First, I need a tight and enthusiastic small team. (The Institute will come later; I didn't mention this word at all.) Just a room or two, and a few volunteer researchers. Surprisingly, there were no questions, and no objections.

This passage may exemplify the general programmatic mindset of many Soviet life-extensionists (such as Komarov, Chebotarev and others), for whom the principal necessary steps were first to set a very general goal, set up a collaborative study group (better an Institute) and then conduct research until the Day of Revelation.

Most Soviet life-extensionist gerontologists, however, were more moderate in their goals, but nonetheless fairly ambitious. Amosov too was much more modest in his practical recommendations. Since the mid-1960s, he published an immensely popular series of works on healthy life extension, circulating in millions of copies, including *Thoughts and Heart* (Mysly i Serdze, 1964), *Thoughts on Health* (Razdumia o Zdorovie, 1977), *The Physical Activity and Heart* (Physicheskaya Activnost i Serdze, 1989), *The Overcoming of*

Aging (*Preodolenie Starenia*, 1996). His summary of this life-long research entitled *An Experiment in the Overcoming of Aging* (an expanded version of *The Overcoming of Aging* of 1996), was completed just before Amosov's death in 2002.[730] While writing in 1996, Amosov was fairly critical of the current abilities of medicine to prolong human life, yet he was still quite optimistic regarding his own health and life-extending regimen:[731]

> Many lives saved can be attributed to the achievements of medicine. Only in our institute [the Institute of Cardiovascular Surgery in Kiev], there were at least 60,000. However, medicine saves the lives of some people, but shortens the lives of others (of the majority!). This sounds paradoxical, but this is so. The scientific-technological (and economic) progress created excellent conditions for people's existence, guarding them against hunger, cold, physical overwork and many other factors that shortened the lives of our distant ancestors. If concomitantly man would adhere to a correct life-style, that is, follow a necessary regimen, that I call "The regimen of limitations and exertions" (including daily gymnastics, jogging and dietary restriction), then death should retreat much farther.

Amosov's immediate aspirations were limited to living a century:[732] "As regards the experiment: I am positive that, with a healthy heart, great exertions will ensure active life until 90 years of age, and for the lucky (or unlucky?) ones – even until 100 years."

To this regimen of intensive gymnastic activity (at least 1000 "movements" per day) and dietary restrictions, Amosov attributed his youthful mental and physical shape far into his 80s, despite his previous tuberculosis and heart conditions. Even in 2001, at the age of 88, he described his physical and mental conditions as excellent. Yet, in the last section "On the course of the experiment and its complications" completed in November 2002 after a myocardial infarction and just before his death (in December 2002), Amosov's conclusion of his experiment is sobering:[733]

> It is so easy to deceive oneself, when one wishes to. And so, now there is a heart attack "out of the blue." Doctors say "It happens." Any gerontologist would say: "Such exertions are not suitable for an old man." And so they told me. I didn't listen. In conclusion, "Amosov made a mistake." … Training is beneficial, but it won't give you much. The greater the age, the less are the hopes from exercise. … Corrections are necessary. Exertions should be reduced… Honestly, I am not guilty. … From the very beginning nine years ago, I have been toiling myself, I did not persuade the aged to imitate me. I advised everybody: "At the maximum – do half of my exertions! Gradually and once again gradually increase them, from three to six months." Luckily, the aged did not follow me, and for the young such exertions are in the normal sports range. … The limits of the exertions were clearly exaggerated. But that's what experiments are for: to test maximal endurance. … And most importantly, there was the elementary ignorance in diagnosing the increased heart valve stenosis. … Everything was done honestly: each half a year I underwent an examination. And everybody was silent, relying on my authority. "He is the chief. He knows everything himself." This is the way it goes.

Thus, to his last days, Amosov remained a life-extensionist, seeking suggestions for a correct centenarian regimen from his own impending destruction, the destruction that appeared to be hastened by his own efforts to avoid it. And so did Metchnikoff on his deathbed.[734]

Other leading Soviet gerontologists too posed somewhat moderate goals and considered a centenarian life-span as a worthy first aim. This was clearly the position of the majority of scientists at the Kiev Institute of Gerontology. As Yuri Konstantinovich Duplenko with the Kiev Institute of Gerontology stated regarding Komarov's immortalist program:[735]

> The successes in the understanding of the material foundations of life's processes, achieved during the

latest period, are able to inspire faith in the unlimited possibilities of influencing these processes for the purpose of life-prolongation. Perhaps, this entailed the appearance of unjustified prognostications about the realization of potential reserves for human life-extension up to 1000 years… and the pronouncement of immortology as a new direction of biomedical science. Such claims remind of the naive hopes of scientists of the distant past and hinder the strengthening of the scientific prestige of the research directed toward a radical increase of the specific life-span.

Equally cautious was the Institute's foremost researcher, the director of its department of physiology, Academician Vladimir Veniaminovich Frolkis (1924-1999).[736] Frolkis absolutely believed that a radical extension of the human life-span is possible and, within his "adaptive-regulatory theory of aging," coined the concept of "Vitauct" or resistance to aging, outlining some of its major processes and suggesting potential ways toward its reinforcement.[737] (At the time, other theories of "regulation" for life-extension were suggested by Alexey Olovnikov, Vladimir Dilman and others.[738]) Yet, Frolkis too struggled against excessive expectations:[739]

> Presently gerontologists are considered not with skepticism, but with hope, even with faith in the achievement of a truly 'cosmic' success in the task of life prolongation. In this field, there should be neither dark pessimism nor unjustified, thoughtless optimism which disorients not only the wide public, but also the scientific circles. Such 'propaganda' devalues the laborious scientific search, the struggle for each year of human life.

The uppermost limit of expectations of the Institute's scientists was summarized by the Institute's director, Chebotarev: "We believe that human beings not only can, but must live more than 100-120 years on the average."[740]

Whatever the level of hopes, during Brezhnev's period and the last years of Soviet orthodoxy, the Soviet gerontologists' support of the Socialist State was almost unanimous (perhaps with a single exception of Zhores Medvedev which will be discussed further on). In Amosov's futurist novel, *Notes from the Future* (1965), where he let his extreme expectations be expressed, the future social systems (as of 1991) are basically the same as the present social systems (of 1970), only somewhat adjusted and improved. "I don't think the world has changed a great deal. … But technology, of course, is completely new." The Soviet Union is essentially the same Soviet Union, only ever stronger and more prosperous:[741]

> Economic insecurity as such has disappeared completely. The state provides for all basic needs of all people… They have become accustomed to prosperity, good health, freedom. … Society supports loafers as well as everybody else. … But habitual loafers are well known, and are listed at the Central Labor Office. The state tries to reeducate them, to inject industriousness into them, not so much through financial rewards as by psychological and sociological means – self-respect, respect of others, and the sense of duty.

In contrast, the United States continues to be ruled by greed, commercialism, unearned luxury and inequality: "Americans are still dollar-conscious, it seems." Furthermore, American scientists lead the world in the most inhumane and dangerous direction.[742] Thus the biologist Karl Schwartzerberg of Columbia University is made to say:[743]

> Man can be perfectly controlled by chemistry – and will be. We are developing a whole set of chemicals which will control people's emotions as well as their body functions. … Happiness will be produced in our laboratories and sold in drugstores at nominal prices. … And then all your socialist and communist notions will become obsolete; people will be perfectly happy even in the most uncomfortable physical

surroundings. We are composing a new Golden Age in our test tubes.

But even the United States gradually moves in the direction of socialism, in Amosov's view the only social system worthy and able to persist in the future. As said by Dr. Barrow, "Any organized society is basically socialistic. Socialism, as a planned economy, is here and nothing can stop it, because it is an efficient form of management. Only, we avoid this word." Japan too will not escape the evolution toward socialism: "Here they have developed unique 'socialism-capitalism' – a purely Japanese sociological phenomenon, rooted deep in Japanese national character," whereby "they have created one of the highest standards of living in the world with a tremendously high educational level."

According to the assertions of Soviet gerontologists of Brezhnev's times and the last years of the Soviet state, in the contemporary period as well, there is no alternative to socialism if wishing to prolong healthy human life. As can be seen, the rhetoric in that sub-period of Soviet history did not change much as compared to Stalin's times. Thus, according to Duplenko (1985):[744]

> Many scientists, especially in capitalist countries, pose the questions: Is it needed? Should human longevity be facilitated? ... Many serious socio-economic problems are posed... the problem of work power, increased social load by the aged ... The pessimistically inclined scientists and politicians do not take into account the close connection between the length of life and the level of public health ... the higher age limit will be accompanied by a significant improvement of the general state of health... therefore we should expect an increased period of creative professional activity. ... In the socialist society, with its indisputable advantages of social arrangement, planned economy and science organization, the most beneficial conditions have been created for the solution of the problem of aging and longevity.

And similarly, though perhaps less emphatically, Frolkis opposed "researchers in the West who depict gloomy perspectives for the society burdened by increasing numbers of the aged" and, in contrast, portrayed an optimistic scenario, possible only in the USSR, where life-extension would signify an improved health and well-being of the citizens, as well as an increased period of their productivity and creativity (1988).[745]

During Brezhnev's period, there was perhaps no stronger propagandist of the socialist system than the director of the Kiev Institute of Gerontology, Dmitry Fedorovich Chebotarev. As the advisor on gerontology and geriatrics of the World Health Organization (1961-1983) and president of the International Association of Gerontology (1972-1975), he did much to promote the image of the Soviet Union as a peace-loving, cooperative country and a leader in the reinforcement of international health care. At home, in the capacity of the President of the Association of Gerontologists and Geriatricians of the USSR (1963-1987) he was undoubtedly one of the most powerful figures in the field, and may well be considered an executive arm of the state. And when something or somebody did not cohere with the state, Chebotarev would make sure the disturbance was dealt with. Thus, during the Ninth International Congress of Gerontology in Kiev in 1972, over which Chebotarev presided, a controversy broke out about the exclusion from the conference of the dissident Russian gerontologist (and a "moderate" life-extensionist) Zhores Alexandrovich Medvedev (b. 1925).[746] Medvedev was blacklisted after the appearance in 1969-1970, in Soviet "self-publication" (*samizdat*) and in the West, of his "subversive" works, especially "The Rise and the Fall of T.D. Lysenko" criticizing authoritarianism in Soviet science, and other publications opposing restrictions on international travel and postal censorship.[747] Consequently, Medvedev was dismissed from his post at the Institute of Medical Radiology in Obninsk, and committed to a psychiatric hospital in 1970. Thanks to a public outcry and appeal by prominent Soviet intellectuals (Peter Kapitsa, Andrey Sakharov and others), he was released and even found a position as a senior scientist at the Institute of Physiology in Borovsk. But clearly, his participation in the Kiev Gerontological World Congress was highly undesirable for the Soviet Program

Committee. Though the American representatives at the conference (Nathan Shock, Leonard Hayflick, and others) insisted on his inclusion as "the first scientist to formulate the error theory of aging," Chebotarev made sure he did not participate on the pretext that "Dr. Medvedev had changed his line of research in recent years and there were so many abstracts submitted by senior Soviet scientists that many had to be omitted in order to keep the program in balance with respect to American and Soviet speakers." Medvedev was eventually forced to move to the UK in 1974. Since then he has continued to defend Western liberties, criticize Russian socialism and authoritarianism, as well as publish works on life-extension.[748] Chebotarev's involvement in this case shows once again that life-extensionists, for the most part, have been loyal supporters of the existing social and ideological system. And Medvedev's fate shows that those few who oppose that system are likely to be forced out or repressed. If they are fortunate, they can find and integrate into a different system that they can support.

During the last years of Soviet orthodoxy, the task of life-extension was perceived as a general social task, but clearly also as a task of particular relevance for many in the Soviet Government. In the early 1980s, the signs of senility of Brezhnev (1906-1982) and of many in the party apparatus were obvious to everybody.[749] Brezhnev's followers in the post of the General Secretary of the Communist Party – Yuri Vladimirovich Andropov (1914-1984) and Konstantin Ustinovich Chernenko (1911-1985) – both died less than a year and a half after entering the post. Whether due to the realization of the general importance of life-extension for the Soviet people, or some private considerations, the development and consolidation of Soviet longevity research culminated in 1980 with the inception of the USSR all-state "Life-Extension Program" – "Programma Prodlenie Zhizni" – that coordinated the activities of over 80 scientific institutes, and was led by the Kiev Institute of Gerontology.[750] The program was a counterpart (though on a much lesser scale) to the all-state "Food Program" approved in May 1982, and the "InterCosmos" program launched in 1978. The "Life-Extension Program" structure (roughly resembling the structure of the American National Institute on Aging established during Richard Nixon's administration in 1974) included: 1) research of fundamental mechanisms of aging at the molecular, cellular, tissue, organ and systemic levels; 2) experimental approaches toward life-extension (including testing substances such as pro-biotic 'gerolact,' cardiac glycosides, vitamins and anti-oxidants, entero-sorbents, macroergics, adreno-reactants, etc); 3) research of age-related pathology; 4) social and hygienic gerontology. Chiefly through the intensive development of the first two parts – basic research of aging and experimental life-extension – it was hoped to achieve "control over the rates of aging" by 2010. The program persisted during the last years of Brezhnev's rule, throughout Andropov's and Chernenko's rule and in the first couple of years of Gorbachev's rule.

The program was extinguished during the "Perestroika" (Reconstruction), but was a substantial element in the Party General Secretary Mikhail Sergeevich Gorbachev's earlier "Uskorenie" (Acceleration) plan of 1985, or the "Program for the Acceleration of Scientific and Technological Progress." Indeed, Gorbachev's revisions of Khrushchev's (third) Program of the Communist Party and his "Main directions for the economic and social development of the USSR for 1986-1990 and for the period up to 2000" – approved by the 27th congress of the Communist Party (February 25 - March 6, 1986, the penultimate congress before the collapse of the Soviet Union in 1991) – removed all mention of building communism or the "material-technological basis of communism" any time soon. Yet, Gorbachev's programmatic tasks were little less ambitious. For the period 1986-2000, it was planned to "double the gross national income," "increase by a factor of 2.3-2.5 the productivity of public labor," "reduce manual labor by 20%," "solve the food and accommodation problems" (i.e. eliminate food deficits and give each family a private apartment).[751] "All these," Gorbachev's program stated, "can be achieved only by a radical technical reconstruction, ... acceleration of the scientific and technological progress, deployment of the newest technology."[752] Life-extensionism fitted perfectly with these aspirations. And so, according to the revised Party Program, "The task of utmost primary importance is *the strengthening of health of the Soviet people, the prolongation of their active life*" (emphasis in the original).[753] This was the final statement of Russian Soviet life-extensionism coherent with communist ideology.

17. New Russia

Great longevity was promised by the Soviet system, but the Soviet system did not have great longevity. The progress did not extrapolate to infinity. With the collapse of the Soviet Union in 1991, Russian economy and science were in shambles, including life-extensionist enterprises.[754] After a peak in life expectancy in 1986 (commonly associated with Gorbachev's anti-alcohol campaign), with the disintegration of the Soviet Union the average life-expectancy plummeted.[755] Many gerontologists left the country.[756] Presently, concomitantly with an apparent economic stabilization, health programs and scientific research have been generally receiving a stronger state support[757] and life-extensionism has been resurfacing.

Life-extensionist, even immortalist aspirations have remained strong in Russia (apparently less so in Ukraine, the former center of Soviet life-extensionism, or other republics of the former Soviet Union).[758] Yet, a dramatic transformation occurred in the movement's social dependence after the collapse of the USSR. In place of the previous central state support, Russian life-extensionists became strongly dependent on donations from new Russian capitalists.[759] (The latter however came in with limited enthusiasm as no immediate profits were expected.) Thus, Prof. Vladimir Anisimov, director of the department of Carcinogenesis and Aging at the Institute of Oncology in St. Petersburg, and president of the Gerontological Society of the Russian Academy of Sciences (who has been focusing on the influence of light and sleep-wakefulness cycles on the life-span), has received state funding, but also received investments from Timur Artemiev, director of the "Evroset" (Euronet) corporation. Evgeny Nudler, who had formerly immigrated to the US and in 2005 returned to Russia as the head of the "Geron" laboratory in Moscow, was financed by Dmitry Zimin, director of the "Beeline" corporation. "Russia2045" movement was founded in 2011 by the director of the New Media Stars corporation, Dmitry Itskov.[760] The movement has been dedicated to the rapid development of regenerative therapies, artificial organs, even of an entire artificial body, as a means to radical life extension.[761]

A public movement "For the Prolongation of Life" (*Za Uvelichenie Prodolzhitelnosti Zhizni*) was established in Yaroslavl in 2006, by the entrepreneur, director of the consulting company "President" Mikhail Batin. The movement, asserting that "The idea of life extension should become the new national idea,"[762] has been dedicated to support basic longevity research and life-prolonging hygienic measures, and has been largely financed out of the founder's own pocket. In July 2012, on Batin's initiative, the establishment of a political party "For the Prolongation of Life" (or "Longevity Party") was proposed.[763] The "Russian Transhumanist Movement," Russia's major grassroots advocacy group for radical life extension, founded in 2003 and presided over by the economist Danila Medvedev, has also largely subsisted on private donations. Other foundations have emerged in an attempt to collect donations from the wealthy to support longevity research, such as "Eternal Youth" (Vechnaya Molodost, operating since 2005).[764] Several peculiar life-extensionist projects sprang up, for the most part commercially driven,[765] but also state and university funded.[766] Several centers for fundamental and applied longevity research have persisted, in particular in Moscow.[767]

Perhaps the most eminent contemporary Russian life-extensionist has been the member of the Russian Academy of Sciences, Vladimir Petrovich Skulachev (b. 1935), since 2002 dean of the Faculty of Bioinformatics and Bioengineering at Moscow University, famous for such statements as "Aging is an atavism which must be overcome" and "Man will live hundreds of years."[768] Vladimir Skulachev, aided in his research by his sons, Maxim, Innokenty and Konstantin, has aimed to achieve this purpose by a super-efficient delivery of antioxidants to the mitochondria via his "new type of compounds (SkQs) comprising plastoquinone (an antioxidant moiety), a penetrating cation, and a decane or pentane linker" (altogether commonly termed "Skulachev's ions").[769] (The concept of such ions originated in 1969.[770]) By the use of such antioxidant-carrying, mitochondria-penetrating ions, Vladimir Skulachev has intended to eliminate oxidative damage entirely and cancel "the program of aging."[771] Being an academician and university professor, he has received some state funding, but a large portion of it came from the Russian nouveau-riche, nay "oligarch," the Russian billionaire aluminium magnate, president of "Russian Aluminium" Oleg Deripaska, and other private sponsors.[772] Furthermore, the Skulachevs have made no secret of their intentions to commercialize the project as intensively as possible. However, the Skulachevs have often stressed that the products, whenever they are perfected, will be eventually made cheap and hence widely accessible. After several years of announcing that the life-extending product is being patented, and therefore little open information appears, it was announced that in December 2009 clinical trials of "Skulachev's ions" began for some ophthalmologic applications. (In September 2012, a product (eye drops) appeared on the market.[773] Yet, the clinical results for ophthalmological and other age-related conditions have still been uncertain.) In any case, the integration of life-extensionism into the new corporate Russia has been completed.

A corresponding change occurred in the ideological justifications of Russian life-extensionism, as many life-extensionists had to adjust to the capitalist ideology of "New Russia." The "prolongation of life" has remained in the program of the Communist Party of the Russian Federation.[774] But the majority of Russian life-extensionists modified or radically changed their ideological allegiance. Thus, the philosopher Igor Vladimirovich Vishev (b. 1933), the author of the concepts of "Homo Immortalis" and "Immortology," began his career at the Department of Scientific Atheism of Moscow University, and since 1968 served as a professor at the Faculty of Marxism-Leninism at Chelyabinsk Polytechnic Institute (in the Southern Urals region). Vishev's views of the 1980s are exemplified by his letter of September 1983 to the General Secretary of the Communist Party, Yuri Andropov. In that letter, Vishev suggested the establishing of a "Scientific Juvenological Center" aimed at investigating "methods of preservation and restoration of youth" and achieving "indefinitely long life" or "practical immortality." Vishev gave the following justifications:[775]

> Such studies, intrinsically valuable in and of themselves, will deal the final blow to religion and mysticism, as man will take the solution of the problem of immortality into his own hands, ... they will increase the attraction of the communist ideal to which we are staunchly dedicated. ... In the face of the imperialist threat of thermonuclear war and universal annihilation ... the proposed research will intensify the struggle for the right to life.

Yet, in his more recent work, *On the Way to Practical Immortality* (Na Puti k Prakticheskomu Bessmertiu, 2002), his tone regarding the communist declarations made in that letter appeared to become somewhat apologetic: "The letter naturally reflected the spirit and the style of that epoch." In the work of 2002, some nostalgia over the communist ideals and the socialist order of life was still felt:

> In fact, many socialist achievements and communist perspectives were lost either entirely or partly, they were proclaimed utopian by their adversaries; the Soviet Union, which undoubtedly had a great potential, was dismembered into several so called 'sovereign' states which were all, without exception, plunged into the profoundest crisis. The nationalist elite that took power in those countries only exacerbated the destructive process.

It seems as if Vishev would still preserve the socialist ideological and social order, would still aspire to the creation of a perpetual human being in a perpetual socialist system, if only he could. Nonetheless, any such mentions of communism, socialism or capitalism, were carefully moved to the background, made almost imperceptible in the bulk of general techno-progressive and scientific-positivist considerations. Now, the radical life-extension was considered by Vishev as a precondition for happiness in "any society," and the communist society was regarded as a "particular case" (cf. the previous dominant statements of Soviet life-extensionists that socialism was intrinsically conducive to longevity, while capitalism was intrinsically conducive to life-shortening). Thus, even such a self-avowed "Bolshevik" as Vishev made ideological adjustments and attempted to integrate into the new social order.

In other authors, in place of "dealing blows to religion and mysticism," Pravoslav Christianity, with its hope of universal salvation, has resurfaced as an ideological foundation for Russian life-extensionism, going back to Fedorov's original propositions. These have been the views of the Pravoslav biophysicist Boris Georgievich Rezhabek (b. 1939) and of the entire "Fedorov's movement" centered at "N.F. Fedorov's Museum-Library" in Moscow.[776]

The majority of life-extensionist authors have now discarded communist ideology altogether. There is perhaps no clearer showcase of the conversion than that of Vladimir Khatskelevich Khavinson (b. 1946), director of the St. Petersburg Institute of Bioregulation and Gerontology and vice president of the Gerontological Society of the Russian Academy of Sciences. His research and development have mainly concerned "peptide bioregulators," such as those produced from the pineal gland/epiphysis (epithalamin) and thymus (thymalin) of animals, intended to regulate the organism into a youthful functioning – yet another ramification of rejuvenative opotherapy. Presently, synthetic analogues have also been under development. Much of this work has been co-authored with Prof. Vladimir Nikolaevich Anisimov (b. 1945), president of the Russian Gerontological Society, and Prof. Vyacheslav Grigorievich Morozov (1946-2007).

As can be seen in one of Khavinson's interviews (2007),[777] Russian national pride emerged as a substantial element of the new longevity research ideology: "Now everybody recognizes," Khavinson asserted, "that in the study of life-extension we are at the forefront of the entire planet." Yet, capitalist ambition became perhaps an even more salient feature. As Khavinson related, the research of "bioregulators" began at the Leningrad Military Medical Academy, directly funded by the USSR Ministry of Defense. "In the USSR," Khavinson recalled, "they did not spare funds for scientific work, whatever we asked they gave us." Now, time has come to capitalize on these developments. Presently, in Khavinson's view, "science is also a business." There now appeared to exist little ideological incentive, except for a bluntly admitted pursuit of profit:

> At one time, representatives of large foreign pharmaceutical companies offered us to sell them the patents on our inventions. Do you know why? So we wouldn't be tangling between the feet of large companies, who do not need novel achievements. It is much more profitable for them to buy out prospective developments from a scientist, so he wouldn't bother them any longer, rather than to invest in the development of a new line of research. We were offered millions of dollars, so I should just close down my research! And do you know what will be the profits from our business in the near future on the European and World scale? Tens of billions of dollars yearly! I will explain: the world population is 6 billion people. One billion comprise the people who care about their health and longevity, plus they can afford to buy our preparations. A prophylactic yearly course of treatment should preferably be given to anybody who reaches the age of 40. The price of the course will be minimum $100. We multiply the $100 by a billion, and we get 100 billion. And the cost of production of our preparations, when we learn to make them synthetically, will be 2-3 or at the most 4 dollars, in the most beautiful packaging.

Surely, now all of us (both former enthusiasts and skeptics of life-extension) would be lining up to pay Khavinson the billions, except, at the present (2014 as of this writing), his preparations are not yet widely available, and it is unclear when they will be. Their exact potential for actual human life-extension also remains to be ascertained.[778] Whatever the immediate prospects, the main point is that Khavinson, a former Soviet military researcher (in the rank of colonel) and communist, has become an outspoken capitalist, having rapidly attuned to the new social and ideological order.

18. Chapter conclusion. The perpetuation of the current social order

Evidently, Russian life-extensionism underwent radical transformations in its ideological motives: from the monarchic government, through the enthusiasm of the emerging Soviet state, Stalin's personality cult, Khrushchev's "spring," Brezhnev's late stage of Soviet orthodoxy, Gorbachev's "Perestroika," and eventually the "new" corporate Russia. And at each stage, life-extensionism blended into the ruling ideological and social regime. Furthermore, stemming from its fundamental aspiration to a prolonged bodily and social equilibrium, life-extensionist philosophy integrally supported the existing ideological and social system, advocating for its endurance. Thus, far-reaching unorthodox scientific projects took their root in a fundamentally conservationist, even stand-pattist, political outlook. Among the ideological justifications for Russian life-extensionism, competition with the United States was an important factor, well throughout the 20th and the beginning of the 21st century.

Chapter 4.

The United States: In the name of capitalism, religion and eugenics – commercial enterprise, basic research and the struggle for funding.
The United Kingdom: philosophical and scientific discussions on life-extension, for and against

1. Chapter summary

At the end of the 19th and the beginning of the 20th century, American life-extensionism was strongly associated with the currently dominant social-ideological patterns, exhibiting three major components: dependence on capitalist enterprise, religious sentiment and advocacy of eugenics. While religious and eugenic rationales later lost in prestige, capitalist considerations have remained prevalent. Since the 1930s, after the loss of popularity of immediate rejuvenation by organotherapy, a shift of emphasis occurred toward basic research of mechanisms of aging and toward increasing institutional consolidation of gerontology. Rejuvenation attempts reemerge with great force since the 1950s, corresponding with contemporary advances in replacement medicine. The US then becomes the world leader in longevity research and practice.

In the UK, the discussions of life-extension in the early 20th century appear to be predominantly literary and philosophical, elaborating on the potential promise and peril of life-extension. Since the 1940s, longevity research in the UK intensifies, in a close, world-leading alliance with American researchers.

2. The American tradition. The field of unlimited possibilities

In the United States, life-extension formed an integral part of the "American Dream," since the foundation of the country. The founder of American life-extensionism was Benjamin Franklin (1706-1790), who stated that "the rapid progress true science now makes, occasions my regretting sometimes that I was born so soon" and that "all diseases may by sure means be prevented or cured, not excepting even that of old age, and our lives lengthened at pleasure even beyond the antediluvian standard" (1780).[779] Franklin also took keen interest in prolonged suspended animation and resuscitation. Early in the 19th century, remedies against old age were sought by Dr. Benjamin Rush (1745-1813), a signatory of the Declaration of Independence, as well as by his pupil Charles Caldwell (1772–1853). Rush's regimental suggestions for increasing longevity were based on puritan values, including industriousness, temperance, mental vigor, equanimity, marriage and healthy parentage. His medical suggestions included the famous "Rush's thunderbolt"[780] – a mixture of calomel and jalap, of great intestinal purgative potency; those who survived it, were said to have long lives ahead of them.

At the end of the 19th - early 20th century, the work of American biologists on the theories of aging was widely recognized as ground-breaking. Thus, the cytomorphological theory of Charles Sedgwick Minot (1852-1914) gave an initial impulse for the extensive study of cellular aging.[781] The elucidation of biological mechanisms of rejuvenation was of particular interest. According to one of America's foremost researchers of rejuvenation, Charles Manning Child (1869-1954), rejuvenation, originating from cell dedifferentiation, widely occurs not only in lower animals and plants, but also in higher animals and man (1915). In lower animals, the processes of regeneration, recovery and rejuvenation can prevail, allowing for a radical prolongation of life. Thus, in the lower invertebrates, "senescence may be retarded or inhibited for a long time and probably indefinitely by the simple means of underfeeding." But in higher animals, the extent of rejuvenation is limited due to the fact of tissue specialization: "In the higher forms death becomes inevitable and necessary because the capacity for rejuvenescence is limited by the greater stability of the substratum."[782] Nonetheless, Child believed that rejuvenation and life-extension in man will be possible, but only after further extensive research into the immensely complex causes and mechanisms of aging, careful to elucidate but not to upset the delicate stability of the organism's "protoplasmic substratum" that emerged during millions of years of evolution.

Another prominent American biologist, Jacques Loeb (1859-1924), posited similar premises on the nature of aging and reached similar conclusions regarding the practicality of rejuvenation and life-extension. Loeb did believe in the potential immortality of certain types of living beings (individual cells and lower animals). He saw aging and death in higher animals and man as due to tissue specialization and deregulation, which can nonetheless be subject to intervention. He suggested several methods for experimental life prolongation, such as reduction of oxygen supply or lowering body temperature in lower animals (1917). Many of these studies were coauthored with Loeb's doctoral student, the future Nobel Laureate in chemistry John Howard Northrop (1891-1987). Loeb, however, was uncertain whether and when a rejuvenative/life-prolonging substance will be found for humans: "There is no end to the substances capable of hastening death. Shall we ever find a substance which will prolong the duration of life? At present we can neither deny nor affirm this possibility." What he was certain of, however, was that the prolongation of human life, to any degree, would only be possible if extensive, well supported research is conducted into the subject: "If this development continues to receive the support it deserves, the time is bound to come when each human being can be guaranteed with a fair degree of probability a full duration of life."[783]

These and many other contemporary American biologists – Alexis Carrel, Raymond Pearl, Herbert Spencer Jennings, Thomas Harrison Montgomery, Edwin Grant Conklin, Gary Nathan Calkins, Thorburn Brailsford Robertson, James Rollin Slonaker – agreed in the major point that aging is related to tissue differentiation and, since differentiation is only a particular state of living beings, death from aging is not a necessary consequence of life. They further agreed that the prolongation of life will be possible thanks to thorough basic research into the mechanisms of aging.

The works of the early 20th century American physicians – Dr. Ignatz Leo Nascher (1863-1944, born in Vienna and brought to New York as an infant), Dr. Malford Wilcox Thewlis (1889-1956) and Dr. Francis Everett Townsend (1867-1960) – set the foundations for clinical geriatrics. In these authors too, the possibility of a significant life-extension was affirmed and extensive further basic and clinical research was proposed as a first precondition to achieve this task. In the writings of the founder of geriatrics (and the author of the term "geriatrics"[784]), Ignatz Leo Nascher, the life-extensionist sentiment is strong. Nascher believed that "as a humanitarian it is [the physician's] duty to prolong life as long as there is life and to relieve distress wherever he may find it" (1914).[785] Nascher recognized the current limitations in the ability of medical science to prolong life, stemming from the imperfect knowledge of mechanisms of aging and of the action of anti-aging interventions (1919):[786]

> Very little is known about the action of serums, vaccines, hormones, internal secretions, electrotherapy, radiotherapy, and other non-medicinal agents upon the senile organism. Organotherapy has not given the same favorable results in senile cases as in younger individuals and the actual value of this class of medicinal agents in the aged is uncertain.

The limitations of knowledge, however, did not imply that the research of aging and longevity should be abandoned, but rather intensified. Yet, Nascher pointed out that, in the present milieu, there are strong economic and social pressures against such research (1914):[787]

> There is, however, a natural reluctance to exert oneself for those who are economically worthless and must remain so, or to strive against the inevitable, though there be the possibility of momentary success, or to devote time and effort in so unfruitful a field when both can be used to greater material advantage in other fields of medicine. Still these objections are paltry when applied to the physician's self-imposed obligation to relieve distress and prolong life.

Despite these pressures, Nascher asserted, longevity research needs to continue, as it not only carries the promise of a significant human life-extension, but also the promise of a deeper understanding of life processes in general. State support, according to Nascher, will be essential for the continuation of this research:

> There is another point of view from which the physician should consider the aged and their diseases, that of the scientist, for here is a most interesting study, presenting problems that are intimately bound up in the grand mystery of life and death. In this direction the French and German investigators are far ahead of their American confreres, not so much in the quality of work done and positive results achieved, as in the quantity, the number of investigators, the many lines of investigation, and the opportunity afforded them to carry on scientific research. There the State takes an interest in scientific work, lending its aid, and there is substantial recognition of work accomplished. The lack of opportunity to carry on research work in this country except at a heavy expense to the individual, or else at the sacrifice of the credit and benefit arising from successful research work, is probably the main reason for the neglect of the scientific study of senility and its diseases.

Yet, the purpose and promise of this research are too great to be abandoned:

In recent years considerable work has been done in blood-pressure investigations, cancer research, arteriosclerosis and other factors related to senility and its diseases. The extent and depth of these investigations, which are really studies into the causes and results of senile changes, and the ever-increasing scope of these investigations, give promise of ultimate success in discovering the fundamental causes of senescence. Perhaps there may be controllable causes or causes which can be minimized so as to defer senility and prolong life to its physiological end. The prolongation of life is after all the aim and goal of the physician's endeavors.

Thus, despite the current hardships in conducting longevity research, significant practical advances in the field are being made, and will continue to be made, provided this research is not depleted of support.

Nascher, Child, Loeb, Carrel, in fact the majority of the early 20th century American longevity researchers, were extremely cautious regarding the immediate prospects for life-extension. Yet, they believed that aging and death are not necessary consequents of all life, but rather the result of tissue specialization, errors in tissue interrelations and failures of regulation. With a more profound understanding of the mechanisms of aging, they believed, potential rejuvenative and life-extending means might be found. Thorough preliminary research, according to them, was imperative for success. But financial support for such research was permanently lacking.[788]

Beside the cautious and underfunded cohort of basic longevity researchers and clinicians specializing in old age, there existed in the beginning of the 20th century a wide-spread "rejuvenation" movement that promised immediate and far reaching results. In the late 1890s-early 1900s, the number of American nostrums for rejuvenation and life-extension was stupendous, manifesting what the historian Stewart Hall Holbrook termed "the golden age of quackery" when referring to general American medical practice (1959). Holbrook demarcated the "golden age" circa 1850 up to 1906, that is, until the ratification of the US "Pure Food and Drug Act."[789] As will be shown shortly, commercial rejuvenation flourished in "the golden age," and continued, though perhaps in a diminished degree, afterwards. In the commercial enterprises, there was no time to wait for a thorough understanding of the aging process, as money needed to be made here and now.

3. The invasion of non-invasive methods

In America of the early 1900s, "over-the-counter" rejuvenative and life-extending nostrums and health improvement systems counted by the dozens. As this work has argued, life-extensionism may have stemmed from a conservationist desire to maintain the "status quo," to preserve the homeostasis for as long as possible, perhaps indefinitely. Paradoxically, this desire to sustain the stability, may explain both the yearning for prolonged existence (life-extensionism) and the aversion to new biomedical interventions ("bio-conservatism" or "therapeutic nihilism"). As mentioned, some of the more notable methods of "rejuvenation" of the early 20th century involved the extirpation of the large intestine to eliminate the seat of putrefactive bacteria, and transplantations of sex glands (mainly from animals) to provide a new source of vital energy. However, having one's intestines removed or testicles transplanted for "life-prolongation" not only carried a tangible risk of complications, but also threatened the current perception of the body. Thus, from the simultaneous desire for perpetuation and the fear of far-reaching intrusion, there may have grown the fascination with "minimally invasive" or "non-invasive" methods of rejuvenation (many of which were "minimally-" or "non-" efficient).

The extent of attraction to "non-invasive" (now also commonly termed "holistic") methods can be appreciated from the works of Morris Fishbein (1889-1976), the editor of the *Journal of the American Medical Association* (from 1924 to 1950) and a sworn enemy of all "quackery." In *The Medical Follies* (1925) and *The New Medical Follies* (1927),[790] Fishbein encapsulated the therapeutic fashions in America at the beginning of the century, and explained the advance of "non-invasive" (holistic) health systems at the retreat of "heroic" (reductionist) medicine:[791]

> Those were days of heavy drugging. ... Homeopathy owed its initial success to the fact that it prescribed small doses of remedies in vast quantities of water, and so did not interfere with the natural tendency of the body to recover. On this tendency – the *vis medicatrix naturae* [the healing power of nature] – all of the cults of history have floated their frail vessels.... Since most of the remedies promoted [in eclecticism] have since been shown to be quite inert or utterly inadequate in the large majority of cases, the vogue of the cult must have rested on the same desire to escape over drugging that promoted homeopathy.

As Fishbein noted, the unorthodox systems succeeded largely because they often did less damage than the conventional methods.[792] Fishbein testified to the rampant flourishing of rejuvenative folk medicine in the US, with "cures" ranging from the "Indian Lukutate" (a laxative alcohol solution containing senna, buckthorn, cascara sagrada, and fruit sugar, and advertised as a "pure extract of Oriental fruits"[793]), to opotherapy or glandular extracts, to magnetic garments. Most of these systems were highly optimistic as to the possibility of combating the infirmities of old age and highly encouraging for the practitioners. Some specifically promised rejuvenation.

The lexicon of unorthodox health systems, provided by Fishbein, is very informative, including over 60 systems.[794] "Vita-o-Pathy," started in the late 19th century by Orrin Robertson of Cass County, Missouri, was perhaps the most eclectic of all, combining some 36 systems:[795]

> Prana-Yama, Zoism, Spiritual Science, Psychic Sacrology, Somnotherapy, Christian Science, Osteopathy, Chiropathy, Divine Science, Botanic, Allopathy, Biopneuma, Prayer Cure, Rest Cure, Diet Cure, Eclecticism, Hydropathy, Magnetism, Phrenopathy, Nervauric Therapeutics, Electro-Therapeutics, Chromotherapy, Vitapathy, Homeopathy, Psychopathy, Magnetic Massage, Faith Cure,

Biochemic System, Therapeutic Sarcognomy, Physio-Medicine, Mechanical Therapy, Suggestive Therapeutics, Auto-Suggestion, Tripsis, Spondylotherapy, Chirothesia.

Special chapters in Fishbein's books were dedicated to Electrode Therapy (starting with Elisha Perkins' (1741-1799) electric "tractors," patented in 1796, with ample future progeny), Homeopathy, Osteopathy, Chiropractic, Electronic Healing (Electronic Resonator of Albert Abrams, 1863-1924), Birth Control, Antivivisectionism, Rejuvenation, Physical Culture, Body Building, the Cult of Beauty, Reduction (slimming diets), Bread (whole grain) and other "Dietary Fads" (such as acidified milk products), Eclecticism, Physical and Electric Therapy, and Psychoanalysis.

    In these therapies, in order to achieve a healing balance, a great emphasis was placed on the regulatory, revitalizing power of the mind or faith healing, insofar as mind and faith were presumably the entities least changed by material "intrusions." A great weight was given to hygienic living (baths, healthful diets, rest, sleep, exercise) alongside a panoply of other minimally invasive treatments (herbal remedies, magnets, massages) – at low production costs and profitable for the independent provider, psychologically unthreatening and affordable for the autonomous middle-class consumer. All these health and longevity systems were a direct offshoot of the American free-market economy. They were all boldly commercial, involving extremely aggressive (and sometimes fraudulent) advertising and marketing techniques, and flourishing under the contemporary slack regulation of medical accreditation and practice. Almost anyone with sufficient means could receive some sort of a medical diploma from a private university and go into the healing and life-extending business.

4. Capitalism, Religion and Eugenics

One of the more successful fin-de-siècle life-extensionist enterprises, emphasizing non-invasive methods, was that of Dr. John Harvey Kellogg (1852-1943), since 1876 head of the Battle Creek Sanatorium in Michigan. Kellogg's 'high-tech' recreation facilities employed state-of-the art hygienic regimentation (exercise and rest), physiotherapy (emphasizing colonic cleansing, light and air baths) and dietetics (recommending probiotic, vegetarian and low-protein diets). Being a devout member of the Seventh-Day Adventist Church, Kellogg's suggestions for life-extension were perfectly fitted to Puritan religious values. A major focus was on enforcing sexual moderation and purity. This, according to Kellogg, was a primary means for conserving the vital energy, since "the development of the sexual instincts perfectly corresponds with the longevity." The task involved the promotion of late marriages and eradication of masturbation.

Closely related to the question of sexual purity, was the correct selection of parents. This too, according to Kellogg, played a crucial role for life-prolongation. When discussing the question "Who may marry?" – Kellogg asserted that "the marriage of diseased persons and kindred violations of the laws of human hygiene have been not unimportant factors in producing this most appalling diminution in the length of human life" (*Plain Facts for Old and Young*, 1879).[796] The term "eugenics" was coined somewhat later, by Sir Francis Galton in 1883, in *Inquiries into Human Faculty and its Development,* and eugenics became entrenched in America after the publication of Charles Benedict Davenport's books *Eugenics: The Science of Human Improvement by Better Breeding* (1910) and *Heredity in Relation to Eugenics* (1911). Both Galton and Davenport expounded on the relation between the selection of appropriate parents and longevity of the offspring.[797] Yet, Kellogg's work of 1879 contained an earlier expression of the same idea. Throughout his career, Kellogg remained a strong supporter of eugenics, as he believed that by measures of "race betterment" human life can be prolonged to twice its current value. In 1915, he wrote:[798]

> The Committee on Conservation of Natural Resources appointed by President [Theodore] Roosevelt [established in 1909, terminated in 1921] did not neglect to consider the greatest of all the national assets – human vitality. This committee pointed out the surprising fact that the average man is only fifty per cent efficient; that we live out less than one half the natural duration of life, that we consume more food than is needed to maintain efficient life, and that one half of all human beings born either die before reaching maturity or fall into the defective, dependent, or delinquent classes. Special study and effort is now being made to prevent this terrible loss to the nation in human vitality and efficiency by the study of methods of race betterment.

Increasing the healthy life-span was, in turn, seen as a primary means for increasing capitalist efficiency and fostering contentment among all the classes of society:

> The ability to do and to endure, to keep on doing what one finds profitable, useful, and agreeable, is the very essence of personal, social, and national well being. ... Statesmen, professional men, business men, leaders in industry and politics, and workers of all classes are asking, how can I increase my efficiency? …. The conservation of health means increased prosperity and happiness. It means, as we have seen, the ability to do more work and to do it better.

In 1906, Kellogg, together with Prof. Charles Davenport and Prof. Irving Fisher (1867-1947, the future president of the American Life Extension Institute) founded the "Race Betterment Foundation," where, among other topics, life-extension was a primary purpose. In the late 19th-early 20th century America, life-extensionism, it would seem, was inseparable from the broader eugenics movement. Moreover, in John

Kellogg's works, life-extensionism primarily derived from the currently dominant socio-ideological fashions – Puritan values (particularly in relation to sex), eugenics, and capitalism – in a peculiarly American blend.

Kellogg's other recommendations for life-prolongation were in perfect alignment with the precepts of his church. He advocated abstinence from alcohol and tobacco, as he believed the human body is a "Living Temple" which becomes corrupted by these poisons.[799] The conduct of life, he preached, should be religiously "industrious" and "simple": "All examples of extraordinary longevity which have been reported have been of persons who had led active, even laborious, lives, and whose habits in diet and in other respects were simple and regular."[800] Diet, according to Kellogg, must be "moderate," "simple," "wholesome" and "vegetarian" as that of the meek laborers and true Christian devotees: "The examples of great longevity are nearly all to be found in the lowly ranks of life, among peasants and laborers, persons of simple and temperate habits. Many of them have lived on a simple diet of bread, milk, and vegetables."[801]

Among the means for life-extension, nothing, according to Kellogg, could compare to a proper dietary regimen, if only it were known and followed: "There is every reason to believe that if these [probiotic] acid formers retained undisputed possession of the intestinal tract the span of human life would be extended very greatly beyond the present age limit and humanity would be saved from a vast multitude of physical, mental and moral disorders and miseries."[802] Man "should have lived to the age of one hundred years at the least"[803] and Kellogg himself intended to live at least a century (although he died 9 years short of the term).[804] Generally, the grand task of radical life-extension, involving the development of and adherence to the correct diet and codified rules of behavior, was, according to Kellogg, perfectly compatible with a Christian worldview: "The greatest problem before the human race during all the ages of its conscious history has been how to defeat old Father Time in his machinations against human life, how to postpone to the latest possible moment the fulfillment of the ancient fiat, 'Dust thou art and unto dust shalt thou return.'"[805]

The search for perfect, health-sustaining and life-prolonging foods led to Kellogg's invention of cornflakes (in 1894), which gave rise to Kellogg's food empire. Kellogg's life-extensionist enterprise was firmly grounded on a commercial basis. The Battle Creek sanatorium was a premium resort for the wealthy, its patients including the US president William Howard Taft, the writer George Bernard Shaw, the industrialist Henry Ford, the inventor Thomas Alva Edison, and more. With the establishment of Kellogg's "Sanitas Food Company" in 1897 to produce whole grain cereals, the commercial element became pervasive, later involving a fierce and prolonged litigious feud between John Harvey and his brother Will Keith over proprietary rights.

Another highly commercially successful (and "minimally invasive") life-extensionist enterprise was that of Webster Edgerly (1852-1926, also known by his pseudonyms, Edmund Shaftesbury and Everett Ralston), a pioneer of the American self-help book genre. In 1876, he founded the "Ralston Health Club" which grew into "Ralston Company." In 1895, he bought an estate intended to house a community of his followers at Ralston Heights, New Jersey, presently Hopewell. And in 1900, he co-founded the "Ralston Purina Company," which later gave rise to "Ralcorp Holdings," presently one of the major American food corporations. At its peak (around 1905), the "Ralstonist" movement boasted of tens, up to hundreds, of thousands of members.

The fascination with super-longevity is well exemplified in the publications of the Two Hundred Years Method (TYM) Club, "founded upon the plan of universal life" by Edgerly in 1889.[806] According to the TYM credo: "If human life were to be prolonged for two hundred years it must be maintained through the continual and ceaseless use of the Natural Laws which are set forth in the methods of this organization." Edgerly listed many examples of super-centenarians and, even though some doubts were cast as to their authenticity, at least some of them were deemed to be authentic. These furnished proof of the attainability of the 200 year mark, since, Edgerly implied, even the lives of super-centenarians, such as Thomas Parr (allegedly 1483-1635), Henry Jenkins (allegedly 1501-1670), Peter Zarten (allegedly 1539-1724) and others, were cut short by pernicious influences that could be avoided. From studying the super-centenarians' life-histories – the majority were said to be "simple people" who lived a "frugal" and "industrious" life – the TYM activists derived a longevity regimen based on Puritan self-discipline.

Religion, capitalism and eugenics were the constitutive elements in Edgerly's life-extensionist philosophy. His "Ten Natural Laws" for life-prolongation were modeled on the Ten Commandments.[807] The "Laws" were clearly influenced by contemporary vitalist chiropractic theories. The 10 "laws" and the corresponding 100 "edicts" included regulations for "vital support" (the right way to stand and walk); poise; respiration; "chest growth"; protecting the "life zone" (chest and torso area containing "vital organs," here one of the curious recommendations was to avoid washing the chest with water, but instead cleanse it with antiseptics); cleansing and exercising the "physical zone" (i.e. the rest of the body, "empty of organs"); drink (for example, recommending "vitalized water" which is "first made pure by distillation; then is given life by exposure to pure bacterial air"); "wholesome" food; strengthening capillary functions; and action ("activities should make sedentary habits impossible"). The regimen particularly emphasized the conservative Christian values of "self-denial" and "self-restraint," cultivation of "energy of the will and determination." Religion, according to Edgerly, was a primary motivation for life-extension. By strictly following the life-extensionist regimen, the proponent was "obeying the mandate of the Creator." "A person of fixed habits," Edgerly maintained, "is necessarily the superior of those whose habits are haphazard. One is nearer to Creator, the other is the waif of circumstance." Furthermore, "the longest lives have been given to men and women who have combined service to God with reasonable habits of living."

Capitalism was another primary constituent of Edgerly's philosophy. According to Edgerly, capitalism is the natural, God-given order of society and, as such, will continue for all posterity: "It has been many times said that, if there were to be an even distribution of the wealth of the world, the same conditions that now exist would soon be re-established, for the wealth-taking class will take wealth wherever it is." It was the task of the followers of the Two Hundred Year Club to be stalwart, long-enduring soldiers of industry, to achieve "progress in health" and a high degree of "material prosperity," while the capitalist society indefinitely continues on its course of progressive development "until the solemn march of time brings humanity to a glorious reward." Beside the theoretical adherence to capitalism, Edgerly's intention to make money is transparent. Much of the Ralston Company's income came from selling the self-help books. Hence, Edgerly devised a deliberate scheme of "steps" and "degrees" within the club hierarchy to promote the book sales. Entering or completing a "degree" basically meant buying a book, and involved a sort of a "pyramid" where those who bring new recruits would receive a discount.

Edgerly's works were concerned not only with life-extension, but also with other concomitant aspects of human development, each aspect constituting a different "step" or area of personal progress. Physical health and longevity were considered just the basis for all the other "great departments of human life": "1. The Physical for the Body. 2. The Mental for the Mind. 3. The Electral for the Nervous Life. 4. The Sexual for the Correlative Nature. 5. The Ethical for the Moral Nature."[808] The discussion of personal betterment could then not be far from the general tenets of Eugenics or "The Science of Human Improvement." Indeed, in the rulings of the Two Hundred Year Club, openly eugenic (selectionist) or racist views are hardly perceptible. Yet, elsewhere Edgerly made some very revealing statements on the subjects, such as "Watermelons are poisonous to most Caucasians"; "The yellow race are seeping in very rapidly" or "Make cordons round the pest sections of the cities."[809]

Yet another branch of (minimally invasive) American life-extensionism was concerned with climate and real estate. Whereas Florida was discovered around 1513 by Juan Ponce de León (1474-1521) in his search for the "Fountain of Youth," California was considered by many "the promised land" of longevity. Since its discovery around 1542 by João Rodrigues Cabrilho (1499-1543), prominently throughout the 19th century, and up to the beginning of the 20th century, there existed an extensive literature commending the life-prolonging qualities of the Californian climate.[810] Thus, for example, the San Francisco physician Marion Thrasher (1842-1926), in *Long Life in California* (1915), asserted that California held records of longevity for virtually all life-forms. Trees there grow the tallest and live the longest, the record-holding Redwood

(Gigantic Sequoia) being over 350 feet tall and 5,500 to 8,000 years old. Animals there grow the largest and live the longest. Children develop the fastest. While, according to Thrasher "the normal limit of man's life is 125 to 150 years," the best chances to reach and exceed it are in California. "Indians here 185 years old live 60 years in other states." California has "by far the lowest death rate … in the United States or in any other country in the world, and the highest average length of life" (India, 20 years; England, 37; Massachusetts, 47; California, 74). Among "the whites" of California, there are "more centenarians here than in any other State or country." The general explanation for this alleged phenomenon was the *mild* and *constant* climate of the "Golden State."[811] (Curiously, life-extensionist authors who lived in colder areas, such as the German Christoph Wilhelm Hufeland, asserted that the sturdy northern climate is more conducive to longevity, which was also supported by ample testimonial and circumstantial evidence. Also, the life-prolonging benefits of the sea-shores versus the mountains diverged radically, depending on the altitude near which the authors lived.) The adjustment of the life extensionists to their particular "milieu" was absolute, even in terms of geography.

There were considerable implications of the belief in the singular life-prolonging qualities of the Californian climate for attracting migrants and real-estate development (also according to Thrasher, in California "money is made without effort").[812] Dr. Peter Charles Remondino of San Diego (1846-1926), in "Climate in its Relation to Longevity" (1890), even argued for a reduction of life-insurance premiums in Southern California, due to its astonishing salubrity.[813] Certainly, the current position of California as a world stronghold of life-extensionism, with a vast host of anti-aging research and development facilities, clubs and clinics, stems from a long tradition.

As in the other cases, capitalism, eugenics and religion figure prominently in Thrasher's philosophy. Without an extensive eugenics program, Thrasher believed, "we cannot hope for a material increase in the number of centenarians." According to Thrasher, eugenic selection will relieve the economic burden imposed by the sick and aged: "Eugenics should be formulated into legal enactments of the state and made as other laws, compulsory requirements. Then would we no longer need almshouses, for all could take care of themselves, nor penitentiaries, for there would be no criminals nor perverts, nor hospitals, for none would be sick." The cause of eugenics was sanctified by religion. Allowing the marriage of and care for "the deformed and the cripple" would be ungodly: "to charge God with all this miserable work would certainly be impeaching His wisdom."[814]

5. The rejuvenators

In contrast to the life-prolonging life styles, that were "minimally invasive" into one's own body, there existed the more "invasive" techniques of rejuvenation. Rejuvenative opotherapy, using glandular preparations, particularly sex gland extracts, to retard aging, flourished in the US since the late 19th century. The modern period in rejuvenative opotherapy can be traced back to the ground-breaking address to the French Biological Society of 1889 by Charles-Édouard Brown-Séquard (1817-1894, whose father, incidentally, was an American).[815] In that address, Brown-Séquard reported the rejuvenating effects of injections of animal testicular extracts on himself, and thereby established the field of therapeutic endocrinology. Yet, in that address, he also admitted that similar preparations had been used by American rejuvenators earlier:[816]

> In the US especially, and often without exact knowledge of my experiments, without the most rudimentary consideration of rules regarding subcutaneous injections of animal materials, many clinicians but more often quacks and charlatans have exploited the deep desire of a large number of individuals and have exposed them to the greatest risks at the least, if they have not done worse.

The use of sex gland preparations for rejuvenation, and the more general association of sexual energy with the life-force and longevity, could, of course, be traced at least two thousand years further back to Chinese Taoist external and internal alchemy (and major European rejuvenators, such as Steinach and Voronoff, were well aware of this tradition). Yet, as regards the modern phase of this method, Brown-Séquard's work, even though pioneering in its own right, testifies that rejuvenative opotherapy and, by extension, therapeutic endocrinology, had been at least partly rooted in the late 19th century American rejuvenation enterprises.

The extent of commercialization of rejuvenative opotherapeutic preparations in that period in America was truly astonishing. According to a contemporary account, by the early 1900s, medical practitioners "administered cardin for heart disease and nephrin for kidney trouble, cerebrin for insanity (save the mark!), and even prostate tissue for prostatism – and with reported good results!"[817] One of the more famous and early operators was Dr. William M. Raphael of Cincinnati, Ohio, since 1858 selling the rejuvenating, potency-enhancing tonics, the "Cordial Invigorant" and "Galvanic Love Powder." In Raphael's sales-pitches, the recipe for the "invigorant" was procured from the Bedouin sheik Ben Hadad "109 years old." Another successful enterprise that flourished toward the end of the 19th century was that of Fred A. Leach, from San Diego, California, owner of the "Packers Product Company," peddling the "Orchis Extract" – "a substance from the testicles of rams." Perhaps the most dramatic story was that of the "Animal Therapy Company," that produced "lymph compounds" and was operated by James M. Rainey and Louen V. Atkins. On September 16, 1910, Rainey shot Atkins dead, after a quarrel over a five-dollar bill.[818] A multitude of other, perhaps less dramatic but nonetheless widely spread, rejuvenative endeavors were operated through mail-orders, walk-in trade at "anatomical educational museums" all across the US, local pharmacies and traveling salesmen. Whatever is one's perception of "quackery," rejuvenative endocrinology clearly received a boost from this unregulated, rampant proliferation of rejuvenating "free-enterprises."

Americans were also among the first to conduct operations on sex glands – vasectomy and sex gland transplantations – for rejuvenative and other therapeutic purposes.[819] Vasoligation was first used for the sterilization of prisoners at the Indiana Reformatory at Jeffersonville in 1899 by Harry Clay Sharp (1870-1940) to fight "sexual over-excitement" and "excessive masturbation" of the inmates. This libido-damping effect was in exact contradiction to Steinach's later rationale for the rejuvenative action of vasoligation (proclaimed in the 1910s-1920s), according to which a major improvement from the operation was believed to be an increase in sexual drive and power. The later adopters of Steinach's rejuvenative vasoligation argued that the loss of sex drive in Sharp's procedure could be explained by accidentally destroying testicular blood vessels during the sterilization, which was carefully avoided in Steinach's procedure.[820] Also in the classical Steinach's vasoligation, usually only one spermatic duct was ligated, to avoid infertility. Yet, Sharp's earlier work also reported an increase in the inmates' vitality, which was attributed to the retention of the seminal fluid following the vasectomy. In fact, Sharp explicitly quoted the studies on testicular extracts by earlier rejuvenators – Drs. Brown-Séquard (1889, France), Poehl (1890, Russia) and Zath (1896, US) – as providing direct theoretical support for the invigorating effects of his operation.[821] One of the staunchest supporters of Steinach's rejuvenative vasoligation in America was Dr. Harry Benjamin (1885-1986, born in Berlin, and settling in New York in 1914). Practicing the operation in New York and San Francisco in the 1920s, Benjamin reported some of the best results of Steinach's vasoligation (65 to 75% "successes" in "revitalization"), obtained on the largest clinical material (hundreds of cases, the sample size comparable to that of Steinach himself).

Rejuvenation by the transplantation of sex glands (male and female) was also practiced in the US. This method of rejuvenation is chiefly associated with the name of the French-Russian Voronoff, whose fame peaked in the 1920s. Yet, earlier (and apparently the first) testicular gland transplantations in humans were conducted in the 1910s by the Americans: Victor Darwin Lespinasse (1911), Levi Jay Hammond and Howard Anderson Sutton (1912), George Frank Lydston (1914) and Max Thorek (1919).[822] In 1919-1920, Dr. Leo Leonidas Stanley (1886-1976) performed sex gland transplantations in hundreds of prisoners of the San Quentin penitentiary, California.[823] At about the same time (1918-1920) John Romulus Brinkley of Kansas (1885-1941), began transplanting goat testicles into humans for rejuvenation (the technique was severely criticized by the transplanter of ape testicles, Voronoff).[824] Despite the constant assaults by the stalwarts of the American Medical Association (such as Morris Fishbein) and even by other rejuvenators (Voronoff) who thought Brinkley went 'too far,' Brinkley was one of the biggest entrepreneurs in Kansas. He owned the "Kansas first, Kansas best" (KFKB) radio station, one of the first to exploit radio advertising, purchased the baseball team "Brinkley Goats," and ran twice for Kansas governor.[825]

As discussed in the previous chapters, sex gland stimulation, supplementation and replacement formed a main-stem of the Western life-extensionist movement of the 1900s-1920s. And contemporary American medical practitioners were at the forefront. Fishbein grouped "rejuvenators," using glandular extracts and transplantations, with faith-healers, hygienists and mild drug and physical therapists. Yet, as it appears, they constituted a distinct and powerful stream within life-extensionism. They did not shun surgical intrusions, yet unlike many conventional surgeons, instead of "subtraction," or in Metchnikoff's terms, "destroying the agents of damage," they put their faith in "supplementation," in "strengthening the noble tissue" through organ and tissue replacements or regeneration. Moreover, their methods were also distinct from the "non-invasive" (holistic) life-extension methods in their reductionist approach, deemphasizing the psychosomatic power and targeting a single component (gland or hormone) in the human machine. The practical failure of the latter methods may have adversely impacted on the general perception of reductionist rejuvenation. Some "rejuvenators" falsely promised immediate and outstanding effects, and for that were often branded as "quacks."[826] But many presented their techniques as the first experimental stages in a research program for the future, establishing the grounds for transplantation surgery and therapeutic endocrinology.

The rejuvenative enterprises in the US followed the contemporary dominant social and ideological fashions, showing the three recurrent components: commercialism, religious sentiment and eugenic beliefs. The examples above show the lengths to which some of the rejuvenators were willing to go to earn a living. The struggle for earnings often involved aggressive (and sometimes fraudulent) advertisements and fierce partnership quarrels. Such a rampant proliferation of rejuvenative nostrums may have been possible 'only in America,' under the contemporary slack regulation and the prevailing spirit of free enterprise. Religious motives were often used in advertising, involving testimonials from (often anonymous) high standing clerics as particularly trustworthy. In the broadcasts of Brinkley's KFKB radio station, for example, religious sermons, gospel music and advertisements of religious items, occupied a central position, in turn lending Brinkley social respectability. As for eugenics, selectionism indeed played a very minor (if any) role in the sex gland supplementation business. Sex gland operations, however, had a major relation to eugenics. Sharp's sterilization procedure was the first of its kind, and hence a defining event in the history of American eugenics, as well as in the history of rejuvenative techniques. To recall, by the mid-1920s, eugenic legislation, including compulsory sterilization of criminals, "feeble-minded" and other "defectives" were passed in 33 states, with estimated 60,000 sterilized victims.[827]

6. The triumph of the medical establishment

The proliferation of rejuvenative free enterprises did not remain unchecked for long. Since the early 1900s, the American medical establishment was beginning to close down on "quackery." The counter-offensive was spearheaded by the journalist Samuel Hopkins Adams (1871-1958), beginning in 1905 with a series of articles in *Collier's Weekly*, tellingly entitled "The Great American Fraud."[828] To stall the dissemination of unproved remedies, in 1906, the first national food and drug regulation laws were passed. In June 1906, President Theodore Roosevelt signed the "Federal Pure Food and Drug Act" which became law on January 1, 1907. The food and drug regulations were further extended and made more stringent in 1938 by President Franklin Delano Roosevelt.[829]

Moreover, in 1910, the famous report by Abraham Flexner (1866-1959) on "Medical Education in the United States and Canada," sponsored by the Carnegie Foundation (founded in 1905 by the steel magnate Andrew Carnegie, 1835-1919) condemned unorthodox and "unscientific" medical practices and teachings. According to the Flexner Report, "the chiropractics, the mechano-therapists, and several others… are unconscionable quacks," while homeopaths, eclectics, physiomedicals and osteopaths are "medical sectarians" thoroughly incompetent and unscientific. Flexner condemned through and through the training institutions of the "medical sects": In case they are rich, they are "mercenary" and "fairly reek of commercialism"; and in case they are poor, their financial condition demonstrates "the utter hopelessness of the future of these schools."[830] In either case, they for the most part need to be closed.

Following the adoption of the Flexner Report's recommendations by the American Medical Association's "Council on Medical Education" (established in 1904), by 1920 the number of "Medicinae Doctor" (MD) degree granting institutions was reduced from 160 as of 1904 to 85, and the number of medical students decreased from more than 28,000 to 14,000.[831] The rejuvenative/life-extensionist movement could not have remained unaffected, as many rejuvenators belonged to either the chiropractic/osteopathic schools (e.g. Edgerly), or homeopathic/drug-eclectic persuasions (e.g. Leach). From 1906 onward through the 1930s, the "Propaganda Department" (renamed in 1925 the "Bureau of Investigation") of the American Medical Association was constantly cracking down on rejuvenative preparations – the "Lukutate" containing "100% Oriental Fruits"; Clayton E. Wheeler's "gland rejuvenation" extracts; Matthew and Herman Richartz's "Ekater-Enger orchic substance", etc. – banning them from mails and celebrating the victories on the pages of the *Journal of the American Medical Association*.[832]

Coinciding with the partial suppression of practical rejuvenative 'folk' medicine in the 1900s-1910s, the "basic-scientific" branch of life-extensionism became reinforced. Since the late 1900s through the 1910s, extensive basic research on aging and experimental life-extension was conducted at the Rockefeller Institute for Medical Research (established in 1901 in New York by the oil magnate John Davison Rockefeller, 1839-1937, under the direction of Simon Flexner, 1863-1946, the older brother of Abraham Flexner of the "Flexner Report"). The life-extension studies at the Rockefeller Institute, most notably those of Alexis Carrel, Jacques Loeb and Raymond Pearl (the latter was funded by the Institute), intensified in the 1910s, as will be discussed in greater detail later on. Also in the 1910s, the field of geriatrics was established by Ignatz Leo Nascher, emphasizing the crucial need for extensive basic studies of "senile changes" that "give promise of ultimate success in discovering the fundamental causes of senescence" and "prolongation of life."[833] The theories of aging and rejuvenation by Charles Sedgwick Minot (1908, 1913) and Charles Manning Child (1915) also emerged after the "Pure Food and Drug Act" of 1906.

Of course, practical rejuvenation did not stop, and basic theoretical and experimental longevity research did not start in 1906. As early as 1859 through 1870, several American authors (such as Stanford Emerson Chaillé and Samuel Prentist Cutler) were fascinated by the "mineralization" theory of aging (a.k.a. accumulation of "alkaline residues," "calcification" or "ossification" theory, accompanied by the suggestion of "vegetable acid cleansing" therapy) proposed by Édouard Robin of the French Academy of Sciences in

1858.[834] Since 1888 through the 1900s, Charles Asbury Stephens theorized on aging and dying as being a result of disruption of cell "cooperation" and even conducted some experiments on cell and tissue cultures (see below). In these early works, however, basic longevity research hardly went beyond pure theorizing. After the "Pure Food and Drug Act" and the "Flexner Report," basic theoretical and experimental longevity research received a much greater weight, displacing the earlier attempts at immediate rejuvenative interventions. This cycle seems to have repeated in the 1920s-1930s. The European boom of endocrine rejuvenative operations of the 1920s (vasectomy and sex gland transplantations) was rapidly reproduced in the US, with Harry Benjamin and John Brinkley rising in notoriety. Yet, as the effectiveness of these new rejuvenative operations fell far behind their promise, a counter-reaction again occurred in the 1930s toward basic theoretical and experimental longevity research, as will be exemplified in the following sections.

7. The life-extensionist as elite scientist – Charles Asbury Stephens (1844-1931)

One of the early proponents of basic longevity research was the man of letters and physician Charles Asbury Stephens. In 1888, funded by an enterprising philanthropist, he established in Norway, Maine, a laboratory, exclusively dedicated to promoting "researches into the causes of old age and death" – apparently among the first in the United States – which he hoped to transform into a large-scale institute and a center of international cooperation for aging and longevity research. The institute funding ceased in 1905 when the philanthropist passed away.[835] Stephens proposed a wide variety of research programs for potential life prolongation, ranging from "concentrating nervous energy" to improving sleep to finding perfect nutrition that would not generate life-clogging waste products.[836] His actual laboratory research, however, seems to have mainly consisted in reproducing the cell culture experiments of the French biologist Émile François Maupas (1888),[837] that is, observing the death of cell colonies and their rejuvenation by cell conjugation, and attempting to postpone the cells' death by modifications of their nutritive medium.

Yet, Stephens' visions went much farther. For Stephens, individual physical immortality, even resurrection, appeared to be distinct possibilities. In *Natural Salvation: Immortal Life On the Earth from the Growth of Knowledge* (1910; first published in 1903),[838] Stephens argued that ultimate destruction is not an inexorable law of biology, not even of physics: the "cell-of-life" is potentially deathless, and material "elements" are virtually unchangeable. The possibilities of indefinite maintenance of human personality and resurrection (similar to the views of his Russian contemporary, Nikolay Fedorov) were derived from the concepts of Lamarckian inheritance and the "Ether of Space."[839] Past personalities were said to be inherited (remembered) in the progeny, as they lie "dormant in the brain of their descendants," and can be reawakened. Furthermore, human thoughts and memories form an imperishable physical trace, an "echo" or "mirror-picture" that are "present in the ether everywhere" and can be recaptured in some distant future.[840] The concepts of life extension, physical immortality, symbolic immortality and resurrection were thus synthesized:[841]

> [We] may possibly know a species of resurrection, if our descendants shall desire to call us up… More than this we cannot yet hope. … Enough, till the grander day comes when our children, transfigured and perfected in their organisms by the growth of knowledge, shall cease to die. But even in that grander day we shall be with them. … And if we have worked for that grand day, they will love us. *Morituri*. But that thought is our compensation, our solace in death.

Stephens recognized the immense technical difficulties involved in the program of radical life-extension or revival, and allocated a few hundred years to accomplish the tasks. Still, dedicated, collective effort, he believed, could substantially accelerate the progress. Thus, Stephens substituted the program of prolonged basic research in place of attempts at immediate practical rejuvenation. Interestingly, the expanded and widely disseminated edition of Stephens' *Natural Salvation* appeared in 1910, after the retreat of "rejuvenative medicine," in line with the apparent general trend toward a strengthening of basic research. Yet, unlike most American longevity researchers of the time, Stephens perceived the promise of basic research to be limitless.

The social and ideological underpinnings of Stephens' work also reflected the currently prevalent socio-ideological fashions. The themes of capitalism, religion and eugenic-type elitism are present in his works, though modified. Stephens' dependence on philanthropic donations from industrial enterprise was apparent. (Notably, practitioners of rejuvenation also depended on entrepreneurship, but in a more direct way, that is, on their own entrepreneurship.) Accordingly, the way to human enhancement and longevity envisioned by Stephens was perfectly attuned to American industrialism. Stephens valorized global "commerce," flourishing under the American leadership, as the foundation for progress and cooperation, whereby "the fruits and goods from every quarter of the planet are brought to our doors; ... the land [is] overspread with wires, which put us in thought-touch with our fellows, thousands of miles away."[842] Notably, not only did Stephens emphasize that life-extensionism is as traditional as human existence, but he also expressed the conservative sentiment that the present pattern of American capitalist economy would be strengthened and perpetuated.

Stephens' attitude to established religions was predominantly negative. Instead of the old creeds, he proposed a new religion, the "The Promethean Faith," the creed in which salvation through faith and supernatural intervention is substituted by salvation through biomedical science. The term "salvation" was understood literally as an eternal rescue from death. According to Stephens, religion based on supernaturalism is a dead and deadening concept, for it displaces human hopes toward the afterworld and curbs the effort to prolong this worldly life. In contrast, the "Promethean faith" is the religion of life preservation, as ancient as "the instinct effort of the protozoon to save itself." Compared with this natural effort at salvation, "the World's five great Creeds are as novelties of yesterday."[843] Yet, Stephens felt obliged to pay homage to the Judeo-Christian religious tradition, and derived from it the inspiration for the grand task of revival: "The strata-ed, human brain-of-man," he wrote, "is our *sheol*" (the storing place for the memories of past generations awaiting revival). Thus, the "Hebraic doctrine of the resurrection" (cf. Maimonides' 13th article of faith in resurrection[844]) obtains, in Stephens' view, a realistic interpretation and becomes a subject of human study and effort. In Stephens' neo-Lamarckist belief in the inheritance of acquired personal traits, "The Hebrew woman's prayer 'Give me children or I die'" underscores a biological truth. Furthermore, Stephens proposed a rationalist explanation for psychic, spiritual or even miraculous occurrences, recorded in religious traditions and taken by Stephens to have a factual basis. As human thoughts and memories form an "echo" in the "ether of space," Stephens proposed that the spiritual phenomena must be redefined using the term "etheric."[845]

Stephens was a self-proclaimed "constructive socialist" (which somewhat clashed with his valorization of "commerce"). He, as in fact the majority of life-extensionists across the world (Fedorov, Metchnikoff, Voronoff, Steinach, Shaw, Bogdanov) perceived the quest for extreme longevity as a common, unifying task of humanity, a quest toward the harmonization and stabilization of the biological and social organism. He wrote:[846]

> What the cell-of-life is in the animal organism the individual citizen is in the nation.... Better, longer, happier life came to the unicell from organization in the multicell. Greatly prolonged life, looking toward deathless life, will come to the human multicell from cooperative organization in the greater, stronger life of united humanity, or that portion of it which can be brought to recognize the truth and act together.

However, more than a slight tint of elitism and racism is present in Stephens' writings:[847]

> The burden of progress and achievement will long rest with the dominant race. Certain of the lower races, like the lower animals, will of necessity be coerced for the general good and for their own good. ... With the Indo-European rests not only the responsibility to do right for the world, but the duty of seeing to it that others do right. This is, in very truth, "the white man's burden," the gravest of all responsibility.

Apparently, the "cooperative organization" Stephens wished to promote was the type of organization entrenched in his own time and place.

8. Basic longevity research. Immortal Soma – Jacques Loeb (1859-1924), Leo Loeb (1869-1959), Alexis Carrel (1873-1944), Raymond Pearl (1879-1941)

In the late 1900s through the 1910s, basic longevity research in the US continued with an increasing intensity. According to Stephens, international, or at least national research cooperation would have to exist in order to achieve success in this area. "Nothing like economic division of labor in scientific research, or in the way of mutual aid," he lamented, "has yet been attempted."[848] Stephens' laboratory was intended to instigate such cooperation, which however did not seem to come to any notable fruition. Yet, Stephens' vision was beginning to be realized later on. A series of concerted, penetrating and well-funded studies at the Rockefeller Institute for Medical Research in New York (established in 1901)[849] concerned potential immortality and life prolongation. The institute became a center of what may be termed the "somatic immortalist" school of longevity research, positing the potential immortality of somatic (body) cells and tissues (as compared to reproductive/"germinal" cells and tissues whose immortality was taken for granted).[850] The institute's founder, John Davison Rockefeller (1839-1937), the famous oil tycoon, the richest man in America and its greatest philanthropist, as a young man allegedly set to himself the goals of making $100,000 and living to 100 years. He exceeded the first goal ten thousand fold, but came about two years short of the second, despite his best efforts.[851] Not surprisingly then, besides combating a wide range of diseases – such as hookworm and yellow fever – a considerable portion of the Rockefeller Institute basic research concerned longevity and immortality.[852]

A principal investigator of these subjects was Jacques Loeb (1859-1924), a German-born Jew of Portuguese and Italian origin, who was educated at the Universities of Berlin, Munich and Strasbourg, and immigrated to the US in 1892.[853] In 1910, he became the director of the Department of Experimental Biology at the Rockefeller Institute. Another principal figure was Alexis Carrel who moved in 1904 from Lyon, France, to Chicago. In 1906 Carrel was attached to the Rockefeller Institute and became its full member in 1912. The third was the American-born Raymond Pearl (1879-1941) who, for most of his career, worked at Johns Hopkins University in Baltimore, but also cooperated at the Rockefeller Institute and was funded by the Rockefeller foundation. Another investigator, though not formally affiliated with the Rockefeller Institute, was Jacques Loeb's brother Leo Loeb (1869-1959), who was educated in Zurich and held positions in the universities of Illinois, Princeton and Washington University in Saint Louis. A "Westward Drift" of life-extensionists can be observed: for Carrel, Jacques and Leo Loeb, America provided richer opportunities than France and Germany.

Jacques Loeb's experiments on parthenogenesis, or the production of organisms, such as sea urchins and even more complex animals like frogs, from the egg without the involvement of sperm, showed that by simple inhibitors (such as lack of oxygen, low temperature, potassium cyanide - KCN, or chloral hydrate in proper concentrations), the life-span of the egg could be prolonged several times. Moreover, by changing the chemical composition of the medium, according to a straightforward protocol, eggs were induced to develop into a complete organism. It was therefore inferred that the conditions for potential immortality are purely physicochemical conditions, subject to manipulation.

According to Jacques Loeb, cellular immortality can easily account for the long life-span of organisms (such as the giant sequoia that may live thousands of years), and it is not the phenomenon of longevity that needs special explaining, but rather why the organisms die at all. Similarly to Metchnikoff, Jacques Loeb believed that "natural death" seldom occurs in nature, and in higher animals and man death is principally due to external damage by "microorganisms or other injuries to vital organs" and particularly the "cessation of respiration."[854] Immortality of higher organisms was, according to Jacques Loeb, well within theoretical possibility, since "the natural death of the metazoa is perhaps a secondary phenomenon due to the cessation of respiratory motions or of the heart beat. ... If respiratory motions and circulation could be maintained indefinitely even the metazoa might be found to be immortal."[855]

Further evidence for potential immortality came from Alexis Carrel's experiments on tissue culture. (Carrel's extensive longevity research is discussed in greater detail earlier, in Chapter One.) Essentially, Carrel claimed the possibility to culture somatic cells (chick fibroblasts) practically indefinitely (1912).[856] Some cultures were said to be maintained for decades, far beyond the life-span of the animals from which they originated. This appeared to be a decisive proof of the potential immortality of somatic cells. In later years, Carrel's results were disputed, particularly with the positing in 1961 of the "Hayflick limit" on the number of cell divisions, causally unaffected by cell culture conditions.[857] Many of Hayflick's criticisms were anticipated by Carrel and his co-workers. Thus, it was strongly affirmed that the conditions of the culture medium, for example the rate of accumulation of the waste products in the medium, do affect the extent of multiplication and growth of cultured tissues: "When the composition of the medium is maintained constant," Carrel claimed, "the cell colonies remain indefinitely in the same state of activity. ... In fact, they are immortal."[858]

Experiments similar to Carrel's were performed by other researchers with the Rockefeller Institute: A.H. Ebeling, R.G. Harrison, M.T. Burrows, M.R. Lewis and others. According to Carrel, cells grown in the serum of younger individuals survive longer and divide for a greater number of times. The introduction of youthful "embryonic juice" was deemed by Carrel necessary for a prolonged cell life. The results obtained by Carrel for cell cultures appeared to be analogous to those obtained by the French researcher Pierre Lecomte du Noüy (1883-1947, an associate member of the Rockefeller Institute from 1920 to 1927) in entire organisms. During WWI and afterwards, du Noüy investigated wound healing and regeneration and attempted to accelerate them.[859] He established that the rate of regeneration in younger individuals is greater than in the older, a phenomenon, he believed, that could serve as an indicator of "physiological age." According to Carrel and his collaborator, Albert Henry Ebeling (1921), du Noüy's findings corroborated the role of the medium for cell and tissue regeneration and longevity:[860]

> The different values of the index of cicatrisation of a wound, ... taken from measurements made by du Noüy,... resembled the decrease in the rate of growth of fibroblasts in function of the age of the animal. This suggested the existence of a relation between the factors determining both phenomena [cell immortality and regeneration].

The question then arose whether the inhibition of cell growth was due to toxic waste products accumulated in the serum of older individuals, or that the serum of older organisms loses some "growth promoting" substance. Carrel decided for the former explanation, i.e. the increase of toxic waste products, since "in high and low concentrations of the serum of young animals, no difference in the rate of multiplication of fibroblasts was observed." In any case, potential cell immortality appeared certain, and could be nullified only by the effects of the surrounding milieu.

The "killer" argument, later proposed by the Hayflick school, that somatic cells that divide indefinitely only appear normal, but are in fact "tumor-like" was known to researchers in the early part of the century, and did not seem to them to diminish, but rather to strengthen the validity of the concept of the potential immortality of somatic cells. Thus, Leo Loeb wrote (1921):[861]

> Tumor cells are, therefore, merely somatic cells which have gained an increased growth energy and at the same time somehow gained, in some cases, the power to escape the destructive consequences of homoiotoxins. The ability of certain tumors to grow in other individuals of the same species has enabled us to prove, through apparently endless propagation of these tumor cells,... that ordinary somatic cells possess potential immortality in the same sense in which protozoa and germ cells possess immortality.

When accepting the potential immortality of somatic cells, the radical life extension of the body, composed of such cells, appeared theoretically possible.[862] Yet, a transition from this theoretical possibility to practical life extension was hardly in sight.

9. Theories of Aging

If the living matter is potentially immortal, why do the higher organisms undergo aging and death? The common answer given in the first quarter of the century by both American and European longevity researchers was that aging and death are the results of differentiation, disharmony, disproportion or discoordination of the various parts of the organism. As Raymond Pearl succinctly explained in *The Biology of Death* (1922):[863]

> The discontinuity of death is not a necessary or inherent adjunct or consequence of life, but is a relatively new phenomenon, which appeared only when and because differentiation of structure and function appeared in the course of evolution. ... Somatic death results from an organic disharmony of the whole organism, initiated by the failure of some organ or part to continue in its normal harmonious functioning in the entire differentiated and mutually dependent system.

In addition to Metchnikoff's proposition of a disharmony between parenchyma and connective tissue, and the disarray in the function of endocrine glands that formed the theoretical basis for the majority of contemporary rejuvenation attempts – a multitude of other disharmonies were postulated, in the US as well as in Europe: disproportions between the nucleus and the cytoplasm, between the cell surface and volume, limitations of growth and differential growth of various tissues, disproportions between the brain weight and the body weight (the "cephalization factor"), inequitable distribution of nourishment between the body center and periphery, etc.[864]

Alongside the various types of disproportion, a series of theories posited that senescence and death ensue because the body is "overburdened" by the products of its own activity which it is unable to assimilate or eliminate: metaplasms (products of the protoplasm that are incapable of assimilation), toxic waste products, various slags, drosses and clinkers (*schlacken* in German, such as "flocculated" protein colloids) produced by the body machinery.[865] Beside Metchnikoff's theory of auto-intoxication by intestinal putrefactive bacteria (e.g. via production of phenol, indole, scatole, ether-sulfuric acid, etc.), other theories of internal intoxication included those of the American zoologists Thomas Harrison Montgomery (1906)[866] who suggested that toxic metabolic products accumulate through faulty excretory processes, and C.M. Child who also believed that senescence arises from an accumulation of toxic substances and that rejuvenation can be achieved by their elimination (1915).[867] Among the more common "poisons" and "slags" there were named the excessive accumulation of minerals (calcium and magnesium) and organic substances (such as lipoids, especially cholesterol).

A number of workers proposed that senescence owed to the accumulation of slag resulting from physico-chemical changes in cell colloids. Protoplasm colloids (especially proteins) were said to undergo spontaneous, entropic changes, tending to decrease their surfaces, hence to condense, flocculate, or precipitate, clogging the cell and leading to its dehydration. The process of colloid condensation was termed "colloid hysteresis" and the separation of large colloid aggregates – "synaeresis." The colloid condensation was characterized by an increased cyclization of chemical bonds, and increased chemical complexity of the aggregates (a forerunner of the cross-linkage theory of aging). This theory was extremely popular in Europe in the 1920s, yet also with quite a few followers in the United States.[868] A number of theories assigned the main cause of senescence to exogenous agents – environmental toxins, cosmic radiation, heavy water, radioactive particles, even gravity – rather than to endogenous damage.[869] These were the forerunners of the theories of aging based on free radical damage and somatic mutation accelerated by radiation and other exogenous mutagens. Thus, the basic longevity researchers were on a reductionist quest to detail the works of the aging mechanism, which was deemed a necessary apprenticeship prior to any attempts at mastery.

Without being able even to list the parts of the mechanism, what hope was there to think of intervention and improvement?

10. Rectifying "Discord" and conserving "Vital Capital"

Socio-biological parallelisms were explicit in many of these theories of deterioration. Jacques Loeb (1919) summarized the "socio-economic" metaphor of aging, including the disparity of supply and demand, the excessive or deficient production or consumption, the accumulation of toxic waste products, and the disaccord between the productive and regulatory units, on the one hand, and harmful or idle units, on the other:[870]

> ...death is not inherent in the individual cell, but is only the fate of more complicated organisms in which different types of cells or tissues are dependent upon each other... one or certain types of cells produce a substance or substances which gradually become harmful to a vital organ like the respiratory center of the medulla, or that certain tissues consume or destroy substances which are needed for the life of some vital organ. The mischief of death of complex organisms may then be traced to the activity of a black sheep in the society of tissues and organs which constitute a complicated multicellular organism.

Thus, postponing the deterioration of the organism was seen as a task parallel to postponing the deterioration of the society. The socio-biological parallelisms between the harmonious functioning of the body and of the society were commonplace in the early 20th century, as for example in Vilfredo Pareto's "Social Functionalism" (1916) or Frederick Winslow Taylor's theory of "Scientific Management" (1911).[871] The life-extensionists added to such parallelisms the element of duration. They emphasized that the equilibrium of the body, as of the society, is vital not only for the systems' efficiency at any given moment, but also for their prolonged maintenance.

The general theoretical question raised in the contemporary basic longevity research was whether the organizational disharmony, seemingly inherent in the "constitution" of the living organism, could be rectified to extend its durability. The "constitutional" disharmony implied grave difficulties in attempting a modification, and a limit to the organism's existence. At the same time, through a better understanding of the mechanism, pathways were hoped to be found to counteract the attrition and achieve longevity, by adjusting or replacing parts of the organism. In Metchnikoff's terms, by "strengthening the noble elements" and attenuating the "primitive, harmful elements," homeostatic organizational stability was hoped to be established.

Many American researchers of aging were coping with the problem of defining the extent to which an edifice seemingly destined to fall by its very structure, could be propped. Thus, Alexis Carrel, the chief proponent of potential immortality, was ambivalent on the possibility and desirability of radical organic modifications for life-extension. On the one hand, he did assert that "the greatest desire of men is for eternal youth" and that, by understanding the mechanisms of aging, methods of intervention can be designed, and this desire can be, to some extent, gratified. "A better knowledge of the mechanisms of physiological duration," Carrel postulated, "could bring a solution of the problem of longevity." Eventually, "a partial reversal of physiological time will become realizable" and "Man's longevity could probably be augmented."[872] At the same time, the spatial disharmony and temporal dissonance (or "heterochrony" – the fact of different subsystems aging at different rates) inherent in the organism, appeared to Carrel too formidable to be ever fundamentally conquered. "Death," Carrel maintained "is the price [man] has to pay for his brain and his personality." Moreover, according to Carrel, the achievement of individual longevity may intensify the disharmonies inherent in the society. Or rather, the achievement of longevity will perpetuate present disharmonies, unless the prolongation of life is coupled with an improvement in the quality of life. Thus Carrel's attitude was an ambivalent mixture of humility in the face of the immense complexity of the aging processes and hopeful determination to continue on the path of research that would eventually lead to an ability to intervene in these processes.

Raymond Pearl (1922) argued, in a similar manner, that even though the understanding of the mechanisms of aging may help to prolong human life, the differentiation and dissonance are inexorably engraved in the human nature, and the life-span is an inherited quantity that can only be manipulated to a very limited degree. And yet, in Pearl's experiments, the life-span of *Drosophila* flies was extended significantly by genetic selection, decreasing the population density, chemical stimulants, or else by lowering the temperature of their environment and reducing their food supply, generally decreasing their "rate of living." Still, the exact extent to which biomedical interventions could become successful, where the life-span's "inherent limitations" would end and life-extension "possibilities" begin, remained uncertain.[873]

There apparently was not much to be done about the "brain to body mass ratio" or "surface to volume ratio" or "nuclear material to cytoplasm ratio." Nonetheless, some mild adjustments were proposed (for example, through diet and exercise). The fundamental, inherent disaccord, first suggested by Metchnikoff and later reiterated by several American researchers, between the "parenchyma," the "putrefactive intestinal bacteria," the "connective tissue" and the "phagocytes," suggested an entire series of potentially harmonizing interventions: from immunotherapy to yogurt diet to colectomy. As regards the accumulation of "colloidal waste products," at the first glance, trying to stop "colloid condensation" appeared as futile as attempting to stop mechanical entropy. Nonetheless, some researchers proposed the application of substances, such as lecithin, as "protective colloids."[874] The biochemist Thorburn Brailsford Robertson (1884-1930) of Berkeley, California, was particularly interested in the disharmonies of "cellular elements" vs. "connective tissue elements" and in the research of "protective colloids" (1913, 1923).[875] Later on, Wilder Dwight Bancroft (1867-1953) of Cornell University, New York, attempted to combat colloid condensation by the use of emulsifiers and hydration-promoting ions such as sulfocyanates.[876] An even stronger objection to the inexorable power of "colloid condensation" was the assertion that living systems cannot be equated to mechanical colloid systems. As one of the pioneers of American gerontology, Albert Ingram Lansing (1915-2002) of Washington University in St. Louis, Missouri, pointed out, "The old organism does not contain old colloids, it contains newly formed colloids of an old character."[877] Insofar as the human body constantly replaces and regenerates all of its materials, and as long as the regenerative ability is strong enough, slag accumulation and degeneration do not prevail. Regeneration, as was believed by Carrel and du Noüy, could be stimulated by the use of "embryonic juice" or glandular extracts. The accumulation of toxins too presented a grave, though in principle not insurmountable problem: the existing systems of toxin elimination, they believed, could be stimulated (e.g. by hormone supplements or "embryonic juice"). Generally, the disharmony in the endocrine gland function appeared to be the most pliable. Yet there was a wide gap between these basic, theoretical suggestions and actually recommending rejuvenative measures using endocrine stimulation, supplementation or replacement.

Notably, in the US in the 1910s-1920s, the majority of biological explanations for the life-span limit were aligned to the capitalist economy model. If aging and death were due to a disaccord among organs and tissues, conversely, life-prolongation meant maintaining the "equilibrium" of the different parts of the body through their respective "supplementation" or "reduction": strengthening the functional parts and attenuating the harmful or interfering parts. The "disequilibrium" was mainly seen in terms of a "struggle" or "competition" of organs and tissues (the idea originating with the German zoologist Wilhelm Roux, in 1881).[878] The metaphors for the body balance, for aging and life-prolongation, were explicitly derived from the capitalist socio-economic order. There were perceived in the body, as in the society, a "division of labor" which could become threatening if some segments over- or under-produce, a "class struggle" between different types of tissues ("noble" or "primitive"), "competition," incompliance to "control." The theories of aging simulated a manufacture: the machinery appeared to wear out and clog, the produce piled up and obstructed traffic, different production units lost coordination, energy supplies dwindled. Attrition seemed to be inherent in such a specialized system, but at the same time, possibilities for a prolonged, perhaps indefinite, homeostatic maintenance were indicated.

Further continuing the capitalist metaphor, the body was said to "inherit" a "vital capital" (the term first used by the "classical liberal" British philosopher Herbert Spencer in 1897).[879] Its currency was the "vital substance" or "vital energy." When the vital credit is spent, the organism goes bankrupt and dies (the idea apparently also imported from Germany).[880] Accordingly, to achieve longevity, the person needed to be thrifty with this capital. Nonetheless, as the capital was of a restricted, "inherited" quantity, frugalness could preserve it only to a limited degree. According to Jacques Loeb (1908, 1916), by lowering the body temperature in cold-blooded animals, their metabolic rate could be reduced and their life prolonged.[881] Life prolongation by lowering the metabolic rate in *Drosophila* flies was achieved by Raymond Pearl (1922, 1924).[882] Yale biologists Thomas Burr Osborne (1859-1929) and Lafayette Benedict Mendel (1872-1935), accomplished the retardation of the metabolic rate and life-prolongation in rats by reducing the animals' food intake (1915, 1917).[883] In 1914, Francis Peyton Rous (1879-1970) of the Rockefeller Institute suggested that a reduced food intake in rodents limits their risk of cancer.[884] In 1929, the "Great Depression" erupted in the US in the wake of the "Prosperity Rush," demonstrating the adverse effects of hyper-activity, over-spending and over-consumption.[885] In the same year, Clive Maine McCay (1898-1967) of Cornell University confirmed the life-prolonging properties of lowering the metabolic activity by reducing food intake. McCay further advanced the life-extending "Calorie Restriction" method after the end of the Depression, in 1935.[886] To the present day, calorie restriction remains the most effective method for life-prolongation in virtually all animal models.[887]

In this line of thought, Raymond Pearl (1922) emphasized that the duration of life is a hereditary quantity, and illustrated his celebrated "Rate of Living" theory of aging in the following terms:[888]

> I believe, ... that environmental circumstances play their part in determining the duration of life, largely, if not in principle entirely, by influencing the rate at which the vital patrimony is spent. If we live rapidly, like Loeb and Northrop's *Drosophila* at the high temperatures, our lives may be more interesting, but they will not be so long. The fact appears to be, ... that heredity determines the amount of capital placed in the vital bank upon which we draw to continue life, and which when all used up spells death; while environment using the term in the broadest sense to include habits of life as well as physical surroundings, determines the rate at which drafts are presented and cashed.

The "conservative" method for life-prolongation was later elaborated by the American physiologist, the author of the theory of homeostasis, Walter Cannon (1871-1945), and the Austro-Hungarian/Canadian endocrinologist, the author of the general adaptation theory of stress, Hans Selye (1907-1982).[889] (See below on the relation of life-extensionism to Cannon's and Selye's concepts of homeostasis.)

Essentially seeing vitality as a bank account, Pearl did not expect any sudden and substantial revenues, only modestly wished the vital capital to last the century quota. In Pearl's perception, the social dynamics are analogous to those of the body. According to him, increased longevity is likely to lead to the "population problem" – the resources being overtaxed by the aged. To him, social resources, as body resources, seemed fixed and non-renewable. Nonetheless, he was ambivalent regarding the limit of life. A limit was posited, but it could be stretched, by manipulating the environment or by eugenics. In this view, biomedical knowledge did not appear futile, but pointed to tangible life-prolonging interventions:[890]

> The sciences and arts of biology, medicine and hygiene [have] the fundamental purpose of learning the underlying principles of vital processes, so that it might ultimately be possible to stretch the length of each individual's life on earth to the greatest attainable degree. ... Well conceived and careful studies... seem likely to yield large returns.

The idea of the "vital capital" thus implied a search for new biological resources and resistance to "overheating." Even more generally, these passages show that American life-extensionism was inseparable from the currently dominant socio-ideological order. Not only did American longevity researchers depend on capitalist enterprise for funding, not only did they advocate the perpetuation of capitalism, but their views seemed inextricable from capitalism even in terms of scientific metaphor. That alliance may seem only obvious and natural, unless recalling the examples where life-extensionism was adjusted to a radically different socio-ideological order (e.g. German fascism or Russian communism).

11. Institutionalization of longevity research and advocacy – The Life Extension Institute

A number of longevity research centers were established in the US in the first quarter of the century: Stephens' fledging laboratory, several laboratories at the unprecedentedly well-funded and comprehensive Rockefeller Institute. Several popular life-extension clubs emerged as well – the "Eternity Club" in Minneapolis, the "Century Society" in Philadelphia, the "One Hundred Year Club" in Seattle, the "Two Hundred Years Method (TYM) Club" in New Jersey – each comprising several up to tens of thousands of members. But perhaps the most authoritative life-extensionist organization, having a sizable impact on health policy, was the "Life Extension Institute," inaugurated in 1913. Its board included the former US president (1909-1913) William Howard Taft as chairman; Irving Fisher, Professor of Political Economy at Yale, as chairman of its Hygiene Reference Board; Eugene Lyman Fisk as its Medical Director; Elmer Ellsworth Rittenhouse, Conservation Commissioner of the Equitable Life Assurance Society, as its President; and Harold Alexander Ley, the industrial contractor, as treasurer (Ley initiated the Institute establishment in 1909). A host of luminaries of science and medicine acted on the Institute's Hygiene Reference Board, including John Harvey Kellogg, William James Mayo, Walter Bradford Cannon, Alexander Graham Bell, Charles Benedict Davenport and many more.[891] The Life Extension Institute was heavily backed by capital, the initial funds approximating $1M, and the board of directors including an impressive cohort of bankers and financial tycoons.[892] The Metropolitan Life Insurance Company played a dominant role. Since its inception in 1863, the company funded extensive demographic comparative longevity studies and issued hygienic recommendations for life-extension, aiming (and often succeeding) at increasing the period of productivity of the insured workers and correspondingly reducing insurance benefit payments.[893] The Life Extension Institute thus obtained the unprecedented support of the political, economic and scientific establishments.[894]

The Life Extension Institute's credo was presented in its programmatic work – *How to Live. Rules for Healthful Living Based on Modern Science* – written by the Institute leaders, Irving Fisher and Eugene Lyman Fisk, and prefaced by William Taft.[895] First published in 1915, by 1919 the book held 15 editions, reaching a total of over 100,000 copies. Irving Fisher believed that human life can be prolonged to at least a century and a half,[896] but in *How to Live*, such aspirations were cautiously veiled. Instead, the work spoke of life prolongation in general. According to Taft (1919):[897]

> The mere extension of human life is not only in itself an end to be desired, but the well digested scientific facts presented in this volume clearly show that the most direct and effective means of lengthening human life are at the same time those that make it more livable and add to its power and capacity for achievement.

This statement resolved the perceived opposition (pointed out in Herbert Spencer and later in Alexis Carrel and many others) between increasing "life quantity" at the expense of "life quality." The life "quantity" and "quality," it was argued, are inseparable concomitants.

The work, and the Institute generally, advocated regular physical examinations, campaigned against "poisons" (alcohol and tobacco) and for the adherence to "The Sixteen Rules of Hygiene" regarding "Air" – advising to "Ventilate every room you occupy; Wear light, loose and porous clothes; Seek out-of-door occupations and recreations; Sleep out, if you can; Breathe deeply"; "Food" – "Avoid overeating and overweight; eat sparingly of meats and eggs; eat some hard, some bulky, some raw foods; eat slowly; drink sufficient water"; "Poisons" – "eliminate thoroughly, regularly and frequently; stand, sit and walk erect; do not allow poisons and infections to enter the body; keep the teeth, gums and tongue clean"; and "Activity" – "Work, play, rest and sleep in moderation; keep serene and whole hearted."[898] The Institute's method was thus succinctly defined: "The method to be used to prolong life is very simple and the same as applied to ordinary machinery – inspection and repair."[899]

How to Live first appeared during WWI, and the Institute's health education pamphlets were distributed in hundreds of thousands of copies at Draft Boards and among the US Army and Navy troops. As the present work argues, the majority of life-extensionist authors were well adjusted to the existing social order and strove to maintain its stability. In the times of war, this meant increasing the fitness and vitality of soldiers and workers of the "home front," to smoothen the operation of the war machine. In the times of peace, this meant increasing the fitness and vitality for constructive activities. As Taft summarized in 1919:[900]

> Thus we have, as a by-product of a terrible and devastating war, the revelation of a great national need and, in consequence, a new and lively interest in human vitality and efficiency. …The principles of individual hygiene which have been applied in the training and guidance of the soldier should also be imprest [sic] upon the civilian. … I therefore commend anew this book to the earnest attention of our people at a time when, as a nation, we are turning from the destructive, life-destroying activities of war to the constructive, life-renewing activities of peace.

Interestingly, the prevalent rhetorical device employed by life-extensionists at the times of peace and prosperity was that the health and life conditions generally improve, and the life-expectancy grows – with a sufficient effort and investment, these trends will only intensify, to effect a radical prolongation of life. Such an attitude, for example, can be seen in *The Philosophy of Long Life* of Jean Finot (1900) who spoke of the ever improving quality of life and the ever increasing "love and universal fraternity."[901] Fisher and Fisk's argument, developed during the War, was almost precisely the opposite: the health conditions worsen, diseases rage and life-expectancy generally declines – therefore the society should enlist all its efforts to combat the existential threat. The various forms of life-extensionism were a product of their time and locale.

The form of life-extensionism represented by the Institute was also distinguished by a strong advocacy of eugenics, understood not just generally as "the science of human improvement" and "healthful living," but in the very specific sense of selective breeding for longevity. To recall, in the first quarter of the century, eugenics was promoted world-wide (in Canada, Australia, Britain, Sweden, Denmark, Norway, Finland, Switzerland, Germany, Austria, France, Brazil, Argentina, Japan, Russia, etc. etc.) and seemed to many a most practicable program for radical life prolongation. In the US, by the mid-1920s, eugenics laws, including forced sterilization and prohibition of marriage of the "defectives," were passed in 33 states.[902] Though, grave difficulties and dangers were perceived in such programs. Raymond Pearl cautioned (1922):[903]

> …almost any breeder of average intelligence, if given omnipotent control over the activities of human beings, could, in a few generations, breed a race of men on the average considerably superior. … But, as a practical person, I am equally sure that nothing of the sort is going to be done by legislative action or any similar delegation of power. Before any sensible person or society is going to entrust the control of its germ-plasm to politics or to science, there will be demanded that science know a great deal more than it now does about the vagaries of germ-plasm and how to control them. Another essential difficulty is one of standards. … what individual or group of individuals could possibly be trusted to decide what it should be? … One cannot but feel that man's instinctive wariness about experimental interferences with his germ-plasm is, in considerable degree, well-founded.

Fisher and Fisk were, however, much more positive about the medical and humanitarian potential of eugenics.[904] According to them, "Eugenics does not require the old Spartan practice of infanticide, nor does Eugenics propose to do violence in any other way to humanitarian or religions feeling." Eugenics, according to these authors, only implies taking heed whom the person marries so as to avoid serious and often fatal health problems in the offspring. The main suggestion was that couples should "not expect physical, mental and moral perfection in any one individual, but look for a majority of sterling traits." From an avid and playful explication of the Mendelian laws of inheritance, using plumage and bean colors, the authors proceeded to "propose to restrict marriages or mating of those unfit to marry" – the "defectives" and "feeble-minded." The results, they hoped, would be truly magnificent: "In the light of modern eugenics we could make a new human race in a hundred years…"

The Life Extension Institute was very attentive to capitalist concerns. Its successes were often measured in monetary terms, and its motto may well have been 'life-extension saves money.' Several American life insurance companies (particularly the Metropolitan) made the Institute's recommendations obligatory, requiring the insured to undergo screening medical examinations, prophylactic measures and health education programs, and consequently gaining large dividends from the resulting decreases in mortality and morbidity. Such practices and campaigns by the Institute and insurance companies were hailed by German life-extensionists of the 1930s, such as Ludwig Roemheld (1933) and Gerhard Venzmer (1937), who advocated similar practices in Germany.[905] Beside the insurance companies, the "Workmen's Compensation Act" of July 1914, according to a contemporary account, "would render the institute of inestimable value to large employers of labor" as it would serve to extend the workers' productivity.[906]

The organization exists to this very day and continues to provide corporate and personal prophylactic examinations, preventive and primary health care and wellness programs. It was consecutively owned by Dun & Bradstreet (1969), Control Data Corporation (1978) and UM Holdings Ltd (1986), with General Motors Corporation as its largest client since 1946. In 1995, the Life Extension Institute acquired the Executive Health Group, its next largest competitor.[907] In 2000, the organization changed its name to "Executive Health Evaluations (EHE) International" which is now the largest (and oldest) preventive health company in America, carrying out millions of examinations yearly, and operating in 42 states.[908]

Thus, Life Extension Institute succeeded in enlisting massive, perhaps unprecedented, industrial, medical-scientific and governmental support for the goal of life-prolongation, blending into the state establishment, and affecting public health policy. Since the beginning of the 1930s, with the general strengthening of the central government, American life-extensionism became further linked with the state, both institutionally and ideologically.

12. The Great Depression, the New Deal, WWII – Projections of socio-biological stability. Walter Cannon (1871-1945)

In the 1930s-1940s, in the United States, general health care improved, becoming more available to the low-income populace.[909] The proportion of the aged population increased and the importance of care for the aged was becoming increasingly recognized.[910] Several social programs were geared toward improving the well-being of the aged. The Townsend Plan (commonly referred to as the "Townsend Crusade" by Dr. Francis Everett Townsend, 1867-1960) called for guaranteed monthly pensions.[911] The Townsend Plan eventually influenced the establishment of old-age pensions within the Social Security Act passed in August 1935 by President Franklin Delano Roosevelt (1882-1945), in the framework of the "New Deal."[912] During WWII, that ultimately ended the "Great Depression," the war effort contributed to the development of surgery, prosthetics and mass production of pharmaceuticals (specifically the mass production of antibiotics).[913] However, the amount of writing on radical life-extension, particularly on rejuvenation, was largely reduced, being of less immediate relevance during the peak of the great depression (1929-1933) and the war.[914] Through the 1930s-1940s, there was no shortage of pessimistic or derogatory writings on rejuvenation attempts.[915] Still the discussion of life-extension did not cease. In the post-depression period especially (the period of the New Deal, 1933-1939) the amount of life-extensionist literature increased. Alongside the more medical and scientific works, there existed idealistic and spiritual works featuring either hard-core spiritualism or assuming a more scientific appearance.[916] "Natural-hygienic" approaches to longevity were widely spread, and the search for "natural," balanced and anti-toxic diets continued.[917] Basic research developed the genetic theory of aging.[918] The seminal "Rate of Living" theory of aging was consolidated by Raymond Pearl in 1928, during the "Prosperity Rush" and just before the "Great Depression." Immediately following it, in 1935, the connection between low energy expenditure (calorie restriction) and longevity was emphasized by the professor of animal husbandry, Clive Maine McCay (1898-1967) of Cornell University, Ithaca, New York.[919] The life-extensionists catered to the changing social situations, advocating for the stability or "equilibrium" within the currently dominant socio-ideological order.

As noted earlier, "Biological Equilibrium" was a key term in the life-extensionist literature since an early time, through the *fin-de-siècle* and onward, figuring prominently in Metchnikoff's, Voronoff's, Steinach's, Bogdanov's and Pearl's writings, in fact in the vast majority of life-extensionists. Yet for these authors, the term somewhat vaguely and qualitatively implied the existence of "deficits" or "excesses" of some cells, tissues or substances that could be correspondingly "replenished" or "eliminated" in order to achieve a "harmony of parts," and hence stability, robust health and longevity.[920] In *The Wisdom of the Body* (1932), the American physiologist Walter Bradford Cannon (Chair of the Department of Physiology at Harvard Medical School) provided more detailed descriptions of various homeostatic steady states – the constancy of blood water and salt contents, blood proteins, glucose and fat levels, oxygen supply, pH and temperature stability, etc. – involving elaborate feedback mechanisms, quantitative thresholds and dose-responses. The term "homeostasis" was coined by Cannon in 1926[921] and was considered by him to be more appropriate to describe stability in a living body than "equilibrium": While thermodynamic "equilibrium" came to signify stagnation and death, "homeostasis" implied a regular flow, or "steady states…complex and peculiar to living beings … a condition which may vary, but which is relatively constant."[922]

According to Cannon (a board member of the American Life Extension Institute), maintaining homeostasis has direct implications for longevity, both bodily and social, as "the stabilizing processes in a body politic [as in an individual organism]… when once discovered and established, might be expected to continue in operation as long as the social organization itself, to which they apply, remains fairly stable in its growth."[923] Later on, the Hungarian/Canadian endocrinologist Hans Selye (the founder of stress physiology) furthered the theory of homeostasis, introducing the concept of the "General Adaptation Response." Emerging from Selye's work on sex hormone replacement, the concept was first published in 1936,[924] and widely disseminated after the publication of Selye's popular book *The Stress of Life* (1956). In Selye, the theory of homeostasis, developing Cannon's views, had clear-cut implications for longevity research as well. Life-extension, Selye believed, could be achieved by a harmonious distribution of stress and energy resources among different parts of the body, maintaining homeostasis.[925]

In Cannon, the parallelism between biological and social equilibrium is explicit. Not only are the regulation and stabilization of the organism analogous to the regulation and stabilization of the society, but "the homeostasis of the individual human being is largely dependent on social homeostasis."[926] According to Cannon, stability is vital for the prolonged survival of both the biological and social body:[927]

> The organism suggests that *stability is of prime importance*. It is more important than economy. ... The organism suggests, furthermore, that the importance of stability warrants a specially organized control, invested by society itself with power to preserve the constancy of the fluid matrix, i.e. the processes of commerce. (Emphasis in the original.)

Yet, insofar as Cannon postulated that the stability of biological organisms depends on the particular structure of those organisms and their particular circumstances, so the stabilization of social organisms too depends on a particular social order and circumstance:[928]

> Various schemes for the avoidance of economic calamities have been put forth ... Communists have offered their solution of the problem and are trying out their ideas on a large scale in Soviet Russia. The socialists have other plans for the mitigation of the economic ills of mankind. And in the United States, where neither communism nor socialism has been influential, various suggestions have been offered for stabilizing the conditions of industry and commerce.

That is to say, the effort toward the stabilization and hence prolonged maintenance of the American social organism needs to be adjusted to the present capitalist order and not to some abstract "Utopias":[929]

> In all such proposals a much greater control of credit, currency, production, distribution, wages and workmen's welfare is anticipated than has been regarded as expedient or justifiable in the individualistic enterprises of the past. ... Among these suggestions are the establishment of a national economic council ... and the assuring of the general public through governmental regulation that in any arrangements which are made its interests will be protected.

This was precisely the philosophy of Franklin Delano Roosevelt's "New Deal" aimed to combat the Great Depression by increasing the governmental regulation of industry and commerce. Though the "New Deal" was enacted in 1933, with Roosevelt's ascent to presidency, its philosophy was widely professed at least as early as 1932, during the presidential campaign. Thus, Cannon's views on the prolonged stabilization of the biological and social organism were exactly aligned to the concurrent dominant socio-ideological paradigm.

Cannon believed that governmental regulation (which he equated with "intelligence") would improve not only economic stability, but health and longevity as well:[930]

> The projection of the schemes, however, is clear evidence that in the minds of thoughtful and responsible men a belief exists that intelligence applied to social instability can lessen the hardships which result from technological advances, unlimited competition and the relatively free play of selfish interests. By application of intelligence to medico-social problems, destructive epidemics such as the plague and smallpox have been abolished; fatal afflictions, e.g. diphtheria and tuberculosis, have been

greatly mitigated and largely reduced …. These achievements all involve social organization, social control, and a lessening of the independence of the individual members.

He dispelled the fears that the central regulation would diminish individual freedom or individual accomplishment. Rather, it would provide the stability needed for free development:[931]

Without homeostatic devices we should be in constant danger of disaster, unless we were always on the alert to correct voluntarily what normally is corrected automatically. With homeostatic devices, however, that keep essential bodily processes steady, we as individuals are free from such slavery – free to enter into agreeable relations with our fellows, free to enjoy beautiful things, to explore and understand the wonders of the world about us, to develop new ideas and interests, and to work and play, untrammeled by anxieties concerning our bodily affairs. The main service of social homeostasis would be to support bodily homeostasis.

Yet, Cannon admitted that the current knowledge of the mechanisms of both social and biological homeostasis, and of the means toward its reestablishment, is very imperfect: "The multiplicity of these schemes is itself proof that no satisfactory single scheme has been suggested by anybody."[932] Hence the need for further extensive research of these mechanisms was emphasized, in order to insure both social and biological longevity.

13. "Further research is needed" – Clifford Cook Furnas (1900-1969)

The importance of "further research" was even more strongly emphasized by Clifford Cook Furnas, the Big Ten champion in the cross country and member of the US Olympic team (1920), in the 1930s a professor of chemical engineering at Yale University, in New Haven, Connecticut, and a corporate consultant, and since the 1940s a pivotal figure in the US military aeronautic programs (Chairman of the Guided Missile Commission - 1952-1953, Chairman of the Defense Science Board - 1961-1965, etc). Though hardly comparable in scientific stature to Cannon's writings, Furnas' *The Next Hundred Years. The Unfinished Business of Science* (1936),[933] written during the "Second New Deal," may be considered as representative of contemporary life-extensionist aspirations. The major ingredients of the early century life-extensionist 'American blend' – capitalism, religion and eugenics – are clearly observable, though significantly modified.

Furnas was apparently not religious. He spoke at length about religious "superstitions" hindering the march of scientific and technological progress. Yet he accepted the status-quo of "truce" where some scientists could be "supernaturalists on Sunday and naturalists on Monday."[934] For Furnas, the preservation of capital was a vital concern. As he succinctly stated, "Personally, I am not ready to overthrow capitalism for I may want to get back my $100 [loaned to others] at the age of 65." Yet, in accord with the prevalent philosophy of the "New Deal," he valorized governmental control of big business and "social cooperation": "To take full advantage of our technological improvements and not have serious social consequences requires a far greater industrial and social cooperation than we care to have." According to Furnas, in a country fully "saturated with capitalism" where "everything we have is owed to someone else," universal increases of capital are impossible.[935] Yet, radical improvements in the standard of living, health and longevity, *are* possible through accelerated development of science and technology. Improved technology will be able to cure diseases, provide unlimited food, mineral resources and energy, while automation will lead to "thousand-fold increases in production." Thus, capitalism will be well able to persist indefinitely.

One of the methods Furnas envisioned to decrease the economic burden, and increase the overall productivity, was by eugenics, namely, "eliminating the undoubtedly unfit," particularly "the short lived defectives." He was well aware of the scientific and social complexities that may be involved in the "battle of eugenics." According to him, standards for selection are difficult to establish, as "the most ardent eugenist never looks askance at his own blood." Eugenic selection may impair the variety and adaptability of the human stock, for we do not know what type of humans the future may need. "There *is* a place for the strong back and weak mind. ... We praise our geniuses but if they were all put in one pen to breed together for a few generations there would be the most terrible crop of neurotics the world has ever seen." Hence, according to Furnas, negative selection is only justifiable for the "undoubtedly unfit": "The only people we can afford to exterminate [via sterilization] are the obviously defective, either physically or mentally. If intelligent eugenists can accomplish this – power to them; if they attempt much more in our present state of limited knowledge the evil they do will live for generations after them." In time, Furnas believed, the crude present methods of eugenic selection (such as sterilization) may be superseded by more refined ways to "breed a super-race by genetic manipulation." "When everything concerning heredity, growth and development is known," he foresaw, "perhaps it will be possible to modify undesirable germ plasm through a period of growth so that a potential moron may grow into at least a moderately good citizen." Yet, he admitted that contemporary science was too far from that level of knowledge. (How imperfect the knowledge was, is exemplified by the fact that Furnas, like the majority of contemporary biologists, believed that a gene is "a minute living blob of protein.") The imperfections of the present state of knowledge and the perfections aspired to in the future, dictated the "great need for research."[936]

Eugenic selection, however, was seen as only one of the possible methods for life-extension. But the "great need for research" was universal for any potential method. Furnas strongly advocated radical life-extension: "Sociologically [death] is unfair and inefficient" – he asserted. Life extension, according to him, will ensure a prolonged period of productivity: "In all discussions of the advantages of longevity it is of course understood that what is desirable is not mere longevity but a lengthening of the period of full physical and mental, especially mental, powers." And there is even "a better reason for postponing death… People should live longer because most of them want to; want to very badly."[937] Yet the means to fulfill that wish are simply not at hand. Hence, 'further research is needed.' Furnas discussed quite a few possible causes for aging suggested by the time: depositions of calcium, accumulation of poisons, maladjustment of hormone secretions, accumulation of "heavy hydrogen," decrease in the percentage of "bound water," etc. "None has been proved, nor has any of them added anything to anyone's life."[938] Yet, with an extensive research, the mechanisms of aging may be clarified, and targets for repair identified:[939]

> It is easier to rationalize than to determine biological facts. We can tell ourselves that death is not so bad and after a while we may make ourselves believe it. The scientists are a long way from making "death take a holiday" and their progress along the road is discouragingly slow. The price of a few battleships given to a foundation for the study of the causes of senility might eventually put a different face on the matter, but almost everyone thinks we need the battleships and almost no one thinks we need the old men. … If you ask a good biologist if we will ever know what senility is and if we do whether we will be able to control it and thus postpone death indefinitely, he will just look at you over the top of his glasses. … The answer is so far beyond the present knowledge that the expert cannot allow himself even to think about it. … Meanwhile we keep dying and he writes theses upon the sex life of the earthworm.

Furnas' bottom line is that capitalism could be preserved indefinitely, with an ever-increasing productivity, and both the worker and the employer could live very long and happy, if only more funds were given to research.

Shortly after the publication of *The Unfinished Business of Science*, Furnas' dreams were beginning to be realized. More "theses" on aging and longevity were beginning to be written and researchers of aging and longevity were beginning to band together, raising the amount of funding, prestige and social leverage.

14. The movement toward consolidation. Cowdry's Problems of Ageing (1939, 1952)

Toward the 1950s there was a marked increase in the institutional organization and cooperation in longevity research and advocacy, in the US and throughout the world. From the beginning of the 20th century, the majority of life-extension advocates (Metchnikoff, Voronoff, Bogdanov, Finot, Fedorov, Stephens, and many others) called for an extensive scientific-institutional collaboration and massive public support of longevity research. Despite these appeals, not much cooperation and support seems to have occurred in their time. The authors (the luckier or better established ones) were able to set up aging research programs in their laboratories, where aging often was only a single subject among many (as in Child's or Carrel's labs) and published nostrums and manifestos that were disseminated in the general public and either adopted into individual health regimens or read for curiosity.

In the 1930s, cooperative, institutional research and public and governmental support were beginning to mount. Notably, in the 1930s-1940s, the institutionalization and support of longevity research were first established almost exclusively within national frameworks. Only in the 1950s some forms of official international cooperation emerged. The founding of The International Association of Gerontology (IAG) in 1950 marked a decisive turn toward international cooperation in the field, and can be considered a watershed event in the history of aging and longevity research. The collaboration proceeded in the following routes: convening of the first national (and later international) conferences, establishment of the first aging research institutes, appearance of the first dedicated journals, organization of the first national (and later international) research associations and government-sponsored agencies.[940]

In the US, the institutionalization proceeded in the same paths. In 1939, *Problems of Ageing – Biological and Medical Aspects* was edited and published by the anatomist, zoologist and cytologist Edmund Vincent Cowdry (1888-1975) of the Washington University School of Medicine in St. Louis, Missouri (with further editions in 1942 and 1952). The volume brought together, for the first time, the works of prominent American longevity researchers, including Edmund Cowdry, Walter Cannon, William MacNider, Clive McCay, Louis Dublin, and others.[941] The authors formed the core of The American Geriatrics Society that was inaugurated in 1942 and of the Gerontological Society of America established in 1945. The latter developed from the American "Club for Research on Aging" set up in 1939 by 24 scientists, including Edmund Cowdry, Clive McCay, Nathan Shock, William MacNider, and Lawrence Frank. These societies established *The Journal of Gerontology* in 1946 and *Journal of the American Geriatrics Society* in 1953, making aging research more visible. Notably, at its inception, the Gerontological Society of America was predominantly concerned with biomedical aspects of aging, with a salient emphasis on basic longevity studies (unlike its more recent "social" emphasis).[942]

In 1940, a Unit on Gerontology was authorized by the US Surgeon General, Thomas Parran (1892-1968), within the US National Institutes of Health, and was directed by Edward Julius Stieglitz (1899-1956).[943] In 1955, Surgeon General Leonard Andrew Scheele (1907-1993) decreed the establishment of 5 Centers of Geriatric research, including the Duke University Center for the Study of Aging and Human Development in Durham, North Carolina, directed by Ewald Busse (1917-2004), and the Gerontology Research Center in Baltimore, Maryland, under the leadership of Prof. Nathan Wetherell Shock (1906-1989). In 1958, the American Association of Retired Persons (AARP) was founded by Dr. Ethel Percy Andrus (1884-1967).[944] After an advocacy campaign conducted by Denham Harman (b. 1916), Robert Neil Butler (1927-2010) and other prominent members of the "Gerontological Society," during the presidency of Richard Milhous Nixon (1913-1994),[945] in 1974, the US National Institute on Aging (NIA) was established, with a clear emphasis on basic biological research.[946]

Edmund Vincent Cowdry played a crucial role in the institutionalization of the American and international research of aging and longevity.[947] (The second crucial figure was the Russian/British physician and physiologist Vladimir Korenchevsky, the founder of British institutional gerontology, whose work will be discussed in the subsequent section on British life-extensionism.) The publication of Cowdry's *Problems of Ageing – Biological and Medical Aspects* in 1939[948] was a seminal event. Preceding it was the conference on aging, held at Woods Hole, Massachusetts, on June 25-26, 1937, funded by the Josiah Macy Jr. Foundation, and jointly sponsored by the Union of American Biological Societies and the National Research Council. Among the 25 articles included in the volume, 15 were contributed by the conference participants. Thus, the Woods Hole conference could vie for priority with the Kiev conference on "The genesis of aging and prophylaxis of untimely aging" (held on December 17-19, 1938) as being the world's first scientific conference dedicated to the subjects of aging and longevity. Both the proceedings of the Kiev conference and Cowdry's *Problems of Aging* were released in the same year, respectively, in January and February 1939. The Kiev conference may have been pioneering in its emphasis on the prolongation of life, while the Woods Hole conference and the subsequent *Problems of Aging* were more cautious and largely concerned with the basic biology of aging. Whatever the world priority, these conferences and volumes were certainly the first of their kind in the USSR and US.

Despite the generally cautious and academic tone, Cowdry's *Problems of Aging* contained strong life-extensionist attitudes, accompanied by an almost universal call for further basic research and funding. Thus, according to the foreword of Lawrence Kelso Frank of the Josiah Macy Jr. Foundation: "Two major questions arise in any consideration of the health-care of the aging: how can we prolong human life and how can we lessen, if not eliminate, those malfunctionings and disturbances which impair the existing lifetime of so many older individuals? ... few definite answers are possible in our present state of knowledge."[949] According to Clive McCay, the pioneer of calorie restriction research, "There is no doubt that the life span can be modified considerably by such factors as nutrition and living regime." In McCay's experiments on rodents, the animal life-span was conclusively extended by reducing the energy content of the diet (caloric intake) with the consequent slower growth. Yet, "Evidence to show specific dietary effects upon the course of development of degenerative diseases in man is difficult to obtain." Some data were available. Thus, "[Lowell] Langstroth (1929) has presented clinical evidence of the beneficial effects of diets richer in vitamins and poorer in purified foodstuffs." Yet, according to McCay, there will be a long road ahead before defining nutritional requirements for healthy life-extension:[950]

Almost nothing is known of the biochemistry of ageing. However, definite advances must be recognized in the discoveries that the life span can be greatly extended by manipulating the diet. Our philosophy need no longer anchor us to the concept of a fixed life span. Evidently there are great gaps in our knowledge of the chemical changes with aging. Little hope of progress in studying the process of ageing, can exist until special institutes of research are established in which whole groups of specialists will devote their lives to cooperative attempts to solve the intricate problems. The field of nutrition probably affords the most promising line of attack but specialists in this field must work side by side with physicists, biochemists, bacteriologists, pathologists, physiologists, histologists and psychologists. Thus far Russia seems to be the only nation that has established such institutes. When these cooperative attacks are made upon the basic problem of age-changes it is likely that the by-products will afford entirely new methods of attacking the diseases of old age such as cancer, arteriosclerosis and those of the heart.

Also according to Walter Cannon's paper, a significant life-extension is possible, and the possibility is derived from the concept of homeostasis: "A routine existence within the limits of easy adaptation may continue indefinitely without revealing any weakness; but exposure to a stress which encroaches on the limits quickly discloses that they have become much restricted." Cannon too indicated the need for further research: "It is well to emphasize also that more evidence is needed on numerous points considered in this chapter, before knowledge of the limitations of homeostatic mechanisms in senescence is quite satisfactory."[951] Cowdry followed Cannon's lead, asserting that "the operation of the balancing mechanisms called homeostatic by Cannon, provides constancy within narrow limits in the temperature of the body, sugar content, acid-base equilibrium and other properties of the blood." Hence, learning to maintain homeostasis will mean learning to extend life. He pointed out the present shortcomings of knowledge (for example, the uncertainty whether blood capillary permeability to proteins increases or decreases with age). But he was nonetheless hopeful: "We may hope to find clues as to the nature of the processes of ageing by balancing local alterations in cells and fibers against changes in fluid environment." Yet again, more research was needed: "The systematic microscopic investigation of the processes of ageing has barely been commenced. Many techniques are available, yet have not been applied. Cooperation between cytologists and biochemists is essential." [952] Yet another pioneering figure in American gerontology, William MacNider, also expected great achievements from further study of stabilizing cellular homeostasis, cell repair and cell resistance to injury.[953]

Life-extensionist aspirations, as well as calls for extending research, continued to be expressed in the next editions of Cowdry's *Problems of Aging*. As Cowdry wrote in the foreword to the third edition of 1952:[954]

> New avenues of investigation have been opened and existing ones have been followed. Only a small part of the information now available, however, has been utilized to enable people to live healthier, more useful and longer lives. What may be regarded as pilot experiments in relatively concise areas of human betterment reveal that if work in this direction is dignified by public approval and support, the following decade will be one of great achievement.

Such "moderate life-extensionist" sentiments seem to characterize the institutionalization of gerontology in the US in the late 1930s through the early 1950s. It appears that by no means was then aging research conceived as a purely academic pursuit or as a predominantly sociological study of the aged. The general objectives of life-prolongation were mentioned, though were somewhat humbled: Reaching a life-span of 100 years appeared in *Cowdry's Problems of Aging* as a reasonable goal. This is the life span that "nature has allotted to man" according to Louis Israel Dublin, chief statistician of the Metropolitan Life Insurance Company.[955] The "moderation" in goals might have been a survival need for the scientists: their aims needed to be both sufficiently modest to maintain responsibility and respectability, yet sufficiently optimistic to justify further efforts. This balanced approach seems to have prevailed in gerontology, though it has been sometimes criticized as insufficiently bold by more "radical" (differently affiliated) life-extension proponents.[956] With the "moderate life-extensionist" motivations, emphasizing the well-being of the growing aged population and striving to combat the increasing incidence of age-related diseases in Western countries, Cowdry and Korenchevsky organized the International Association of Gerontology in 1950,[957] laying down the foundations for the institutionalization of gerontology as an international scientific discipline.

15. Consolidation continues

The founding of the International Association of Gerontology (IAG) in 1950 may indicate a victory of the movement for international cooperation. With the Cold War at its peak, the IAG provided a collaborative platform, including representatives from both the "Western" and "Eastern" blocs, from both developed and developing countries.[958] From its foundation, the Association saw itself as a potential supplicant and subsidiary to the recently established United Nations Educational, Scientific and Cultural Organization (UNESCO, founded in 1945), the World Health Organization (WHO, founded in 1948), the International Labour Organization (ILO, incorporated into the United Nations system in 1946), and the Council for the Coordination of International Congresses of Medical Sciences (CCICMS, founded in 1949).[959] The International Association of Gerontology was, of course, not always an ideal family, and ideological struggle sometimes erupted (as in the 1972 IAG World Congress in Kiev, Ukraine). Still, the participation of both the US as well as the USSR and other countries of the "Eastern bloc" in the Association was indispensable.

The IAG funding was initially supplied mainly by private foundations: the Nuffield Foundation endowed by Lord Nuffield (the industrialist William Richard Morris, 1877-1963) from the UK, and the Josiah Macy Jr. Foundation from the US (established in New York in 1930 in honor of the Sea Captain and patron of science Josiah W. Macy, Jr, 1838-1876). Another major philanthropist, supporting cooperation in gerontology, was the Chemical Industries of Basel – Chemische Industrie Basel, CIBA. (Though funded by the Swiss chemical concern (one of the precursors of Novartis) which was then directed by Robert Käppeli (1900-2000), the foundation was incorporated in 1949 in London, where it was directed by Sir Gordon Ethelbert Ward Wolstenholme (1913-2004) and presided over by the Nobel prize winning physiologist Edgar Douglas Adrian (1889-1977), president of the Royal Society and Master of Trinity College, Cambridge.[960]) Since 1954, the CIBA Foundation sponsored international colloquia on endocrinology, aging and longevity, as well as on scientific, technological and medical futurism, such as the *Man and His Future* colloquium held in 1963 in London, which included some of the world's foremost biologists and where the topic of radical life-extension was central (this work will be discussed later on).[961] State support for gerontological research was, however, gradually mounting, for example, through the National Institutes of Health in the US or through various national ministries of health.

Since the time of the IAG establishment, the US has provided most of the funding for aging research, and has since become the undisputed world leader in longevity research and advocacy. It appears that prior to 1945, aging and longevity research was more evenly distributed among several Western countries. Thus, according to Nathan Shock's bibliography,[962] in 1930, the number of publications (scientific journal articles and books) on theories of aging in the US, UK, France, Germany, Russia, and Italy was 1 or 2 for each country. In 1945, the amount of publications in the US was 5, and one or zero in each of the latter countries (perhaps for quite an obvious reason of war devastation in Europe). In 1960, the number of publications in the US was 15, and 1 or 2 in the UK, France, Russia, West Germany, Italy, Netherlands, and Argentina each. Presently, in the US, the expenditures on health generally, and on longevity research in particular, by far exceed any other country, in both relative and absolute terms.[963] Nonetheless, according to many longevity research advocates, the expenditures are still far from being adequate.[964] Increasing the funding is now a major drive for American grass-roots life-extensionist organizations and public initiatives,[965] as it was for individual life-extensionists since the early 20th century. In all the advocacy efforts for expanding longevity research, since then to the present, the adaptation to the existing economic and state structures has been indispensable.

Since the 1930s, state support has been deemed essential for the progress of longevity science. The 1930s are well known for the unprecedented strengthening of state governments and state ideologies, most notably in the USSR (after the end of the "New Economic Policy" in 1929 and the instatement of Stalin's "personality cult") and Germany (after the assent of National Socialism to power in 1933). Though in a milder form, governmental control was increased under Franklin Delano Roosevelt's "New Deal" (1933-1939) in the US as well. Not surprisingly then, contemporary researchers felt that "the maintenance [of an institute for aging and longevity research] would require moral and financial state support" (1937).[966] The relationship between longevity research and the state might have been a two-end bargain: the research required state ideological endorsement and material support for its continuation, while state ideologies enlisted the hope for life-extension to increase their appeal.

Hence, contrary to the common perceptions of its "radicalism" and "infringement," life-extensionism might be seen as a fundamentally conservationist enterprise. It was based on the desire to preserve the "equilibrium" or "identity" – both bodily and social. Generally, the life-extensionist works have expressed the desire to sustain the individual and social body in an ideal working order. And the order the life-extensionists wished to sustain was, in most cases, the one they lived in. Several cases of social dissension of life-extensionists can be cited.[967] However, these examples might be an exception rather than the rule. The majority of the life-extensionist authors writing in the 1930s through the 1950s seem to express the desire to perfect and/or perpetuate the social order in which they were embedded. The Soviet scientists propagandized socialism as a necessary condition for the pursuit of longevity (probably without any choice to say otherwise, but also very likely in a sincere belief), the German scientists upheld National Socialism, and the Americans supported Roosevelt's capitalism.[968]

The strong ties of longevity research with prevalent state regulation and ideology may have played a decisive role in the institutionalization of gerontology, a field permeated by 'moderate life-extensionist' aspirations. By the 1950s, a gradual movement occurred worldwide toward first national and then international institutional cooperation in the field. Since the 1950s, the US assumed the leading position in the collaborative effort. Since that time, life-extension research and practice, on an international scale, largely meant American life-extension research and practice. As will be exemplified below, the American achievements were trailblazing.

16. The 1950s-1960s. The evolution of rejuvenation methods: From organotherapy to replacement medicine. The cycle of hopefulness

It is difficult to establish precise periods in the history of the life-extensionist intellectual movement. Should its new (modern) stage be associated with the establishment of the International Association of Gerontology in 1950? Is it to be traced from the 'official' advent of the era of molecular biology (1953)? Was the emergence of Denham Harman's "Free radical theory of aging" a turning point in longevity theory and practice (1956)? Or did the publication of Alex Comfort's *Biology of Senescence* (1956) or Robert Ettinger's *Prospect of Immortality* (1964), herald a new period of public awareness? Even without any specific 'breaking point,' the 1950s may mark a transition in this history: In the period c.1930-1950, earlier theories of aging and rejuvenation practices were thoroughly and critically reevaluated, and foundations for generically new ones established. By the 1950s, the hopes for immediate, far reaching rejuvenation through "organotherapy," so rampant in the 1920s, were virtually abandoned. Instead, great expectations for future life-extension were pinned on continuous basic longevity research, informed by the explosive scientific and technological advances of the mid-century: replacement medicine, molecular biology and cybernetics. Thus, a trend can be observed toward and after the 1950s: a marked decrease of interest in immediate rejuvenation and increase of interest in biological theories of aging that endeavored first to fully elucidate its mechanisms and thereafter pinpoint targets for intervention.

It is particularly difficult to segment the history of life-extensionism in the first half of the 20th century due to the thematic continuity of research in this period. It is easier to see a transition phase from the early 1950s onward. Figures 1 and 2 (at the end of this work) show the amount of scientific publications worldwide in two areas traditionally associated with the life-extensionist thought: Rejuvenation and Theories of Aging, based on Nathan Shock's comprehensive *Classified Bibliography of Gerontology and Geriatrics* (1951, 1957, 1963).[969] As Figure 1 shows, the discussion on rejuvenation arose immediately after WWI (prior to 1920 there were virtually no publications on the subject) and reached a peak in 1928 (27 publications). After that year, the discussion gradually subsided, and since 1940 reached a plateau of 2-4 publications yearly (with a slight resurgence in 1952). In the 1930s-1940s, there were relatively few new generic interventions proposed for rejuvenation. The works of the 1930s largely elaborated and critically assessed the propositions that had arisen in the 1900s-1920s, most notably, hormone replacement therapies. The enthusiasm of the 1920s regarding these interventions, their promise and effectiveness, became considerably mitigated.[970] The initial motivations to find far-reaching rejuvenative means were substituted by more modest ones.

As Metchnikoff asserted, the science and art of life-extension consist in eliminating "damaging elements" while strengthening or replenishing "noble elements" of the body, in order to achieve a harmonious state of stable equilibrium.[971] These basic concepts have continued to appear in the popular life-extensionist literature up to the present time, however, the understanding of what constitutes an equilibrium, precisely what elements are harmful or beneficial, and how these can be respectively eliminated or replenished, have undergone dramatic transformations. The anticipated effectiveness of the interventions also ranged widely. After a refractory period of the late 1940s-early 1950s, toward the end of the 1950s new hopes for practical rejuvenation began to emerge.

The publicity gradually shifted to anti-oxidants, after Denham Harman's seminal work of 1956 delineating the "free radical theory of aging."[972] Since 1956, anti-oxidant treatments have assumed the function earlier associated with "anti-toxic" sera in eliminating age-related "waste products" or "damaging elements" and come to dominate the field of longevity research and advocacy for the next half a century. The suggested forms and effects of supplementation or "replenishing the noble elements" – a prevalent idea in virtually all rejuvenation attempts – also metamorphosed continuously. (The anti-oxidants too came to be seen as "supplements.") The rejuvenative opotherapy or organotherapy of the 1900s-1920s, using animal organ extracts or derivatives as the chief means of supplementation, gave birth to synthetic hormone replacements developed during the 1930s-1950s.[973] In the late 1950s-early 1960s, hormonal therapy appears to resume its prominence in the anti-aging arsenal, when synthetic hormones cheapened and became more widely available. Thus, the declining levels of 17-ketosteroids, and their particular forms, such as Dehydroepiandrosterone (DHEA), Androsterone and Estrone, were said to play a crucial role in aging by the American inventor of oral contraceptives, Gregory Pincus (1955). Pincus and his group sought to develop "a regime of steroid administration designed to restore the urinary pattern to that of normal young men."[974] In the 1960s, in the US, thyroxine (a thyroid hormone) was popularized for anti-aging by Charles Brusch and Murray Israel, estrogen and progesterone by Robert Wilson, and thymosin (a thymus hormone) by Allan Goldstein.[975] Hormone levels of the hypothalamus (such as dopamine and growth hormone-releasing hormone) were assigned a crucial role in the onset of aging by the American researchers William Donner Denckla and Caleb Finch (c. 1975), who referred to the hypothalamus as a hormonal "aging clock."[976]

Since the 1960s, supplementation with anti-oxidants and hormones became big business. Corporate interests can be easily observed in the related American life-extensionist literature. Thus, in Durk Pearson and Sandy Shaw's *Life Extension. A Practical Scientific Approach* (1982), large pharmaceutical companies were hailed as the main flag-ships of progress, while the Food and Drug Administration was posed as the main bureaucratic snag and small firms were implicated as the major source of charlatanism.[977] Capitalism and struggle for funding thus remained dominant motifs in American life-extensionism.

Not only hormones, but entire organs became targets for supplementation and replacement. The timeline of transplantation medicine was continuous[978] and should necessarily include the original rejuvenation attempts of the 1920s-1930s through sex gland transplantation (Serge Voronoff, Harry Benjamin), "artificial revival" by resuscitation devices and embryonic/juvenile cell transplants and tissue transplants (Alexis Carrel).[979] However, in the 1950s, transplantation surgery advanced drastically, following a deeper understanding of immune rejection and tolerance mechanisms. The development of immunosuppressive means (drugs and radiation), that would enable human organ transplants "to take," became possible thanks to the studies conducted since the 1950s.[980] In the 1950s-1960s, transplantations were conducted for a great array of organs, from head to toe, mainly by American surgeons. There were transplanted the human kidney (1954), heart valve (1955), bone-marrow containing adult stem cells (1956). There followed the transplantation of the liver (1963), lung (1963), hand (1964), pancreas (1966), and heart (1967). Head was transplanted in a monkey in 1963.[981] In the 1960s, experiments with testicular and ovarian transplants continued as well, though apparently in a rather diminished scope and without any practical human applications.[982] Moreover, in the late 1950s-early 1960s, the possibilities of growing desired organs and tissues outside of the body by directing the differentiation of "totipotent" cells or regenerating tissues within the body, were being raised (Philip Siekevitz, 1958).[983] Reconstructive/cosmetic surgery also expanded dramatically after World War II.[984]

In the 1950s, the advances in prosthetics, bionics and resuscitation technology paralleled those in live tissue transplantation. Landmarks in the field included the artificial heart valve development (1951), the first successful cardiac pace-making electro-stimulators (1952), the heart and lung machine (1953), and the artificial kidney (1955). In 1956, hyperbaric oxygen pressure chambers were successfully applied as an aid in cardiac surgery and resuscitation. (Resuscitation apparatuses may just be considered as large artificial organs, whose size would in time diminish.) In the late 1950s-early 1960s there emerged the first practical artificial hip replacements (1962), the first prototypes of biosensors and artificial blood (1962), a computer-controlled arm (1963), and synthetic skin (1965).[985] The timeline of prosthetics/artificial organ development, similarly to live transplants, appears to be rather continuous and extensive.[986] Yet, even though the first prototypes had emerged earlier; in the 1950s through the early 1960s the fields of live transplantations and prosthetics markedly expanded in application, scope and publicity, with the US in the lead.[987] Notably, live transplantation technology converged with artificial organ and resuscitation technology, with developments in one field encouraging developments in others. Thus, for example, John Heysham Gibbon (1903-1973) of

Jefferson Medical College, Philadelphia, performed in 1953 the first successful open-heart bypass graft operation, aided by the heart-lung machine that he developed. Another converging stream was the advancement in tissue and organ preservation, particularly by cryo-preservation, making the prolonged storage of transplantable organs possible.[988]

The advances of the 1950s in live tissue transplantation and prosthetics were so momentous that many, and mainly in the US, began to speak about the possibility of replacing *all* worn-out body tissues and organs (including parts of the nervous/brain tissue) through biological or bionic replacement parts. Some researchers cautioned that transplantations could impair the delicate body balance.[989] Still, the anticipation of unlimited tissue replacements was rampant. This anticipation was expressed, for example, by the American transplantation surgeons Robert Brittain and Richard Lillehei in 1963 and the biochemical engineer Lee Browning Lusted in 1962.[990] The potential for the repair of the human machine by mechanical aids and resuscitation devices, by transplants, organ and tissue cultures, tissue regeneration and tissue preservation, appeared to the life-extensionists of the early 1960s to be virtually unlimited. Thus, famously, the American physicist Robert Ettinger (1918-2011) in *The Prospect of Immortality* (1964) envisioned a complete "overhaul and rejuvenation" of the entire body that will be achieved by the replacement therapy of the future, following the body's indefinite preservation by freezing ("cryo-preservation" or "cryonics"), and thus attaining virtual immortality.[991] By 1964, there existed in the US at least 3 grass-roots societies dedicated to setting up "the freezers," though the first "patient" – the psychologist Dr. James Hiram Bedford (b. 1893) – was cryo-preserved only in 1967. Ettinger's "prospect of immortality" was a projection of contemporary technologies.[992] These popular themes that emerged with great force in the late 1950s-early 1960s – the unlimited potential of organic and bionic replacements, tailor-made culture of organs and tissues for transplantation, and cryonics – have continued to be discussed and developed in life-extensionist literature up to the present time.[993]

A distinction can thus be observed in the attitudes toward potential rejuvenative treatments before and after the 1950s. Prior to the 1930s, there was much interest in practical rejuvenation through opotherapeutic or organotherapeutic means (most notably, through hormone replacements, sex gland stimulation and transplantation, blood and sera transfusions). After that, in the 1930s through 1940s, the discussion of immediate rejuvenation seems to have been (at least temporarily) laid to rest (as shown by the numbers of relevant publications, see Figure 1 and Table 1 below). Since the 1950s, the focus shifted to the direct descendants of organotherapy: live organ and bionic transplants. However, unlike the earlier rejuvenation attempts that sought immediate and far-reaching effects (which in later assessments appeared exaggerated),[994] the full life-extending potential of organ transplants was hoped to be realized in a more distant future.

17. Theories of Aging. A new longevity research paradigm based on contemporary scientific advances: the advent of molecular biology and cybernetics

The 1950s may also represent a demarcation period with regard to theories of aging, which were transformed by contemporary scientific and technological advances, particularly by the emergence of molecular biology and cybernetics. Here too, the US was in the lead.

The development trend is shown by the pattern of publications on theories of aging, which was markedly different from that on rejuvenation (Figure 2). According to Nathan Shock's *Classified Bibliography*, since 1900 through 1923 the yearly number of theoretical publications was relatively stable (2-4 yearly). Since 1924 through 1946 it grew in bulk (to about 7-8 per year on the average, with sporadic fluctuations), then notably subsided, reaching a minimum of 1 or 2 publications in 1950-1951. Since 1952 the numbers again started steeply increasing, reaching 24 publications in 1960.[995] According to these figures, it can be inferred that by the 1950s the issue of practical human rejuvenation became deemphasized. In contrast, publications on theories of aging rose to prominence in the 1950s (Figures 1 and 2, Table 1). The theoretical works were infused by the premise that from a deeper understanding of mechanisms of aging, new life-prolonging interventions may eventually emerge (though likely after a very considerable delay).

The theories of aging developed in the 1920s through 1940s gave way to generically new ones in the 1950s. No single cause, such as hormonal imbalance or auto-intoxication, was any longer believed to be the main cause of aging, but perhaps only a single (and minor) one.[996] New damaging agents were being sought. In the 1920s-1940s, a dominant theory of aging was the entropic transformation and condensation of cell colloids (mainly proteins), forming flocculates and precipitates that clog the cell machinery. Yet, there was a great uncertainty of the colloids' molecular structure.[997] Some surmises were advanced regarding "cyclization" as an underlying chemical cause of colloid condensation, yet the colloids were essentially seen as blobs of matter that "congest" or "dissolve," with molecular composition and mechanisms unknown. A possible genetic basis of aging and longevity was also proposed in the 1920s and continued to be investigated through the 1940s.[998] Yet it was described mainly in terms of population genetics, without reference to molecular mechanisms. There were also discussions of "cell memory" yet again without knowing its molecular works.[999]

The advent of molecular biology in the late 1940s-early 1950s changed the theoretical perception dramatically, as the precise molecular structure of proteins and genetic material became known.[1000] The new theories of aging of the late 1950s were aligned to the new paradigm of molecular biology, attempting to identify causes of aging at the molecular level.[1001] Foremost, there appeared the seminal work "Aging: A Theory based on Free Radical and Radiation Chemistry" (1956) by the former Californian Shell Oil chemist, Denham Harman (b. 1916, working consecutively in Berkeley, Stanford and Nebraska universities). This work played a pivotal role both in practical attempts at rejuvenation and in theoretical notions of aging. The theory fully incorporated concepts from molecular biology, such as preventing DNA damage that was said to be caused by oxidative free radicals. Based on the theory, Harman proposed the application of "anti-oxidants" or "easily reduced compounds" that would neutralize the free radicals. Such compounds, he suggested, can be used as "chemical means of prolonging effective life" and to "slow down the aging process." The theory was mainly derived from earlier industrial developments (the prevention of food spoilage and rubber aging by anti-oxidative agents), radiation chemistry and radiobiology (namely, the anti-radiation protection which boomed after the first deployments of nuclear energy by the US military in 1945 and electrical power industry in 1951).[1002] The theory gave rise to the wide use of anti-oxidant supplements, building on the infrastructure of the vitamin industry which was by then firmly established.[1003]

There also emerged in the 1950s the "Cross-linkage theory of aging" advanced by the Finnish/American researcher Johan Bjorksten (1907-1995, working in Madison, Wisconsin), in his influential study "A common molecular basis for the aging syndrome" (1958). The theory was a direct descendant of the "colloidal condensation" theory of aging. According to Bjorksten's theory, "crosslinks" between biomolecules (proteins, nucleic acids, etc.) produce rigid agglomerates that block cellular functions.[1004] Later on, agglomerates of cross-linked, lipid-containing "lipofuscin" (or "age pigment") were increasingly seen as a major type of molecular debris leading to senescence.[1005] Based on the theory, Bjorksten sought to employ proteolytic enzymes or free radical bursts (oxido-reductive depolymerization) that would be able to break molecular crosslinks and thus restore tissue flexibility and molecular turnover to prolong life. He also sought to use chelating agents that would be able to remove heavy metals, a major cause of cross-linkage, and thus prevent the formation of the clogging aggregates in the first place. "Gradually removing crosslinked aggregates faster than they are formed," he insisted, "could be our last remaining challenge before we reach the ultimate future, where death is not caused by aging, but only by accident, suicide, or violence."[1006] A special interest of Bjorksten's was the use of "chalones" – growth-inhibiting hormones or polypeptides of animal origin, active in very small doses – which he believed could be employed in the treatment of cancer and auto-immune diseases and in wound repair. These lines of research have been continued by life-extensionists to this very day.[1007]

Another ground-breaking theory of aging that fully incorporated the molecular-biological perspective was the "Somatic mutation theory" proposed in 1958-1959 by the Italian/American radiation biologist Gioacchino Failla (1891-1961) and the Hungarian/American nuclear physicist Leó Szilárd (1898-1964, the author of the concept of nuclear chain reaction of 1933).[1008] The theory suggested that aging is caused by random DNA damage in somatic cells and that the extent of damage is enhanced by radiation. DNA damage and its exacerbation by radiation were also the underlying concepts in the work of another prominent American researcher of aging, Bernard Louis Strehler (1925-2001, from the NIH Gerontology Research Center, Baltimore, Maryland), as presented in 1959.[1009] The "Somatic mutation theory" is considered to be the first truly stochastic theory of aging.[1010] A later stochastic theory was the "error catastrophe theory" proposed in 1963 by the British/American biochemist Leslie Eleazer Orgel (1927-2007, Cambridge University, UK - Salk Institute for Biological Studies, La Jolla, California), which emphasized the accumulation of errors in the protein-synthesis apparatus (particularly RNA and enzymes necessary for DNA transcription). As Orgel suggested, "the inheritance of inadequate protein-synthesizing enzymes can be as disastrous as the inheritance of a mutated gene."[1011] It was also argued that inaccurate protein synthesis and inaccurate DNA synthesis are coupled phenomena.[1012]

Several American researchers of the 1960s – Ronald Hart, Richard Setlow, George Sacher, Richard Cutler, Bernard Strehler, and others – actively investigated enzymatic mechanisms of DNA repair, as it was believed that an improved DNA repair system can protect the stability of the human genome against mutations and thus radically increase the human life span.[1013] In the 1960s, Roy Lee Walford (1924-2004, from the University of California at Los Angeles) developed the "auto-immune" or "immunologic" theory of aging, where the increased attack on the organism by its own immune cells was supposed to be due to somatic DNA aberrations (1960, 1969). Among the life-extension methods investigated by Walford, there were underfeeding (calorie restriction), cooling (hypothermia) and hibernation (induced by such means as chlorpromazine and marijuana derivatives, and other hibernation induction triggers), and anti-inflammatory diets and medications.[1014] In 1961, Leonard Hayflick (b. 1928, at the time working at the Wistar Institute of Anatomy and Biology, Philadelphia) posited a limit to the number of cell divisions, suggesting that cellular aging is genetically predetermined.[1015] The molecular biological view of mechanisms of aging converged with contemporarily emerging mathematical theories of mortality[1016] and evolutionary theories of aging that were highly conscious of molecular mutagenesis and sought to identify genetic targets for anti-aging interventions.[1017]

The development in the late 1940s of Cybernetics (Norbert Wiener, 1948), Information Theory (Claude Shannon, 1949)[1018] and the emergence of the first digital electronic computers (John von Neumann's works on computer architecture, and the Electronic Numerical Integrator And Computer – ENIAC constructed in 1946) were other momentous influences on changing theories of aging. After the advent of cybernetics and information theory, more rigorous quantitative descriptions of homeostasis or deregulation due to aging processes were sought, and analogies between automata and human organisms suggested.[1019] Since the 1950s, Information Theory has been widely applied in biomedicine, in diagnosis in particular, especially for the assessment of aging and age-related diseases.[1020] According to Bernard Strehler (1960), the purpose of applying Information Theory in aging and longevity research was to determine whether "the deterioration of organisms with time" is due to "the disorganization of a stored collection of directions (information)" or "the disruption of information transmitting or decoding machinery." Another goal was to find "the special problems encountered in the design of self-replicating machines" and "self-replicating and self-repairing machines," particularly referring to DNA replication and DNA repair.[1021] Special cybernetic consideration was given to increasing the "reliability" of a living "automaton," for example, through redundancy of its parts, continuous replacement of parts, or incorporating quality monitors over the working components and replacing those components that show signs of "wearing out."[1022]

In more futuristic visions, computers figured as harbingers and models of extreme longevity, since, in the words of the British Artificial Intelligence pioneer Donald MacCrimmon MacKay (1922-1987, the founder of the Department of Communication and Neuroscience at the University of Keele, England), "such a machine could be everlasting, because repair men could plug in any part as soon as it needed replacement" (1963).[1023] The concept of "cyborgization" (the term coined by the American electronics engineer Manfred Clynes and psycho-pharmacologist Nathan Kline in 1960), or indefinitely maintainable man-machine synergy, was advanced. The possibilities of using the rapidly accelerating computer technology for life-extension appeared unrestricted. As it was suggested in the 1960s, the computers could be used for massive biological data analysis and thus discovering effective life-extending interventions. They would calculate and prescribe the balance of billions of biochemical interactions. They would preserve and expand the human personality. They would enable the unlimited production and communication for future long-lived generations. These possibilities, raised in the early 1960s, have continued to be raised to the present day.[1024]

With respect to futurism, two other crucial advances in science and technology, though not immediately related to biomedicine, shaped the vision of life-extension: namely, nuclear research (epitomized by the "atom-splitting" Manhattan project of 1939-1947) and space exploration (Sputnik, 1957; the first man in space, 1961; the Apollo program, 1961-1975). As mentioned, radiation research was a basic for Denham Harman's "Free radical" theory of aging (1956), and the massive data collection on anti-oxidant protection against radiation sickness has served to justify the dietary supplementation with anti-oxidants for life-extension.[1025] More radically, attaining unlimited nuclear (and ideally thermonuclear) energy was seen as a likely concomitant to attaining increased individual longevity. This was the view of the British pioneer of population genetics, John Burdon Sanderson Haldane (1892-1964, consecutively with Oxford, Cambridge and London universities, and since 1956 working on statistics, genetics and biometry in Calcutta and Orissa, India). In Haldane's forecast, "rational animals of the human type cannot achieve the wisdom needed to use nuclear energy unless they live for several centuries" and the long-lived individuals will be likely resistant to radiation (1963).[1026] An alternative to this scenario, according to Haldane, is that a nuclear war will wipe out the human race or reduce it to barbarism. According to Robert Ettinger, unlimited energy (to be supplied through nuclear fission or nuclear fusion), availability of matter (to be ensured by earth and space exploration), and unlimited organization (or labor to be provided by robotics) are the three components that will supply the unlimited wealth necessary to sustain a population with potentially boundless longevity and size.[1027]

Space exploration, in such predictions, will play an indispensable role in preventing overpopulation and deficit of resources that may arise due to increased population longevity, by providing new settlement areas and mining capabilities. And conversely – as suggested by Haldane and the American science fiction writer Robert Anson Heinlein (1958)[1028] – in order to permit space travel and settlement, human beings will likely be genetically modified (to adapt to new living conditions) and extremely long-lived (to last the journey). The potential for progress, exemplified by computer, nuclear and space research, seemed unlimited, and life-extensionism fitted well into the paradigm.

Perhaps of more immediate significance was the fact that nuclear research and space exploration were then perceived (and continue to be perceived) by many as a model for how longevity research must be pursued.[1029] "The splitting of the atom" or "sending man into space" were government-sponsored projects, involving tens of thousands of scientists and billions of dollars of investments, producing results that may have seemed science fiction just a few years earlier. And it is through massive participation (based on competition and/or organized effort) and massive investments that longevity research was ever hoped to succeed.

Though a "Manhattan-type" project to combat aging remains a dream, as it was a dream in the 1950s, nevertheless, significant steps were taken in the 1950s toward institutionalization, collaboration, public recognition and support of this endeavor. This might have been a partial reaction to the war, as the unprecedented horror of massive slaughter may have produced a corresponding unprecedented desire and drive not only to prevent future wars, but to cooperatively postpone suffering and death altogether. The works of at least two prominent life-extensionist authors – the American biochemist Linus Carl Pauling (1901-1994, working at the California Institute of Technology, the recipient of the Nobel Prize in Chemistry in 1954 and the Nobel Peace Prize in 1962) and the British gerontologist, poet and political philosopher, Alexander Comfort (1920-2000, conducting research at the University College, London) – make the connection between pacifism and life-extensionism explicit.[1030]

As the present data suggest, life-extensionism was able to survive under any current socio-economic and ideological conditions. Yet it appears to have thrived somewhat better under peaceful and prosperous conditions. It is of note that the discussions on rejuvenation rose to prominence immediately after WWI (Figure 1, Table 1). The number of publications on rejuvenation began steadily declining since 1930, during the period of world upheavals (the world economic crisis of 1929-1933, the rise of personality cults, WWII, the post-war impoverishment in Europe). On the other hand, the number of basic longevity studies began steeply rising since 1953, following the post-War reconstruction (Figure 2). Thus, the increased interest in life-prolongation may correspond to periods of peace, stability and prosperity, raising the aspirations to have 'more of the good life.' Yet, survival was a persistent feature at any time, adapting to both the prevalent social-ideological and scientific paradigms. Thus, after WWII, any mention of eugenics seems to have disappeared from the American life-extensionist literature (in line with the common disfavor of the subject). Not many references to religion could then be found either. Yet capitalism and the untiring pursuit of funding, both governmental and private, remained dominant. As the above examples show, American life-extensionist works also inseparably fitted into the currently dominant scientific paradigms.[1031]

18. British Allies – Literary and philosophical life-extensionism: The optimistic vs. the pessimistic view. The reductionist vs. the holistic approach

As exemplified in the previous sections, since the 1950s the US has dominated the field of longevity research. Yet, in that period, the contribution of British researchers was also very considerable. This does not seem to have always been so. From an early date and well up to the 1940s, most of the discussions of longevity in Britain seem to have been decidedly theoretical, philosophical or literary, rather than experimental-scientific or practical. Literary fiction and philosophical treatises, being the primary venues of discussion, were chiefly occupied by envisioning the promise and peril of extending longevity.

Britain had an enduring tradition of philosophical life-extensionism, even of outspoken immortalism. Its leading proponents included Roger Bacon (1214-1294),[1032] Francis Bacon (1561-1626),[1033] William Godwin (1756-1836)[1034] and William Winwood Reade (1838-1875).[1035] At the end of the 19th century, the novelist and philosopher Samuel Butler (1835-1902) was a chief proponent of the Lamarckian theory of aging that sought "the principle underlying longevity" in the "lateness in the average age of reproduction" (Samuel Butler, *Essays on Life, Art and Science*, 1889).[1036] The positivist philosophy of Herbert Spencer (1820-1903) too was highly compatible with life-extensionism.[1037] Spencer's dictum that "the highest conduct is that which conduces to the greatest length, breadth, and completeness of life," could serve as a life-extensionist motto, and it did serve as such in Albert Dastre's writings.[1038] The Spencerian parallelism between a harmonious body and a harmonious society, and the notion of progress as achieving a long-lasting social and biological "equilibrium" reappeared in many life-extensionist authors. According to Spencer, there are two mutually exclusive attitudes to life prolongation, the optimistic and the pessimistic:[1039]

> By those who think life is not a benefit but a misfortune, conduct which prolongs it is to be blamed rather than praised…Those who, on the other hand, take an optimistic view, … are committed to opposite estimates; and must regard as conduct to be approved that which fosters life in self and others, and as conduct to be disapproved that which injures or endangers life in self or others…. Legislation conducive to increased longevity would, on the pessimistic view, remain blameable; while it would be praiseworthy on the optimistic view.

Spencer and other British philosophers sympathetic with life-extension favored the optimistic view, and so did several prominent English fiction writers, such as George Bernard Shaw (1856-1950). The tradition of radical life-extensionism has recently been continued by British philosophers, such as John Harris, Max More and (the Swedish born) Nick Bostrom – the founders of contemporary transhumanism.[1040]

The pessimistic view as to the possibility and desirability of life prolongation existed as well, and even appears to have been dominant in Britain, detailing why life-extension should not be pursued. Here too, the main venues of discussion were philosophical treatises and works of literary fiction. The tone was set by Thomas Robert Malthus (1766-1834), in his dispute with the proponents of indefinite life-extension, William Godwin and Nicolas Condorcet (Malthus, *An Essay on the Principle of Population, as it Affects the Future Improvement of Society with Remarks on the Speculations of Mr. Godwin, M. Condorcet, and Other Writers*, 1798).[1041] Contrary to a common belief, the essay was not exclusively concerned with dangers of overpopulation *per se*, derived from the hypothesized inherent geometric growth of the population that cannot be supported by the inherent arithmetic growth of food supplies. The underlying subject was rather the opposition to the idea of unlimited human perfectibility and, in the first place, the human ability to achieve indefinite prolongation of life. According to Malthus, the suppositions of indefinite life-extension overlook inherent limits of human development that can by no means be overcome. Thus, according to Malthus, the main "fallacy" of the "conjecture concerning the organic perfectibility of man, and the indefinite prolongation of human life" is that it "infers an unlimited progress from a partial improvement, the limit of which cannot be ascertained." Thus, concerning the upper limits of human endurance, Malthus pointed out that a driven mind can enable a person to run twenty miles, but not a hundred. Another example of a "limit" was furnished by the breeding of animals: Even though sheep breeders favor small heads and short legs in their sheep, "the head and legs of these sheep would never be so small as the head and legs of a rat." By analogy, partial improvements in longevity cannot lead to unlimited improvements. For Malthus, a sworn enemy of the idea of evolution ("organic perfectibility"), transformations from legged to leg-less creatures, and vice versa, appeared entirely impossible.[1042] Malthus' main thesis was that increases in food production would always lag after increases in population, leading to perpetual threats of starvation and making sustained human development impossible (the theoretical supposition that proved wrong in practice time and again). Increased longevity, according to him, would only increase the population burden. He noted "the very great additional weight that an increase in the duration of life would give to the argument of population" (the supposition that also proved wrong).[1043] Even while foreseeing an unsupportable increase in population, the use of birth control, the means to "prevent breeding," was rejected by Malthus out of hand, as "promiscuous concubinage" or "something else as unnatural" that would "destroy [the] virtue and purity of manners." Ultimately, Malthus suggested, if human beings did not suffer and die, "Providence must necessarily be at an end, and the study will even cease to be an improving exercise of the human mind."

The Malthusian logic has persisted in the intellectual discourse on life extension, in Britain and other English-speaking countries. The opposition to the very idea of increasing longevity has been widespread, even among physicians and researchers of aging,[1044] using quite a typical (and standardly refutable) set of ethical arguments.[1045] Yet, the opposition to the pursuit of life-extension appears to have been the strongest in literary circles, most prominently in Britain. Since an early date, the representation of this pursuit was usually highly negative.[1046] Several positive literary treatments can be cited, particularly Bernard Shaw's works *Man and Superman* (1903)[1047] and *Back to Methuselah* (1921).[1048] Following Shaw's enthusiasm, there was in the 1930s a sprinkle of optimistic, positive literary representations. James Hilton's *Lost Horizon* (1933)[1049] romanticized life-extension achieved by the residents of Shangri-La. However, Western technology is largely irrelevant, even antagonistic, to this quest. (The only hint at the life-extending method practiced in Shangri-La is that it involved "drug-taking and deep-breathing exercises," as well as the assumption that "the atmosphere" of the Himalayan valley is "essential.") Aldous Huxley's novels seem to exhibit a balanced attitude to life extension, yet strongly tending toward its negative representation.[1050] William Olaf Stapledon's approach is also balanced, yet tending toward the positive.[1051] Some authors expressed their strong opposition to aging and dying, yet paradoxically without any apparent implications for a practical pursuit of life extension.[1052] Generally, the prevailing attitude to the pursuit of life-extension in English literature, since the *fin-de-siècle* to the present, has been hostile.[1053]

Still, "the optimistic view" of life prolongation was present in English fiction. From the quest for the Holy Grail and King Arthur's suspended animation at Avilion in Thomas Malory's *Morte Darthur* (1470),[1054] life-extension has been a noble literary theme. In the early twentieth century, Bernard Shaw's *Back to Methuselah* was the foremost optimistic literary treatment of the subject. Shaw's representation sharply contrasted with earlier *fin-de-siècle* and contemporary dystopian treatments of extreme longevity in English literature, which presented its attainment as leading to eternal boredom, stagnation, inequality, exploitation and overpopulation. In contrast, Shaw's work provided a set of arguments why life-extension should be actively pursued. Its main benefit would be to allow for the accumulation of wisdom necessary for the survival of the human society.[1055] This belief accorded with Shaw's humanistic, socialist and pacifist worldview.

A critical trait in Shaw's views concerning the prolongation of life seems to be its idealistic, holistic, anti-reductionist and anti-technological proclivity. Shaw's holistic vision of life-extension employed contemporary concepts of "eugenics,"[1056] "psycho-regulation,"[1057] "energy field,"[1058] and "Neo-Lamarckism" (the latter had a strong relation to the work of the contemporary Austrian Neo-Lamarckist rejuvenator Paul Kammerer[1059]). Shaw did not seem to believe that any technological devices or material adjustments generally, are of a decisive value for life extension. Devout as he was to vegetarian diet, hygiene and sanitation, so was he opposed to vaccination and vivisection.[1060] According to him, the power of the will, the Yoga-like meditations, the direct control of the mind over the body, over inheritance, over the Life Force, are more important for longevity than pharmacology. In *Back to Methuselah*, Shaw mocked the possibility of creating a healing "lozenge" (in the context ironic of Metchnikoff's diets). According to Shaw, the effects of art and science can never compare to those he believed can be produced by a meditative psyche. The homunculus (Galatea) created by art and science (*Back to Methuselah*, Part V) is only a frail parody of a human being. In contrast, a Yogi-like mastermind, by directing and accumulating the vital energy, can succeed in altering the body and sustaining it indefinitely.

Thus, Shaw's views of life-extension seem to have been rather outside the field of reductionist biology, and were more attuned to a 'holistic' hygienic approach. This is hardly surprising, as Shaw himself, as well as other British life-extensionist philosophers and men of letters of the time (such as James Hilton) were outside the field of reductionist science. As discussed in the previous chapters, in Germany and France, after a surge of interest in reductionist rejuvenation of the 1920s, a reaction occurred in the 1930s toward a more holistic, hygienic approach to life-extension. In Russia, rejuvenation techniques remained popular through the 1930s and 1940s. In the US, in the 1930s, there occurred a shift of emphasis from the attempts at immediate rejuvenation toward basic research of mechanisms of aging (still a reductionist approach, yet less ambitious). In contrast, in Britain, reductionist rejuvenation did not seem to be very popular in the first place (with a few exceptions, which will be discussed in the next section). The literary discussions seem to have dominated over actual research of aging or practice of rejuvenation. Was British conservatism stalling the encroachments of rejuvenators, rather preferring to commit attempts at life-extension to paper instead of one's own body? Was it a manifestation of an aristocratic, "gentlemanly" disdain for the practice of mechanical "tinkering" and crude "butchering"?[1061] Whatever the explanation might be, the battle between reductionist and holistic approaches that was fought by Shaw, has continued to be waged on the pages of English science fiction well to the present.[1062]

There appears to be a paradox in Shaw's attitudes to life-extension. Namely, Shaw's suggestions that humanity should emulate Methuselah, coexisted with his advocacy of euthanasia or "humane killing." Thus, in 1922, in the year *Back to Methuselah* was first performed, Shaw also wrote on "Imprisonment": "If people are fit to live, let them live under decent human conditions. If they are not fit to live, kill them in a decent human way." Furthermore, "if we kill incurable criminals we may as well also kill incurable invalids." In a later revision of that essay made in 1934, the terms were even harsher. While in 1922 Shaw suggested that "[capital punishment] may be extended to criminals of all sorts," in the 1934 version, he proposed extending it to "social incompatibility of all sorts."[1063] In 1933, in the preface to *On the Rocks*, Shaw further spoke of "the necessity for killing." Moreover, the contemporary surge of the "movement … for the extermination" in Nazi Germany and Communist Russia, in the British Empire and the United States, and in India among Hindus and Moslems, altogether indicated "a growing perception that if we desire a certain type of civilization and culture we must exterminate the sort of people who do not fit into it."[1064] How could any of these statements be compatible with the desire to prolong human life?

One possible explanation might be that Shaw spoke in fashion. In the early 1930s, the time of the world-wide "growing perception" of the "political necessity for killing," other British "humanistic" thinkers expressed their support for euthanasia. For example, the British philosopher and mathematician Bertrand Russell (1872-1970), suggested in 1934 killing people in cases of "hydrophobia," "pneumonic plague," "congenital idiocy" and other cases which "cannot be useful to society." Also "criminals condemned to long terms ought to have the option of euthanasia." But Russell was consistent; according to him, the prolongation of life was undesirable both for the very sick and the very old. Thus, in "The Menace of Old Age" (1931) he was greatly worried by the prospect that "every increase in medical skill is bound to make the world more and more conservative." Hence, he proposed "to prevent all researches calculated to prolong the life of the very old." And in "How to grow old" (1944), he maintained that "in an old man ... the fear of death is somewhat abject and ignoble."[1065] In 1935, Bertrand Russell, Bernard Shaw and Herbert George Wells (1866-1946), were among the co-founders of the British "Voluntary Euthanasia Society."[1066] While Russell may have been consistent, why would Shaw join that society, when he advocated "the extension of human life to three hundred years"?[1067] And why would Wells join it, who dreamt of a "Utopia" where people enjoy "peace, power, health, happy activity, length of days and beauty" (1923)?[1068] One may suspect that, as the valuation of life may have generally eroded in the 1930s (cf. the contemporary events in Germany, Russia, etc., mentioned by Shaw), the authors blended into a prevailing thought pattern. Indeed, Shaw's pro-euthanasia statements became more harsh and pronounced in the 1930s. Yet in 1922, they were harsh enough. Wells too, hoped for the "euthanasia of the weak and sensual" at least as early as 1901.[1069] The fashion for euthanasia may have started earlier and only intensified in the 1930s.[1070] Yet, the question remains regarding the very principal possibility of any reconciliation of life-extensionism with euthanasia.

The coexistence of these notions may not be entirely paradoxical. As proposed by Metchnikoff, the art of life-extension consists in strengthening or replenishing "noble elements" while attenuating or destroying "harmful elements" in the body, in order to maintain a harmonious state of body equilibrium.[1071] This general assumption might have been shared by most life-extensionists, if only they knew exactly what "elements" and in which quantities were "noble" or "harmful," and exactly how they could be respectively replenished or destroyed. Moreover, the metaphorical parallelism between the harmonious functioning of the body and the harmonious functioning of the society was virtually ubiquitous in life-extensionist writings in the first half of the twentieth century. It would then appear logical that the possibility of eliminating "harmful elements" in the society should have been also raised. But in fact it was very seldom raised. When speaking of the society, life-extensionist authors rather preferred to focus on "strengthening the noble elements": enhancing human creativity, physical and intellectual growth, apparently in the hope that the positive, "noble" elements would eventually overwhelm the "harmful" ones. They propagandized the conservation of the current social orders, but often turned a blind eye to the repressions and persecutions these orders committed. More likely than not, they were aware of the atrocities occurring at their time, but preferred to 'stay out of it' (after all, not all scientific or sociological treatises always have to relate to atrocities). Rather, they preferred to cultivate their own garden (a 'fool's paradise'?).

Shaw was one of the very few life-extensionist authors who chose to confront the problems of evil and extermination headlong, and to apply the socio-biological metaphor all the way (and not surprisingly waded himself into a deep mire.) Even his pro-euthanasia treatise on "Imprisonment" was predicated on the premise that "life is the most precious of all things, and its waste the worst of crimes." Yet, essentially, "the community has a right to put a price on the right to live in it. That price must be sufficient self-control to live without wasting and destroying the lives of others." People, according to Shaw, should "justify their existence" and give an account of whether they "pull their weight" in the society. If they don't "pull their weight" they can be "disposed of" (1922).[1072] "The essential justification for extermination," he continued in the preface to *On the Rocks*, "is always incorrigible social incompatibility and nothing else" (1933).[1073] Thus, the general life-extensionist desire to maintain social harmony and equilibrium, protecting "our civilization," is present in these works as well, even if it would involve killing those incompatible with the "harmony."

Shaw vehemently denied any malice on his part: "I produce a terrified impression that I want to hang everybody. In vain do I protest that I am dealing with a very small class of human monsters." Killing was proposed for the "incurably ferocious," for "the small number of dangerous or incorrigibly mischievous human animals. ... all hopeless defectives, from the idiots at Darenth to the worst homicidal maniacs at Broadmoor." (Caring for the "idiots" as an exercise in communal survival did not seem to occur in this reasoning.) For all such cases, the solution was that "You kill or you cage: that is all." Yet, Shaw expected that, if the exterminations are carried out on absolute "necessity," the numbers of the killed would be small:

"I should be surprised if, even in so large a population as ours, it would ever be thought necessary to extirpate one criminal as utterly unmanageable every year; and this means, of course, that if we decide to cage such people, the cage need not be a very large one." (Shaw much preferred killing to caging, as presumably the former was less "cruel," even though in no way "deterring.") He warned against the possibility that the killings might be carried on too far, as he believed himself to be "the last person to forget that Governments use the criminal law to suppress and exterminate their opponents" (1922).[1074] In the preface to *On the Rocks*, he further emphasized the immense dangers of abusing the "powers of life and death" and stressed that "there is a large field for toleration here" (1933).[1075] And even earlier, in *Man and Superman*, Shaw noticed that the desire to eradicate vices and cruelties, often led to even worse vices and cruelties, even to killing in holy indignation: "The very humanitarians who abhor [tortures and vices] are stirred to murder by them"(1903).[1076] In short, euthanasia was not to be applied easily.

Non-violent solutions were proposed by Shaw as well. Even the apparently "incorrigible" can be reformed by education: "You can exterminate any human class not only by summary violence but by bringing up its children to be different" (1933).[1077] The possibility of "therapeutic treatment" was also raised: "a considerable case can be made out for at least a conjecture that many cases which are now incurable may be disposed of in the not very remote future by either inducing the patient [or "incurable pathological case of crime"] to produce more thyroxin or pituitrin or adrenalin or what not, or else administering them to him as thyroxin is at present administered in cases of myxoedema." Shaw also noticed the existence of "an unconquerable repugnance to resort to killing" and of "individuals who are resolutely and uncompromisingly opposed to slaying under any provocation whatever." Furthermore, Shaw pointed out, "It may be argued that if society were to forgo its power of slaying, and also its practice of punishment, it would have a strong incentive to find out how to correct the apparently incorrigible." He even considered the possibility that the "party" of the "unconquerable repugnance" to any form of killing may prevail. "But unless we rule killing out absolutely, persons who give more trouble than they are worth will run the risk of being apologetically, sympathetically, painlessly, but effectually returned to the dust from which they sprung" (1922).[1078] Many life-extensionists would wish they were not "returned to the dust," and would rather uphold the proposition of the founder of geriatrics, Ignatz Leo Nascher, "to prolong life as long as there is life"[1079] (at least, as far as one can tell, no open proponent of life-extension was ever implicated in killing, including Shaw). Yet, the "party" or "form" of life-extensionism advocated by Shaw, at least on Shaw's own presumptions (notably, a very rare form), was compatible with euthanasia. And there have been many "forms" and "parties" in life-extensionism.

Even in *Back to Methuselah*, a life-extensionist manifesto through and through, the possibility of extermination is looming. There, the "long-lived" are greatly disturbed by the ferocity, cruelty, destructiveness and stupidity of the "short-lived." As stated by one of the "long-lived": "Therefore I say that we who live three hundred years can be of no use to you who live less than a hundred, and that our true destiny is not to advise and govern you, but to supplant and supersede you. In that faith I now declare myself a Colonizer and an Exterminator."[1080] Shaw's sympathies were clearly with the serene and powerful "long-lived," rather than with the pesky and savage "short-lived." (In a similar fashion, in H.G. Wells' *Men Like Gods* (1923), the Utopians, with all their "peace, … length of days and beauty" pose a potential threat to ordinary humanity. According to one of the Utopians, "From what I know of your people and their ignorance and obstinacies it is clear our people would despise you; and contempt is the cause of all injustice. We might end by exterminating you."[1081]) Shaw and Wells may have wished to induce the "short-lived" to reform their savage ways and progress toward the higher, "long-lived" human forms. Still, the threats of selectivity and of the extermination of the non-select by the select are very disturbing.

The form of life-extensionism represented by Shaw, incorporating killing as a part of the program, did not appear to produce much sympathy. Shaw considered *Back to Methuselah* to be his highest, most significant literary contribution, but apparently it was not often recognized as such.[1082] In fact, it played right into the prevailing dystopian, "pessimistic" English literary-fictional representation of the pursuers of radical life extension. The introduction of the "long-lived" as a potential "exterminator" immediately conjures up the monstrous images of the arrogant and egotistic "immortals" that have been so frequent in English fiction, from Bram Stoker's *Dracula* (1897) to Joanne Kathleen Rowling's *Harry Potter* series (1997-2007). The immortal monster prolongs its life at the expense or at the disregard of all the others, and the only way to stop it is to drive a wooden stake through its heart or smash the utensil into which its personality is uploaded. (The monstrous image is at least as old as Medea who used the power of rejuvenation for murder, in Ovid's *Metamorphoses*, 8 AD.[1083])

Shaw might not have sided with Dracula, but rather with the good doctor Van Helsing who fought the vampire. Shaw was enraged by "whoever enlarges his life ... at the expense of the lives of others," particularly "these two bragging predatory insects, the criminal and the gentleman" (1922).[1084] Once again, "The political necessity for killing [a dangerous human being] is precisely like that for killing the cobra or the tiger: he is so ferocious or unscrupulous that if his neighbors do not kill him he will kill or ruin his neighbors" (1933).[1085] Furthermore, Shaw exhibited a great uncertainty regarding the kinds of people that would be "incorrigibly" incompatible with the society and for whom imprisonment or killing would be justified. Thus, he noted that "most prisoners [are] no worse than ourselves." He was uncertain about "who is to be trusted with the appalling responsibility of deciding whether a man is to live or die, and what government could be trusted not to kill its enemies under the pretence that they are enemies of society." He proposed that "certain lines have to be drawn limiting the activities of governments, and allowing the individual to be a law unto himself." "Freedom of speech and conscience" must be secured, including the toleration of "sedition and blasphemy" as conditions for social evolution. Furthermore, he doubted whether "it would ever be thought necessary to extirpate one criminal as utterly unmanageable every year." Yet, despite all these reservations and stipulations, Shaw believed that "Such responsibilities [deciding whether a man is to live or die] must be taken" by the government (1922).[1086] And thus the possibility of selecting those fit "to live or die" remained.

While Shaw's work on "Imprisonment" concerned mundane and current matters, and *Back to Methuselah* presented a more futuristic, hypothetical, even hyperbolic case, the problem of selectivity in the extension (or termination) of life remains critical. Not just in the realm of science fiction and philosophy, a certain degree of selectivity was indicated in the works of several of Shaw's contemporary German and American life-extensionist scientists and physicians (see above). And in science fiction and philosophical treatises, the problem of selectivity has been absolutely central, one of the crucial points in the ethical debate "for and against" life extension, perhaps even the most vital one. Its centrality should be anticipated. "Overpopulation" and "shortage of resources," often presumed to be the result of life-extension, can be in principle corrected by birth control or by agricultural/technological means (it may be unlikely that people will learn to increase the human life-span, but not learn to increase the yield of crops). "Boredom" is not very likely either, as most people (unless they are suicidal) may find the task of staying alive for any period of time not extremely boring. But the problem of selectivity, or as it has been sometimes termed "inequitable access to life-extending technologies" seems to be a very tangible one, seeing the current inequalities in the access to health care. The frequent literary-fictional representation of the pursuers of immortality as self-serving monsters, contemptuous of the lives of everybody else, may not be so easily dismissed as a manifestation of "apologism" – the desire to rationalize, justify or apologize for our short life-spans, a part of Aesop's good old "sour grapes" syndrome, vilifying what we cannot yet have. The existing instances of inequality and exploitation allowed the fiction writers to foresee their "radical extension." The British bioethicist John Harris argued in "Immortal Ethics" (2003) that the impossibility to grant the good of life-extension to all people should not prevent granting it to some people.[1087] But what if life-extension is granted to some select few, *because it is taken from everyone else*? Notably though, Harris did add that "the principle [of global justice] requires that strenuous and realistic efforts be made to provide the benefits of the technology justly and as widely as possible."

After WWII several philosophical and science fictional works in English-speaking countries began to consider the possibility of absolutely universal, non-selective distribution of life-extending technologies, or at least a universal *choice* to use such technologies. Thus, for example, in *Bug Jack Barron* by the American writer Norman Spinrad (1969, first serialized in England), the pursuer of immortality, the billionaire Benedict Howards, is still an egotistic monster, killing children to use their glands for the "immortality treatment." He is thwarted. But the ultimate hope expressed in the book is that the immortality treatment should be made available "for everyone without murder."[1088] Such hopes, however, have been rather seldom expressed in English fiction. Usually, the immortality-aspiring monster is killed, and the pursuit of immortality is left at that. The popularity of the dystopian scenario is not surprising: it is exciting and cathartically purges the reader of the fear of death. But it could hardly contribute to the popularity of life-extensionism in Britain. The cause was not helped by the fact that even such a well-expressed life-extensionist author as Shaw, presented himself as a champion of selectivity in life-extension (or extermination). It may be "rational" on capitalist assumptions to provide life-extending commodities to the people who are most able to pay for them, or on military socialist assumptions, grant them to those considered most "productive," "deserving," or "socially compatible" – under the "impossibility" of providing them to all. But still there might be an incapacitating, "unconquerable repugnance," unaffected by any rationalizations, to their supply to the select, while letting the rest die. In the later part of the century, it seems to have been increasingly realized that a more desirable and sympathetic course might be to strive to provide life-extending prospects to all from the outset. As said by a (catholic) protagonist in Poul Anderson's science fiction novel *The Boat of A Million Years* (1989, US), "can't you see the dying down of war, arms races, terrorism, despotism, given such a prospect before everyone?"[1089] And as stated in James Halperin's novel *The First Immortal* (1998, US), "Let us make victory against death a prospect for all, lest it become a prospect for none."[1090]

Yet, the hope for universal life-extension does not seem to have been expressed in English fiction of Shaw's time. Rather, the pursuit of extreme longevity was chiefly depicted as egotistic, destructive, exhaustive, unjust or boring (notice the supposition of many conflicts due to life extension in some dystopian works, and boredom in others). These predominantly negative representations may have stemmed either from rationalization and "apologism" or genuine fear of social calamities that life-extension would bring. In any case, the dominant negative literary perception may have been one of the causes for the relative lack of interest in practical or scientific life-extensionism in Britain well until the 1940s. If radical life-extension was most commonly demonized and defeated even in science fiction, who would wish to pursue it in real life? And so, for a long period, the discussions of life-extension in Britain mainly remained in the realm of literary fiction and philosophy, and even there were for the most part negative.

19. British longevity research: practice, theory and programs

Even though traditionally the prevalent areas of discussion of life-extension in Britain seem to have been rather literary and philosophical (at least until the 1940s), there was still some interest in the subject in scientific and popular scientific literature. Hygienic regimens for life-extension enjoyed perhaps the greatest and most sustained interest, being possibly perceived as less 'farfetched' than speculations about radical and immediate rejuvenation. One of the earliest published works in this tradition was *Art of longevity, or, A diaeteticall institution* (1659) by Edmund Gayton (1608-1666).[1091] At about the same time, in 1666, Dr. John Smith urged:[1092]

> Let none give over their patients when they come overburdened with the infirmities of Age, as though they were altogether incapable of having any good done unto them. Those that are negligent toward their Ancient Friends, are very near of kin to those inhuman Barbarians and Americans, who both kill and devour them.

Lending his authority to the hygienic tradition, William Harvey (1578-1657) examined the famous super-centenarian, Thomas Parr, who was said to have lived to the age of 152 years and 9 months (1483-1635). Harvey appreciated Parr's moderate and industrious habits of life and expressed the conviction that "had nothing happened to interfere with the old man's habits of life, he might perhaps have escaped paying the debt due to nature for some little time longer."[1093] The physician Thomas Sydenham (1624-1689) admonished that the patients should not "have so worn themselves out in youth as to have brought on a premature old age."[1094] Yet, for Sydenham, aging appears to be a rather negligible concern as compared to acute and epidemic diseases. Another British pioneer of the study and care of the elderly and prolongevist was Sir John Floyer (1649-1734), who advocated personalized therapy adjusted to the aged patient's "constitution" and whose favorite prescriptions were moderation and cold or hot baths.[1095] A few popular works on longevity were published in the 19th century, such as those of Sir John Sinclair (1754-1835) and Prof. John Abernethy (1764-1831), undergoing several editions.[1096] At the turn of the twentieth century, the physician and public health activist Sir James Crichton-Browne (1840-1938), continued to argue for the improvement of health care for the aged, and for the "prevention of senility." He opposed fads, stating that "there is no shortcut to longevity. To win it is the work of a lifetime."[1097]

Rejuvenation techniques seem to have been much less popular in England than in France, Austria, Germany, Russia or the US. Yet, some interest in rejuvenation can be cited. Among the renowned early seekers of the "elixir of life" were the chemist Robert Boyle (1627-1691) and the physician George Thomson (1619-1677).[1098] The surgeon John Hunter (1728-1794) believed that human life can be sustained far beyond contemporary boundaries and worked on gland grafting,[1099] body preservation by embalming and freezing, and resuscitation.[1100] In the early 1900s, based on Metchnikoff's theory that the large intestine is the chief source of bacterial putrefaction and intoxication leading to senescence, the Scottish surgeon William Arbuthnot Lane (1856-1943) bravely extirpated the large intestines, to eliminate the cause of aging once and for all. Metchnikoff supported the logic behind such interventions, but cautioned against their wide application, pointing out the need to improve the surgical technique and generally advocating minimally invasive methods (1910). According to Metchnikoff, the number of operations performed by Lane was rather limited, little more than 50, with 39 cases reported, 9 of which (23%) resulting in death.[1101]

Voronoff's rejuvenative sex gland transplantation, that was a craze in France in the 1920s, was prohibited by law in England.[1102] Steinach's vasoligation was permitted, though not nearly as widely used as in Austria, Germany, France, Russia or the US. One of Steinach's prominent followers in Britain was Norman Haire (1892-1952),[1103] yet the number of operations he performed could be hardly compared to the hundreds of operations done by Eugen Steinach in Austria, Peter Schmidt in Germany or Harry Benjamin in the US. The Steinach procedure was chiefly known in England by its effects on a literary figure, the poet William Butler Yeats (1865-1939). The rejuvenative operation was performed on Yeats by Haire in 1934, and Yeats' tempestuous poetry following the operation often evoked a greater interest than the operative technique itself (see Richard Ellmann's *W.B. Yeats's Second Puberty*, 1985).[1104]

The work of the British physician Sir Geoffrey Langdon Keynes (1887-1982, brother of the economist John Maynard Keynes) on blood transfusion in the 1910s, was a basis for Alexander Bogdanov's research on rejuvenation by blood transfusion. Before establishing the laboratory and eventually the Institute for Blood Transfusion in Moscow, which were explicitly dedicated to finding means for rejuvenation, in 1921 Bogdanov traveled to England to study Keynes' experience, as well as to acquire literature and equipment.[1105] The world search for rejuvenative replacement therapy later received a further great boost by the work of the British immunologist Peter Medawar (1915-1987, the Nobel Laureate in medicine for 1960 and Director of the British National Institute for Medical Research since 1962). In 1951, he suggested the use of immuno-suppressants (such as cortisone) to prevent tissue rejection, thus making tissue replacements feasible.[1106] (See the previous section on the relation of life-extensionism to transplantation research.) In 1963, Medawar dispelled the doubts about the potential use of replacement/transplantation therapy as a cure or palliative for senile deterioration: "I don't think it will be possible to repudiate the idea of using replacement therapy until we have a theory of ageing, which we haven't got at present."[1107] During WWII and shortly afterwards, a group of scientists with the National Institute for Medical Research in London – Alan Sterling Parkes, Christopher Polge, Audrey Smith and James Lovelock – achieved breakthroughs in tissue cryopreservation by introducing cryoprotectants such as glycerol and dimethyl sulfoxide.[1108] Their work facilitated transplantation surgery and reproductive technology, as well as established the basis for cryonics, making a stride toward implementing John Hunter's original vision.

Still, well until the 1940s, the discourse on life-extension in British scientific and popular-scientific literature seems to have been almost exclusively theoretical, with a strong emphasis on the evolutionary theories of aging. The German life-extension advocate Gerhard Venzmer (1937) quoted Darwin as postulating the maximal human life-span of 180-200 years.[1109] I was unable to find the source of this quote from Darwin. However, Charles Darwin was indeed somewhat interested in longevity. He pondered the question "What animals can be budded and rendered of great age as must be inferred from what Mr. [Thomas Andrew] Knight says?"[1110] Darwin posited the hereditary basis of longevity. He claimed that "Seeing how hereditary evil qualities are, it is fortunate that good health, vigour, and longevity are equally inherited." Long-lived and healthy parents were found by Darwin to produce long-lived and healthy offspring. He further examined the relation between the expenditure of reproductive elements ("gemmules") and aging, and between the conservation of the reproductive elements and the "repair of wasted tissues" (*The Variation of Animals and Plants under Domestication*, 1875).[1111] These ideas were directly related to Edwin Ray Lankester's work *On Comparative Longevity in Man and the Lower Animals*, indicating mutual influence (1870).[1112] Charles Darwin's interest in longevity might have been partly inherited from his grandfather, the physician Erasmus Darwin, who wrote on the subject extensively.[1113] The British founder of eugenics, Sir Francis Galton (Erasmus' grandson and Charles Darwin's second cousin) also expounded on the relation between heredity and longevity of the offspring (1883).[1114] Further correlations between longevity and heredity were established by the statisticians Karl Pearson, Mary Beeton and George Yule of the University College, London, at the turn of the 20th century.[1115] Since Darwin, evolutionary theories of aging have been a mainstay of British longevity research, from Alfred Russel Wallace (1870, 1889)[1116] through George Parker Bidder (1925, 1932), Julian Sorrell Huxley (1932), John Burdon Sanderson Haldane (1941), Peter Brian Medawar (1952), John Maynard Smith (1962), Thomas Burton Loram Kirkwood (1977) and Michael Robertson Rose (1979).[1117] The theories expounded on the emergence of senescence in the course of evolution, either as a result of direct selection for aging (senescence appearing as an evolutionary advantage to 'clear the space' for new life forms) or indirect selection (evolved body structures making immortality impossible) or as a result of "evolutionary neglect" (evolution "not caring" for the survival of organisms past their reproductive age). Yet, apparently, there were few implications for practical human life extension.

Evolution was also an important concern in the work of Alex Comfort (1920-2000), a researcher of the biology of senescence at the Department of Zoology, University College, London, as well as a poet, novelist and political philosopher, perhaps best known for his life-affirming, pacifist poetry, as in *A Wreath for the Living* (1942), sexological treatises, such as *The Joy of Sex* (1972), and anarchistic views (despite his good social and financial standing) as in *Writings Against Power and Death* (1994). According to Comfort, the evolutionary theory of senile deterioration as a result of "declining efficiency of the evolutionary pressure toward survival at different ages" is most plausible "insofar as any general theory of senescence is justified." Still, basic laboratory research of aging, he believed, deserved a much greater emphasis. "The gerontologist, with the prolongation of human life in mind," Comfort asserted, "is interested in something which is not, as such, of interest to the evolutionary 'demon.'"[1118]

In Comfort's summative and programmatic book *The Biology of Senescence* (1956), it can be observed how molecular biological considerations of the process of aging, and laboratory research generally, shifted into focus. Comfort, perhaps one of the first among longevity researchers, began to consider "genes as catalytic molecules" and "somatic mutations."[1119] Comfort opposed earlier generalist non-molecular "fundamentalist theories of aging" (wear and tear, mechano-chemical deterioration of cell colloids, inherent changes in specific tissues, such as nervous, endocrine, vascular and connective tissue), and the "epiphenomenalist group" (theories of toxicity, accumulation of 'metaplasm' or cell metabolites, effects of gravity, heavy water or cosmic rays). Neither was he particularly fond of general developmental theories that "stress the continuity of senescence with morphogenesis" or the "operation of an Aristotelian entelechy," theories of fixed quantity reactions, critical volume-surface relations, depletion, and cessation of somatic growth.[1120] Rather than multiplying theoretical generalizations on the nature of aging, Comfort proposed an extensive program of laboratory data collection that would alone make any future interventions into senescence conceivable. The program would include a continuous collecting of data on the distribution of senescence in various species and evidence of their maximum longevity, the influence of genetic constitution on longevity (mainly presented in population genetics terms), the relation of growth and senescence, cell and endocrine senescence.

The corollary of Comfort's work was a call for further research and an outline of possible future research directions: "study of the phylogeny of senescence in vertebrates, study of the correlations and the experimental modification of growth and development in populations where the life-span can be concurrently measured, and study of tissue-environment relationships through the creation of age chimeras."[1121] According to Comfort, the purpose of this research is human life-extension:[1122]

> Insofar as biology is more than a branch of idle curiosity, its assignment in the study of old age is to devise if possible means of keeping human beings alive in active health for a longer time than would normally be the case – in other words, to prolong individual life. People now rightly look to 'science' to provide the practical realization of perennial human wishes which our ancestors have failed to realize by magic – or at least to investigate the prospect of realizing them. Under the influence of the study which is necessary to fulfill such wishes, the character of the wish itself generally changes in the direction of realism, so that most people today would incline to prefer the prospect of longevity, which may be realizable, to a physical immortality which is not, and, *pari passu*, 'potentielle Unsterblichkeit' [potential immortality] is already disappearing from the biological literature. An analogous process can be seen in the psychology of individual growing-up.

In Comfort's belief, only after an extensive research program to achieve a complete understanding of senescence, "a radical interference with the whole process of ageing… will give us 150 or 200-year lifespans."[1123] Thus, the hope for immediate rejuvenation in the present was as much as absent and, instead, extensive basic biological research was called for to provide life-prolonging means in the future. As no single "magic bullet" appeared possible at the time, the task of a life-extensionist scientist, Comfort believed, was to collect all data and all theory available, in all research avenues conceivable, prior to even beginning to think about an actual intervention.

Among many others, Alex Comfort kindled the interest in longevity research in his colleague at the Zoology Department of the University College, London, the geneticist John Maynard Smith (1920-2004). Comfort started his life-extension experiments on *Drosophila* flies in the early 1950s at Maynard Smith's laboratory, but the experiments were not very successful. As Maynard Smith testified: "Jean Clarke was the technician there and she was very quickly convinced that no fly looked after by Alex would live for more than one week. So we more or less took over because we felt sorry for him."[1124] Unlike Comfort, Maynard Smith succeeded in prolonging the lives of *Drosophila* flies by reproduction-diminishing brief heat exposures, sterilizing radiation and hybridization between inbred lines. He also wrote on evolutionary theories of aging.[1125]

Basic biological laboratory research of aging seems to have developed quite late in the UK. It was not until the Russian-born Vladimir Korenchevsky (1880-1959) organized the British Society for Research on Aging in 1939 and, with the grant of £3,000 from Lord Nuffield (William Richard Morris, 1877-1963, owner of the Morris Motor Company), established in 1944 the Gerontological Research Unit in Oxford, that British aging research became institutionalized.[1126]

Korenchevsky's views were deeply rooted in the Russian life-extensionist tradition. He was a student of Nikolay Kravkov, Russia's foremost life-extensionist pharmacologist, rejuvenator and resuscitator. For brief periods Korenchevsky worked with Elie Metchnikoff (in 1908) and Ivan Pavlov (in 1910), specializing in toxicology and anti-toxic treatments, and pathophysiology, particularly, the pathophysiology of the aged. After the Russian Socialist Revolution of 1917, Korenchevsky worked in the Soviet Military Medical Academy. In 1919, he joined the counterrevolutionary White Army in Sevastopol, Crimea, and after its defeat he fled to England in the same year. Having obtained British citizenship, he became a leading, organizing figure of the Russian academic diaspora: In 1921 he organized the "Russian Academic Association" in London, and in 1928 he was elected president of the congress of Russian Academic Organizations in Belgrade, Serbia.[1127] Thus, the import of the Russian life-extensionist tradition may have been considerable, particularly the emphasis on a pro-active approach to the treatment of aging and the striving toward a collaborative organization of aging and longevity research (the aspirations pronounced by Kravkov, Metchnikoff, Bogdanov and many other Russian life-extensionists).

Vladimir Korenchevsky played an instrumental role in the national and international organization of gerontological research. Hailed by Nathan Shock as the "apostle of an international forum in gerontology," he "spread the gospel of gerontology,"[1128] and worked to provide continuous communications and funds for researchers of aging worldwide. In 1939, Korenchevsky organized the "British Club for Research on Aging" (renamed "The British Society for Research on Aging" in 1946) that became the launching platform for the institutionalization of gerontology in Britain.[1129] Later on, mainly funded by Lord Nuffield, Korenchevsky extended the organizational efforts internationally, helping to establish gerontological associations in other countries. Largely thanks to these efforts, by the end of the 1940s, gerontological societies emerged in the US and Argentina (1945), Sweden and Denmark (1946), Switzerland and Netherlands (1947), Belgium, Czechoslovakia, Finland and Spain (1948), Australia, Canada and Italy (1949), with initial memberships ranging from 6 (Czechoslovakia) to 217 (USA, 1945).[1130] These national associations, emerging in the 1940s, were in fact initially designated by Korenchevsky as "Branches of the British Club for Research on Aging." Following the efforts of Korenchevsky and his ally, the American gerontologist Edmund Vincent Cowdry, in 1950 the International Association of Gerontology was established and the First World Congress of Gerontology was held in Liège, Belgium. These events marked the establishment of gerontology as an international, institutional scientific discipline.[1131]

Through the late 1940s-early 1950s, Korenchevsky propagandized the "conditions desirable for the rapid progress of gerontological research" and its central aim of "not only a longer life but a stronger one" (1946). According to Korenchevsky, "the 'elixir of life' appears to consist mainly in elimination of all abnormal causes of aging and not in the use of stimulating compounds only" (such as hormones and vitamins).[1132] In Korenchevsky's view, "present day old age is a kind of illness"[1133] and "the final solution of all basic gerontological problems depends entirely on biological and medical research in this field" (1952).[1134] In his vision, the treatment of senility should involve "Rejuvenative, Preventive and Eliminative" interventions (1950).[1135] Korenchevsky waxed skeptical about the "rejuvenative" treatments by "stimulating factors" – hormones and vitamins – that were proposed earlier. His own experiments with these substances failed to produce "a true anti-aging effect on the whole organism." Any single factor or a group of factors, he assumed, would be "unlikely to reverse all processes of senescence."

Nonetheless, he remained generally optimistic of the prospects of rejuvenation: "The future treatment of senility will be manifold," he foresaw, "and the so-called 'elixir of life' will contain many chemical substances." He put greater faith in "preventive and eliminative treatment of senility," that is, the "prevention or elimination of all [age-related] diseases." "The power of science," he asserted, "is very great, and the progress of research sometimes becomes unexpectedly rapid. For example, the recent discoveries of the decholesterolization by, and some other effects of, choline, inositol and methionine and the decalcifying properties of the anti-stiffness factor opened new promising vistas for research on treatment of arteriosclerosis and angina pectoris, both number one killers of old persons" (1950). Korenchevsky believed "it is improbable that all [complicating diseases, secondary and primary causes of aging] can be eliminated, especially the primary causes, the elimination of which would make immortality possible." Nonetheless, "elimination of some so far unknown secondary causes of aging might become practicable when their nature and action are investigated by gerontology."[1136]

In Korenchevsky's view, the gerontological research has to be "manifold" and involve "all branches of science and medicine." The specific requirements for the progress of gerontology were the following (1946):[1137]

1) Just as in the case of the splitting of the atom, co-operative research work of scientists and medical men of all civilized countries is necessary. 2) Large funds must be available for financing gerontological research in all its manifold aspects. 3) The establishment of experimental and clinical Institutes for Research on Aging in major countries. 4) The formation of groups of research workers in gerontology and geriatrics. 5) Large long-term grants to existing scientific and medical laboratories and research hospitals who will agree to start gerontological or geriatrical research work.

Several central features can be observed in Korenchevsky's life-extensionist program: the rejection of earlier far-reaching rejuvenation attempts and instead expressing "moderate life-extensionist" aspirations, bold enough to stimulate investment, but modest enough to maintain scientific responsibility. And above all, the program involved the demand for a radical expansion of basic biological research of aging and for large scale cooperation and financial support for that purpose. These features appear to be very similar to the contemporary American life-extensionist programs. In fact, the establishment of the "British Club for Research on Aging" was almost precisely simultaneous with the institutionalization of aging and longevity research in other countries: the publication of Cowdry's *Problems of Aging* in 1939 in the US, the inauguration of Bürger's *Zeitschrift für Altersforschung* (Journal for Aging Research) in 1938 in Germany, Bogomolets' *Conference on the Genesis of Aging and Prophylaxis of Untimely Aging* in 1938 in the USSR, and the establishment of Binet's "French Society of Gerontology" in 1939 and Carrel's "French Foundation for the Study of Human Problems" in 1941. These concurrent events may indicate that the institutionalization of aging and longevity research in the late 1930s was an international trend, and perhaps a result of international competition. Thus, allusions to successes of Russian institutional longevity science and calls for their reproduction were then made by Americans (Clive McCay) and British researchers (J.B.S. Haldane). Korenchevsky acted as an exile counterweight to the Soviet Russian scientific establishment, while drawing on his native tradition. German life-extensionists (Venzmer, Roemheld) wished to emulate and surpass the achievements of American life-extensionists. Russian longevity scientists (Bogomolets, Nagorny) posed their research in direct confrontation to the studies done in "capitalist" and "fascist" countries, while citing evidence from those studies extensively. French researchers (Carrel, Binet) competed with and drew on both American and British experiences. Yet, paradoxically, aspirations for international cooperation in the field were at the same time expressed throughout. Out of this paradoxical blend of competition and striving for cooperation, after WWII international aging and longevity research emerged as an internationally associated discipline.

As this work has argued, in particular countries, life-extensionism was nourished by and gave support to the dominant ideological and socio-economic paradigms. Korenchevsky's work in the UK exemplifies the dependence on the philanthropy of big business, accompanied by a sustained aspiration toward increased public support. Such aspirations and dependencies are very similar, perhaps identical, to those of the contemporary American life-extensionists. Not surprisingly, representatives from the US and UK became the closest partners and the strongest forces in the International Association of Gerontology, that was initially mainly funded by British and American private sources (the British Nuffield foundation and the American Josiah Macy Jr. foundation) and later also by public sources.

Several large scale projects in the UK now continue Korenchevsky's legacy of organizational efforts toward cooperation and advocacy for longevity research, in close partnership with American scientists. Korenchevsky's "British Society for Research on Ageing" exists until the present day. Until 2014, it has been chaired by Prof. Richard Faragher of the University of Brighton, who has also been chair of the International Association of Biomedical Gerontology.[1138] Together, the British Society for Research on Ageing, the British Geriatrics Society and the British Society of Gerontology, form the British Council for Ageing. Though hardly radical in their goals, the common purpose of these organizations is to "improve longevity" and "provide informed opinions and influence policy making with respect to older people and research in to the ageing process."[1139] In 2000, the International Longevity Centre – United Kingdom was established, dedicated to "addressing issues of longevity, ageing and population change." The organization became one of the major branches of the International Longevity Center – US, established in 1990 by the American gerontologist Robert Neil Butler.[1140] On May 10-11, 2010, a conference was held by the Royal Society in London, entitled "The New Science of Ageing." The conference focused on pharmacological and genetic control of aging and experimental life-extension, and brought together some of the foremost American and British researchers. Several prestigious British conferences related to life-extension have been held since.[1141]

Among the less "main-stream" projects, in 1992, the British Longevity Society was established by Marios Kyriazis (born in 1956 in Cyprus). Kyriazis has been a staunch supporter and provider of anti-aging medicine, which he defined as a "branch of medical science and clinical medicine, aimed at treating some of the underlying processes of ageing and at alleviating or postponing any age-related ailments, with the ultimate goal of extending the healthy lifespan of humans" (2006). He also authored the concept of "Extreme Lifespans through Perpetual-equalising Interventions (ELPIs, 2010)." According to this concept, radical life extension will be inevitable thanks to the ever increasing complexity and equilibration (optimization) of life. Human beings will be able to accelerate the progress toward a "transhuman" stage of "global integration" or "developmental singularity" characterized by "extreme life-spans." In practical terms, the inching toward "extreme life-spans" would be facilitated by "hormesis" (challenging external stimulations) and "environmental enrichment" (constant intellectual stimulation aided by "nutritional factors" and "physical exercise") that would stimulate anti-aging defense systems of the body. The hypothesis, however, acknowledged the possibility that "overstimulation may cause deleterious results." Thus Kyriazis combined transhumanist and singularitarian philosophy with the practice of anti-aging medicine.[1142] Generally, the "anti-aging" movement flourishes in the UK.[1143]

Since 2002, the SENS project (Strategies for Engineered Negligible Senescence) has been led by Dr. Aubrey David Nicholas Jasper de Grey of Cambridge (b. 1963). Up-to-date, this has been perhaps the most ambitious life-extensionist endeavor in the UK, perhaps in the world, even though its support has mainly come from the United States (see below).[1144] The SENS project synthesizes various approaches toward "curing aging": biological replacements, maintaining homeostasis, repairing molecular damage by genetic engineering, eliminating metabolic waste products by enzymatic catalysis and other methods. SENS suggests testing specific interventions against "The Seven Deadly Things" or seven types of pathogenic damage accumulating with age: cell loss or atrophy, nuclear mutations and epimutations, mutant mitochondria, cellular senescence, extracellular cross-links, extracellular junk, and intracellular junk. In the SENS program, upon a successful elimination of the agents of damage, human senescence and death from aging are expected to be entirely eliminated and the only remaining cause of death will be external accidents, with a death rate comparable to that of present day youth. Provided massive scientific and public support for life-extension research, according to Aubrey de Grey, an "escape velocity" could be reached, when new life-extending technologies would arrive in time to save those who would die without them. De Grey's call to arms to fight aging is perhaps best summarized in his slogan: "Wake Up – Aging Kills!"[1145] As Aubrey de Grey acknowledged, "SENS is in a very real sense a revival of the "rejuvenation" theme that was of diminishing prominence over the course of the 20th century."[1146]

20. Man and His Future

Since Korenchevsky's efforts, the cooperation in aging and longevity research has continued to increase, in an almost inseparable alliance of British and American life-extensionists. In 1963, British and American scientists once again joined forces in an attempt to envision and encourage the future evolution of humanity. They met to discuss it in the colloquium entitled *Man and His Future* held that year in London and sponsored by the Chemical Industries of Basel (Chemische Industrie Basel, CIBA).[1147] At that conference, longevity was considered among the primary parameters of human development. The colloquium enlisted some of the world's foremost scientists, mostly from the United States and United Kingdom – Nobel laureates, founders of entire scientific fields, heads of academic institutions. The United States was represented by Joshua Lederberg, Hermann Muller, Hilary Koprowski, Albert Szent-Györgyi, Gregory Pincus, Carleton Coon, Hudson Hoagland, and others. On the British side, there were Francis Crick, J.B.S Haldane, Peter Medawar, Julian Huxley, Alan Parkes, Alex Comfort, Donald MacKay, Colin Clark, Jacob Bronowski, and others. The representation from other countries was very limited, including John Brock (UK/South Africa), Artur Glikson (Israel), and Brock Chisholm (Canada).[1148] In their discussions, human biological enhancement generally and life-extension in particular, were the central items on the agenda.

The authors expounded on the prolongation of life through eradication of infectious diseases (Hilary Koprowski, Julian Huxley); through diets "as a means of prolonging life" (John Brock); or even through "a radical interference with the whole process of ageing" (Alex Comfort). They spoke about life-extension via replacement of worn-out tissues: by organ transplants (Peter Medawar), hormone replacements (Gregory Pincus), or robotic parts (Donald MacKay). Life extension could also be achieved through eugenic "germinal choice" or "consciously directed genetic change" (Hermann Muller, Francis Crick) by using what Muller termed "nano-needles." Alternatively, it could be done through "*euphenics*, the engineering of human development" using artificial organs, synthesis of specific proteins (hormones, enzymes, antigens and structural proteins), and genetic modifications of animals "as sources for spare parts" (Joshua Lederberg). Humans could evolve into long-lived, radiation-resistant beings, and "centenarians, if reasonably healthy, would generally be cloned, if this is possible" (J.B.S. Haldane).[1149] Alongside the technological solutions, long life was also hoped to be achieved through "natural means" (Jacob Bronowski). Albert Szent-Györgyi summarized the general hopeful attitude: "Present experience gives a ray of hope that we may get closer to the solution of three major human problems (prevention of cancer, extension of youth and dealing with over-population)." Francis Crick appeared to be even more optimistic than Szent-Györgyi. In responding to Szent-Györgyi's notice of the immense complexities involved in attempting to reduce and manipulate life processes, Crick said, "I agree with you that it is all very complicated, but not as complicated as you're trying to frighten us with."[1150]

In case the problem of over-population, commonly associated with increasing longevity, would emerge, solutions would be at hand. One effective solution would be the use of contraceptives, as proposed at the conference by the inventor of the oral contraceptive pill, Gregory Pincus, and by the founder of cryobiological preservation, Alan Parkes.[1151] Potential shortages of resources due to overpopulation did not appear to the authors of *Man and His Future* to pose an insurmountable problem either. According to Colin Grant Clark – director of the Agricultural Economics Research Institute, Oxford – agricultural productivity, even at the present level, will be more than sufficient to prevent food deficits. Malthus' pessimistic theoretical assumptions, Clark argued, have been and will continue to be proved wrong: "Historical, geographical, and anthropological evidence – most of which, to be fair, was not available to Malthus – alike show how untrue his proposition was." "Our land requirements," Clark calculated, "using the best agricultural methods now available – though great further improvements will be possible" are 1800 square meters/person or 5.5 persons/hectare (allowing for an average food requirement of "500 kilograms per person per year or 1,370 grams per person per day").[1152] Such an agricultural productivity will be well capable of sustaining 45 billion people. "The world," Clark estimated, "has the equivalent of 6,600 million hectares of good agricultural land." With the addition of potential agricultural land in the wet tropics of Africa, Latin America and Asia, "we must have 8,200 million hectares in all, capable of giving a diet containing meat and dairy products on a North American scale to 45,000 million people." A shortage of mineral and energy resources was also not to be expected with the use of proper technology:[1153]

> Finally, let me mention our requirements of minerals. They are being reduced by technical improvements and our descendants will not have to worry about mineral fuels, since they will have all the energy they want in the form of nuclear power, solar batteries, and cars driven by batteries on hydrogen, if necessary. The only exception is aluminium, but we could produce this from ordinary clays if supplies of bauxite ran out. If we assume a world population of 45,000 million consuming minerals at the present North American rate per head, the supplies of these minerals available in the top 1,500 metres of the earth's crust would keep them going for some multiple of 10^5 years (10^8 in the case of aluminium). So it looks like being a long time before our descendants have to dig deeper into the earth – or search in other planets outside it – for minerals.

Furthermore, according to Artur Glikson, an Israeli architect and ecologist, natural resources can be long sustained by an "ecological integration" with the environment, for example, by creating "an artificial landscape, in which fertility and water are preserved by actively fulfilling the rule of return observed in nature." According to Donald MacKay, a pioneer of artificial intelligence research, machine intelligence will play an ever greater role in an efficient distribution of resources for prolonged biological and social stability.[1154]

Indeed, no one at the meeting spoke of immediate rejuvenation or of immediate and universal elimination of scarcity. These tasks were left for the future. Some consideration was given to possible disruptions of social stability due to rapidly developing and increasingly complicated technologies. Yet, according to the physical anthropologist Carleton Stevens Coon (US), social stability will be maintained through "games," "laws," "rituals," "education," or training an "elite."[1155] According to Hudson Hoagland and Brock Chisholm, human intelligence must be improved to help us resolve any difficulties that may arise.[1156] Throughout *Man and His Future*, the society was assumed to be capable of stable, indefinite continuation. Despite the far-reaching visions, no immediate (or even distant) interventions into the current social structure were suggested.

Man and His Future is a remarkable document, which may be considered a milestone of British and American scientific futurism.[1157] It shows that the issue of life-extension, and even radical life-extension, was far from being on the fringe of scientific and intellectual discourse, but occupied some of the world's leading scientists and in many instances motivated their work. It shows that far reaching imagination is difficult to eliminate from the scientific pursuit. *Man and His Future* is also remarkable in that it presented a clearly defined set of goals for human development, life-extension being one of the pivotal. The participants in that conference of 1963 outlined prospective directions for future development of science that are still followed today by many. Thus, in considering the present day literature of the transhumanist intellectual movement – the movement that derived from Julian Huxley's conceptual term "transhumanism" for the incessant evolutionary transformation of the human species – it may be observed that the long-term goals posited now, are very much the same as those proposed 50 years ago, even with regard to specific technologies. Without daring to enter into any quantitative and qualitative technological comparisons, and without venturing into any feasibility assessments, past or future, the thematic similarities are nonetheless striking. Now, as 50 years ago, the fields of reproductive control, tissue and organ transplantation, cryobiology, genetic engineering and developmental engineering, micro-machining and artificial intelligence, are seen as guarantors of a radical improvement and prolongation of life.[1158] As the (mainly British and American) transhumanist literature emphasizes, the life-expectancy continues to rise, agricultural and industrial productivity continues to increase, computing power continues to multiply, reproductive technology and biological engineering are ever more widely employed. No limit is foreseen to their capabilities. *Man and His Future* demonstrates that the fascination with these technological capabilities did not start now, as sometimes presumed by those who tend to better remember the immediate past, which appears to them much more eventful than anything that had happened earlier. It started at least half a century ago, and as this work has endeavored to demonstrate, many of these hopes and developments can be traced at least a century further back.

21. The present time: Life extension programs in the US

In the US, the organizational efforts in gerontology that began in the 1930s by Cowdry, in cooperation with Korenchevsky, now continue on an ever larger scale. A list of currently existing American life-extensionist grass-roots organizations, clubs, institutes and associations would be extensive indeed. Just a few examples of the larger or more radical ones may be mentioned here. An immense organizational effort to promote the research of aging was made by Dr. Robert Neil Butler (1927-2010), the founding director of the NIH National Institute on Aging (1975-1982). In 1990, Butler grounded the International Longevity Center – US, with the expressed mission "to educate individuals on how to live longer and better." This was one of several research support programs spearheaded by Butler.[1159] Presently, the Global Alliance of the International Longevity Centers, including 12 member states, has been said to be "the only non-governmental organization capable of carrying out global research and education projects on population aging."[1160] In 1970, the American Aging Association was founded thanks to the efforts of Dr. Denham Harman, the father of the "free-radical theory of aging." To the present, the association espouses the "drive to conquer aging" and is referred to by some as the "American Anti-Aging Association."[1161] The Gerontology Research Group, founded in 1990 and led by Dr. Leslie Stephen Coles of the University of California, Los Angeles, has been "dedicated to the quest to slow and ultimately reverse human aging within the next 20 years."[1162] The Trans-NIH Geroscience Interest Group, founded in 2011 by Dr. Felipe Sierra, Director of the Division of Aging Biology of the National Institute of Aging, has been devoted to address the biology of aging – "the largest single risk factor for most chronic diseases."[1163] Several research projects related to life-extension have been conducted at the US Defense Advanced Research Projects Agency (DARPA).[1164] Many other organizations have been dedicated to gerontological research and care, with varying levels of ambition. Even though far-reaching life-extensionist goals have been seldom expressed in these organizations, many of them historically derive from life-extensionist enterprises and are life-extensionists' natural allies.[1165] Increasing the social and material support of life-extensionist organizations has been an ongoing struggle.

Several currently active life-extensionist organizations are explicitly commercial, such as the Life Extension Foundation,[1166] the American Academy of Anti-Aging Medicine (A4M),[1167] as well as other organizations for anti-aging medicine,[1168] yet they have weighed heavily in support of longevity research.[1169] Several companies have been dedicated to developing technologies for regenerative and rejuvenative medicine,[1170] established by prominent researchers of aging, such as Michael West,[1171] Leonard Guarente,[1172] Cynthia Kenyon,[1173] Anthony Atala,[1174] Craig Venter,[1175] George Church,[1176] and others.[1177] Subsidiaries dedicated to research and development for life-extension have been founded by large corporations, including Google.[1178] The development of this emerging industry has not always been smooth sailing and incentives for its development and accessibility of its products have been sought.[1179]

Several classes of potential therapeutics have been promoted as particularly promising, including telomerase activators,[1180] as well as potential activators of sirtuins and several other purported "longevity genes,"[1181] other forms of "gene therapy" and RNA interference therapy,[1182] cell cycle regulation (inducers of cell death and cell regeneration),[1183] stem cells and their products,[1184] modulators of mitochondrial energetics,[1185] and an additional wide variety of dietary supplements,[1186] and other potential longevity-promoting substances.[1187] The new biotechnological approaches to life extension may be seen as supplementary to the "age-honored" hygienic and regimental ways, such as avoidance of poisons, exercise, reasonable diet and good rest.[1188] The practical implementation of the various pathways to life-extension yet remains to be thoroughly investigated.

Several research centers have been dedicated to finding means of postponing aging.[1189] Several institutional longevity research programs and consortia have been formed.[1190] Dr. Aubrey de Grey's SENS project (Strategies for Engineered Negligible Senescence, the term coined by de Grey in 2002) has been perhaps the highest aspiring and motivated life-extensionist project in the world, raising unprecedented public interest (largely in the US).[1191] There have also been several advocacy groups, loosely associated into life-extensionist community networks.[1192] Several public figures have expressed their support of radical life-extension.[1193] Practically the entire Transhumanist social movement – including virtually all of its numerous groups advocating for the improvement (enhancement) of human capabilities through an ethical use of technology – has openly upheld the cause of radical longevity.[1194]

Various particular methods toward radical life extension have been advocated. Thus, many far-reaching hopes and fears have been raised regarding nanotechnology and nanomedicine,[1195] investigating the use of nano-particles as therapeutic agents and even discussing potential uses of nano-robots for tissue repair.[1196] Cryonics has been a widely proclaimed (though not as widely practiced) "insurance policy" to preserve the body until the emergence of effective reparative technologies.[1197] Indefinitely sustainable man-machine synergy or "cyborgization"[1198] has been investigated as yet another path to radical life extension.[1199] Here a major topic of discussion has been the "technological singularity" – chiefly understood as an explosion of technological capabilities that would make some kind of "cybernetic immortality" possible. The leader of the "singularitarian" school has been the inventor and industrialist Ray Kurzweil.[1200] The concept of "technological singularity" had a long history, that can be traced back to the ideas of Vernor Vinge, Pierre Teilhard de Chardin, Vladimir Vernadsky, and others.[1201] The concept now has many supporters, mainly in the US and mainly in the high-tech and computer industry.[1202] The topics related to indefinitely maintainable man-machine symbiosis have included artificial intelligence,"[1203] "robotic organs,"[1204] "cybernetic immortality" via "mind uploading"[1205] and related subjects such as "memory preservation,"[1206] "brain mapping"[1207] and "quantified self."[1208]

The safety of any such potential interventions – either bio-technological, nano-technological or info-technological – has been one of the primary concerns.[1209] Another great concern derives from the very diversity of proposed approaches. Which approach or which combination of approaches is more promising or cost-effective for bringing the maximal possible prolongation of life, in the best of health, to the largest numbers of people (hopefully all), in the shortest time? A comprehensive quantitative evidence-based and integrative "roadmap" to life extension, that would reliably evaluate, prioritize and forecast the current technological trends, as well as the potential social implications and determinants of emerging life extending technologies, has been desired.[1210]

The cultural and socio-political diversity among proponents of life extension has also been broad. Even though great many life-extensionists are pronouncedly technophilic, materialistic or atheist (rather preferring to be termed "agnostics"), there have been quite a few life-extensionist communities of "spiritual" orientation.[1211] Religion is still a force in American life-extensionism, as it was in the *fin-de-siècle*, though it is now much less emphasized.[1212] Both in the US and globally, life-extensionist beliefs have been expressed by proponents of Christianity,[1213] Islam,[1214] Judaism,[1215] Hinduism,[1216] Buddhism,[1217] Taoism[1218] and other religions.[1219] Thus, life-extensionism can be once again seen as a unifying pursuit, which however does not diminish cultural and social variation.

Considering the social and material support for life-extensionism, despite the hopes often expressed for increased government funding, the more radical life-extensionist enterprises still largely depend on capitalist philanthropies, such as those of the billionaires John Sperling,[1220] Larry Ellison[1221] and Peter Thiel.[1222] An interesting development occurred with the establishment of the Maximum Life Foundation – "The Aging Research Foundation for Life-extension and Longevity" – in 1999 by the venture capitalist David Kekich (b. 1943). One of the foundation's central projects has been the Manhattan Beach Project. It has been hoped to be on the scope of the original "Manhattan Project" that developed the atomic bomb, except that the Manhattan Beach Project will be life-preserving rather than life-destroying. Another difference will be that the "Manhattan Beach Project" – driven by libertarian capitalists – will be entirely privately owned and will endeavor to keep its prospective developments as close to the private owners and as far away from governmental coercion as possible. The numerous advocates for the appropriation of longevity research by the private sector have put in question the role of governmental regulation for accelerating the development of innovative medical technologies. In their expositions, such regulation is mostly inhibitive. They have further pointed out the reluctance of the government to invest in longevity research. Whatever the justifications, the dominant involvement of private enterprise in American longevity research is undeniable.

As the Manhattan Beach project summary stated, "Entrepreneurs and investors will own the vast majority of the equity in the enterprises. The emerging longevity sector is expected to exceed one trillion dollars. It will spawn a whole new breed of billionaires who will get rich by extending our lives." The summit of the Manhattan Beach Project, on November 13-15, 2009, at Manhattan Beach, California, according to the sympathetic account of the libertarian journalist Ronald Bailey, witnessed the beginnings of "Immortality Incorporated" (another proposed name was "MaxLife Capital").[1223] This was planned to be a public life extension research company that would invest specifically in companies that research technologies to stop and reverse aging. In the meantime, the company would still seek to raise $5 million before going public. The company, called "Age Reversal Inc." (ARI) was eventually launched in 2010. Yet, very soon Kekich found that "investment bankers and investors didn't want to take a risk on a company without operating history and without the prospect of short-term earnings" and "the harsh reality is, most people are more concerned with how much they can make and how quickly – and how much they can lose."[1224]

Capitalism and the quest for funding still seem to be prevalent elements in American Life-extensionism, as they have been since the late 19th century.

22. Chapter conclusion. The US leads the world of longevity research, practice and advocacy

It is a striking demographic feature of the present day life-extensionism that the fascination with prospective life-preserving technologies is prevalent with the young and relatively well-to-do men, many of whom have studied exact and biomedical sciences. The majority of life-extensionists have been citizens of affluent Western countries, while the representation from the Developing World has been minimal.[1225] The representation from the United States has been by far the largest, somewhat remotely followed by the United Kingdom, as it was since the inception of the International Association of Gerontology in 1950. Nonetheless, currently the representation from the "developing world" and a wider massive involvement in the subject have been gradually growing.[1226]

Yet, in the recent period, wealthy westerners have represented perhaps the widest cohort of contemporary life-extensionists. It is therefore hardly surprising that, in such a milieu, capitalism has been highly prized and widely praised. It is the increased competition and "decentralization" of biomedical industry, it has been commonly asserted, that would make the rapid development of life-extending technologies possible. Several prominent members of the American "pro-market" libertarian political movement have been explicit life-extensionists, including the 1984 and 2008 Libertarian Party presidential candidate Mary Jane Ruwart, the 2008 Libertarian Party Vice Presidential candidate Wayne Allyn Root, as well as the anti-aging entrepreneurs Durk Pearson and Sandy Shaw, professor of law at the University of Tennessee Glenn Reynolds, the science editor of *Reason* magazine Ronald Bailey, and the inventor and industrialist Ray Kurzweil.[1227] Life extension has also played a part, though perhaps less pronounced, in the politics of both the Democratic and Republican parties.[1228]

In the "market-oriented" view, the poor are not entirely forgotten. In this view, thanks to competition and ever expanding markets, life-extending technologies are hoped to eventually become cheaper and more widely accessible. It is often hoped that new technologies will so immensely increase the "pie" of world wealth, that everyone will have their fill. Provisions for the actual distribution of the "pie" are mostly left in silence. In particular, the mechanisms for the universal distribution of future life-extending technologies remain obscure. The improvements of health and longevity in disadvantaged communities (if anyone cares to remember them at all) will 'just get there' in the future by their participation in the global market and by cheapening the prices, certainly not from any special allocations (or relocations) of the present wealth. It must be added, however, that a social-democratic stream in life-extensionism and transhumanism has also been strong, chiefly drawing from the experience of European social democracies.[1229] With the strengthening of the global "social justice movements" in 2011-2012, life-extensionism was increasingly linked with the issue of social justice.[1230]

While those closer to the private sector have favored the "libertarian solution" for life-extension; those closer to established academic or governmental institutions have preferred lobbying the government for increased longevity research funding. A notable initiative of the latter kind has been the "Longevity Dividend Campaign" started in 2006, which however does not seem so far to have yielded any outstanding changes in funding policy.[1231] Commercial and academic/institutional developers of life-extending methods are often found to be in opposition to each other (that is, of course, unless academic scientists rapidly translate their research to the private sector).[1232] Thus, once again, life-extensionists can be seen as advocates for the continuation of the existing socio-ideological pattern of which they form a part. And in case the patterns diverge, as in the opposition between "free-markets" vs. "governmental programs," life-extensionists will advocate for the perpetuation of the pattern in which they are integrated the closest.

The advocates of "free markets" seem to have been gaining the upper hand so far. Up-to-date, proponents of the futurist "Upwinger," "Extropian," "Transhumanist/Humanity Plus" or "Singularitarian" intellectual movements have seemed to advocate for a radical change (enhancement) of human nature and society, almost beyond recognition: attaining unlimited energy, matter availability, space expansion, robotic labor, cognitive enhancement through symbiosis with artificial intelligence, in short, going to the next stage of human and social evolution, with unlimited capabilities and wealth. Similar aspirations were expressed in the US and other parts of the Western world at least since the *fin-de-siècle*. Unlimited health enhancement and radical life extension have been the first items on the agenda of the current futurist movements. In fact, these movements are now perhaps the only ones that openly embrace the cause of radical longevity. Still, even when speaking about changing the human nature and society beyond recognition, the underlying primary desire is to preserve *ourselves* and make *our society* more durable.

Markets and democracies, hence, have figured prominently in contemporary Western discussions of our post-human future. There is perhaps no stronger advocate of competitive "decentralization," of "the movement toward democracy and capitalism" during our evolutionary transcendence, than Ray Kurzweil, president of Kurzweil Technologies Inc.[1233] One may wonder how much unplanned "decentralized" activity takes place in Kurzweil Technologies, and whether careful planning is deemed undesirable for that particular company. But, on a global scale, "decentralized/free markets" have been recurrently hailed by Kurzweil as the guarantors of human survival. Kurzweil has been echoed by many voices praising the progressive and philanthropic role of markets and high-tech entrepreneurship.[1234] Such an ideological preference might be expected as, in a capitalist milieu, to doubt the necessity of free markets for progress would be sheer blasphemy. Thus in this case also, life-extensionism can be seen as an expression of a desire to preserve a prevailing socio-ideological paradigm.

General conclusion. Life-extensionism as a pursuit of constancy

As this work has argued, in different national contexts, different ideological schemes – secular humanism or religion, elitism or egalitarianism, idealism or materialism, socialism or capitalism, liberalism or totalitarianism – appear to have yielded different justifications for the necessity of life prolongation and longevity research and to impact profoundly on the way such goals were conceived and pursued. As the works of the proponents of human enhancement and longevity exemplify, the authors adapted to particular national ideological environments and served as agents for their continuation. Several conclusions can be drawn from these examples of adaptation of life-extensionism to the specific socio-ideological environments.

First, these adaptations may question the claims of a particular ideology for supremacy in the promotion of life-extension and life-enhancement. Any claims that atheism, capitalism or hedonism is more conducive to the pursuit of longevity, can be countered by historical examples where religion, socialism or asceticism, were the foundations. No ideological system seems to have a monopoly, however strongly it asserts that it constitutes the rock-solid ground for this pursuit. It may be that, rather than providing such a foundation, political ideologies enlist the hope for life extension to increase their appeal. Life extension may thus represent a cross-cultural value, yet often involving antagonistic social theories and political movements.

Secondly, in the authors under consideration, the goal of life extension has been associated with a striving for stability and equilibrium, desiring to stabilize and thus perpetuate the current state of the body or personality, and the present social system. In this sense, life-extensionism may be a fundamentally conservative (or conservationist) enterprise. Therefore, the impression that life-extensionism represents a form of utopianism, a fringe or revolutionary movement, or an advocacy of a radical change of human nature – should be rejected or accepted only with profound reservations. Historically, the proponents of radical life extension may have envisioned no greater change to human nature than the extent to which maintenance of an ancient edifice changes the nature of that edifice, or the extent to which the (often high-tech) restoration and conservation of an old work of art make it a forgery. The life-extensionists may indeed have strived for a perfected society, which one might call a "utopia," but that "utopian" society, they hoped, would uncannily resemble the one they lived in, with all or most of its institutions intact and all the near and dear ones alive and around.[1235] The life-extensionist movement may have been profoundly anti-revolutionary, if only for the simple reason that opposing the existing social system would nullify public support for longevity research. After a revolution has won, the life-extensionists may side with the winner (either opportunistically or in a firm belief, or both).

The adaptability of life-extensionists to the changing social patterns was rapid, but paradoxically, once an adaptation had been established, it would appear in their writings that the new pattern would continue indefinitely, or only with very minor modifications. Thus, in Russia, after the socialist revolution, the life-extensionists swiftly changed their rhetoric from praising rural patriarchy, absolutism and Pravolslav Christianity, to exalting socialism, atheism and state regulation and planning that, according to them, were the only long-viable constructs. In Germany, life-extensionists swiftly adapted to the philosophy of preserving the "national body" in line with the prevalent fascist ideology. In France, with the strengthening of traditionalist and religious sentiments, life-extensionists promulgated conceptions adjusted to the trend, adopting "holistic" systems for life-extension that contained strong religious and traditional cultural elements. And in American life-extensionism, capitalism, religion and eugenics had been dominant before WWII, in line with the common popularity of these notions. After the war, with eugenics losing favor and religion underemphasized in scientific pursuits, capitalism remained dominant. As these instances suggest, rather than speaking of life-extension generally, it may be necessary to speak of the extension of particular "life-forms" or "life-patterns" – personal or social – the forms that are being perpetuated or fixated upon.

We may perhaps want to select or at least discuss the patterns that we indeed wish to perpetuate. In other words, we may consider what social practices, ethical precepts or power structures may be involved in the pursuit of life-extension, or what would be the form of society in which we would wish to live long. An undesirable yet unchangeable pattern may be a dystopian prospect indeed. The question may still be raised regarding the form of society that is most conducive to longevity research or to actually increasing human longevity.

The desire to preserve constancy was difficult to fulfill, as changes in general and deteriorative changes in particular were difficult to resist. This might be one of the reasons why radical life-extensionism has not become entrenched in the public mind. The task of maintaining constancy, equilibrium or homeostasis, is daunting and goes against too many odds. Yet, the human desire to maintain constancy, in spite of all change, has been persistent as well, and in this regard life-extensionism is nothing exceptional. The inevitability of change has been acknowledged in many conceptions of social organization. But the desire for constancy and stability has been acknowledged as well. Thus according to Hegel's classical conception of the "Zeitgeist," or the "Spirit of the Age," prolonged periods of stability are not tolerated, but subverted by internal oppositions. "Periods of happiness," Hegel wrote in *The Philosophy of History* (1837) "are blank pages in [the History of the World], for they are periods of harmony – periods when the antithesis is in abeyance." Civilizations are always changing, manifesting the development and realization of Spirit. But Spirit itself does not change. "Spirit is immortal; with it there is no past, no future, but an essential *now* ... the present form of Spirit comprehends within it all earlier steps. ... what Spirit is it has always been essentially; distinctions are only the development of this essential nature."[1236] Moreover, according to Hegel, as far as human capacity for reasoning goes, constancy is always sought for. As Hegel stated in the *Encyclopedia of the Philosophical Sciences* (1830), "reflection is always seeking for something fixed and permanent, definite in itself and governing the particulars."[1237] Also in Marxism, owing a great deal to Hegelianism, change and constancy are pervasive concerns. In Marxism, "social formations" constantly change, being subverted by economic developments and class struggle; that is, until the "social formation" of communism will be reached, where there will be no class struggle and which will presumably continue indefinitely.[1238] Even much earlier, transformative change has been a fundamental concept of Taoism, the "Book of Changes" being one of the most venerated texts of Chinese philosophy. Yet, Taoism also envisions the attainment by the society of the state of "Taiping" – the state of Great Peace – that will enjoy stability and harmony for eternity.[1239] (Ironically, the historical Kingdom of Taiping (1850-1864) was one of the most turbulent and violent in Chinese history.)

The recognition of the inevitability of change and the desire for perpetuation are also present in many works of cultural history, particularly the works on the history and sociology of science. Thus, in Thomas Kuhn, paradigms constantly shift, being subverted by anomalies. Yet Kuhn also points out the resilience of established paradigms. "And at least part of that achievement always proves to be permanent" (1962).[1240] (Or else, logically, the paradigmatic belief in paradigm shifts may itself pass.) Bruno Latour speaks of "reference" as "our way of keeping something *constant* through a series of transformations"(1999).[1241] Steven Shapin speaks of "broad European changes in attitudes to knowledge in general and to the relations between knowledge and social order." Yet, immediately afterwards, he notices "a state of *permanent crisis* affecting European politics, society and culture" (1996).[1242] Michel Foucault, in *The Order of Things* (1966), speaks of "fixism" and "evolutionism" as "two simultaneous requirements," and "these two requirements are complementary, and therefore irreducible."[1243] Stability and fixity are intrinsically related to order. And the striving for stable equilibrium has been related to rationalism.[1244]

Further confirming the inexorable presence of both the concepts of change and of constancy, the American social historian Peter Burke (2005) defines "modernity" as "the assumption of fixity" and "post-modernity" as "the assumption of fluidity," "the collapse of the traditional idea of structure," "destabilization and decentering." In a "modern" or "modernizing" discourse, the rhetoric of progressive change is persistent, but it may conceal a deep-seated desire for fixity and stability. And in a "post-modern" discourse, "change" is a commendation in and of itself, and "constancy" or "stability" are commonly either ignored or vilified. Yet the desire to preserve constancy cannot be easily rejected. Burke points out numerous attempts "to freeze the social structure," to "resist change." "Such activities," Burke suggests, "surely deserve a place in any general theory of social change."[1245]

On a continuum between the desire for absolute change and the desire for absolute constancy, the life-extensionists would seem to stand closer to the pole of constancy. Indeed, without some notion of constancy, the concept of life-extension, even of survival, would be meaningless. Consider such cases as the atoms of a decomposing human body merging with the Universe, or human life being transformed into the life of grave worms, as discussed by Jean Finot in *The Philosophy of Long Life* (1900).[1246] Many boundaries are "transcended" in such "transformations," but one can hardly speak of "life-extension." If extinction is determined by "a critical rate of long-term environmental change beyond which extinction is certain"[1247] (notice, *any* change), then life-extensionists would wish to be as far from this rate of change as possible.[1248] Or else, they would wish to design the technological armor that would make us impervious to such changes. Without work invested in maintaining constancy, spontaneous deteriorative change may be expected. Thus, in the sense of Burke's definition, life-extensionism is a very "modern" endeavor. The rhetoric of progressive change is emphasized, but not just any change for the change's sake, but only such change that would serve to perpetuate some existing structure. In the words of the protagonist of Giuseppe di Lampedusa's *The Leopard* (1960) "If we want things to stay as they are, things will have to change."[1249] And in the words of Lewis Carroll, "it takes all the running you can do, to keep in the same place" (1871).[1250]

The question may arise: what is it exactly that the life-extensionists would endeavor to fix? For a religious person, the answer might be easier. The things that require preservation might be the eternal, imperishable soul maintained in a robust temple of the body, and a God-decreed social order and way of life. But for a materialist, believing in the contingent and temporal construction of physical objects, the answer may be much more difficult. Should the preservation efforts be directed to some arbitrary structure, archetype, memory, connections, the Zeitgeist? Hegel's classical notion of the "zeitgeist" has now been generally discarded. As succinctly stated by the Austrian-British art historian Ernst Gombrich in his book *In Search of Cultural History* (1969), in place of the Hegelian search for expressions of the universal spirit of the age, there comes the search for connections within the surrounding culture, since "any event and any creation of a period is connected by a thousand threads with the culture in which it is embedded."[1251] And thus, an historian may observe distinctions between specific social and ideological environments or "embedding cultures" (for example, French liberalism followed by conservatism, German fascism, Russian communism or American capitalism), even without providing exact definitions, and may detect specific adaptations.

With reference to life-extensionism, the adaptations can be clearly perceived. Even the very terms for life-extensionism have varied according to period and context – internal alchemy, gerocomia, macrobiotics, rejuvenation, experimental gerontology, anti-aging, prolongevity, life-extensionism, immortalism, transhumanism – as befits the circumstance of political correctness. And the terms for progress, within which life-extensionism has been commonly embedded, have changed as well: in place of the somewhat archaic "meliorism" and the somewhat ominously sounding "progressivism," now the more popular terms are "making the world a better place," "sustained human development" or "continuous evolution." An apparent persistent opposition in life-extensionist methodologies, between what has been termed here "reductionist" vs. "holistic" approaches to life-extension, emphasizing, respectively, targeted repairs of the human machine vs. psychosomatic effects, also seems to have undergone shifts of terminological fashions. These were manifested in the dichotomies of "mechanism" vs. "vitalism," "materialism" vs. "idealism," "invasive/artificial therapeutics" vs. "non-invasive/natural hygiene." The respective terms are not entirely synonymous. Moreover, "reductionist" and "holistic" methods for life-extension were often combined by the proponents. Yet, similarities and continuities between the respective terms can nonetheless be observed.

But perhaps the most salient adaptations were to what might be termed the "dominant socio-ideological order" – "liberalism" or "conservatism," "fascism," "communism" or "capitalism" – whose prevalence in the specific countries and periods under consideration is apparent. Though these "dominants" may seem similar to Hegelian manifestations of the "Spirit of the Age," they are rather expressions of the "interconnected embedding culture," not something "essential" but rather categorical and contingent. The adaptations of life-extensionism to particular contexts took various forms. These included the rhetorical support of the ruling socio-ideological order (if only to ensure that the research is not shut down by the authorities), and the positing of metaphorical socio-biological parallels between the workings of the body and the society in which the authors lived. Moreover, specific research projects were favored as compatible with the ruling socio-ideological order, for example, the strengthening of "holism" in France in line with the strengthening of "traditionalism"; the favoring of "natural" longevity regimens in Germany in line with the nationalist romanticizing of "natural ways"; the reductionist engineering of the human body in the Soviet Union in line with the professed ideology of optimal social planning and engineering a "new man"; or the proliferation of rejuvenative nostrums in America under the spirit of free enterprise. But variation was only a part of the adaptation process; another was conservation. The life-extensionists did not simply "adapt" to the changing socio-ideological conditions, but sought to conserve the adaptation, sought to make their relations to the environment or "embedding culture" stable. I argue that the support of the existing ruling regime, whatever it may be, may derive from the nature of life-extensionism that seeks stability and constancy.

The life-extensionists' inherent desire for constancy has stood in stark contrast to "apocalyptic" beliefs. Such beliefs have been present throughout the century, and they had been recently intensifying. There has existed an extensive literature expecting (and accepting) "the world as we know it" to end anytime soon.[1252] In morbid excitement, the prophets of the apocalypse have often stressed the great corruption of humanity, expressing what the novelist John Updike termed "a smug conviction that the world was doomed" (1972).[1253] Insofar as humanity was seen as inherently corrupt and self-destructive, a thorough "cleansing" appeared to be in order, through an all-out war between "the sons of light" and "the sons of darkness," separating the bad "weeds" from the good "wheat." Often the apocalyptists, both secular and religious, have had some very strong convictions about who the "sons of darkness" and the "weeds" are.

Historically, the life-extensionists have exhibited none of this attitude. They might find it difficult to distinguish and separate between the "weeds" and the "wheat" and would request a longer life-span to figure it out. Until then, the entire societal and personal 'bundle' may need to be conserved. Alternatively, many life-extensionists appear to have realized the existence of corruption and exploitation in the current society, whose perpetuation would be highly undesirable. But at the same time they also extrapolated on the manifestations of creativity, benevolence and justice, as well found in the current society, and considered them to be worth preserving indefinitely. Thus they would follow the ancient Talmudic command that "the sins will cease" but not "the sinners."[1254] While the "apocalyptic" view largely assumed that human attempts to resist catastrophic changes are destined to failure, the life-extensionists, even though recognizing existential threats (and the threat of senescent death in the first place), valorized our ability to defend ourselves.[1255] Whatever the explanation or underlying motives, life-extensionism appears to be a profoundly anti-revolutionary, anti-catastrophic, anti-apocalyptic ideology. As the American author William Bailey emphasized in his bibliography on *Human Longevity from Antiquity to the Modern Lab* (1987), "Death be not proud, this heartening literature opposes Armageddon and the perniciousness of nature to say that we can extend life and enjoy many another springtime."[1256]

The questions still remain considering what exactly the life-extensionists would desire to maintain constant, and whether anything at all can be maintained constant. An answer may be again suggested by Taoism. As the great teacher of Taoist immortalists, Lao Tse said of the Tao (the Way or Course): "How still it was and formless, standing alone, and undergoing no change. … Great, it passes on in constant flow."

Moreover, human beings should "possess the attributes of the Tao."[1257] No wonder then that in Taoism, radical life-extension and conservation of order have always been all-pervasive aspirations. If Heraclites could not "step twice into the same river; for other waters are continually flowing in"; Lao Tse could, for the course of the flow may remain constant. It seems as if Walter Cannon's concept of homeostasis, described in *The Wisdom of the Body* (1932), follows directly Lao Tse's notion of the Tao. In Cannon, some constancy is provided by what he terms "the interesting fact that we are separated from the air which surrounds us by a layer of dead or inert material." Yet, for Cannon, the organism is not entirely separated from the environment, but related to it through a constant flow of materials and energy: "the internal, proximate environment of the cells is made favorable by keeping the fluids on the move and constantly fresh and uniform."[1258] A modern textbook definition of homeostasis would say the same. The materials may be exchanged, but the course and the form of their flow remain constant: "life is characterized by a continuing flow of material and energy, and a steady state is reached if all possible disturbing factors remain constant."[1259]

A ramification of this idea can be found in the philosophy of Arthur Schopenhauer (1788-1860) who postulated that "the dead body is a mere excrement of a constant human form."[1260] That is to say, all materials in the human body are being incessantly replaced, and only their form or arrangement is constant. Schopenhauer's "pessimistic philosophy," calling for reconciliation with death, was hardly compatible with life-extensionism. Metchnikoff took great pains to refute it, among other reasons, because he did not believe that the "form" or "ideal," either of an individual or of a species, can exist without a material substrate (1903).[1261] Yet, the particular idea of Schopenhauer's regarding the constancy of the human form, despite the material replacements, was approvingly cited by Max Bürger, the proponent of "biorhesis" or stable biological flow (1947).[1262] Indeed, various "replacement therapies" – ranging from hormones through vitamins and minerals to stem cells, artificially grown organs and bionic prostheses, while maintaining the constancy of the human form – have constituted the core of life-extensionist methodologies throughout the century, mainly in its "reductionist" and "materialistic" branches. In the "holistic" and "idealistic" branches, some essential core of human personality was believed to be able to directly control the body and resist bodily changes.

The idea of maintaining the constancy of structure and function through a continuous replacement of material components can be traced even further back to the "Paradox of the Ship of Theseus" first mentioned by Plutarch (c. 46-120 CE).[1263] The great quandary was whether a ship, all of whose parts are replaced, will retain its identity. For the life-extensionists this has been a vital issue and their implicit (and often explicit) answer has been that it would indeed remain essentially the same, since its structure and function would be preserved. And if some new components were to be added to the "Theseus' Ship" to improve and prolong its performance, it would be essentially the same as well, since a major part of its structure and function would remain constant. However, if it were to break down into components, even though the materials would remain, the constancy of form would be lost.[1264]

Science fiction related to life-extension had a field-day with this paradox. Among many examples, in Stanislaw Lem's "Do you exist, Mr. Johns?" (1955),[1265] the protagonist has all his biological components replaced by artificial ones, and the company that produced them claims ownership over him. The cybernetic human vehemently defends his right to an identity (that is, being identical to the former biological human), since his personal memories are uniquely his own and are only preserved in a different substrate. Even as a biological entity, all of the materials in his body were being replaced in a very short time. By analogy, he is now no more the property of the company than he was the property of the grocer who formerly supplied him with food.

Beyond science fiction, for practicing life-extensionist scientists, the "Theseus' Ship Paradox" has been a practical concern, as it informed the search for replacement therapies. And the paradox has been resolved in a similar, positive manner. Thus Aubrey de Grey wrote in *Ending Aging* (2007):[1266]

> I emphasized … that the body is a machine, and that that's both why it ages and why it can in principle be maintained. I made a comparison with vintage cars, which are kept fully functional even 100 years after they were built, using the same maintenance technologies that kept them going 50 years ago when they were already far older than they were ever designed to be.

The main point is that for the life-extensionists, the possibility of maintaining the constancy of form has been certain.

The same valorization of the constancy of form recurs in Ray Kurzweil's *The Singularity Is Near: When Humans Transcend Biology* (2005). Yet, instead of the classical notion of the "form," Kurzweil uses the term "pattern." Kurzweil is a world-renowned expert in pattern recognition, having pioneered several ground-breaking developments in optical character recognition, speech recognition, stock-market pattern recognition and more. The entire world, according to Kurzweil, consists of "patterns of information": patterns of matter and energy, biological and social patterns. And there are precise procedures in information theory to determine the extent to which various patterns, material, biological or social, are similar or different (for example through the use of entropy and mutual information).[1267] Hence, for Kurzweil, the maintenance of specific patterns is a very tangible and practical task. Insofar as orderly information patterns are constantly maintained by biological systems, such patterns can be similarly (perhaps even better) maintained by machines. A dedicated life-extensionist, Kurzweil suggests the constancy of an information pattern, comprising the human body and mind, as an underlying concept for indefinite survival:[1268]

> My body is temporary. Its particles turn over almost completely every month. Only the pattern of my body and brain have continuity. … Knowledge is precious in all its forms: music, art, science, and technology, as well as the embedded knowledge in our bodies and brains. Any loss of this knowledge is tragic. … Death is a tragedy. It is not demeaning to regard a person as a profound pattern (a form of knowledge), which is lost when he or she dies. That, at least, is the case today, since we do not yet have the means to access and back up this knowledge.

By perfecting the means of preserving our "information patterns" and eventually "backing them up," human life can be preserved indefinitely: "We are now approaching a paradigm shift in the means we will have available to preserve the patterns underlying our existence." And further down the line, "As we move toward a nonbiological existence, we will gain the means of 'backing ourselves up' (storing the key patterns underlying our knowledge, skills, and personality), thereby eliminating most causes of death as we know it."[1269] Thus the conservation of the existing patterns is an explicit goal. Kurzweil admits that he cherishes all his memories and never discards any memorabilia, since they constitute the unique pattern of his personality. Furthermore, one might suggest, Kurzweil's valorization of "democracy and capitalism"[1270] is a part of the overall program to conserve the "pattern" in which he exists.

It should be noted, however, that, in Kurzweil, the conserved pattern is not perceived as entirely stagnant. Rather, the underlying metaphor is that of a continuous growth, building new structures on the existing foundation or "core" and including the already existing building blocks. Kurzweil uses the analogy of an old computer file that may be preserved in a new computer, yet with many new files added to it. Similarly, our general "mind file" will be conserved, yet augmented with new extensions. The "core pattern" is not entirely unchangeable either, but it changes slowly and gradually. Still, the continuity of the pattern is maintained: "You change your pattern – your memory, skills, experiences, even personality over time – but there is a continuity, a core that changes only gradually." Technological enhancement will not radically modify this "core": "that's just a surface manifestation. My true core changes only gradually."[1271] Kurzweil uses as an epigraph to his discussion of longevity the statement by the American computer scientist Vernor Vinge (1993) that the technologically enhanced human "would be everything the original was, but vastly more."[1272] Still, it would be "everything the original was." Thus, the underlying desire for constancy is affirmed once again.

The assumption of constancy may also answer the frequently raised question: 'Why would we want to prolong life?' If human life is an absolute value now, and its value will remain the same tomorrow or in a hundred years, then all the efforts to preserve human life at any moment and for any period of time are justified. It is important to bear those philosophical considerations in mind when researching the historical motivations for life-extensionism.

Finally, and paradoxically, out of the desire for constancy, novelty arises. It is easy to dismiss the pursuit of life-extension as a "pipe dream." Yet, many examples show that the scientific contributions of life-extensionist researchers have been considerable, and often pioneering: the first attempts at therapeutic endocrinology, blood transfusion, transplantation, cell and tissue therapy, probiotic diets, general hygiene, and more.[1273] These developments may have been not just due to 'aiming high' and in the process bound to achieve at least some results, even though most often falling short of the original aspirations. Rather, the scientific advances made by the life-extensionists may be the product of their underlying conservative bent on stability and perpetuation. As the stability of the internal milieu could not be achieved by contemporary medical technology, innovative interventions were sought. Consider, for example, such late 19th-early 20th century developments as Nikolay Pirogov's plaster casts to fixate the bone (c. 1870), Porfiry Bakhmetiev's preservation of animals by freezing (c. 1900), or Auguste Lumière's introduction into biomedicine of film and auto-chrome plates to safeguard images of the body (c. 1900). All these can be viewed as technological novelties employed in the service of maintaining constancy. And if some methods of maintaining constancy failed, such as the reductionist "endocrine" rejuvenation – new methods of maintaining homeostasis would emerge, such as improved (and still reductionist) replacement techniques or more systemic, holistic or hygienic approaches.

Many life-extensionist scientists spoke explicitly about their desire to maintain constancy through novel technological means. Thus, the British pioneer of X-ray crystallography, John Desmond Bernal (1901-1971), asserted that new technological extensions of existing human capabilities will lead to an indefinite extension of life and a greater "fixity" of human personality (1929): "This capacity for indefinite extension might in the end lead to the relative fixity of the different brains; and this would, in itself, be an advantage from the point of view of security and uniformity of conditions." At any rate, the technologically modified man may have a better chance for self-preservation than an unmodified one, even at an early and imperfect stage of the modifying technology: "But though it is possible that in the early stages a surgically transformed man would be at a disadvantage in capacity of performance to a normal, healthy man, he would still be better off than a dead man."[1274] The Russian pioneer of neurophysiology, Ivan Pavlov, spoke about the fundamental drive of science toward equilibration of living systems (1923): "There will be a time, even though a remote one, when mathematical analysis, based on natural science, will encompass, by magnificent mathematical equations, all existing equilibria."[1275] And as one of the foremost twentieth century life-extensionists, the French-American pioneer of organ transplantation and tissue engineering, Alexis Carrel contended (1935): "Science has supplied us with means for keeping our intraorganic equilibrium, which are more agreeable and less laborious than the natural processes. …the physical conditions of our daily life are prevented from varying."[1276] Thus the desire for fixity and equilibration may have been a pervasive motive in life-extensionism and often a source of new developments in biomedical science and technology. As the "philosopher of long life," Jean Finot asserted (1900), the primary purpose of biomedical advances is not to change, but to "preserve and greatly strengthen existing life."[1277]

Supplemental Materials

Figures 1&2.

Publications on Rejuvenation (1) and Theories of aging (2). 1900-1960.

Based on the analysis of Nathan Shock's *Classified Bibliography of Gerontology and Geriatrics*, Stanford, 1951, 1957, 1963.

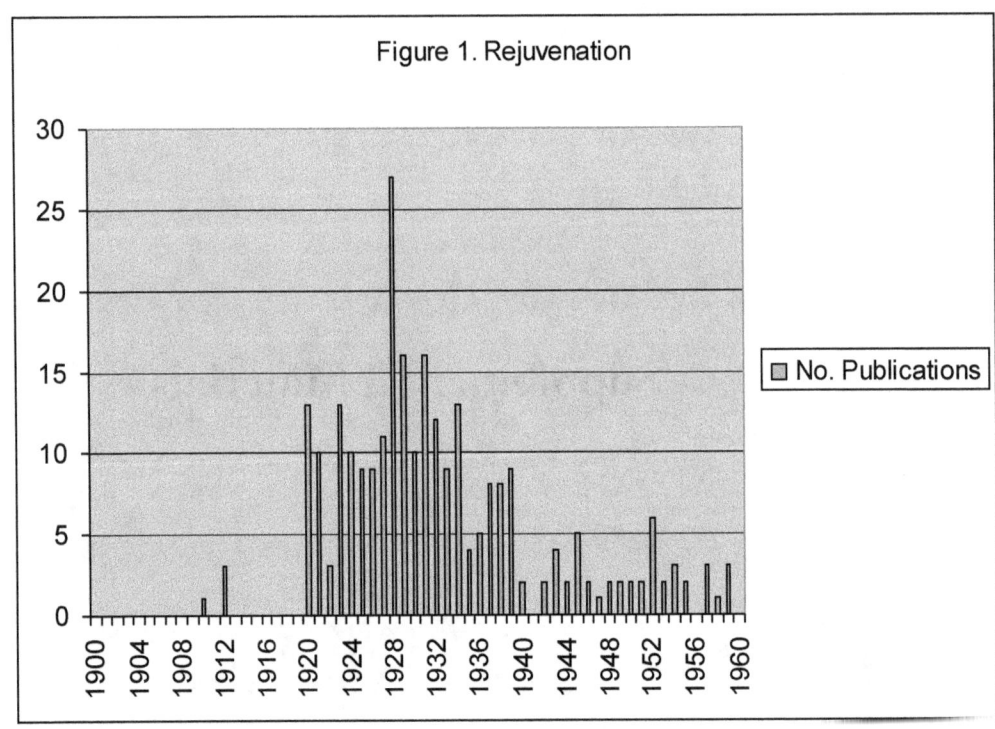

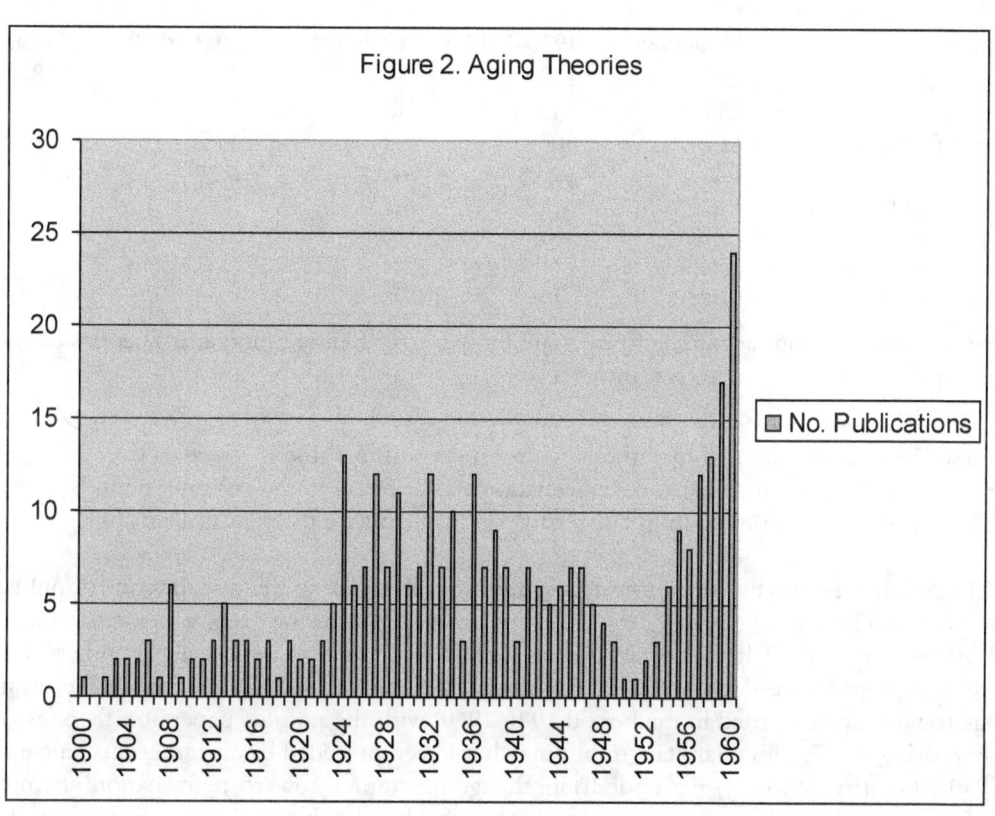

Table 1.

Distribution of publications on Rejuvenation and Theories of Aging according to periods

Periods	1900-1929	1930-1939	1940-1945	1946-1960	Total
Total	239	174	49	146	608
Rejuvenation	125	94	15	31	265
Theories of Aging	114	80	34	115	343
Ratio Rejuvenation/Theories	1.1	1.175	0.44	0.27	0.77

Hypothesis H0: Distributions according to rows are the same for Rejuvenation and Theories of Aging.
Hypothesis H1: The "null" hypothesis is incorrect.

Calculating the χ^2 criterion for the table of conjunction (A. Buhl, P. Zofel, SPSS Version 10, Addison-Wesley, New York, 2001), the null hypothesis with respect to the Table is rejected (P=0.05). This signifies that the interest in Theories of Aging and Rejuvenation was different in the different periods.
(I thank Dr. David Blokh of Ben Gurion University, for his help with the statistical analysis.)

The differences in ratios demonstrate a significant difference in the distributions of publications on Rejuvenation and Theories of Aging in the different periods. As can be seen, a dramatic shift of interest occurred in the period 1940-1945, corresponding with WWII. In that period, the number of studies on theories of aging began exceeding those on rejuvenation by a factor of 2.3 (a reversal of the relative weight ratio). This trend only intensified in the period 1946-1960, with the ratio of papers on theories of aging to rejuvenation rising to 3.7. This quantitative shift may have been preceded by an earlier qualitative shift in the period 1930-1939, that is to say, the underlying change of attitude toward rejuvenation attempts. As the efficacy of contemporary rejuvenation techniques was found to be thoroughly wanting, a recoil may have occurred toward the more basic, theoretical approach. The influence of the war may have also been considerable, as attempts at far-reaching rejuvenation may have been of less immediate relevance during the war. The reversal of the relative weight ratio may have been also partly related to the emergence of the first national institutional gerontolological associations in 1938-1939 in the US, UK, Russia, Germany and France.

Main thematic classification

The topics are arranged according to Chapter and Section number, I – Introduction, C – Conclusion, SN – Supplemental Materials and Notes. To find specific terms, please use a "search in text" option.

Rejuvenation methods – surgical, organotherapeutic and pharmacological – R: 1.6, 1.8, 1.9, 1.10, 1.11, 1.12, 1.15, 1.16, 1.18, 2.2, 2.3, 2.10, 2.11, 3.10, 3.11, 3.13, 3.17, 4.5, 4.16, 4.17, 4.19, 4.21, SN

Life extending holistic systems and regimens – L: 1.7, 1.15, 1.17, 1.18, 2.2, 2.5, 2.6, 3.9, 4.3, 4.4, 4.11, 4.18, 4.19, SN

Theories of aging and longevity – T: 1.2, 1.6, 1.13, 1.14, 1.15, 2.7, 2.9, 2.12, 3.11, 3.12, 3.16, 3.17, 4.2, 4.7, 4.8, 4.9, 4.10, 4.12, 4.14, 4.16, 4.17, 4.19, 4.19, SN

Institutional organization and social milieu of life-extensionists – O: 1.2, 1.3, 1.4, 1.5, 1.15, 1.16, 1.17, 2.4, 2.5, 2.6, 2.7, 2.8, 2.9, 2.12, 3.2, 3.3, 3.4, 3.5, 3.11, 3.12, 3.14, 3.15, 3.16, 3.17, 3.18, 4.2, 4.4, 4.6, 4.11, 4.12, 4.13, 4.14, 4.15, 4.18, 4.19, 4.20, 4.21, 4.22, SN

Life-extensionist philosophy – P: I.1., I.2, 1.1, 1.11, 1.13, 1.14, 1.16, 1.17, 1.18, 1.19, 2.1, 2.9, 2.12, 2.13, 3.1, 3.6, 3.7, 3.8, 3.14, 3.16, 3.18, 4.1, 4.2, 4.7, 4.12, 4.13, 4.18, 4.20, 4.21, 4.22, C, SN

Name Index

This section lists over 700 authors and actors mentioned in the main text.
Overall, more than 2,000 authors have been cited in the References and Notes Section (not listed here).

Abbas II, Abderhalden E; Abernethy J; Abrams A; Abrikosov AI; Adams SH; Adenauer K; Adrian ED; Alexander II; Alexander III; Allendy R; Amosov NM; Anderson P; Andreev FA; Andropov YV; Andrus EP; Anichkov NN; Anisimov VN; Anna Ioanovna; Anokhin PK; Aravandinos A; Artemiev T; Ascheim S; Aslan A; Atkins LV; Avicenna;

Belokopytova O; Benjamin H; Bürger M; Bacon F; Bacon R; Bailey R; Bailey W; Bakhmeteff B; Bakhmetiev PI; Balsamo G (Cagliostro A); Bancroft WD; Bardach M; Baruch B; Basilevich IV; Batin M; Baudet R; Bedford JH; Beeton M; Behring EA; Bekhterev VM; Belkin RI; Bell AG; Bender H; Benetato G; Benjamin H; Berdiaev NA; Bergson HL; Bernal JD; Bernard C; Bert P; Besredka AM; Biancani E; Biancani H; Bichat MFX; Bidder GP; Bier A; Binet L; Biot R; Bjorksten J; Blok A; Blum V; Boerhaave H; Bogdanov AA (Malinovsky AA); Bogomolets AA; Bogomolets OA; Bordet J; Bostrom N; Botkin SP; Bourlière F; Boyle R; Brehmer W; Brezhnev LI; Brinkley JR; Brittain R; Brock JF; Bronowski J; Broussais FJ; Brown-Séquard CE; Brusch C; Bryukhonenko SS; Buffon GL; Bulankin IN; Bulgakov MA; Burke P; Burrows MT; Busquet P; Busse E; Butenandt A; Buteyko KP; Butler RN; Butler S; Bykov KM;

Cabrilho JR; Caldwell C; Calkins GN; Calment J; Cannon WB; Canstatt C; Capra F; Card T; Cardenal L; Carnegie A; Carrel A; Carroll L; Carton P; Casabianca J; Casanova G; Catherine II; Cavazzi F; Cazalis H; Ceaușescu N; Chaillé SE; Chantereine J; Chaplin C; Charcot JM; Chebotarev DF; Chechulin SI; Chekhov AP; Chekhov MP; Chernenko KU; Child CM; Chisholm B; Chizhevsky AL; Church GM, Churchill WLS; Clark CG; Clarke J; Clifford C; Clynes M; Cohausen JH; Coles LS; Collin R; Comfort A; Comte A; Condorcet N; Conklin EG; Conti L; Coon CS; Corman L; Cornaro L; Cornil L; Coudert F; Cowdry EV; Crichton-Browne J; Crick FHC; Cutler R; Cutler SP;

d'Herelle F; Dakin H; Damsky A; Danilevsky VY; Dartigues L; Darwin CR; Darwin E; Dastre A; Davenport CB; de Gaulle C; de Grey ADNJ; de León JP; de Morant GS; Delore P; Demikhov VP; Denckla WD; Denis JB; Deripaska O; Derviz GV; Descartes R; di Lampedusa GT; Dilman VM; Dinaburg AG; Dior C; Domagk GJP; Doppler K; Doriot J; Dostoevsky FM; Driesch HAE; du Laurens A; du Noüy PL; Dublin LI; Dumontpallier VAA; Duplenko YK; Durand-Fardel CL; Dzerzhinsky FE;

Ebeling AH; Edgerly W (Shaftesbury E/Ralston E); Edison TA; Ehrlich P; Elberg VA; Elisaveta Petrovna; Ellison L; Ellmann R; Empedocles; Engels F; Ettinger RCW;

Fahrenkamp K; Failla G; Faragher R; Faure M; Fedorov NF; Fedorovitz L; Ferreyrolles P; Feynman RP; Filatov VP; Finch C; Finot J (Finkelstein J); Finsen N; Fischer JB; Fishbein M; Fisher GF; Fisher I; Fisk EL; Flamel N; Fleming A; Flexner A; Flexner S; Flourens MJP; Floyer J; Folbort GV; Ford H; Foucault M; Frank LK; Franklin B; Frenkel ZG; Freud S; Friedrich II; Friedrich III; Frolkis VV; Frumusan J; Funk C; Furnas CC; Fürstenberg A;

Galen; Galimard P; Galton F; Gardner TS; Gatzanyuk MD; Gayton E; Geibel E; Gengou O; Gibbon JH; Glikson A; Godwin W; Goethe JW; Goetze-Claren W; Goldstein A; Goldstein M; Gombrich EHJ; Gorbachev MS; Gorbunova GP; Gorev NN; Gorky M (Peshkov AM); Greeff JH; Gruman GJ; Guéniot A; Guarente LP; Guillerey M; Gurwitsch AG;

Haberlandt G; Hahnemann CFS; Haire N; Haldane JBS; Haller A; Halperin J; Hammond LJ; Harman D; Harms JW; Harrington A; Harris J; Harrison RG; Hart R; Hartmann M; Harvey W; Hayflick L; Hegel GWF; Heinlein R; Henry IV; Heraclites; Herodotus; Hesiod; Hilton J; Hippocrates; Hirsch M; Hirsch S; Hirschfeld M; Hoagland H; Holbrook SH; Horth O; Hufeland CW; Hunter J; Huxley A; Huxley JS; Huxley T;

Ibn Saud A (Abdulaziz Al-Saud); Ilyinsky I; Ischlondsky NE; Israel M; Itskov D; Ivanov PK;

Jaccoud S; Jenkins H; Jennings HS; Jerne NK; Jottras P; Jughashvili EG;

Kahn G; Kammerer P; Kanaev PP; Kapitsa PL; Käppeli R; Karpovich PV; Kauders O; Kavetsky RE; Kekich DA; Kellogg JH; Kellogg WK; Kenyon C; Keynes GL; Keynes JM; Khalatov SS; Khavinson VKh; Khrushchev NS; Kirkwood TBL; Kislovsky DA; Klimaszewski W; Kline N; Knight TA; Knipper-Chekhov O; Koch R; Kogan AB; Kogan GM; Köhler G; Kolb K; Koltsov NK; Komarov LV; Koprowski H; Korenchevsky VG; Korschelt E; Kotsovsky DA; Kovrigina MD; Krauze NI; Kravkov NP; Krimsky A; Krylov N; Krylov P; Kuhn TS; Kuliabko AA; Kuprevich VF; Kuprianov M; Kurchakov NN; Kurzweil R; Kustria DK; Kyriazis M;

Lacassagne A; Laignel-Lavastine M; Lakatos V; Lamarck JB; Lane WA; Langstroth L; Lankester ER; Lansing AI; Lao Tse; Last E; Latour B; Laval P; Lavoisier A; Le Compte H; Leach FA; Leake C; Lebedinsky NG; Lechler KL; Lederberg J; Lederer E; Leipunsky A; Lem S; Lenin VI (Ulyanov-Lenin VI); Leontovich AV; Lepeshinskaya OB; Leprince A; Leriche R; Lespinasse VD; Lessius L; Lévi L; Levy-Lenz L; Lewis MR; Ley HA; Lichtenstern R; Lickint F; Liebig J; Liek E; Lillehei R; Lindbergh C; Liozner LD; Lister J; Loeb J; Loeb L; Lomonosov MV; Lorand A; Louis XIV; Louis XV; Louis XVI; Lovelock J; Lubishev AA; Lumière A; Lumière L; Lusted LB; Lydston GF; Lysenko TD;

Mac Auliffe L; MacKay DM; MacNider WDB; Macy JWJr; Maimonides; Malory T; Malthus TR; Mann T; Mareuse J; Margueritte P; Maria Feodorovna; Marinesco G; Martiny M; Martynov AV; Marx KH; Masing-Delic I; Maspero H; Matthes K; Mauclaire P; Maugham S; Maupas E; Maximov AA; Maxwell JC; Mayakovsky VV; Mayo WJ; Mayzlish RM; McCay CM; McGrady PM; Medawar PB; Medvedev DA; Medvedev ZA; Medvedeva NB; Melnikov L; Melville H; Mendel LB; Mengele J; Merle P; Metalnikov SI; Metchnikoff E (Mechnikov II); Mettenheimer C; Michael I; Michalowsky IO; Michurin IV; Milstein C; Minot CS; Mittelstedt P; Miyagawa Y; Montgomery TH; More M (O'Connor MT); Morgagni GB; Morgan TH; Morozov VG; Morris WR (Lord Nuffield); Mühlmann MS (Milman MS); Muller HJ;

Nagorny AV; Nascher IL; Needham J; Nemes-Nagy Z; Nemilov AV; Neumann J; Niehans P; Nietzsche FW; Nikitin VN; Nikolay I; Nixon RM; Nordau M; Northrop JH; Nudler E; Nyiszli M;

Olovnikov AM; Orbeli LA; Orgel LE; Orwell G (Blair EA); Osborne TB; Ovid;

Paul I; Paracelsus (Hohenheim PATB); Pareto V; Parhon CI; Parkes AS; Parr T; Parran T; Pasternak B; Pasteur L; Pauling LC; Pavlov IP; Pearl R; Pearson D; Pearson K; Pech E; Perevodchikov IN; Perkins E; Pétain P; Petit R; Petrova MK; Pettenkofer MJ; Pettinari V; Pincus G; Pirogov NI; Pliny the Elder; Plutarch; Poehl AV; Polezhaev LV; Polge C; Pope Pius XII (Pacelli EMGG); Poucel J; Pravdina LI; Preobrazhensky B; Proctor RN; Protasov AP; Ptukha M; Puchkov NN; Puchkov NV;

Raab W; Rainey JM; Raphael WM; Rapoport IA; Rayleigh J; Reade WW; Reichstein T; Remondino PC; Retterer E; Reynolds G; Rezhabek B; Richartz H; Richartz M; Ries W; Rittenhouse EE; Robertson O; Robertson TB; Robin E; Robinson GP; Rockefeller JD; Roemheld L; Roentgen W; Romeis B; Roosevelt FD; Roosevelt T; Root WA; Rosanov VN; Rose MR; Rosenberg CE; Rostand J; Rothschild H; Rous FP; Roux W; Rowling JK; Rubner M; Rush B; Russell B; Ruwart MJ; Ruzicka V (Růžička V);

Sacher G; Sakaki Y; Sakharov AP; Sand K; Sardou G; Sarkizov-Serazini IM; Sauerbruch EF; Savoire C; Schapiro B; Scheele LA; Schenk EG; Schiller F; Schlüter O; Schlomka G; Schmalhausen II; Schmidt AA; Schmidt P; Schopenhauer A; Schubert R; Schultz EA; Sechenov IM; Seiler BW; Selye H; Semmelweis I; Setlow R; Severzov AN; Shannon C; Shapin S; Sharp HC; Shaw GB; Shaw S; Shereshevsky NA; Shipachev VG; Shock NW; Siekevitz P; Sierra F; Sigaud C; Sinclair J; Skapier J; Skulachev MV; Skulachev VP; Slonaker JR; Smith A; Smith J; Smith JM; Smuts J; Sokolov N; Solntzev VI; Spemann H; Spencer H; Spengler O; Speransky AD; Sperling J; Spinrad N; St. Bernadette; St. Germain; Stalin IV (Jughashvili IV); Stanley LL; Stapledon WO; Steidle L; Steinach E; Stephan P; Stephens CA; Steudel J; Stieglitz EJ; Stocker S; Stoff H; Stoker B; Strasburger E; Straus R; Strazhesko ND; Strehler BL; Studitsky AN; Sutton HA; Svedberg T; Swanson G; Sydenham T; Szabo I; Szent-Györgyi A; Szilárd L;

Taft WH; Tarkhanov IR (Tarkhan-Mouravov IR); Taylor FW; Termen LS (Theremin L); Thaddea S; Thewlis MW; Thiel PA; Thomson G; Thooris A; Thorek M; Thrasher M; Timiriazev CA; Tirala L; Tito

JB; Tkachuk V; Tokin BP; Tolstoy LN; Tomilin SA; Tournaire R; Townsend FE; Trotsky LD (Bronshtein LD); Tsiolkovsky KE; Tuke H; Turovets IM; Tushnov MP; Tyndall J;

Ullmann E; Updike J;

Valedinsky I; Van Helmont JB; Vannier L; Variot G; Vasilenko VH; Vasiliev LL; Veatch R; Venzmer G; Verigo AB; Vermehren F; Vernadsky VI; Verschuer O; Verzár F; Villanova A; Vinge V; Vinogradov V; Vishev IV; Vishnevsky AV; Vlasov S; Vobly K; Voit C; Vorobiev VP; Voronoff SA; Voronoy YY; Voronzova MA; Voroshilov CE; Voynar AO (Voynard AI);

Wagner-Jauregg J; Waldeyer HW; Walford RL; Wallace AR; Weindling P; Weismann FLA; Wells HG; West MD; Wheeler CE; Wiener N; Wiles P; Wilhelm II; Wilhelm O; Wilson R; Winter P; Wollman E; Wolstenholme GEW;

Yeats WB; Yudina ND; Yule G;

Zath Dr; Zalmanov AS; Zamyatin EI; Zarten P; Zavadovsky BM; Zavarzin AA; Zbarsky BI; Zimin D; Zondek B; Zondek H; Zybelin SG

References and Notes

[1] Valuable sources on the early history of aging research and care for the elderly (though for the most part omitting the authors' aspirations to radical life prolongation) include:
Joseph T Freeman, "The History of Geriatrics," *Annals of Medical History*, 10, 324-335, 1938; Frederic D. Zeman, "Life's Later Years: Studies in the Medical History of Old Age," *Journal of Mount Sinai Hospital*," 16, 308-322; 17, 53-68, 1950; Sona Rosa Burstein, "Gerontology: a Modern Science with a Long History," *Post Graduate Medical Journal* (London), 22, 185-190, 1946; Sona Rosa Burstein, "The foundations of geriatrics," *Geriatrics*, 12, 494-499, 1957; Sona Rosa Burstein, "The historical background of Gerontology," *Geriatrics*, 189-193, 328-332, 536-540, 1955; Gerald J. Gruman, "An Introduction to Literature on the History of Gerontology," *Bulletin of the History of Medicine*, 31, 78-83, 1957; Mirko D. Grmek, *On Ageing and Old Age, Basic Problems and Historic Aspects of Gerontology and Geriatrics*, Monographiae Biologicae, 5, 2, Den Haag, 1958; Johannes Steudel, "Zur Geschichte der Lehre von den Greisenkrankheiten," (The history of the study of the diseases of old age), *Sudhoffs Archiv für Geschichte der Medizin und der Naturwissenschaften*, 35, 1-27, 1942; Paul Lüth, *Geschichte der geriatrie: Dreitausend Jahre Physiologie, Pathologie und Therapie des alten Menschen* (The history of geriatrics, three thousand years of physiology, pathology and therapy of the aged), Enke, Stuttgart, 1965; Pat Thane, "Geriatrics," pp. 1092-1115, in *Companion Encyclopedia of the History of Medicine*, Edited by W.F. Bynum and R.S. Porter, Routledge, London and NY, 2001; V.N. Anisimov, M.V. Soloviev, *Evoluzia Concepcy v Gerontologii* (The Evolution of Concepts in Gerontology), Aesculap, Saint Petersburg, 1999; David Boyd Haycock, *Mortal Coil. A Short History of Living Longer*, Yale University Press, New Haven and London, 2008; Lucian Boia, *Forever Young: A Cultural History of Longevity from Antiquity to the Present*, translated from French by Trista Selous, Reaktion Books, London, 2004. The latter two works do consider the aspirations to radical life extension, yet emphasize the time frame and national contexts other than the present work, i.e. give relatively little space to the first half of the 20th century (about 40 pages each); in the modern period they focus on the US/UK, and do not consider many authors and socio-ideological and scientific aspects examined in the current study.

[2] The term "life-extensionism" has been increasingly used in on-line health forums and books about longevity since about 2000, though it was also used in the 1980s and 1990s.
There were other kinds of "extensionism" in the US as early as the 19th century, such as "slavery extensionism," "church extensionism" and "university extensionism."
But the earliest reference to "Life-extensionism" that I could find was in 1929, in the journal *Eugenics: A Journal of Race Betterment*. The reference was by Dr. Clarence Gordon Campbell (1868-1956), president of the American Eugenics Research Association, addressing and criticizing the largely "euthenic"/"environmentalist" views of Dr. Eugene Lyman Fisk (1867-1931), director of the American Life Extension Institute (established in 1913). It is likely, though, that the term is even older. As Campbell's article stated:
"the life extensionists ... might adopt the Irish toast: "May you live to be a hundred, and then be hanged for rape.""
(C.G. Campbell, "Eugenics and euthenics," *Eugenics: A Journal of Race Betterment*, 2(9), 21-25, September 1929.)

[3] One of the earliest representations of rejuvenation and life-extension, as well as one of the earliest known works of literature, is the Sumero-Babylonian *Epic of Gilgamesh*, a story about the hero's struggle with death (the most complete version has been dated from c. 1300 BCE to 650 BCE, but the story possibly originated as early as about 3000 BCE).
According to the *Epic of Gilgamesh*:
"There is a plant like a thorn with its root [deep down in the ocean], Like unto those of the briar (in sooth) its prickles will scratch [thee], (Yet) if thy hand reach this plant, [thou'lt surely find life (everlasting)]" (*The Epic of Gilgamesh*, translated by R. Campbell Thompson, 1928, Tablet 11. "The Flood, lines 268-270, The magic gift of restored youth," reprinted at *Sacred Texts*, http://www.sacred-texts.com/ane/eog/index.htm).

The plant has been sometimes likened to box-thorn and dog-rose.

There are striking parallels between the description of the immortalizing plant and the story of the extremely long-lived Utnapishtim in the epic of *Gilgamesh* and the biblical stories (with the composition sometimes dated c. 1300 BCE to 450 BCE) about the "tree of life" and about the extreme longevity of antediluvian patriarchs (*Genesis 2:9, 3:22-24, 5:1-32*).

In the *Rigveda*, one of the earliest known Vedic collections of India (c. 1700-1100 BCE), the entire Book 9 is composed of hymns praising the immortality-giving "Soma" plant (*The Hymns of the Rigveda*, translated by Ralph T.H. Griffith, 1896, reprinted at http://www.sacred-texts.com/hin/rigveda/). Recent identifications of "Soma" range from fly-agaric, ephedra and cannabis, to sacred lotus, heather and honey.

In the ancient Indian epic of the *Ramayana* (often dated c. 400 BCE, and sometimes purported to relate to events occurring 4,000 and even 5000 BCE), the monkey king Hanuman uses the *Sanjeevani* plant (translated as "One that infuses life" and commonly identified as the lycophyte *Selaginella bryopteris*, growing at the Dunagiri (Mahodaya) mountain in the Himalayas) to revive Rama's younger brother Lakshman, severely wounded by Ravan. (*Ramayan of Valmiki*, Translated Into English Verse by Ralph T. H. Griffith, 1870-1874, Book 6, Canto CII "Lakshman Healed," reprinted at http://www.sacred-texts.com/hin/rama/index.htm.)

In the *Avesta*, the sacred text of the Iranian Zoroastrian religion (with estimated dates of origin ranging from 1200 BCE to 200 BCE), during the rule of the mythical king Jamshid (Yima), people knew no disease, aging and death. ("Avesta: Venidad. Fargard 2. Yima (Jamshed) and the deluge," translated by James Darmesteter, from *Sacred Books of the East*, American Edition, 1898, http://www.avesta.org/vendidad/vd2sbe.htm.) The legendary "cup of Jamshid" was said to be a container for the elixir of immortality and at the same time a means for information retrieval (scrying/remote viewing). According to Ferdowsi's (940-1020, CE) *Shah Nameh*, Jamshid became proud and his reign of prosperity and longevity was terminated by the demonic king Zahhak. (Ferdowsi, *The Epic of Kings*, Translated by Helen Zimmern, 1883, "The Shahs of Old," http://www.sacred-texts.com/neu/shahnama.txt.)

In one of the earliest known Egyptian medical papyruses, "The Edwin Smith Surgical Papyrus" (commonly dated to the period of the New Kingdom of Egypt, c. 1500 BCE), there is a "Recipe for Transforming an Old Man into a Youth." The recipe involved the external use of bruised and dried *hemayet*-fruit (with recent identifications varying from fenugreek to almond). The remedy would not only have a cosmetic anti-aging effect – remove wrinkles, beautify the skin, remove blemishes, disfigurements, and "all signs of age" – but it would also have a true rejuvenating effect, as it would remove "all weaknesses which are in the flesh" (James Henry Breasted (Translator and Editor), *The Edwin Smith Surgical Papyrus*, The University of Chicago Press, Chicago, Illinois, 1930, XXI9-XXII10, pp. 506-507).

And in yet another ancient Egyptian medical papyrus, *The Ebers Papyrus* (c. 1500-1600 BCE), there are anti-aging cosmetic remedies to prevent the graying of hair (for example by the use of honey, onion water, donkey liver and crocodile fat), and to stimulate hair growth (for example by the use of flaxseed oil, gazelle excrements and snake fat). Actual treatment of aging was also mentioned:

"When you examine a person ... whose heart is weak as when old age comes upon him, you say: 'This is an accumulation of diseased juices,' the person should not arrogantly dismiss the disease or trust in weak remedies."

(H. Joachim (Translator and Editor), *Papyrus Ebers. Das Älteste Buch Über Heilkunde* (The Ebers Papyrus, The Oldest Book on Medicine), Georg Reimer, Berlin 1890, pp. 105-107, 43-44.)

The legendary chief minister to the Egyptian pharaoh Djoser, and the reputed builder of the first step pyramid, Imhotep (c. 2650-2600 BCE), too was said to be skilled in the art of rejuvenation.

According to Chinese legend, the Yellow Emperor, Huangdi (Huang-ti), fabled to have originated many fields of Chinese culture around 2600-2700 BCE, also possessed the secret of immortality. Do not confuse him with China's first historical emperor and seeker of immortality, Qin Shi Huang-di, 259-210 BCE. (See "Imhotep", "Huang-ti," "Shih huang-ti," *Encyclopedia Britannica*, Deluxe Edition CD, London, 2000.)

In ancient American legends, some of the immortality-giving, rejuvenating and healing plants included cocoa, cactus, aloe (octli), ayahuasca (caapi jungle vine), manioc and maize. (*Larousse World Mythology*, edited by Pierre Grimal, Gallery Books, NY, 1989, pp. 463, 479; Tamra Andrews, *Nectar and Ambrosia: An Encyclopedia of Food in World Mythology*, ABC-CLIO, Santa Barbara, CA, 2000; http://www.scribd.com/doc/70492442/The-Encyclopedia-of-Psychoactive-Plants.)

[4] The word "alchemy" apparently took root in Europe in the 12th century. The first alchemical text translated from Arabic into Latin was presumably done by Robert of Chester in 1144 and was entitled *Liber de compositione alchimiae* (The book of alchemical composition). This was allegedly a translation from Arabic into Latin of an epistle of the Egyptian-Greek-Christian alchemist Marianos to the Arab alchemical adept, the Umayyad prince Khalid ibn Yazid (665-704 CE).

(*Alchemy Academy Archive*, January 2002, "Maryanos," http://www.levity.com/alchemy/a-archive_jan02.html.)

The world "al-kimia" is of Arabic origin (originating with Khalid ibn Yazid?), "al" being the Arabic definite article, and the etymology of "kimia" being very uncertain, with hypotheses ranging from the Greek "Khemeioa" (appearing c. 296 CE. in the decree of the Roman Emperor Diocletian banning the "old writings" of Egyptian "makers" (counterfeiters) of gold and silver; "Khemia" ("the land of black earth," the old name of Egypt); "khymatos" (pouring/infusing in Greek); "khymos (the Greek word for juice), etc.

(Douglas Harper, *Online Etymology Dictionary*, 2012 http://www.etymonline.com/index.php?search=alchemy; *Alchemy Academy Archive*, June 2006, "Diocletian's Edict against alchemy," http://www.levity.com/alchemy/a-archive_jun06.html.)

One of the founding figures of alchemy is considered to be Abu Mūsā Jābir ibn Hayyān (also known as Jabir in Arabic and Geber in Latin, c. 721-815) whose theory of elements profoundly influenced both the Islamic and European (Latin-Christian) alchemy. In one of his treatises Jabir stated:

"If you could take a man, dissect him in such a way as to balance his natures [qualities] and then restore him to life, he would no longer be subject to death. ... This equilibrium once obtained, they will no longer be subject to change, alteration or modification and neither they nor their children ever will perish."

(Quoted in Gerald Joseph Gruman, *A History of Ideas about the Prolongation of Life. The Evolution of Prolongevity Hypotheses to 1800*, Transactions of the American Philosophical Society, Volume 56 (9), Philadelphia, 1966, "Arabic Alchemy: The Missing Link?" p. 60,)

And of course, practices definable as "alchemical" had existed from Egypt to China hundreds of years before the appearance of the word "alchemy."

[5] The term "gerocomia" ("gerocomica" or "gerontocomia" from the Latin "care for the aged") was used since the time of Galen (Aelius/Claudius Galenus, c. 129-217 AD, Galen, 5th Book, *De tuenda Sanitate. Gerontocomia* (5th book *On the Preservation of Health. Gerontocomia*), quoted in Sir John Floyer [1649-1734], *Medicina gerocomica, or, The Galenic art of preserving old men's healths*, J. Isted, London, 1725, p.107).

Another influential work on gerocomia was Gabriele Zerbi [1445-1505], *Gerontocomia, scilicet de senium cura atque victu* (1489, "Gerontocomia, or, care and nutrition for old age," written in Rome upon request of Pope Innocent VIII, 1432-1492).

[6] The term "gerontology" (the study of aging) was coined in Elie Metchnikoff, *Études sur la Nature Humaine* (Etudes on the Nature of Man), Masson, Paris, 1903. The Russian edition used here is I.I. Metchnikoff, *Etudy o Prirode Cheloveka* (Etudes On the Nature of Man), Izdatelstvo Academii Nauk SSSR (The USSR Academy of Sciences Press), Moscow, 1961 (1915, first published in 1903), Ch. 12 "Obshy Obzor I Vyvody" (General review and conclusions), p. 242.

[7] The term "anti-aging" was first widely used in the 1920s-1930s, but mainly in chemical engineering, particularly with regard to the protection of rubber, using "anti-agers, age resisters, or anti-oxidants" (George Oenslager, Ch. 13. "Chemical and Engineering Advances in the Rubber Industry," in Sidney D.

Kirkpatrick (Ed.), *Twenty-Five Years of Chemical Engineering Progress*, American Institute of Chemical Engineers, NY, 1933, p. 181).

A Canadian patent "for a process of treating rubber or similar material ... in the presence of anti-ageing material" was issued in 1917 (*The Canadian Patent Office Record*, September 1917, p. 2778).

The term was then very seldom used in the field of medicine, as for example in the cursory mention of "anti-aging treatment" in "The Cry for Youthfulness," *The Urologic and Cutaneous Review*, vol. 32, 1928, p. 40.

The term "anti-aging" seems to have become noticeable in medical discourse only in the late 1940s, with the appearance of such expressions as "anti-aging program" (Charles Ward Crampton, *Live Long and Like It*, 1948, p. 16); "anti-aging work" (*Science Digest*, vol. 21, 1947, p. 30), "anti-aging science" (*Nation's Business*, vol. 37, 1949, p. 76), "anti-aging factors," *Conference on Problems of Aging, Transactions of the Tenth and Eleventh Conferences, 1948-1949*, Josiah Macy Jr. Foundation, NY, 1950, p. 168).

One of the earliest scientific reports on "anti-aging" substances as well as "gerontotherapeutics" that I could find, was made in 1948 by Thomas Samuel Gardner (1908-1963) from Rutherford, New Jersey, the original home base of the Becton-Dickinson medical technology company. (Thomas Samuel Gardner, "The design of experiments for the cumulative effects of vitamins as anti-aging factors," *Journal of the Tennessee Academy of Sciences*, 23(4), 291-306, 1948.)

As Gardner stated, "In conclusion it would not be an exaggeration to suggest that longevity research work may be the most profitable field of human well-being, as well as from the commercial returns, of all of the present fields of medical investigations" (p. 302).

Interestingly enough, T.S. Gardner's work inspired the philanthropist Paul Glenn to establish in 1965 the "Glenn Foundation," one of the earliest American sponsors of basic research into the molecular biology of aging and age-related diseases (http://web.mit.edu/science/alumniandfriends/profiles/glenn.html).

[8] The term "prolongevity" or "prolongevitism" was coined by the medical historian Gerald Joseph Gruman (1926-2007, University of Massachusetts), in *A History of Ideas about the Prolongation of Life. The Evolution of Prolongevity Hypotheses to 1800*, Transactions of the American Philosophical Society, Volume 56 (9), Philadelphia, 1966, reprinted as Gerald J. Gruman, *A History of Ideas About the Prolongation of Life (Classics in Longevity and Aging)*, Springer Publishing Company, New York, 2003.

[9] It seems, the term "Immortalism," referring to the advocacy of radical life extension, took hold in the late 1960s, after the inauguration of the journal *The Immortalist* in 1967 by the Immortalist Society, led by the American founder of cryonics Robert Ettinger.

The term gained in popularity after the publication of Alan Harrington philosophical-apologetic book, *The Immortalist*, in 1969.

But the term appeared in English as early as the 18th century (though since then until the 1960s it mainly referred to the belief in the immortality of the soul).

Thus in 1774, the book-seller, historian and philosopher William Creech (1745-1815) of Edinburgh, Scotland, wrote:

"The Immortalist, therefore, is the only philosopher, who can render the parts of Christianity consistent and homogeneous. For, if the essence of man be not evanescent and fluctuating, but immaterial and permanent; if the body be no more than a mere instrument of action and vehicle of perception, then is essential identity ascertained; then the idea of accountability remains; and the man is as much punishable in a new body for crimes committed in the old one, as a robber is amenable in a new suit of cloaths, for a theft committed in those he formerly wore."

(William Creech, "Review of New Publications. Institutes of Natural and Revealed Religion, vol. III. Containing a view of the Doctrines of Revelation. By Joseph Priestley LL.D. F.R.S.," *The Edinburgh Magazine and Review by a Society of Gentlemen*, Volumes 1-2, November 1774, p. 714.)

[10] The term "Transhumanism" appeared in varying contexts since the second half of the 20th century. Sources mentioning this term include:

Pierre Teilhard de Chardin, "From the Pre-Human to the Ultra-Human: The Phases of a Living Planet (written in 1950 and first published in 1951) and "The Essence of the Democratic Idea: A Biological Approach," (written in 1949, but at the time unpublished), in Teilhard de Chardin, *The Future of Mankind*, translated by Norman Denny, Harper & Row, New York, 1959;

Julian Huxley, "Transhumanism," in *New Bottles for New Wine*, Chatto & Windus, London, 1957, pp. 13-17;

Fereydun M. Esfandiary (a.k.a. FM-2030), *Are You a Transhuman?* Warner Books, New York, 1989, and in FM-2030, "New Concepts of the Human," 1966, quoted in Nick Bostrom, "The Transhumanist FAQ. What is a transhumanism and the transhuman?" 1999, 2003;

Max More, "Transhumanism: Toward a Futurist Philosophy," 1990, 1996.

[11] Jean-Marie Robine and Michel Allard, "The oldest human," *Science*, 279 (5358), 1834-1835, 1998.

[12] For 2011, the PubMed database search on "life extension" yielded about 2,000 results, and only about 20 (~1%) contained "life-extension" or "life-span extension" in the title of the article. Apparently, in the vast majority of cases, "life" and "extension" are considered separately.

[13] For example, Max Fogiel (Ed.), *The Biology Problem Solver. A Complete Solution Guide to Any Textbook* (Research and Education Association, Piscataway, New Jersey, US, 1990, republished in 2001) contains about 800 problems ("for undergraduate and graduate studies") relating to all areas of biology and physiology. Yet the single problem even remotely related to aging is that on the menopause. The term "death" is scarcely ever mentioned and is not indexed. And many more such examples can be cited.

[14] *Natural History of Pliny* (translated by John Bostock and H.T. Riley), Henry G. Bohn, London, 1855, vol. 2, Book 7, Ch. 2. "The wonderful forms of different nations," p. 134, Ch. 49. "The greatest length of life," p. 200.

Herodotus (c. 484-425 BCE) spoke of the extremely long-lived "macrobii" and "Hyperboreans" in Herodotus, *The Histories* (edited by A.D. Godley, 1920), sections 3.17-25, 4.13.1, reprinted at the Perseus Project, Tufts University, Boston MA, http://www.perseus.tufts.edu/.

The extremely long-lived Hyperboreans are also mentioned in the *Pythian Odes* of Pindar (c. 518 - 438 BC) (Pindar, *Odes*, "Pythian Ode 10," edited by Diane Svarlien, 1990, Perseus Project, http://www.perseus.tufts.edu).

Hesiod spoke of the Hyperboreans in the "Catalogues of Women" (*Hesiod, The Homeric Hymns, and Homerica, by Homer and Hesiod*, Project Gutenberg, http://www.gutenberg.org/files/348/348-h/348-h.htm#2H_4_0024).

[15] *Hufeland's Art of Prolonging Life*, Edited by Erasmus Wilson, Lindsay & Blakiston, Philadelphia, 1867, pp. IX-X, originally Christoph Wilhelm Hufeland, *Makrobiotik; oder, Die Kunst das menschliche Leben zu verlängern*, first published in Jena in 1796.

[16] The scientists considered in this work, include (in rough chronological order):

In the period 1890-1930 – the Russian/French immunologist Elie Metchnikoff, the French physiologist Charles-Édouard Brown-Séquard, the Russian/French surgeon Serge Voronoff, the French philosopher Jean Finot, the Austrian physician Eugen Steinach, the French/American biologist Alexis Carrel, the Russian physician and politician Alexander Bogdanov, the German/American biologist Jacques Loeb, the American biologists Charles Stephens and Raymond Pearl, and the British playwright Bernard Shaw.

In the period 1930-1950 – the French biologist and inventor Auguste Lumière, the German physicians Ludwig Roemheld, Gerhard Venzmer and Max Bürger, the Romanian gerontologists Dimu Kotsovsky and Constantin Ion Parhon, the Swiss physicians Paul Niehans and Fritz Verzár, the Russian physiologists Alexander Bogomolets and Alexander Nagorny, the American physiologists Edmund Cowdry, Clive McCay and Walter Cannon, and the Russian/British physician Vladimir Korenchevsky.

In the period 1950-1980 – the Russian scientists Vasily Kuprevich, Lev Komarov and Nikolay Amosov, and the American scientists Linus Pauling, Robert Ettinger, Denham Harman, Johan Bjorksten and Bernard Strehler.

And in 1980-2012 – the British Aubrey de Grey, the Russian Vladimir Skulachev, the Americans Ray Kurzweil and Michael West, and many others.

[17] Charles E. Rosenberg, "Holism in twentieth-century medicine," in Christopher Lawrence, George Weisz (Eds.), *Greater Than the Parts: Holism in Biomedicine, 1920-1950,* Oxford University Press, Oxford, 1998, p. 335.

[18] Gerald Joseph Gruman, *A History of Ideas about the Prolongation of Life. The Evolution of Prolongevity Hypotheses to 1800*, Transactions of the American Philosophical Society, Volume 56 (9), Philadelphia, 1966, reprinted as Gerald J. Gruman, *A History of Ideas About the Prolongation of Life (Classics in Longevity and Aging)*, Springer Publishing Company, New York, 2003.

[19] Herbert Spencer, *The Data of Ethics*, Ch. 3. "Good and Bad Conduct," § 9, Williams and Norgate, London, 1879, reprinted in the Online Library of Liberty, http://oll.libertyfund.org.

[20] Robert Veatch, *Death, Dying, and the Biological Revolution. Our Last Quest for Responsibility*, Yale University Press, New Haven CT, 1977, Ch. 8 "Natural death and public policy," pp. 293-305; John Harris, "Immortal Ethics," presented at the International Association of Biogerontologists (IABG) 10th Annual Conference "Strategies for Engineered Negligible Senescence," Queens College, Cambridge, UK, September 17-24, 2003, reprinted in *Strategies for Engineered Negligible Senescence: Why Genuine Control of Aging May Be Foreseeable* (Aubrey de Grey, Ed.), *Annals of the New York Academy of Sciences*, 1019, 527-534, June 2004.

Further on the ethics of radical life extension, see Christine Overall, *Aging, Death, and Human Longevity: A Philosophical Inquiry* (University of California Press, Berkeley, CA, 2003), finding "the usual arguments against seeking immortality" unconvincing (p. 153). See also Frida Fuchs-Simonstein, *Self-evolution: The Ethics of Redesigning Eden* (Yozmot, Tel Aviv, 2004), arguing that "It seems improbable to find ethical objections to increasing life spans, or even immortality, when it means saving someone's life" (p. 181).

[21] Richard P. Feynman, "What Is and What Should be the Role of Scientific Culture in Modern Society," presented at the Galileo Symposium in Florence, Italy, in 1964, in Richard P. Feynman, *The Pleasure of Finding Things Out: The Best Short Works of Richard P. Feynman*, Perseus Books, NY, 1999, p. 100.

[22] *Merriam-Webster Dictionary*, in *Encyclopedia Britannica*, Deluxe Edition CD, London, 2000. Similar definitions of adjustment, adaptation, balance and equilibrium, are given in the Free Online Dictionary, http://www.thefreedictionary.com/.

[23] Gerald Joseph Gruman, *A History of Ideas about the Prolongation of Life. The Evolution of Prolongevity Hypotheses to 1800*, Transactions of the American Philosophical Society, Volume 56 (9), Philadelphia, 1966.

[24] Ilia Stambler, "Life extension – a conservative enterprise? Some fin-de-siècle and early twentieth-century precursors of transhumanism," *Journal of Evolution and Technology*, 21(1), 13-26, 2010, http://jetpress.org/, http://jetpress.org/v21/stambler.htm.

[25] Mike Jay and Michael Neve (Eds.), *1900: A Fin-de-siècle Reader*, Penguin Books, London, 1999, pp. ix-xvii.

[26] On the quantum leap in Therapeutic Activism at the end of the 19th – beginning of the 20th century, see Roy Sydney Porter, *The Greatest Benefit to Mankind: A Medical History of Humanity*, W.W. Norton & Company, NY, 1998 (1997), Ch. XIV "From Pasteur to Penicillin," Ch. XVII "Medical Research," Ch. XIX "Surgery"; and W.R. Albury, "Ideas of Life and Death," in *Companion Encyclopedia of the History of Medicine*, Edited by W.F. Bynum and Roy Porter, Routledge, London and NY, 2001, pp. 253-254. For the best available history of early life-extensionism, see Gerald J. Gruman, *A History of Ideas about the Prolongation of Life. The Evolution of Prolongevity Hypotheses to 1800*, Transactions of the American Philosophical Society, Volume 56 (9), Philadelphia, 1966.

[27] Jean Finot, *The Philosophy of Long Life* (translated by Harry Roberts), John Lane Company, London and New York, 1909, pp. 77-78, first published in French as *La Philosophie de la Longévité*, Schleicher Freres, Paris, 1900.

[28] Jean Finot, *The Philosophy of Long Life*, 1909 (1900), pp. 25-31.

[29] Joseph T Freeman, "The History of Geriatrics," *Annals of Medical History*, 10, 324-335, 1938.

[30] *Nicolas Flamel. Alchimia,* translated by G.A. Butuzov, Peterburgskoe Vostokovedenie (St. Petersburg Oriental Studies), St. Petersburg, 2001, p. 77.

[31] Mirko Dražen Grmek, *On Ageing and Old Age, Basic Problems and Historic Aspects of Gerontology and Geriatrics, Monographiae Biologicae,* 5, 2, Den Haag, 1958, p. 46.

[32] Descartes, *Discourse on the Method of Rightly Conducting the Reason, and Seeing Truth in the Sciences,* translated by John Veitch, The Open Court Publishing Company, Chicago, 1903, Part 6, pp. 66-67 (first published in 1637, translated in 1850, online at http://www.gutenberg.org/files/59/59-h/59-h.htm; Gerald J. Gruman, *A History of Ideas about the Prolongation of Life. The Evolution of Prolongevity Hypotheses to 1800,* 1966, p. 78.

[33] Sergey Ryazanzev, *Thanatologia* (Thanatology), The East-European Institute of Psychoanalysis, Saint-Petersburg, 1994, pp. 331-332;
The Complete Memoirs of Jacques Casanova de Seingalt 1725-1798, translated by Arthur Machen, London, 1894, reprinted in Project Gutenberg, http://www.gutenberg.org/ebooks/2981.

[34] Marie-Jean-Antoine-Nicolas Caritat, Marquis de Condorcet, Outlines of an historical view of the progress of the human mind (a posthumous work by M. de Condorcet, first published in 1795 as Esquisse d'un tableau historique des progrès de l'esprit humain, translated from the French, the English translator is unnoted in this edition), Printed by Lang & Ustick, Philadelphia, 1796, "Tenth Epoch. Future Progress of Mankind," pp. 290-291, reprinted at the Online Library of Liberty, http://oll.libertyfund.org; Gerald J. Gruman, A History of Ideas about the Prolongation of Life. The Evolution of Prolongevity Hypotheses to 1800, Transactions of the American Philosophical Society, Volume 56 (9), Philadelphia, 1966, p. 87.
Condorcet's book ends with these words:
"And how admirably calculated is this view of the human race, emancipated from its chains, released alike from the dominion of chance, as well as from that of the enemies of its progress, and advancing with a firm and indeviate step in the paths of truth, to console the philosopher lamenting the errors, the flagrant acts of injustice, the crimes with which the earth is still polluted? It is the contemplation of this prospect that rewards him for all his efforts to assist the progress of reason and the establishment of liberty. He dares to regard these efforts as a part of the eternal chain of the destiny of mankind; and in this persuasion he finds the true delight of virtue, the pleasure of having performed a durable service, which no vicissitude will ever destroy in a fatal operation calculated to restore the reign of prejudice and slavery. This sentiment is the asylum into which he retires, and to which the memory of his persecutors cannot follow him: he unites himself in imagination with man restored to his rights, delivered from oppression, and proceeding with rapid strides in the path of happiness; he forgets his own misfortunes while his thoughts are thus employed; he lives no longer to adversity, calumny and malice, but becomes the associate of these wiser and more fortunate beings whose enviable condition he so earnestly contributed to produce" (pp. 292-293).

[35] Auguste Comte, *The Catechism of Positive Religion* (Translated by Richard Congreve), John Chapman, London, 1858, pp. 97, 104, 224-225 (first published in French as *Catéchisme positiviste ou Sommaire exposition de la religion universelle,* Paris, 1852 – The Positivist Catechism or Summary Exposition of the Universal Religion).

[36] Broussais' methods of conserving the vital energy built on a long tradition: In the pre-chemotherapeutic era, many techniques employed by physicians/medicine-men for the conservation of the "vital heat" – such as drug-sedation, starving, blood-letting, freezing, purging, and even incantation – were designed to overcome stimulation and quiet the person down.
This view was fundamentally opposed to Bichat's idea of internal stimulation as a means to counter the threat of destruction by the environment. In more practical medical terms, the importance of internal stimulation was supported by the English physician John Brown (1810-1882), the proponent of physiological excitation.
The French historian of medicine Charles Daremberg poignantly noted the old conflict between the Stimulation and Relaxation schools (1870):

"All of [John] Brown's patients were destined to become athletes. All of Broussais's were supposed to be reduced to the state of diaphanous bodies. One left Brown's care with a ruddy complexion, Broussais's as a pale and a winding sheet. For Brown stimulation was the remedy, for Broussais irritation was the ill."
(Charles Daremberg, *Histoire des sciences médicales*, 1870, quoted in Georges Canguilhem, "John Brown's system: An Example of Medical Ideology," in *Ideology and Rationality in the History of the Life Sciences*, translated from French by Arthur Goldhammer, Cambridge MA, The Massachusetts Institute of Technology Press, 1988, article note 11, p. 49.)
On Bichat's theory, see W.R. Albury, "Ideas of Life and Death," in *Companion Encyclopedia of the History of Medicine*, Edited by W.F. Bynum and R.S. Porter, Routledge, London and NY, 2001, pp. 253-254.

[37] J.B. Lamarck, *Zoological Philosophy*, translated by Hugh Elliot, Macmillan and Co., London, 1914 (originally published in 1809), p. 193.

[38] Marie Jean Pierre Flourens, *On Human Longevity and the Amount of Life upon the Globe translated from the French by Charles Martel*, Bailliere, London, 1855 (first published in French in 1854), pp. 54, 75; S.E. Chaillé, "Longevity," *New Orleans Medical and Surgical Journal*, 16, 417-424, 1859, p. 419; S.P. Cutler, M.D. Holly Springs, Miss, "Physiology and Chemistry of Old Age," *New Orleans Journal of Medicine*, 23(1), 96-104, 1870; M.F. Legrand, "Peut-On Reculer le Bornes de la Vie Humaine?" (Can we reverse the limitations of human life?), *L'Union Médicale*, 13 (5), 65-71, 1859 – reprinted in Geraldine M. Emerson (Ed.), *Benchmark Papers in Human Physiology, Vol. 11, Aging*, Dowden, Hutchinson and Ross, Stroudsburg PA, 1977, pp. 28-47.

[39] Pat Thane, "Geriatrics," in *Companion Encyclopedia of the History of Medicine*, Edited by W.F. Bynum and R.S. Porter, Routledge, London and NY, 2001, pp. 1092-1115.

[40] Some of the notable French physicians working on the pathology of aging, following Durand-Fardel and Charcot, were Henri Cazalis (1840–1909), who coined the maxim "the man is as old as his arteries"; Jules Boy-Teissier (1858-1908), physician at the Hopital St. Marguerite, a home for the aged in Marseille, who published in 1895 *Lectures on the Diseases of the Aged*; and Yves Delage (1854-1920) who emphasized the role of tissue differentiation in aging, and the role of heart disease in senile pathology (1903). Delage was also suggested to be the forerunner of nuclear transfer experiments (Jean-Claude Beetschen, Jean-Louis Fischer, "Yves Delage (1854-1920) as a forerunner of modern nuclear transfer experiments," *International Journal of Developmental Biology*, 48, 607-612, 2004). Further see Stearns, P.N, *Old age in European society: the case of France*, Holmes & Meier, New York, 1976.
Of special note is the "mineralization" theory of aging (a.k.a. accumulation of "alkaline residues," "calcification" or "ossification" theory) proposed by Édouard Robin of the French Academy of Sciences in 1858, which considered lactic acid and "vegetable acids" as possible means to dissolve the "mineral matters" and thus prolong life. This might be considered one of the first modern scientific theories of aging. (S.E. Chaillé, "Longevity," *New Orleans Medical and Surgical Journal*, 16, 417-424, 1859, p. 419; S.P. Cutler, M.D. Holly Springs, Miss, "Physiology and Chemistry of Old Age," *New Orleans Journal of Medicine*, 23(1), 96-104, 1870.)
The theory was presented by Édouard Robin in 1858 to the French Academy of Sciences in the address "Sur le causes de la vieillesse et de la mort sénile" (On causes of aging and senile death). Excerpts are quoted in M.F. Legrand, "Peut-On Reculer le Bornes de la Vie Humaine?" (Can we reverse the limitations of human life?), *L'Union Médicale*, 13 (5), 65-71, 1859, reprinted in Geraldine M. Emerson (Ed.), *Benchmark Papers in Human Physiology, Vol. 11, Aging*, Dowden, Hutchinson and Ross, Stroudsburg PA, 1977, pp. 28-47.

[41] Claude Bernard's statement *"La fixité du milieu intérieur est la condition de la vie libre, indépendante"* appears in his *Leçons sur les Phénomènes de la Vie Communs aux Animaux et aux Végétaux* (Lectures on the phenomena of communal life in animals and plants), J.-B. Baillière, Paris, 1878, edited by Albert Dastre, vol. 1, Lecture 2 "Les Trois Formes de la Vie" (The three forms of life), Subsection 3 "La vie constante" (The constant life), p. 113; J.M.D. Olmstead, *Claude Bernard, Physiologist*, Harper & Brothers, NY, 1938, p. 254.

[42] Metchnikoff coined both the term "gerontology" ("the study of aging") and "thanatology" ("the study of death") in the *Etudes On the Nature of Man* in 1903. (Elie Metchnikoff, *Etudy o Prirode Cheloveka* (Etudes On the Nature of Man), Izdatelstvo Academii Nauk SSSR (The USSR Academy of Sciences Press), Moscow, 1961 (first published in 1903), Ch. 12 "Obshy Obzor I Vyvody" (General review and conclusions), p. 242.)

[43] Alfred I. Tauber, "Metchnikoff and the phagocytosis theory," *Nature Reviews: Molecular Cell Biology*, 4, 897-301, 2003; Alfred I. Tauber, *The Immune Self: Theory or Metaphor?* Cambridge University Press, Cambridge, 1994, p. 16.

[44] One of Metchnikoff's earliest works on aging was: Metchnikoff, "La Dégénérescence sénile" (The senile degeneration), *Année biologique*, 3, 249-267, 1897.

Yet, Metchnikoff traced the beginning of publicity of his aging and longevity research to his presentation on April 22, 1901, at the Manchester Literary and Philosophical Society, where he "laid out a program of investigations aimed to unravel the problem of aging, the problem that had seemed almost intractable" (Elie Metchnikoff, "Borba so Starcheskim Pererozhdeniem" (The struggle against the degeneration of senescence), in *I.I. Metchnikoff. Sobranie Sochineniy* (Collected Works), Eds. N.N. Anichkov and R.I. Belkin, The USSR Academy of Medical Sciences, Moscow, 1962, vol. XV, pp. 346-350).

[45] Theories of aging were proposed before Metchnikoff, such as the depletion of the "vital heat" and "vital moisture" (Francis Bacon, England, 1623), the depletion of the "vital force" (Christoph Wilhelm Hufeland, Germany, 1796), "tissue mineralization" (Édouard Robin, France, 1858), "tissue differentiation" (August Weismann, Germany, 1882), production of "clinkers of metabolism" (Hugo Eisig, Germany, 1887), accumulation of "metaplasms" (Max Kassowitz, Austria, 1889).

Yet, Metchnikoff's was arguably the first theory of aging and longevity based on meticulous histological observations and on a model of physiological behavior of living tissues.

[46] Elie Metchnikoff, *Etudy o Prirode Cheloveka* (Etudes On the Nature of Man), Izdatelstvo Academii Nauk SSSR (The USSR Academy of Sciences Press), Moscow, 1961 (1903). The first French edition, *Études sur la Nature Humaine*, was published in Paris (Masson) in 1903. The Russian translation used here was done by Elie Metchnikoff and his wife Olga. The book is also available in English (*The Nature of Man: Studies in Optimistic Philosophy*, translated by P.C. Mitchell, Putnam, NY, 1903). Unless otherwise specified, all the excerpts quoted in the present work are in my translation.

[47] Elie Metchnikoff, *Etudy Optimisma* (Etudes of Optimism), Nauka, Moscow, 1988 (1907). It is also available in the original French, *Essais optimistes* (Paris, 1907) and in English, *The Prolongation of Life: Optimistic Studies*, translated by P.S. Mitchell, published by Heinemann (London, 1907) and Putnam (New York, 1908).

[48] Elie Metchnikoff, "Avtobiographia" (autobiography), pp. 297-298, vol. xvi; "Rasskaz o tom, kak i pochemu ya poselilsia za granizey" (The story of how and why I emigrated, 1909), pp.38-45, vol. xiv, in *I.I. Metchnikoff. Sobranie Sochineniy* (Collected Works), Eds. N.N. Anichkov and R.I. Belkin, The USSR Academy of Medical Sciences, Moscow, 1962; Olga Metchnikoff, *Zhizn Ilyi Ilyicha Mechnikova* (The Life of I.I. Metchnikoff), Literatura, Moscow, 1926.

[49] Metchnikoff, "The story of how and why I emigrated" (1909), in *I.I. Metchnikoff. Collected Works*, 1962, vol. xiv, p. 43.

[50] Metchnikoff's mother, nee Emilia Nevachovitz was from a Jewish merchant family, daughter of Lev Nevachovitz, a renowned Ukrainian writer who converted to Protestantism. His father, Ilya Ivanovich Metchnikoff, was a retired Russian Imperial Guard officer, who allegedly squandered most of Emilia's dowry on gambling.

Despite this rather disadvantageous background of relatively scarce financial means and Jewish ancestry, Metchnikoff was able to make a brilliant scientific career. After his graduation at Kharkov University in 1864 (at the age of 19), his post-graduate studies of marine fauna embryology in Germany and Italy became possible thanks to the 1600 ruble scholarship procured on the recommendation of the Russian surgeon Nikolay Pirogov – an exceptional opportunity for a person with such a disadvantaged background.

(Elie Metchnikoff, "Ilia Ilyich Metchnikoff" (Curriculum Vitae), 1888, pp. 292-296, vol. xvi; "Avtobiographia" (Autobiography, 1908), pp. 297-298, vol. xvi; "Rasskaz o tom, kak i pochemu ya poselilsia za granizey" (The story of how and why I Emigrated, 1909), pp.38-45, vol. xiv, in *I.I. Metchnikoff. Sobranie Sochineniy* (Collected Works), Eds. N.N. Anichkov and R.I. Belkin, The USSR Academy of Medical Sciences, Moscow, 1962.)

[51] Metchnikoff, "Borba so Starcheskim Pererozhdeniem" (The struggle against the degeneration of senescence"), in *Metchnikoff, Collected Works*, vol. xv, 1962, pp. 349-350.

[52] Metchnikoff, "The story of how and why I emigrated," in *I.I. Metchnikoff. Collected Works*, 1962, vol. xiv, pp. 44-45.

[53] Peter Wiles, "On Physical Immortality, II" *Survey*, 57, 142-161, 1965.

[54] Elie Metchnikoff, *Sorok Let Iskania Razionalnogo Mirovozzrenia* (Forty Years in Search of a Rational Worldview), 1914, in I.I. Metchnikoff, *Academicheskoe Sobranie Sochineniy* (Elie Metchnikoff. Academic Collected Works, Ed. G.S. Vasezky), Academia Medizinskikh Nauk SSSR (The USSR Academy of Medical Sciences), Moscow, 1954, vol. 13, pp. 9-22.

[55] Henri Bergson's attitude to life-extension may be diametrically opposed to Metchnikoff's. Bergson doubted the desirability of individual life-extension, as he believed that "everything is *as if* this death had been willed, or at least accepted, for the greater progress of life in general." Nonetheless, he did assume the great capacity of the spirit, of life, of the "vital impetus," to resist the degradation of matter.
He even claimed that humanity may eventually overcome death:
"the whole of humanity, in space and in time, is one immense army … able to beat down every resistance and clear the most formidable obstacles, perhaps even death."
Yet, this "spiritual resistance" to death apparently had little to do with Metchnikoff's "materialistic" position, striving to maintain the human machine by removing or adding its material "elements." In fact, Bergson directly attacked the "mechanistic" views of Metchnikoff, Charles Sedgwick Minot and Félix Alexandre Le Dantec who explained aging by "the constant accumulation or loss of a certain kind of matter." In turn, Metchnikoff was entirely unable to accept Bergson's beliefs in the intangible "intuition," "soul" and "will."
(Henri Bergson, *Creative Evolution*, authorized translation by Arthur Mitchell, Henry Holt and Company, NY, 1911, pp. 17-18, 246-247 note 1, 271, first published in French in 1907.)

[56] Stephen Kunitz, "Medicine, Mortality, and Morbidity," in *Companion Encyclopedia of the History of Medicine*, Edited by W.F. Bynum and R.S. Porter, Routledge, London and NY, 2001, pp. 1693-1711.

[57] As stated by Metchnikoff:
"It has been long noted that aging is very similar to disease. Therefore it is not surprising that human beings feel a strong aversion to aging. … Undoubtedly, it is a mistake to consider aging as a physiological phenomenon. It makes as much sense to accept aging as a normal phenomenon, because everybody ages, as it makes sense to accept childbirth pain as normal, because only very few women are spared it. In both cases, we deal, of course, with pathological and not with purely physiological phenomena. Inasmuch as people endeavor to mitigate or eliminate the pains of a woman in labor, it is as natural to endeavor to eliminate the evils brought by aging."
(Elie Metchnikoff, *Etudy o Prirode Cheloveka* (Etudes on the Nature of Man), The USSR Academy of Sciences Press, Moscow, 1961 (1903), Ch. X. "Vvedenie v nauchnoe izuchenie starosti" (An introduction to the scientific study of aging), p. 201.)

[58] Metchnikoff's theory was conflicted on the existence of a life-span limit. According to Metchnikoff, natural death is a rare occasion, but it does occur. He gave an example of *Palingenia virgo*, a mayfly species who, after their metamorphosis, live happily and actively only for one day and, having performed their reproductive function, die for no apparent reason. Metchnikoff found no pathological causes for their demise. Their rapid death is, so to speak, programmed into their existence.

This led Metchnikoff to hypothesize a similar phenomenon for human beings. At the present, he argued, people do not live up to their longevity potential and are killed prematurely by violent causes. However, there might be a point in their life when death will occur naturally, as it is programmed. He, rather arbitrarily, set this limit to the natural life-span at approximately 150 years. At that point, people are supposed to become "full of years," completely lose the desire to live and acquire a new, yet unknown "death instinct," a sense of complete life fulfillment – as he believed happens in mayflies. (To recall, according to Freud, the "death instinct" is present at any stage of life.)

This position seems paradoxical. Alongside the belief in potential immortality, alongside the elaborate methodologies for life-extension, Metchnikoff eventually strove to reach a point in life when life-extension is no longer desirable. Earlier in *The Etudes on the Nature of Man*, he ardently argued with pessimistic philosophers, Arthur Schopenhauer (*The World as Will and Idea*, 1844), Eduard Hartmann (*The Philosophy of the Unconscious*, 1870) and Philipp Mainländer (*The Philosophy of Redemption*, 1872), who became saturated with life and ready to abandon the pursuit of it much earlier. Quite coherently with the pessimistic philosophy, Mainländer committed suicide at the age of 34 (1841-1876).

But for Metchnikoff, a life-span of 150 years versus 70 meant a world of difference. At 150, pessimistic philosophies were supposed to start making sense. In this regard, Metchnikoff might have been not entirely consistent with his own teachings. Yet, until that hypothetical limit is reached, the pursuit of life extension and the pursuit of immortality follow the same path.

(Elie Metchnikoff, *Etudy o Prirode Cheloveka* (Etudes on the Nature of Man), The USSR Academy of Sciences Press, Moscow, 1961 (first published in 1903.)

[59] Metchnikoff's exact statement was:

"We saw that, during aging, there occurs a struggle between noble elements and phagocytes ["low/primitive elements"], and that the vitality of the former is for the most part diminished, whereas the latter, on the contrary, show increased activity. Therefore, it would seem that the means to use in the struggle against pathological aging should be, on the one hand, the strengthening of the most valuable elements of the organism, and on the other, the attenuation of the aggressive onslaught of the phagocytes. I must point out to the reader from the beginning that this problem is not yet solved, but its solution does not involve anything impossible. It is a scientific question, like many others."

(Elie Metchnikoff, *Etudy o Prirode Cheloveka* (Etudes on the Nature of Man), The USSR Academy of Sciences Press, Moscow, 1961 (1903), Ch. X. "Vvedenie v nauchnoe izuchenie starosti" (An introduction to the scientific study of aging), pp. 201-202.)

[60] Tjalma, W.A., et al., "Prophylactic human papillomavirus vaccines: the beginning of the end of cervical cancer," *International Journal of Gynecological Cancer*, 14, 751-761, 2004.

[61] Peter Baldry, *The Battle Against Bacteria. A Fresh Look*, Cambridge University Press, Cambridge, 1976 (1965); Paul de Kruif, *Microbe Hunters*, Harcourt, Brace, Jovanovich, NY, 1926; Thomas Hausler, *Viruses vs. Superbugs: a solution to the Antibiotics Crisis?* Translated by Karen Leube, Macmillan, London, 2006.

[62] James C. Riley, *Rising Life Expectancy: A Global History*, Cambridge University Press, Cambridge, 2001.

[63] Robert Baille Pearson, *The Prolongation of Life Through Diet*, Health Inc, Denver Colorado, 1941; Linus Pauling, *How To Live Longer And Feel Better*, W.H. Freeman and Company, NY, 1986; Roy L. Walford, *Beyond the 120-Year Diet: How to Double Your Vital Years*, Four Walls Eight Windows, New York, 2000.

[64] Herbert Spencer, *The principles of Ethics. Vol. 2*, Ch.1. "Animal ethics," 1897, reprinted in The Online Library of Liberty, http://oll.libertyfund.org/.

[65] In Japan, the pro-biotic dietary product ("yakult") was developed in the 1930s by Minoru Shirota (1899-1982).

The term "probiotic" was apparently first introduced only in 1953 by the German bacteriologist Werner Georg Kollath (1892-1970).

Yet, the assumed action of "probiotic" dairy products was precisely as that first proposed by Metchnikoff for "fermented" dairy products.

(On the origin of the term "probiotics" see Young Jin Baek and Byong H. Lee, Ch. 12, "Probiotics and prebiotics as bioactive compounds in dairy products," in Young W. Park (Ed.), *Bioactive Components in Milk and Dairy Products*, Wiley-Blackwell, Hoboken NJ, 2009, pp. 287, 290-291.)

[66] Arnold Lorand, *Old Age Deferred. The Causes of Old Age and its Postponement by Hygienic and Therapeutic Measures*, 6th edition, Translated with additions by the Author from the Third German Edition, 1926, F.A. Davis Company, Philadelphia, 1926 (first published by F.A. Davis in 1910), pp. 455-456.

[67] Arnold Lorand, *Old Age Deferred*, 1926 (1910), p. 178.

[68] Jean Finot, *The Philosophy of Long Life*, 1909 (1900), pp. 43-44.

[69] V.N. Nikitin, "I.I. Metchnikoff i problema dolgoletia" (I.I. Metchnikoff and the problem of longevity), in *Collected Works of I.I. Metchnikoff* (Ed. N.N. Anichkov and R.I. Belkin), The USSR Academy of Medical Sciences, Moscow, 1962, vol. 15, p. 389.

[70] In the words of Lao-Tse, the great teacher of Taoist immortalists (c. 6th century BCE), "For regulating the human in our constitution and rendering the proper service to the heavenly, there is nothing like moderation."

The passage continues, "It is only by this moderation that there is effected an early return (to man's normal state). That early return is what I call the repeated accumulation of the attributes (of the Tao). With that repeated accumulation of those attributes, there comes the subjugation (of every obstacle to such return). Of this subjugation we know not what shall be the limit; and when one knows not what the limit shall be, he may be the ruler of a state."

(*Lao-Tse. The Tao Teh King. The Tao and Its Characteristics*, Translated by James Legge, 1880, Section 59.1-2, reprinted in Project Gutenberg, http://www.gutenberg.org/ebooks/216.)

Moderation, particularly moderation in diet, has remained the absolute life-extensionist consensus for centuries. (See Steven Shapin, Christopher Martyn, "How to live forever: lessons of history," *British Medical Journal*, 321, 1580-1582, 2000.)

It has never been agreed on, however, what exactly a "moderate" measure is. Some rules of thumb for a correct "moderate" diet were sometimes proposed. Thus, for example, the Flemish priest Leonardus Lessius (1554-1623) proposed in his *Treatise of Health and Long Life* (*Hygiasticon*, 1613) the "Seven Rules for the better Discovery of this right Measure":

1) The meals should be less than would render the person "incapable or unfit for his mental employments";
2) The meals taken should not produce "dullness or heaviness of disposition";
3) If changing the diet, "it is to be done cautiously and by Degrees";
4) "For those, who are much advanced in Years, and for those also, that are of weak complexions, twelve or fourteen Ounces of Food a-Day are judged sufficient";
5) Any type of food may be good in small measure: "Any Sort of Food that is common to one suits agreeably enough with hale Constitutions, if so be not too much of it be taken at one Time";
6) The diet must be simple and uniform: "Every man should above all Things forbear Variety of Dishes, and the luxurious Artfulness of Cookery";
7) Appetites should not be excessively aroused by the "Fancy of Imagination."

But even these rules were partly disputed, for example, by George Cheyne (1671-1743) in his *Essay on Health and Long Life* (1724) and by Lessius' English translator, the apothecary Timothy Smith.

(*A Treatise of Health and Long Life, with the Sure Means of Attaining It. In 2 Books. The first By Leonard Lessius. The*

Second by Lewis Cornaro, Translated into English by Timothy Smith, London, C. Hitch, 1743, pp. 14-32.)
And yet, despite the uncertainties regarding the exact "moderate" measure, the importance of moderation, of consuming less than people usually do, has been emphasized throughout by most authors.
Perhaps one of the very few dissenters from this consensus was the famous French lawyer, physician and gastronome Jean Anthelme Brillat-Savarin (1755-1826). In his *Physiologie du goût* (*The Physiology of Taste*, 1825), Brillat-Savarin spoke of the "Longevity of Gourmands" and claimed:
"I am happy, I cannot be more so, to inform my readers that good cheer is far from being injurious, and that all things being equal, gourmands live longer than other people. This was proved by a scientific dissertation recently read at the academy, by Doctor Villermet [the hygienist Louis René Villermé, 1782-1863]. ... Those who indulge in good cheer, are rarely, or never sick. ... as all portions of their organization are better sustained, nature has more resources, and the body incomparably resists destruction."
(Jean Anthelme Brillat-Savarin, *The Physiology of Taste; or, Transcendental Gastronomy* (translated by Fayette Robinson), Lindsay & Blakiston, Philadelphia, 1854, pp. 194-196, first published in 1825, http://www.gutenberg.org/cache/epub/5434/pg5434.html.)

[71] Roy Sydney Porter, *The Greatest Benefit to Mankind: A Medical History of Humanity*, W.W. Norton & Company, NY, 1998 (1997), Ch. XIX "Surgery," pp. 600-601.

[72] Elie Metchnikoff, "Mirosozerzanie i Medizina" (The worldview and Medicine), first published in *Vestnik Evropy* (The European Bulletin) in 1910, in *I.I. Metchnikoff, Akademicheskoe Sobranie Sochineniy* (The Academic Collection of Works), Edited by G.S. Vasezky, The USSR Academy of Medical Sciences, Moscow, 1954, vol. 13, pp. 204-205.

[73] Elie Metchnikoff, "Dolgovechnost ili prodolzhitelnost zhizni" (Longevity or the Span of Life), in *Collected Works of I.I. Metchnikoff* (Ed. N.N. Anichkov and R.I. Belkin), The USSR Academy of Medical Sciences, Moscow, vol. 15, 1962, p. 361.

[74] Bogomolets disputed Metchnikoff's tenet of the non-functionality of connective tissue buildup and saw such tissues as the sustaining "root of the organism."
(Alexander A. Bogomolets, *The Prolongation of Life*, translated by Peter V. Karpovich and Sonia Bleeker, Essential Books, Duel, Sloan and Pearce, NY, 1946, pp. 82-88.)

[75] See, for example, David Hamilton, *The Monkey Gland Affair*, Chatto and Windus, London, 1986.

[76] August Weismann, *Über die Dauer des Lebens* (On the Duration of Life), G. Fischer, Jena, 1882; August Weismann, *Über Leben und Tod* (On Life and Death), G. Fischer, Jena, 1884.

[77] The notion of aging as due to differentiation was espoused by Max Kassowitz (1889), Jules Boy-Teissier (1895), Joseph Delboeuf (1896), Alexander Pawlinoff (1896), Moisey Mühlmann (1900), Max Verworn (1903), Yves Delage (1903), Herbert Spencer Jennings (1906), Richard von Hertwig (1906), Charles Sedgwick Minot (1908), Peter Kaschtschenko (1914), Waldemar Schleip (1915), Charles Manning Child (1915), Rudolf Ehrenberg (1923), Alexis Carrel (1929), Eugen Korschelt (1924), Nikolay Plavilschikoff (1925), Max Hartmann (1926), Georg Perthes (1927), Aldred Scott Warthin (1928), Dimu Kotsovsky (1929,1931), Edmund Vincent Cowdry (1942) and many others.
Reviews of the early 20th century theories of aging are given in Dimu Kotsovsky, "Das Alter in der Geschichte der Wissenschaft" (Aging in the history of science), *Isis*, 20(1), 220-245, November 1933; Max Bürger, *Altern und Krankheit* (Aging and Disease), Zweite Auflage, Georg Thieme, Leipzig, 1954 (first published in 1947); Alexander Comfort, *The Biology of Senescence*, Butler & Tanner, London, 1956; Mirko D. Grmek, *On Ageing and Old Age, Basic Problems and Historic Aspects of Gerontology and Geriatrics*, Monographiae Biologicae, 5, 2, Den Haag, 1958; Alfred Cohn and Henry Murray, "Physiological Ontogeny I.. The Present Status of the Problem," *The Quarterly Review of Biology*, 2(4), 469-493, 1927; Albert I. Lansing, "General Physiology," in *Cowdry's Problems of Ageing, Biological and Medical Aspects*, edited by Albert I. Lansing, The Williams and Wilkins Company, Baltimore, 1952 (first published in January 1939), pp. 3-23; Raymond Pearl, *The Biology of Death*, J.B. Lippincott Company, Philadelphia, 1922 (first delivered as a series of lectures at the

Lowell Institute in Boston in December 1920); Charles Manning Child, *Senescence and Rejuvenescence*, University of Chicago Press, Chicago, 1915; Auguste Lumière, *Sénilité et Rajeunissement* (Senility and Rejuvenation), Bailliere J.B.& Fils, Paris, 1931; Alexander Vasilievich Nagorny, *Starenie I Prodlenie Zhizni* (Aging and Life-extension), Sovetskaya Nauka, Moscow, 1950, first published as Nagorny A.V., *Problema Starenia i Dolgoletia* (The Problem of Senescence and Longevity), Kharkov State University, Kharkov, 1940.

[78] Besides Henri Cazalis (1840-1909), the heart was attributed the primary importance in senescence by Samuel Basch (1837-1905), Henri Huchard (1844-1910), Otto Josue (1869-1923), Oskar Klotz (1878-1936), Felix Marchand (1846-1928), Émile Demange (1846-1904, published in 1887) and Nikolay Anitchkoff (1885-1964, published in 1912).

[79] The role of the brain was emphasized by the Russian neurophysiologists Moisey Samuilovich Mühlmann (1867-1940, published in 1900, 1911, 1926) and Ivan Petrovich Pavlov (1849-1936), the German physiologists Oskar Vogt (1870-1959), Moritz Ribbert (1855-1920), Robert Rössle (1876-1956), and Jürgen Wilhelm Harms (1885-1956), the Italian Ugo Cerletti (1877-1963), and the Romanian Georges Marinesco (1863-1938).

[80] Deterioration of the "middle brain" was emphasized by the German/Jewish pathologists Samson Hirsch (1926, 1929) and Friedrich Levy (1933).

[81] Generally, one of the earliest references to biological human "enhancement" that I could find was in Arnold Lorand, *Human Intelligence: Its Enhancement by Hygienic and Therapeutic Means: A Guide to Rational Thinking*, translated by Philipp Fischelis, F. A. Davis Company, Philadelphia, 1927.

[82] Arnold Lorand, *Old Age Deferred*, 1926 (1910), pp. 457-458, 51, 445-447.

[83] Arnold Lorand, *Old Age Deferred*, 1926 (1910), p. v.

[84] Morris Fishbein, *The Medical Follies. An Analysis of the Foibles of Some Healing Cults, Including Osteopathy, Homeopathy, Chiropractic, and the Electronic Reactions of Abrams, with Essays on the Antivivisectionists, Health Legislation, Physical Culture, Birth Control, and Rejuvenation*, Boni and Liveright, New York, 1925, pp. 169, 234.

[85] Metchnikoff's views on the "limit" to the human life and possibilities of the prolongation of life, are summarized by the following statements:
"It is possible that, in the human species, the limit of life is not as constant as in mayflies, and therefore it is impossible to constrain it by any number. In most cases it should be much higher than 100 years and only in exceptional cases it could reach below this limit. …
It is possible that the feeling of saturation with life, as strange as it may sound to us, is nothing else than the instinct of natural death that develops in well preserved aged people who reach 140-180 years."
(Elie Metchnikoff, *Etudy o Prirode Cheloveka* (Etudes on the Nature of Man), The USSR Academy of Sciences Press, Moscow, 1961 (1903), XI "Vvedenie v nauchnoe izuchenie smerti" (Introduction to the Scientific Study of Death), pp. 227, 229-230.)
And more specifically, on the "normal limit" of life:
"It is very difficult to establish the limit of 'normal' life for a human being who has been undergoing chronic poisoning, leading to premature death. Only hypothetically one can determine an average limit of such a life at 100-120-145 years, considering also that this rule must have many exceptions."
(Elie Metchnikoff, "Predislovie ko Vtoromu Izdaniu *Sorok Let Iskania Razionalnogo Mirovozzrenia* (Introduction to the Second Edition of 'Forty Years in Search of a Rational Worldview'), 1914, in I.I. Metchnikoff, *Academicheskoe Sobranie Sochineniy* (Elie Metchnikoff. Academic Collected Works, Ed. G.S. Vasezky), Academia Medizinskikh Nauk SSSR (The USSR Academy of Medical Sciences), Moscow, 1954, vol. 13, p. 21.)
Or else:
"Cases of living to 150 years and more are very rare, but old people who lived to 100 years and slightly more were nothing unusual."
(Elie Metchnikoff, "Dolgovechnost ili prodolzhitelnost zhizni" (Longevity or the Span of Life), in *Collected*

Works of I.I. Metchnikoff (Ed. N.N. Anichkov and R.I. Belkin), The USSR Academy of Medical Sciences, Moscow, vol. 15, 1962, p. 359.)

Ultimately, about a year before his death, Metchnikoff concluded:

"When the sorrows and worries of the present moment, engulfed by the world war, will be long gone and archived, the problems of life and death will preserve their dominant significance.

Let us hope that the works of our institute, in which I will no longer be able to participate, will broadly contribute that people in the future will be given the opportunity to reach the normal limit of life, longer than it is now."

("Rech I.I. Mechnikova na Prazdnovanii ego Semidesyatiletia" (Speech of I.I. Metchnikoff at his 70th anniversary), in *Collected Works of I.I. Metchnikoff* (Ed. N.N. Anichkov and R.I. Belkin), The USSR Academy of Medical Sciences, Moscow, vol. 15, 1962, p. 371.)

[86] Sigmund Freud, *Beyond the Pleasure Principle*, Translated by James Strachey, Liveright, New York, 1976 (1920), p. 38.

[87] Mirko Dražen Grmek, *On Ageing and Old Age, Basic Problems and Historic Aspects of Gerontology and Geriatrics*, *Monographiae Biologicae*, 5, 2, Den Haag, 1958, pp. 3, 44, 53-54, 59.

[88] Gerald J. Gruman, *A History of Ideas About the Prolongation of Life: The Evolution of Prolongevity Hypotheses to 1800*, 1966, Ch. V "Taoist Prolongevitism in Practice. Sexual Techniques," pp. 45-47.

[89] Johann Heinrich Cohausen, *Tentaminum physico-medicorum curiosa decas de vita humana theoretice et practice per pharmaciam prolonganda* (Curioous physico-medical theoretical and practical attempts to prolong the decades of human life by pharmacological means), M.A. Fuhrmannum, Osnabrück, 1714 (1699).

The book discusses over 260 life-prolonging elixirs from mineral, plant and animal sources.

[90] Johann Heinrich Cohausen, *Hermippus Redivivus or the Sage's Triumph over Old Age and the Grave, Wherein a method is laid down for prolonging the life and vigour of man, including a commentary upon an ancient inscription, in which this great secret is revealed, supported by numerous authorities. The whole is interspersed with a great variety of remarkable and well attested relations*, J. Nourse, London, 1748.

The manuscript was first published in Latin in 1742 as *Hermippus redivivus, sive exercitatio physico-medica curiosa de methodo rara ad CXV annos prorogandae senectutis per anhelitum puellarum, ex veteri monumento romano deprompta, nunc artis medicae fundamentis stabilita, et rationibus atque exemplis, nec non singulari chymiae philosophicae paradoxo, illustrate et confirmata* (Hermippus revived, or the curious physico-medical exercise of a rare method to postpone old age to 115 years by the breath of young maidens, taken from an old Roman tomb, now establishing the foundations of the art of medicine, with reasons and examples, and also singular chemical-philosophical pardoxes), John Benjamin Andrae & Henry Hort, Frankfurt am Main, 1742.

[91] Christoph Wilhelm Hufeland, *Makrobiotik; oder, Die Kunst das menschliche Leben zu verlängern*, Sechste verbesserte Auflage, A.F. Macklot, Stuttgart, 1826 (Macrobiotics or the art of prolonging human life, the 6th improved edition), first published in Jena, in 1796.

[92] Brown-Séquard, C.E. "Des effets produits chez l'homme par des injections sous-cutanées d'un liquide retiré des testicules frais de cobaye et de chien" (Effects in man of subcutaneous injections of freshly prepared liquid from guinea pig and dog testes), *Comptes Rendus des Séances de la Société de Biologie*, Série 9, 1, 415-419, 1889, reprinted and translated in Geraldine M. Emerson (Ed.), *Benchmark Papers in Human Physiology, Vol. 11, Aging*, Dowden, Hutchinson and Ross, Stroudsburg PA, 1977, pp. 68-76.

[93] C.E. Brown-Séquard, "Des effets produits chez l'homme par des injections sous-cutanées d'un liquide retiré des testicules frais de cobaye et de chien" (Effects in man of subcutaneous injections of freshly prepared liquid from guinea pig and dog testes), *Comptes Rendus des Séances de la Société de Biologie*, Série 9, 1, 415-419, 1889, in *Benchmark Papers in Human Physiology*, Vol. 11, 1977, pp. 68-76; Newell Dunbar (Ed.), *The "Elixir of Life." Dr. Brown-Sequard's own account of his Famous Alleged Remedy for Debility and Old Age, Dr. Variot's Experiments, and Contemporaneous Comments of the Profession and the Press*, J.G Cupples Company, Boston, 1889.

[94] Quoted in Eugene Steinach and John Loebel, *Sex and Life; Forty Years of Biological and Medical Experiments*, Faber, London, 1940, pp. 49-50.

[95] Victor Alphonse Amédée Dumontpallier, "Remarques au sujet de la communication de M. Brown-Séquard" (Remarks regarding the communication of M. Brown-Séquard), *Comptes Rendus des Séances de la Société de Biologie*, Série 9, 1, 419, 1889, reprinted in *Benchmark Papers in Human Physiology, Vol. 11*, 1977, p. 78.

[96] C.E. Brown-Séquard, "Effects in man of subcutaneous injections of freshly prepared liquid from guinea pig and dog testes," reprinted in *Benchmark Papers in Human Physiology, Vol. 11*, 1977 (1889), p. 76.

[97] C.E. Brown-Séquard, "Du Role Physiologique et Thérapeutique d'un suc extrait de testicules d'animaux d'après nombre de faits observes chez l'homme" (The physiological and therapeutic role of animal testicular extract based on several experiments in man), *Archives de Physiologie Normale et Pathologique*, 1(5), 739-746, 1889, reprinted in *Benchmark Papers in Human Physiology, Vol. 11*, 1977, p. 101, 104-105.

[98] Thus regarding the claims of the French homeopathic doctor, Mériadec Conan, Brown-Séquard asserted:
"M. Conan is occupied only with methods of treating illness. My communications to the Society have had the object not illness; I have had in mind only the changes caused in man by age…"
In addition, Conan's methods of extract preparation were said to be unsound:
"He destroyed all the organic properties of these parts before using them. In fact, for six days he subjected them to desiccation in a stove at 70°. After that, no one will find that what I did resembles in any way that which the homeopathic doctor did."
(C.E. Brown-Séquard, "Correspondance Imprimée. Remarques de M. Brown-Séquard a l'egard de la reclamation de M. Conan" (Printed Correspondence and Comments, regarding the claims of M. Conan), reprinted in Geraldine M. Emerson (Ed.), *Benchmark Papers in Human Physiology, Vol. 11, Aging*, 1977, p. 87.)

[99] C.E. Brown-Séquard, "The Physiological and Therapeutic role of animal testicular extract," 1889, reprinted in *Benchmark Papers in Human Physiology, Vol. 11*, 1977, p. 100.

[100] C.E. Brown-Séquard, "The Physiological and Therapeutic role of animal testicular extract," 1889, reprinted in *Benchmark Papers in Human Physiology, Vol. 11*, 1977, pp. 93-94, 101-102.

[101] *A Treatise of Health and Long Life, with the Sure Means of Attaining It. In 2 Books. The first By Leonard Lessius. The Second by Lewis Cornaro,* Translated into English by Timothy Smith, London, C. Hitch, 1743; Christoph Wilhelm Hufeland, *Makrobiotik oder Die Kunst, das menschliche Leben zu verlängern*, first published in 1796, translated into English as *The Art of Prolonging Life*, Edited by Erasmus Wilson, J. Churchill, London, 1859, Lindsay & Blakiston, Philadelphia, 1867.

[102] There has been a vast, ancient tradition advising on sexual moderation in order to achieve life-extension, from Aristotle's *On Length and Shortness of Life* and *On Youth, Old Age, Life and Death, and Respiration*, through Taoist physicians, to the Renaissance and early modern hygienists (Luigi Cornaro, Leonardus Lessius, Christoph Wilhelm Hufeland, and others). Some (rare) recent examples of this attitude are Edwin Flatto's *Warning: Sex May Be Hazardous to Your Health* (1977) or David Pratt's *Sex and Sexuality* (2002).
(Aristotle, *On Length and Shortness of Life*, Aristotle, *On Youth, Old Age, Life and Death, and Respiration*, translated by G.R.T. Ross, in *The Complete Works of Aristotle: The Revised Oxford Translation*, Princeton University Press, Princeton, 1984, vol. 1, pp. 740-744, 745-763; Gerald J. Gruman, *A History of Ideas about the Prolongation of Life. The Evolution of Prolongevity Hypotheses to 1800*, Transactions of the American Philosophical Society, Volume 56 (9), Philadelphia, 1966; Edwin Flatto, *Warning: Sex May Be Hazardous to Your Health*, 2nd ed., Arco, New York, 1977; David Pratt, *Sex and Sexuality*, 3.6. "Pleasure at a Price," New York, 2002, http://davidpratt.info/sex3.htm.)
This tradition appears to be born out by the findings of the consistently higher longevity among monks.
(Jenner B, "Changes in average life span of monks and nuns in Poland in the years 1950-2000," *Przegląd Lekarski*, 59(4-5), 225-229, 2002; De Gouw HW, et al, "Decreased mortality among contemplative monks in The Netherlands," *American Journal of Epidemiology*, 141(8), 771-775, 1995; Marc Luy, "Sex differences in mortality - time to take a second look," *Zeitschrift für Gerontologie und Geriatrie*, 35(5), 412-429, 2002; Marc

Luy, *Klosterstudie* - The Cloister Study, http://www.klosterstudie.de/.)

There have also been indications about the longer life span of eunuchs.

(Hamilton JB, Mestler GE, "Mortality and survival: comparison of eunuchs with intact men and women in a mentally retarded population," *Journal of Gerontology*, 24(4), 395-411, 1969; John P. Phelan, Michael R. Rose, "Why dietary restriction substantially increases longevity in animal models but won't in humans," *Aging Research Reviews*, 4(3), 339-350, 2005.)

A series of studies showed the costs of reproduction for longevity in animal models.

(E.g. Rolff J, Siva-Jothy MT, "Copulation corrupts immunity: a mechanism for a cost of mating in insects," *Proceedings of the National Academy of Sciences USA*, 99(15), 9916-9918, 2002; Wayne A. Van Voorhies, "Production of sperm reduces nematode lifespan," *Nature*, 360, 456-458, 1992.)

More generally, according to the "Disposable Soma" theory of aging, energy expenditures on reproduction come at the cost of energy expenditures on the maintenance of the body.

(Tom Kirkwood, *Time of Our Lives. The Science of Human Aging*, Oxford University Press, Oxford, 1999.)

And in the "Phenoptosis" or "programmed aging" theory, sex is considered as a trigger of the 'self-destruct' mechanism in animals.

(Skulachev, V.P., "Aging is a specific biological function rather than the result of a disorder in complex living systems: biochemical evidence in support of Weismann's hypothesis," *Biochemistry*, Moscow, 62(11), 1191-1195, 1997.)

Yet, presently, even when discussing the "disposable soma theory," the "phenoptosis theory" or the "costs of reproduction," sexual moderation in humans is hardly ever recommended. Instead, there is a common popular belief that "sex is good for longevity."

"Sex is good," for example, according to Dr. Mark Stibich http://longevity.about.com/od/lifelongrelationships/p/sex_longevity.htm; Dr. Michael Roizen http://www.scribd.com/doc/24835112/Longevity-Sex; or the "sensual product designer" Anne Enke http://www.anneofcarversville.com/superyoung/sex-and-longevity-health-benefits-of-loving-sex.html; and many more such examples of the popular stance can be added.

A single most widely cited study in this approach is George Davey Smith, Stephen Frankel, and John Yarnell's "Sex and death: are they related? Findings from the Caerphilly cohort study," *British Medical Journal*, 315, 1641-1644, 1997, which correlated between high sexual activity and lower mortality in a group of aged men.

The majority of authors now heavily emphasize the possibility and desirability of sex for the aged, the benefits of sex with a constant partner, its positive role for the production of stress-reducing hormones, and for the improvement of self-image and connection.

The terms "exhaustion" and "moderation," that had been prevalent among the earlier hygienists, are now hardly ever present, and at any rate the possibility of a "threshold" or "tradeoff" in sexual activity with reference to human longevity is hardly ever considered.

(See for example, *Normal Aging: Reports from the Duke Longitudinal Study, 1955-1969*, edited by Erdman Palmore, Duke University Press, Durham NC, 1970, pp. 266-303; Nathan W. Shock, et al., *Normal Human Aging: The Baltimore Longitudinal Study of Aging*, NIH Publication, 1984, pp. 164-165; Thomas H. Walz, Nancee S. Blum, *Sexual Health in Later Life*, Lexington Books, Lexington MA, 1987.)

The works of the early 20th century rejuvenators – Steinach, Voronoff and others, who foregrounded and praised the vigorous sexual activity of the aged – may have been instrumental in this shift from an "ascetic" to a "hedonistic" view of the role of sex for healthy longevity.

[103] Quoted in David Hamilton, *The Monkey Gland Affair*, Chatto and Windus, London, 1986, p. 2.

[104] Serge Voronoff, *Rejuvenation by Grafting*, translated by Fred F. Imianitoff, George Allen and Unwin Ltd, London, 1925. Other popular works by Voronoff were: *The Conquest of Life*, translated by G. Gibier Rambaud, Brentano's Ltd, London, 1928; *The Sources of Life*, Bruce Humphries, Boston, 1943; *Greffes*

Testiculaires, Librairie Octave Doin, Paris, 1923. Interesting archival materials on Serge Voronoff can be found at the website of Aaron Voronoff (descendant of Serge Voronoff's brother Jacques) entitled *Interstitial Immortality* (accessed 2012, http://voronoff.wordpress.com/).

[105] Serge Voronoff, George Alexandrescu, *La greffe testiculaire du singe à l'homme* (Testicular grafts from apes to men), Doin, Paris, 1930.

[106] V.N. Anisimov, M.V. Soloviev, *Evoluzia Concepcy v Gerontologii* (The Evolution of Concepts in Gerontology), Aesculap, Saint Petersburg, 1999 (reprinted in the *Russian Biomedical Journal,* www.Medline.ru, vol.3), "Perspectivy razvitia gerontologii v Rossii" (Perspectives of the development of gerontology in Russia, http://www.medline.ru/public/art/tom3/geront/pt1-1.phtml#pt1p1).

[107] For example, Hakim and Papalois' *History of Organ and Cell Transplantation* (2003) makes no mention of Voronoff, in fact, no mention of sex gland transplantation or opotherapy generally. (Nadey S. Hakim, Vassilios E. Papalois (Eds.), *History of Organ and Cell Transplantation*, Imperial College Press, London, 2003.) In Roy Porter's *The Greatest Benefit to Mankind* (W.W. Norton & Company, NY, 1997), in the section on "Transplants," Voronoff is absent.

From several works of medical history, it may appear as if the timeline of transplantation research ends by 1910, with skin graft attempts (among the first, reported in 1869 by the French Jean Casimir Félix Guyon, and in the same year by the Swiss Jacques Reverdin), bone grafts (done by the Scottish William MacEwen in 1880), cornea transplants (conducted by the Czech Eduard Zirm in 1906), and the development of anastomosis (blood vessels suturing) and transplantation experiments in animals by the French/American Alexis Carrel (performed between 1902-1910, earning Carrel the Nobel Prize in 1912, though a priority dispute existed between Carrel and Mathieu Jaboulay of Lyon and Emerich Ullmann of Vienna).

Then, it may appear, after a 30-40 years gap, the time line resumes into being, after the British/Brazilian/Lebanese Nobel laureate in medicine Peter Brian Medawar's suggestion of 1951 to use immuno-suppression (by cortisone) to prevent tissue rejection. Medawar's suggestion was then followed by the first successful transplantations of the kidney (by the Americans John Hartwell Harrison and Joseph Edward Murray, 1954), the lung (by the American James Daniel Hardy, 1963), the heart (by the South African Christian Neethling Barnard, 1967), and so forth.

The works of several life-extensionists and rejuvenators (discussed in this and the following sections of the present study) appear to have been largely left out of such historical accounts. The histories exclude, for example, Voronoff's and Steinach's sex gland transplantations of the 1910s-1920s, Sergey Bryukhonenko, Sergey Chechulin and Vladimir Demikhov's work on the "revival of the organism" in the USSR in the 1920-1930s, the implantation of embryonic (especially placental) tissues by the Russians Vladimir Filatov and Nikolay Krauze, and by the Germans Max Bürger and Georg Weitzmann in the 1930-1940s, and the Cellular Therapy by the Swiss Paul Niehans in the 1930s-1940s. All these works were done with the explicit aim of rejuvenation and life-prolongation, but they may have seemed too unorthodox to include in the time line.

Furthermore, William H. Schneider's inclusive *Quality and Quantity: The Quest for Biological Regeneration in Twentieth-Century France* (Cambridge University Press, Cambridge, 2002) makes no specific mention either of Metchnikoff or Voronoff whose rejuvenation experiments were the ultimate hype in France in the 1920s and fitted perfectly into the French "Regeneration" paradigm in the aftermath of WWI.

In the rare cases when Voronoff's work is discussed – as in David Hamilton's *The Monkey Gland Affair* (Chatto and Windus, London, 1986) – the verdict is ruthless. Armed with the later knowledge of tissue compatibility, Hamilton maintains that "Voronoff was wrong there is no doubt" and his experiments "could not have possibly worked" (p. 64.)

Thus, Voronoff's role as a pioneer of live tissue transplantation in humans, deriving from a life-extensionist pursuit, still needs to be acknowledged.

[108] Gerald J. Gruman, *A History of Ideas about the Prolongation of Life*, 1966, p. 6.

[109] Voronoff, *The Sources of Life*, 1943, Ch. "The Struggle Against Old Age," p. 65.
[110] Voronoff, *The Conquest of Life*, 1928, p. 201.
[111] Voronoff, *The Conquest of Life*, 1928, p. 17.
[112] Sergey Metalnikov, *Problema Bessmertia v Sovremennoy Biologii* (The Problem of immortality in modern biology), Petrograd, 1917.
Later editions: Sergey Metalnikov, *Problema Bessmertia I Omolozhenia v Sovremennoy Biologii* (The Problem of Immortality and Rejuvenation in Modern Biology), Slovo, Berlin, 1924; Serge Metalnikov, *La Lutte contre la Mort* (The struggle against death), Gallimard, Paris, 1937.
[113] Voronoff, *The Conquest of Life*, 1928, pp. 23, 24, 26.
[114] Voronoff, *The Conquest of Life*, 1928, pp. 32, 40.
[115] Voronoff, *The Conquest of Life*, 1928, p. 199.

The major problem with Voronoff's organ replacement operations, as it is presently understood, is that animal transplants would be rejected by the human host as incompatible. Today, the grave problem of invariable rejection of animal organ transplants by the human host is hoped to be overcome by the use of transgenic animals whose organs are made more "human-like" by genetic engineering, or by sophisticated immunosuppression. (Scott McCartney, *Defying the Gods: Inside the New Frontiers of Organ Transplants*, Macmillan, New York, 1994; Robert Michler, "Xenotransplantation: Risks, Clinical Potential, and Future Prospects," *Emerging Infectious Diseases*, 2(1), 64-70, 1996; Dooldeniya M, Warrens A, "Xenotransplantation: where are we today?" *Journal of the Royal Society of Medicine*, 96, 11-117, 2003.)

Another grave concern is the transmission of known and unknown pathogens via the animal-to-human transplants. This threat is now hoped to be prevented by strict surveillance.

The fear of latent pathogens is now so widespread that some authors even accused Voronoff's ape grafts of originating AIDS. Presumably, it took some 50 years for the threat to materialize. (Gebhardt, D.O.E, "Origin of AIDS," *Lancet*, 338, December 28, 1991, p. 1604.)

To recall, AIDS was also blamed on Hilary Koprowski's polio vaccination in the Belgian Congo in 1957-1960, and on the global eradication of smallpox in 1977.

(Brian Martin, "Peer review and the origin of AIDS - a case study in rejected ideas," *BioScience*, 43, (9), 624-627, 1993, http://www.uow.edu.au/~bmartin/pubs/93bs.html; http://www.righto.com/theories/polio.html; Pearce Wright, "Smallpox vaccine 'triggered AIDS virus'", May 11, 1987, http://www.wanttoknow.info/870511vaccineaids; Roy Porter, *The Greatest Benefit to Mankind: A Medical History of Humanity*, W.W. Norton & Company, NY, 1998 (1997), p. 491.)

Voronoff, however, acted on the state-of-the-art knowledge of his time, according to which sex gland transplantation from apes (mainly chimpanzees) was entirely plausible. He mounted a formidable defense of his method, describing in great detail the surgical technique (self-admittedly owing to Carrel's novel blood vessels suturing method), animal experiments and human trials. He set out to demonstrate by histological experiments that the transplanted sex glands are not "reabsorbed" (rejected) by the organism, but continue to live, whether transplanted between subjects of the same animal species, or very close species (Voronoff, *Rejuvenation by Grafting*, 1925, p. 29).

A firm evolutionist, Voronoff believed that humans and apes are evolutionary close enough to allow transplantation, their sex glands appearing similar both morphologically and immunologically. "But that man and the apes should possess similar serum reactions is a far more significant fact; for it shows that, not only is there a striking anatomical resemblance between the two, but that there exists an undoubtable biological affinity" (Voronoff, *Rejuvenation by Grafting*, 1925, pp. 29, 36). Thus, no tissue rejection was expected. And, of course, only apparently healthy specimens were selected for transplantation.

The evolutionary affinity between Man and Apes was then perceived to be even closer by the Russian/French/Soviet zoologist Ilya Ivanovich Ivanov (1870-1932) who, between 1910 and 1930, attempted man-ape crossings, aiming to produce specimens of unprecedented hybrid vigor.

(Kirill Rossiianov, "Beyond Species: Il'ya Ivanov and His Experiments on Cross-Breeding Humans with Anthropoid Apes," *Science in Context*, 15, 277-316, 2002.) Reports about such crossing experiments have cropped up as late as 1988 in the UK; one is mentioned in John Harris' *Clones, Genes and Immortality*, 1998. (John Harris, *Clones, Genes and Immortality*, Oxford University Press, Oxford, 1998, pp.183, 296 note 14.)

Voronoff provided no evidence of a radical extension of life by the grafting operations, though he offered ample data on rejuvenating effects: restoring daily functioning, enhancing the "capacity to work," increasing "physical energy" and improving "mental capacity." These, and not exclusively or primarily the improvements of sex function, were the objectives. "The testicular graft," Voronoff emphasized, "is not an aphrodisiac." Patients' case histories testified to reinvigorating effects, and numerous "before and after" pictures showed enfeebled elderly men turning into sturdy, manly individuals. (Voronoff, *Rejuvenation by Grafting*, 1925, pp. 45-50.)

Voronoff took great pains to rule out auto-suggestion as a cause of these effects, relying on 'suggestion free' animal experiments and on the long-lasting results in humans that were allegedly unaffected by mood and consciousness shifts.

Besides the autosuggestion, the stimulating effects in humans could be well explained by an action of reabsorbed hormones, i.e. "embedding opotherapy," or by a non-specific stimulation due to the intrusion (cf. criticisms of Brown-Séquard methods).

Voronoff discarded such possibilities in humans, based on animal experiments where the grafted glands were "far from being absorbed" and where the effects of sex tissue transplantations were specific.

There was also quite a distinct possibility of physical exhaustion due to over-stimulation. Thus, "[Case 4] at the end of two years and seven months, required a new graft, having been over-prodigal of the vital energy supplied him by the old one." According to Voronoff, the patient's own wasteful habits, not the operation, was responsible for the exhaustion: "In the graft science has given to humanity a means of increasing a vital capital; it remains with man that this capital should not be recklessly squandered" (Voronoff, *Rejuvenation by Grafting*, 1925, pp. 50-51).

In summary, Voronoff presented his method as an initial stage for a research program for the future development of rejuvenative/regenerative/replacement therapy, yet even the current results were presented by him as little short of miraculous.

[116] Voronoff was first married to Marguerite Barbe (from 1897 to 1910), daughter of Ferdinand de Lesseps, the developer of the Suez Canal; then to Evelyn Bostwick (1919-1921), daughter of Jabez Bostwick, a founding partner of the Standard Oil Company, and then to the noblewoman Gertrude Schwartz (since 1931).

[117] Voronoff, *The Sources of Life*, 1943, p. 64.

[118] A literary expression of this popular stance is given in Mikhail Bulgakov's *The Heart of a Dog* (1925), where Prof. Preobrazhensky, performing sex gland transplants for rejuvenation, was directly modeled after Voronoff. Preobrazhensky's wealthy clients, who were able to pay for the rejuvenative operations, are clearly ridiculed as undeserving. One of Preobrazhensky's test subjects, the "low class" Sharikov attacks the professor:

"Of course. How can we be your comrades? We didn't learn in universities, we didn't live in apartments with 15 rooms, with bathrooms." Or else, "One spreads himself in seven rooms, he has forty pairs of trousers. Another is wandering around, looking for sustenance in garbage bins."

(Mikhail Bulgakov, *Sobachie Serdze* (The Heart of a Dog), in *Mikhail Bulgakov. Povesti, Rasskazy, Felietony, Ocherki* (Novels and Sketches), Literatura Artistike, Kishinev, 1989 (1925), pp. 192, 204.)

[119] Robert Wilson, *Feminine Forever*, M. Evans & Co., NY, 1966.

[120] Eugen Steinach and John Loebel, *Sex and Life; Forty Years of Biological and Medical Experiments*, Faber, London, 1940, pp. 56, 213.

[121] Voronoff, *Rejuvenation by Grafting*, 1925, pp. 31-32.

[122] Grant JA, "Victor Darwin Lespinasse: a biographical sketch," *Neurosurgery*, 39, 1232-233, 1996; Dirk Schultheiss and Rainer M. Engel, "G. Frank Lydston (1858–1923) revisited: androgen therapy by testicular implantation in the early twentieth century," *World Journal of Urology*, 21(5), 356-363, 2003; Max Thorek, "The present position of testicle transplantation in surgical practice: A preliminary report," *Endocrinology*, 6, 771-775, 1922.

[123] Voronoff, *Rejuvenation by Grafting*, 1925, p. 32-34.

[124] In this English edition of Voronoff's *Rejuvenation by Grafting*, some names appear to have been misspelled. Here are the full names of the prominent testicle grafters in animals and humans mentioned by Voronoff and the dates of their work:
John Hunter (1770), Arnold Adolf Berthold (1849), Jean Marie Philippeaux (1858), Paolo Mantegazza (1864), Eugen Steinach (1911), Albert Pézard (1911), Boris Zavadovsky (1921), Knud Sand (1922), Alexander Lipschütz (1924), Moritz Nussbaum (1880, 1905), Johann Meissenheimer (1922), Charles Claude Guthrie (1910), Richard Mühsam (1925-1926), Erwin Payr (1906), Placide Mauclaire (1922), Robert Lichtenstern (1916), Max Thorek (1912), H. Jeanée (1925, 1928), Eugen Enderlen (1921), Levi Jay Hammond and Howard Anderson Sutton (1912), Victor Darwin Lespinasse (1911), Robert Tuttle Morris (1914), George Frank Lydston (1912), Charles Morgan McKenna (1920), Leo Leonidas Stanley (1918), George David Kelker (1918), Paul Lissmann (1914), Arthur Gregory (1922), E. Mariotti (1919), R. Falcone (1920).
(Further see Thomas Schlich, *The Origins of Organ Transplantation: Surgery and Laboratory Science, 1880-1930*, University of Rochester Press, NY, 2010.)

[125] Richard Ellmann, *W.B. Yeats's Second Puberty*, Library of Congress, Washington, 1985, p. 7; Paul E. Stepansky, *Freud, Surgery, and the Surgeons*, Routledge, London, 1999, p.137. Freud underwent the procedure in 1923 and Yeats in 1934.

[126] Eugen Steinach, *Verjüngung durch experimentelle Neubelebung der alternden Pubertätsdrüse* (Rejuvenation by experimental revitalization of the aging puberty gland), Springer, Berlin, 1920; Eugen Steinach and John Loebel, *Sex and Life; Forty Years of Biological and Medical Experiments*, Faber, London, 1940.

[127] Eugen Steinach. *Verjüngung durch experimentelle Neubelebung der alternden Pubertätsdrüse* (Rejuvenation by experimental revitalization of the aging puberty gland), Springer, Berlin, 1920, p. 13.
Some of the central points regarding the operation are described on pp. 8, 42, 52, 54, 58, 60.

[128] The first male to female "sex reassignment surgery" was performed on Lili Elbe (b. Einar Wegener, 1882-1931) in 1930 in Dresden, Germany, by Magnus Hirschfeld and Kurt Warnekros – Steinach's followers.

[129] The nomination for "rejuvenation" was done by Steinach's Japanese follower Yasusaburo Sakaki (1870-1929) of Kyushu Imperial University, http://www.med.kyushu-u.ac.jp/psychiatry/cn3/pg28.html. See The Nomination Database for the Nobel Prize in Physiology or Medicine, 1901-1953, http://nobelprize.org/nobel_prizes/medicine/nomination/database.html.

[130] Steinach's first report of his rejuvenation research was "Untersuchungen über die Jugend und über das Alter" (Studies of Youth and Old Age) presented to the Academy of Sciences in Vienna on December 5, 1912. It is reproduced in full in Eugen Steinach, *Verjüngung durch experimentelle Neubelebung der alternden Pubertätsdrüse* (Rejuvenation by experimental revitalization of the aging puberty gland), Springer, Berlin, 1920, pp. 61-63.
Even as early as 1912, having only started the animal experiments, Steinach asserted that this research could lead to a "prolongation of the period of active life" and a "limited prolongation of the life-span."

[131] The historical connection between the rejuvenation procedures and the wide spread of vasectomy and tube ligation may be emphasized.
As of 1996, according to the National Institute of Child Health and Human Development of the US National Institutes of Health (NIH), worldwide "A total of about 50 million men have had a vasectomy – a

number that corresponds to roughly 5 percent of all married couples of reproductive age. In comparison, about 15 percent of couples rely on female sterilization for birth control. Approximately half a million vasectomies are performed in the United States each year. About one out of six men over age 35 has been vasectomized, the prevalence increasing along with education and income." (http://web.archive.org/web/20061001131154/http://www.nichd.nih.gov/publications/pubs/vasectomy_safety.cfm.)

According to other reports, as of 2005, it was estimated that more than 200 million women world wide had been sterilized by tube ligation. By that time, in the US, about 600,000 women underwent sterilization by tube ligation yearly. (*World Contraceptive Use 2005*, United Nations Department of Economic and Social Affairs; New York, 2005, http://www.un.org/esa/population/publications/contraceptive2005/WCU2005.htm; http://apps.who.int/rhl/fertility/contraception/lxhcom/en/index.html; http://www.rhtp.org/contraception/iud/alternative.asp.)

It should also be noted that, contrary to the original expectations of the rejuvenators, vasectomy does not appear to have a great effect on longevity.

According to Giovannucci E, Tosteson TD, Speizer FE, Vessey MP, Colditz GA, "A long-term study of mortality in men who have undergone vasectomy," *New England Journal of Medicine*, 326 (21), 1392-1398, 1992, after vasectomy, mortality somewhat decreases in the short term, but in the longer term remains approximately the same.

As the authors summarize:

"Vasectomy was associated with reductions in mortality from all causes (age-adjusted relative risk, 0.85; 95 percent confidence interval, 0.76 to 0.96) and mortality from cardiovascular disease (relative risk, 0.76; 95 percent confidence interval, 0.63 to 0.92). Vasectomy was unrelated to mortality from all forms of cancer (relative risk, 1.01; 95 percent confidence interval, 0.82 to 1.25). Among men who had a vasectomy at least 20 years earlier, the procedure had no relation to mortality from all causes (relative risk, 1.11; 95 percent confidence interval, 0.92 to 1.33) or that from cardiovascular disease (relative risk, 0.85; 95 percent confidence interval, 0.63 to 1.16). However, mortality from cancer was increased in men who had a vasectomy at least 20 years earlier (relative risk, 1.44; 95 percent confidence interval, 1.07 to 1.92). The excess risk of cancer in these men was due primarily to lung cancer. None of the observed associations were confounded by smoking habits, body-mass index, alcohol consumption, or educational level."

The authors conclude:

"Vasectomy is not associated with an increase in overall mortality or mortality from cardiovascular disease. Our study also found no increase in overall mortality from cancer after vasectomy, but there was an apparent increase in the risk of cancer 20 or more years after vasectomy that requires further study."

[132] Steinach and Loebel, *Sex and Life*, 1940, p. 102.

[133] Steinach and Loebel, *Sex and Life*, 1940, pp. 148-149.

[134] Pauchet, V, "Revitalisation par sympathicectomie chimiques des organes sexuel" (Revitalization by chemical sympathectomy of sexual organs), *Journal de médecine*, Paris, 49, 801-803, 1929; Pauchet V, "La rejeunissement de l'organisme par la méthode de Doppler. Technique Opératoire (Rejuvenation of the organism by Doppler's method. Operation technique), *Clinique*, Paris, 25, 327-332, 1930; Berri HD, Silvestre AL, "La simpatectomia quimica. Método de Doppler. Resultados obtenidos en una observación (Chemical sympathectomy. Doppler's method. Observation results), *Semana Médica Buenos Aires*, Argentina, 46, 1199-1201, 1939.

A comprehensive bibliography on rejuvenation is given in Nathan W. Shock (Ed.), *Classified Bibliography of Gerontology and Geriatrics*, Stanford University Press, Stanford CA, 1951 (reprinted in 1980), Section "Rejuvenation," subsections "Sex Glands and Organs" and "Sympathectomy," pp 74-80.

[135] "Vie et rajeunissement: Une nouvelle méthode générale de traitement et mes expériences de

rajeunissement de Bologne et de Paris. Par le Dr. Francesco Cavazzi. Paris: Gaston Doin & Cie, 1934" (Life and rejuvenation. A new general method of treatment and my rejuvenation experiments in Bologna and Paris. By Dr. Francesco Cavazzi), *Journal of the American Medical Association*, 103(24), 1881, 1934 (review).

[136] Lebedinski, N.G, "Stimulierung ("Verjüngung") des alternden Säugetierorganismus durch Zerreissen und Zerdrucken des Hodengewebes" (Stimulation or "rejuvenation" of aging mammals by tearing and squashing the testis tissue), *Latvijas Biologijas Biedribas Raksti*, Riga, 1, 121-122, 1929.

[137] Durand-Fardel, *Traité clinique et pratique des maladies des vieillards* (Clinical and practical treatise on the diseases of the aged), 1854, quoted in Joseph T Freeman, "The History of Geriatrics," *Annals of Medical History*, 10, 324-335, 1938, and in Ignatz Leo Nascher, *Geriatrics, the Diseases of Old Age and their Treatment. Including Physiological Old Age, Home and Institutional Care, and Medicolegal Relations*, P. Blakiston's Son & Co, Philadelphia, 1914, p. 41.

[138] Marie Jean Pierre Flourens, *On Human Longevity and the Amount of Life upon the Globe translated from the French by Charles Martel*, Bailliere, London, 1855 (first published in French in 1854), pp. 54, 75; S.E. Chaillé, "Longevity," *New Orleans Medical and Surgical Journal*, 16, 417-424, 1859, reprinted in Geraldine M. Emerson (Ed.), *Benchmark Papers in Human Physiology, Vol. 11, Aging*, Dowden, Hutchinson and Ross, Stroudsburg PA, 1977, pp. 28-35.

[139] S.E. Chaillé, "Longevity," *New Orleans Medical and Surgical Journal*, 16, 417-424, 1859; S.P. Cutler, M.D. Holly Springs, Miss, "Physiology and Chemistry of Old Age," *New Orleans Journal of Medicine*, 23(1), 96-104, 1870; M.F. Legrand, "Peut-On Reculer le Bornes de la Vie Humaine?" (Can we reverse the limitations of human life?), *L'Union Médicale*, 13 (5), 65-71, 1859, reprinted in Geraldine M. Emerson (Ed.), *Benchmark Papers in Human Physiology, Vol. 11, Aging*, Dowden, Hutchinson and Ross, Stroudsburg PA, 1977, pp. 28-47. The "mineralization" theory was presented in 1858 by Édouard Robin to the French Academy of Sciences in the address "Sur le causes de la vieillesse et de la mort sénile" (On causes of aging and senile death).

[140] Elie Metchnikoff, *Etudy o Prirode Cheloveka* (Etudes On the Nature of Man), The USSR Academy of Sciences Press, Moscow, 1961 (1915, first published in 1903), p. 178.

[141] Jean Finot, *The Philosophy of Long Life* (translated by Harry Roberts), John Lane Company, London and New York, 1909, p. 273, first published in French as *La Philosophie de la Longévité*, Schleicher Freres, Paris, 1900.

[142] The idea of creating a "homunculus" was, of course, not new. As said in Goethe's *Faust*, "What wise or stupid thing can man conceive that was not thought in ages passed away?" A description of creating a homunculus follows: "A great design at first seems mad; but we henceforth will laugh at chance in procreation, and such a brain that is to think transcendently will be a thinker's own creation." This feat is hoped to be achieved through a proper combination of material components: "Through mixture - for on mixture all depends - Man's substance gently be consolidated" (Johann Wolfgang Goethe, *Faust* [1808-1832], Part 2, Act 2, Scene 2, ln. 6809-6810, 6867- 6870, 6850-6851, translated by George Madison Priest, Franklin Center, PA, The Franklin Library, 1979, http://einam.com/faust/index.html). Another famous (negative) literary treatment of the theme is Mary Wollstonecraft (Godwin) Shelley's *Frankenstein; or, The Modern Prometheus* (1818, http://www.gutenberg.org/files/84/84-h/84-h.htm). Finot does review the earlier lore on synthesizing "homunculi" or building human-like "automata": from the myth of Prometheus, through legends of the Golem, the "androids" allegedly constructed by Albertus Magnus and René Descartes, Paracelsus' alchemical theories, stories of "conjuring" homunculi by Count Johann Ferdinand Kueffstein in 1775 (as recorded by the Count's servant Joseph Kammerer, then published in the Freemasons' Almanac, *Die Sphinx*, in Vienna in 1873, and later retold in Somerset Maugham's novel *The Magician* in 1908) and others, Baron Wolfgang von Kempelen's mechanical chess-player (first advertised in 1769 and later proved to conceal a man), and more.

However, the novel element in Finot's teachings seems to be the assertion that the quest to create a "homunculus" may transcend the realm of legends, scary literary fantasies, occult sciences and curiosity chambers, and may gain in feasibility due to the progress of modern biology.

According to Finot, the inspiration and hope come, first of all, from the works on "plastidules" or "fine granulations linked together by very slender filaments" (what we may today call "micro-organelles") as the "first basis of life" or life's "elementary" components – based on the works of Ferdinand Julius Cohn, Thomas Huxley, Otto Bütschli, Eduard Strasburger, Georg Wetzel, Carl Heitzmann, Ernst Haeckel, Claude Bernard, Karl Baer, August Weismann, Charles Darwin and others. Some plastidules were believed to be immortal, and their composition appeared to be subject to manipulation.

The works on "parthenogenesis" or creating "living cells by the help of unfecundated eggs" – by Jacques Loeb, Edmund Beecher Wilson, Thomas Morgan, Martin Fischer, Eugène Bataillon, Yves Delage, Alfred Mathieu Giard, Louis-Félix Henneguy and others – further strengthened the assurance that life can be purposively manipulated through chemistry and physics.

"Organic synthesis" – as performed by Marcellin Berthelot, Justus Liebig, Adolphe Würtz, Leon Lilienfeld, William Perkin, Paul Schützenberger, Paul Sabatier, et al. – reinforced Finot's optimism even more. Indeed, Finot wondered, "How does animal chemistry produce fatty or albumenoid bodies? How indeed! We know nothing, and we shall know nothing for many years" (p. 266). Yet despite the current limitations of knowledge, the progress of science is presumed to be limitless:

"The triumphs of the human brain are unlimited. It would thus be as unjust to attempt to fix bounds for the evolution of chemical synthesis, as it would be bold to assign in advance any limit to physical discoveries" (p. 267).

Finally, the construction of human-like automata or "simulacra of living beings," represents another line of research into the "artificial creation of life." Finot valorized "the artificial creation of living matter" over "making miraculous automata," valorized biology over mechanics. The fascination with "mechanical" models is, according to him, the lot of "simpler," less educated people, and the creation of life directly from inert matter appeared to him less promising than manipulating biological "plastidules" that already exist. Yet, according to Finot, the creation of such automata is a powerful line of advancement.

The conclusion Finot draws – possibly for the first time expressed so candidly, succinctly and enthusiastically – is that "modern science, without giving [the dream of artificial creation] more importance than it deserves, yet slowly travels towards its solution. It is still for ever preoccupied with Homunculi, although it never speaks of them" (Jean Finot, *The Philosophy of Long Life*, 1909 (1900), Ch. VII "Life as an Artificial Creation," pp. 249-279).

[143] The term "transhumanism" has been commonly attributed to Julian Huxley's work of 1957 (Julian Huxley, "Transhumanism," in *New Bottles for New Wine*, Chatto & Windus, London, 1957, pp. 13-17, http://www.transhumanism.org/index.php/WTA/more/huxley; Nick Bostrom, "A History of Transhumanist Thought," *Journal of Evolution and Technology*, 14(1), 1-30, 2005, http://jetpress.org/volume14/bostrom.html).

And the term "transhuman" is even older, appearing, for example, in Robinson Jeffers' poem "New Year's Dawn" (1947) which speaks of "transhuman beauty" alongside "inhumanism" and "inhuman beauty" in "The Double Axe. Part II. The Inhumanist" – 1942-1947 (Robinson Jeffers, *The Double Axe and Other Poems*, Random House, New York, 1948, pp. 142, 81).

Yet, the coherent "transhumanist" intellectual movement took shape only in the later part of the 20th century (in fact promoting views quite the opposite to Jeffers' "inhumanist" ecologism).

Prominent thinkers of the modern transhumanist intellectual movement include:

Fereydun M. Esfandiary (*Up-Wingers. A Futurist Manifesto*, 1973), Robert Ettinger (*Man into Superman*, 1972), Eric Drexler (*Engines of Creation: The Coming Era of Nanotechnology*, 1986), Ray Kurzweil (*The Singularity Is Near. When Humans Transcend Biology*, 2005), such philosophers as Max More, Tom Morrow, and Natasha Vita-

More (founding the Extropy Institute in 1988), Nick Bostrom, David Pearce, and James Hughes (founding the World Transhumanist Association in 1998), John Harris, Vernor Vinge, Ramez Naam, Giulio Prisco, Ronald Bailey and others.

Many contemporary transhumanist scientists believe that sentient/immortal life will result from research into Artificial Intelligence (Marvin Minsky, Hans Moravec, Ray Kurzweil, Eliezer Yudkowsky, Ben Goertzel and others). The Singularity Institute for Artificial Intelligence, founded in 2000, has investigated this possibility.

Other scientists hope to manipulate / enhance / create / immortalize living matter through nanotechnology (including Robert Freitas, Ralph Merkle, and Christine Peterson – leaders of the Foresight Institute, founded by Eric Drexler in 1986).

Research in biotechnology "proper" is yet another of the converging roads (the so-called Nano-Bio-Info-Cogno technologies or NBIC convergence) that will, according to transhumanist philosophy, ultimately lead to human enhancement, life extension and the artificial creation of life.

(The term "NBIC" was introduced in *Converging Technologies for Improving Human Performance: Nanotechnology, Biotechnology, Information Technology and Cognitive Science, NSF/DOC-sponsored report*, Edited by Mihail C. Roco and William Sims Bainbridge, National Science Foundation, Kluwer Academic Publishers, Dordrecht, Netherlands, 2003.)

Transhumanism is defined primarily as "The intellectual and cultural movement that affirms the possibility and desirability of fundamentally improving the human condition through applied reason, especially by developing and making widely available technologies to eliminate aging and to greatly enhance human intellectual, physical, and psychological capacities" (Nick Bostrom, "The Transhumanist FAQ. A General Introduction," published by the World Transhumanist Association (currently Humanity Plus) in 2003, http://humanityplus.org/learn/transhumanist-faq/).

Many contemporary biologists advocating radical life-extension (such as Aubrey de Grey) are sympathetic to transhumanism, and the vast majority of transhumanists are ardent advocates of radical life-extension. One can even argue that the radical prolongation of healthy human life is the single most important objective of transhumanism. Transhumanism is now perhaps the only intellectual movement that openly supports this goal. After life-extension, the artificial creation of life (either through biotechnology, nanotechnology or artificial intelligence) is the next item on the agenda.

In juxtaposing Finot's teachings (of 1900) with those of present-day transhumanists, I would like to suggest that Jean Finot may be considered a true pioneer of transhumanism, expressing concerns and aspirations generic to transhumanist philosophy, more than half a century before the term "transhumanism" emerged, and almost a century before "transhumanism" formed into a recognizable intellectual movement.

Finot seems to have been well acquainted with contemporary scientific trends and based his optimistic forecasts on these trends. And this is the argumentative strategy that many contemporary transhumanists employ. Consider, for example, the evolutionary proximity of Finot's areas of interest with those of modern transhumanists: "organic synthesis" vs. "nano-technology," "automata" vs. "artificial intelligence," manipulation of "plastidules" vs. "biotechnology."

Another tenet that Finot emphasized was "transformation," to the grotesque points when he affirmed that human life does not end in the grave, but transforms itself into the life of worms, or when he foresaw a future divide between "homunculi" and "man-monkeys" (p. 275). And "transition" and "transformation" are in the very definition of "Transhumanism."

Thus, Jean Finot, who seems to have been almost entirely forgotten by scholars (even in transhumanist circles), deserves recognition, both as an author of an original, consistent life-extensionist philosophy, and as a major *fin-de-siècle* precursor of present-day transhumanism.

(Ilia Stambler, "Life extension – a conservative enterprise? Some fin-de-siècle and early twentieth-century precursors of transhumanism," *Journal of Evolution and Technology*, 21(1), 13-26, 2010, http://jetpress.org/v21/stambler.htm.)

[144] Albert Dastre, *La Vie et la Mort* (Life and Death), Flammarion, Paris 1907 (first published in 1903), Livre 1, "En Marge de la science. Les doctrines générales sur la vie et la mort. Leurs transformations successives" (Book 1. The margins of science. General doctrines of life and death. Their successive transformations), pp. 1-51.

Dastre's beliefs were deeply rooted in scientific materialism, or rather "Unicism."

Dastre reviews recent animist views of Paul-Émile Chauffard, Alexander von Bunge and Georg Eduard von Rindfleisch, and their antecedents, Aristotle, Saint Thomas and Georg Ernst Stahl, positing the existence of an immortal and reasonable soul.

He considers the neo-vitalist views of Rudolf Heidenhain, Armand Gautier and Johannes Reinke, and the "old" vitalists, Paracelsus, Jan Baptist van Helmont, Paul Joseph Barthez, Theophile Bordeu, Georges Cuvier and Marie François Xavier Bichat, for whom the "vital force" was a kind of a "soul of a second, inferior degree of majesty."

And finally, he presents the "materialist" views of René Descartes, Giovanni Alfonso Borelli, Archibald Pitcairne, Stephen Hales, Daniel Bernouilli, Franciscus Sylvius le Boë, Herman Boerhaave, Antoine Lavoisier, Thomas Willis, Charles Darwin and Ernst Haeckel.

Dastre uses the terms "materialism," "mechanism," "unicism" and "monism" quite interchangeably, and subsumes them all under the heading of "physico-chemical doctrine of life." "Materialistic unicism" is the preferred term, which implies the identification of life with physical and chemical forces.

"Materialistic Unicism" recognizes that the operation of living beings is different from inanimate matter. For example, the diffusion of substances through a living membrane is different from the diffusion through an inanimate membrane. This, however, does not require some special "vital force," or involvement of a "thinking soul," only a different, more complex "arrangement" of physicochemical forces.

Dastre further equates between "reduction," "materialism" and the "physicochemical doctrine" (Dastre, pp. 11, 27). He clearly sides with these doctrines and discards any ability of a "thought," "soul" or "vital force," irreducible to physicochemical phenomena, to exercise any influence on life-processes. (According to his testimony, this was the majority position among the biologists of his time.)

A chief reason why Dastre discards the "vital force" as a valid concept in biological research generally, and in relation to life-extension in particular, is that the "physicochemical mechanisms" are subject to human intervention: matter can be added or removed. The vital force is, in contrast, "capricious," "a willful genie," it cannot be measured or even adequately described, let alone manipulated.

[145] A life-extensionist of Metchnikovian persuasion, Dastre believed that aging and death are the result of the disharmony of human nature, of imbalanced internal and external environment. This disharmony, however, can be repaired based on increasing biological knowledge.

He reaffirmed Metchnikoff's view that human beings need to strive to a limit of life-fulfillment, to a natural death "after a long and healthy life, exempt from mortal accidents, a natural and desirable event, a fulfilled need," and emphasized how yet remote we are from achieving this goal.

In some respects, however, Dastre was more far-reaching and persistent than Metchnikoff. According to Dastre, even though the attainment of the phase of life-fulfillment may dispel the fears of death for the individual who reaches that stage, death will still remain a tragedy, it will be a tragedy for the close ones whom the person leaves behind.

Moreover, Dastre perceived no 'written in stone' law that there should be a life limit at all. People may "remain forever in full health and guarded from disease," he wrote, as "science will vanquish the evil of disease." Not only disease, "death, 'the last enemy to be vanquished,' to use to St. Paul's expression, will be defeated by the power of science."

(Albert Dastre, *La Vie et la Mort* (Life and Death), 1907 (1903), pp. 337, 348-349.)

[146] Sergey Metalnikov, *Problema Bessmertia i Omolozhenia v Sovremennoy Biologii* (The Problem of Immortality and Rejuvenation in Modern Biology), Slovo, Berlin, 1924 (1917).

[147] Claude Bernard, *Leçons sur les Phénomènes de la Vie Communs aux Animaux et aux Végétaux* (Lectures on the phenomena of communal life in animals and plants), J.-B. Baillière, Paris, 1878, edited by Albert Dastre, vol. 1, p. 113, subsection "La vie constante" (the constant life).

[148] Benno Romeis, *Altern und Verjüngung. Eine Kritische Darstellung der Endokrinen "Verjüngungsmethoden", Ihrer Theoretischen Grundlagen und der Bisher Erzielten Erfolge*, Verlag von Curt Kabitzsch, Leipzig, 1931 (Aging and Rejuvenation. A Critical Presentation of Endocrine "Rejuvenation Methods," Their Theoretical Foundations and Up-to-Date Successes).

[149] Auguste Comte, *The Catechism of Positive Religion* (translated by Richard Congreve), John Chapman, London, 1858 (first published in French as *Catéchisme positiviste ou Sommaire exposition de la religion universelle*, Paris, 1852 – The Positivist Catechism or Summary Exposition of the Universal Religion), pp. 498-450.

Yet other points of Comte's teachings may have resonated strongly with Carrel. According to Comte, health improvement needs to derive from a moral imperative, and to serve society:

"As a matter of fact, the precepts of health can secure active obedience only when they are rested on moral grounds. This is true equally of the individual and the society. It is easily verified by the fruitlessness of the efforts made by our physicians in Western Europe to regulate common diet. They have been fruitless ever since the old religious precepts lost their hold.

Men will not generally submit to any practical inconvenience on the ground of their own mere personal health, – each one is left, on this ground, to judge for himself. And we are often more sensible of actual and certain annoyance than of distant and doubtful advantages.

We must call in an authority superior to all individual judgment, to be able to prescribe, even in unimportant points, rules which shall have any real efficacy. Such rules will then rest on a view of the needs of society, which shall admit of no hesitation as to obedience."

(Comte, *The Catechism of Positive Religion*, 1858 (1852), p. 50.)

Carrel would fully subscribe to these notions of the "common good" and collective moral imperative.

[150] Alexis Carrel, "Le Rôle Futur de la Médecine" (The future role of medicine), in *Médecine Officielle et Médecines Hérétiques* (Official Medicine and Heretical Medicines), Plon, Paris, 1945, pp. 1-9.

[151] Auguste Lumière, *Sénilité et Rajeunissement* (Aging and Rejuvenation), Librairie J.-B. Baillière et Fils, Paris, 1932.

[152] Alexis Carrel, *Man, The Unknown*, Burns & Oates, London, 1961 (1935).

[153] Bruno Salazard, Christophe Desouches, and Guy Magalon, "Auguste and Louis Lumière, inventors at the service of the suffering," *European Journal of Plastic Surgery*, 28(7), 441-447, 2006.

[154] Auguste Lumière, *La Renaissance de la Médecine Humorale*, Deuxie edition (The Renaissance of Humoral Medicine, second edition), Imprimerie Léon Sézanne, Lyon, 1937 (previously published in 1935), pp. 218-219.

[155] There is in Lyon an organization named the "Lumière Institute" even now, yet it serves exclusively as a "Living Museum of the Cinema" (http://www.institut-lumiere.org/; http://www.institut-lumiere.org/english/frames.html).

[156] Auguste Lumière, *Sénilité et Rajeunissement* (Aging and Rejuvenation), Librairie J.-B. Baillière et Fils, Paris, 1932, p. 137.

[157] Auguste Lumière, *Les Horizons de la Médecine* (The horizons of medicine), Albin Michel, Paris, 1937, p. 215.

[158] *Le Petit Comtois*, Novembre 15, 1940, http://www.angelfire.com/biz2/rlf69/CR/lumiere.html.

[159] *Pour la Memoire: Contre rapport. Les dix affaires qui ébranlèrent le monde universitaire lyonnais (1978-1999)* (For the memory: Counter-report. The ten affairs that shook the academic world of Lyon, 1978-1999), issued by

Cercle Marc Bloch, Golias, Ras l'Front, Sos-Racisme, Lyon, December 6, 1999, Ch. VII, "Affaire Lumière" (the Lumière Affair, http://www.angelfire.com/biz2/rlf69/CR/lumiere.html).

[160] Auguste Lumière, *Les Horizons de la Médecine* (The horizons of medicine), Albin Michel, Paris, 1937, Ch. 5 "Les entraves a l'essor de la médecine resultant de l'ostracisme qui frappe les novateurs" (The barriers to the progress of medicine as a result of ostracism striking innovators), pp. 234-246.

[161] Gerald J. Gruman, *A History of Ideas about the Prolongation of Life*, 1966, "The Alchemists," pp. 49-67; Roy Porter, *The Greatest Benefit to Mankind: A Medical History of Humanity*, W.W. Norton & Company, NY, 1998 (1997), Ch. 3 "Antiquity," Ch. 5 "The Medieval West," Ch. 6 "Indian Medicine," Ch. 7 "Chinese Medicine"; Noga Arikha, *Passions and Tempers: A History of the Humours*, Ecco, New York, 2007.

[162] According to Lumière, humoralism became as good as extinct by the mid 19th century. He fully agreed with the Swiss physician Sigismond Jaccoud (1830-1913) who, in the *Ancient and Modern Humoralism* (*De l'humorisme ancien compare à l'humorisme moderne*, 1863), claimed that the modern offshoot of humoralism consisted almost entirely in the collection of data on the chemical composition of blood, without any unifying theory.

While "ancient Humoralism," Lumière affirmed, was "a collection of hypotheses without facts," modern humoralism became reduced to a "collection of facts without hypotheses." And even this "collection of facts" on the composition of blood became undermined by the prevalence of the "solidist" paradigm.

(Auguste Lumière, *La Renaissance de la Médecine Humorale*, Deuxie edition (The Renaissance of Humoral Medicine), Imprimerie Léon Sézanne, Lyon, 1937, pp. 23-25.)

Lumière may have exaggerated the lack of interest in body fluids in contemporary biomedicine and too quick to eulogize the demise of humoralism. In fact, by the end of the 19th century, the dispute raged in immunology between the "humoralist" school attributing immune properties to blood "serum" (the approach represented by Robert Koch) versus the "cytological" school (represented by Elie Metchnikoff) that perceived cells as the main agents of immunity.

(Alfred I. Tauber, "Metchnikoff and the phagocytosis theory," *Nature Reviews: Molecular Cell Biology*, 4, 897-301, 2003; Alfred I. Tauber, *The Immune Self: Theory or Metaphor?* Cambridge University Press, Cambridge, 1994.)

Nonetheless, apparently by the time of Lumière's writing, the "cytological-solidist" approach completely won out. Lumière cited the authoritative and comprehensive 22-volume *Nouveau Traités de Médecine* (New Medical Treatises, 1924, edited by Georges-Henri Roger, Fernand Widal and Pierre Joseph Teissier) that did not include a single chapter on "humors."

The main blame for the "exclusive triumph of solidism" and "the complete abandonment of humoralism" chiefly resided with the "reign of anatomical pathology" and the "advent of microbiology, under the influence of certain authors such as [Rudolph Carl] Virchow [1821-1902] in Germany and [Ferdinand Ritter] Hebra [1816-1880] in Austria."

(Auguste Lumière, *La Renaissance de la Médecine Humorale*, Deuxie edition (The Renaissance of Humoral Medicine), Imprimerie Léon Sézanne, Lyon, 1937, pp. 27-28.)

[163] Auguste Lumière, *La Vie, La Maladie et La Mort, Phénomènes Colloïdaux* (Life, Disease and Death as Colloidal Phenomena), Masson & Cie, Libraires de L'Académie de Médecine, Paris, 1928. This work appeared in German translation by Otto Einstein, *Leben, Krankheit und Tod als Kolloid Erscheinungen*, Franckh'sche Verlagshandlung, Stuttgart, 1931.

The subject was reiterated in *La Renaissance de la Médecine Humorale*, Deuxie edition (The Renaissance of Humoral Medicine), Imprimerie Léon Sézanne, Lyon, 1937 (1935); and in "La Médecine Humorale et ses Résultats" (Humoral Medicine and its Results), in *Médecine Officielle et Médecines Hérétiques* (Official Medicine and Heretical Medicines), "Présences" Plon, Paris, 1945, pp. 53-62.

[164] Lumière lamented that, however important the biological colloids are, there is an almost complete ignorance of their molecular structure. (Auguste Lumière, *La Vie, La Maladie et La Mort, Phénomènes Colloïdaux* (Life, Disease and Death as Colloidal Phenomena), 1928, pp. 63-71.)
He considered some surmises regarding "cyclization" as an underlying chemical cause of colloid (protein) condensation, and the hypothesis that "open" aminoacid chains represent the active forms of proteins that are needed for growth, whereas the "closed chains" are chemically inert and only needed for body "maintenance" – the hypothesis advanced by Georges Bohn in France, Amé Pictet in Switzerland, Thomas Burr Osborne and Lafayette Benedict Mendel in the US.
Based on that theory, by diminishing the dietary amount of "open" amino-acid chains, such as lysine, the growth of animals could be arrested and their life-span prolonged.
However, Lumière did not believe that "cyclization" corresponds either to flocculation or precipitation, because proteins in a healthy and stable "colloidal" state are inert to coloring reagents, while the "dead" flocculates can easily react with the coloring substances. In contrast, in "cyclization," the inert, "closed-chain" proteins lose their chemical reactivity for the coloring.
Lumière thought that flocculation is just a "mechanical reunion of granules" and that the "closed" and "open" protein chains are just the result of random "amassment" of amino-acids, or even that the chains are imaginary constructs altogether, and that "nothing authorizes us to suppose that, in an acyclic body, the constituent atoms of a molecule attach to each other to form a string."
(Auguste Lumière, *Sénilité et Rajeunissement* (Aging and Rejuvenation), Librairie J.-B. Baillière et Fils, Paris, 1932, pp. 111-114.)
Only with the advent of molecular biology in the late 1940s-early 1950s, thanks to the research of William Lawrence Bragg, John Kendrew, Max Perutz (1950), Linus Pauling, Robert Corey (1950, 1951) and others, the precise molecular structure of proteins became known, including the exact laws of protein polypeptide chain formation.
Yet, in fact, the kind of research conducted by Lumière has continued to the present (even though the contributions of Lumière and his predecessors and contemporaries are hardly ever acknowledged).
Thus, the Hematoxylin and Eosin (H&E) staining has been used since the late 19th century, through Lumière to the present, to detect amyloid aggregates (protein/colloidal flocculates in Lumière's terms), which have been implicated in a great number of degenerative diseases, from arteriosclerosis to Alzheimer's. Some recent discoveries in the field of protein/colloidal "stabilization" are very much in line with Lumière's legacy (e.g. Silvestre Alavez, et al., "Amyloid-binding compounds maintain protein homeostasis during ageing and extend lifespan," *Nature*, 472, 226-229, 2011).
[165] Georges Marinesco, *Mécanisme colloidal de la sénilité* (Colloidal mechanisms of aging), Officina tipografica Giannotta, Catania, 1913; Georges Marinesco, "Méchanisme chimico-colloidal de la sénilité et le problème de la mort naturelle" (Chemical-colloidal mechanisms of aging and the problem of natural death), *Revue Scientifique de la France et de l'étranger*, Paris, 1, 673-679, 1914; Vladislav Růžička, "Beitrage zum Studium der Protoplasmahysteresis und der hysterischen Vorgänge (Zur Kausalität des Alterns) I. Die Protoplasmahysteresis als Entropieerscheinung" (Contributions to the study of protoplasm hysteresis and hysteretic processes as a cause of aging. I. Protoplasm hysteresis as an entropic phenomenon), *Archiv für Mikroskopische Anatomie und Entwicklungsmechanik*, Bonn, 101, 459-482, 1924.
[166] The major efforts of the Lumière Institute were directed to combating such diseases as asthma, tuberculosis, cancer and arteriosclerosis, but also a host of other infectious and chronic diseases. The self-reported rate of therapy success was impressive. As Lumière reported in 1945, since the inception of the Institute in 1930, 65% of "cures," 21% of "great amelioration," and 8.5% of "mild amelioration" were effected in about 12,000 patients.
He did not, however, specify the diseases or any patient demographics. Presumably, the numbers referred to cases of "latent chronic infections, chronic intoxications and auto-intoxications, organic and endocrine

dysfunctions, anaphylactic states, bacterial impregnations, stases [blood flow obstructions], etc." Thus, "over 50,000 years of suffering have been eliminated."

(Auguste Lumière, "La Médecine Humorale et ses Résultats" (Humoral Medicine and its Results), in *Médecine Officielle et Médecines Hérétiques* (Official Medicine and Heretical Medicines), "Présences" Plon, Paris, 1945, pp. 53-62.)

[167] According to Lumière's studies, Magnesium Hyposulfite (MgS_2O_3) exerts a triple role:

First, it "modifies the form of precipitations, rendering them completely inoffensive when introduced into the blood stream." Secondly, it "dissolves protein flocculates" that clog the cell operation. And finally, it "attenuates [anesthetizes] the sensitivity of endovascular sympathetic nerve terminals."

Desensitization (or "protection from a shock by a smaller shock") was a key concept in Lumière's theory. Anaphylactic (allergic) shock served as the model. By introducing insoluble "sediment" particles in "sub-threshold" concentrations, it appeared possible to "accustom" the organism to the substance and thus make it resistant to much higher concentrations that would otherwise produce shock and pathology.

Since the "precipitates" and "flocculates" could essentially be any indeterminate small blobs of matter that mechanically "press on vascular sympathetic nerve terminals" (thus producing painful shock and "deregulation" of blood-supply), by introducing one kind of precipitate in small doses it appeared possible to desensitize the body to large toxic doses and also to other toxic substances.

This theory essentially recapitulated the first principle of homeopathy ("like cures like"), and of body "tempering" or "hardening" by exposures to moderately stressful conditions. But Lumière attributed its first scientific elaboration and application to Metchnikoff's pupils, Jules Bordet and Alexandre Besredka at the turn of the 20th century. He saw himself just as a person who empirically tests and develops the tradition and points the direction for further research.

According to Lumière, two processes can preserve the organism against shock: first, "the accustoming of vascular endothelium to excitation by any precipitate"; and second, the "desensitization through a specific antigen."

In true anaphylactic shock, the precipitates (agglomerates) are produced by specific antibody-antigen reactions, and, in this case, mechanical "accustoming" and antigenic "desensitization" are simultaneous, in fact, synonymous processes.

Thus, in Lumière holistic vision, different diseases could produce similar symptoms, due to a general "humoral imbalance," and, accordingly, a single ("desensitizing") substance could be used to treat several diseases.

(Auguste Lumière, *Les Horizons de la Médecine*, 1937, pp. 42-47; Auguste Lumière, *La Vie, La Maladie et La Mort, Phénomènes Colloïdaux*, 1928, pp. 115-171.)

[168] Arthur Lorber and Timothy M. Simon, "Applications and Implications of Gold Therapy," *Gold Bulletin*, 12 (4), 149-158, 1979; Alan K. Matsumoto, et al., "Rheumatoid Arthritis Treatment. Intramuscular gold," The Johns Hopkins Arthritis Center, 2010, http://www.hopkins-arthritis.org/arthritis-info/rheumatoid-arthritis/rheum_treat.html; Loo C, et al., "Nanoshell-enabled photonics-based imaging and therapy of cancer," *Technology in Cancer Research and Treatment*, 3 (1), 33-40, 2004.

[169] Auguste Lumière, *La Renaissance de la Médecine Humorale* (2nd edition of 1937), pp. 245-258; Auguste Lumière, *Le Horizons de la Médecine* (1937), pp. 178-189.

[170] Auguste Lumière, *Sénilité et Rajeunissement* (Aging and Rejuvenation), Librairie J.-B. Baillière et Fils, Paris, 1932, p. 139.

Hippocrates' prescription of "exertion, food, drink, sleep, sexual activity, in moderation" appears in *Epidemics* (Book 6, Section 6.2), in *Hippocrates*, Volume 7, Edited and Translated by Wesley D. Smith, Loeb Classical Library, Cambridge, Massachusetts, 1994, p. 263.

[171] During strenuous physical exertion or excesses in nutrition, the turbulence of the humors is said to increase, facilitating the colloids' "flocculation" and "precipitation." In counteraction, rest and sleep stabilize

the colloids and allow the "leukocytes a sufficient time to purge the blood, through digestion or diapedesis [cell migration], from the precipitates produced by the exercise."

Moderate physical and mental exertions produce a "desensitizing" influence or "train" the organism against higher exertions, similarly to the "desensitizing" chemicals, thus maintaining the "vital equilibrium" (Auguste Lumière, *Sénilité et Rajeunissement*, 1932, pp. 133-141).

[172] Continuous (and moderate) physical and mental activity was considered by Lumière to be vital for longevity, and was related to a generally optimistic, motivated worldview, continued interest in life and social contribution of the aged. Lumière showcased the importance of the psychological or motivational factors by his own example: his creative scientific activities flourished quite late in life (in his 60s and 70s), after he was sufficiently well established and educated to pursue them. (Auguste Lumière, *Sénilité et Rajeunissement*, 1932, pp. 133-141.)

[173] Auguste Lumière, *Sénilité et Rajeunissement* (Aging and Rejuvenation), Librairie J.-B. Baillière et Fils, Paris, 1932, pp. 150-152.

[174] Auguste Lumière, *Sénilité et Rajeunissement* (Aging and Rejuvenation), Librairie J.-B. Baillière et Fils, Paris, 1932, pp. 141-142.

[175] Lumière asserted the possibility of "rejuvenation" of colloids:

"It is already possible, though only in a little measure, to realize this condition [of colloidal rejuvenation] by means of several procedures that employ the same principle: the impregnation of the organism with substances excreted by sex glands. These are the methods of Brown-Séquard, Voronoff, Steinach and Doppler. The ability of the glandular products to increase the power of growth, accompanied by a certain degree of rejuvenation, is very real and proved beyond any doubt."

(Auguste Lumière, *Sénilité et Rajeunissement* (Aging and Rejuvenation), Librairie J.-B. Baillière et Fils, Paris, 1932, pp. 150-152.)

Lumière expressed similar assertions earlier, in *Life, Disease and Death as Colloidal Phenomena* (1928):

"Can we hope to increase the duration of human life and how can we achieve this? *A priori*, it seems that two ways are open to us to attain that goal: a) The hindering of the evolution of colloids toward flocculation. b) The stimulation of cell proliferation. Regarding the first condition, presently, we have no means to intervene in the normal process of colloidal dying, we can accelerate the process, but not inhibit it. We are not so completely disarmed with regard to the second condition, and the study of the rhythm of growth allows us to explain some, still precarious, facts of rejuvenation."

(Auguste Lumière, *La Vie, La Maladie et La Mort, Phénomènes Colloïdaux* (Life, Disease and Death as Colloidal Phenomena), Masson & Cie, Libraires de L'Académie de Médecine, Paris, 1928, pp. 95-96.)

This work appeared in German translation by Otto Einstein, *Leben, Krankheit und Tod als Kolloid Erscheinungen* (Franckh'sche Verlagshandlung, Stuttgart, 1931). The German edition seems to have mistranslated the latter sentence (p. 55). While the original says, "Nous ne sommes pas aussi complètement désarmés vis-à-vis de la seconde condition..." (We are not so completely disarmed with regard to the second condition), it was translated "Auch gegenüber der zweiten Forderung sind wir ziemlich hilflos" (Also against the second challenge we are quite helpless"). The idea of the possibility of rejuvenation may have seemed too farfetched to the German translator.

It should also be noted that Lumière's interest in aging and rejuvenation does not seem to be mentioned at all in his biographies, not even in the very few works that mention his interest in medicine, e.g. Bruno Salazard, Christophe Desouches, and Guy Magalon, "Auguste and Louis Lumière, inventors at the service of the suffering," *European Journal of Plastic Surgery*, 28(7), 441-447, 2006.

[176] Auguste Lumière, *La Vie, La Maladie et La Mort, Phénomènes Colloïdaux* (Life, Disease and Death as Colloidal Phenomena), Masson & Cie, Libraires de L'Académie de Médecine, Paris, 1928, pp. 73-115.

[177] Serge Voronoff, *The Sources of Life*, Bruce Humphries, Boston, 1943, pp. 17-29, 36; Serge Voronoff, *The Conquest of Life*, translated by G. Gibier Rambaud, Brentano's Ltd, London, 1928, pp. 121-145; Serge

Voronoff, *Rejuvenation by Grafting*, translated by Fred F. Imianitoff, George Allen and Unwin Ltd, London, 1925, pp. 57-68.

[178] Auguste Lumière, *Sénilité et Rajeunissement* (Aging and Rejuvenation), Librairie J.-B. Baillière et Fils, Paris, 1932, pp. 94-97.

[179] During his work on sex hormone replacement, Selye observed that a wide variety of dissimilar toxins or stimulants ("diverse nocuous agents") produce the same response in the body (e.g. the same changes in the adrenal, thymico-lymphatic and intestinal systems). And the three stages of the organism's reaction were the "alarm," "resistance" and "exhaustion."

Selye described his discovery of the "non-specific response" as a profound epiphany and as his groundbreaking contribution to medical science, running counter to the entrenched "specialist" paradigm in medicine where each disease was said to be caused by a specific agent and manifested by specific symptoms. Yet, Lumière spoke of the vital importance of "non-specific" responses in therapy earlier:

"This conception [of humoral destabilization through flocculation of colloids] allows us to understand medical mysteries that would remain impenetrable without it. This notion explains to us why a single agent can cause diverse afflictions; why a multiplicity of essentially different agents can generate the same disease; why a single remedy can cure multiple distinct afflictions; why many completely different medications can cure the same syndrome; and why the major symptoms of acute maladies present a remarkable similarity. This is because all these phenomena share one common factor – the flocculate."

(Auguste Lumière, *La Renaissance de la Médecine Humorale* (1935, 2nd edition 1937), p. 267; Hans Selye, *The Stress of Life*, McGraw-Hill, NY, 1956, pp. 14-43.)

[180] Auguste Lumière, *Les Horizons de la Médecine* (1937), p. 60.

[181] Auguste Lumière, *La Vie, La Maladie et La Mort, Phénomènes Colloïdaux* (Life, Disease and Death as Colloidal Phenomena), Masson & Cie, Libraires de L'Académie de Médecine, Paris, 1928, p. 66-67.

[182] Auguste Lumière, *Sénilité et Rajeunissement* (Aging and Rejuvenation), Librairie J.-B. Baillière et Fils, Paris, 1932, p. 91.

[183] Nobelprize.org – The Official Web Site of the Nobel Prize, http://nobelprize.org/nobel_prizes/medicine/laureates/1912/carrel.html.

[184] Carrel's biographies include:
Robert Soupault, *Alexis Carrel. 1873-1944*, Librairie Plon, Paris, 1952; William Edwards and Peter Edwards, *Alexis Carrel: Visionary Surgeon*, Charles C. Thomas, Springfield IL, 1974; David M. Friedman, *The Immortalists. Charles Lindbergh, Dr. Alexis Carrel and Their Daring Quest to Live Forever*, HarperCollins, NY, 2007; Etienne Lepicard, *A Bio-Medical Response to the Crisis of the 1930s: the Construction of Man, The Unknown by Alexis Carrel, 1935*, PhD dissertation, Hebrew University, Jerusalem, 2000.

[185] Alexis Carrel, "Results of transplantation of the blood vessels, organs, and limbs," *Journal of the American Medical Association*, 51, 1662-1667, 1908; Alexis Carrel, "Further Studies on Transplantation of Vessels and Organs," *Proceedings of the American Philosophical Society*, 47(190), 677-696, Nov. 1908, http://www.jstor.org/stable/983837.

[186] Alexis Carrel, "On the permanent life of tissues outside of the organism," *Journal of Experimental Medicine*, 15, 516-528, 1912.

[187] Until the 1930s, the antagonistic relations or disequilibrium of differentiated somatic tissues were seen as the chief culprits of senescence, but senescence was not considered to be inherent in the tissue cells *per se*.

This conviction was greatly reinforced by Carrel's finding of the cultured cells' "potential immortality." This tenet, however, was disputed in the 1960s, with the publication of "Hayflick's limit," positing that normal somatic cells can only undergo a limited number of divisions, and thus inescapably age and die.

Carrel's findings of cells' apparent immortality were explained by Leonard Hayflick as due to the introduction of fresh cells into the culture, perhaps even intentionally, as alleged by the historian of biology Jan Anthony Witkowski.

Yet, Hayflick too well recognized the reality of "cellular immortality" of some cell types and extensively worked with immortal "tumor-like" cell lines (such as HeLa). The equipment from Hayflick's lab, and even Hayflick's own cells, were used in the 1990s-2000s in Michael West's experiments on the "immortalization" of somatic cells using the telomerase mechanism.

Thus, in the course of the century, the concept of "potential cell immortality" seems to have gone a full circle.

(Leonard Hayflick, "The limited *in vitro* lifetime of human diploid cell strains," *Experimental Cell Research*, 37, 614-636, 1965; Witkowski, JA, "Dr. Carrel's immortal cells," *Medical History*, 24 (2), 129-142, 1980; Michael D. West, *The Immortal Cell. One Scientist's Quest to Solve the Mystery of Human Aging*, Doubleday, NY, 2003, p. 121.)

[188] Alexis Carrel and Charles Lindbergh, *The Culture of Organs*, Paul B. Hoeber, NY, 1938; Alexis Carrel, *Man, The Unknown*, Burns & Oates, London, 1961 (1935); William Edwards and Peter Edwards, *Alexis Carrel: Visionary Surgeon*, Charles C. Thomas, Springfield IL, 1974; Robert Soupault, *Alexis Carrel: 1873-1944*, Librairie Plon, Paris, 1952; David M. Friedman, *The Immortalists. Charles Lindbergh, Dr. Alexis Carrel and Their Daring Quest to Live Forever*, HarperCollins, NY, 2007.

[189] Alexis Carrel and Charles Lindbergh, *The Culture of Organs*, Paul B. Hoeber, NY, 1938, pp. 219-221.

[190] "Science: Physical Immortality," *Time*, November 30, 1925; "Lindbergh seeks the secret of life," *Sunday Express*, July 11, 1937.

[191] Alexis Carrel and Charles Lindbergh, *The Culture of Organs*, Paul B. Hoeber, NY, 1938, p. 221.

[192] Cf. the presently existing "Mousery Database," http://mouserydatabase.com/. Data on Carrel's "Mousery" at the Rockefeller Institute can be found at the Rockefeller University Archives, NY http://www.rockarch.org/collections/ru/, http://www.rockefeller.edu/.

[193] Gerhard Venzmer, *Lang leben und jung bleiben!* (Live long and stay young), Franckh'sche Verlagshandlung, Stuttgart, 1941, pp. 18-19; Alexis Carrel, "Further Studies on Transplantation of Vessels and Organs," *Proceedings of the American Philosophical Society*, 47 (190), 677-696, Nov. 1908, pp. 689-690, http://www.jstor.org/stable/983837?seq=13.

In the latter article, Carrel discussed both the possibilities of inhibiting body metabolism by freezing (now this approach is termed "cryopreservation") and its complete temporary arrest by chemical poisons, such as calcium chloride (currently this approach is termed "chemopreservation").

[194] Cited in David M. Friedman, *The Immortalists. Charles Lindbergh, Dr. Alexis Carrel and Their Daring Quest to Live Forever*, HarperCollins, NY, 2007, p. 249.

[195] Legends abound to the present, regarding the yogic suspended animation, associated with the mystical state of "Samadhi" or extreme mental concentration – "the one-pointedness of mind," that allegedly enabled the adepts to survive adverse (e.g. airless) conditions or preserved them far beyond the normal human life-span. (Ernst Muldashev, *Ot kogo my proizoshli?* (Who did we descend from?), Argumenty I Fakty Print, Moscow, 2002, p. 96.)

[196] Lee MW, Deppe SA, Sipperly ME, Barrette RR, Thompson DR, "The efficacy of barbiturate coma in the management of uncontrolled intracranial hypertension following neurosurgical trauma," *Journal of Neurotrauma*, 11 (3), 325-331, 1994; Kees H. Polderman, "Application of therapeutic hypothermia in the ICU: opportunities and pitfalls of a promising treatment modality," *Intensive Care Medicine*, 30, 556-575, 2004.

[197] David M. Friedman, *The Immortalists. Charles Lindbergh, Dr. Alexis Carrel and Their Daring Quest to Live Forever*, HarperCollins, NY, 2007.

[198] Alexis Carrel, *Man, The Unknown*, Burns & Oates, London, 1961 (1935), "Inward time," pp. 130-153.

[199] Alexis Carrel, *Man, The Unknown*, 1935, p. 144. That was indeed the case in the US at that time. Though, in the US, the white men's life-expectancy at birth increased from about 47 in 1900 to about 60 by 1930, the life expectancy at age 50 remained exactly the same – about 71; by 2000 the life-expectancy rose to 75 and 78 for the ages of birth and 50, respectively. (*Human Mortality Database*, http://www.mortality.org.)

[200] Alexis Carrel, *Man, The Unknown*, Burns & Oates, London, 1961 (1935), p. 147.

[201] Carrel wrote:

"Owing to hygiene, athletics, alimentary restrictions, beauty parlours, and to the superficial activity engendered by telephone and automobile, all are more alert than in former times. At fifty, women are still young. Modern progress, however, has brought in its train counterfeit money as well as gold. When their visages, lifted and smoothed by the beauty surgeon, again become flabby, when massage no longer prevails against invading fat, those women whose appearance has been girlish for so many years look older than their grandmothers did at the same age.

The pseudo-young men, who play tennis and dance as at twenty years, who discard their old wife and marry a young woman, are liable to softening of the brain, and to diseases of the heart and the kidneys. Sometimes they die suddenly in their bed, in their office, on the golf-links, at an age when their ancestors were still tilling their land or managing their business with a firm hand."

(Alexis Carrel, *Man, The Unknown*, Burns & Oates, London, 1961 (1935), "Inward time," p. 144.)

"False youth" was apparently a common cultural concern in the US at the time, as evidenced by Noel Coward's song "The Tots," written in 1927 and revised in 1955, perplexing over the question "What's going to happen to the children when there aren't any more grownups? … Having been injected with some rather peculiar glands Darling mum's gone platinum and dances to all the rumba bands..."

[202] Alexis Carrel, *Man, The Unknown*, Burns & Oates, London, 1961 (1935), p. 143.

[203] Carrel exemplified the inherent disharmony in the human organism leading to death:

"In the course of life, the tissues undergo important alterations. … Such changes occur at various rates, according to the organs. Certain organs grow old more rapidly than others. But we do not know as yet the reason for this phenomenon. …"

"Abnormally vigorous organs in a senile organism are almost as harmful as senile organs in a young organism. The youthful functioning of any anatomical system, either sexual glands, digestive apparatus, or muscles, is very dangerous for old men. Obviously, the value of time is not the same for all tissues. This heterochronism shortens the duration of life. If excessive work is imposed on any part of the body, even in individuals whose tissues are isochronic, ageing is also accelerated. An organ which is submitted to over-activity, toxic influences, and abnormal stimulations, wears out more quickly than the others. And its premature senility brings on the death of the organism."

These disharmonies arose as evolutionary adaptations, and hence are extremely difficult to rectify:

"The length of life is conditioned by the very mechanisms that make man independent of the cosmic environment and give him his spatial mobility. … If the volume of the organic medium were much greater, and the elimination of waste products more complete, human life might last longer. But our body would be far larger, softer, less compact. It would resemble the gigantic prehistoric animals. We certainly would be deprived of the agility, the speed, and the skill that we now possess."

(Alexis Carrel, *Man, The Unknown*, Burns & Oates, London, 1961 (1935), "Inward time," pp. 141-142, 148.)

[204] Alexis Carrel, *Man, The Unknown*, Burns & Oates, London, 1961 (1935), pp. 147-148.

[205] Alexis Carrel, *Man, The Unknown*, Burns & Oates, London, 1961 (1935), pp. 144-145.

[206] Alexis Carrel, *Man, The Unknown*, Burns & Oates, London, 1961 (1935), pp. 50-51.

[207] The institutionalization proceeded from the inauguration of the world's first "Institute for the Study and Combat of Aging" in Romania in 1933 by Dimu Kotsovsky, to the organization of the world's first conference on Senility and Longevity in 1938 in the USSR by Alexander Bogomolets, to the foundation of the *Zeitschrift für Altersforschung* (Journal for Aging Research) in Germany in 1938 by Max Bürger, to the establishment of the "British Club for Research on Aging" in 1939 by Vladimir Korenchevsky, to the collaborative publication of *The Problems of Ageing* in the US in 1939 by Edmund Vincent Cowdry, to the formation of the "Fondation Française" by Carrel in 1941.

[208] Alexis Carrel, *Man, The Unknown*, Burns & Oates, London, 1961 (1935), pp. 172-186.

[209] Alexis Carrel, *Man, The Unknown*, 1961 (1935), pp. 178.
[210] Alexis Carrel, *Man, The Unknown*, 1961 (1935), pp. 177-181.
[211] Alexis Carrel, *Man, The Unknown*, 1961 (1935), p. 185.
[212] Alexis Carrel, *Man, The Unknown*, 1961 (1935), pp. 232-241.
[213] Alexis Carrel, *Man, The Unknown*, 1961 (1935), p. 234.
[214] Alexis Carrel, *Man, The Unknown*, 1961 (1935), p. 236.
[215] Alexis Carrel, *Man, The Unknown*, 1961 (1935), p. 234.
[216] Alexis Carrel, *Man, The Unknown*, 1961 (1935), pp. 145-146.
[217] Alexis Carrel, "Le Rôle Futur de la Médecine" (The future role of medicine), in *Médecine Officielle et Médecines Hérétiques* (Official Medicine and Heretical Medicines), Plon, Paris, 1945, pp. 1-9.
[218] Alexis Carrel, *Man, The Unknown*, 1961 (1935), pp. 117-123.
[219] Alexis Carrel, *Man, The Unknown*, 1961 (1935), pp.117-123.
[220] Alexis Carrel, *Man, The Unknown*, 1961 (1935), p. 139.
[221] Alexis Carrel, *Man, The Unknown*, 1961 (1935), p. 121.
[222] Alexis Carrel, *Man, The Unknown*, 1961 (1935), p. 122-123. Carrel's interest in faith healing was undoubtedly further encouraged by his wife, Anne de la Motte, whom he met in 1910 at his annual pilgrimage to Lourdes and married in 1913. This aristocratic woman of great religious devotion was rumored to have (and certainly was very interested in) psychic powers: dowsing skills, telepathy and clairvoyance.
[223] Carrel attested to the contemporary interest in miraculous healing:
"Miraculous cures seldom occur. Despite their small number, they prove the existence of organic and mental processes that we do not know. They show that certain mystic states, such as that of prayer, have definite effects. They are stubborn, irreducible facts, which must be taken into account. The author knows that miracles are as far from scientific orthodoxy as mysticity. The investigation of such phenomena is still more delicate than that of telepathy and clairvoyance. But science has to explore the entire field of reality."
"He has attempted to learn the characteristics of this mode of healing, as well as of the ordinary modes. He began this study in 1902, at a time when the documents were scarce, when it was difficult for a young doctor, and dangerous for his future career, to become interested in such a subject. Today, any physician can observe the patients brought to Lourdes, and examine the records kept in the Medical Bureau. Lourdes is the centre of an International Medical Association, composed of many members. There is a slowly growing literature about miraculous healing. Physicians are becoming more interested in these extraordinary facts."
(Carrel, *Man, The Unknown*, 1961 (1935), p. 122.)
[224] Alexis Carrel, *Man, The Unknown*, 1961 (1935), p. 111.
[225] Alexis Carrel, *Man, The Unknown*, 1961 (1935), p.123. Carrel attempted to explain the power of prayer:
"Certain spiritual activities may cause anatomical as well as functional modifications of the tissues and the organs. These organic phenomena are observed in various circumstances, among them being the state of prayer. Prayer should be understood, not as a mere mechanical recitation of formulas, but as a mystical elevation, an absorption of consciousness in the contemplation of a principle both permeating and transcending our world" (p. 121).
[226] Alexis Carrel, *Man, The Unknown*, 1961 (1935), pp. 7-11.
[227] Alexis Carrel, *Man, The Unknown*, 1961 (1935), pp. 8, 9.
[228] *Médecine Officielle et Médecines Hérétiques* (Official Medicine and Heretical Medicines), Plon, Paris, 1945.
[229] J. Poucel, "La Médecine Naturiste" (The Naturist Medicine), in *Médecine Officielle et Médecines Hérétiques* (Official Medicine and Heretical Medicines), Plon, Paris, 1945, pp. 159-182.
[230] Alexis Carrel, "Le rôle futur de la Médecine" (The future role of medicine), in *Médecine Officielle et Médecines Hérétiques* (Official Medicine and Heretical Medicines), Plon, Paris, 1945, pp. 1-10.

[231] Auguste Lumière, "La Médecine Humorale et ses Résultats" (Humoral Medicine and its results), in *Médecine Officielle et Médecines Hérétiques* (Official Medicine and Heretical Medicines), Plon, Paris, 1945, pp. 53-62.

[232] Hippocrates' most general recommendation for the prolongation of life was:
"Speaking generally, all parts of the body which have a function, if used in moderation and exercised in labours to which each is accustomed, become thereby healthy and well-developed, and age slowly; but if unused and left idle, they become liable to disease, defective in growth, and age quickly."
(*Hippocrates. With an English Translation by Dr. E.T Withington*, William Heinemann, London, 1959 (first printed in 1928), Vol. 3, "On joints," p. 339.)

[233] Pierre Galimard, "La tradition hippocratique et la médecine des correspondances" (The Hippocratic tradition and the medicine of correspondences), in *Médecine Officielle et Médecines Hérétiques* (Official Medicine and Heretical Medicines), 1945, pp.117-139.

[234] Combined, the "neo-Hippocratic" typological paradigm included the following four elemental biotypes:
1) Endodermic/ short asthenic [slender]/ lymphatic [phlegmatic]/ digestive/ round/ atoni-plastic [having little tonus and much plasticity]/ negroid and vegetarian;
2) Mesodermic/ short sthenic [robust]/ sanguine [blood]/ respiratory/ cubical/ toni-plastic/ nomadic and carnivorous;
3) Blastodermic/ tall sthenic/ biliary [yellow bile]/ muscular/ curved/ toni-aplastic [much tonus and little plasticity]/ organized and gregarious;
4) Ectodermic/ tall asthenic/ nervous or atrabiliary [black bile]/ cerebral/ flat/ atoni-aplastic/ passive and pensive.
(Marcel Martiny, "Nouvel Hippocratisme" (New Hippocratism), in *Médecine Officielle et Médecines Hérétiques* (Official Medicine and Heretical Medicines), Plon, Paris, 1945, pp. 141-158.)
These sorts of typologies were also highly popular in contemporary Germany, Switzerland and Italy, e.g. Carl Jung, *Psychologische Typen* (Psychological Types, 1921); Ernst Kretschmer, *Körperbau und Charakter* (The body buildup and character, 1921); Nicola Pende, *La Biotipologia Umana* (Human Biotypology, 1924).
In Nazi Germany, popular proponents of constitutional and racial "typology" ("typenlehre"/"konstitutionslehre") based on inherent hormone levels, were Walter Stemmer, Gerhard Venzmer, and others.
It was for example claimed that members of the "Dinaric race" (a.k.a. Adriatic or Epirotic race prevalent in South-Eastern Europe) are characterized by strong pituitary activity (basically high secretion of human growth hormone) which was responsible for their large noses and prominent chins, as well as for their presumed aggressiveness, persistence and greed (*kauflust*). In contrast, people with little activity of the pituitary have small noses, unobtrusive chins and otherwise "graceful," "regular" and "pretty" facial features and are characterized by "indifference, indulgence, tractability, patience, trust, satisfaction, clumsiness, self-doubt and indecisiveness."
(Gerhard Venzmer, *Deine Hormone – Dein Schicksal? Von den Triebstoffen unseres Lebens* (Your Hormones – Your Fate? The Fuel of Our Life), Franckh'sche Verlagshandlung, Stuttgart, 1940 (1933), XVIII "Wirkstoffe und Wesensart; Hormone und Rasse, das "Horoskop" der Zukunft" (Active substances and character; hormones and the race, the "horoscope" of the future), pp. 178-194; Walter Stemmer, *Klinik der weiblichen Geschlechtshormone: ein Buch für die Praxis*, Enke, Stuttgart, 1933 (The clinic of the female sex hormones. A book for the practice); Walter Stemmer, *Die elemente des psychischen. Ein Beitrag zur allgemeinen Seelenkunde und zur psychosomatischen Medizin auf Grund einer vergleichenden Psychologie des Menschen und der Tiere*, Hippokrates, Stuttgart, 1953 (Elements of the psychical. A contribution to general psychology and psychosomatic medicine based on comparative psychology of people and animals.)

[235] There are, according to Martiny, three levels of intervention or "therapeutic modalities."

The first is the "strong" intervention: acting by destruction, such as surgical ablation of diseased parts, or "topographic actions of release" for example through purging, bleeding, skin irritation, sudation, etc. Such strong interventions act on a single part of the biological unity, but create a general, non-specific response of the entire organism, redirecting the physico-electric vital force, and restoring the organic equilibrium through a "global humoral mechanism."

The second is the "intermediate" modality, based on Hippocrates' principle *"contraria contrariis curantur"* (opposite cures opposite), encompassing "all of the classical [official/modern] medicine." Examples include chemical treatments of infectious diseases. The introduction into therapy of heavy metals (bismuth, arsenic or antimony), those "renovations of Paracelsian medicine," along with alkaloids and other plant derivatives (never forgetting the value of entire plants), synthetic compounds, such as barbiturates and sulfamides – represent precious, though often dangerous, means in the therapeutic arsenal of the "intermediate modality." According to Martiny, only by recognizing the qualitative and quantitative specificity of each organism, and by establishing a limit between an active dose and toxic dose, can the benefits outweigh the risks. Such a dosimetry, according to him, is presently rudimentary and inadequate.

In contrast, the third – "weak" – therapeutic modality is based on Hippocrates' principle *"similia similibus curantur"* (like cures like), reintroduced by Hahnemann's homeopathy in 1810. The chief concepts of the "weak modality" are sensitivity and desensitization. According to Martiny, the advent of homeopathy opened the possibilities for vaccine therapy, serum therapy, auto-hemotherapy, opotherapy, hormone therapy, vitamin therapy, as well as a variety of other "desensitizing" or "tempering" treatments of the kind proposed by Auguste Lumière.

In discussing the "weak modality," Martiny appropriates terminology from "quantum" or "wave" mechanics. He speaks of "micro-dose" therapeutic catalysts working as "biological resonators," reinforcing the "energetic weakening of biological components," harmonizing or stabilizing them and reestablishing their normal rhythm.

The power of the psyche is said to be paramount for such a "reenergizing." Accordingly, mental hygiene, as well as proper professional and social orientation, are crucial for Neo-Hippocratic medicine, as it moves from the material into the spiritual domain.

The "weak modality" is valorized over the "strong" and "intermediate" ones. While the "strong" modalities may completely disregard or even destroy individuality, considering people as automata or elements in a troop, "intermediate" modalities threaten to diminish individuality. In contrast, in the "weak modality," the individuality is preserved intact.

(Marcel Martiny, "Nouvel Hippocratisme" (New Hippocratism), in *Médecine Officielle et Médecines Hérétiques* (Official Medicine and Heretical Medicines), Plon, Paris, 1945, pp. 141-158.)

[236] Vannier surveys the application of the principle "like cures like" from Paracelsus' theory of "effigies" in the 16th century through Oswald Crollius' "Royal Chemistry" and "Doctrine of Signatures" to Athanasius Kirscher's "Symbolism" in the 17th century, and the proposal of the homeopathy doctrine by Samuel Hahnemann in 1810.

By seeking "symbolic analogies" ("sympathies" or "affinities") between poisons and remedies, these authors hoped to discover the means that would "absorb" or "expel" the poisons or "excess matters," and thus "purify" the body and induce it into a stable equilibrium.

(Leon Vannier, "La Tradition Scientifique de l'Homœopathie" (The scientific tradition of Homeopathy), in *Médecine Officielle et Médecines Hérétiques* (Official Medicine and Heretical Medicines), Plon, Paris, 1945, pp. 63-116.)

[237] J. Poucel, "La Médecine Naturiste" (The Naturist Medicine), in *Médecine Officielle et Médecines Hérétiques* (Official Medicine and Heretical Medicines), Plon, Paris, 1945, pp. 159-182.

[238] J. Poucel, "La Médecine Naturiste" (The Naturist Medicine), in *Médecine Officielle et Médecines Hérétiques* (Official Medicine and Heretical Medicines), Plon, Paris, 1945, pp. 168, 172.

[239] Anne Harrington, *Reenchanted Science. Holism in German Culture from Wilhelm II to Hitler*, Princeton University Press, Princeton, New Jersey, 1996, p. 186; Christopher Lawrence and George Weisz, "Medical Holism: The Context," in Christopher Lawrence, George Weisz (Eds.), *Greater Than the Parts: Holism in Biomedicine, 1920-1950*, Oxford University Press, Oxford, 1998, p. 18; Robert Proctor, *Racial Hygiene: Medicine under the Nazis*, Harvard University Press, Harvard University Press, Cambridge, MA, 1988, p. 234.

[240] For example, the American author Marilyn Ferguson, in her famous manifesto of the "New Age" movement, *The Aquarian Conspiracy: Personal and Social Transformation in Our Time* (Putnam, NY, 1980, Ch. 8, "Healing ourselves," pp. 241-277), speaks of a "revolution," of "a new paradigm" in health care, based on "holistic" approaches. She seems to make little notice of how traditional these approaches really are.

In fact, her table juxtaposing the "assumptions of the old paradigm in medicine" with those of "the new paradigm of health" is in many details identical to the table given by Poucel comparing "Classical [orthodox]" medicine with "Naturist Medicine." Thus, for example, according to Ferguson, in the old paradigm, "body [is] seen as machine in good or bad repair," "body and mind are separate" and "mind is secondary factor in organic illness"; while in the new paradigm, "body [is] seen as dynamic system, context, field of energy with other fields," "mind is primary or coequal factor in *all* illness."

And according to Poucel, in the "classical" view, the individual is composed of "compartments," disease is seen as a "lesion of an organ" and "the mind and body are treated separately, as if in duality." The "Naturist medicine," on the other hand, teaches the "principle of unity," "dynamism and harmony of the ensemble," conjoined physical and spiritual education, it refers to "a concrete personality, his familial and social milieu, his cosmic ambiance," to the "whole organism, including the mental sphere," etc.

(J. Poucel, "La Médecine Naturiste" (The Naturist Medicine), in *Médecine Officielle et Médecines Hérétiques* (Official Medicine and Heretical Medicines), Plon, Paris, 1945, pp. 159-182.)

[241] Of course, in the 1930s-1940s, "natural" and "holistic" health systems existed not only in France, but also in Germany, the US, Britain, Russia, and elsewhere. Hence, establishing "world leadership" is very difficult. It is equally difficult to establish "world priorities."

Thus, going further back, in the early 1800s, several "natural" health systems were proposed at about the same time in several countries: Samuel Christian Hahnemann's "homeopathy" in Germany (c. 1810), Per Henrik Ling's "medical gymnastics" in Sweden (c. 1813), Mary Baker Eddy's "Christian Science" in the US (c. 1821). Even earlier, healing by "animal magnetism" was promulgated by Franz Mesmer in Germany, Austria and France almost simultaneously (around 1775). And the lines of descent can be traced further back and across the world. Yet, in the 1930s-1940s, in terms of dissemination and institutionalization of "natural" and "holistic" systems for health and longevity, France seems to be at the forefront.

[242] Among many examples, in the late 19th century, the British sinologist Herbert Allen Giles spoke of "Chinese doctors generally, whose ranks are recruited from the swarms of half-educated candidates who have been rejected at the great competitive examinations, medical diplomas being quite unknown in China." Taoist alchemy and striving for immortality were considered part of the "black arts," "weird stories of Taoist devilry and magic." And Taoism generally, "originally a pure system of metaphysics, it is now but a shadow of its former self, and is corrupted by the grossest forms of superstition borrowed from Buddhism, which has in its turn adopted many of the forms and beliefs of Taoism, so that the two religions are hardly distinguishable one from the other."

(Herbert A. Giles, *Strange Stories from a Chinese Studio*, Thomas de la Rue, London, 1880, vol. 2, pp. 293, 322, vol. 1, pp. XXIX, 13, reprinted in Internet Archive, www.archive.org. Further see J. J. Clarke, *The Tao of the West: Western Transformations of Taoist Thought*, Routledge, London, 2000.)

[243] Henri Maspero, *La Chine antique* (Ancient China), Presses Universitaires de France, Paris, 1965 (1927); Henri Maspero, "Les Procédés de 'nourrir le principe vital' dans la religion taoiste ancienne" (Procedures for nourishing the vital principle in ancient Taoist religion), *Journal Asiatique*, 229, 177-252, 353-430, 1937; Henri Maspero, *Le Taoïsme et les religions chinoises* (Taoism and Chinese religions), Musée Guimet, Paris, 1971 (1950,

published posthumously).

Comprehensive accounts of Taoist prolongevity are given in Gerald J. Gruman, *A History of Ideas about the Prolongation of Life. The Evolution of Prolongevity Hypotheses to 1800*, Transactions of the American Philosophical Society, Volume 56 (9), Philadelphia, 1966, pp. 28-56; Livia Kohn, *Daoism and Chinese Culture*, Three Pines Press, Cambridge MA, 2001, pp. 42-60, 136-152.

[244] Ilza Veith, "Acupuncture therapy – past and present. Verity or delusion," *Journal of the American Medical Association*, 180, 478-484, 1962; Ilza Veith, "Acupuncture in traditional Chinese medicine. An historical review," *California Medicine*, 118 (2), 70-79, 1973.

[245] G. Soulié de Morant, "Acuponcture, énergie vitale et électricité cosmique" (Acupuncture, vital energy and cosmic electricity), in *Médecine Officielle et Médecines Hérétiques* (Official Medicine and Heretical Medicines), Plon, Paris, 1945, pp.183-204.

[246] Zero electric potential (sustained in enclosed spaces, buildings and large cities) is said to be highly detrimental. There, however, seems to be some discrepancy in the findings about the benefits of "positive" and "negative" ambient charges. The negative charge is associated with nourishment and repair. However, it is said that "inhabitants of villages at net positive atmosphere are generally unscathed by tuberculosis and cancer and escape epidemics."

Furthermore, it is unclear whether and how an internal negative charge is related to an external positive charge. The finding about the benefits of the "net positive atmosphere" also seems to be at odds with the therapeutic use of negative air ionization, as practiced by Alexander Chizhevsky and Auguste Lumière (and is still used today).

(G. Soulié de Morant, "Acuponcture, énergie vitale et électricité cosmique" (Acupuncture, vital energy and cosmic electricity), in *Médecine Officielle et Médecines Hérétiques* (Official Medicine and Heretical Medicines), Plon, Paris, 1945, pp.188, 191.)

[247] In the 1960s, the American electro-physiologists, Stephen Smith and Robert Becker, used electromagnetic fields to induce tissue regeneration.

One method for such induction was by implanting a bimetallic (silver-platinum) wire into the animal limb. Becker reported that "a major amount of limb regeneration [results] if the electrical implant is oriented so that the end of the stump is made negative" (with the negative platinum wire at the distal end of the limb), thus reconfirming the nourishing and healing properties of the negative charge.

Becker further hypothesized that "electromagnetic energy was used by the body to integrate, interrelate, harmonize, and execute diverse physiological processes." However, no reference to the forerunning ideas of de Morant or other French researchers was made. In fact, in Becker's *Electromagnetism and Life* (1982), in the sections on acupuncture and on the electromagnetic effects of natural ecological systems, the earliest reference is from 1962.

(Robert O. Becker, Andrew A. Marino, *Electromagnetism and Life*, State University of New York Press, Albany, 1982, pp. 45, 196-206; Robert O. Becker, Gary Selden, *The Body Electric. Electromagnetism and the Foundation of Life*, Quill, NY, 1985, pp. 150-152.)

[248] G. Soulié de Morant, "Acuponcture, énergie vitale et électricité cosmique" (Acupuncture, vital energy and cosmic electricity), in *Médecine Officielle et Médecines Hérétiques* (Official Medicine and Heretical Medicines), Plon, Paris, 1945, p. 196.

[249] Pierre Winter, "Que devrait être une médecine traditionelle?"("What must be the traditional medicine?"), in *Médecine Officielle et Médecines Hérétiques* (Official Medicine and Heretical Medicines), Plon, Paris, 1945, pp. 293-332.

[250] The fusion between religion and medicine originated in Egypt, where the immensely powerful priesthood appropriated the responsibility for the physical preservation of the king and his subjects. (Pierre Winter, "Que devrait être une médecine traditionelle?"("What must be the traditional medicine?"), in *Médecine Officielle et Médecines Hérétiques* (Official Medicine and Heretical Medicines), Plon, Paris, 1945, pp. 293-332.)

(Notice the role of religion for the stabilization and perpetuation of the existing social order, coalescing with the goals of life-extension. In this context, one can argue that the preoccupation of the Egyptians with life-prolongation and immortality, was inseparably linked to their static cosmology, to their obsession with preservation, balance and constancy in all spheres; and furthermore that their pioneering technologies, from pyramid construction to embalming and surgery, emerged for the purpose of preserving constancy. It can also be recalled that within the Hindu "Trimurti" (Trinity) – Brahma the creator, Shiva the destroyer and Vishnu the preserver – the god of Ayurveda, or the science of long life, including the branch of Rasayana or rejuvenation alchemy, was Vishnu the preserver, incarnated as Dhanvantari.)

Winter further mentions the priestly-medical traditions of the Assyrians; the Chinese; the Greeks (the Pythagoreans, the priests of Asclepius, and ultimately Hippocrates who acknowledged the spirit-body connection and "synthesized earlier efforts," but presented a "vulgarized" version of the science that had been "reserved only for the initiates"); the ancient Nordic tribes (about whom data is scarce, but suggests the existence of a priestly-shamanic/medical caste); and finally medieval Christian, particularly Catholic, metaphysicians, such as Albertus Magnus, Thomas Aquinas and Roger Bacon, and Islamic and Jewish medical and religious thinkers: Avicenna, Averroes, and Maimonides.

(Tibetan Buddhism could be added to the list. The 17th century Tibetan Buddhist medical treatise *The Blue Beryl* promises the adept the power to "dispel human suffering, overcome diseases, aging and death," to achieve extreme longevity and eventually immortality "in the body of light," but only if the practitioner is a worthy vessel, following the true teachings of Buddha, and rejecting the heretical tenets of the mystic Buddhist school of Bhutan and the pre-Buddhist Tibetan religion of Bön taught by Shenrab. Jokers and disputers are also excluded from receiving the teaching.

See Parfionovich, Yu. M. (Ed.), *Atlas Tibetskoy Mediziny. Svod Illustraziy k Tibetskomu medizinskomu traktatu XVII veka "Goluboy Beryl"* (Atlas of Tibetan medicine. Collection of illustrations to the 17th century Tibetan medical treatise "The Blue Beryl"), Act, Moscow, 1998, book 4, folium 77, pp. 538-543; Book 2, folium 20, pp. 160-167.)

[251] Bill Halls, "Catholicism under Vichy: a study in diversity and ambiguity," in Harry Roderick Kedward, Roger Austin (Eds.), *Vichy France and the Resistance: Culture and Ideology*, Taylor & Francis, NY, 1985, pp.133-146; George Weisz, "A moment of synthesis: medical holism in France between the wars," in Christopher Lawrence, George Weisz (Eds.), *Greater Than the Parts: Holism in Biomedicine, 1920-1950*, Oxford University Press, Oxford, 1998, pp. 68-93; Etienne Lepicard, *A Bio-Medical Response to the Crisis of the 1930s: the Construction of Man, The Unknown by Alexis Carrel, 1935*, PhD dissertation, Hebrew University, 2000; Pierre Merle, "Guérisons rationnellement inexplicables"(Rationally inexplicable cures), in *Médecine Officielle et Médecines Hérétiques* (Official Medicine and Heretical Medicines), Plon, Paris, 1945, pp. 255-291.

[252] Cf. Jesus healing a paralyzed man at the pool of Bethesda (Beth-Hesda or Beth-Zatha – the house of mercy) in Jerusalem (John 5:1-15) and the healing of the blind at the pool of Siloam (Brechat Hashiloach – 'the sending pool' – in Jerusalem, receiving water from the Gihon Spring, John 9:1-12), among some 20 miraculous cures mentioned in the Gospels, such as the cures of leprosy, dropsy, palsy, bleeding, dumbness, possession, etc. Cf. Jesus reviving the dead: Jairus' daughter (Matthew 9:18-26, Mark 5:21-43, Luke 8:40-56), Widow's son at Nain (Luke 7:11-17) and Lazarus (John 11:1-44). Cf. also Jesus' commandment "Heal the sick, raise the dead, cleanse lepers, cast out demons. You received without paying; give without pay" (Matthew 10:8, http://www.biblegateway.com/passage/?search=Mt%2010:8;&version=ESV).

[253] Miraculous healing has been engrained in the Christian lore since the time of early saints (St. Paul, St. Jude, St. Cosmas and Damian, St. Anthony, St. Ambrose, St. Simon, etc. etc.).
Overall it was estimated that there have been more than 10,000 named Roman Catholic saints, most of whom were said to be able to produce miraculous healing.
(*Catholic Encyclopedia*, "Miracles" http://www.newadvent.org/cathen/10338a.htm; "Canonization" http://www.newadvent.org/cathen/02364b.htm;
http://en.wikipedia.org/wiki/Faith_healing;
http://en.wikipedia.org/wiki/List_of_saints;
http://en.wikipedia.org/wiki/List_of_early_Christian_saints.)
By the turn of the 21st century, "The Catholic Church has officially recognized 67 miracles and some 7,000 inexplicable cures since the Blessed Virgin Mary appeared in Lourdes in February 1858, as attested in the book "The Doctor in the Face of Miracles" ("Il medico di fronte ai miracoli"), written by the Italian Doctors Association" [By Salvino Leone, EDB, Bologna, 1996].
("How Lourdes Cures Are Recognized as Miraculous," *ZENIT Daily Dispatch*, 11 Feb. 2004, http://www.ewtn.com/library/MARY/ZLURDCUR.HTM.)
Cf. also the examples of super-longevity of venerated Christian saints. The following partial list includes some alleged persons, places and dates of life:
Saint Servatius (Tongeren, current Belgium, 8-384 AD, 297 years), Saint Shenouda (Egypt, 348-466, 118 years), Saint Llywarch Hen (Wales, 350-500, 150 years), Saint Kevin of Glendalough (Ireland, 498-618, 120 years), Scolastica Oliveri (Bivona, Italy, 1448-1578, 130 years), Theodosius of Caucasus (a Pravoslav Saint, Stavropol, 1841-1948, 107 years).
(Based on http://en.wikipedia.org/wiki/Longevity_myths.)
Thus, Merle's presentation on "Rationally inexplicable cures" was in line of a very powerful church tradition.

[254] Miraculous cures are posited if "no curative agent can explain the cure," if the disease is organic and not simply psycho-somatic, and if the cure is sudden and immediate, without a period of convalescence.
It might be additionally noted that the tradition of documenting miraculous cures has been well established in the Church since St. Augustine (354-420AD). In the *City of God*, St. Augustine speaks of "miracles which were wrought that the world might believe in Christ, and which have not ceased since the world believed," providing detailed testimonial accounts of miraculous cures of blindness, fistulae, cancer, gout, etc.
More generally, in Saint Augustine, healing and longevity, even immortality in the physical world, are central precepts. "In the Resurrection," he writes, "the substance of our bodies, however disintegrated, shall be entirely reunited," "the flesh shall then be spiritual, and subject to the spirit, but still flesh, not spirit."
Such a body, combining the carnal and the spiritual, will be forever preserved as at 30 years of age, "the age of the fullness of Christ." The body will undergo a thorough reconstruction to attain ideal proportions, whereby "all bodily blemishes which mar human beauty in this life shall be removed in the Resurrection, the natural substance of the body remaining, but the quality and quantity of it being altered so as to produce beauty."
(St. Augustine, *City of God*, translated by Marcus Dods, T&T Clark, Edinburgh, 1886, Book XXII, Chapters 8, 18-21, reprinted in Christian Classics Ethereal Library, http://www.ccel.org/ccel/schaff/npnf102.iv.XXII.html.)

[255] René Allendy, "La médecine et les agents impondérables" (Medicine and imponderable agents), in *Médecine Officielle et Médecines Hérétiques* (Official Medicine and Heretical Medicines), Plon, Paris, 1945, pp. 205-232.
Interestingly, René Allendy's doctoral thesis in medicine concerned life-extension in the alchemical tradition and was apparently one of the first modern works on this subject:

René Allendy, *L'Alchimie et la médecine: études sur les théories hermétiques dans l'historie de la médecine* (Alchemy and medicine: studies of hermetic theories in the history of medicine), Paris, 1912.

[256] Thus, according to the founding president of the American Association for Ayurvedic Medicine, a leading proponent of holistic therapy, Deepak Chopra, "the physical world arose from the quantum field, which is the source of all matter and energy," and this "unified field is inside ourselves, anchoring us to the timeless world with every breath, every thought, every action."

He distinguishes between the concepts of the "physical body, a frozen anatomical sculpture" and "quantum mechanical body, a river of intelligence constantly renewing itself." The former is subject to reductionist mechanical tinkering; the latter is related to the energy and information field of the universe and is subject to the power of mental "awareness."

According to Chopra, the efforts of reductionist experimental life-extension, based on the assumption that "technical ingenuity will solve any problem" are unlikely to produce tangible results any time soon. They "have far to go before achieving any benefit for cells outside glass dishes." And that is because "like other reductionist models, the genetic view of aging ignores life as a whole."

In contrast, the holistic approach promises high and immediate returns. By directly harnessing the power of awareness, the individual may draw nourishing energy from the universe, and may make "quantum leaps" into a youthful state of metabolic equilibrium.

In other words, meditations and affirmations, "setting longevity as a goal," will instantaneously make you young. "Our inner intelligence," Chopra concludes, "can create a permanent state of ageless body and timeless mind," a state "beyond change and death."

(Deepak Chopra, *Ageless Body, Timeless Mind. The Quantum Alternative to Growing Old*, Harmony Books, New York, 1993, pp. 226-229, 280, 287-288, 314-315.)

[257] Since 1934, the Institute published *Revue de Cosmobiologie* (Review of Cosmobiology) – apparently the world's first scientific journal in the field – with Maurice Faure as the editor in chief. It also organized the first Congress of Cosmobiology in Nice, on June 2-7, 1938, under Maurice Faure's presidency.

[258] Sigmund Freud, *Beyond the Pleasure Principle*, Translated by James Strachey, Liveright, New York, 1976 (first published in 1920), p. 38.

[259] Guillerey also notes that the "human being is composed of the body and the spirit" and calls on "psychotherapy" to address both the "psychic" and the related "physical" disharmonies.

The arsenal of psychotherapy includes both the "passive, analytical, regressive" methods, and "active, synthetic, progressive" ones. Guillerey favors "active methods" over passive psychoanalysis, and introduces his own "active" technique of "guided imagination" (rêverie dirigée). In that procedure, the patient confronts problems in his mind and imaginatively solves them, thus exercising an "effort of concentration," learning to "mobilize energy," eventually "reintegrating" and "reconstructing" his organism, and achieving the "interior unification of equilibrium."

(M. Guillerey, "Médecine Psychologique" (Psychological Medicine), in *Médecine Officielle et Médecines Hérétiques* (Official Medicine and Heretical Medicines), Plon, Paris, 1945, pp. 233-254.)

[260] Rene Biot, "Vers l'unité de la médecine" (Toward the unity of medicine), in *Médecine Officielle et Médecines Hérétiques* (Official Medicine and Heretical Medicines), Plon, Paris, 1945, pp. 333-342.

[261] Jottras, "Analyse scientifique et médecine humaine" (Scientific analysis and humane medicine), in *Médecine Officielle et Médecines Hérétiques* (Official Medicine and Heretical Medicines), Plon, Paris, 1945, pp. 29-52.

[262] Rémy Collin, "Existe-t-il une doctrine officielle?" (Is there an official doctrine?), in *Médecine Officielle et Médecines Hérétiques* (Official Medicine and Heretical Medicines), Plon, Paris, 1945, pp. 11-28.

[263] René Rémond, *Les Droites en France* (The right-wingers in France), Aubier, Paris, 1982, pp. 473, 493.

[264] According to George Weisz, after WWII, the French Neo-Hippocratic movement lost much of its popularity and prestige; it "had clearly lost its relevance for the medical profession" (George Weisz, "A

moment of synthesis: medical holism in France between the wars," in Christopher Lawrence, George Weisz (Eds.), *Greater Than the Parts: Holism in Biomedicine, 1920-1950*, Oxford University Press, Oxford, 1998, p. 86).

[265] Sergey Metalnikov, *Problema Bessmertia i Omolozhenia v Sovremennoy Biologii* (The Problem of Immortality and Rejuvenation in Modern Biology), Slovo, Berlin, 1924 (1917); S. Metalnikov, *Immortalité et rajeunissement dans la biologie moderne* (Immortality and rejuvenation in modern biology), Bibliothèque de Philosophie Scientifique, Flammarion, Paris, 1924; Sergey Metalnikov, *La Lutte Contre La Mort* (The Struggle against Death), Gallimard, Paris, 1937.

[266] Sergey Metalnikov, *Problema Bessmertia i Omolozhenia v Sovremennoy Biologii* (The Problem of Immortality and Rejuvenation in Modern Biology), Slovo, Berlin, 1924 (1917), pp. 170, 144, 138, 171-172; Sergey Metalnikov, *La Lutte Contre La Mort* (The Struggle against Death), Gallimard, Paris, 1937, pp. 229, 203, 151, 149 (hereafter referred to as "Metalnikov, 1924" and "Metalnikov, 1937").

[267] Metalnikov, 1924, pp. 172-173; Metalnikov, 1937, pp. 238-239.

[268] Auguste Lumière, *La Vie, La Maladie et La Mort, Phénomènes Colloïdaux* (Life, Disease and Death as Colloidal Phenomena), Masson & Cie, Libraires de L'Académie de Médecine, Paris, 1928, p. 92.

[269] Metalnikov, 1924, p. 171; Metalnikov, 1937, p. 235.

[270] Quoting data from the American "Metropolitan Life Insurance Company" (from 1909 to 1927), Metalnikov points out that their investment of $32 million in health education and hygiene, yielded a decrease in mortality and corresponding economy of $75 million.

In France, between 1880 and 1931, the average duration of life increased by 14 years (from 44 to 58), which was equivalent to saving 300,000 lives each year or 600 million years of additional life for the 42 million inhabitants for the entire period.

Social status is considered as a crucial factor for longevity. According to Metalnikov, it is difficult to decide whether the poor or the rich live longer, or more precisely, whether extreme longevity is more common among the poor or the rich (as the former are plagued by scarcities and the latter by excesses). Yet, statistics indicate that in the higher classes the average mortality is much lower. (Metalnikov, 1937, pp. 158-160.)

[271] Metalnikov, 1937, p. 230.

[272] Alexandre Guéniot, *Pour Vivre Cent Ans. l'Art de Prolonger ses Jours* (To Live a Century. The Art of Prolonging the Days), Librairie Baillière et Fils, Paris, 1936 (1931), pp. 20-23, 61-63, 68-87, 166-167, 172-173, 189-193, 204-206.

[273] Guéniot emphasized vegetables over meat, and was apparently among the first to point out the importance of vitamins for postponing aging. Vegetable sources, he believed, could provide sufficient vitamin contents. To recall, major vitamins were not mass produced or even synthesized before 1935.

Guéniot valorized concerted hygienic measures over reductionist interventions:

"It appears that the prolongation of life until its natural limit of about a hundred years cannot result from any *unique* procedure or act, from *any panacea*. No operation, no graft, however advantageous it may be, no elixir, no blood transfusion, nor anything else, presently has this power. These diverse agents, and the benefits one can derive from them, have only temporary and limited effects. Undoubtedly, the grafts, the renovation and invigoration that they often produce, raise high expectations; perhaps they are the panacea of the future. But today, the attainment of a centenarian life results from a *concert of [hygienic] causes* that assure a regular evolution of the organism, as well as a normal duration of life" (emphasis in the original).

And these are the causes of longevity which Guéniot indicated:

The hereditary vital force; temperance in eating and drinking; abstinence from sexual abuse; ample respiration of pure air; daily friction-massage of the entire body; habitual physical exercises that increase the energy and the suppleness of movement; finally, the rational use of sleep and rest needed to repair all functional expenditures.

"To live to be a hundred, the intervention of this *ensemble* of measures is indeed indispensable. As facts testify, three or four factors sometimes suffice: heredity, temperance, abstinence from any excess, pure air, etc." (Alexandre Guéniot, *Pour Vivre Cent Ans. l'Art de Prolonger ses Jours* (To Live a Century. The Art of Prolonging the Days), Librairie Baillière et Fils, Paris, 1936 (1931), pp. 166-167.)

[274] Harry Roderick Kedward, Roger Austin (Eds.), *Vichy France and the Resistance: Culture and Ideology*, Taylor & Francis, NY, 1985; René Rémond, *Les Droites en France* (The right-wingers in France), Aubier, Paris, 1982.

[275] François Bourlière, "Species differences in potential longevity of vertebrates and their physiological implications," in Bernard L. Strehler (Ed.), *The Biology of Aging. A Symposium Held at Gatlinburg, Tennessee, May 1-3, 1957, Under the Sponsorship of the American Institute of Biological Sciences and with support of the National Science Foundation*, Waverly Press, Baltimore, 1960, pp.128-132; François Bourlière, "Est-il réellement possible d'envisager un traitement biologique de la senescence? (Is it really possible to envision a biological treatment of senescence?) *Gazette médicale de France*, 65, 925-926, 1958; Léon Binet, *Gérontologie et gériatrie. La lutte contre les années* (Gerontology and geriatrics. The struggle against the years), Presses Universitaires de France, Paris, 1961; Léon Binet & François Bourlière (Eds.), *Précis de Gérontologie* (Precis of Gerontology), Masson e Cie., Paris, 1955; Paul Hirschmann, *Vivre Longtemps et Rester Jeune* (To live long and stay young, originally appeared in Yiddish translation by Dr. Noah Gris, entitled *Lebn Lang un Blaybn Yung*), Oyfsnay, Paris, 1975, pp. 156-176.

Binet and Bourlière's cautious attitude to rejuvenation is exemplified in their authoritative *Précis de Gérontologie* (1955). The authors recommend a wide range of anti-aging treatments, including calorie restriction, physical exercise balanced with rest, a wide range of hormonal ("opotherapeutic") supplements (such as sex and thyroid hormones), vitamins, mineral supplements (iodine, calcium, iron, glutamic acid, royal jelly, etc.), and "general stimulating products" – chicken embryo extracts, biogenic/tissue therapeutic stimulators developed by Vladimir Filatov (1875-1956, Ukraine), the anti-reticular cytotoxic serum of Alexander Bogomolets (1881-1946, Ukraine), and the "orthobiotic serum" developed at the Pasteur Institute, Paris, by Michel Bardach.

All the above-mentioned treatments were developed for the explicit purpose of "rejuvenation." Yet, recognizing their limited efficiency, Binet and Bourlière designate them as "stimulating products," and "therapeutics" capable of "correcting certain functions and metabolic perturbations," and "making the life of the aged more comfortable."

"Rejuvenation," however, is decried through and through. The authors emphasize very strongly that the treatments are no "fountain of youth" and "do not authorize us to pronounce that hackneyed word – rejuvenation."

(Léon Binet, François Bourlière, "Problèmes biologiques généraux posés par la sénescence de l'organisme" (General biological problems of organism's aging), Léon Binet, Claude Bétourné, "Aspects thérapeutiques du problème de la sénescence" (Therapeutic aspects of the problem of senescence), in Léon Binet & François Bourlière (Eds.), *Précis de Gérontologie* (Precis of Gerontology), Masson e Cie., Paris, 1955, pp. 34, 539-545.)

[276] Michel Bardach (1899-1960?) was a son of the Ukrainian bacteriologist Jacob Bardach (1857-1929), Metchnikoff's student and collaborator in Odessa, Ukraine.

"Bardach's serum" developed in the 1940s was very much like "Bogomolets' serum" developed in the 1930s in Ukraine, using "stimulating cytotoxic sera." And both owed to the studies of "cytotoxic sera" which were started in the early 1900s by Elie Metchnikoff and his assistant at the Pasteur Institute, Alexandre Besredka (1870-1940, an emigrant from Odessa working at the Pasteur Institute since 1893).

Further on Michel Bardach's "orthobiotic serum" see:

Ilana Löwy, "'The Terrain is All': Metchnikoff's Heritage at the Pasteur Institute, from Besredka's 'Antivirus' to Bardach's 'Orthobiotic Serum'," in Christopher Lawrence, George Weisz (Eds.), *Greater Than the Parts: Holism in Biomedicine, 1920-1950*, Oxford University Press, Oxford, 1998, pp. 257-282; M. Bardach, P. Goret, L. Joubert, "Le Système physiologique du serum conjonctif et sa stimulation par le serum de Bogomoletz" (The physiological system of connective tissue and its stimulation by Bogomolets' serum),

Recueil de Médecine Veterinaire, 124, 337-363, 1948; M. Bardach, E.J. Sobieski, M. Tosquelles, "Premièrs Résultats obtenus avec le sérum orthobiotique en médecine humaine" (The first results obtained with the orthobiotic serum in human medicine), *Archives Hospitalières*, 9, 269-275, 1949.

Other potentially life-prolonging immunological treatments were developed in the 1920s-1930s at the Pasteur Institute. One was the Bacillus Calmette-Guérin (BCG) vaccine prepared by Albert Calmette and Camille Guérin from weakened live bovine tuberculosis bacillus, *Mycobacterium bovis*. It was initially used against tuberculosis and by now has also been tried against other diseases, from leprosy to diabetes and cancer.

Another was Alexandre Besredka's "Antivirus Therapy" made from filtrates of old bacterial cultures, primarily *streptococci* and *staphylococci*. It was supposed to effect local immunization, and was used against a host of infectious diseases, from boils to eye infection to anthrax.

These treatments, however, were not initially intended as "stimulating" or "rejuvenating" agents. (Speil C, Rzepka R, "Vaccines and vaccine adjuvants as biological response modifiers," *Infectious Disease Clinics of North America*, 25 (4), 755-772, 2011; Ilana Löwy, "'The Terrain is All': Metchnikoff's Heritage at the Pasteur Institute, from Besredka's 'Antivirus' to Bardach's 'Orthobiotic Serum'," in Christopher Lawrence, George Weisz (Eds.), Greater Than the Parts: Holism in Biomedicine, 1920-1950, Oxford University Press, Oxford, 1998, pp. 257-282.)

[277] To exemplify Rostand's life-extensionist views, in a preface to Robert Ettinger's *The Prospect of Immortality* (1964), Rostand confirmed:

"we don't have long to wait before we shall know how to freeze the human organism without injuring it. When that happens, we shall have to replace cemeteries by dormitories, so that each of us may have the chance for immortality that the present state of knowledge seems to promise. ... we must begin; the job will be done some day, and for every day that we put it off untold thousands are going to an unnecessary grave."
(Preface by Jean Rostand de l'Academie Française, in Robert Ettinger, *The Prospect of Immortality*, Doubleday, NY, 1964, pp. 9-10.)

Rostand earlier spoke about the "perspectives of rejuvenation" in Jean Rostand, *La biologie et l'avenir humain* (Biology and the human future), Albin Michel, Paris, 1950, pp. 15-23.

[278] In tracing the origin of the modern life-extensionist movement to *fin-de-siècle* France, it should be of course noted that Metchnikoff, Finot and Voronoff were Russian/Jewish immigrants, and Brown-Séquard was the son of an American father and French mother, and was born in Mauritius.

[279] The Claude Bernard Center of Gerontology (Centre Claude Bernard de Gerontologie, Paris), founded by Francois Bourlière in 1956, existed until the dissolution of Claude Bernard's Association in 2003. But the Gerontology Research Unit of the French National Institute of Health and Medical Research (Institut National de la Santé et de la Recherche Médicale - INSERM, Unit 118), founded by Bourlière in 1972, exists to the present.

The French Gerontological Society (Société Française de Gérontologie) established in 1939, among the earliest in the world, by Alphonse Baudouin, Léon Binet and Francois Bourlière, exists as well (http://www.sfgg.fr/). With about 4000 members as of 2011, it was the second largest organization in the IAGG – the International Association of Gerontology and Geriatrics, after the American Gerontological Society (5335 members). (IAGG Member Societies and Council Members, http://www.iagg.info/organization/council; http://www.sfu.ca/iag/links/member.htm.)

On June 5-9, 2009, France for the first time hosted the (19th) IAGG World Congress of Gerontology and Geriatrics, with 5,800 participants from 91 countries. The congress theme was "Longevity, Health and Wealth."

Another major gerontological organization is the National Foundation of Gerontology (Fondation Nationale de Gérontologie), which was established in 1967 and collaborated with the International Longevity Center (US) since 1989 (http://www.fng.fr/).

Yet another is Institut de la Longévité et du Vieillissement (Institute of Longevity and Aging – Scientific Interest Group), created in 2002 at the initiative of Professor Étienne-Émile Baulieu, as a partnership between the French Ministry of Scientific Research, CNRS (Centre national de la recherche scientifique - National Center of Scientific Research), INED (Institut national d'études démographiques - The National Institute for Demographic Studies) and INSERM (Institut National de la Santé et de la Recherche Médicale - National Institute of Health and Medical Research). (http://www.gis-longevite.cnrs.fr/.)

In 2010, Global Forum for Longevity and a longevity research program have been originated by AXA insurance company, headquartered in Paris (http://longevity.axa.com/fr/index.php; https://gallery.axa-research.org/en/actualite/news/discover-book-knowledge-life-risks.htm).

[280] Anti-aging products (sometimes involving stem cells) have been, for example, developed by Laboratoires Inneov, in Nantes, the joint venture established in 2002 by the Swiss food concern Nestle and the French cosmetics concern L'Oréal (http://www.loreal-finance.com/eng/news/active-cosmetics-40.htm; http://www.inneov.com/). These, however, have been mainly concerned with cosmetic products, driving the goals of life-extension far into the background.

Several societies have been dedicated to the research and promotion of anti-aging medicine, including the French Society of Anti-Aging Medicine (Association Française d'Anti-Aging, www.fsaam.com) which has been led by Dr. Claude Dalle of Paris; and the French Society of Medicine and Physiology of Longevity (Société Française de Médecine et de Physiologie de la Longévité, www.sfmpl.org) led by Dr. Christophe de Jaeger. Christophe de Jaeger has also been the founding president of the European Institute of Aging in Paris (l'Institut Européen du Vieillissement, http://www.iev-medecine.org/) and of the European Medical Center for Health and Longevity (Centre Médical Européen Santé et Longévité, also in Paris, http://www.cemesal.org/).

The European Society of Preventive, Regenerative and Anti-Aging Medicine (ESAAM), a sustaining member of the World Anti-Aging Academy of Medicine (WAAAM) was, as of 2007, presided over by the French professors Jean-Jacques Legrand and Antoine Lorcy. As of 2010, it was chaired by Prof. Christos C. Zouboulis, head of the Laboratory for Biogerontology at Charité Universitaetsmedizin, Berlin, Germany. ESAAM's official journal is *Rejuvenation Research*, with Aubrey de Grey as the editor in chief.

(Sources: http://www.waaam.org/member_organizations.php; http://www.esaam-org.eu/; http://www.ecopram.com/; http://www.wosiam.org/index.php?rub=about-WOSIAM; http://www.euromedicom.com/index.php?rub=41; http://web.archive.org/web/20070618101046/www.esaam.com/index.html?pg=board.)

Still, the use of vitamins and dietary supplements in France is among the lowest in the Western countries (~25% of the population as compared to over 50% in the US). (Skeie G, et al., "Use of dietary supplements in the European Prospective Investigation into Cancer and Nutrition calibration study," *European Journal of Clinical Nutrition*, 63 (Supplement 4), 226-238, 2009; The Nielsen Global Report, March 2009, http://en-us.nielsen.com/content/nielsen/en_us/insights/consumer_insight/issue_16/half_of_americans.html.)

[281] The lesser life-extensionist organizations have included:
Objectif Longévité (The Longevity Objective, http://objectif.longevite.over-blog.net/); Pour l'extension radicale de l'espérance de vie (For the radical extension of life-expectancy, http://plus2vie.blogspot.com/); Médecine anti-âge (Anti-age medicine, http://www.medecine-anti-age.com/); Vivre 1000 ans (To live a thousand years, http://vivre1000ans.free.fr; http://www.mfoundation.org/files/sens/index-fr.htm); Immortalité Biologique (Biological immortality, http://immortalite.org; http://sites.google.com/site/immortalitebiologique/); La Longévité Française (The French Longevity, http://www.lalongevitefrancaise.com/; http://www.longevite.fr/); HEALES (Healthy Life Extension Society, though incorporated in Brussels, Belgium, it has many French activists, http://heales.org/FRENCH/); Transhumanistes: Technoprog - l'Association Française Transhumaniste (The French Technoprogressive - Transhumanist Association, http://transhumanistes.com/); Longévité &

Santé (Longevity and Health, http://www.longevite-sante.org/).

[282] Jean-Marie Robine and Michel Allard, "The oldest human," *Science*, 279 (5358), 1834-1835, 1998.

[283] According to the CIA World Factbook, as of 2010, the life-expectancy in France ranked 12th in the word, 13th in 2011, and 15th in 2012. (The CIA World Factbook, https://www.cia.gov/library/publications/the-world-factbook/geos/fr.html.)

According to other sources, in 2011 and 2012 France was in the world's best 10 in terms of life expectancy (6th and 8th place respectively). (http://countryeconomy.com/demography/life-expectancy/france.)

In any case, the value has been one of the highest in the world.

[284] Bruno Simini, "Serge Renaud: from French paradox to Cretan miracle," *The Lancet*, 355 (9197), 48, 2000.

[285] Laure Lapasset, Ollivier Milhavet, Alexandre Prieur, Emilie Besnard, Amelie Babled, Nafissa Aït-Hamou, Julia Leschik, Franck Pellestor, Jean-Marie Ramirez, John De Vos, Sylvain Lehmann, Jean-Marc Lemaitre, "Rejuvenating senescent and centenarian human cells by reprogramming through the pluripotent state," *Genes and Development*, 25, 2248-2253, November 1, 2011.

[286] Johann Wolfgang Goethe, *Faust* [1808-1832], translated by George Madison Priest, Franklin Center, PA, The Franklin Library, 1979, "The First Part of the Tragedy," lines 70-71, http://einam.com/faust/index.html.

[287] Johann Heinrich Cohausen, *Hermippus Redivivus or the Sage's Triumph over Old Age and the Grave, Wherein a method is laid down for prolonging the life and vigour of man, including a commentary upon an ancient inscription, in which this great secret is revealed, supported by numerous authorities. The whole is interspersed with a great variety of remarkable and well attested relations*, J. Nourse, London, 1748.

The book was first published in Latin in 1742: *Hermippus redivivus, sive exercitatio physico-medica curiosa de methodo rara ad CXV annos prorogandae senectutis per anhelitum puellarum, ex veteri monumento romano deprompta, nunc artis medicae fundamentis stabilita, et rationibus atque exemplis, nec non singulari chymiae philosophicae paradoxo* (Hermippus revived, or the curious physico-medical exercise of a rare method to postpone old age to 115 years by the breath of young maidens, taken from an old Roman tomb, now establishing the foundations of the art of medicine, with reasons and examples, and also singular chemical-philosophical pardoxes), Frankfurt am Main, John Benjamin Andrae and Henry Hort, 1742.

[288] Johann Bernhard Fischer, *De Senio Eiusque Gradibus et Morbis* (On Old Age, its Degrees and Diseases), Erfordiae, 1754.

[289] The edition mainly used here is: Christoph Wilhelm Hufeland, *Makrobiotik; oder, Die Kunst das menschliche Leben zu verlängern*, Sechste verbesserte Auflage, A.F. Macklot, Stuttgart, 1826 (Macrobiotics or the art of prolonging human life, the 6th improved edition).

Apparently, the term "Makrobiotik" was added to the title only in the third edition, published in 1805 in Berlin by Ludwig Wilhelm Wittich. A variant of the 1805 edition used the term "Makrobiot" instead.

The first two editions were just entitled *Die Kunst das menschliche Leben zu verlängern* (the art of prolonging human life), and were published in 1796 and 1799 in Jena for Gotthold Ludwig Fiedler, Academische Buchhandlung.

The book has been translated into virtually all European and many other languages, including English and Russian, also consulted here. In English: *The Art of Prolonging Life*, Edited by Erasmus Wilson, J. Churchill, London, 1859, Lindsay & Blakiston, Philadelphia, 1867. And in Russian: *Iskusstvo Prodlevat Chelovecheskuyu Zhizn (Macrobiotika)*, translated by P. Zablotsky, E. Pratz Typography, St. Petersburg, 1852.

The book's epigraph was taken from Goethe's play "Egmont" (1787):

„Süßes Leben! schöne freundliche Gewohnheit des Daseins und Wirkens, von dir soll ich scheiden?"

(Sweet life! The beautiful friendly habit of existence and work. Must I depart from thee?)

[290] *Faust. A Tragedy Translated from the German of Goethe with notes by Charles T Brooks*, Ticknor and Fields, Boston, 1856 (Part 1, Scene 6, "Witches' Kitchen," lines 2345-2377, http://einam.com/faust/index.html).

[291] Famous "Steinachian" rejuvenators included Harry Benjamin in the US, Norman Haire in the UK (Haire performed the Steinach operation on the poet William Butler Yeats in 1934), and Yasusaburo Sakaki in Japan (in 1922 Sakaki nominated Steinach for a Nobel Prize for the "work on transplantation of reproductive glands and particularly on rejuvenation").

Steinach's, Voronoff's and various other forms of operations for "endocrine rejuvenation" were performed at the time by Knud Sand in Denmark; Gregorio Marañón and León Cardenal in Spain; Francesco Cavazzi and Vincenzo Pettinari in Italy; Siegfried Stocker, Karl Kolb, and Paul Niehans in Switzerland; Ottmar Wilhelm in Chile; Constantin Ion Parhon and Georges Marinesco in Romania; Vladislav Ruzicka in Czechia (both Marinesco and Ruzicka were mainly interested in the effects of the operations on the colloidal system); Ivan Osipovich Michalowsky, Naum Efimovich Ischlondsky and Boris Mikhailovich Zavadovsky in Russia.

On the Steinach operation performed on W.B. Yeats and Sigmund Freud, see Richard Ellmann, *W.B. Yeats's Second Puberty*, Library of Congress, Washington, 1985, p. 7; Paul E. Stepansky, *Freud, Surgery, and the Surgeons*, Routledge, London, 1999, p.137.

On Steinach's 11 nominations for the Nobel Prize, see the Nomination Database for the Nobel Prize in Physiology or Medicine, 1901-1953 (http://nobelprize.org/).

On prominent rejuvenators in the German-speaking world, see Benno Romeis, *Altern und Verjüngung. Eine Kritische Darstellung der Endokrinen "Verjüngungsmethoden", Ihrer Theoretischen Grundlagen und der Bisher Erzielten Erfolge*, Verlag von Curt Kabitzsch, Leipzig, 1931 (Aging and Rejuvenation. A Critical Presentation of Endocrine "Rejuvenation Methods," Their Theoretical Foundations and Up-to-Date Successes); Heiko Stoff, *Ewige Jugend. Konzepte der Verjüngung vom späten 19. Jahrhundert bis ins Dritte Reich*, Böhlau Verlag, Köln, 2004 (Eternal Youth. Concepts of Rejuvenation from the late 19th century until the Third Reich, 2004), pp. 69-72.

[292] Vermehren F., "Stoffwechseluntersuchungen nach Behandlung mit Glandula thyreoidea an individuen mit und ohne Myxoedem" (Metabolic examinations after treatment with thyorid glands in individuals with and without myxoedema), *Deutsche Medizinische Wochenschrift*, 43, 1037, 1893.

[293] Benno Romeis, *Altern und Verjüngung. Eine Kritische Darstellung der Endokrinen "Verjüngungsmethoden", Ihrer Theoretischen Grundlagen und der Bisher Erzielten Erfolge*, Verlag von Curt Kabitzsch, Leipzig, 1931 (Aging and Rejuvenation. A Critical Presentation of Endocrine "Rejuvenation Methods," Their Theoretical Foundations and Up-to-Date Successes), "III. Reaktivierungsversuche mit endokrinen Organpräparaten" (Reactivation experiments with endocrine organ preparations), pp. 1944-1965.

[294] Selmar Ascheim was stripped of all posts in 1935 and fled to France in 1937. Bernhard Zondek fled to Stockholm in 1933 and then moved to Palestine in 1940. Hermann Zondek (Bernhard's older brother), fled to Palestine in 1934 and died in Jerusalem in 1979. After 1938, Eugen Steinach had to remain in Switzerland, and Otto Kauders immigrated to the US.

One of the more tragic stories is that of the Jewish-Latvian rejuvenator Naum Lebedinsky (b. Odessa 1888 – d. Riga 1941), head of the Institute of Comparative Anatomy and Experimental Zoology at the University of Latvia, Riga (from 1922 to 1941). After the Nazi occupation of Latvia in July 1941, he was refused by the German High Command to leave the country and committed suicide together with all his family.

[295] Steinach especially prided himself on being the first to determine the role of physiological sex hormones in the human organism in general, and in its reactivation or "rejuvenation" in particular. Even though Steinach valorized "vasoligation" over "transplantation" – the future of rejuvenation, he believed, belongs to synthetic hormone supplements. Compared to surgery, such hormones are cheaper and more readily available to the public, and their therapeutic concentrations can be precisely established.

The road to such synthetic therapies was paved by the success of the German biochemist Adolf Butenandt in synthesizing female sex hormones (estrone in 1929 and progesterone in 1935), as well as by the

Croatian/Swiss Leopold (Lavoslav) Ruzicka's and the Polish/Dutch Ernst Laqueur's successful synthesis of male sex hormones ("androsterone" and "testosterone" from 1931 to 1935).

More impetus was given by the extraction of insulin (1921) by the Canadian Frederick Banting and Charles Best for the treatment of (senile) diabetes ("after applying [on the pancreas] a surgical procedure corresponding in principle to my [Steinach's] vasoligature").

Writing in 1940, Steinach felt that his pioneering efforts in endocrinology and endocrine rejuvenation, starting in the early 1900s, were vindicated:

"The programme which I outlined has been fulfilled. Today the effective agent can be administered in exact doses as a drug, and it can produce the same effect as physiological testicular hormone."

(Eugene Steinach and John Loebel, *Sex and Life; Forty Years of Biological and Medical Experiments*, Faber, London, 1940, pp. 56, 213.)

[296] Benno Romeis, *Altern und Verjüngung. Eine Kritische Darstellung der Endokrinen "Verjüngungsmethoden", Ihrer Theoretischen Grundlagen und der Bisher Erzielten Erfolge*, Verlag von Curt Kabitzsch, Leipzig, 1931 (Aging and Rejuvenation. A Critical Presentation of Endocrine "Rejuvenation Methods", Their Theoretical Foundations and Up-to-Date Successes, hereafter referred to as "Romeis, 1931").

[297] Romeis, 1931, p. 1808.

[298] In Steinach's theory, the vasoligation destroys sperm-producing cells of the testis, leading to a compensatory "proliferation" or "regeneration" of its interstitial cells responsible for the heightened production of sex hormones, which in turn leads to the general "reinvigoration," "revitalization" or "rejuvenation" of the body via increasing the blood flow.

Romeis finds little anatomical evidence for this theory. He does not exclude the possibility of cell regeneration. In fact, in 1921, he was among the first to show that spermatic epithelium can regenerate from "non-differentiated cells" found even in degenerating spermatic canals.

Yet, he finds that "until now, not in a single case, neither after the ligation of the epididymis nor after the ligation of the vas deferens, neither in animals nor in people, was there evidenced a 'proliferation of sex glands' as Steinach presumed."

By and large, the morphology of senile animal testicles was found to be very similar to that of mature ones and to those who have undergone the "rejuvenating operation," thus presenting little evidence for "cell regeneration."

Romeis does accept that "despite the failures, there is a series of observations on people and animals showing that the intervention does lead to a 'refreshing' of underlying organs and functions." Yet, Romeis suggests, its mechanism of action is not the compensatory cell regeneration as posited by Steinach.

Rather, the vasoligation leads to a partial destruction of the testicular tissue, releasing "hormone-containing" and other "degradation products" that are "reabsorbed" and act as "irritants" on the body. (Romeis does not go far in specifying the mechanisms of such an "irritation.")

As there is only a limited amount of tissue to be destroyed and reabsorbed, Romeis argues, the younger and more intact patients will have greater amounts of it, and will therefore exhibit a stronger and more lasting "reinvigoration."

Romeis concludes that the results of the "reinvigorating" interventions are determined chiefly by three factors: the amount of intact testicular tissues, the conductivity of the spermatic canals, and the reactivity of the organism.

In other words, with highly atrophied testicular tissues (that is, in cases of advanced senescence), no success is to be expected.

In summary, according to Romeis, there might be "some effects" to Steinach's operation, yet the results are intractably confounded and inconclusive and, in any case, the operation produces no "true rejuvenation" whatsoever.

(Romeis, 1931, pp. 1809-1873.)

[299] In cases when vasoligation did not succeed in "refreshing" the organism, Steinach suggested "albugineatomy" – the surgical removal of the Tunica Albuginea layer of connective tissue covering the testicles. The goal was to "reduce pressure" on the testicular tissue and allow more room for its regeneration.

As in the case of vasoligation, the positive findings of "general reinvigoration" could not be explained by cell "regeneration" but again by the destruction and subsequent "resorption" of the present testicular tissue after the operation. Since there is only a limited amount of tissue to be destroyed, the activation effects can only be temporary.

This conclusion is strengthened by the findings from Lebedinsky's method of "rejuvenation" that straightforwardly smashed and tore parts of dogs' testicles to achieve a temporary "reinvigoration" by the release of the products of destruction.

Lakatos' modified version of Steinach's albugineatomy attempted to minimize the impairment of the testicular tissue by making a small incision.

In animal experiments, the testicles increased in size, but, as Romeis argued, not because of their regeneration, but because of a "reflexive vasodilation" in response to the trauma. The clinical results on two human subjects could only be described as "poor" (as Romeis slyly adds, Lakatos presumably reported not the worst of his cases).

Regarding Ullmann's "decortication" of testicles, Romeis finds only "superficial protocol data."

The single patient, in Ullmann's report, successively underwent Steinach's double-sided vasoligation, Doppler's phenolization of spermatic cords, injections with testicular extracts – Testikulin and Testogan, and then the removal of the testicular cortex. (Romeis wonders how many other "rejuvenating" operations the patient would undergo.)

The destruction of testicular parenchyma was most likely responsible for whatever temporary and minor "reinvigorating" improvements there might have been. In summary, "decortication is an example of a theoretically as well as experimentally insufficient method, with a premature representation."

(Romeis, 1931, pp. 1874-1878.)

[300] The verdict is similarly severe for Michalowsky's method of ligation of testicles themselves (and not just spermatic ducts as in the original Steinach's procedure). This method was developed by Ivan Osipovich Michalowsky (1877-1937), head of the Institute of Histology at Smolensk University, Russia.

(Do not to confuse him with Ivan Petrovich Michalowsky (1877-1929), head of the Department of Physiology at Tashkent Medical Institute, Uzbekistan, who conducted far reaching experiments on rejuvenation and revival by blood transfusions and even attempted to revive recently mummified persons.)

Contrary to I. O. Michalowsky's claims, no regeneration was noticeable on the micrographs. The same patients also underwent other operations (vasoligation, prostatectomy); therefore it was difficult to specify the effects of a particular intervention. Clinical descriptions such as "the patient immediately recovered and was released in full health" seemed questionable. Thus, Michalowsky's procedure too did not appear to justify the hopes pinned on it.

The "Phenolization" (sympathico-diaphtheresis) of testicular arteries, suggested by Karl Doppler of Vienna, offered no greater hope. Doppler's underlying idea was that by irritating the spermatic cord with toxic phenol, the sympathetic (blood-vessel constricting) nerves of the testis' blood vessels would be damaged, consequently increasing the blood flow into the testis.

(The idea that the destruction of sympathetic vasoconstrictive nerves can increase local blood flow, originated with the French physician René Leriche in 1913. But instead of Leriche's surgical dissection of the sympathetic nerve, Doppler practiced chemical sympathectomy.)

Doppler believed that Steinach's vasoligation too involved damaging sympathetic nerves and therefore its effects could be explained by an increased blood flow to the testicle. According to Romeis's review, clinical trials of Doppler's sympathectomy showed only a very partial and short-lasting success. Moreover, the "reinvigoration" could be well ascribed to the patients' "auto-suggestion."

Romeis deals another blow to the "diathermy of sex glands," involving an electrical heating of the tissue with a subsequent increase in blood flow (hyperemia), first performed by Walter Kolmer and Paul Liebesny in Vienna in 1920.

Romeis describes the clinical trials of the diathermy as "too scant and not especially promising." The invigorating effects might be again due to a "degradation" and "resorption" of the testicular tissue.

The stimulation was less marked after "diathermy" than after Steinach's vasoligation, because in diathermy less tissue was destroyed and more degradation products were removed through the spermatic duct with ejaculation. The retention of sperm, due to vasoligation, was seen as beneficial, as more "degradation products" remained active in the body. In diathermy, the resulting "tonification" of the organism lasted only for a few weeks.

Even less effective and more dangerous was the "rejuvenating" irradiation of the testicles by X-rays (as performed by Harry Benjamin of New York). In 10 cases, sterilizing radiation doses were administered and in only 2 cases there was some "subjective and moderate improvement" noticeable.

Diathermy of female ovaries was conducted as well, together with the application of ovarian extracts and diathermy of the pituitary. Based on the histological structure of atrophied senile ovaries, in Romeis's view, "nothing is to be expected from the heating of these organs after the menopause." The same goes for the irradiation of the ovaries and/or pituitary.

And generally, according to Romeis, the question of how the irradiation of endocrine glands influences the aging processes was at the time fully unresolved.

(Romeis, 1931, pp. 1879-1898.)

[301] The method of sex gland tissue transplantation was chiefly associated with the name of Voronoff, but was also practiced by many others (Harms, Steinach, Lichtenstern, Lydston, Stocker, Kolb, Thorek, Sand, Wilhelm, Pettinari, Parhon, et al.), to improve sexual function specifically and fight aging generally.

The three alternative mechanisms of action were surmised to be either the gradual disintegration of the transplant, leading to the functional "reactivation" of the host's sex glands by the released breakdown products – a sort of homeopathic or autocatalytic/allosteric activation (Harms' theory); or continuous production of sex hormones by the intact, living transplant (Voronoff's theory); or the gradual disintegration of the transplant and resorption of active, hormone-rich breakdown products into the bloodstream, without any restorative action on the sex glands of the host (the view subscribed to by Romeis).

In fact, the sex gland transplantation is the one rejuvenation technique that Romeis actually favors, since "in senile animals of both sexes, the implantation of young or mature sex glands in many cases achieves a 'refreshing' of the entire organism. Single failures do not alter this determination."

In human studies, however, the results were far from conclusive.

Voronoff's summative report of 1930 on 475 cases of male sex gland transplants from apes to humans is said to be "extremely sketchy and of little demonstrative power." Voronoff claimed to have produced improvements of metabolism, such as a decrease in blood cholesterol and sugar and an increase in the general metabolic turnover, but gave no direct evidence for these claims. Only cases of "lasting improvements" were selected for the report, and cases of "short lasting" improvement were designated as failures.

In such a presentation, 90% of cases of "premature aging" (55-70 year old subjects) and about 75% of senile patients (70-85 years old) were successfully treated. Though, on the face of it, the results were remarkable, unfortunately, no specific protocols or symptoms were provided, hence "such incomplete and sporadic observations do not carry the weight of scientific evidence."

The results of other researchers were of a lesser scope and less consistent. Therefore, Romeis comes to the conclusion that "Voronoff's operation, besides the psychological influence on the patients, did not fulfill the expectations associated with it."

The results of ovary transplantations in females (from living human donors, animals such as goats and apes, or corpses) were even less encouraging.

The time of observation was short and the effects even shorter (a few months at the most). Such short lasting effects could only be explained by the breakdown of the transplants and resorption of working substances. According to some reports, after ovary transplantation, younger females showed general signs of "invigoration" and were able to ovulate. In climacteric and senile women, however, no evidence of either ovulation or regeneration was present, and functional improvements were negligible.

Romeis concludes that "until now, the combat of climacteric phenomena or age-related diseases with the help of ovary transplants affords no greater success than other methods."

Nonetheless, Romeis states that "it would be a mistake to deny on these grounds any effectiveness to the transplants, and to condemn the intervention as objectively useless."

Yet, as in the case of Steinach's vasoligation, the promises of sex gland transplants are shown to be exaggerated. The effects are found to be inconsistent, confounded and relativized by an immensity of contributing factors: the patients' age, time of resorption, origin of materials, size of transplanted pieces, place of transplantation, etc. etc.

The mechanisms of action of these procedures were far from being clear, and required a thorough further investigation.

The situation was no clearer when several endocrine glands (thyroid, pituitary and sex glands) were transplanted simultaneously. Evidently, the transplants did not "take in" and did not continue their excretory function in the body, but were degraded and reabsorbed. But it was unclear whether the degradation products effected any "reviving" action on the corresponding organs of the host.
(Romeis, 1931, pp.1898-1943.)

[302] C.E. Brown-Séquard, "Du role physiologique et thérapeutique d'un suc extrait de testicules d'animaux d'après nombre de faits observes chez l'homme," *Archives de Physiologie Normale et Pathologique*, 1, 739-746, 1889 (The Physiological and therapeutic role of animal testicular extract based on several experiments in man), reprinted in Geraldine M. Emerson (Ed.), *Benchmark Papers in Human Physiology, Vol. 11, Aging*, Dowden, Hutchinson and Ross, Stroudsburg PA, 1977, pp. 92-105, quote from p. 100; quoted in Romeis, 1931, p. 1946.

[303] Romeis, 1931, p. 1948.
Further on the history of rejuvenating and therapeutic uses of sex hormones, see:
Chandak Sengoopta, *The Most Secret Quintessence of Life. Sex, Glands, and Hormones, 1850-1950*, The University of Chicago Press, Chicago and London, 2006, Ch. 3 "Sexuality, Aging and the Gonads: New Physiology and Old Values," pp. 69-115, Ch. 5 "The New Hormones in the Clinic," pp. 153-204; Roger Gosden, *Cheating Time. Science, Sex and Aging*, Macmillan, NY, 1996, Ch. 7 "The gland grafters," pp. 150-175, Ch. 8 "Hormones come of age," pp. 176-204; Victor Cornelius Medvei, *The History of Clinical Endocrinology. A Comprehensive Account of Endocrinology from Earliest Times to the Present Day*, The Parthenon Publishing Group, Lancaster and New York, 1993, Ch. 16-18, pp. 159-240; Thomas Schlich, *The Origins of Organ Transplantation: Surgery and Laboratory Science, 1880-1930*, University of Rochester Press, NY, 2010, Part 2. "The Success of Organ Transplantation as a Concept," pp. 23-164, Part 3, "The Failure of Organ Transplantation in Practice," pp. 165-240.

[304] The new experiments with "testicular preparations" of the 1920s did not demonstrate much progress as compared to the earlier ones of the 1890s-1900s.
Thus, the results of Vladimir Korenchevsky's studies with "testicular emulsion" (1925) actually achieved "much worse results than after Brown-Séquard's injections." Animal studies by Wilhelm Lahm (1924) using

new commercial hormonal preparations (Neosex, Novotestal, etc.) were of a tiny scope and highly indeterminate with regard to weight and dose responses.

"Similarly and often even worse grounded" were the effects of other commercially available testicular preparations. "Auto-suggestion" was difficult to rule out, though apparently "not all the effects" could be ascribed to it.

Romeis expected some progress from synthesizing and applying specific sex hormones, rather than just sex gland extracts of indeterminate content.

Such studies were then beginning to be done – by Casimir Funk, Ernst Laqueur, Adolf Butenandt, Leopold (Lavoslav) Ruzicka, B. Harrow and A. Lejwa, S. Loewe and H.E. Voss, C.R. Moore, and others. But specific sex hormone isolation and synthesis were then at the cradles, and their anti-aging effects were only beginning to be explored (a single example was Harry Benjamin's study of 1930, using male sex hormones isolated from urine by the method of Casimir Funk).

Thus, it was too early to trust in the "rejuvenating" properties of male sex hormones, either natural or synthetic.

The first attempts to reactivate female animals using ovarian preparations were conducted in 1925 by Eugen Steinach, Heinrich Heinlein and Berthold Paul Wiesner.

These and other animal experiments – by E. Laqueur, W. Hohlweg and H. Kun – were partial and incomplete (only the best cases were reported, crucial data, such as the dose, age and number of animals, were not included).

"All these cases do not concern central phenomena of aging," Romeis lamented, "and presently only little is known about the influence of ovarian hormones on these phenomena." Human trials were as much as non-existent, except perhaps for a single case report by Harry Benjamin of 1928.

Generally, female hormone preparations, such as Progynon, were said to achieve successes comparable to ovary transplantation. However, the study of their anti-aging properties was only incipient and required a radical expansion and profound elaboration.

Pituitary hormone preparations, especially the gonadotropic hormones inherently linked to the sex gland function (i.e. the follicle-stimulating hormone – Prolan A, and the luteinizing hormone – Prolan B) were also considered as possible anti-aging supplements (by B. Zondek, S. Ascheim, B.P. Wiesner and S. Belawenetz).

For these substances, the results were also inconclusive:

"The contradictions that presently emerge between the observations of different authors call for further investigation. In particular, it needs to be determined how the observed effects divide between Prolan A and Prolan B. Experiments on the influence of the anterior pituitary extracts on the aging phenomena in humans, presently do not exist."

(Romeis, 1931, pp.1950, 1959, 1965.)

[305] Romeis, 1931, pp. 1966.

[306] Romeis, 1931, pp. 1966.

[307] According to Romeis, it would be completely unjustified to believe that the effects of rejuvenative interventions are restricted only to the sexual sphere or act exclusively as sexual stimulants. Rather, the rejuvenation methods affect the entire organism.

Still, the mechanism of such a general organic influence of the rejuvenating hormones was elusive:

The effects of the endocrine rejuvenating techniques could be mediated by whole-body regulating centers, such as the pituitary or the central nervous system, as surmised by Brown-Séquard.

The hormonal supplements could act by improving the general blood flow and thereby increasing the nourishment and buildup of the body, as suggested by Steinach.

The improved blood supply could remedy the dehydration and thickening of tissues that were believed to be the primary aging processes, according to Max Bürger and Georg Schlomka.

The hormonal secretions could also increase cell permeability, combating the cell dehydration and the subsequent loss of function incumbent on aging.
Still, the elucidation of these mechanisms required a vast further research.

[308] Romeis, 1931, pp. 1966-1968.

[309] Nathan W. Shock (Ed.), *A Classified Bibliography of Gerontology and Geriatrics*, Stanford University Press, Stanford CA, 1951; *Supplement One 1949-1955*, 1957; *Supplement Two 1956-1961*, 1963 (reprinted in 1980 by Stanford University Press), Subcategory: "Rejuvenation."

[310] Heiko Stoff, *Ewige Jugend. Konzepte der Verjüngung vom späten 19. Jahrhundert bis ins Dritte Reich* (Eternal Youth. Concepts of Rejuvenation from the late 19th century until the Third Reich), Böhlau Verlag, Köln, 2004 (hereafter referred to as "Heiko Stoff, *Ewige Jugend*, 2004).

[311] Heiko Stoff, *Ewige Jugend*, 2004, p. 13.

[312] As Stoff notes, the distinction between "consumerism" and "productivism" originated in the work of the American sociologist Lawrence Birken, *Consuming Desire. Sexual Science and the Emergence of a Culture of Abundance, 1871-1914*, Ithaca, London, Cornell University Press, 1988 (Heiko Stoff, *Ewige Jugend*, 2004, p. 16).

[313] Heiko Stoff, *Ewige Jugend*, 2004, p. 22.
Stoff draws here on Bruno Latour's notion of technology as an agent of stability. (Bruno Latour, *Pandora's Hope. An Essay on the Reality of Science Studies*, Harvard University Press, Cambridge, MA, 1999.)

[314] Heiko Stoff, *Ewige Jugend*, 2004, p. 23.

[315] Paul Weindling, *Health, Race and German Politics between National Unification and Nazism. 1870-1945*, Cambridge University Press, Cambridge, 1993, p. 554; Robert N. Proctor, *The Nazi War on Cancer*, Princeton University Press, Princeton, 1999, pp. 249-250.

[316] Numerous examples of National Socialist sacrificial rhetoric are provided at the "German Propaganda Archive" of Calvin College, Grand Rapids, Michigan, USA (http://www.calvin.edu/academic/cas/gpa/).
Here are some quotes from Nazi leaders:
"Sacrificing one's own life for the community is the crown of all sacrifice. Adolf Hitler" (#12, 15 - 21 March 1942, *Wochenspruch der NSDAP - Die Nationalsozialistische Deutsche Arbeiterpartei* (Weekly Quotation Posters of the National Socialist German Workers' Party –the NSDAP).
"Better to sacrifice life than loyalty" (#43, 20-26 October 1940, Weekly Quotation Posters of the NSDAP).
"The people are willing to bear any burden, even the heaviest, to make any sacrifice, if it leads to the great goal of victory" ("Nation, Rise Up, and Let the Storm Break Loose," 18 February 1943, Joseph Goebbels).
"They [men who gave their lives for the idea of Germany] hoped that by sacrificing their individual lives, a German community worthy of their lives would grow" (13 June 1936, Rudolf Hess).
And this is what a 5th grade biology textbook for girls taught (Marie Harm, Hermann Wiehle, *Lebenskunde für Mittelschulen. Fünfter Teil. Klasse 5 für Mädchen*, Hermann Schroedel Verlag, Halle, 1942), pp. 168-173:
"The drive for maintaining the species is stronger than the instinct for self preservation. Plants sacrifice themselves for their seeds. Most insects die when they have reproduced. The female rabbit defends her young against hawks, often at the cost of her own life. A fox risks its life to secure food for its young. The life of the individual can be sacrificed to assure the continuation of the species. (The law of the species is stronger than that of the individual!)"
The textbook emphasized Hitler's dictum:
"He who loves his people proves it only by the sacrifices he is prepared to make for it" (Hitler, *Mein Kampf*). (http://www.calvin.edu/academic/cas/gpa/textbk01.htm.)

[317] Another alleged version of the battle-cry of Friedrich II the Great (1712-1786, at the battle of Kolin in 1757) is "Ihr verfluchten Racker, wollt ihr denn ewig leben?" ("Rogues, do you wish to live forever?"). See Christopher Duffy, *Frederick the Great: A Military Life*, Routledge, London, 1988, p. 128 (originally published in 1985).

[318] A comprehensive history of racial hygiene and Nazi medicine is given in Paul Weindling, *Health, Race and German Politics between National Unification and Nazism. 1870-1945*, Cambridge University Press, Cambridge, 1993. The work includes accounts of the most monstrous human experiments conducted on the "inferiors" to test survival techniques for the "superiors" (as revealed in the Nuremberg trials of Nazi medical war criminals).

[319] Robert N. Proctor, *The Nazi War on Cancer*, Princeton University Press, Princeton, 1999.

[320] Robert N. Proctor, *The Nazi War on Cancer*, Princeton University Press, Princeton, 1999, p. 278.

[321] Hufeland thus expressed the non-identity of the pursuit of health and the pursuit of longevity:
"This art [of prolonging life], however, must not be confounded with the common art of medicine or medical regimen; its object, means, and boundaries, are different. ... The medical art endeavors, by corroborative and other remedies, to elevate mankind to the highest degree of strength and physical perfection; while the macrobiotic proves that here even there is a maximum, and that *strengthening, carried too far, may tend to accelerate life, and consequently, to shorten its duration*."
(*Hufeland's Art of Prolonging Life*, Edited by Erasmus Wilson, Lindsay & Blakiston, Philadelphia, 1867, pp. IX-X, emphasis added, first published in 1796.)

[322] The popularity of the Nietzschean concept of the *Übermensch* in Nazi Germany is well known and anti-democratic tendencies of Nietzschean philosophy have been suggested (e.g. Abir Taha, *Nietzsche, Prophet of Nazism: The Cult of the Superman – Unveiling the Nazi Secret Doctrine*, Author House, Bloomington, Indiana, 2005).
Indeed, in Nietzsche, the denigration of the "low," the "weak" and the "mediocre" is ubiquitous.
The Russian fin-de-siècle immortalist Nikolay Fedorov (1829-1903) was among the first to point out the inherently elitist nature of Superhumanity as advocated by Nietzsche, and considered Nietzsche as a true representative of European "Petty Aristocracy," a mouthpiece of militarized Germany, a "philosopher of the Dark Kingdom" (Fedorov, N.F., *Sobranie Sochineniy v Chetyrekh Tomakh* - N.F. Fedorov, Collected works in four volumes, Progress, Moscow, 1995, vol. 2, pp. 118-141).
Fedorov thus summarized his critique of the Nietzschean conception of Superhumanity:
"Superhumanity can either be the greatest of vices, or the greatest of virtues. It is undoubtedly a vice of satanic origin when it consists in the elevation of one or several persons above their equals, that is, above their fathers and brothers. It becomes the greatest vice, when it appropriates immortality as a privilege, when it exalts itself above all, above the deceased and those yet living. Superhumanity in this sense (as a privilege to immortality) is a vice not only moral, but intellectual. ...
But Superhumanity is also the highest virtue, when it consists in the fulfillment of the natural duty of sentient beings to unite, to transform the blind, irrational force of nature that spontaneously creates and destroys, into a force governed by reason."
(Fedorov, "Sverhchelovechestvo, kak porok i kak dobrodetel" (Superhumanity as a vice and as a virtue), in N.F. Fedorov, *Collected works*, Moscow, 1995, vol. 2, p. 135.)
I would further argue that Nietzschean philosophy is hardly compatible with the general task of life extension.
According to Nietzsche, prolonged self-preservation is the lot of mediocrity, vainly attempting to perpetuate the current perceptions of personhood and current social patterns. "Nothing will endure until the day after to-morrow," he wrote in *Beyond Good and Evil*, "except one species of man, the incurably mediocre" (*The complete works of Friedrich Nietzsche in eighteen volumes, edited by Oscar Levy and first published in 1909-1911*, Russell & Russell, New York, 1964, vol. 12, p. 237).

The Superman, in contrast, will indomitably march onto his tragic end. "Tragedy is a tonic" Nietzsche maintained in *The Will to Power* (Nietzsche, 1964, vol. 15, p. 287).

As Fedorov suggested:

"The [Nietzschean] prophecy of the 'culture of a tragic worldview' is the preparation for the future 'world catastrophe,' that is, world demise; and if all humanity must sometime die (and who doubts the inevitability of such a demise?), then its highest goal is to unify into a whole, and as a whole, imbued by the tragic realization, march forward to the forthcoming downfall."

(Fedorov, "Filosof Chernogo Czarstva" (The Philosopher of the Dark Kingdom), in Fedorov 1995, vol. 2, p. 120.)

Nietzsche does often speak of "life affirmation" and "life enhancement" and, in *The Will to Power*, refers to death as a "foolish physiological fact," opposing the dominant Christian assumption that "one should live in such a way that *one may have the will to die at the right time!*" (Nietzsche's emphasis, 1964, vol. 15, p. 338). For a brief period (around 1876) he was a follower of Cornaro's hygienic regimen (Stanley Hall, *Senescence, the last half of life*, D. Appleton & Company, New York, 1922, p. 22).

But nowhere in his writings does Nietzsche seem to overtly set longevity as a goal for the Superhuman. On the contrary, in *Thus Spake Zarathustra*, he treats the pursuit of longevity with utter contempt:

"What matter about long life! What warrior wisheth to be spared!" He makes his contempt even more explicit when he claims: "I love those who do not wish to preserve themselves, the down-going ones do I love with mine entire love: for they go beyond."

(Nietzsche 1964, vol. 11, pp. 53, 244.)

For Nietzsche, strength is by no means equivalent to longevity:

"The strong are, after all, weaker, less wilful, and more absurd than the average weak ones. They are *squandering* races. *"Permanence"* in itself, can have no value: that which ought to be preferred thereto would be a shorter life for the species, but a life *richer* in creations."

(Nietzsche 1964, *The Will to Power*, vol. 15, p. 304, emphasis in the original.)

Thus, according to Nietzsche, the ecstatic momentous enhancement of life is to be preferred over a long (and presumably conservative and boring) self-preservation.

[323] Paradigmatically, Adolf Hitler's *Mein Kampf* (My Struggle, 1923) strongly insisted on operating the German society along "natural" principles, that is, according to the laws perceived in the animal kingdom:

"When men have lost their natural instincts and ignore the obligations imposed on them by Nature, then there is no hope that Nature will correct the loss that has been caused, until recognition of the lost instincts has been restored. Then the task of bringing back what has been lost will have to be accomplished" (Vol. 2, The National Socialist Movement, Ch. 2 "The State").

"For as soon as the procreative faculty is thwarted and the number of births diminished, the natural struggle for existence which allows only healthy and strong individuals to survive is replaced by a sheer craze to 'save' feeble and even diseased creatures at any cost. And thus the seeds are sown for a human progeny which will become more and more miserable from one generation to another, as long as Nature's will is scorned" (Vol. 1, A Retrospect, Ch. 4 "Munich").

"But then Nature abhors such intercourse [between individuals of different species] with all her might; and her protest is most clearly demonstrated by the fact that the hybrid is either sterile or the fecundity of its descendants is limited. In most cases hybrids and their progeny are denied the ordinary powers of resistance to disease or the natural means of defence against outer attack" (Vol. 1, Ch. 11 "Race and People").

"When a State is composed of a homogeneous population, the natural inertia of such a population will hold the Stage [sic] together and maintain its existence through astonishingly long periods of misgovernment and maladministration. It may often seem as if the principle of life had died out in such a body-politic; but a time comes when the apparent corpse rises up and displays before the world an astonishing manifestation of its indestructible vitality" (Vol. 2, Ch. 1 "Weltanschauung and Party").

(Adolf Hitler (1889-1945), *Mein Kampf* (first published in 1923, translated into English by James Murphy, the translation was first published on March 21st, 1939 by Hurst and Blackett, NY, Reprinted in Project Gutenberg of Australia, http://gutenberg.net.au/ebooks02/0200601.txt.)

[324] Ludwig Roemheld, *Wie verlängere ich mein Leben?* (How do I prolong my life?) Ferdinand Enke Verlag, Stuttgart, 1941 (first presented in 1933), hereafter referred to as Ludwig Roemheld, 1941 (1933).

[325] Ludwig Roemheld, 1941 (1933), p. 3.

[326] Ludwig Roemheld, 1941 (1933), p. 6.

[327] According to Roemheld, in Germany, the life-expectancy increased from 37 in 1880 to 57 in 1933. In England, it increased from 43 in 1880 to 56 in 1930. And in Denmark, it rose from 43 in 1844 to 61 in 1925.

[328] Notably, several prominent contemporary German life-extensionists were physicians at resorts in the Baden-Württemberg area in Southern Germany: Roemheld was the chief sanatorium physician in Gundelsheim in Baden-Württemberg. Julius Hermann Greeff was a physician at Bad Cannstatt (a district of Stuttgart, Baden-Württemberg's capital, and one of Europe's largest mineral spa resorts). And Gerhard Venzmer was also a physician in Stuttgart (see below).

[329] Ludwig Roemheld, 1941 (1933), p. 4.

[330] Ludwig Roemheld, 1941 (1933), p. 6.

[331] According to Roemheld, the suppression of infectious diseases became possible primarily thanks to the "Three great German deeds: the discovery of anti-diphtheria serum by [Emil Adolf von] Behring, the discovery of Tuberculosis bacillus by Robert Koch and the discovery of *Spirochaeta pallida*, the cause of syphilis, by [Fritz Richard] Schaudinn."

Contributions from other nations are not mentioned.

According to Roemheld, not only hygiene, but also clinical medicine contributed in a high degree to the prolongation of the human life-span. "The fabulous successes of surgery," thanks to refined operation techniques and antiseptics, often permitted to "tear the patients away from the threshold of death."

Internal medicine is also considered to be a crucial factor for life-extension. Roemheld rages against the popular perception of internal medicine as a "stepchild" in the medical family, a specialty doing nothing more than to "overwhelm the body with poisons." He points out that most severe, life-shortening diseases belong to the care of internal medicine, and the field undergoes rapid progress.

The life-prolonging means in the arsenal of internal medicine include prophylaxis of heart and kidney diseases through special dietetic measures (such as salt-free nutrition), removal of seats of infection (tonsils, teeth), etc.

Another example of the power of internal medicine is the discovery of insulin that dramatically reduced the suffering and prolonged the lives of millions of diabetic patients. Here too, the contribution of German scientists is said to be paramount. Roemheld claims that insulin was developed by "the Americans," "based on German research" by George Ludwig Zuelzer.

(Actually, the recipients of the Nobel Prize of 1923 for insulin discovery were the Canadians Frederick Grant Banting and Charles Herbert Best, and the priority dispute over insulin discovery also involved the Romanian Nicolae Paulescu, the Americans Israel Simon Kleiner and Ernest Lyman Scott, the Polish-German-Jewish Oscar Minkowski, and others. But Roemheld's point is that Germany has all the priority.)

Similarly, the use of Strophanthin (or Ouabain, a cardiac glylcoside) against congestive heart failure or cardiac arrhythmia (still used today) "prolonged the life of heart patients for years and decades, as well as made their suffering bearable." Roemheld thanks the "practical physician Fraenkel from Badenweiler" for the discovery of Strophanthin. (Albert Fraenkel, 1864-1938, was a German Jew, a fact which Roemheld omits. Fraenkel was stripped of his professorship in 1933 and of his medical license in 1938.)

Notably, no "rejuvenation" surgeries or supplements are listed among the life-prolonging triumphs of medicine.

(Ludwig Roemheld, 1941 (1933), pp. 8-9.)

[332] Ludwig Roemheld, 1941 (1933), p. 9.
[333] Ludwig Roemheld, 1941 (1933), p. 10.
[334] Ludwig Roemheld, 1941 (1933), p. 11.
[335] Ludwig Roemheld, 1941 (1933), p. 11.
In Roemheld, the general factors determining human longevity are almost identical to those proposed by Hufeland: "the innate quantity of vital power," "firmness of organization of the vital organs," and the rates of "consumption" and "renovation" ("restoration" or "regeneration") of the vital force and of the organs (*Hufeland's Art of Prolonging Life*, Edited by Erasmus Wilson, Lindsay & Blakiston, Philadelphia, 1867 (1796), pp. 48-49).
[336] According to Roemheld, quite consistent with the general fascist stance, the children that undergo the "hardening" will be more "useful and healthy" and will have an advantage in longevity over the "feeble and spoiled" ones.
The youth need to be strengthened by a "natural life-style, sports, moderate simple nutrition (particularly avoiding excesses of meat), avoidance of unnatural stimulants of the body and mind, successful suppression of the sex drive until the time of manhood."
Early sex-life, before proper sexual development, is said to be detrimental to longevity, and its dangers need to become a subject of sex education. Marriage, on the other hand, "undoubtedly prolongs life through the orderly regulation of behavior" (Ludwig Roemheld, 1941 (1933), pp. 11-12)
[337] Similar recommendations were given in Wladislaw Klimaszewski, *Gründliche Gesundung. Vollkraft, Erfolg, Verjüngung*, 7. Auflage, München, 1937 (Fundamental health improvement: Full power, Success, Rejuvenation, 7th edition, first published in 1925).
[338] A series of other measures for life-extension is proposed.
Through the joint efforts of the medical authority and each and every individual, the German people are called to battle against life-shortening diseases, the "dangerous enemies," with "science providing the weapons."
In the first line of combat are "diseases of the constitution" aggravated by "hereditary burdens" – emaciation and obesity, gout and diabetes. "It is the task of the controlling physician to regulate the life-style of the patients so as to achieve prevention."
A wide arsenal of means is recommended to combat infectious diseases (which were at the time still the main cause of death in Germany). The first rule is to "avoid infection, unless it is impossible due to professional duty."
For cancer, the earliest possible detection and treatment are said to be of crucial importance for extending the life-expectancy.
For gastro-intestinal diseases, moderate nutrition, rich in fiber and vitamins, is recommended.
Roemheld refers to the controversy raging in Germany regarding the life-prolonging properties of the so-called "raw" (uncooked, unprocessed, "natural") foods (*Rohkost*), mainly consisting of vegetarian dishes. He admits that there might be some benefits from such a diet for some people, and it can be enjoyed for some periods of time. Yet, he believes, human beings are no vegetarians by nature, and the faddist claim that "moderate consumption of meet shortens life is an unproven fable."
Generally, any "one-sidedness" in diet can be dangerous. Roemheld recalls Paracelsus' old slogan "the dose makes the poison" and points to the importance of a personalized approach to diet (e.g. bread can be a poison for a diabetic and meat can be poisonous for sufferers of gout and arteriosclerosis).
Excessive consumption of salt and strong spices is to be avoided. The dangers of "luxury foods"/"stimulants"/"enjoyment poisons" (*genussmittelen* or *genussgifte*) are emphasized. For example, excessive alcohol consumption is warned against, yet there is little harm in "a little glass of wine to improve appetite." "Of all the different stimulants, coffee is the least detrimental for longevity." But the "best drinks are milk and yogurt."

Regarding "nervous deterioration" and "diseases of the heart and of the circulation system," it is noted that these diseases take one of the highest death tolls in civilized countries. And yet, against these diseases, the prevention via a "rational life-style" and "avoidance of damage" works most successfully.

Several particular preventive measures are recommended: first and foremost, exercise (though the dangers of exhaustive, excessive high-performance sports are noted), the "hardening" (Abhärtung) by cool air and water baths, the "diaphragmatic" (abdominal) breathing, sufficient rest, sleep and equanimity.
(Ludwig Roemheld, 1941 (1933), pp. 11-23, 26-32.)

[339] Ludwig Roemheld, 1941 (1933), p. 23-24.

[340] As Roemheld attests, a great variety of rejuvenating nostrums were widely promulgated in contemporary Germany: "today iodine becomes famous, tomorrow garlic or some other chemical preparation."

The danger of such fads may outweigh their benefits. Slimming treatments, for example, may not be entirely ineffectual, "when conducted under the control of a physician, these are life-prolonging treatments." However, uncontrolled, improper self-medication with slimming endocrine preparations (thyroid, lipolysin, etc.) put the consumers in great danger.

Roemheld attests that "since the discoveries of Steinach, sex hormones have played a great role" among such nostrums. "Almost daily, new preparations are recommended by enterprising companies to strengthen sexual function, especially in men, and to postpone aging." The uncontrolled use of such preparations is strongly discouraged.
(Ludwig Roemheld, 1941 (1933), p. 25.)

[341] Ludwig Roemheld, 1941 (1933), p. 26.

[342] Gerhard Venzmer, *Lang leben und jung bleiben!* (Live long and stay young), Franckhsche Verlag, Stuttgart, 1937 (republished in 1941), hereafter referred to as "Venzmer, 1937."

[343] Venzmer, 1937, p. 13.

[344] Venzmer, 1937, pp. 24, 26.

[345] Venzmer, 1937, pp. 45-46.

According to Venzmer, the exact value of a "natural" life-span is very difficult to determine. Yet, based on growth-to-lifespan time ratios in animals, "the crown of creation should live to be 195! [15 years until reaching sexual maturity multiplied by 13]." This estimate, in Venzmer's view, agrees with the maximal life-span of 180-200 years assigned to man by Hufeland and Darwin.

These calculations of the maximal human life-span of about 200 years were suggested by the renowned German physiologist Eduard Pflüger (1829-1910), based on his studies of ratios of the periods of growth to life-span in animals. The German physiological chemist August Pütter (1879-1929), on the other hand, determined a "natural" human life-span to be in the area of 100 years, based on the maximal human life spans currently observed.
(Eduard Friedrich Wilhelm Pflüger, *Über die Kunst der Verlängerung des menschlichen Lebens* (On the art of prolonging human life), Strauss, Bonn, 1890; August Fr. Rob Pütter, "Altern und Sterben" (Aging and Dying), *Virchows Archiv für pathologische Anatomie und Physiologie*, 261 (2), 393-424, 1926.)

"There is a variety of opinions about the 'natural' duration of human life," Venzmer notes, "perhaps it is even impossible to precisely define how old a person 'must really be.' It is likely that there are no rigid physiological laws for the duration of human life; rather, it is sufficient to know that this duration broadly depends on hereditary structure and predisposition to a long or short life, on the power of internal defenses, on the degree of damage from the environment, on the height of civilization, on the well being and personal life-style of the individual. …

We need not lament this gap in our knowledge. Rather, we have a much more valuable insight, namely, that life can really be prolonged, as modern experiences unequivocally demonstrate"
(Venzmer, 1937, pp. 45-46).

Regarding the supposition of the 200 year human life-span by Darwin (Venzmer, 1937, pp. 43, 47), the source of this quotation from Darwin could not be found. Darwin was indeed somewhat interested in life extension (see the subsequent section on "British Longevity Research"). The subject is discussed by Charles Darwin in his *Notebook E: [Transmutation of species, 1838-1839]*, and in *The Variation of Animals and Plants under Domestication*, 2nd Edition, John Murray, London, 1875, vol. 1, pp. 455, 476, vol. 2, pp. 56, 376 (reprinted in *Darwin Online* http://darwin-online.org.uk/). However, no reference by Charles Darwin to particular values of maximum human longevity could be found.

It is also possible (though less likely) that Venzmer refers here to Charles Darwin's grandfather, the physician Erasmus Darwin (1731-1802) who wrote about life-extension in *The Temple of Nature; or, the Origin of Society. A Poem, with Philosophical Notes*, T. Bensley, London, 1803, Canto II. 1.4. Note 7. II "Means of preventing old age" and further in *The Botanic Garden*, 1789, Part II, l. 387, and in *Zoonomia*, 1796, Vol. 1. Sect. XXXVII. IV. "Old age and death from inirritability," V. "Art of producing long life" (Reprinted in *Project Gutenberg*, http://www.gutenberg.org/ebooks/26861). However, no specific values for maximal human longevity could be found in Erasmus Darwin either.

Hufeland's *Makrobiotik*, on the other hand, unequivocally posited a "natural" human life-span of 200 years (*Hufeland's Art of Prolonging Life*, Edited by Erasmus Wilson, Lindsay & Blakiston, Philadelphia, 1867 (1796), p. 109).

[346] Venzmer, 1937, pp. 7-8.

[347] Venzmer, 1937, p. 8.

[348] On the strong ties between German and American eugenicists, see Edwin Black, *War Against the Weak: Eugenics and America's Campaign to Create a Master Race*, Thunder's Mouth Press, New York, 2004 (2003).

[349] Venzmer, 1937, pp. 84-85.

[350] Venzmer, 1937, pp. 101-103.

[351] Venzmer, 1937, pp. 106-107.

[352] Venzmer, 1937, p. 125.

[353] Soldier Mittelstedt's case was earlier discussed by Hufeland in *Macrobiotics* (*Hufeland's Art of Prolonging Life*, 1867 (1796), Ch. 5, "Instances of Longevity," pp. 92-93; by George M. Gould and Walter L. Pyle in *Anomalies and Curiosities of Medicine* (1896, Ch. 8 "Longevity," http://www.gutenberg.org/files/747/747-h/747-h.htm) and several other authors.

[354] Robert Koch (1843-1910) and Emil Adolf von Behring (1854-1917) broke the grounds in microbiology and created the first anti-bacterial sera.

Ignaz Semmelweis (1818-1865) pioneered antiseptic obstetrics and thus helped to radically reduce child mortality (though an Austro-Hungarian, he could be annexed to the German tradition).

Max Joseph von Pettenkofer (1818-1901) established the field of public hygiene.

Carl von Voit (1831-1908) founded the "new science of nutrition."

[355] The tradition of German longevity research, referred to by Venzmer, was extensive indeed: Christoph Wilhelm Hufeland (1762-1836) was one of the earliest and most influential modern theorists and practitioners of life-extension hygiene.

August Weismann (1834-1914) proposed the inherent immortality of the living matter and suggested tissue differentiation to be the primary cause of senescence and death.

Otto Bütschli (1848-1920) and David von Hansemann (1858-1920) hypothesized the existence of a particular "life enzyme" which originated in sex glands and provided the organism with youthful vitality. Accordingly, the exhaustion of the "life enzyme" resulted in senescence.

Eduard Pflüger (1829-1910) wrote extensively "On the Art of Prolonging Human Life" (*Über die Kunst der Verlängerung des menschlichen Lebens*, Strauss, Bonn, 1890). He posited a "Growth Drive" (*wachstumstrieb*) whose diminishment leads to aging and death. According to Pflüger's calculations of the periods of growth versus the entire life-course, the human life should be in the vicinity of 200 years.

A similar concept was considered by Richard von Hertwig (1850-1937) and by August Pütter (1879-1929). Pütter, however, believed the "normal" life span should be in the range of 100 years.

Max Rubner (1854-1932) related the life-span to the sum total of metabolic energy expenditure.

Emil Abderhalden (1877-1950) developed the colloidal flocculation theory of aging. (Abderhalden, the president of the German Academy of Natural Scientists "Leopoldina" from 1931 to 1950, was a crucial figure in the longevity research during the Nazi period, and will be discussed in greater detail further on.)

[356] Mortality statistics by Hermann Nothnagel and Georg Wolff suggested points of socio-demographic intervention.

Rules for healthy, life-prolonging (mixed) nutrition were developed by Carl von Noorden, Adolf Bickel and many others.

Trail-blazing work was done by German scientists on the use of vitamins (particularly vitamins C and A) to combat infectious and age-related diseases. (The leading proponent of vitamin C therapy in the US, the Nobel laureate Linus Pauling, recognized the works of the Nazi German physicians Roger Korbsch and Hans Ertel as pioneering in this area. See Linus Pauling, *How To Live Longer And Feel Better*, W.H. Freeman and Company, NY, 1986, p. 112.)

[357] Indeed, there were discussions of centenarians and supercentenarians ever since ancient Greece, as in Herodotus (c. 484-425 BC), *History* (B. i., c. 163); and in the Roman Empire, as in Cicero (106 BC-43 BC), *De Senectute* (On Old Age, section 19); Valerius Maximus (worked c. 14 AD-37 AD), *Factorum ac Dictorum memorabilium* (Memorable deeds and sayings, B. viii. c. 13); Lucian (c. 125 AD-180 AD), *De Macrobiis* (On Long-lived Men, though of disputed authorship), and Pliny the Elder (Gaius Plinius Secundus, 23 AD-79 AD), *Naturalis Historia* (The Natural History, published about 77-79 AD) among other sources. (Quoted in *Natural History of Pliny*, translated by John Bostock and H.T. Riley, Henry G. Bohn, London, 1855, vol. 2, Book 7, Ch. 49. "The greatest length of life," and note 41 *ibid*, p. 200.)

And the discussions of centenarians continued all through the beginning of the 20th century, at times including some considerations of their life-style and diet. Thus the 19th century English physician Charles Watkyns de Lacy Evans wrote:

"On reviewing nearly two thousand reported cases of persons who lived more than a century, ... we notice great divergence both in habits and diet, but in those cases where we have been able to obtain a reliable account of the diet, we find one great cause which accounts for the majority of cases of longevity, *moderation in the quantity of food.*"

(Charles Watkyns de Lacy Evans, *How to Prolong Life*, Sawyer, London, 1910 (1879), Ch. 4, p. 93, emphasis in the original; also quoted in F.C. Havens, *The Possibility of Living 200 Years*, The Two Hundred Company, San Francisco, California, 1896, pp. 184-185.)

The lives of centenarians were a topic of interest in the works of such 19th century authors as Helen Densmore, Sir Henry Thompson, Sir James Crichton-Browne, Wilhelm Ebstein and Herbert Spencer.

Yet, it seems, none of these early works had included any specific demographic and epidemiological data on the centenarians and in most cases the estimates of age were not very reliable.

It appears that one of the earliest scientific works that did include reliable demographic-epidemiological data on centenarians was that of Dr. Julius Hermann Greeff of Bad Cannstatt at Stuttgart.

The study was started in the late 1920s and published in 1933. It examined 124 centenarians (81 women and 43 men), selected according to official state statistical records, with the purpose to characterize the factors that may have contributed to their longevity.

The data included areas of habitat, distinctions between men and women, data on the centenarians' parents and children, and exact details about the subjects' health status, habits, occupations and diets. This was apparently one of the first such dedicated studies in the world.

(J. H. Greeff, "Hundertjährige" (Centenarians), *Archiv für Rassen- und Gesellschafts-Biologie* (Archive for Racial and Social Biology), 27 (3), 241-270, 1933.)

It appears that the only former cognate study, that Greeff cited, was published only slightly earlier, in 1929, by Dmitri Michaykoff, Professor at the University of Sofia, Chairman of the General Council of Statistics of the Kingdom of Bulgaria (Dmitri Michaykoff, *Les centenaires en Bulgarie* (The centenarians in Bulgaria), Imprimerie de L'Etat, Sofia, 1929).

The first concerted team expedition to study centenarians (in Abkhazia, Georgia) was led slightly later, in 1937, by the Soviet-Russian Professor Ivan Basilevich from Kiev, Ukraine. (I.V. Basilevich, "Syndrom normalnoy starosti" (The syndrome of normal aging), pp. 255-308, in *Starost* (Aging), Kiev, 1938, see the next chapter on Russian life-extensionism.)

[358] Thus, according to Greeff, the long-lived were characterized by high fertility (on the average, 5 children for men and 6 for women). All women but one, and all men but two, were married (and these two men lived the longest of all).

The dietary habits varied greatly, with preferences ranging from oatmeal gruel to bacon, but "moderation" was noted as a part of most diets. Yet, among all the centenarians examined, both men and women, there was not a single vegetarian. The vast majority had a "liking for fat." Greeff hypothesized that fat was needed to retain some "vitamin reserve" necessary for longevity.

None of the women consumed tobacco, but all men, with a single exception, smoked. The vast majority, both men and women, also exhibited a "significant enjoyment of alcohol."

Most people "even at the highest age were excellent walkers. Thus we see that walking, without a doubt, is one of the best means to preserve human health" (p. 264).

Generally, Greeff dispelled the fears of "over-aging":

"Is the great experience of a long human life worth nothing? The existence of many old people signifies a healthy national power and a high quality of hygiene and medical art."

He fully agreed with the Austrian banker, hygienist and parliamentary Alfred von Lindheim (1836-1913) who stated in *Saluti Senectutis* (Healthy Old Age, 1909) that "Long life is a great benefit for the individual, for the state and for the entire humanity" (p. 258).

(J. H. Greeff, "Hundertjahrige" (Centenarians), *Archiv fur Rassen- und Gesellschafts-Biologie* (Archive for Racial and Social Biology), 27 (3), 241-270, 1933.)

[359] Fahrenkamp's research on the psychosomatic regulation of the cardio-vascular system extended from the 1920s to the 1940s.

(Karl Fahrenkamp, *The Psychophysical Interaction in Diseases Associated with Hypertension*, Hippocrates Verlag, Stuttgart - Berlin, 1926; Karl Fahrenkamp, *Kreislauffürsorge und gesundheitsführung* (Care of the circulation system and healthy life-style), Hippokrates-Verlag, Stuttgart, 1941.)

[360] Though not mentioned by Venzmer directly, the Danzig physician Erwin Liek (1878-1935) was one of the most influential proponents of the "natural" and "holistic" approaches to longevity and one of the fiercest detractors of rejuvenative surgery.

Liek attacked sympathectomy and other methods of surgical rejuvenation as "aberrant ways of surgery" (*Irrwege der Chirurgie*, 1929). Rather, Liek generally considered the effects of sympathectomy to be due to the "interrelation of the Psyche – Vascular System" (Romeis, *Altern und Verjüngung* (Aging and Rejuvenation), 1931, p. 1895).

According to the historian Robert Proctor,

"[Liek] is widely reviled today as the 'father of Nazi medicine.' ... As founding editor of *Hippokrates*, a magazine of general health interest with strong ties to homeopathy and the natural foods movement, [Liek] helped to usher in a broader and more holistic medicine of the sort embraced by many Nazi leaders – medical men like Kurt Klare, Karl Kötschau, Walter Schultze, and Ernst Günther Schenck, but also high-placed politicos like Heinrich Himmler, Julius Streicher, Rudolf Hess, and even Hitler himself."

(Robert Neel Proctor, *The Nazi War on Cancer*, Princeton University Press, Princeton, 1999, pp. 22.)

Liek is cited as a chief health authority by Roemheld (*Wie verlängere ich mein Leben?* 1941 (1933), p. 4).

[361] An exemplary quote is from the "Dedication" of Goethe's *Faust*:
"Naught, yet enough had I when but a youth, ... Give impulse its unfettered dower, ... Oh, give me back my youth again! ... Youth, my good friend, you need most in the fight ... Age makes not childish, as one oft avers; It finds us still true children merely."
(Johann Wolfgang Goethe, *Faust* [1808-1832], translated by George Madison Priest, Franklin Center, PA, The Franklin Library, 1979, "Dedication," http://einam.com/faust/index.html, quoted in Venzmer, 1937, p. 204.)

[362] Venzmer, 1937, p.72.

[363] Heinrich Wilhelm Gottfried von Waldeyer-Hartz, "Über Karyokinese und ihre Beziehungen zu den Befruchtungsvorgängen" (On caryokinesis and its relations to fertilization processes), *Archiv für mikroskopische Anatomie und Entwicklungsmechanik*, 32, 1-122, 1888;

Heinrich Wilhelm Gottfried von Waldeyer-Hartz, "Über einige neuere Forschungen im Gebiete der Anatomie des Centralnervensystems" (On some new research in the area of the anatomy of the central nervous system), *Deutsche medizinische Wochenschrift*, Berlin, 17, 1213-1218, 1244-1246, 1287-1289, 1331-1332, 1350-1356 - 1891.

HWG Waldeyer authored several pioneering works on sex cell development, that Venzmer might refer to, such as *Eierstock und Ei* (Ovary and Egg), 1870, or *Die Geschlechtszelle* (The sex cells), 1879.

Yet, insofar as Venzmer, like many other contemporary German writers, generally does not mention the cited authors' first names, it is also possible that he refers here to Leonhard Waldeyer, author of "Zur Frage der Reaktivierung von senilen menschlichen Ovarien" (On the question of reactivation of senile human ovaries), *Zentralblatt für Gynäkologie*, 58, 2882-2891, Dec. 8, 1934.

[364] Anne Harrington, *Reenchanted Science. Holism in German Culture from Wilhelm II to Hitler*, Princeton University Press, Princeton, New Jersey, 1996, Ch. 6 "Life Science, Nazi Wholeness, and the "Machine" in Germany's Midst," Section "The 'Jew' as Chaos and Mechanism," p. 181.

It should be noted that recent psychometric studies indicated that Jews on the average indeed tend to score somewhat worse on visuo-spatial (holistic-perceptual) abilities and somewhat better on analytical (reductionist-perceptual) abilities. (Gregory Cochran, Jason Hardy, Henry Harpending, "Natural History of Ashkenazi Intelligence," *Journal of Biosocial Science*, 38 (5), 659-693, 2006.) This would not seem to be a good reason to eliminate the entire analytical approach (or its proponents for that matter). Still, in the Nazi discourse, the analytical/mechanistic approach was thoroughly depreciated, including the reductionist/analytical/mechanistic "rejuvenation."

Notably, however, the Nazi valorization of "wholeness" over "mechanism" (present in Venzmer's book as well) appears to be rather feigned. The book includes many extremely "mechanistic" descriptions, illustrating the 'checks and balances' of the human organism. Also many techniques of reductionist rejuvenation (most notably hormone supplements) are still included in the "macrobiotic" arsenal.

Only the notions of "Jews" and "Rejuvenation" are excluded in an ostensibly political-rhetorical attempt to get rid of both in one stroke.

[365] Interestingly, in 1975, a book was published in Yiddish in France by Paul Hirschmann with almost precisely the same title as Venzmer's – *Lebn Lang un Blaybn Yung* – Live Long and Stay Young.

Generally, it is structured very similarly to Venzmer's work, including chapters about theories of aging, comparative longevity, suggestions on diet and exercise to combat age-related diseases. However, in Hirschmann's work, Jews and rejuvenation techniques figure prominently.

(Paul Hirschmann, *Vivre Longtemps et Rester Jeune* (Live long and stay young, originally appeared in the Yiddish translation by Dr. Noah Gris, entitled *Lebn Lang un Blaybn Yung*), Oyfsnay, Paris, 1975, Ch. 10, "Der kampf kegen freezeitiger elterung bemeshech fun doyres" (The struggle against premature aging through generations), pp. 156-176.)

[366] Venzmer, 1937, pp. 147-148.

[367] Venzmer, 1937, "Die bekämpfung vorzeitigen alterns durch wirkstoffe" (The struggle with premature aging by active substances), pp. 147-156.

[368] Venzmer was the author of a very popular book on endocrinology: Gerhard Venzmer, *Deine Hormone – Dein Schicksal? Von den Triebstoffen unseres Lebens* (Your Hormones – Your Fate? The Fuel of Our Life), Franckh'sche Verlagshandlung, Stuttgart, 1940 (first published in 1933), including an extensive discussion on the endocrinological determination of the character type and race (Ch. XVIII, "Wirkstoffe und Wesensart; Hormone und Rasse, das 'Horoskop' der Zukunft" – "Active substances and character; hormones and the race, the 'horoscope' of the future," pp. 178-194).

(Gerhard Venzmer, *Deine Hormone – Dein Schicksal?* 1940 (1933), pp. 185-186, 192-193, and further on the subject in Gerhard Venzmer, *Dein Kopf – dein Charakter* (Your Head – Your Character), Francksche Buchhandlung, Stuttgart, 1934 – *Deutsches Ärzteblatt*, 1934, Nr. 14.)

Yet, in *Your Hormones – Your Fate?* rejuvenative operations on endocrine organs are thoroughly rejected. They are "not only unsafe, but also involve great material and conceptual difficulties" (Venzmer, *Deine Hormone*, 1940, p. 65.)

Also, in *Lang Leben und Jung Bleiben*, rejuvenative operations on endocrine glands are emphatically dismissed. They may be useful only for "a few single cases" and represent a "bygone chapter in medicine."

Furthermore, Venzmer believed that self-medications with at the time widely available and aggressively advertised "hormonal rejuvenation means" are "doomed to failure" and can "discredit hormone therapy." Their consumption is "money thrown away."

Nonetheless, hormonal preparations, when administered by a "learned physician" and in accordance to "personal constitution and hereditary predisposition" can go a long way in combating "untimely aging." These may include sex hormone preparations for men and women, pancreatic, thyroid, pituitary and adrenal extracts, their combinations, with possible additions of vitamins (vitamins C and A are recurrently mentioned) and minerals (Zink, Iron, Calcium, Iodine, etc.).

(Venzmer, 1937, pp. 147-156.)

[369] Venzmer, 1937, p. 152.

[370] Venzmer, 1937, p. 150.

[371] Venzmer, 1937, p. 124.

[372] *Hufeland's Art of Prolonging life,* Edited by Erasmus Wilson, Lindsay & Blakiston, Philadelphia, 1867, p. 18.

[373] Edwin Black, *War Against the Weak: Eugenics and America's Campaign to Create a Master Race*, Thunder's Mouth Press, New York, 2004 (2003), p. 299.

[374] Gerhard Venzmer, *Erbmasse und Krankheit. Erbliche Leiden und ihre Bekämpfung* (Heredity and disease. Hereditary suffering and its elimination), Franckh'sche Verlagshandlung, Stuttgart, 1940 (1933), pp. 39-41.

[375] Venzmer, 1937, p. 80.

[376] Advice on nutrition occupies the central place. Venzmer delineates the bitter controversies in nutrition politics:

"Unfortunately, one cannot write about nutrition without making some part of humanity his bitter enemies. A fanatic of a 'correct nutrition' makes no concessions, for him 'his' way of nutrition is the only right one, and whoever thinks otherwise is an idiot, and possibly also a 'traitor to national health' or a 'paid agent of certain industries.'"

Venzmer ridicules the excessive zeal of partisans of particular dietary programs:

"It is a shame how the fanatics of particular forms of diet, with every bite they take, constantly worry whether the food they consume is the 'right one.' They eventually feed themselves so 'healthy' that, through sheer anxiety, they cannot enjoy the 'health' that they buy so dearly" (Venzmer, 1937, p. 127).

Any "one-sidedness" in diet is said to be detrimental. The rampant contemporary German obsession with "raw" foods and vegetarian foods (Hitler's famous dietary preference) is disputed. An analysis of the structure of human teeth and intestines leads to the conclusion that "an ordinary person is no vegetarian by nature." Another argument against the refusal of meat is that it can cripple German economy.

Venzmer's own recommendation is "mixed" nutrition, as diverse as possible, or "alternate" nutrition with different types of foods emphasized on particular days of the week (*Wechselkost*), e.g. "fruit day," "vegetable day," "cereal day," "milk day," etc. (recipes are included). Foods rich in vitamins and dietary fiber ("ballast materials" or cellulose) are recommended. And, of course, the most important rule is moderation in the consumption of food and drink, which is said to be the key to life-extension.

377 Moderation is also said to be the key factor in the consumption of the "means of enjoyment" or "luxury foods" (*genussmittelen*), such as alcohol, tobacco or coffee.

Venzmer emphasizes the dangers posed by alcoholism and tobacco addiction to the national health and longevity. The danger of alcoholism was then perceived to be so great that the "Law for the Prevention of Genetically Diseased Offspring" stipulated sterilization in particular cases of severe alcoholism. Venzmer fully sympathizes with the anti-tobacco campaign led by Dr. Fritz Lickint, including restrictions of tobacco sales, cultivation of "nicotine-free" or "nicotine poor" tobacco, and massive propaganda efforts explaining the dangers of tobacco abuse.

Yet, according to Venzmer, if used in moderation, the "means of enjoyment" may cause no great harm. "Among the German centenarians, examined by Dr. J.H. Greeff, almost everybody, both men and women, both in youth and in old age, appreciated a 'good drop' of alcohol, and all the men, with a single exception, were not averse to smoking" (Venzmer, 1937, p. 140).

"Whoever can do without these means of enjoyment, which in the right measure can lighten our existence, let him be praised. However, a person who takes a gentle drop of alcohol with friends, who after a joyful meal takes in a liquor, or smokes a cigar or a cigarette, who dispels morning or after-work fatigue with a cup of coffee – needs not fear that he shortens the years allotted to him" (Venzmer, 1937, p. 147).

378 Additional hygienic advice concerns exercise, particularly regarding the "hygiene of the cardiovascular system" – gymnastics, breathing exercises, "tempering" air and water baths. "Disturbances of circulation" are seen as formidable "national diseases" gravely undermining national stamina. Yet, at the same time, it is emphasized that these diseases are the most susceptible to preventive hygienic measures, often neglected because of common sloth and inertia.

Activity, according to Venzmer, needs to be balanced by "healthy sleep" for the "restoration of the nervous system." Rest, generally, is said to be a mighty means toward life-extension, that can neutralize the detrimental effects of the exhaustively hectic life-style in "civilized societies." "Free time," "holidays," "hobbies" are all means for life-extension.

Further hygienic recommendations relate to the combat of major age-related diseases, such as "menopause in men and women," obesity, gout, diabetes, arthritis, osteoporosis, hearing and sight impairment, paradontosis, gastrointestinal disturbances, prostate enlargement, arteriosclerosis, dementia, cancer, etc. The recommendations are quite general, and summarized in the discussion of cancer:

"Man should avoid irritants of any kind, … avoid chronic inflammation, enjoy luxury foods with reserve, not eat very hot spices, neither swallow up his meals, but chew carefully, ensure regular bowel activity, use rational nutrition containing as little "denatured" products as possible, and including accompanying substances in the amounts and compositions necessary for normal metabolism and healthy balance of fluids…. Mixed nutrition is most beneficial and should not contain too much animal fat or carbohydrates, neither excessive amounts of salt and spices. Finally, appropriate body movement should enhance blood circulation, and aid in the removal and excretion of poisonous waste products of metabolism" (Venzmer, 1937, pp. 186-187).

Skin and dental care, massages, cosmetics and general cleanliness too play an important role in "graceful aging," in "becoming old, while remaining young." Tooth decay is linked to a variety of diseases, and the importance of dental hygiene is strongly stressed. Skin care is mentioned as a veritable contributor to general health and longevity, and cosmetic care as highly important for the self-esteem and enjoyment of life of the aged.

[379] Venzmer, 1937, "Ärger und Verstimmung als Verkürzer der Lebenserwartung" (Anger and dissatisfaction shorten the life-expectancy), pp. 88-93.

[380] Venzmer, 1937, "Altern und Seele" (Aging and the soul), pp. 197-207.

[381] Venzmer, 1937, p. 206.

[382] Sigismund Thaddea, "Die Probleme des Alterns beim schaffenden Menschen. Die Gesundheitsführung – Ziel und Weg," *Monatsschrift des Hauptamtes der NSDAP, des Sachverständigenbeirates und des Nationalsozialistischen Ärztebundes*, e.V. Berlin, 1940, Teil I (The problem of aging for creative people. Health guidance – the goal and the way, Monthly of the Main Office of the National Socialist German Workers' Party, Expert Advisory Council and National Socialist Medical Association, Part 1, Berlin, 1940, p. 463). Quoted in Brigitte Michel, *Gesundheitsförderung und Prävention im Alter (Health Promotion and Prevention in Old Age)*, Dissertation, Medizinische Fakultät Charité - Universitätsmedizin Berlin, 2006, Ch. 6, pp. 78-79, http://www.diss.fu-berlin.de/diss/receive/FUDISS_thesis_000000002467.

[383] J. Steudel, "Zur Geschichte der Lehre von den Greisenkrankheiten" (On the history of the study of the diseases of old age), *Sudhoffs Archiv für Geschichte der Medizin und der Naturwissenschaften*, 35, 1-27, 1942 (quote from p. 26).

[384] W. Klimaszewski, *Gründliche Gesundung. Vollkraft, Erfolg, Verjüngung*, 7. Auflage, München, 1937 (Fundamental health improvement: Full power, Success, Rejuvenation, 7th edition, first published in 1925), "Nationalökonomie – Volkswirtschaft" (National economy – People's economy), p. 120, "Daurende Vollkraft" (Durable full power), p. 131, "Verjüngung" (Rejuvenation), p. 178.

[385] *Zeitschrift für Altersforschung. Organ für Erforschung der Physiologie und Pathologie der Erscheinungen des Alters* (Journal for Aging Research: Periodical for the research of physiology and pathology of the phenomena of aging), Dresden-Leipzig, Band 1, Heft 1 (volume 1, issue 1) 1938. Notably, the journal's first volume is catalogued as of 1939. However, its first introductory issue is dated as of July 1938. Hereafter, the establishment of the journal will be noted as of 1938.

[386] Nathan W. Shock, "Physiological Aspects of Aging in Man," *Annual Review of Physiology*, 23, 97-122, 1961; Watanabe S, "Outline of gerontology in Japan and some aspects of life span and causes of death in the Japanese," *Journal of Gerontology*, 14, 299-304, 1959; Stephen J. Cutler and Jon Hendricks, "Gerontology," in *Macmillan Reference USA's Encyclopedia of Aging*, NY, 2002; Nobuo Yanagisawa, "Geriatrics and Longevity Sciences in Japan," *Internal Medicine*, 36(10), 667-668, 1997.

[387] Dimu A. Kotsovsky, "Gerontology in Eastern Europe," *Review of Eastern Medical Sciences*, 13, 7-16, 1958.

[388] Bürger M, Abderhalden E, "Zur Einführung" (Introduction), *Zeitschrift für Altersforschung. Organ für Erforschung der Physiologie und Pathologie der Erscheinungen des Alters* (Journal for Aging Research: Periodical for the research of physiology and pathology of the phenomena of aging), Dresden-Leipzig, 1(1), 1-2, 1938.

[389] Max Bürger, *Altern und Krankheit* (Aging and Disease), Zweite Auflage (2nd Edition), Veb Georg Thieme, Leipzig, 1954 (1947); Werner Ries, *Max Bürger (1885-1966). Internist, Pathophysiologe, Altersforscher. Ausgewählte Texte* (Max Bürger: Internist, Pathophysiologist, Researcher of Aging. Selected writings), Johann Ambrosius Barth, Leipzig, 1985.

[390] Max Bürger, *Altern und Krankheit* (Aging and Disease), 1954, "Makrobiotik," pp. 279-282.

[391] The most famous showcases of basic biomedical research during the Nazi period included the discovery of sulphanilamides (the "sulfa drugs") – the first successful anti-bacterial pharmaceuticals – by Gerhard Domagk (around 1936); the invention and development of electron microscopy by Ernst Ruska (around 1937); and the synthesis of sex hormones by Adolf Butenandt (around 1935). All these scientists received Nobel Prizes. Domagk was awarded the Nobel Prize in Medicine and Butenandt in Chemistry in 1939, but could receive the prizes only after the war, as the Nazi government prohibited the acceptance. Ruska was awarded the Nobel Prize in Physics in 1986.

Butenandt was apparently more sympathetic to the Nazi regime than Domagk. While Butenandt joined the NSDAP (*Die Nationalsozialistische Deutsche Arbeiterpartei* - the National Socialist German Workers' Party) in 1936, Domagk was in 1939 for some time detained by the Gestapo (*Geheime Staatspolizei* - Secret State Police).

[392] Zahler H., "Die Auffrischung greisenhafter Hunde mittels Hoden wirkstoffen und ihre Wirkung auf Hoden und Prostata," Aus dem Laboratorium der Chirurgischen Klinik der Charité. [Geh. Rat Prof. Dr. Sauerbruch] (The refreshing of senile dogs by means of active substances of the testes and their effects on the testes and prostate. From the laboratory of the Charité Surgical Clinic under the direction of Privy Councilor, Prof. Dr. Sauerbruch), *Virchows Archiv für pathologische Anatomie und Physiologie und für klinische Medizin*, 305 (1), 65-107, 1939.

[393] Julius Wagner-Jauregg, *Über die menschliche Lebensdauer. Eine populär-wissenschaftliche Darstellung* (On the duration of human life. A popular scientific presentation), Deutscher Alpenverlag, Innsbruck, 1941 (published posthumously).

Wagner-Jauregg thought that "changes in hereditary structure" (mutations, *erbgefügsänderungen*) are the worst enemies of longevity, even though they may be important for the "adjustment to changing environmental relations." Yet their effects on the life-span are mostly negative: mutations are "mainly accompanied by lesser vitality, that is, lesser resistance against damaging influences, lesser fertility and shortening of life" (p. 54).

According to Wagner-Jauregg, "The farthest form of inbreeding, that is, the mating between non-blood-related individuals of the same race, surely provides the most beneficial prospects in relation to fertility and longevity." Hybrid vigor (*Wuchern/luxuriren der bastarde*), according to this fascist philosophy, can in principle produce exceptionally long-lived individuals, yet their future offspring will have a "tendency to restoring the normal state" and the "fertility of such excessive formations is low" (pp. 82, 76).

The importance of early marriages and early child-bearing for the longevity of the offspring was emphasized (p. 79).

Wagner-Jauregg worried that the increasing number of the long lived should not become a "burden for the younger layers of the nation." Yet ultimately he asserted that life-extension is inseparably connected with increasing vitality and fertility, insofar as mutations are at the same time damaging to vitality, fertility and longevity.

Therefore, "if people wish a heightened vitality and greater fertility, they should reconcile with the burden of longevity" (p. 85).

[394] According to the views of Hans Driesch, as expressed in his article "Leben, Tod und Unsterblichkeit" (Life, Death and Immortality, 1926), vitality is determined by "entelechy, an immaterial whole-making and wholeness-ensuring nature factor" which "operates soul-like."

This notion developed the classical Aristotle's conception where entelechy is a purpose-oriented force, transforming the potential into actual existence.

Though Driesch acknowledges that the body can be damaged by mechanical injury or chemical intoxication, these do not necessarily lead to deterioration and death, as the living organism has the ability to "restitute" itself, and the extent of restitution is determined by the force of the "entelechy."

Death, according to Driesch, can result either from an active working of the entelechy (upon completion of its "program" or when "it knows that regeneration is no longer possible"), or from a passive expiring of the entelechy from the matter.

Driesch does accept the concept of potential physical (material) immortality, as found in unicellular organisms, which he understood as an indefinite maintenance of the "complex of relations" or the "form" of the living being. In fact, he sees the potential material immortality of protozoa as a proof of the existence of immaterial entelechy, giving the living matter the ability to regenerate, despite mechanical or chemical injuries, in the apparent absence of any mechanical "wear and tear."

(Hans Driesch, "Leben, Tod und Unsterblichkeit" (Life, Death and Immortality), in *Leben, Altern, Tod* (Life, Aging, Death), Senckenberg-Bücher II. Hugo Bermühler Verlag, Berlin, 1926, pp. 69-81.)

These views were later fervently supported by Max Bürger. (Max Bürger, *Altern und Krankheit* (Aging and Disease), Zweite Auflage (2nd Edition), Veb Georg Thieme, Leipzig, 1954, first published in 1947.)

[395] *Leben, Altern, Tod* (Life, Aging, Death), Senckenberg-Bücher II. Hugo Bermühler Verlag, Berlin, 1926, included the following works:

Eugen Korschelt (Marburg, 1858-1946), "Die Lebensdauer der Tiere und Pflanzen" (The duration of life in animals and plants), pp. 3-17; S. Hirsch (Frankfurt am Main), "Über Alterserscheinungen und Tod" (On aging phenomena and death), pp. 18-39; Jürgen W. Harms (Tübingen, 1885-1956), "Verjüngung und Verlängerung des Lebens" (Rejuvenation and prolongation of life), pp. 40-56; Max Hartmann (Berlin, 1876-1962), "Tod, Fortpflanzung und Verjüngung" (Death, reproduction and rejuvenation), pp. 57-68; Hans Driesch (Leipzig, 1867-1941) "Leben, Tod und Unsterblichkeit" (Life, Death and Immortality), pp. 69-81.

[396] In emigration, Hirsch continued his research at the Department of Pathology at Brussels University. In 1950 he was one of the representatives for Belgium at the First International Congress of Gerontology, in Liege, Belgium, that took place on July 10-12, 1950. The congress was a defining moment in the institutionalization of gerontology as an international discipline. The conference chair and the first president of the International Association of Gerontology was the Belgian physiologist Lucien Brull (1876-1972).

In 1954, Hirsch was one of the first to relate aging research to cybernetics, in the presentation: Hirsch S, "Senescence, entropy, and cybernetics: a clarification of basic concepts in gerontological research," in *The 3rd Congress of the International Association of Gerontology. London. 1954. Old Age in the Modern World*, E&S Livingstone Ltd., London, 1955, pp. 622-627.

In 1956, Hirsch summarized his research of aging in "Was bedeutet Altern? Einige grundsätzliche Bemerkungen zur heutigen medizinischen und biologischen Alternsforschung" (What does aging signify? Some fundamental remarks on today's medical and biological research of aging), *Die Medizinische*, Stuttgart, F.K. Schattauer, 7(1), 1-8, January 1956.

[397] Eugen Korschelt (1858-1946) retired in 1928 at the age of 70. (*Korschelt-Festband: Eugen Korschelt zu seinem 70. Geburtstage im Jahre seines Abschiedes vom Lehramt, gewidmet v. seinen dankbaren Schülern u. Freunden* (Korschelt compendium: Dedicated to Eugen Korschelt's 70th anniversary in the year of his retirement from teaching, from thankful students and friends), Akademische Verlagsgesellschaft, Leipzig, 1928.)

Jürgen Wilhelm Harms (1858-1956) published textbooks on animal reproduction (*Die Fortpflanzung der Tiere*) and practical zoology (*Zoobiologie für Mediziner und Landwirte*) as late as 1954.

[398] Hans Driesch, *Parapsycholgie. Mit einem Nachwort von Hans Bender*. 3 Auflage (Parapsychology, with an afterword by Hans Bender, 3rd edition, Kindler Verlag, München, 1975, Geist und Psyche (Spirit and Psyche series), first published in 1932, p.101.

[399] In 1933, Driesch was forced to retire from his post as a professor and director of philosophical seminars at the University of Leipzig. According to one of Driesch's loyal disciples, the American pioneer of paranormal research, Joseph Banks Rhine (1895-1980), Driesch's worldview stood in stout opposition to "scientific materialism," "the Nazi movement" and the "false healing message of communism."

(Joseph Banks Rhine, "Über Hans Driesch" (On Hans Driesch), in Hans Driesch, *Parapsycholgie*, 3rd edition, Kindler Verlag, München, 1975, p. 9.)

(J.B. Rhine established a "Para-psychological laboratory" at Duke University, in Durham, North Carolina, in the early 1930s. He coined the term "Extra-sensory perception" in 1935 and established in 1937 the *Journal of Parapsychology*, which has been published by Duke University Press, since then to the present.)

Driesch was equally acclaimed by his German follower in the research of the paranormal, the professor of psychology at Freiburg, Hans Bender (1907-1991). Bender claimed in 1975 that Driesch's parapsychological research was "under political suspicion and persecution ... during the national-socialist regime."

(Hans Bender, "Zur entwicklung der Parapsychologie von 1930-1950" (On the development of parapsychology, from 1930 to 1950), in Hans Driesch, *Parapsycholgie,* 3rd edition, Kindler Verlag, München, 1975, p. 146.)

However, Hans Bender himself joined the NSDAP in 1937, and from about 1941 to 1944 lectured on parapsychology at the Reich University of Strasbourg (Reichsuniversität Strassburg). He was also actively involved in the research of the paranormal at the Paracelsus Institute (Paracelsus-Institut) in Strasbourg (c. 1943-1944).

(Ernst Klee, *Das Personenlexikon zum Dritten Reich. Wer war was vor und nach 1945*, 3. Auflage, Fischer Taschenbuch Verlag, Frankfurt am Main, 2011 (Persons lexicon of the Third Reich. Who was who before and after 1945), 3rd edition, p. 37; Frank-Rutger Hausmann, *Hans Bender (1907-1991) und das "Institut für Psychologie und Klinische Psychologie" an der Reichsuniversität Strassburg 1941-1944* (Hans Bender, 1907-1991, and the "Institute for Psychology and Clinical Psychology" at the Reich University of Strasbourg, 1941-1944), Ergon Verlag, Würzburg, 2006, pp. 108-110.)

Further on the Nazi involvement with the occult and the paranormal, see Nicholas Goodrick-Clarke, *The Occult Roots of Nazism: Secret Aryan Cults and Their Influence on Nazi Ideology: The Ariosophists of Austria and Germany, 1890-1935*, The Aquarian Press, Wellingborough, UK, 1985; Peter Levenda, *Unholy Alliance: A History of Nazi Involvement With the Occult*, Continuum, New York, 2002, and other works.

Still, Driesch's studies of the paranormal were indeed disapproved by some Nazi officials, leading to his retirement.

[400] The work at Bender's "Paracelsus Institute" in Strasbourg was funded by the Alsatian publisher Friedrich Spieser (1902-1987). Another active figure in the Paracelsus Institute in Strasbourg was the painter, poet and astrologer Thomas Ring (1892-1983).

See Frank-Rutger Hausmann, *Hans Bender (1907-1991) und das "Institut für Psychologie und Klinische Psychologie" an der Reichsuniversität Strassburg 1941-1944* (Hans Bender, 1907-1991, and the "Institute for Psychology and Clinical Psychology" at the Reich University of Strasbourg, 1941-1944), Ergon Verlag, Würzburg, 2006, pp. 14, 108-110.

The main research areas at the Institute included: magnetopathy (magnet therapy), mineral waters (including radium), eye diagnosis (iridodiagnosis), diagnosis by radioesthesia (dowsing or wand divination) and medical astrology.

A major focus was on the diagnostic and therapeutic methods developed by Dr. Matthias Leisen (1879-1940) and his daughter Katharina Vanselow-Leisen. According to Matthias Leisen's theory, diseases originate due to excessive accumulation of minerals (*schlacken*), mainly metal elements, in the tissues. These mineral deposits were hoped to be detected via special dowsing devices, the so-called Leisen's forks (*Leisengabeln*) made of glass filled with water containing various corresponding diluted metal elements.

After diagnosing a disease presumably due to an excess mineral, the so-called "Leisen's cure" (*Leisenkur*) was applied. The cure consisted of a tee extracted from a plant containing the corresponding mineral in diluted quantities.

The diluted minerals in the tea were supposed to draw away the excessive mineral in the tissues by the homeopathic principle "like dissolves like." In this way both acute conditions, such as cramps and functional disturbances, and chronic age-related diseases, such as rheumatism, arthritis, liver insufficiency, diabetes, heart disease and even cancer, were hoped to be cured.

For example, to combat Rheumatism (presumably due to an excess of Lithium) a tea was made from the Yarrow plant (*Achillea millefolium*) presumably containing diluted Lithium. This method has been applied to the present time.

(Frank-Rutger Hausmann, *Hans Bender*, 2006, pp. 118-120; Peter Germann, "Die Leisen Kur: Therapie schlackenbedingter Krankheiten" (Leisen's cure: The Therapy of diseases due to mineral slags), *Heilpraxis*

Magazin 8. Jahrgang, Nr. 1, Februar 2004, http://www.natuerlich-gesund-online.info/artikel_2004/leisen_kur.php;
Katharina Vanselow-Leisen, L. Feist, *Die Leisen-Kur: Zur Therapie schlackenbedingter Krankheiten* (Leisen's cure: On the Therapy of diseases due to mineral slags), Lorber-Verlag, Bietigheim-Bissingen, 2007.)

[401] The "Paracelsus Institute" established in 1935 in Nuremberg, Bavaria, was sponsored by the Gauleiter of Franconia Julius Streicher (1885-1946) and was initially led by the oncologist Wilhelm von Brehmer (1883-1958).

One of the institute's major purposes was to achieve "disease free, natural aging." This was hoped to be done with the aid of proper nutrition and exercise; mineral supplements (employed both as "building substances" and "catalysts"); fostering the revitalizing power of the mind and the mind-body unity; and by the suppression of "cancer pathogens" (presumably the microorganisms termed *Siphonosporae polymorphae*, allegedly developing in an alkaline environment).

The cancer suppression involved anti-cancer vaccination; stimulation of mesenchymal (immune-cell producing) tissues by plant glycosides; manipulation of blood acidity by a diet of fruits and vegetables, herbal extracts and acid vapor inhalations – the so-called Säuretherapie (acid therapy) first proposed by Sigmund von Kapff (1864-1946) around 1908, etc.

(*Paracelsus-Institut: Gründung und Aufgaben des Paracelsus-Institutes*, Verein Deutsche Volksheilkunde, Paracelsus-Institut, Nürnberg, 1936 [*The Paracelsus Institute, The Foundation and Tasks of the Paracelsus Institute*, published by The Association of German Folk Medicine, Paracelsus Institute, Nuremberg, 1936, pp. 26, 55].)

[402] At the Paracelsus Institute, aging and therefore predisposition to cancer were associated with increased blood alkalinity (blood pH was measured within the body by the specially designed "haemo-ionometer").

Very curiously and contrastingly, at about the same time, aging and cancer were also associated with increased acidity. For example, combating acidity and increasing tissue alkalinity, using alkaline drinks and baths, was recommended as an anti-aging measure by the Soviet researcher Olga Lepeshinskaya (O.B. Lepeshinskaya, "Starost i borba s neu" (Aging and the struggle with it), *Nauka I Zhizn* (Science and Life), 7, 11-13, 1951).

Brehmer himself admitted that "Two major directions stand opposed to each other … the proponents of alkalinization of blood" and those who "aspire to acidification of blood" (Wilhelm von Brehmer, "Vortrag" (Presentation at the opening of the Paracelsus Institute), *Paracelsus-Institut*, 1936, p. 28).

With specific reference to cancer, Brehmer's association of cancer with alkalinity appears to be in direct opposition to the association of cancer with acidity suggested by the German Nobel Prize winning biochemist Otto Heinrich Warburg (1883-1970). According to Warburg's observations, cancer cells increasingly produce energy via glycolysis (glucose fermentation) rather than via oxygen respiration, thus producing an increased amount of lactic acid (the so called "Warburg effect"). The increased acidity, in turn, has been suggested to drive cancer progression and invasion. Could von Brehmer's opposition be related to the fact that Warburg was half-Jewish?

(See Otto Warburg, Franz Wind, Erwin Negelein, "The metabolism of tumors in the body" (From the Kaiser Wilhelm Institut für Biologie, Berlin-Dahlem, Germany, Received for publication, April 29, 1926), *The Journal of General Physiology*, 8, 519-530, 1927. The concept was originally published in German in O. Warburg, K. Posener, E. Negelein, "Ueber den Stoffwechsel der Tumoren" (The metabolism of tumors), *Biochemische Zeitschrift*, 152, 319-344, 1924.)

Since then, Warburg's view appears to have won over Brehmer's. The association of cancer with acidity has been acknowledged by a large number of researchers, and several treatments aimed at decreasing the tissue acidity have been suggested to prevent the spread of cancer (yet careful not to induce general alkalosis).

(See, for example, Ian F. Robey and Natasha K. Martin, "Bicarbonate and dichloroacetate: Evaluating pH altering therapies in a mouse model for metastatic breast cancer," *BioMed Central (BMC) Cancer*, 11, 235, 2011.)

[403] The "Paracelsus Institute" founded in 1951 in Bad Hall, Upper Austria, has been a leader in the research of rejuvenative iodine therapy and balneology or mineral baths research. (http://www.paracelsus-badhall.at/en/balneology.)

[404] On various additional Paracelsus' institutes, clinics and societies, see Peter Dilg, "Paracelsus-Forschung Gestern und Heute" (Paracelsus Research. Yesterday and Today), in Peter Dilg, Hartmut Rudolph (Eds.), *Resultate und Desiderate der Paracelsus-Forschung* (The results and goals of Paracelsus research), Franz Steiner Verlag, Stuttgart, 1993, pp. 9-24.

[405] Paul Weindling, *Health, Race and German Politics between National Unification and Nazism. 1870-1945*, Cambridge University Press, Cambridge, 1993, pp. 559, 579.

[406] Ute Deichmann, Benno Müller-Hill, "The fraud of Abderhalden's enzymes," *Nature*, 393, 109-111, 1998; Michael Kaasch, "Sensation, Irrtum, Betrug? – Emil Abderhalden und die Geschichte der Abwehrfermente," *Acta Historica Leopoldina*, 36, 145-210, 2000 (Sensation, Error, Fraud? – Emil Abderhalden and the history of defense enzymes); Mir Taher Fattahi, "Emil Abderhalden (1877-1950): Die Abwehrfermente. Ein langer Irrweg oder wissenschaftlicher Betrug?" (The defense enzymes. A prolonged error or scientific fraud?), MD dissertation, Ruhr-Universität Bochum, Germany, 2005.

The principle of the Abderhalden reaction was that blood sera from different organs were incubated with different tissues, and then tested for products of protein degradation (small peptides and amino-acids).

The presence of the products of protein degradation (and hence the presence of "protective proteolytic enzymes" in the serum) was determined either by changes in the light polarization of the serum or, most commonly, by a reaction with ninhydrin which yielded purple color of different intensity: the stronger the color the greater, supposedly, was the amount of breakdown products. The amounts of the protein breakdown products were believed to reflect the state of tissue degradation.

From the very beginning of its application, the Abderhalden reaction was widely disputed as regards its specificity, reproducibility of its results and possible mechanisms.

See, for example, Donald D. Van Slyke, Mariam Vinograd-Villchur, J.R. Losee, "The Abderhalden reaction," *Journal of Biological Chemistry*, 23 (1), 377-406, 1915; J. Bronfenbrenner, "The mechanism of the Abderhalden reaction. Studies on immunity. I," *The Journal of Experimental Medicine*, 21, 221-238, 1915; James W. Jobling, A.A. Eggstein, William Petersen, "Serum proteases and the mechanism of the Abderhalden reaction. Studies on ferment action. XX," *The Journal of Experimental Medicine*, 21, 239-249, 1915.

Jobling et al. thus summarized their critique of the Abderhalden reaction:

"Again, the idea is advanced [by Abderhalden] that as a result of tissue destruction or of infection, lytic bodies or proteases are found which are capable of splitting the infecting organism (as in tuberculosis) or the cell (as in carcinoma). Now it is a well established fact (reaction of [Ernst] Freund and [Gisa] Kaminer) that in carcinoma we have actually the reverse of this process; i.e., the blood of the carcinoma patient has lost the power to dissolve carcinoma cells normally possessed by the serum. Yet the Abderhalden theory is based on the diametrically opposite supposition without any experimental basis in its support."

It should be noted, however, that it is today well recognized that, in many cases, particularly at earlier stages of cancer progression, the immune defense system can indeed recognize and attack cancer cells.

(E.g. Pettit SJ, Seymour K, O'Flaherty E, Kirby JA, "Immune selection in neoplasia: towards a microevolutionary model of cancer development," *British Journal of Cancer*, 82, 1900-1906, 2000; Toubi E, Shoenfeld Y, "Protective autoimmunity in cancer" (Review), *Oncology Reports*, 17(1), 245-251, 2007; Bhardwaj N, "Harnessing the immune system to treat cancer," *Journal of Clinical Investigation*, 117(5), 1130-1136, 2007, Review.)

Furthermore, the presence of proteases, particularly Matrix-Metalloproteases is now a common marker for malignant, metastatic tumors.

(Mignatti P, Rifkin DB, "Biology and biochemistry of proteinases in tumor invasion," *Physiological Reviews*, 73, 161-185, 1993; Duffy MJ, "Proteases as prognostic markers in cancer," *Clinical Cancer Research*, 2, 613-

618, 1996; Roy R, Yang J, Moses MA, "Matrix metalloproteinases as novel biomarkers and potential therapeutic targets in human cancer," *Journal of Clinical Oncology*, 27(31), 5287-5297, 2009, Review.)

Thus, Abderhalden may be well considered as one of the pioneers of cancer immunodiagnosis and immunotherapy. Furthermore, the current methods of cancer diagnosis according to the level of proteases may be well considered as direct descendants of Abderhalden's original idea of the association between cancer and proteolysis.

[407] George Wolf, "Emil Abderhalden: His Contribution to the Nutritional Biochemistry of Protein," *Journal of Nutrition*, 126, 794-799, 1996; Werner Ries, *Max Bürger (1885-1966). Internist, Pathophysiologe, Alternsforscher. Ausgewählte Texte* (Max Bürger: Internist, Pathophysiologist, Researcher of Aging. Selected writings), Johann Ambrosius Barth, Leipzig, 1985, "Emil Abderhalden" pp. 146-148; Hanson H, "Die Bedeutung Emil Abderhaldens für die Entwicklung der Physiologie des Eiweiss in den letzten 50 Jahren" (The significance of Emil Abderhalden for the development of protein physiology in the last 50 years), *Wissenschaftliche Zeitschrift der Martin-Luther-Universität Halle-Wittenberg. Mathematisch-naturwissenschaftliche Reihe*, R.3, 319-324, 1953/1954; Schoen E, "Emil Abderhalden zum Gedächtnis. Das soziale Wirken Abderhaldens," *Nova Acta Leopoldina, Neue Folge*, 14, 178-179, 1952 (In memory of Emil Abderhalden. The social works of Abderhalden); Schlüter O, "Emil Abderhalden zum Gedächtnis. Abderhalden und die Leopoldina" (In memory of Emil Abderhalden. Abderhalden and the Leopoldina), *Nova Acta Leopoldina, Neue Folge*, 14, 147-154, 1952.

As an example of the positive attitude, the East German gerontologist Werner Ries claimed in 1985 that Abderhalden's work "will be a lasting legacy of a great humanist. ... Thanks to him, the Academy [Leopoldina] was sustained through the difficult years of the world economic crisis, fascism and the war and retained its significant place in the international scientific community" (Werner Ries, *Max Bürger*, 1985, "Emil Abderhalden," p. 147).

Ries particularly honored Emil Abderhalden's *Gedanken eines Biologen zur Schaffung einer Völkergemeinschaft und eines dauerhaftes Friedens* (Thoughts of a biologist on the creation of international community and lasting peace, Rascher, Zürich, 1947).

[408] George Wolf, "Emil Abderhalden: His Contribution to the Nutritional Biochemistry of Protein," *Journal of Nutrition*, 126, 794-799, 1996; Werner Ries, *Max Bürger (1885-1966). Internist, Pathophysiologe, Alternsforscher. Ausgewählte Texte* (Max Bürger: Internist, Pathophysiologist, Researcher of Aging. Selected writings), Johann Ambrosius Barth, Leipzig, 1985, "Emil Abderhalden" pp. 146-148; Hanson H, "Die Bedeutung Emil Abderhaldens für die Entwicklung der Physiologie des Eiweiss in den letzten 50 Jahren" (The significance of Emil Abderhalden for the development of protein physiology in the last 50 years), *Wissenschaftliche Zeitschrift der Martin-Luther-Universität Halle-Wittenberg. Mathematisch-naturwissenschaftliche Reihe*, R.3, 319-324, 1953/1954; Schoen E, "Emil Abderhalden zum Gedächtnis. Das soziale Wirken Abderhaldens," *Nova Acta Leopoldina, Neue Folge*, 14, 178-179, 1952 (In memory of Emil Abderhalden. The social works of Abderhalden); Schlüter O, "Emil Abderhalden zum Gedächtnis. Abderhalden und die Leopoldina" (In memory of Emil Abderhalden. Abderhalden and the Leopoldina), *Nova Acta Leopoldina, Neue Folge*, 14, 147-154, 1952.

[409] Emil Abderhalden, Max Bürger, "Zur Einführung" (Introduction), *Zeitschrift für Altersforschung*, Band 1, Heft 1 (vol. 1, issue 1), Juli 1938, pp. 1-2.

[410] Friedrich Vogel, Arno G. Motulsky, *Human Genetics. Problems and Approaches*, 3rd Edition, Springer-Verlag, Berlin, 1997, Section 14.3.1 "Races," pp. 610-616, Section 17 "Differences in IQ and achievement between ethnic groups," pp. 705-710.

Presently, the American National Institutes of Health mandate the inclusion of ethnic minorities in all clinical trials. See *NIH Policy and Guidelines on the Inclusion of Women and Minorities as Participants in Research Involving Human Subjects*, October, 2001, http://grants.nih.gov/grants/funding/women_min/women_min.htm.

[411] Paul Weindling, *Health, Race and German Politics between National Unification and Nazism. 1870-1945*, Cambridge University Press, Cambridge, 1993, p. 562.

[412] The "experiments" included:
High-altitude experiments; Freezing experiments; Malaria experiments; Mustard ("lost") gas experiments; Sulfanilamide experiments; Bone, muscle, and nerve regeneration, and bone transplant experiments; Seawater experiments; Epidemic jaundice experiments; Typhus ("spotted fever") and other vaccine experiments; Poison experiments; Incendiary bomb experiments; Sterilization experiments; Skeleton collection; experiments on Tubercular Polish nationals; Euthanasia (the secret killing of the aged, insane, incurably ill, deformed children, and others, beginning in asylums in Germany and later in the concentration camps and occupied territories); Phenol (gas edema) experiments; Phlegmon experiments (to test treatments for sepsis and related diseases); Polygal experiments (to test the effectiveness of polygal, a blood coagulant, for the treatment of wounds).

The "freezing experiments" were charged against Hermann Becker-Freyseng, Karl Brandt, Rudolf Brandt, Karl Gebhardt, Siegfried Handloser, Joachim Mrugowsky, Helmut Poppendick, Oskar Schroeder, Wolfram Sievers, and Georg August Weltz. Becker-Freyseng, K. Brandt, Gebhardt, Mrugowsky, Poppendick, and Weltz were acquitted; R. Brandt, Handloser, Schroeder, and Sievers were convicted. The "regeneration experiments" were charged against Karl Brandt, Rudolf Brandt, Fritz Fischer, Karl Gebhardt, Siegfried Handloser, Herta Oberheuser, and Paul Rostock. The charge against R. Brandt was withdrawn. K. Brandt, Handloser, and Rostock were acquitted; Fischer, Gebhardt, and Oberheuser were convicted.

Generally, out of these, only Weltz and Rostock were acquitted on all charges, the rest were sentenced to death, 10 to 20 years or life in prison.

(The Harvard Law School Library. Nuremberg Trials Project. A Digital Document Collection. "The Medical Case. Case No. 1," http://nuremberg.law.harvard.edu/php/docs_swi.php?DI=1&text=medical; Nuremberg Military Tribunals. Indictments. Office of Military Government for Germany (US), Nuremberg, 1946, at Military Legal Resources, http://www.loc.gov/rr/frd/Military_Law/NT_major-war-criminals.html.)

[413] Gerhard Venzmer, *Lang Leben und Jung Bleiben!* (Live long and stay young), Franckhsche Verlagshandlung, Stuttgart, 1937 (republished in 1941), "Einflüsse der Erblichkeit und der persönlichen Lebensführung auf das Altwerden" (Influences of heredity and personal life style on aging), pp. 78-87.

[414] *The Future of Aging: Pathways to Human Life Extension* (Gregory M. Fahy, Michael D. West, L. Stephen Coles and Steven B. Harris, Eds.), Springer, NY, 2010.

[415] Of additional note, in this regard, is the fact that Europe's largest medicinal herbs plantation was established at the Dachau concentration camp. This fact is emphasized by Robert Proctor (Robert N. Proctor, *The Nazi War on Cancer*, Princeton University Press, Princeton, 1999, pp. 158, 265).

[416] It might be also interesting to note that transplantation, hypothermia and cryobiology research rapidly expanded in the United States immediately after the war. It might be of interest to trace the involvement of German physicians and biologists transferred to the US after the war in the framework of the project "Paperclip" in these particular fields. But here too I have very little data for speculations.

[417] Werner Ries, *Max Bürger (1885-1966). Internist, Pathophysiologe, Alternsforscher. Ausgewählte Texte* (Max Bürger: Internist, Pathophysiologist, Researcher of Aging. Selected writings), Johann Ambrosius Barth, Leipzig, 1985, hereafter referred to as "Werner Ries, *Max Bürger*, 1985."

[418] Many members of the Schlomka family now live in Israel, many of whom are Holocaust survivors. The fate of Georg Schlomka is unclear, except that he continued to publish well into the 1960s, predominantly in Swiss and East German journals (such as *Deutsche Gesundheitswesen* - German Healthcare).

[419] Bürger M, Schlomka G, "Ergebnisse und Bedeutung chemischer gewebsuntersuchungen für die Altersforschung" (The results and significance of chemical examination of tissues for aging research), *Klinische Wochenschrift*, 7, 1944, 1928.

[420] Max Bürger, "Zum Geleit!" (Foreword), *Deutsche Zeitschrift für Verdauungs- und Stoffwechselkrankheiten, Februar 1949*, quoted in Werner Ries, *Max Bürger*, 1985, p. 32.

[421] Quoted in Werner Ries, *Max Bürger*, 1985, p. 38.

[422] Max Bürger, *Altern und Krankheit* (Aging and Disease), Zweite Auflage (2nd Edition), Veb Georg Thieme, Leipzig, 1954 (1947), "Makrobiotik," pp. 279-282.

[423] Vitalism had been traditionally strong in Germany. The founding fathers of German life-extensionism – Johannes Heinrich Cohausen (1665-1750) and Christoph Wilhelm Hufeland (1762-1836) – were steadfast adherents of vitalism.

Thus, Hufeland postulated that "the duration of life in a being will be proportioned to the *innate quantity of vital power* ... the more intensively a being lives, the more will its life lose its extension" (*Hufeland's Art of Prolonging Life*, Edited by Erasmus Wilson, Lindsay & Blakiston, Philadelphia, 1867 (1796), pp. 49, 50-51).

And Cohausen thought that "growth ... is owing to the strength of the vial flame" (Johann Heinrich Cohausen, *Hermippus Redivivus or the Sage's Triumph over Old Age and the Grave*, J. Nourse, London, 1748 (1742), p. 75).

Other 18th-19th century German pioneers of aging research, such as Johann Bernhard Fischer (1685-1772), Burkhard Wilhelm Seiler (1779-1843) and Carl Friedrich Canstatt (1807-1850) also subscribed to the vitalist views.

Thus, Canstatt (1839), deeply influenced by the philosopher Friedrich Schelling (1775-1854), spoke of "the gravitation of life to organic individual formation" as the "Egoistic organizing principle" and "the gravitation of individuals to the Universe" as "the Cosmic principle." Accordingly, aging and disease result from "the struggle between these two poles striving for exclusivity." Moreover, the process of aging, according to Canstatt, is analogous to the involution of life into lower forms. (Quoted in Johannes Steudel, "Zur Geschichte der Lehre von den Greisenkrankheiten" (The history of the study of the diseases of old age), *Sudhoffs Archiv für Geschichte der Medizin und der Naturwissenschaften*, 35, 1-27, 1942, quote on p. 21.)

The influence of the vitalistic-idealistic philosophy of Arthur Schopenhauer (1788-1860) on aging research was strongly felt, for example, his postulate that "the dead body is a mere excrement of a constant human form." That is to say, all materials in the human body are being incessantly replaced, and only their form or arrangement is constant (quoted in Max Bürger, *Altern und Krankheit*, 1956 (1947), p. 41).

Quotes from Schopenhauer abounded in the scientific texts on aging (e.g. in S. Hirsch, "Über Alterserscheinungen und Tod" (On aging phenomena and death), in *Leben, Altern, Tod* (Life, Aging, Death), Hugo Bermühler Verlag, Berlin, 1926, pp. 38-39).

In contrast to Schopenhauer's pessimism and resignation to mortality, the vitalistic teachings of Albert Schweitzer (1875-1965) on "vital spirituality," "reverence for life," "clinging to life" under all conditions, positively supported the efforts for life-prolongation. (Albert Schweitzer, *An Anthology*, Beacon Press, Boston, 1947, pp. 218, 255.)

In the works of the twentieth century's strongest proponents of vitalism – Hans Driesch and Max Bürger – the concept of "entelechy" or "vital factor" implies quite a rigid limit to the life-span:

As the operation of the "vital factor" can be likened to a pre-planned construction enterprise, when the construction program is complete, the "vital force" has no more room to operate and its action ceases.

Nevertheless, the vitalist perception did offer several theoretical possibilities for life prolongation. Basically, the force of entelechy could be enhanced by manipulating body structure.

First of all, the life force can be conserved by diminishing activity.

Secondly, the structure can be reduced – for example, by dissolving structure or by amputation – to "free the room" for a continued action of the "vital force." In this way, Bürger writes, "the catastrophic end can be postponed either by dissolving structure or by a forced regeneration after amputation."

Evidence for the renewal of the life force's constructive activity upon diminution of structure, appeared to be given by Max Hartmann's experiments showing that amputations could prolong the life of protozoa virtually unlimitedly. Further evidence was given by the low differentiation (structuredness) of immortal germ cells, and by the fact of regeneration induction after injury.

In this view, according to Bürger, during a moderate exercise, some body structures become partly worn out, the life force receives a "new room" to operate and rebuilds the lost structures even stronger than before.

Thirdly, and perhaps most importantly, according to Bürger, the immaterial "vital force" could be directly affected by another "immaterial" entity – the mind.

(Max Bürger, *Altern und Krankheit* (Aging and Disease), 1954, "Das Altern im Lichte der vitalistischen Autonomielehre" (Aging in the light of the theory of vitalistic autonomy), pp. 39-41.)

[424] Max Bürger, *Altern und Krankheit* (Aging and Disease), 1954, "Das Altern im Lichte der vitalistischen Autonomielehre" (Aging in the light of the theory of vitalistic autonomy), pp. 39-41.

[425] Max Bürger, "Die Bedeutung des Alternsvorgangs für die Klinik" (The significance of the aging processes for the Clinic," *Zeitschrift für Altersforschung*, vol. 1, issue 1, 1939, pp. 3-8, the article was written on March 27, 1938).

Bürger claimed in that article that "According to [his] conception, individual organs age simultaneously and harmoniously, while each individual organ and tissue follows the law of aging which is the same for the entire organism."

And further on, "The law of the synchronous aging of organs can be disrupted by pathological conditions, where individual organ systems age prematurely, particularly those organ systems that are under special stress."

In other words, the (unspecified) "normal" life span is determined by the working off of the initial (unspecified) amount of the vital force, i.e. "normal" aging. However, this "normal," "synchronous" aging process is often greatly accelerated by disease or overwork of particular organ systems, bringing the organism "out of synchronization" and thereby shortening the "natural" span of life.

[426] Max Bürger, *Altern und Krankheit*, 1954, "Makrobiotik," pp. 279-282

[427] As of 2010, the German Society of Gerontology and Geriatrics (DGGG) included 1,100 members and was led by Manfred Gogol, Clemens Tesch-Roemer and Thomas Klie. The society publishes *Die Zeitschrift für Gerontologie und Geriatrie* (Journal of Gerontology and Geriatrics). (http://www.dggg-online.de/; http://www.iagg.info/organization/council.)

[428] Research institutes, involved in longevity studies, have included the Leibniz Institute for Age Research in Jena (http://www.fli-leibniz.de/); Fraunhofer Institute for Cell Therapy and Immunology IZI, in Leipzig (http://www.izi.fraunhofer.de/; http://www.izi.fraunhofer.de/eng/content/arbeitsgeb/stemcell.htm); Translation Center for Regenerative Medicine at the University of Leipzig (http://www.trm.uni-leipzig.de/); and others.

Several German research groups on systems biology have been involved in aging research, as listed by JenAge – Jena Center for Systems Biology of Aging (http://info-centre.jenage.de/ageing/german-systems-biology-initiatives-in-ageing-research.html).

The EU IDEAL Aging consortium (Integrated Research on Developmental Determinants of Aging and Longevity - http://www.ideal-ageing.eu/) has included the research group on Aging and Human Immunology at the University of Tübingen (http://www.ideal-ageing.eu/participants/participants_item/t/university_of_tubingen).

The GerontoShield consortium "pursues a systems biology-driven approach to understand the molecular processes affected during immunosenescence to derive strategies to overcome them in order to establish immune interventions tailored for the elderly" and involves several German universities (http://www.gerontoshield.de/partners/).

[429] Renowned centers of cryopreservation research, especially in reproductive medicine, include Universitotsfrauenklinik at Cologne and the Institute of Reproductive Medicine at Munster University. There are also more radical cryonics-related groups (http://www.biostase.de/).

[430] In 2004, the European Union (EU)-Integrated Project Genetics of Healthy Aging (GEHA) was launched, with 25 partners (24 from Europe plus the Beijing Genomics Institute of China), with the purpose "to identify genes involved in healthy aging and longevity, which allow individuals to survive to advanced old age in good cognitive and physical function and in the absence of major age-related diseases" (Franceschi C, ..., Vaupel JW, "Genetics of healthy aging in Europe: the EU-integrated project GEHA (Genetics of Healthy Aging)," *Annals of the New York Academy of Sciences*, 1100, 21-45, 2007; http://www.mucosa.de/geha/projekt.html; http://www.geha.unibo.it/).
In 2013, some summative results from this project were published, including some genetic correlates of longevity. (Marian Beekman, ... Claudio Franceschi and the GEHA consortium, "Genome-wide linkage analysis for human longevity: Genetics of Healthy Aging Study," *Aging Cell*, 12(2):184-93, 2013 http://onlinelibrary.wiley.com/doi/10.1111/acel.12039/full.)
This project has been led by the German (American born) demographer James Walton Vaupel, the founding director of the Max Plank Institute for Demographic Research in Rostock (since 1996). Vaupel's life-extensionist views are explicit, arguing for the "plasticity of longevity" and projecting a massive increase in the proportion of people living beyond a century.
(Jim Oeppen and James W. Vaupel, "Broken Limits to Life Expectancy," *Science*, 296(5570), 1029-1031, 2002; http://user.demogr.mpg.de/jwv/; http://www.demogr.mpg.de/; http://discovermagazine.com/2003/nov/cover; Vaupel JW, "The future of human longevity – how important are markets and innovation?" *Science Aging Knowledge Environment*, 2003(26), PE18, 2003; Vaupel JW, "Biodemography of human ageing," *Nature* 464(7288), 536-542, 2010; Christensen K, Doblhammer G, Rau R, Vaupel JW, "Ageing populations: the challenges ahead," *Lancet*, 374(9696), 1196-11208, 2009.)

[431] Several projects have been initiated in Germany in bioinformatics of aging and longevity, intended to "decipher" the aging process and calculate the most effective life-extending interventions based on available data bases. A leading group is the Institute for Biostatistics and Informatics in Medicine and Ageing Research – IBIMA University of Rostock (lead by Prof. Dr. Georg Fuellen, http://www.ibima.med.uni-rostock.de/ ; http://www.ibima.med.uni-rostock.de/IBIMA/symposium/2013/). Another expansive project is Denigma – Digital Enigma to Decipher the Aging Process (http://www.denigma.de/).

[432] Some resources on the expanding field of biomedical engineering in Germany:
http://www.mastersportal.eu/disciplines/30/bio-biomedical-engineering.html
http://www.zdf.de/ZDFmediathek/beitrag/video/2061668/Ewig-jung---bald-nicht-mehr-alt

[433] The German Society of Anti-Aging Medicine, German Hormone Society, and German Society of Hemotoxicology, are affiliated member organizations of the World Anti-Aging Academy of Medicine (WAAAM) and of the European Society of Preventive, Regenerative and Anti-Aging Medicine (ESAAM). The German Society of Anti-Aging Medicine has been chaired by Prof. Dr. med. Bernd Kleine-Gunk, of Euromed Clinic, Fürth. As of 2010, the European Society of Preventive, Regenerative and Anti-Aging Medicine (ESAAM) was presided over by the German (Greek born) Professor Christos C. Zouboulis, Chair of the Departments of Dermatology and Immunology, Dessau Medical Center and Head of the Laboratory for Biogerontology, Dermato-Pharmacology and Dermato-Endocrinology at the Institute of Clinical Pharmacology and Toxicology, Charité Universitaetsmedizin, Berlin.
(Sources: http://www.gsaam.de/; http://www.esaam-org.eu/; http://www.ecopram.com/; http://web.archive.org/web/20070618101046/www.esaam.com/index.html?pg=board;

http://www.wosiam.org/index.php?rub=about-WOSIAM;
http://www.euromedicom.com/index.php?rub=41; http://www.waaam.org/member_organizations.php.)

[434] Small scale (and radical) life-extensionist organizations have included the German Transhumanist Association (http://www.detrans.de/; http://www.detrans.de/links.html); Forever - Magazin für Physische Unsterblichkeit (Magazine for physical immortality, http://www.physischeunsterblichkeit.de/) and a few others.

[435] Nathan W. Shock, *The International Association of Gerontology. A Chronicle. 1950 to 1986*, Springer Publishing Company, NY, 1988, pp. 5-8; E.V. Cowdry, "V. Korenchevsky, father of gerontology," *Science* (NY), 130, 1391-1392, November 20, 1959.

[436] The Japanese journal *Acta Gerontologica Japonica* (*Yokufuen Chosa Kenkyu Kiyo*), established in 1930, was apparently the world's first journal on aging and longevity, predating Bürger's and Kotsovsky's journals. Pioneers in the study of Japanese centenarians and some of the leading contributors to *Acta Gerontologica Japonica* were Matataro Matsumoto (1865-1943, a founder of Japanese psychology) and his students Izuo Terasawa and Kakusho Tachibana (Kakusho Tachibana, "Trends in gerontology and problems of the aged in Japan," *The Gerontologist*, 6(4), 215-217, 1966).

(On the later development of gerontology in Japan, see Naoko Muramatsu, Hiroko Akiyama, "Japan: Super-Aging Society Preparing for the Future," *The Gerontologist*, 51(4), 425-432, 2011.)

In the 1930s, prominent Japanese researchers of metabolic changes and nutritional requirements of the aged were H. Obata of Fukuoka, H. Yamagata and K. Yamamoto of Tokyo. In 1930, Minoru Shirota of Kyoto (1899-1982) developed the pro-biotic (lactobacillic) drink Yakult, inspired by Metchnikoff's ideas about the life-prolonging effects of fermented dairy products.

In the 1920s-1930s, "natural" approaches to longevity flourished in Japan. In 1927, Katsuzo Nishi (1884-1959), founded a peculiar system for increasing health and longevity (Nishi-Shiki or Nishi-Kai). In Nishi's theory, longevity was largely determined by the activity of blood capillaries. Hence, his system was mainly predicated on "capillary therapy," a set of exercises (*undō*) designed to boost capillary blood flow. Nishi was also an Aikido master, a student of the Aikido founder Morihei Ueshiba (1883-1969). Hence the principle of the "harmony of the life-force" (Ai-Ki) was central to his teachings. Nishi's "Six laws of health" involved exercises some of which were incorporated into Aikido calisthenics ("aiki-taiso"). The healing system of "Reiki" was founded in 1922 by Mikao Usui (1865-1926). In Reiki, longevity was hoped to be achieved through channeling cosmic energies.

Another prominent contemporary Japanese longevity advocate was George Ohsawa (born Nyoichi Sakurazawa, 1893-1966). In 1928, he wrote his first books on "Macrobiotics," the term he borrowed from Hufeland. However, Ohsawa's life-extension system was largely based on the traditional Eastern concepts of balancing the "yin" and "yang" in the body through appropriate foods, rather than on Hufeland's rules. From 1929 to 1935, Ohsawa lived in France, cooperating with George Soulié de Morant in introducing acupuncture and other forms of Far-Eastern (Taoist/Zen) medicine to the West. Today, Ohsawa's "macrobiotic diet" is practiced widely throughout the world.

(Michio Kushi and Alex Jack, *The Macrobiotic Path to Total Health: A Complete Guide to Preventing and Relieving More Than 200 Chronic Conditions and Disorders Naturally*, Ballantine Books, NY, 2003; Michio Kushi and Alex Jack, *The Book of Macrobiotics: The Universal Way of Health, Happiness, and Peace*, Japan Publications, Tokyo-New York, 1994.)

[437] In Fascist Italy (1922-1943), extensive longevity studies were conducted, including rejuvenation experiments (by Francesco Cavazzi, Vincenzo Pettinari, Giacondo Protti), basic research of the morphology and physical chemistry of aging (by Nicola Pende, M. Crepet, A. Ferrara), and studies of the influence of environmental factors on longevity (by A. Campani, L. Gabbano, A. Niciforo) and others.

[438] In Fascist Spain under Francisco Franco, among the most highly honored longevity researchers were Academician Leon Cardenal (1866-1951) who conducted extensive rejuvenation experiments and

Academician Antonio de Gregorio Rocasolano (1873-1941) who developed the physical-chemical theory of aging. (The peak of their scientific achievement, however, was during the Monarchy and the Second Republic, before the Spanish civil war of 1936 and the establishment of Franco's dictatorship in 1939.)

[439] In the Hungarian Kingdom, which during Miklós Horthy's regency (1920-1944) leaned heavily toward fascism, one of the foremost researchers of aging and longevity was Istvan Szabo (1899-1972). Szabo was one of the primary contributors to the *Zeitschrift für Altersforschung* and focused on "heterochronous" and "synchronous" aging processes. In the late 1930s, he worked to establish an "Old Age Institute" in Hungary, but his plans apparently did not come to fruition. (Istvan Szabo, "Purposes and aims of a projected Old Age Institute," *Altersprobleme*, 1(2), 49-50, 1937.)

[440] For the discussion of contemporary French life-extensionism, see the previous chapter, particularly as regards the work of Auguste Lumière and Alexis Carrel.

[441] In the 1930s-1940s, the Scandinavian (Nordic) countries – Denmark, Sweden, Norway and Iceland – were noted for their high longevity (~66-67 years as of 1939, the highest in the world) and for their laws protecting racial eugenic purity. Yet, it seems, few outstanding longevity researchers and advocates operated at the time in these countries, with a few notable exceptions.

The same can perhaps be said for the Baltic countries – Lithuania, Latvia and Estonia, as well as Finland (the longevity was also relatively high, about 55-60 years in the late 1930s, and eugenic laws were passed in Finland and Estonia, and were on the verge of passing in Latvia and Lithuania). Yet, there too were few famous longevity advocates. Still, several prominent researchers can be mentioned:

Thus, the Danish physician Knud Sand (1887-1968), working from 1925 to 1957 as a professor of forensic medicine at the University of Copenhagen, became world-famous in the 1920s as a daring practitioner of endocrine rejuvenation operations. The pathologist and eugenicist Oluf Thomsen (1878-1940, also of Copenhagen) was mainly interested in the genetics of longevity.

In Sweden, the winner of the 1926 Nobel Prize in Chemistry, Theodor Svedberg (1884-1971) of Uppsala University, worked on the colloidal theory of aging. Carl Gustaf Santesson (1862-1939), professor of pharmacology at Stockholm University, was a strong proponent of temperance as a means to prolong life.

The founders of Finnish gerontology were professors Werner Oswald Renkonen (1872-1951) and Eeva Jalavisto (1909-1966) of Helsinki, focusing on the genetics of longevity.

In Latvia, perhaps the most ambitious proponent of life-extension and rejuvenation was the Jewish (Russian born) Professor Naum Lebedinsky (1888-1941), working as the head of the Institute of Comparative Anatomy and Experimental Zoology at the University of Latvia in Riga, from 1922 to 1941, until driven to suicide by the Nazis.

All of these researchers (except for Renkonen and Jalavisto) were members of the Institute for the Study and Combat of Aging, with the headquarters in Romania (see below).

[442] Romania was definitely a leader in Eastern Europe. Its leadership is demonstrated by the establishment of trailblazing institutions for longevity research, such as those of Kotsovsky and Parhon (see below). Still, particular longevity studies were conducted in the 1920s through the 1940s in other East European countries:

In Bulgaria, in the 1930s-1940s, research groups of Dimitr Krilov (Krilow) and Jivko Lambrev conducted local demographic studies of longevity. Bulgaria was famed for a high proportion of centenarians among its peasants (though these high numbers were mainly due to flawed birth registration). Krilov's and Lambrev's groups, however, did not develop into institutes. Neither did the laboratory of Mikhail Popov at the University of Sofia, who conducted extensive studies on biological stimulants (their doses and duration of activity) between the years 1925-1950.

In Czechoslovakia, in the 1920s, Vladislav Ruzicka (Růžička, 1870-1934) of Prague was a world renowned researcher of basic (colloidal) processes of aging and a renowned eugenicist – the founder of the Institute of Biology at Prague University and president of the Czech Eugenics Society. (Do not confuse him with

Lavoslav Růžička (1887-1976), the Swiss-Croatian synthesizer of male sex hormones.) Vladislav Ruzicka too was unable to establish a gerontological institute.

There was some longevity research conducted in Yugoslavia in the 1920s-1930s. There, a notable theoretician of life-extension was the Russian-born neurologist and psychiatrist Nikolay Krainsky, the author of the "Law of Conservation of Energy applied to Psychical Activity" (1897). In the 1920s-1930s, he worked as a professor of psychology at the University of Belgrade and cooperated with other European researchers of aging. Yet, in terms of institutionalized research in Yugoslavia, the Problems of Ageing and Longevity were included into the working program of the Yugoslav Academy of Sciences and Arts only in 1958.

In Poland, gerontological research became institutionally associated in 1973, at the initiative of Jerzy Piotrowski.

[443] Elie Metchnikoff, *Etudy o Prirode Cheloveka* (Etudes On the Nature of Man), The USSR Academy of Sciences Press, Moscow, 1961 (First published in 1903), Ch. X. "Vvedenie v nauchnoe izuchenie starosti" (An introduction to the scientific study of aging), p. 198.

[444] Georges Marinesco, *Mécanisme colloidal de la sénilité* (Colloidal mechanisms of aging), Officina tipografica Giannotta, Catania, 1913 (in the same year published in Bucharest, Berlin, Paris and Vienna); Georges Marinesco, "Méchanisme chimico-colloidal de la sénilité et le problème de la mort naturelle" (Chemical-colloidal mechanisms of aging and the problem of natural death), *Revue Scientifique de la France et de l'étranger*, Paris, 1, 673-679, 1914; Georges Marinesco, *Sénilité et rajeunissement* (Aging and rejuvenation), J. Mällo, Tartu, 1925.

[445] Dimu A. Kotsovsky, "Gerontology in Eastern Europe," *Review of Eastern Medical Sciences*, 13, 7-16, 1958.

[446] The list of the Institute's "corresponding honorary members and collaborators" (all professors except for only a few "Doctors") included:

Emil Abderhalden (Halle), Boris Babkin (Montreal), Heinrich Jakob Bechhold (Frankfurt am Main), Jan Bělehrádek (Brno), Harry Benjamin (New York), Theodor Brugsch (Halle), Pol Bouin (Strasbourg), Hans Berger (Jena), H. Bitterling (Cologne), Max Bürger (Leipzig), Luigi Castaldi (Cagliari), Francesco Cavazzi (Bologna), Charles Manning Child (Chicago), Francis Albert Eley Crew (Edinburgh), W. Crozier (Cambridge, Massachusetts), Hans Joachim Deuticke (Bonn), A. del Castillo (Buenos Aires), Karl Doppler (Vienna), M. Ernest (London), Casimir Funk (Paris), M. Fischer (Cincinnati), Julius Hermann Greeff (Cannstatt), Max Hartmann (Berlin), Jürgen Wilhelm Harms (Jena), Willy Hellpach (Heidelberg), Alfonso Luis Herrera (Mexico), Günther Hertwig (Rostock), E. Hernandez (Paris), A. Hill (London), Hélan Jaworski (Paris), H. Johnson (Pittsburg), Rudolf Kafemann (Königsberg), D. Kellner (Budapest), Herbert Koch (Leipzig), Eugen Korschelt (Marburg), Dimu Anatoli Kotsovsky (Kishinev), Wladislas Kopaczewski (Paris), C. Krause (Giessen), Nikolay Krainsky (Belgrade), Dimitr Krilow (Sofia), Naum Lebedinsky (Riga), Wladimir Lepeschkin (Vienna), I. Lepsi (Kishinev), S. Levi (Turin), Alexander Lipschütz (Santiago), Auguste Lumière (Lyon), Arnold Lorand (Nice), Leo Loeb (St. Louis), Georges Marinesco (Bucharest), Sergey Metalnikov (Paris), L. Müller (Erlangen), Anton Nemilov (Leningrad), H. Obata (Fukuoka), F. Orthner (Ried), Constantin Ion Parhon (Bucharest), A. Pawlowsky (Soroca), Raymond Pearl (Baltimore), Boris Perott (London), Giacondo Protti (Venice), O. Polimanti (Perugia), L. Polew (Rabat, Morocco), A.H. Roffo (Buenos-Aires), Benno Romeis (Munich), Antonio Grigorio Rocasolano (Saragossa), R. Roosen (Cologne), Robert Rössle (Berlin), Oswald Spengler (Munich), H. Stieve (Berlin), Richard Shope (Princeton), Knud Sand (Copenhagen), Carl Gustaf Santesson (Stockholm), Hans Spemann (Freiburg), Eugen Steinach (Vienna), B. Stempel (Prague), Theodor Svedberg (Uppsala), István Szabó (Budapest), Oluf Thomsen (Copenhagen), H. Theilhaber (Munich), E. Wedekind (Munich), Berthold Paul Wiesner (Edinburgh), Ottmar Wilhelm (Concepción), Herbert Spencer Jennings (Baltimore), Arthur Jores (Hamburg), H. Yamagata (Tokyo).

[447] Nathan W. Shock (Ed.), *A Classified Bibliography of Gerontology and Geriatrics*, Stanford University Press, Stanford CA, 1951; *Supplement One 1949-1955*, 1957; *Supplement Two 1956-1961*, 1963 (reprinted in 1980 by Stanford University Press).

[448] *Universitas*, vol. 6, 1951, p. 1279.

[449] Dimu Kotsovsky, "Das Alter in der Geschichte der Wissenschaft" (Aging in the history of science), *Isis*, 20(1), 220-245, November 1933.

[450] Dimu A. Kotsovsky, "Gerontology in Eastern Europe," *Review of Eastern Medical Sciences*, 13, 7-16, 1958.

[451] Kotsovsky, D., "Über Beschleunigung und Hemmung des Alterns beim Menschen" (On the acceleration and inhibition of aging in humans), *Wiener Klinische Wochenschrift*, 55, 307-308, 1942.

[452] Kotsovsky, D., "Altersreaktion (Neue Wege zur Erforschung der Alternsdynamik). I. Acidität als Index der Altersreaktion. II. Altersreaktion, festgestellt durch das ausfallen des Eiweisses. III. Altersreaktion, festgestellt durch das Ausfällen des Hühnereiweisses mit Hilfe einer 2%igen Tanninlosung im Laufe von 12 Tagen unter verschiedenen photochemischen Bedingungen," *Schweizerische Zeitschrift für Biochemie*, 1(2/3), 90-93, (4/5), 159-168, 1941-1942 (The aging reaction: New ways for researching the dynamics of aging. I. Acidity as an index of the aging reaction. II. Aging reaction determined by the precipitation of proteins. III. Aging reaction determined by the precipitation of chicken proteins with a 2% solution of Tannin during 12 days under different photochemical conditions).

Kotsovsky's experiments with young and old chicken proteins precipitated with tannin (a plant polyphenole solution), as well as with saltpeter/nitric acid and hydrochloric acid, showed that proteins from young chickens produced less precipitate and their rate of precipitation was faster. Thus, age could be estimated simply by looking at the amount of precipitated protein and its rate of change in the test tube. Furthermore, when mixing old and young proteins, the mixture behaved similarly to young proteins.

Another crucial index in the "aging reaction" was acidity. According to Kotsovsky's experiments with agave juice, young plants were more acidic and the old ones were more alkaline. These findings agreed with those of the contemporary German physician, head of the Paracelsus Institute, Wilhelm von Brehmer, who claimed that "ionic values of blood increase from youth to old age slowly from the acidic to the alkaline line. Thus we can also understand why cancer is predominantly a disease of old age" (according to Brehmer, alkaline blood favors the development of cancer). Thus, in these authors, aging was apparently associated with increasing alkalinity.

(*Paracelsus-Institut: Gründung und Aufgaben des Paracelsus-Institutes*, Verein Deutsche Volksheilkunde, Paracelsus-Institut, Nürnberg, 1936 [*The Paracelsus Institute, The Foundation and Tasks of the Paracelsus Institute*, published by The Association of German Folk Medicine, Paracelsus Institute, Nuremberg, 1936, presentation by Wilhelm von Brehmer, p. 26].)

Other researchers, however, associated increasing alkalinity with youth, energy and healing. Thus, the contemporary Soviet researcher Olga Lepeshinskaya, showed that the flocculation and precipitation of proteins (colloidal hysteresis) could be slowed down and even reversed by adding alkaline liquid medium, such as baking soda (sodium bicarbonate – $NaHCO_3$).

(O.B. Lepeshinskaya, "Starost i borba s neu" (Aging and the struggle with it), *Nauka I Zhizn* (Science and Life), 7, 11-13, 1951.)

Even now, despite the apparent significance of the acid-alkaline balance for the aging process, the exact recommendations for establishing this balance have remained very unclear, with uncertain empirical grounds (see, for example, Theodore A. Baroody, *Alkalize or Die: Superior Health Through Proper Alkaline-Acid Balance*, Eclectic Press, Waynesville NC, 1999).

[453] Kotsovsky D, "Rassen-Hygiene und Altersbekämpfung," *Monatsberichte. Institut für Altersforschung und Altersbekämpfung Organ für Internationale Altersforschung und Altersbekämpfung*, 1, 2-13, Aug-Sept. 1936 (Racial hygiene and the combat of aging).

⁴⁵⁴ Dimu Kotsovsky, "Allgemeine vergleichende biologie des Alters. Genese des Alters" (General Comparative biology of Aging: The genesis of aging), *Reviews of Physiology, Biochemistry and Experimental Pharmacology*, 1(31), 132-164, 1931 (First presented in August 1929, at the XIII International Congress of Physiologists, Boston, USA).
Twenty years later, Kotsovsky continued to emphasize the rejuvenative power of sleep. (Dimu Kotsovsky, "Schlaf als Verjüngung" (Sleep as Rejuvenation), *Hippokrates. Zeitschrift für Praktische Heilkunde*, 23, 689-692, 12 Dec. 1950.)

⁴⁵⁵ Building on earlier studies (by Max Verworn, Emil Kraepelin, Ludimar Hermann, Gustav Embden, Max Offner, Clifton Fremont Hodge, Angelo Mosso, Nathaniel Kleitman, Wilhelm Weichardt, and others) that demonstrated that fatigue leads to body intoxication, energy depletion, weakened immunity and whole body equilibrium impairment, Kotsovsky concluded that rest is vitally important for life extension:
"Periods of rest…contribute to the general life-prolongation, as the resting state diminishes tissue disintegration and energy expenditure, and opposes the action of harmful metabolic products accumulating during the awake state."
Further to emphasizing the "need for a longer general rest," Kotsovsky discussed various means of artificial sleep induction – by light, scent, mechanical and electrical stimulation in animals and hypnosis in humans – and the lower susceptibility of people to such manipulations. Despite the obvious regenerative ability of sleep, Kotsovsky maintained, the understanding of sleep mechanisms was very deficient and had not obtained a sufficient attention of the researchers of aging.
(Dimu Kotsovsky, "Allgemeine vergleichende biologie des Alters. Genese des Alters" (General Comparative biology of Aging: The genesis of aging), *Reviews of Physiology, Biochemistry and Experimental Pharmacology*, 1(31), 132-164, 1931.)
Yet, even by then, some initial attempts were made to reinforce recovery processes, for example using the anti-fatigue anti-toxic serum ("*antikenotoxin*") and oxygen therapy against fatigue, suggested in 1904 by the chemist Wilhelm Weichardt of the University of Erlangen, Germany. (See Anson Rabinbach, *The Human Motor: Energy, Fatigue, and the Origins of Modernity*, University of California Press, Berkeley, CA, 1992, pp. 142-145.)

⁴⁵⁶ Kotsovsky D, "Die Psycho-Hygiene und das Problem des Alters und der Lebensverlängerung" (Psycho-hygiene and the problem of aging and life-extension), *Zeitschrift für psychische hygiene*, 6, 8-10, 1933; Kotsovsky D, "La problème de la vieillesse en sociologie," *Altersprobleme*, 1 (2), 76-87, 1937 (The problem of senescence in sociology, first published in *Revue Internationale de Sociologie*, Nr. 5-6, May-June 1932).

⁴⁵⁷ Kotsovsky D, "Über die Hormonwirkung auf die Altersentwicklung der Pflanzen" (Hormone effect on the aging of plants), *Der Biologe*, 11, 276-278, 1942.

⁴⁵⁸ Kotsovsky D, "Über Beschleunigung und Hemmung des Alterns beim Menschen" (On the acceleration and inhibition of aging in humans), *Wiener Klinische Wochenschrift*, 55, 307-308, 1942.

⁴⁵⁹ Kotsovsky D, "Lebensdauer und Arbeit" (The duration of life and work), *Naturwissenschaftliche Rundschau*, 12, 511, 1951.

⁴⁶⁰ Kotsovsky D, "Zur Bekämpfung des Antagonismus der Altersklassen: Positivum und Negativum des Alters" (On the combat against the antagonism between age classes: the positive and the negative in aging), *Praxis*, 47, 650-653, 1958.

⁴⁶¹ Dimu Kotsovsky, *Tragödie des Genius. Genialität - Altern - Tod. Prinzipielle Untersuchung* (The Tragedy of Genius. Genius, Aging, Death. Principal Explorations), München, 1959; Dimu Kotsovsky, *Das Altersproblem in der Geschichte. Versuch einer bio-sozialen Synthese. The Problem of Aging in History. Attempt of a Biosocial Synthesis*, München, 1960; Dimu Kotsovsky. *Dostojewski, Tolstoj und der Bolschewismus. Über die historische Verantwortung des Schriftstellers. – Dostoyevsky, Tolstoy i Bolshevism. Ob istoricheskoy otvetstvennosti pisatelia.* (Dostoevsky, Tolstoy and Bolshevism. On the historical responsibility of writers), München, 1960.

[462] According to Kotsovsky's analysis, the "romantic, emotional, extrovert group of geniuses" (including poets, sculptors, painters and composers) lived the shortest (about 60 years), and their fate was "often tragic." In contrast, the "classical, rational, introvert" geniuses lived some 8-10 years longer (about 68-70) and their fate was "seldom tragic."
The explanation consisted in a difference in energy expenditures: "erratic, irregular, irrational and unsparing" of the romantics, and "rhythmic, rational and sparing" of the classics.
(Dimu Kotsovsky, *Tragödie des Genius. Genialität - Altern - Tod. Prinzipielle Untersuchung* (The Tragedy of Genius. Genius, Aging, Death. Principal Explorations), München, 1959, p. 371-375.)

[463] Dimu Kotsovsky, *Tragödie des Genius. Genialität - Altern - Tod. Prinzipielle Untersuchung* (The Tragedy of Genius. Genius, Aging, Death. Principal Explorations), München, 1959, p. 371.

[464] Dimu Kotsovsky, *Das Altersproblem in der Geschichte. Versuch einer bio-sozialen Synthese. The Problem of Aging in History. Attempt of a Biosocial Synthesis*, München, 1960, p. 48-54.

[465] Dimu Kotsovsky, *Das Altersproblem in der Geschichte. Versuch einer bio-sozialen Synthese. The Problem of Aging in History. Attempt of a Biosocial Synthesis*, München, 1960, pp. 43-44, 35.

[466] Dimu Kotsovsky. *Tragödie des Genius. Genialität - Altern - Tod. Prinzipielle Untersuchung* (The Tragedy of Genius. Genius, Aging, Death. Principal Explorations), München, 1959, p. 62.

[467] Dimu Kotsovsky, *Das Altersproblem in der Geschichte. Versuch einer bio-sozialen Synthese. The Problem of Aging in History. Attempt of a Biosocial Synthesis*, München, 1960, pp. 2, 49.

[468] Dimu Kotsovsky, *Das Altersproblem in der Geschichte. Versuch einer bio-sozialen Synthese. The Problem of Aging in History. Attempt of a Biosocial Synthesis*, München, 1960, pp. 53-54.

[469] Dimu Kotsovsky, *Das Altersproblem in der Geschichte. Versuch einer bio-sozialen Synthese. The Problem of Aging in History. Attempt of a Biosocial Synthesis*, München, 1960, p. 49.

[470] Dimu Kotsovsky, *Das Altersproblem in der Geschichte. Versuch einer bio-sozialen Synthese. The Problem of Aging in History. Attempt of a Biosocial Synthesis*, München, 1960, pp.39-42.
To support the importance of freedom for longevity, Kotsovsky claims that "animals in captivity live shorter than in their free, natural state." The only example he provides is that of gorillas that live 20-25 years in the zoo, and "almost twice as long in the wild, when they are well cared for and do not need to fight for existence."
The fact is, however, that, unlike animals in captivity, animals in the wild almost never live to an old age. This fact has formed the basis for the majority of modern evolutionary theories of aging. And so, if Kotsovsky was wrong, would this diminish the deontological value of freedom?
Still, in support of Kotsovsky's assumption, prison inmates were reported to have worse health and higher mortality, even to age more rapidly, than the free population, possibly due to substandard nutrition, bad hygiene and abuses.
See Susan J. Loeb, Azza Abudagga, "Health-related research on older inmates: an integrative review," *Research in Nursing and Health*, 29, 556-565, 2006.
On evolutionary theories of aging, see Peter B. Medawar, *An Unsolved Problem of Biology*, H.K. Lewis, London, 1952; Peter B. Medawar, "The definition and measurement of senescence," in *CIBA Foundation Colloquia on Ageing, Vol. 1, General Aspects* (Wolstenholme G.E.W. and Margaret P. Cameron, Eds.), J. & A. Churchill Ltd, London, 1955, pp. 4-15; George C. Williams, "Pleiotropy, natural selection, and the evolution of senescence," *Evolution*, 11, 398-411, 1957; Vladimir Frolkis, *Aging and Life-Prolonging Processes*, Springer-Verlag, Wien, 1982; Vladimir Frolkis, *Starenie i Uvelichenie Prodolzhitelnosti Zhisni* (Aging and Life-prolongation), Nauka, Leningrad, 1988; Michael R. Rose, *Evolutionary Biology of Aging*, Oxford University Press, Oxford, 1991; Tom Kirkwood, *Time of Our Lives. The Science of Human Aging*, Oxford University Press, Oxford, 1999.)

[471] Kotsovsky asserts that "the terror and tragedy are that the revolutionary ideas of totalitarianism, messianism and 'police state'… were expressed in an artful form by such geniuses of Russian thought as

Dostoevsky, Kropotkin, Lev Tolstoy and Berdiaev."

Dostoevsky was chiefly held responsible for propagating the idea of a messianic role of the Russian people, hence "Russian chauvinism that nourishes Russian imperialism and constitutes a threat to humanity."

Tolstoy "by word and by deed was destabilizing the social order which gave him the opportunity to maintain his large family." Plus, in his private life, Tolstoy was often tyrannical to his family and treacherous to his followers, thus giving a bad example for the Russian people.

Even Nikolay Berdiaev (an apostle of anti-Communism, who was deported from Soviet Russia in 1922) shared the blame for fostering Russian chauvinism, a major foundation for Russian Communism.

Kotsovsky also denounces anti-Semitism as a sin (particularly Dostoevsky's anti-Semitism). However, some anti-Semitic notes seem to surface. Thus, regarding the participation of Jews in the Russian revolution, Kotsovsky slyly refers the reader to "materials that can be found in the Museum of the Bund [the Jewish Socialist party of Lithuania, Poland and Russia] in New York," which seems to imply that the evidence for the responsibility of the Jews for the Revolution is actually quite considerable.

(Dimu Kotsovsky, *Dostojewski, Tolstoj und der Bolschewismus. Über die historische Verantwortung des Schriftstellers. – Dostoyevsky, Tolstoy i Bolshevism. Ob istoricheskoy otvetstvennosti pisatelia* (Dostoevsky, Tolstoy and Bolshevism: On the historical responsibility of writers), München, 1960, pp. 9, 79-80, 100, 126, 138, 147.)

[472] Constantin I. Parhon, M. Goldstein, *Les sécrétions internes; pathologie et physiologie* (Internal secretions: pathology and physiology), Maloine, Paris, 1909.

[473] Nathan W. Shock, *The International Association of Gerontology. A Chronicle – 1950-1986*, Springer Publishing Company, NY, 1988; V.N. Anisimov, M.V. Soloviev, *Evoluzia Concepcy v Gerontologii* (The Evolution of Concepts in Gerontology), Aesculap, Saint Petersburg, 1999, reprinted in the *Russian Biomedical Journal*, www.Medline.ru, vol.3, "Perspectivy razvitia gerontologii v Rossii" (Perspectives of the development of gerontology in Russia); Leake, C.D., "Russian and Iron Curtain Proposals for Geriatric Therapy," *Geriatrics* 14(10), 670-673, 1959, reprinted in Geraldine M. Emerson (Ed.), *Benchmark Papers in Human Physiology, Vol. 11, Aging*, Dowden, Hutchinson and Ross, Stroudsburg PA, 1977; The Institute of Gerontology and Geriatrics "Ana Aslan," http://www.ana-aslan.ro/index_en.htm.

[474] The Institute of Gerontology and Geriatrics "Ana Aslan," http://www.ana-aslan.ro/index_en.htm.

[475] The institute in Kiev was led by Professors Nikolay Gorev (the first director) and Dmitry Chebotarev (the second director), and was established under the decree of Clement Voroshilov (1881-1969, then president of the Supreme Soviet), and of the USSR Minister of Health Maria Kovrigina.

[476] Quoted in Dimu A. Kotsovsky, "Gerontology in Eastern Europe," *Review of Eastern Medical Sciences*," 13, 7-16, 1958.

[477] Leake, C.D. "Russian and Iron Curtain Proposals for Geriatric Therapy," *Geriatrics* 14(10), 670-673, 1959, reprinted in Geraldine M. Emerson (Ed.), *Benchmark Papers in Human Physiology, Vol. 11, Aging*, Dowden, Hutchinson and Ross, Stroudsburg PA, 1977; Ana Aslan, "Recent experiences on the rejuvenating action of Novocain (H3), together with experimental, clinical, and statistical findings," *Therapiewoche*, 8, 10-19, 1957; Ana Aslan, "A new method for the prophylaxis and treatment of aging with Novocain – eutrophic and rejuvenating effects," *Therapiewoche*, 7, 14-22, 1956; Patrick M. McGrady, *The Youth Doctors*, Coward-McCann, NY, 1968; Joel Kurtzman and Phillip Gordon, *No More Dying. The Conquest of Aging and the Extension of Human Life*, Tarcher Inc, Los Angeles, 1976.

[478] Among other effects, there were recorded an increase of albumins and decrease of globulins, and improvement of lipoprotein balance (an increase in "Alpha" High Density Lipoprotein, now colloquially referred to as "good cholesterol," and a decrease in "Beta" Low Density Lipoprotein - the "bad cholesterol").

[479] Ana Aslan, *Research on novocain therapy in old age, in English translation; a collection of seven papers from Die Therapiewoche, 1956-1957*, Consultants Bureau, New York, 1959; Ana Aslan, *Gerovital H3: Neurotropic,*

antidepressive, vitaminic, eutrophic and regenerative factor in the treatment of ageing and trophic disturbances, National Institute of Gerontology and Geriatrics, Bucharest, 1977.

[480] In 1982, "Gerovital" was banned in the US by the American Food and Drug Administration as an unproved remedy. For a summary of Gerovital's efficiency or rather lack thereof, see Ostfeld A, Smith CM, Stotsky BA, "The systemic use of procaine in the treatment of the elderly: a review," *Journal of the American Geriatrics Society*, 25(1), 1-19, January 1977.

[481] Patrick M. McGrady, *The Youth Doctors*, Coward-McCann, NY, 1968, pp. 181-192; Neica L, Aldulea N, "Ana Aslan, The Woman Who Defeated Time," *Bulletin of the Transilvania University of Brasov, Series 6: Medical Sciences*, 6(51), 161-166, 2009; "Ana Aslan" bio at the Institute of Gerontology and Geriatrics "Ana Aslan," http://www.ana-aslan.ro/index_en.htm.

[482] Paracelsus (Phillippus Aureolus Theophrastus Bombastus von Hohenheim, 1493-1541), the renowned alchemist and occultist, and the founder of medical chemistry (iatrochemistry), believed that human life can be prolonged well beyond half a millennium or even "as long as Methuselah" and sought the *quinta essentia* that would have "the power to change us, to renew us, and to restore us" and that would "transform the body, removing its harmful parts, its crudity, its incompleteness, and transform everything into a pure, noble and indestructible being."

(*Paracelsus. Selected Writings*, edited by Jolande Jacobi, Princeton University Press, Princeton, 1995 (1951), "Man and Works. Alchemy. Art of Transformation," p. 148; further see *The Hermetic and Alchemical Writings of Paracelsus*, Edited by Arthur Edward Waite, James Elliott and Co., London, 1894, vol. II, Part II, Hermetic Medicine, Book 8 "Concerning the elixirs," pp. 69-76, "A book concerning long life," pp. 108-123; James Campbell Brown, *A History of Chemistry from the Earliest Times* (2nd edition, J.& A. Churchill, London, 1920, first published in 1913), Ch. 11 "The Protest of Paracelsus," pp. 108-117.)

Paracelsus suggested quite a few elixirs for rejuvenation. See John Uri Lloyd, *Pharmaceutical preparations. Elixirs, their history, formulae, and methods of preparation. Including Practical processes for making the popular elixirs of the present day, and those which have been official in the old pharmacopoeias. Together with a résumé of unofficial elixirs from the days of Paracelsus*, Second Edition, Robert Clarke and Company, Cincinnati, 1883, pp. 5-16, 149.

One of the more famous was Paracelsus' "Elixir Proprietatis" (proprietary elixir), containing myrrh, aloe and saffron. "The Paracelsus' Elixir Proprietatis" was to be found in pharmacopoeias and pharmacies as late as the end of the 19th century, usually concocted by the following inexpensive formula: Aloes, Myrrh and Saffron – 3 troy ounces each, in 2 pints of alcohol. Judging from Paracelsus' 48 year life-span, the elixirs were not particularly effective. The prevailing version of his death, however, was that he was murdered, and against such a cause of death surely no pharmacological remedies would avail.

Another Swiss master of rejuvenation was Albrecht von Haller (1708-1777), the renowned initiator of physiology. (Albrecht von Haller spent 17 years in the German University of Göttingen, thus he might perhaps be better described as a Swiss-German. Paracelsus too, though Swiss-born, traveled and worked all across Europe, and was as much at home in Germany as in Switzerland.)

In the 8th volume of Albrecht von Haller's monumental *Elementa Physiologiae Corporis Humani* (Physiological Elements of the Human Body, 1757-1766, published after Haller's return to Switzerland), a large section "Decrementum" is dedicated to aging and longevity.

(Alberto v. Haller, *Elementa Physiologiae Corporis Humani. Tomus Octavus*, Lausannae, Sumptibus Societatis Typographicae. M. DCC. LXXVIII, "Sectio III. Decrementum," pp. 68-124.)

In that section, Haller treated of age-related deterioration of various organs and humors; comparative longevity of plants, animals and humans; possibilities of rejuvenation (synonymous with regeneration of organs); and causes of longevity (including climate and diet). A list of super-centenarians, some of whom were said to live nearly two centuries and even more, was included to show that such tremendous life-spans are reasonably attainable. Furthermore, Haller, a devout Christian, cited the longevity of antediluvian patriarchs as a fact. He was unsure about the causes for the dramatic decline in human longevity after the deluge (perhaps it was the abandonment of a vegetarian diet, perhaps a "total mutation" in the air, seasons, weather, earth, etc.). Haller believed that "this is a problem insoluble due to the paucity of data," nonetheless under proper environmental, physical and spiritual conditions, Methuselah's longevity were feasible. Haller's "Acid elixir" (Elixir Acidum Halleri, composed of a mixture of sulphuric acid and alcohol – of uncertain safety) was still to be found in the German Pharmacopoeia of 1872.

Haller apparently was influenced in his pursuit of longevity and rejuvenation by his teacher, the Dutch physician Herman Boerhaave (1668-1738), yet another pioneer of physiology (introducing, among other notions, the concept of hemodynamic equilibrium). In *Elementa Chemiae* (1724), Boerhaave presented five different processes for making Paracelsus' Elixir Proprietatis, and highly recommended its efficacy: "Thus we obtain an acid, aromatic medicine, of great use in the practice of physic; for when externally applied, it cleanses and heals putrid, sinuous, and fistulous old ulcers, defends the parts from putrefaction, and preserves them by a true embalming" (quoted in Lloyd, 1883, pp. 12-13).

Boerhaave was also famous for his attempt to rejuvenate Amsterdam's burgomaster by the ancient art of "Gerocomia": ordering him to sleep between two young persons (after King David's example, as described in 1 Kings 1:2). This attempt is related in Hufeland's *Makrobiotik; oder, Die Kunst das menschliche Leben zu verlängern* (Macrobiotics or the art of prolonging human life), A.F. Macklot, Stuttgart, 1826 (1796), p. 7. Curiously, the burgomaster lay between "two young persons" – "zwischen zwei jungen Leuten" (lying between two girls might have been culturally inappropriate). Boerhaave further extensively treated on aging and longevity in his *Aphorismi de cognoscendis et curandis morbis* (Aphorisms on the Recognition and Treatment of Diseases, Leiden, 1709).

Boerhaave fostered the interest in these subjects in yet another of his famous pupils and assistants, the Dutch-Austrian physician Gerard van Swieten (1700-1772). Aging and longevity were among the major topics of van Swieten's five-volume commentary on Boerhaave's "Aphorismi," but perhaps van Swieten's most renowned apology of healthy life-extension was his *Oratio de senum valetudine tuenda* (Oration on the care of health of the aged, published in 1778, first presented in 1763).

Another crucial figure in the succession of rejuvenators from Paracelsus to Boerhaave was the Dutch (Flemish) physician Jan Baptist van Helmont (1579-1644), the author of "pneumatic"/gas chemistry and one of the most prominent philosophers of vitalism. Van Helmont's elixir (as presented in the London Pharmacopoeia of 1770, though of uncertain authenticity) consisted of "Any fixed alkaline salt [such as potassium carbonate, K_2CO_3], Socotrine Aloes, Saffron, Myrrh, Sal Ammoniac [ammonium chloride, NH_4Cl], Mountain wine [vinum album/white wine]" which one needed to "macerate without heat, for a week or longer, then filter through paper" (Lloyd, 1883, p. 100).

Boerhaave traced the line of succession in the development of elixirs further back from Paracelsus, and gave a special credit to the 15th century (rather legendary) German Benedictine monk Basil Valentine. Boerhaave wrote:

"M.C. Clerc thinks there are indications of chemical medicines in Thaddeus the Florentine, who lived in the thirteenth century, in Albertus Magnus, Friar [Roger] Bacon, and Isaac Hollandus. Helmont has taken pains to show that Basil Valentine was prior to Paracelsus by a hundred years. ... [Basil Valentine] would seem to have been the first who applied chemistry to medicine; for after every preparation he never fails to give some medicinal use thereof. Paracelsus, Helmont, the elder [Nicolas] Lemery, and many others of modern fame, owe a great part of what is valuable in them to this author; so that it is not without reason that he is judged the father of the modern chemisis and the founder of the chemical pharmacy" (quoted in Lloyd, 1883, p.12).

[483] As *Encyclopedia Britannica* summarizes (Deluxe Edition CD, 2000), "Switzerland's traditional neutrality and its laws of political asylum have made the country a magnet for many creative persons during times of unrest or war in Europe. The mid-19th century was such a period, as were the 1930s and '40s, when the rise of fascism caused a number of German, Austrian, and Italian writers such as Thomas Mann, Stefan George,

and Ignazio Silone to seek harbour in Switzerland" ("Switzerland: Cultural Life"), "Despite being surrounded by Nazi and fascist enemies, Switzerland survived as the only democratic state in central Europe" ("Swiss Neutrality").

[484] Nathan W. Shock (Ed.), *A Classified Bibliography of Gerontology and Geriatrics*, Stanford University Press, Stanford CA, 1951; *Supplement One 1949-1955*, 1957; *Supplement Two 1956-1961*, 1963 (reprinted in 1980 by Stanford University Press).

[485] Between 1939 and 1944, the life-expectancy in Switzerland was about 64-65 years, exceeded only by Scandinavian countries – Sweden, Norway and Iceland – where the life-expectancy was approximately 66-67 years.

Even recently, Switzerland has retained one of the highest ranks in life-expectancy. For the period 2005-2010, with the world average life-expectancy of about 67.9 years, Switzerland was estimated to be in the second place (about 81.8 years), only very slightly surpassed by Japan (82.7 years). (Noteworthy, in the 1930s and early 1940s the life-expectancy in Japan ranged between 47 and 50 years – about twenty years lower than Switzerland.)

(Source: Compiled by Earth Policy Institute from United Nations Population Division, *World Population Prospects: The 2010 Revision*, CD-ROM Edition, New York, April 2011.)

With slight variations, Switzerland ranks among the highest in life-expectancy according to most recent estimates.

(E.g. *The Human Mortality Database*, http://www.mortality.org/; World Bank, http://data.worldbank.org/indicator/SP.DYN.LE00.IN?order=wbapi_data_value_2009+wbapi_data_value+wbapi_data_value-last&sort=desc; *CIA – The World Factbook*, https://www.cia.gov/library/publications/the-world-factbook/rankorder/2102rank.html.)

[486] J. Robine and F. Paccaud, "Nonagenarians and centenarians in Switzerland, 1860–2001: a demographic analysis," *Journal of Epidemiology and Community Health*, 59(1), 31-37, January 2005.

[487] In this regard, the "Endowment Theory," suggesting a greater valuation on the objects the person already has, may be relevant. (Richard Thaler, "Towards a positive theory of consumer choice," *Journal of Economic Behavior and Organization*, 1, 39-60, 1980.)

Simply put, when the quality of life is satisfactory, life extension or the preservation of the present well-being may be desired, while under adverse conditions, hopefulness may diminish.

The underlying principle might thus be interpreted as "The more one has, the more one wants."

A contrary argument may be raised, based on the "Denial defense mechanism" – the theory expounded by Anna Freud. (Anna Freud, *The Ego and the Mechanisms of Defense* (originally published in German in 1936, translated in 1937 by Cecil Baines), Karnac Books, London, 1993, Ch. 7 "Denial in Word and Act," p. 83.)

According to the "denial" theory, adverse circumstances may induce overly hopeful expectations, when the person is unable to cope with calamity.

However, even if assuming that far reaching hopes (viz. life-extension) may be a reaction to strenuous social conditions, the subject then may not have the necessary resources to act on or disseminate his/her optimism.

Indeed, a recent evolutionary model argued that when resources are abundant, there emerge altruistic behaviors directed toward the prolongation of life for other members of the group. As the study stated, "weak local competition (abundant resource, low crowding), favor survival (e.g., nest defense) and fecundity (e.g., nurse workers) altruism, which are "r-strategies" that increase a social group's growth rate."

(J. David Van Dyken, Michael J. Wade, "Origins of altruism diversity I: the diverse ecological roles of altruistic strategies and their evolutionary responses to local competition," *Evolution*, 66(8), 2484-2497, April 16, 2012, http://onlinelibrary.wiley.com/doi/10.1111/j.1558-5646.2012.01630.x/abstract.)

[488] Eugene Steinach and John Loebel, *Sex and Life; Forty Years of Biological and Medical Experiments*, Faber, London, 1940, p. 233.

[489] S. Stocker, "Reimplantation der Keimdrüsen beim Menschen" (Reimplantation of genital glands in humans), *Correspondenz-Blatt für Schweizer Aerzte*, Basel, 7, 193-224, Feb. 12, 1916.

[490] K. Kolb, "Ueber einen verjungungsversuch bei einer ziege" (On a rejuvenation experiment in a she-goat), *Verhandlungen der Schweizerischen naturforschenden Gesellschaft*, 10, 311, 1922.

[491] Paul Niehans, "Les glandes endocrines et les méthodes de rajeunissement," *Schweizerische Medizinische Wochenschrift*, 10, 255-264, 1929 (Endocrine glands and methods of rejuvenation); Paul Niehans, *La sénescence et le rajeunissement*, Vigot Frères, Paris, 1937 (Aging and rejuvenation); Paul Niehans, *20 Jahre Zellular-Therapie*, Verlag Urban & Schwarzenberg, Berlin-München-Wien, 1952 (Twenty Years of Cell Therapy).

The historian Paul Lüth attributes to the surgeon Hermann Küttner of Breslau (Wroclaw, Poland) the priority in applying cell therapy by injection in 1912.

(Paul Lüth, *Geschichte der geriatrie: Dreitausend Jahre Physiologie, Pathologie und Therapie des alten Menschen* (The history of geriatrics, three thousand years of physiology, pathology and therapy of the aged), Enke, Stuttgart, 1965, p. 227; Küttner, H., "Die Verimpfung anstelle der Transplantation hochwertiger Organe," Breslauer Chirurgische Gesellschaft (Inoculation instead of transplantaiton of highly valuable tissues, first presented at the Breslau Surgical Society), *Zentralblatt für Chirugie*, 1, 390-397, 1912).

Nonetheless, the first wide application of cell therapy can be undoubtedly attributed to Paul Niehans.

[492] Niehans's method consisted in the following:

"Organs are taken from domestic animals, under veterinary control. The intervention is painless for the animals; a sudden blow on the head replaces anesthesia. The operation and the dissolution of organs into cells (in Ringer solution) proceed under sterile conditions. Until now, I have used cells from practically all the organs from fetuses, youthful or adult animals, for medical purposes …

A diseased organism takes in the cells, serving for its healing, exceptionally well. Even multiple intramuscular injections containing foreign proteins, whether in little children or the elderly, are extremely well tolerated, when the cells or cell cultures are fresh and germ-free, and when the conserved cells are injected without delay. With over 3000 injections of about 10 cm^3 of cell suspension, I have only seldom observed a slight reaction, and never any anaphylactic shock effects or even excitation.

Whether the injected fresh cells travel to their targets or are degraded on the spot by the organism to produce the building blocks necessary for its healing, is a question that must be answered by a physiologist and not by a physician. One thing is clear: Cell Therapy promotes the regeneration of damaged cells, until reaching a necessary quantitative activity threshold. Beyond that physiological threshold, there is no activity; therefore the introduced cells can cause no damage."

(Paul Niehans, *20 Jahre Zellular-Therapie* (Twenty Years of Cell Therapy), Verlag Urban & Schwarzenberg, Berlin-München-Wien, 1952, p. 1.)

[493] Paul Niehans, *20 Jahre Zellular-Therapie* (Twenty Years of Cell Therapy), Verlag Urban & Schwarzenberg, Berlin-München-Wien, 1952, p. 15.

[494] Paul Niehans, *20 Jahre Zellular-Therapie* (Twenty Years of Cell Therapy), Verlag Urban & Schwarzenberg, Berlin-München-Wien, 1952, pp. 1, 8-9.

[495] Anthony Atala, "Extending life using tissue and organ replacement," *Current Aging Science*, 1(2), 73-83, 2008; Mariusz Z. Ratajczak, et al, "Hunt for Pluripotent Stem Cell – Regenerative Medicine Search for Almighty Cell," *Journal of Autoimmunity*, 30(3), 151-162, 2008; Alan Trounson, "New perspectives in human stem cell therapeutic research," *BioMed Central - BMC Medicine*, 7(29), 1-5, 2009.

The first therapeutic use of adult stem cells via bone marrow transplantation is commonly traced to the late 1950s, to the works of the French Doctor Georges Mathé (1957-1959) and the American Doctor Edward Donnall Thomas (1956-1957).

(Thomas ED, Lochte HL, Lu WC, et al., "Intravenous infusion of bone marrow in patients receiving radiation and chemotherapy," *New England Journal of Medicine*, 157 (11), 491-496, 1957;

Mathé G, Jammet H, Pendic B, et al., "Transfusions et greffes de moelle osseuse homologue chez de humains irradiés à haute dose accidentellement" (Transfusions and grafts of homologous bone marrow in humans after accidental high dosage irradiation), *Revue francaise des études cliniques et biologiques*, 4, 226-238, 1959.)

[496] Paul Niehans Clinic, http://www.paulniehans.ch/clinic.htm.

[497] Another contemporary user of "nucleic acid therapy" (RNA and DNA injections) was the English physician Max Odens. (Max Odens, "Prolongation of the Life Span in Rats," *Journal of the American Geriatrics Society*, 21 (10), 450-451, 1973; http://www.cryonet.org/cgi-bin/dsp.cgi?msg=23318; http://www.cryonet.org/cgi-bin/dsp.cgi?msg=7556; http://www.sens.org/node/1972.)

[498] Porteus MH, et al., "A look to future directions in gene therapy research for monogenic diseases," *Public Library of Science - PLoS Genetics*, 2 (9), 1285-92, 2006; Woods NB, et al., "Gene therapy: therapeutic gene causing lymphoma," *Nature*, 440 (7088), 1123, April 2006; Rattan SI, Singh R, "Progress and prospects: gene therapy in aging," *Gene Therapy*, 16(1), 3-9, 2009.

As Rattan and Singh pointed out: "Whether genetic redesigning can be achieved in the wake of numerous and complex epigenetic factors that effectively determine the life course and the life span of an individual still appears to be a 'mission impossible.'"

[499] Patrick M. McGrady, *The Youth Doctors*, Coward-McCann, NY, 1968, pp.59-122.

[500] Patrick Michael McGrady (1932-2003) would later also play a significant part in the life-extension movement, having co-authored with Nathan Pritikin *The Pritikin Program for Diet and Exercise* (1979, emphasizing "low fat" and "high fiber/low sugar" diet, http://www.pritikin.com/), and having founded the cancer information center CANHELP (1983, still active, http://www.canhelp.com/).

[501] This story about Niehans's birth also appears in "Niehans, Chirurgie ohne Messer" (Niehans: Surgery without a knife), *Der Spiegel*, No. 13, pp. 32-41, March 27, 1957, http://www.spiegel.de/spiegel/print/d-41120750.html

[502] Paul Niehans Clinic Switzerland, http://www.paulniehans.ch/clinic.htm.

[503] The royal jelly (or "bee's milk"), allowing the queen bee to dramatically surpass the life-span of a working bee (2-5 years vs. 20-50 days respectively), was propagandized as a life-extending supplement in the 1950s by the French professor Rémy Chauvin (1913-2009).

(Chauvin R, Louveaux J, "Etude macroscopique et microscopique de la gelée royale" (Macroscopic and microscopic studies of the royal jelly), *Apiculteur*, 100, 33-43, 1956; Chauvin R, "La gelée royale. Composition biochimique" (The royal jelly, biochemical composition), *Apiculteur*, 100, 44-55, 1956; Chauvin R, "Sur un principe de la gelée royale d'abeilles, actif sur la glycémie des mammiferes" (On a principle of bees' royal jelly, affecting glycemia in mammals), *Comptes rendus des séances de l'Academie des Sciences*, 243, 1920-1921, 1956.)

The "chicken embryo" therapy or "embryotherapy" was promoted since 1945 by the Yugoslav-born French rejuvenator Ivan Popov.

(Ivan Popov, *Stay Young. The Secrets of a World-famous Youth Doctor*, Grosset & Dunlap, NY, 1975, Ch. 19 "Modern Organic Clinical Procedures for Revitalization I: Embryotherapy," pp. 232-242.)

But, of course, "chicken embryo" therapy has been known since ancient times, from the Far East (balut, maodan) to the Near East (chicken embryo dishes).

(See Fred Rosner, *Medicine in the Bible and the Talmud*, Ktav, Hoboken NJ, 1995 (1977), "Therapeutic Efficacy of Chicken Soup," pp. 136-139; Jerry Hopkins, *Asian Aphrodisiacs: From Bangkok to Beijing*, Periplus, Singapore, 2007, p. 153.)

[504] Alan Harrington's *The Immortalist* (1969) exemplifies the attitude perceiving religion as inherently harmful to life-extension:

"Religious orthodoxy was invented to give everyone a chance to earn life everlasting. ... The false gods to whom the immortality-hunter formerly bowed will be reduced to artifacts. ... Our mission will be simply, first, to attack death and all of its natural causes, and, second, to prepare for immortality."
(Alan Harrington, *The Immortalist*, Celestial Arts, Millbrae, California, 1977, first published in 1969, pp. 92, 273.)
For the atheist immortalist, the unlimited prolongation of life (that is, life extension that is not constrained by any particular/arbitrary/ordained date) is logically equivalent to the pursuit of physical immortality, hence religion is inimical to both.
[505] James Joseph Walsh, in *Old-Time Makers of Medicine*, makes a thorough case for the support of medical science by the Catholic Church and by the Papal Office during the Middle Ages, including such examples as the patronage by Abbot Desiderius (Pope Victor III, 1026-1087), the work of Pope John XXI the physician (1215-1277), the medical research of Cardinal Nicolas Cusanus (1401-1464), the studies of the Jesuit priest Athanasius Kircher (1602-1680) and great many other clerics and medical scholars.
(James Joseph Walsh, *Old-Time Makers of Medicine. The Story of The Students And Teachers of the Sciences Related to Medicine During the Middle Ages*, Fordham University Press, NY, 1911; James Joseph Walsh, *The Popes and Science*, Fordham University Press, NY, 1908; James Joseph Walsh, *Priests and Long Life*, J. F. Wagner, New York, 1927.)
Perhaps the most extreme case of 'interest in medicine' was that of Pope Innocent VIII (1432-1492) who was said to drink the blood of boys for rejuvenation.
(Mirko D. Grmek, *On Ageing and Old Age, Basic Problems and Historic Aspects of Gerontology and Geriatrics*, *Monographiae Biologicae*, 5, 2, Den Haag, 1958, pp. 45-46.
The source of this story is in Stefano Infessura (1435-1500), *Diario della Città di Roma* (Diary of the City of Rome, Forzani, Roma, 1890, pp. 275-276.)
In more recent times, the encyclicals of Pope John Paul II (1920-2005) called to "reaffirm the culture of life." In the 1995 encyclical, *Evangelium Vitae*, John Paul II issued the dictum to use all modern biomedical means to prevent death, that "life may be always defended and promoted."
(Ioannes Paulus PP. II, *Evangelium Vitae, On the Value and Inviolability of Human Life*, March 25, 1995, http://www.vatican.va/holy_father/john_paul_ii/encyclicals/documents/hf_jp-ii_enc_25031995_evangelium-vitae_en.html.)
And his own struggle with death to the last days provides a testimony to following this principle (despite some journalistic allegations to the contrary, e.g. Jeff Israely, "Was John Paul II euthanized?" *Time*, September 21, 2007, http://www.time.com/time/world/article/0,8599,1664189,00.html).
Indeed, a recent study showed that religious people tend to be more likely to refuse a "do-not-resuscitate" status than non-religious.
(Maria A. Sullivan, Philip R. Muskin, Shara J. Feldman, Elizabeth Haase, "Effects of Religiosity on Patients' Perceptions of Do-Not-Resuscitate Status," *Psychosomatics*, 45, 119-128, April 2004.)
The Pope Benedict XVI (b. 1927) took radical life extension very seriously. On April 3, 2010, in the Easter Vigil Homily, Benedict XVI said the following:
"Man's resistance to death becomes evident: somewhere – people have constantly thought – there must be some cure for death. Sooner or later it should be possible to find the remedy not only for this or that illness, but for our ultimate destiny – for death itself. Surely the medicine of immortality must exist. Today too, the search for a source of healing continues. Modern medical science strives, if not exactly to exclude death, at least to eliminate as many as possible of its causes, to postpone it further and further, to prolong life more and more."
Indeed, these statements were followed by poignant concerns:

"Would that be a good thing? Humanity would become extraordinarily old, there would be no more room for youth. Capacity for innovation would die, and endless life would be no paradise, if anything a condemnation."

Yet, the fact that the Pope acknowledges, and even somewhat sympathizes with the possibility of radical life extension is unprecedented.

(*Easter Vigil, Homily of His Holiness Benedict XVI, Saint Peter's Basilica, Holy Saturday, 3 April 2010*, http://www.vatican.va/holy_father/benedict_xvi/homilies/2010/documents/hf_ben-xvi_hom_20100403_veglia-pasquale_en.html.)

Thus, the interest of Pope Pius XII in rejuvenating treatments seems to be not unusual.

[506] *Pontifical Academy for Life. Declaration on the Production and the Scientific and Therapeutic Use of Human Embryonic Stem Cells* (signed by the Academy President, Prof. Juan de Dios Vial Correa), Vatican, August 25, 2000, http://www.vatican.va/roman_curia/pontifical_academies/acdlife/documents/rc_pa_acdlife_doc_20000824_cellule-staminali_en.html; *The Catholic Encyclopedia* "Abortion," http://www.newadvent.org/cathen/01046b.htm; *Teachings of the Catholic Church on Abortion*, http://www.priestsforlife.org/magisterium/bishops/09-02-26-scranton-holy-communion.htm; at *Priests for Life*, http://www.priestsforlife.org/.

[507] Verzár immigrated to Switzerland in 1930. Prof. Francois Bourlière explains (and cites Prof. Otto Lowei in testimony) that the possible cause for Verzár's emigration was that, as "Rector Magnificus" of the University of Debrecen, Hungary, Verzár was overswept by administrative work and outraged by the students' ultra-nationalist and anti-Semitic movement that he had to suppress. Verzár himself was not Jewish, but a descendant of a Reformed Protestant family. In 1929 he was appointed Presbyter of the Reformed Church in Debrecen.

(Francois Bourlière, "Chapter 1. Fritz Verzár: The man and his work," in *Perspectives in Experimental Gerontology*, Nathan Wetherill Shock (Editor), Charles C. Thomas, Springfield, IL, 1966, pp. 5-18.)

[508] Verzár's institute was the world's second or third in the field, depending on whether counting Kotsovsky's institute. The Russian/British gerontologist and life-extensionist Zhores Medvedev considered Verzár's institute even as the world's first and claimed that Verzár "had the greatest impact on the development of experimental gerontology in Europe."

The East-German gerontologist Werner Ries furthermore honored Verzár as "father of experimental gerontology."

(Zhores A. Medvedev, "The past and the future of experimental gerontology," *Archives of Gerontology and Geriatrics*, 9, 201-213, 1989; Werner Ries, *Max Bürger (1885-1966). Internist, Pathophysiologe, Alternsforscher. Ausgewählte Texte*, Johann Ambrosius Barth, Leipzig, 1985, "Fritz Verzár" pp. 155-157.)

Further on Verzár's legacy, see:

Herman Le Compte, "All you need is…sense (to live long and happily)," *Acta Gerontologica et Geriatrica Belgica*, 9(1), 12-21, 1971; L.V. Komarov, H. Le Compte, "People can and must live not by decades but by centuries," *Acta Gerontologica et Geriatrica Belgica*, 10(2-3), 87-97, 1972; Imre Zsolnai (Zs.)-Nagy, "Fritz Verzár was born 120 years ago: A personal account," *Archives of Gerontology and Geriatrics*, 43 (1), 1-11, July 2006; Nathan W. Shock, *The International Association of Gerontology. A Chronicle. 1950 to 1986*, Springer Publishing Company, NY, 1988, p. 40.

[509] Fritz Verzár, "Veränderungen der thermoelastischen Kontraktion von Sehnenfasern im Alter," *Helvetica Physiologica Acta*, 13, 64-67, 1955 (Changes in thermoelastic contraction of tendon fibers in aging); Fritz Verzár, "Das Altern des Kollagens," *Helvetica Physiologica Acta*, 14, 207-221, 1956 (The aging of collagen); Fritz Verzár, "Das Altern des Kollagens in der Haut und in Narben," *Schweizerische medizinische Wochenschrift*, 91, 1234-1236, 1961 (The aging of collagen in the skin and scars).

The emergence in the 1950s of the "Cross-linkage theory of aging" is now mainly associated with the work of the Finnish/American researcher Johan Bjorksten, especially with his seminal study "A common

molecular basis for the aging syndrome" (*Journal of the American Geriatrics Society*, 6, 740-748, 1958).
The theory was self-admittedly inspired by Bjorksten's work in the 1940s on cross-linking in thin protein gels, as chief chemist for Ditto Inc, a subsidiary of Eastman Kodak.
(Johan Bjorksten, "Recent developments in protein chemistry," *Chemical Industries*, 48, 746-751, 1941; Johan Bjorksten, "Chemistry of duplication," *Chemical Industries*, 49, 2, 1942.)
Several gerontologists, however, give Verzár the credit for developing the "cross-linkage" theory.
Thus, according to Werner Ries, "[Verzár's] studies of the biology of aging culminated in the discovery of cross-links in collagen" (Ries, 1985, p.157).
And according to Nathan Shock,
"[Verzár] was primarily responsible for developing the cross-linking theory of aging… In the area of biology [c. 1966] the study of connective tissue as a model for aging based on the formation of inter- and intramolecular cross-links was emphasized in many papers. …
The cross-linking theory of cellular aging dominated the field of cellular gerontology at this time. Although Dr. Verzár played front and center stage at this drama, other investigators such as [Johan] Bjorksten, [Robert Austin] Milch, [Francis Marott] Sinex, and others from the United States, as well as [Milos] Chvapil from Czechoslovakia and [Joseph] Balo and [Ilona] Banga from Hungary, provided an international scope to the idea that cross-linking at the molecular level played a key role in aging at the cellular level."
(Nathan W. Shock, *The International Association of Gerontology*, 1988, pp.113-115.)
Bjorksten, however, strongly asserted his priority:
"F. Verzár, in 1955 proposed the theory of crosslinkages in aging, on the basis of his work with collagen in tendons and skin. A year later, after his first publication my prior work was brought to his attention by Dr. K. H. Gustavson in Stockholm. Dr. Verzár wrote me a very polite letter of apology. We had occasions to collaborate later and became good friends."
(Johan Bjorksten, *Longevity, 2: Past, Present, Future*, Jab Publishing, Charleston SC, 1987, "Re-discoveries of the Crosslinkage Theory of Aging," p. 104.)
The search for agents capable of breaking the "cross-links" has been continued by life-extensionists to the present. See, for example, Aubrey de Grey and Michael Rae, *Ending Aging. The Rejuvenation Breakthroughs That Could Reverse Human Aging in Our Lifetime*, St. Martin's Press, NY, 2007 (Ch. 9, "Breaking the Shackles of AGE," pp. 165 199).

[510] Herman Le Compte, "All you need is…sense (to live long and happily)," *Acta Gerontologica et Geriatrica Belgica*, 9(1), 12-21, 1971; L.V. Komarov, H. Le Compte, "People can and must live not by decades but by centuries," *Acta Gerontologica et Geriatrica Belgica*, 10(2-3), 87-97, 1972 (p. 93).

[511] In an interview, Verzár empathically exclaimed:
"Niehans? Cured the Pope? Yes… cured him of constipation. Hah! But that's all. And with what? With mashed potatoes! No, I will not discuss Niehans with you. Either one discusses serious things or things that are not serious. Niehans is *not* serious."
(Patrick M. McGrady, *The Youth Doctors*, Coward-McCann, NY, 1968, p. 85.)

[512] Herman Le Compte, "All you need is…sense (to live long and happily)," *Acta Gerontologica et Geriatrica Belgica*, 9(1), 12-21, 1971 (quote from p. 18).

[513] Francois Bourlière, "Ch. 1. Fritz Verzár: The man and his work," in *Perspectives in Experimental Gerontology*, Nathan Wetherill Shock (Editor), Charles C. Thomas, Springfield, IL, 1966, pp. 5-18 (quote from p. 18).

[514] Anna Ioanovna (1693-1740), niece of Peter I the Great, ruled 1730-1740.

[515] Elisaveta Petrovna (1709-1762), daughter of Peter I, ruled 1741-1762. Elisaveta's sponsorship also made possible the establishment of the University of Moscow by Mikhail Lomonosov in 1755 and the foundation of the Academy of Fine Arts in St. Petersburg by Ivan Shuvalov in 1757 (Russia's first).

[516] J. Steudel, "Zur Geschichte der Lehre von den Greisenkrankheiten" (On the history of the study of the diseases of old age), *Sudhoffs Archiv für Geschichte der Medizin und der Naturwissenschaften*, 35, 1-27, 1942, pp. 15-18.

[517] S.A. Tomilin (Kiev), "Statistica Dolgovechnosti" (Statistics of Longevity), p. 252, in *Starost. Trudy Konferenzii po Probleme Geneza Starosti I Profilaktiki Prezhdevremennogo Starenia Organisma. Kiev 17-19 Decabria. 1938*, Izdatelstvo Akademii Nauk USSR, Kiev, 1939 (Aging. Proceedings of the Conference on the Problem of the Genesis of Aging and Prophylaxis of the Organism's Untimely Aging, Kiev, December 17-19, 1938, Publication of The Ukrainian Soviet Socialist Republic Academy of Sciences, Kiev, 1939).

[518] Catherine II the Great (1729-1796) ruled 1762-1796. Like Fischer, she was also German, born Sophie Friederike Auguste von Anhalt-Zerbst-Dornburg. Catherine's "enlightenment" was commonly paraded by her friendly correspondence with the French philosophers Denis Diderot, François-Marie Arouet (Voltaire) and Charles-Louis de Montesquieu.

[519] Paul I (1754-1801) ruled 1796-1801. Maria Feodorovna (1759-1828), born Duchess Sophie Marie Dorothea Auguste Louise von Württemberg, was the mother of the Czars Alexander I and Nikolay I.

[520] Mikhail Batin, *Lekarstva ot Starosti. Argumenty Nauchnogo Iimmortalisma* (Cures for Aging. Arguments of Scientific Immortalism), Obshestvennaya Organizazia Za Uvelichenie Prodolzhitelnosti Zhizni (The public organization for increasing the life-span), Izdatelstvo I.V.Balabanov, Yaroslval, 2007, p. 49.

Maria Fedorovna's chief occupation, both before and after Paul I's death, was patronage over charities, including the introduction of measures against the high mortality in charitable institutions for children, invalids and the aged.

[521] Alexander II (1818-1881) ruled 1855-1881. After his assassination by "Narodnaya Volya" – "The People's Will" revolutionaries, his son Alexander III (1845-1994) ruled from 1881 to 1894.

[522] Nikolay II (1868-1918) was the last Russian emperor, ruling from 1894 to 1917.

[523] Ivan Romanovich Tarkhanov (1846-1908) was a world foremost authority on the theory of aging, being the first to propose the role of the "power of heredity" associated with the "physical substrate" of the nucleus in determining the life span. (I.R. Tarkhanov, "Dolgoleite zhivotnykh, rasteniy i ludey" (The longevity of animals, plants and men), *Vestnik Evropy*, v. 5, 1891, p. 159.)

He was also a world foremost authority on rejuvenation by glandular extracts, recognized as such by leading European rejuvenators. Thus, Arnold Lorand submitted to Tarkhanov's authority, and Steinach acknowledged Tarkhanov as an inspiration for his own life-long research. The monograph *Organotherapy* (1905), co-authored by Tarkhanov and Poehl, became a defining work in the field.

(Aleksandr Vasil'evich Von Poehl, J. Von Tarchanoff, *Rational Organotherapy. Translated from the Russian*, Blakiston's Son & Co., Philadelphia, 1906; *Enziklopedicheskiy Slovar Brokgauza I Efrona*, St. Petersburg, 1890-1907 (The encyclopedic dictionary of Brockhaus and Efron), Articles "Tarkhanov," "Poehl," http://www.vehi.net/brokgauz/index.html; Walter E. Dixon, "A note on the physiological action of Poehl's spermine," *Journal of Physiology*, 25(5), 356-363, August 1900.)

Dr. Poehl's pharmacy, founded by Alexander's father Wilhelm, had the status of a "Supplier of the Royal Court" since 1871. The chemist Dmitry Ivanovich Mendeleev (1834-1907) was a frequent visitor and Alexander Poehl's close friend. The pharmacy has been preserved in St. Petersburg to the present.

[524] "Bakhmetiev, Porfiriy Ivanovich," *Bolshaya Sovetskaya Enziklopedia* – The Great Soviet Encyclopedia in 30 volumes, Sovetskaya Enziklopedia, Moscow, 1969-1978, http://bse.sci-lib.com/; http://slovari.yandex.ru/; Bakhmetiev P.I., "Kak ya nashel anabioz u mlekopitayushikh" (How I discovered anabiosis in mammals), *Priroda* (Nature - Moscow), 1 (5), 606-622, May 1912; Kulagin N., "Pamiati P.I. Bakhmetieva" (In Memory of P.I. Bakhmetiev), *Priroda*, 2 (10), 1127-1130, October 1913, http://www.ammonit.ru/text/405.htm.

Of interest is the article by Bakhmetiev, entitled "Rezept dozhit do XXI veka" (A recipe to live until the 21st century), published in *Jurnal Estestvoznania I Geografii* (Journal of Natural Sciences and Geography), 8, 1901, self-cited in Bakhmetiev, *Priroda*, October 1912, p. 617. The purpose of Bakhmetiev's studies of anabiosis is thus made explicit.

[525] Negovsky V.A., *Ot Smerti k Zhizni* (From Death to Life), Znanie, Moscow, 1964; Blokhin N.N., *O Dostizheniakh I Putiakh Razvitia Medizinskoy Nauki* (On the achievements and ways of development of medical science), Znanie, Moscow, 1968; Leontiev A.N., Luria A.R., Markosian A.A (Eds.), *Chelovek* (Human being), Pedagogika, Moscow, 1975, Ch. "Ozhivlenie Organisma" (The revival of the organism), pp. 169-175.

[526] The works of the Russian physiologists (or rather founders of physiology) Ivan Mikhailovich Sechenov (1829-1905, descending from a family of noblemen) and Ivan Petrovich Pavlov (1949-1936, born in the family of a priest and educated at a seminary) are of special note in relation to the pursuit of life-extension.

These scientists established the fundamentals of homeostatic neuro-humoral regulation and suggested techniques for the prolongation of functional activity, such as Sechenov's "alternate activity" or Pavlov's "stimulation-inhibition" cycles.

Perhaps even more furthering the life-extensionist pursuit, they affirmed that the biomedical intervention into organisms can be far reaching, and the possibilities of artificially keeping the organism alive are limited only by the current level of biological knowledge and technology.

Sechenov admired alchemy and considered himself as an heir to the alchemical tradition:

"It is terrible to think what would have become of mankind if the rigid mediaeval guardians of public opinion had succeeded in burning and drowning as sorcerers and evildoers all those [alchemists] who worked hard at imageless ideas and who were unconsciously creating chemistry and medicine."

(Sechenov, "Reflexes of the Brain," first published in 1863, p. 32, in I. Sechenov, *Selected Physiological and Psychological Works*, Translated from the Russian by S. Belsky, Edited by G. Gibbons, Foreign Languages Publishing House, Moscow, 1956.)

Furthermore, Sechenov believed that the "final aim of life" is the "preservation of the individual" and that "the preservation of life at any given moment is achieved through continuous transformations."

"This follows from the well-known fact that in all organisms preservation of the integrity of the body and of life is achieved not by the immutability of things established, but by continuous partial decay and restoration of the elements of the body. As long as the development of the organism is positive, i.e., as long as growth is in the ascendancy, the process of creation predominates over that of decay; in maturity the two processes are balanced; in old age, or in the period of decline, the process of decaying prevails."

(I. Sechenov, "The elements of thought," 1878, p. 279, in I. Sechenov, *Selected Physiological and Psychological Works*, 1956.)

Since the late 1880s to the end of his days, Sechenov worked on the problems of energy replenishment and balance (mainly O_2/CO_2 balance, where Sechenov determined the role of saturated and unsaturated hemoglobin in breathing). He furthermore wished to achieve "indefatigability" by rationing periods of rest and activity.

In *The Physiological criteria for determining the duration of the working day* (1895), Sechenov suggested alternate activity, i.e. alternately exercising different systems of the body, as a way toward perfect recreation and "indefatigability."

(Ivan Mikhailovich Sechenov, *Avtobiographicheskie Zapiski* (Autobiographic Notes, written in 1904, and first published in 1907), Izdatelstvo Academii Nauk SSSR, Moscow, 1945, pp. 143-145, 170-171.)

However, despite these elements compatible with life-extensionism, in Sechenov the attainment of radical longevity is hardly considered.

Pavlov, on the other hand, was an explicit life-extensionist, confessing half a year before his death:

"I very, very much want to live long… at least to 100 years… and even longer."

(Quoted in Alexander Vasilievich Nagorny, *Starenie I Prodlenie Zhizni* (Aging and Life-extension), Sovetskaya Nauka, Moscow, 1950, pp.167, 212.)

[527] Elie Metchnikoff, "Ilia Ilyich Metchnikoff" (Curriculum Vitae), 1888, pp. 292-296, "Avtobiographia" (Autobiography, 1908), pp. 297-298, vol. xvi; "Rasskaz o tom, kak i pochemu ya poselilsia za granizey" (The story of how and why I Emigrated, 1909), pp.38-45, vol. xiv, in *I.I. Metchnikoff. Sobranie Sochineniy* (Collected Works), Eds. N.N. Anichkov and R.I. Belkin, The USSR Academy of Medical Sciences, Moscow, 1962.

[528] Among the admirers of Fedorov were the apostles of the "Russian Idea" such as Vladimir Soloviev and Nikolay Berdiaev, as well as the founders of "Russian Cosmism" such as Konstantin Tsiolkovsky, Vladimir Vernadsky and Alexander Chizhevsky.
(N.A. Bediaev, "Religia Voskreshenia. Philosophia Obshego Dela N. Fedorova" (The Religion of Resuscitative Resurrection. "The Philosophy of the Common Task" of N. Fedorov), *Russkaya Mysl*, 7, 1915, pp. 76-120; English translation by Fr. S. Janos, 2002, http://www.berdyaev.com/berdiaev/berd_lib/1915_186.html.)

[529] According to Fedorov, human beings must endeavor to a perfect, coherent society that would be indefinitely maintained by mutual aid. Such a society will outgrow the "infantile" concept of a "Superman," and there will be in it no "egoism or altruism," no "mastery or slavery," only the "relatedness" and brotherly love of all humankind.

In such a society, individual death, "the last enemy to be overcome," will be vanquished by regulating and purifying the internal body environment (to prevent intrinsic death) and by controlling the external environment (to prevent extrinsic accident). The latter goal involved the colonization of the entire earth surface and space exploration, and was a source of inspiration for Fedorov's pupil, the rocket pioneer Konstantin Tsiolkovsky.

According to Fedorov, physical immortality will be attained by all, with no exceptions. Moreover, achieving immortality only by one future generation, while all the past ones remain disintegrated, seemed to Fedorov incompatible with universal justice, with Christian compassion. Therefore, humanity needed to work toward the resurrection of all who have ever lived.

How was this even theoretically possible? First of all, according to Fedorov, human remains could be identified "according to vibrations coming from within the earth" using "atmosphere as an optical instrument," and then "re-synthesized, atom by atom." Each individual was believed to leave a trace in the universe, which could be recollected. Secondly, Lamarckian inheritance was invoked, suggesting the preservation of personal traits in the progeny. "The power of reproduction" that eternally sustains life on earth, was hoped to be directed to the revival of the ancestors.

(Fedorov N.F., "Vnutrennaya reguliazia" (Internal regulation), *Sobranie Sochineniy v Chetyrekh Tomakh* (N.F. Fedorov. Collected works in four volumes), Progress, Moscow, 1995., vol. 3, pp. 136-138.)

A selection of Fedorov's works is available in English: Fedorov N.F., *What Was Man Created For? The Philosophy of the Common Task: Selected Works*, Koutiassov E. and Minto M., (Eds.), Lausanne, Switzerland, Honeyglen, 1990.

[530] Fedorov N.F., *Sobranie Sochineniy v Chetyrekh Tomakh* (N.F. Fedorov. Collected works in four volumes), Progress, Moscow, 1995, vol. 3, "Vopros o Zaglavii" (Question of the title), p. 74.

[531] Fedorov N.F., *Sobranie Sochineniy v Chetyrekh Tomakh* (N.F. Fedorov. Collected works in four volumes), Progress, Moscow, 1995, vol. 3, "Vnutrennaya Regulazia" (Internal Regulation), p.136.

[532] The enduring "new Soviet man" has since been often jokingly referred to as "Homo soveticus longevus." On the history of the term "Homo soveticus," see Aleksandr Zinovyev, *Homo Sovieticus*, Grove-Atlantic, NY, 1986.

[533] It has been sometimes argued that Russian socialism is a direct continuation of Russian Czarist imperialism, mainly implying a continuation of messianic and totalitarian undercurrents. (Dimu Kotsovsky. *Dostojewski, Tolstoj und der Bolschewismus. Über die historische Verantwortung des Schriftstellers. – Dostoyevsky, Tolstoy i Bolshevism. Ob istoricheskoy otvetstvennosti pisatelia* (Dostoevsky, Tolstoy and Bolshevism. On the historical responsibility of writers), München, 1960.)

However, the radical transformations of social institutions after the revolution, such as the eradication of the rule of the high aristocracy of land owners and the suppression of religion, may qualify as paradigm shifts in the social and ideological order to which the life-extensionists had to adjust.

[534] Bogdanov A.A., *Tectologia – Vseobshaya Organizazionnaya Nauka* (Tectology – The Universal Science of Organization), L.I. Abalkin et al. (eds.), in 2 volumes, Economica, Moscow, 1989 (first published in 1913-1928).

There is a particular emphasis on life-extension in Vol. 1, Ch. 4, Section 1, "Kolichestvennaya i strukturnaya ustoychivost" (Quantitative and structural stability), pp. 206-216; and Vol. 2, Ch. 5, Section 5, "Tectologia borby so starostiu" (Tectology of the struggle with aging), pp. 78-89.

According to Bogdanov, system stability can be threatened both by increasing complexity and heterogeneity, threatening to tear the system apart, and by increasing simplification and homogeneity, threatening to reduce the system into nothingness.

[535] Bogdanov advocated blood transfusion as a principal means for enhancing individual viability and longevity.

The Moscow Institute for Blood Transfusion was established to research and realize this possibility, after studying state of the art techniques of the Czech surgeon Yan Yansky (1873-1921), of the Americans William Lorenzo Moss (1876-1957) and Winifred Ashby (1879-1975) and of the British Geoffrey Langdon Keynes (1887-1982).

The idea of blood transfusion was nothing new. It was attempted by the Italian alchemist and mathematician Gerolamo Cardano and the German alchemist and physician Andreas Libavius in the 16th century. Animal blood was transferred to humans by the English Richard Lower in 1657 and by the French Jean-Baptiste Denis in 1667.

The identification of blood with vitality, and its use for rejuvenation were even older. The Florentine physician and philosopher Marsilio Ficino (1433-1499) prescribed drinking young people's blood mixed with sugar. Pope Innocent VIII (1432-1492) was said to drink the blood of boys for rejuvenation (this pope tried Ficino for heresy).

(Mirko D. Grmek, *On Ageing and Old Age, Basic Problems and Historic Aspects of Gerontology and Geriatrics*, Monographiae Biologicae, 5, 2, Den Haag, 1958, pp. 45-46.)

The most notorious of all was the Hungarian/Slovak countess Elisabeth Bathory (1560-1614), bathing in the blood of peasant girls whom she ordered to kill – the real life basis for vampire stories. Bathory, being a royalty, could not be executed, and spent the rest of her days under "house arrest" in the castle of Cachtice, Slovakia.

(Bathory's crimes are related in her trial records by her judge, Count György Thurzó (1564-1616), see Susanne Kord, *Murderesses in German Writing, 1720-1860: Heroines of Horror*, Cambridge University Press, Cambridge, 2009, p. 56; Michael Farin, *Heroine des Grauens: Wirken und Leben der Elisabeth Báthory in Briefen, Zeugenaussagen und Phantasiespielen* (The heroine of horror: the acts and life of Elisabeth Bathory, in letters, testimonies and fantasy plays), P. Kirchheim, München, 1989.)

The social implications of those "rejuvenation attempts" are clear: the Oppressor saps the vitality of the Oppressed. Bogdanov radically transformed the social metaphor: blood transfusion now became the highest manifestation of collectivism, of the blood brotherhood of all humankind.

Nowadays, the life-saving abilities of blood transfusion can be hardly doubted. But in the 1920s, this was still an experimental and rather unsafe intervention. Bogdanov, who conducted most dangerous experiments on himself, died in one of them in 1928, apparently due to an infection of the donor blood or imperfect knowledge of blood type compatibility.

[536] Bogdanov A.A., *Borba za zhiznesposobnost* (The Struggle for viability), Moscow, 1927, Ch. 1, reprinted at http://www.bogdinst.ru/vestnik/v13_01.htm.

Bogdanov A.A., *Statyi A. A. Bogdanova, opublikovannie v Vestnike Mezhdunarodnogo Instituta A. Bogdanova*

(Articles by A.A. Bogdanov, reprinted in the Bulletin of the International Alexander Bogdanov Institute) 13(1), 14(2), 2003. Bogdanov Institute, Yekaterinburg, http://www.bogdinst.ru/bogdanov/index.html.

[537] According to Bogdanov, the new Proletarian Culture was required to create "a new human type, organizationally complete, free of the earlier narrow-mindedness born of the fragmentation of the human being in specialization, free of the individual seclusion of the will and feeling born of the economic disparity and struggle."

(Alexander Bogdanov, "Vvedenie k knige V.O. Lichtenstand *"Goethe,"* (Introduction to V.O. Lichtenstand's "Goethe"), Moscow, 1920.)

Under socialism infused by the "Proletarian Culture," Bogdanov hoped, human beings will become well adjusted and harmoniously developed, and hence long lived. Slightly modifying Marx and Engels's dictum that "the free development of each is the condition for the free development of all" – Bogdanov believed that the sustained viability of each must become the condition for the sustained viability of all, and vice versa.

(Bogdanov A.A., *Statyi A. A. Bogdanova, opublikovannie v Vestnike Mezhdunarodnogo Instituta A. Bogdanova* (Articles by A.A. Bogdanov, reprinted in the Bulletin of the International Alexander Bogdanov Institute) 13(1), 14(2), 2003, Bogdanov Institute, Yekaterinburg, http://www.bogdinst.ru/bogdanov/index.html.)

[538] The ideas of Leon Trotsky (Lev Davidovich Bronshtein, 1879-1940) regarding the harmonizing role of the new socialist culture were also highly compatible with life-extensionism. Trotsky hoped:

"Man at last will begin to harmonize himself in earnest. ... Emancipated man will want to attain a greater equilibrium in the work of his organs and a more proportional developing and wearing out of his tissues, in order to reduce the fear of death to a rational reaction of the organism towards danger. ... All the arts – literature, drama, painting, music and architecture will lend this process beautiful form."

(Leon Trotsky, *Literature and Revolution*, 1924, Ch. 8, "Revolutionary and Socialist Art," reprinted at http://www.marxists.org/archive/trotsky/1924/lit_revo/ch08.htm.)

And later, in 1932, Trotsky continued to preach harmonization:

"Once he has done with the anarchic forces of his own society man will set to work on himself, in the pestle and the retort of the chemist. ... Socialism will mean a leap from the realm of necessity into the realm of freedom in this sense also, that the man of today, with all his contradictions and lack of harmony, will open the road for a new and happier race."

(Leon Trotsky, "The Future of Man" (Conclusion of Trotsky's *Speech on Russian Revolution* (In Defense of October) delivered at Copenhagen in November 1932, *Fourth International*, 8 (7), 223, July-August 1947, http://www.marxists.org/history/etol/newspape/fi/vol08/no07/trotsky2.htm.)

[539] The following theory was offered by Bogdanov for the revitalizing power of blood mixing:

"The "conjugation" of blood, as well as the conjugation of cells, has this property that, even without the exact determination of the weakest components, it typically supports them. This occurs because the "weakest components" in Organism A are not the same as in B and vice versa. By the exchange of vital resources, they share exactly what each element needs, therefore it is the weak components that are supported.

And if there is only a little deficit, as usually happens in prolonged processes of deterioration, then even the smallest support can have a radical significance, allowing the organism to fully utilize its own resources for its own restoration, which was previously hindered by chain functional disarray."

(Alexander Bogdanov, "O Fiziologicheskom Kollektivizme" (On Physiological Collectivism), Moscow, 1922, reprinted in *Statyi A. A. Bogdanova, opublikovannie v Vestnike Mezhdunarodnogo Instituta A. Bogdanova* (Articles by A.A. Bogdanov, reprinted in the Bulletin of the International Alexander Bogdanov Institute), 13(1), 14(2), 2003, Bogdanov Institute, Yekaterinburg, http://www.bogdinst.ru/vestnik/v14_05.htm.)

Curiously, in the USSR, concurrently with Bogdanov's "physiological collectivism," there appeared studies underwritten by what one can only term "physiological separatism," such as E.O. Manoylov, "Popytka raspoznavania chelovecheskikh ras po krovi" (An attempt to distinguish human races by blood), *Vrachebnoe Delo* (The medical work), Kharkov, No. 15-17, pp. 1152-1154, 1925. The work claimed to demonstrate stronger processes of blood oxidation among the Jews as compared to the Russians, and apparently had some implications for maintaining the stability of ethnic blood lines.

[540] Alexander Bogdanov, *Krasnaya Zvezda* (The Red Star), Pravda, Moscow, 1989 (1908), part 2, section 5, http://lib.ru/RUFANT/BOGDANOW/red_star.txt.

[541] Koltsov believed that if the laws of inheritance were known to exploiters (e.g. Russian or American slave-owners), the first thing they would do would be to "eliminate all the rebellious people, all who do not comply with the hard conditions of slavery" and then breed castes of strong physical workers, agile artisans, "decorative oddities" (freaks), etc.

(N.K. Koltsov, *Uluchshenie Chelovecheskoy Porody* (The Improvement of the Human Breed), Vremia, Petrograd, 1923, pp. 8-19, http://oleg-devyatkin.livejournal.com/60218.html.)

However, the eugenic enhancement of the party elite was considered desirable. Thus, referring to the low birth rate among the members of the All-Union Communist (Bolshevik) Party (VKPb), Koltsov wondered "What would we say about a horse or cattle breeder that every year castrates his most valuable producers, preventing their procreation?"

(N.K. Koltsov, "Noveyshie popytki dokazat nasledstvennost blagopriobretennykh priznakov" (New attempts to prove the heredity of acquired characteristics), *Russky Evgenicheskiy Zhurnal* (The Russian Eugenics Journal), 2(2-3), Moscow, GIZ, 1924, p. 15. See Maria Malikova, "NEP kak opyt sozialno-biologicheskoy hybridizazii" (The New Economic Policy as an experiment in socio-biological hybridization), *Otechestvennye Zapiski*, 28(1), 175-192, 2006, http://www.el-history.ru/node/360.)

Koltsov's optimistic views on the possibility of "avoiding aging and restoration of youth" and on the "positive results" of rejuvenative operations are presented in the following works:

N.K. Koltsov, "Smert, starost, omolozhenie" (Death, aging, rejuvenation), in N.K. Koltsov (Ed.), *Omolozhenie. Sovremennie Problemy Estestvoznania* (Rejuvenation, a special volume in Modern Problems in Natural Sciences), vol. 2, 1923, pp. 5-27; N.K. Koltsov, "Predislovie Redaktora" (Editor's Introduction) and N.K. Koltsov, "Noveyshaya Amerikanskaya literatura v oblasti operativnogo omolozhenia cheloveka" (The newest American literature on operative human rejuvenation), in N.K. Koltsov (Ed.), *Omolozhenie. Sovremennie Problemy Estestvoznania* (Rejuvenation, the second special volume on Rejuvenation in Modern Problems in Natural Sciences), vol. 16, 1924, pp. 5-10, 124-147, quotes from pp. 7-8.

Beside Koltsov's own, the latter two volumes include articles by Elie Metchnikoff, Serge Voronoff, Harry Sharp, Peter Schmidt, and Benno Romeis.

Koltsov also wrote on rejuvenation in N.K. Koltsov, *Omolozhenie Organizma po Metodu Steinacha* (The organism's rejuvenation by Steinach's method), Vremia, Peterburg, 1922.

[542] A.S. Zalmanov, *Tainaya Mudrost Chelovecheskogo Organizma* (The Secret Wisdom of the Human Body), 2nd Edition, Leningrad, 1991, including Zalmanov's "Secrety I Mudrost Tela" (The secret and wisdodm of the body, 1958), "Chudo Zhizni" (The miracle of life, 1960) and "Tysyachi putey k vyzdorovleniu" (Thousands of ways toward healing, 1965). (http://polbu.ru/zalmanov_secretwisdom/.)

[543] Maximow A., "Der Lymphozyt als gemeinsame Stammzelle der verschiedenen Blutelemente in der embryonalen Entwicklung und im postfetalen Leben der Säugetiere," *Folia Haematologica*, Leipzig, 8, 125-134, 1909 (The lymphocyte as a common stem cell for different blood elements in embryological development and post-fetal life of mammals).

An earlier precursor of the stem cell concept was the supposition of 1875 by the German-Jewish pathologist Julius Friedrich Cohnheim (1839-1884) about the origin of cancer from residual embryonic cells. (Julius Cohnheim, "Congenitales, quergestreiftes Muskelsarkom der Niere" (Congenital sarcoma of striated muscle of the kidney), *Virchows Archiv für pathologische Anatomie und Physiologie und für klinische Medizin*, 65, 64-69, 1875.)

[544] A forceful account of the persecutions of educated professionals is given in Yakov Lvovich Rapoport, *Na Rubezhe Dvukh Epoch. Delo Vrachey 1953 Goda (On the Verge of Two Epochs, The Doctors' Case 1953)*, Kniga, Moscow, 1988.

The most famous cases of repressions were "the sabotage trials" of Donbass' engineers (1928), the arrest and death of the geneticist Nikolay Vavilov (1887-1943) and the massive practice of imprisonment of scientists and engineers in the so called "sharaga's" or "scientific labor camps" (especially during 1941-1945).

The last years of Stalin's rule (1948-1953) were notorious for the agitation against cybernetics in the press, the "Doctors case," the "Pavlov's" campaign in physiology, and the thriving of Lysenkoism.

(See Yuri Isaacovich Stetzovski, *Istoria Sovietskikh Repressiy* (The History of Soviet Repressions), Znak-SP, Moscow, 1997.)

[545] The Russian archiving projects "Voennaya Literatura" (Military Literature, http://militera.lib.ru/) or "Chronos" (http://www.hrono.ru) contain many such heroic calls to sacrifice personal convenience or even life for the community. During WWII ("The Great Patriotic War"), "No counting sacrifices" was a common directive (Volkogonov D.A., "Triumf i Tragedia" (Triumph and Tragedy), *Octyabr*, 8, 1989, p. 62).

In the time of war, such calls might be easier to understand, but they emerged even before Hitler's assault on the Soviet Union and its entering the World War on June 22, 1941. Thus, on June 27, 1940, there went into effect the Decree of the Presidium of the USSR Supreme Soviet on the "Transition to an eight hour working day, seven day working week, and the prohibition of voluntary resignation of workers and office employees from works or offices." According to the decree, the working day for all Soviet workers and office employees was increased to 8 hours, and the working week was increased from 6 days to 7 days (with Sunday as "the day of rest"). Voluntary resignation and change of work place were prohibited for all workers (except for special cases). Violators of this prohibition (both workers and their managers) were liable to an imprisonment for 2 to 4 months. Unjustified absences from work were punished by compulsory work for 6 months, with 25% reduction of salary. The decree was signed by the President of the USSR Supreme Soviet Mikhail Kalinin, and Secretary of the Presidium of the USSR Supreme Soviet Alexander Gorkin. The justification for the decree was that "We need to exert all the efforts for the further development of industry, for the strengthening of our state." (The decree is reprinted in *Archiv Chronos* (*Chronos Archive*) http://www.hrono.info/dokum/194_dok/19400626bkpb.html; http://www.hrono.ru/dokum/194_dok/194007318chas.html.)

[546] *The Great Soviet Encyclopedia* (1978) gave the following figures:

"The highest life expectancy is in Sweden (74 years), in the country that has not participated in wars for 150 years ... The average life-expectancy in Russia in 1896-97 was 32 years, in the USSR in 1926-27 – 44 years, in 1958-59 – 69 years, in 1970-71 – 70 years. This is the result of the increasing prosperity of the people, improving the conditions of their work, way of life, rest and nutrition."

("Prodolzhitelnost Zhizni Cheloveka" – Human Life-expectancy, *Bolshaya Sovetskaya Enziklopedia* – The Great Soviet Encyclopedia in 30 volumes, Sovetskaya Enziklopedia, Moscow, 1969-1978, http://bse.sci-lib.com/.)

The Soviet gerontologist Vladimir Nikolayevich Nikitin pointed out in 1963 that the mortality rate in Russia decreased from 30.2 per 1000 in 1913 (just before WWI) to 7.5 in 1956 (a four-fold decrease). Thus, "the greatest decrease in mortality in that period took place in the USSR, under the conditions of the socialist socio-economic formation."

(A.V. Nagorny, V.N. Nikitin, I.N. Bulankin, *Problema Starenia I Dolgoletia* (The Problem of Aging and Longevity), Gosudarstvennoe Izdatelstvo Medizinskoy Literatury, Moscow, 1963, Ch. XXI. V.N. Nikitin, "Problema Udlinenia Zhizni" (The problem of life-extension), pp. 687-688.)

The Russian dissident-emigrant gerontologist Zhores Medvedev found in 1985 that "Life expectancy in the USSR declined from 71 years in 1964 to 70 years in 1970, and to an estimated 68 years in 1983. Decline was due to increase in infant mortality and increase in mortality among middle aged and elderly, and reflects public health problems" (precisely coinciding with Brezhnev's rule, 1964-1982). (Zhores A. Medvedev, "Negative Trends in Life Expectancy in the USSR, 1964-1983," *Gerontologist*, 25 (2) 201-208, 1985.)

However, in 1986, with the average life-expectancy at birth of 69.4 years, the USSR still ranked among the 30 countries with the highest life-expectancy (this high value has been often attributed to Gorbachev's anti-alcohol campaign of 1985).

This value and rank are based on the World Development Indicators Database of the World Bank and the Nation Master Database (www.worldbank.org/data/; http://www.nationmaster.com/graph/hea_lif_exp_at_bir_tot_yea-life-expectancy-birth-total-years&date=1986).

The State Statistics Committee of the Russian Federation (Gosudarstvenny Komitet Rossiyskoy Federazii po Statistike) gives slightly higher figures for 1986-1987 – 70.13 years (http://www.gks.ru/bgd/regl/B08_16/IssWWW.exe/Stg/html1/02-09.htm; http://www.gks.ru/free_doc/2005/b05_13/04-26.htm).

And so does the Human Mortality Database (69.96 years, http://www.mortality.org), further improving Soviet Russia's rank.

The level of 69-70 years was maintained until 1990 (i.e. until the collapse of the Soviet Union) and then began to decrease dramatically. By 1994, the life expectancy in Russia dropped to 63.85 years (57.42 years for men, and 71.08 years for women, according to the Russian Federation Statistics Committee).

In the decade from 1995 to 2005 the average life-expectancy oscillated around 65 years, and since 2005 began to recover slightly. Yet, as of 2011, after some 20 years of capitalism and democracy, according to the CIA World Factbook, with the estimated average life-expectancy of 66.3 years, Russia was ranked in the 162nd place among the 222 countries of the world (or 132nd among the 192 United Nations member states) or thereabouts. In 2014, it was estimated to be in the 151st place, with about 70 years life expectancy (https://www.cia.gov/library/publications/the-world-factbook/geos/rs.html).

The situation in other republics of the former Soviet Union is not much better, sometimes even worse. As of June 15, 2011, the estimated life expectancy in Georgia was ~77.12 (64th place among 222 countries of the world), Moldova ~71.37 years (136th), Belarus – 71.20 (139th), Ukraine – 68.58 (150th place), Tajikistan – 66.03 (165th place), according to the CIA World Factbook.

Further on Russian and Soviet life-expectancy and its varied estimates, see Graziella Caselli, Guillaume J. Wunsch, Jacques Vallin (Eds.), *Demography: Analysis and Synthesis*, Elsevier Academic Press, Burlington, MA, 2006, vol. 4, Ch. 118, S. Ivanov, A. Vishnevski and S. Zakharov, "Population Policy in Russia," pp. 415-416; Nicholas Eberstadt, Nick Eberstadt, *The Tyranny of Numbers. Mismeasurement and Misrule*, AEI Press, La Vergne TN, 1995, pp. 127-129; Michael Ryan, "Life expectancy and mortality data from the Soviet Union," *British Medical Journal*, 296, 1513-1515, 1988; V.A. Borisov (Ed.), *Naselenie Mira. Demograficheskiy spravochnik* (Population of the World. Demographic Factbook), Mysl, Moscow, 1989, "Smertnost i prodolzhitelnost zhizni (Death rate and life expectancy), pp. 195-239.

[547] Ivan Mikhailovich Sarkizov-Serazini, *Put k Zdoroviu, Sile I Dolgoy Zhizni* (The Way to Health, Strength and Long Life), Fizkultura i Sport, Moscow, 1987 (reprints of Prof. Sarkizov-Serazini's writings from the 1930s-1950s).

[548] Zakhary Grigorievich Frenkel, *Udlinenie Zhizni I Deiatelnaya Starost* (The Prolongation of Life and Active Aging), Izdatelstvo Academii Medicinskikh Nauk SSSR (The USSR Academy of Medical Sciences Press), Moscow, 1949 (first published in 1945).

[549] Loren Raymond Graham, *Science in Russia and the Soviet Union: a Short History*, Cambridge University Press, Cambridge, 1993.

[550] Iosif Vissarionovich Stalin, *Polnoe Sobranie Sochineniy v 16 Tomakh* (Iosif Vissarionovich Stalin, Complete Collected Works in 16 volumes), vol. 14, Petrograd, St. Petersburg, 2003, http://petrograd.biz/stalin/vol-

14.php; "Perepiska Stalina s Materiu I Rodnymi" (Stalin's correspondence with his mother and relatives), in *Velikie Vlastiteli Proshlogo* (Great Rulers of the Past), Moscow, 2004, http://www.vlastitel.com.ru/stalin/reform/mama.html.

[551] Zhores A. Medvedev, Roy A. Medvedev, *The Unknown Stalin*, translated by Ellen Dahrendorf, I.B. Tauris & Co., London, 2006, p. 4.

[552] I.V. Basilevich, "Syndrom normalnoy starosti" (The syndrome of normal aging), pp. 255-308, S.I. Berulava, "Resultaty issledovania glubokikh starikov Abkhazii na tropicheskie bolezni" (Results of the examination of persons of great age in Abkhazia with regard to tropical diseases), pp. 341-344, in *Starost. Trudy Konferenzii po Probleme Geneza Starosti I Profilaktiki Prezhdevremennogo Starenia Organisma. Kiev 17-19 Decabria. 1938*, Izdatelstvo Akademii Nauk USSR, Kiev 1939 (Aging. Proceedings of the Conference on the Problem of the Genesis of Aging and Prophylaxis of the Organism's Untimely Aging, Kiev, December 17-19, 1938, Publication of The Ukrainian Soviet Socialist Republic Academy of Sciences, Kiev, 1939 – hereafter referred to as "*Starost*, 1938"); Zakhary Grigorievich Frenkel, *Udlinenie Zhizni I Deiatelnaya Starost* (The Prolongation of Life and Active Aging), Izdatelstvo Academii Medicinskikh Nauk SSSR (The USSR Academy of Medical Sciences Press), Moscow, 1949.

[553] I.V. Basilevich, "Syndrom normalnoy starosti" (The syndrome of normal aging), in *Starost, 1938*, pp. 255-308. The research of the Caucasus centenarians continued throughout the Soviet history (e.g. Georgi Z. Pitskhelauri, *The Longliving of Soviet Georgia*, Translated and Edited by Gari Lesnoff, Human Sciences Press, NY, 1982, particularly Ch. 3 "Soviet Georgia: A Republic of Longliving People," pp. 35-44).
Since 1937, Georgia has remained an epicenter of study of exceptional longevity also by Western researchers, alongside other regions of fabled longevity, such as the Hunza valley in Pakistan; Vilcabamba in Ecuador; Sardinia, Italy; Okinawa, Japan; Loma Linda, California; and the Nicoya Peninsula, Costa Rica.
(See Alexander Leaf, *Youth in Old Age*, McGraw-Hill Book Company, NY, 1975; Dan Buettner, *The Blue Zones: Lessons for Living Longer From the People Who've Lived the Longest*, National Geographic, 2010.)

[554] Ivan Mikhailovich Sarkizov-Serazini, *Put k Zdoroviu, Sile I Dolgoy Zhizni* (The Way to Health, Strength and Long Life), Fizkultura i Sport, Moscow, 1987 (reprints of Prof. Sarkizov-Serazini's writings from the 1930s-1950s, first published in 1957).

[555] Ivan Mikhailovich Sarkizov-Serazini, *Put k Zdoroviu, Sile I Dolgoy Zhizni* (The Way to Health, Strength and Long Life), Fizkultura i Sport, Moscow, 1987, p. 6 (hereafter referred to as "Sarkizov-Serazini, 1987").

[556] Ivan Mikhailovich Sarkizov-Serazini, "Ko vsem nauchnim rabotnikam" (To all scientific workers), *Fizicheskaya Kultura* (Physical Culture), April 1923, reprinted in Sarkizov-Serazini, 1987, pp. 7-8.

[557] Sarkizov-Serazini's celebrity friends included, among others, the biologist Ivan Michurin, the poet Andrey Bely, the writer Maximilian Voloshin, the actress Maria Ermolova, the painter Alexander Gerasimov, etc.

[558] Ivan Mikhailovich Sarkizov-Serazini, "Nasha Rodina – Strana silnikh i krepkikh ludey" (Our Motherland – the country of strong, robust people), in Sarkizov-Serazini, 1987, pp. 35-37.

[559] Sarkizov-Serazini, 1987, p. 45.

[560] Sarkizov-Serazini, 1987, pp. 37-38.

[561] Sarkizov-Serazini, 1987, pp. 40-41.

[562] Ivan Mikhailovich Sarkizov-Serazini, "Starost i prichiny prezhdevremennogo starenia" (Aging and causes of premature aging, written in 1956), in Sarkizov-Serazini, 1987, pp. 56-57.

[563] Ivan Mikhailovich Sarkizov-Serazini, "O dolgozhiteliakh" (On the long-lived), in Sarkizov-Serazini, 1987, p. 50.

[564] Alexander Vasilievich Nagorny, *Starenie I Prodlenie Zhizni* (Aging and Life-extension), Sovetskaya Nauka (Soviet Science), Moscow, 1950, pp. 154-155.

[565] Alexander Vasilievich Nagorny, *Starenie I Prodlenie Zhizni* (Aging and Life-extension), Sovetskaya Nauka, Moscow, 1950, pp. 160-161.

[566] Alexander Vasilievich Nagorny, *Starenie I Prodlenie Zhizni* (Aging and Life-extension), Sovetskaya Nauka, Moscow, 1950, pp. 217-219.

[567] Folbort, G.V., Semernina, A.V., "Izmenenie rabotosposobnosti centralnikh elementov, obespechivayushikh vyshuyu nervnuyu deatelnost, pri starenii u sobak" (Changes in the working capacity of the central elements responsible for the high nervous activity, in aging dogs), in *Starost*, 1938, p. 204.

[568] Zakhary Grigorievich Frenkel, *Udlinenie Zhizni I Deiatelnaya Starost* (The Prolongation of Life and Active Aging), Izdatelstvo Academii Medicinskikh Nauk SSSR (The USSR Academy of Medical Sciences Press), Moscow, 1949, pp. 4, 81 (first published in 1945); quoted in Nagorny, 1950, pp. 185-186, Nikitin, 1963, p. 695.

[569] Folbort, G.V., Semernina, A.V., "Izmenenie rabotosposobnosti centralnikh elementov, obespechivayushikh vyshuyu nervnuyu deatelnost, pri starenii u sobak" (Changes in the working capacity of the central elements responsible for the high nervous activity, in aging dogs), in *Starost*, 1938, p. 205.

[570] Elie Metchnikoff, *Etudy o Prirode Cheloveka* (Etudes on the Nature of Man), Izdatelstvo Academii Nauk SSSR (The USSR Academy of Sciences Press), Moscow, 1961 (1903), Ch. 8. "Popytki filosofskich system borotsia s disharmoniami chelovecheskoy prirody" (Attempts of philosophical systems to combat the disharmonies of human nature), pp. 141-165.

[571] Maxim Gorky [b. Alexey Maximovich Peshkov, 1868-1936], "Chelovek" (The Human Being), 1898, az.lib.ru/g/gorxkij_m/text_0017.shtml

572 Alexander Vasilievich Nagorny, *Starenie I Prodlenie Zhizni* (Aging and Life-extension), Sovetskaya Nauka, Moscow, 1950, pp. 202, 219.

[573] Quoted in Alexander Vasilievich Nagorny, *Starenie I Prodlenie Zhizni* (Aging and Life-extension), Sovetskaya Nauka, Moscow, 1950, pp.167, 212.

[574] This statement was made by Pavlov to the Soviet ambassador in the UK, Ivan Maisky, during Pavlov's visit to London, in July 1935, during the 2nd. International Neurological Congress, and was related to the press. (Ivan Mikhailovich Maisky, *Vospominania Sovetskogo Diplomata* (Memoirs of a Soviet Diplomat), Nauka, Moscow, 1971, p. 308, reprinted at http://militera.lib.ru/memo/russian/maisky_im1/03.html.)

[575] Ivan Petrovich Pavlov, "O Perspektivakh Raboty v 1935 godu" (Work perspectives in the year 1935), in I.P. Pavlov, *Polnoe Sobranie Sochineniy* (Complete Works, 2nd Edition), Izdatelstvo Academii Nauk SSSR (The USSR Academy of Sciences Press), 1951, vol 1, p. 15 (reprinted at http://bibnav.feb-web.ru/ru/text/pavlov_pss_v1_1951/), an excerpt from the newspaper article "Academic I.P. Pavlov O Sovetskoy Rodine" (Academician I.P. Pavlov about the Soviet Motherland), *Izvestia*, No. 157, July 6, 1935.

[576] Karl Marx and Frederick Engels, *Manifesto of the Communist Party*, London, 1848, The concluding statement of Part II. "Proletarians and Communists" (http://www.anu.edu.au/polsci/marx/classics/manifesto.html).
Perhaps one of the best examples for the sacrificial demand explicitly made on Soviet citizens was the decree of June 27, 1940, of the Presidium of the USSR Supreme Soviet on the "Transition to an eight hour working day, seven day working week, and the prohibition of voluntary resignation of workers and office employees from works or offices" – essentially introducing a system of industrial 'serfdom.'
(The Decree is reprinted in *Archiv Chronos - Chronos Archive*, http://www.hrono.info/dokum/194_dok/19400626bkpb.html; http://www.hrono.ru/dokum/194_dok/194007318chas.html.)

[577] Trofim Denisovich Lysenko, "O nasledstvennosti i ee izmenchivosti" (On heredity and its variability, 1943, p. 436), "O putiakh upravlenia rastitelnimi organizmami" (On the ways of governing plant organisms, 1940, p. 324), in T.D. Lysenko, *Agrobiologia: Raboty po voprosam genetiki, selekzii i semenovodstva* (Agrobiology: Works on problems of genetics, selection and semenology), Gosudarstvennoe Izdatelstvo Selskokhozyaystvennoy Literatury, Moscow, 6th edition, 1952, first published in 1949, http://imichurin.narod.ru/lysenko/Index.html; quoted in Alexander Vasilievich Nagorny, *Starenie I Prodlenie*

Zhizni (Aging and Life-extension), Sovetskaya Nauka, Moscow, 1950, p. 191.

[578] Alexander Vasilievich Nagorny, *Starenie I Prodlenie Zhizni* (Aging and Life-extension), Sovetskaya Nauka, Moscow, 1950, pp. 191, 218.

[579] Evgeny Ivanovich Zamyatin, *Mi* (We, 1920), Znamia, Moscow, 1988, 5-6, reprinted at http://az.lib.ru/z/zamjatin_e_i/text_0050.shtml, Entry 3.

[580] Vladimir Vladimirovich Mayakovsky, *Klop* (The Bedbug), Pravda, Moscow, 1988 (1929), reprinted at http://ru.wikisource.org/, Acts 5-9.

[581] Mikhail Afanasievich Bulgakov, *Diavoliada. Povesti, Rasskazy, Felietony, Ocherki* (Diaboliada. Novels, Short Stories, Feuilletons, Essays), Literatura Artistike, Kishinev, 1989, "Sobachie Serdze" (The heart of a dog), pp. 152-224.

[582] Mikhail Afanasievich Bulgakov, *Diavoliada. Povesti, Rasskazy, Felietony, Ocherki* (Diaboliada. Novels, Short Stories, Feuilletons, Essays), Literatura Artistike, Kishinev, 1989, "Rokovye Yaytza" (The fatal eggs), pp. 96-152.

[583] The idea of using electromagnetism for revival and prolongation of life originated at least as early as the 18th century – cf. the stimulation of "animal electricity" in dead frogs by the Italian physician Luigi Aloisio Galvani (1737-1798, starting these experiments around 1771); and the attempts of Luigi Galvani's nephew, Giovanni Aldini (1762-1834) to revive human corpses by electric stimulation (performed around 1798-1802).

The theme has been exploited in science fiction at least since the early 19th century (cf. Mary Wollstonecraft Shelley, *Frankenstein; or, The Modern Prometheus*, 1818).

[584] Irene Masing-Delic, *Abolishing Death. A Salvation Myth of Russian Twentieth-Century Literature*, Stanford University Press, Stanford CA, 1992, p. 5.

[585] It appears that "The Heart of a Dog" reached truly massive popularity in the Soviet Union only after the production of the story-based movie in 1988, following the first Soviet publication in 1987, at the time of Perestroika, when the dominant Soviet ideology was increasingly questioned.

Similarly, until the late 1980s, Zamyatin's "We" was published and popular in the West, perhaps also among a very small portion of Russian dissidents. It became known to the wide Soviet-Russian readership only after 1988.

[586] Peter Wiles, "On Physical Immortality," *Survey*, 56, 125-143, 1965; Peter Wiles, "On Physical Immortality, II," *Survey*, 57, 142-161, 1965.

One of the earliest Western works on Soviet life-extensionism seems to be Edward Podolsky's biomedical survey, *Red Miracle: the Story of Soviet Medicine* (1947), which can now be considered as an historical testimony. (Edward Podolsky, *Red Miracle: the Story of Soviet Medicine*, Beechhurst Press, NY, 1947, Part 2 "The Miracles of Soviet Medicine," pp. 119-220.)

[587] Thus, for example, most comprehensive accounts of Soviet science, such as Loren Graham's *Science in Russia and the Soviet Union* (1993) including the extensive appendix on Soviet Biological Sciences, Medicine and Technology or David Joravsky's *The Lysenko Affair* (1970) – listing hundreds of scientists and officials – never mention Alexander Bogomolets or his work on rejuvenation (despite the fact that Bogomolets headed the Ukrainian Academy of Sciences for 15 years and enjoyed Stalin's and Khrushchev's personal patronage), neither relate to the works of other prominent life-extensionists, including university department chairs, such as Alexander Nagorny, Moisey Mühlmann, and Vladimir Filatov, and Academicians, such as Nikolay Amosov, Alexander Kuprevich and Vladimir Skulachev. Nikolai Krementsov's *The Cure* (2002) mentions Bogomolets only briefly, without reference to rejuvenation.

(Loren Raymond Graham, *Science in Russia and the Soviet Union: a Short History*, Cambridge University Press, Cambridge, 1993; Nikolai Krementsov, *The Cure: A Story of Cancer and Politics from the Annals of the Cold War*, University of Chicago Press, Chicago, 2002; David Joravsky, *The Lysenko Affair*, Harvard University Press, Cambridge MA, 1970.)

Even in discussing relatively well known Soviet scientific fields and their leaders, such as physiology and Ivan Pavlov, or genetics and Nikolay Koltsov, the fact that both Pavlov and Koltsov were life-extensionists is never mentioned.

(E.g. Mark Adams, "Science, Ideology, and Structure: The Kol'tsov Institute, 1900-1970," in Linda Lubrano and Susan Gross Solomon (Eds.), *The Social Context of Soviet Science*, Westview Press, Boulder Colorado, 1980, pp. 173-204; Loren R. Graham, "Reasons for Studying Soviet Science: The Example of Genetic Engineering," in Linda Lubrano and Susan Gross Solomon (Eds.), *The Social Context of Soviet Science*, 1980, pp. 205-240; Babkin, B.P., *Pavlov: A Biography*, University of Chicago Press, Chicago, 1949; David Joravsky, *Russian Psychology: A Critical History*, Blackwell, Oxford, 1989; Alex Kozulin, *Psychology in Utopia: Toward a Social History of Soviet Psychology*, MIT Press, Cambridge MA, 1984.)

Furthermore, the life-extensionist teleology is not generally considered a part of Marxist philosophy.

(E.g. George Novack, *Polemics in Marxist Philosophy*, Monad Press, NY, 1978, or Martin Jay, *Marxism and Totality: The Adventures of a Concept from Lukacs to Habermas*, University of California Press, Berkeley, CA, 1986.)

[588] O.B. Lepeshinskaya, "Starost i borba s neu" (Aging and the struggle with it), *Nauka I Zhizn* (Science and Life), 7, 11-13, 1951; Alexander Vasilievich Nagorny, *Starenie I Prodlenie Zhizni* (Aging and Life-extension), Sovetskaya Nauka (Soviet Science), Moscow, 1950; Ivan Mikhailovich Sarkizov-Serazini, *Put k Zdoroviu, Sile I Dolgoy Zhizni* (The Way to Health, Strength and Long Life), Fizkultura i Sport, Moscow, 1957, e.g. Sarkizov-Serazini, "Nasha Rodina – Strana bogatirey i silnikh ludey" (Our Motherland – the country of strong, robust people, written c. 1951), pp. 8-26.

[589] Yakov Lvovich Rapoport, *Na Rubezhe Dvukh Epoch. Delo Vrachey 1953 Goda* (On the Verge of Two Epochs. The Doctors' Case 1953), Kniga, Moscow, 1988, "'Zhivoe veshestvo' i ego konez. Otkrytie O.B. Lepeshinskoy i ego sudba" (The 'living substance' and its end. The discovery of O.B. Lepeshinskaya and its fate), pp. 250-270 (reprinted at http://lib.ru/MEMUARY/1953/rapoport.txt);

O.B. Lepeshinskaya, "Starost i borba s neu" (Aging and the struggle with it), *Nauka I Zhizn* (Science and Life), 7, 11-13, 1951 (http://publ.lib.ru/ARCHIVES/N/"Nauka_i_jizn"'/_"Nauka_i_jizn"'_1951_.html); Lepeshinskaya, O. B., *Proiskhozhdenie Kletok iz Zhivogo Veshestva* (The Origin of Cells from Living Substance), Pravda, Moscow, 1951 (http://www.oldgazette.ru/lib/lepesh/index.html).

[590] A.V. Nagorny, "K voprosu o faktorakh, obuslovlivayushchikh dlitelnost zhizni" (On the question of factors determining the duration of life), in *Starost*, 1938, pp. 156-172, citations from pp. 168-169.

[591] Frederick Engels, *Anti-Dühring. Herr Eugen Dühring's Revolution in Science*, translated by Emile Burns, Progress Publishers, Moscow, 1947 (1877), Part I. Philosophy. VIII. "Philosophy of Nature. The Organic World, Conclusion" (http://www.marxists.org/archive/marx/works/1877/anti-duhring/ch06.htm).

[592] Porfiry Korneevich Ivanov, "Istoria I Metod Moey Zakalki," Leningrad, 1951 (The history and method of my tempering).

The article is reprinted in Yuri Zolotarev, *Novoe – Nebyvaloe. Opyt Dela Uchitelia Ivanova. V pomosh vypolnyaushim sovety "Detki"* (The new and uncommon. The experience of the work of the teacher Ivanov. In aid to the followers of the advice of the "Child"), Izdatelstvo KSP, Moscow, 1995, p. 59.

The book will be hereafter referred to as "Ivanov, 1995."

"The Child" – "Detka" – was the nickname of Ivanov and of his health-improvement system.

[593] Porfiry Korneevich Ivanov, "Moya Zakalka" (1958), reprinted in Ivanov, 1995, p. 293.

[594] Porfiry Ivanov, "Nizko klanyayus i proshu vas ludi," 1978 (I bow to you deeply and ask you people), reprinted in *Sistema Ivanova* (Ivanov's System, http://sistemaivanova.narod.ru/portal/).

[595] Porfiry Ivanov, "Moy podarok molodezhi" (My present to the youth, 1983), reprinted in *Sistema Ivanova* (Ivanov's System, http://sistemaivanova.narod.ru/portal/).

[596] Vladimir Ivanovich Vernadsky, *Nauchnaya Mysl kak Planetnoe Yavlenie* (Scientific thought as a planetary phenomenon), Nauka, Moscow, 1991 (written c. 1937-1938), reprinted in *Biblioteka Maxima Moshkova Lib.Ru*, http://lib.ru/FILOSOF/WERNADSKIJ/mysl.txt.

[597] Konstantin Eduardovich Tsiolkovsky, *Volia I Vselennaya* (The Will and the Universe), 1928, http://www.kirsoft.com.ru/freedom/KSNews_30.htm.

[598] Alexander Leonidovich Chizhevsky, "Teoria Kosmicheskikh Er" (The Theory of Cosmic Eras), in *Aeriony i zhizn. Besedy s Ziolkovskim* (Aero-ions and life. Conversations with Tsiolkovsky), Mysl, Moscow, 1999 (1920), p. 670, reprinted at http://www.fantclubcrimea.info/5-ciolkov.html.
Chizhevsky's practical work on life-extension through irradiation and exposure to negative air ions (aerions) will be referred to later on.

[599] Porfiry Korneevich Ivanov, "Istoria I Metod Moey Zakalki," Leningrad, 1951, in Ivanov, 1995, p. 57.

[600] Porfiry Ivanov, "Nizko klanyayus i proshu vas ludi," 1978 (I bow to you deeply and ask you people), reprinted in *Sistema Ivanova* (Ivanov's System, http://sistemaivanova.narod.ru/portal/).

[601] Porfiry Korneevich Ivanov, "Istoria I Metod Moey Zakalki," Leningrad, 1951, in Ivanov, 1995, pp. 52, 54

[602] Andrey Levshinov, *Sistemy Ozdorovlenia Zemli Russkoy* (Health Improvement Systems of the Russian Land), Olma Press, Moscow, 2002, pp. 145-158.

[603] Porfiry Korneevich Ivanov, "Istoria I Metod Moey Zakalki," Leningrad, 1951, in Ivanov, 1995, pp. 60-61.

[604] Sergey Vlasov, "Experiment dlinoyu v polveka" (An experiment lasting half a century), *Ogonek*, No. 8, 1982, http://uchitel-ivanov.narod.ru/experiment.html.

[605] The advice of Ivanov's health and longevity system "Detka" - "The Child" included:
1) "Twice a day bathe in cold, natural water"; 2) Associated with the bathing, "stand on the earth or on the snow." "Breathe several times through the mouth and ask nature for health"; 3) "Do not drink alcohol and do not smoke"; 4) Complete fasting (no food and drink) at least once a week, and at least for 24 hours; 5) Walking barefoot, breathing and meditating; 6) "Love the surrounding nature. Do not spit around and do not spit out anything. Get used to it – this is your health"; 7) "Greet all people"; 8) "Help people as much as you can"; 9) "Defeat greed, sloth, self-complacency, covetousness, fear, hypocrisy, pride"; 10) "Free your head from thoughts about diseases, weaknesses, death. This is your victory"; 11) "Do not separate the thought from the deed. It is good to read, but the most important thing is to do"; 12) "Relate to others and pass on the experience of this Deed, but do not boast and do not condescend. Be modest" (http://uchitel-ivanov.narod.ru/detka.html).

[606] A.V. Nagorny, V.N. Nikitin, I.N. Bulankin, *Problema Starenia I Dolgoletia* (The Problem of Aging and Longevity), Gosudarstvennoe Izdatelstvo Medizinskoy Literatury, Moscow, 1963. Ch. XXI. V.N. Nikitin, "Problema Udlinenia Zhizni" (The problem of life-extension), pp. 682-752.

[607] *Bolshaya Sovetskaya Enziklopedia* (The Great Soviet Encyclopedia in 30 volumes), Sovetskaya Enziklopedia, Moscow, 1969-1978, "Kravkov Nikolay Pavlovich," http://bse.sci-lib.com/.

[608] In the 1990s, the Vitamins Institute was commercialized and starved of government funding, virtually bringing its operation to a halt. (Lavrov, B.A., *Ocherki po Istorii Sovetskoy Vitaminologii (1917-1967)* (Essays on the history of Soviet Vitaminology, 1917–1967), Medizina, Moscow, 1980; Lyubarev, A.E., "Kratkaya istoria vitaminnogo instituta" (A brief history of the Vitamins Institute), 2003, http://lyubarev.narod.ru/science/VNIVI.htm.)

[609] O.B. Lepeshinskaya, "Starost i borba s neu" (Aging and the struggle with it), *Nauka I Zhizn* (Science and Life), 7, 11-13, 1951, http://publ.lib.ru/ARCHIVES/N/.
Lepeshinskaya's studies of the effects of the alkaline-acid balance on aging and longevity were sometimes criticized, even ridiculed, as in Yakov Lvovich Rapoport, *Na Rubezhe Dvukh Epoch. Delo Vrachey 1953 Goda* (On the Verge of Two Epochs. The Doctors' Case 1953), Kniga, Moscow, 1988, "'Zhivoe veshestvo' i ego

konez. Otkrytie O.B. Lepeshinskoy i ego sudba" (The 'living substance' and its end. The discovery of O.B. Lepeshinskaya and its fate), pp. 250-270 (reprinted at http://lib.ru/MEMUARY/1953/rapoport.txt).

Yet, the problem of the influence of increasing acidity or alkalinity on the aging processes appears to be a significant and persistent one, even though complex and even confusing. The therapeutic use of acids or alkali as rejuvenating means has a very long history.

Acidic remedies can be traced back to the use of hydrochloric acid by the American physicians Burr Ferguson and Walter B. Guy in the 1930s, Elie Metchnikoff's "acidified milk products" in the early 1900s, Édouard Robin's "lactic acid" and "vegetable acids" in the mid 19th century, the "Acid elixir" developed by the 18th century Swiss physician Albrecht von Haller ("Elixir Acidum Halleri" – a mixture of sulphuric acid and alcohol of uncertain concentration and method of preparation), perhaps even to Galen's "theriac" or "the universal antidote" that may have contained Salicylic acid.

On the other hand, alkaline mineral waters and bicarbonate drinks have been also used externally and internally since at least as early as the time of Galen. At the end of the 19th century, the German-Baltic-Swiss chemist Gustav von Bunge stressed the importance of body alkalinity. In the 1920s-1930s, attempts of tissue alkalinization through various types of diet were made by health practitioners across the world: the Americans William Howard Hay, DeForest Clinton Jarvis, Edgar Cayce and Alfred W. McCann, the Russian-Indian Helena Ivanovna Roerich, the Austrian Franz Xaver Mayr, the Japanese Hirotaro Nishizaki, the Swedish-German Ragnar Berg, and many others.

Lepeshinskaya seems to adhere to this line of the "alkalizing" tradition. Recently, notable western proponents of body "alkalinization" have been the American authors Robert O. Young, Theodore Baroody and Ray Kurzweil, the Italian Tullio Simoncini, and many others.

Yet, the role of the acid-alkaline balance in aging appears to be very complex and any particular measures for its correction appear to be very uncertain.

Thus, for example, according to Baroody, certain acidic foods and food additives (such as hydrochloric acid and fruit vinegars) can be "alkaline forming" (improving digestion and preventing the formation of acids in the stomach). Baroody also makes a distinction between the possibilities of increasing alkalinity "in the tissues" vs. "in the blood." Thus the dichotomy between the "benefits" vs. the "costs" of increasing alkalinity appears to be deconstructed.

Furthermore, the very possibility of a sustained, absolute, significant and general pH shift in the buffered body medium appears to be problematic. The benefits of Lepeshinskaya's alkalizing treatments were equally uncertain.

(See John Uri Lloyd, *Pharmaceutical preparations. Elixirs, their history, formulae, and methods of preparation*, Second Edition, Robert Clarke and Company, Cincinnati, 1883, p. 99; *Three Years of HCL Therapy, As Recorded in articles in The Medical World*, With Introduction by Henry Pleasants, Jr., AB, MD, FACP, Associate Editor, Published by W. Roy Huntsman, Philadelphia, PA, 1935, http://www.townsendletter.com/Dec2006/HCl1206.htm; Richard I. Levin, "Theriac Found? Nitric Oxide-Aspirin and the Search for the Universal Cure," *Journal of the American College of Cardiology*, 44 (3), 642-643, 2004, http://content.onlinejacc.org/cgi/content/full/44/3/642; Ray Kurzweil and Terry Grossman, *Fantastic Voyage. Live Long Enough to Live Forever*, Plume, NY, 2005, "The importance of being alkaline," pp. 45-48; Theodore A. Baroody, *Alkalize or Die: Superior Health Through Proper Alkaline-Acid Balance*, Eclectic Press, Waynesville NC, 1999; Robert O. Young, *The pH Miracle*, Grand Central Life & Style, NY, 2010; Gabe Mirkin, "Acid/Alkaline Theory of Disease Is Nonsense," *Quackwatch*, January 11, 2009, http://www.quackwatch.org/01QuackeryRelatedTopics/DSH/coral2.html.)

[610] V.N. Nikitin, "Problema Udlinenia Zhizni" (The problem of life-extension), in A.V. Nagorny, V.N. Nikitin, I.N. Bulankin, *Problema Starenia I Dolgoletia* (The Problem of Aging and Longevity), Gosudarstvennoe Izdatelstvo Medizinskoy Literatury, Moscow, 1963, pp. 682-752.

[611] Petrova, M.K., "Proishozhdenie starosti i profilaktika prezhdevremennogo starenia" (The origin of aging and prophylaxis of untimely aging), *Nauka I Zhizn* (Science and Life), 11-12, 16-22, 1944; Petrova, M.K., *O roli funkzionalno oslablennoy kory golovnogo mozga v vozniknovenii razlichnikh patologicheskikh prozessov v organizme* (The role of functionally weakened brain cortex in the emerging of various pathological processes in the organism), Medgiz, Leningrad, 1946.

[612] Some of the first works of Alexander Gavrilovich Gurwitsch (1874-1954, Crimean Simferopol-Moscow-Leningrad) on the growth-promoting ultraviolet "mitogenic radiation" were:

Gurwitsch A., "Über Ursachen der Zellteilung" (On causes of cell division), *Wilhelm Roux' Archiv für Entwicklungsmechanik der Organismen*, 52, 167-181, 1922;. Gurwitsch A., "Die Natur des spezifischen Erregers der Zellteilung," (The nature of specific stimulants of cell division), *Wilhelm Roux' Archiv für Entwicklungsmechanik der Organismen*, 100, 11-40, 1923; Gurwitsch A., "Physikalisches über mitogenetische Strahlen" (Physical aspects of mitogenic rays), *Wilhelm Roux' Archiv für Entwicklungsmechanik der Organismen*, 103, 490-498, 1924.

His later works included:

A.G. Gurwitsch, *Teoria biologicheskogo Polia* (The theory of biological field), Sovetskaya Nauka (Soviet Science), Moscow, 1944; A.G. Gurwitsch, *Mitogeneticheskoe izluchenie* (Mitogenic radiation), Medgiz, Moscow, 1945 (3rd edition, first published in 1934).

A thorough review of Soviet "mitogenic radiation" research until the 1930s, including its role in the research of fatigue and aging, was given in Semen Yakovlevich Salkind, "Mitogeneticheskie luchi i ich znachenie v medizine" (Mitogenic rays and their significance for medicine), in *Problemy teoreticheskoy i prakticheskoy mediziny* (Problems of Theoretical and Practical Medicine), Biomedgiz, Moscow-Leningrad, 1936, reprinted in *Medznanie*, 2007 (http://www.medznanie.ru/lib/problema.html; http://www.medznanie.ru/lib/problema114.html).

Regarding yet another powerful line of research, on negative air charge, in the very first presentation of his method of negative "air ionization" in 1919, Alexander Leonidovich Chizhevsky (1897-1964, Kaluga-Moscow) stated:

"With a common access to the "ionization" of dwellings, man of the future, using this method, will be able to conduct planned struggle for his longevity."

(Chizhevsky, A.L., "Ionizatsia vozdukha, kak fiziologicheski activniy factor atmosfernogo electrichestva. Experimentalnie issledovania (Air ionization as a physiologically active factor of atmospheric electricity. Experimental Studies), Doklad (Presentation), Kaluga, 1919, quoted by Chizhevsky in his book, *Aeroionifikatsia v narodnom hozyaystve*, Stroyizdat, Moscow, 1989 (Aero-ionization in people's economy, 2nd edition, first published in 1960), Ch. 1. "Istoria problemy aeroionifikazii i atmosfernoe elektrichestvo" (The history of the problem of aero-ionization and atmospheric electricity), p. 36.)

The work of Alexey Osipovich (Iosifovich) Voynar (1904-1963, Donetsk, Ukraine), Leonid Leonidovich Vasiliev (1891-1966, Leningrad), Alexander Bronislavovich Verigo (1893-1953, Leningrad), and Alexander Alexandrovich Lubishev (1890-1972, Crimean Simferopol-Leningrad-Kiev-Ulyanovsk), too was intimately related to healing, rejuvenation and life-extension via electromagnetic influences.

(A.O. Voynar, L.L. Vasiliev, A.L. Chizhevsky, "Starenie i Omolozhenie Organizma v Svete Electrochimii" (Aging and rejuvenation of the organism in the light of electrochemistry), *Trudy Centralnoy Nauchno-Issledovatelskoy Laboratorii Ionizatsii - CNILI "Problemy Ionifikazii,"* Moscow, vol. 3, pp. 381-390, 1934 (Proceedings of the Central Scientific Laboratory for the Study of Ionization – "Problems of Ionization," directed by Chizhevsky, having branches in several cities with the headquarters in Moscow);

A.A. Lubishev, A.G. Gurwitsch, *Dialog o Biopole – Sbornik* (A dialogue on the biofield – collected works), Ulyanovsk, 1998.)

Further on the work of Soviet electro- and photo-physiologists, see for example, Helios Foundation, http://chizhevskiy.ru/; Cycles Research Institute http://cyclesresearchinstitute.org/cycles-research/general-chizhevsky.shtml; V.L. Voeikov, L.V. Beloussov, "From Mitogenic rays to Biophotonics," in *Biophotonics and Coherent Systems* (L.V. Beloussov, F.-A. Popp, V.L. Voeikov and R. van Wijk, eds.), Springer, NY, 2007, http://www.scribd.com/doc/60996082/V-S-Martynnyuk-Biophotonics-and-Coherent-Systems-in-Biology-pdf.

[613] "Raboty na Russkom I Ukrainskom Yazykakh" (Works in Russian and Ukrainian Languages), in *Starost. Trudy Konferenzii po Probleme Geneza Starosti I Profilaktiki Prezhdevremennogo Starenia Organisma. Kiev 17-19 Decabria. 1938*, Izdatelstvo Akademii Nauk USSR, Kiev 1939 (Aging. Proceedings of the Conference on the Problem of the Genesis of Aging and Prophylaxis of the Organism's Untimely Aging, Kiev, December 17-19, 1938, Publication of The Ukrainian Soviet Socialist Republic Academy of Sciences, Kiev, 1939), pp. 457-466.

[614] Mark Mirsky, "Yu. Yu. Voronoy – A Pioneer of clinical kidney transplantation," *Agapit. The Ukrainian Historical and Medical Journal*, No. 14-15, pp. 10-15, 2004, available at The National Museum of Medicine of Ukraine, Kiev, http://web.archive.org/web/20070125075033/http://www.histomed.kiev.ua/agapit/ag1415/eng/page03.php.htm.

[615] V.N. Nikitin, "Problema Udlinenia Zhizni" (The problem of life-extension), in A.V. Nagorny, V.N. Nikitin, I.N. Bulankin, *Problema Starenia I Dolgoletia* (The Problem of Aging and Longevity), Gosudarstvennoe Izdatelstvo Medizinskoy Literatury, Moscow, 1963, pp. 682-752.

Evgeny Alexandrovich Schultz, "O Molodenii," *Novie Idei v Biologii, Sb. 3. Smert I Bessmertie*, Obrazovanie, St. Petersburg, 1914 ("On Rejuvenation," in *New Ideas in Biology*, a series edited by Professors Vladimir Alexandrovich Wagner and Evgeny Alexandrovich Schultz, Vol. 3, Death and Immortality), pp. 136-148. Beside Schultz's own, the volume includes articles by August Weismann, Charles Sedgwick Minot and Richard Hertwig.

[616] Filatov, V.P., *Tkanevaya Terapia* (Tissue Therapy), Gosizdat Uzbekskoy SSR, Tashkent, 1943.

[617] Paul U. Unschuld, *Huang Di Nei Jing Su Wen: Nature, Knowledge and Imagery in an Ancient Chinese Medical Text*, University of California Press, Berkeley, 2003, p. 286. According to Unschuld, the medical treatise *Huang Di Nei Jing Su Wen (The Inner Canon of the Yellow Emperor*, composed ca. 200 BCE) which is commonly considered as the primary text of traditional Chinese medicine, contained very little information on pharmacological agents, and much more on theory (yin and yang, the vessels, the five elements, Tao, ways of the immortals, etc.), unlike the Mawangdui manuscripts that contained extensive information on opotherapy.

[618] *An English translation of the Sushruta samhita, based on original Sanskrit text, Edited and published by Kaviraj Kunja Lal Bhishagratna*, Calcutta, 1907, 1911, 1916, Vol. 2, Chikitsasthanam (Therapeutics), Ch. 26, p. 512.

[619] James Joseph Walsh, *Old-Time Makers of Medicine. The Story of The Students And Teachers of the Sciences Related to Medicine During the Middle Ages*, Fordham University Press, NY, 1911, pp. 67-68, footnote 1, reprinted in Project Gutenberg, http://www.gutenberg.org/etext/20216; Fred Rosner, *Medicine in the Bible and the Talmud*, Ktav, Hoboken NJ, 1995 (1977), "Therapeutic Efficacy of Chicken Soup," pp. 136-139.

[620] *Bolshaya Sovetskaya Enziklopedia* (The Great Soviet Encyclopedia in 30 volumes), Sovetskaya Enziklopedia, Moscow, 1969-1978, "Proteinoterapia" (protein therapy) http://bse.sci-lib.com/article093541.html, "Tkanevaya Terapia" (tissue therapy) http://bse.sci-lib.com/article110899.html; M.P. Tushnov, "Lizatoterapia, ee teoreticheskoe obosnovanie i prakticheskoe primenenie" (Lysotherapy, its theoretical foundation and practical application), N.A. Shereshevsky, "Vorposy lizatoterapii" (Problems of lysotherapy), O.A. Stepun, "Spezificheskie i nespezificheskie factory organoterapii" (Specific and non-specific factors in organotherapy), in *Problemy teoreticheskoy i prakticheskoy mediziny* (Problems of Theoretical and Practical Medicine), Biomedgiz, Moscow-Leningrad, 1936, reprinted in *Medznanie* (2007) http://www.medznanie.ru/lib/problema.html.

[621] Claude Bernard's exact statement was:

"Ces deux opérations de destruction et de rénovation, inverses l'une de l'autre, distinctes dans leur nature, sont absolument connexes et inséparables : elles sont la condition l'une de l'aulre. Les phénomènes de destruction fonctionnelle sont euxmêmes les instigateurs et les précurseurs de la rénovation matérielle."
(Originally published in Claude Bernard, *Cours de physiologie générale au Muséum d'Histoire naturelle: Des phénomènes de la vie communs aux animaux et aux Végétaux* (Course in General Physiology at the Museum of Natural History: The Phenomena of Communal Life in Animals and Plants), *Revue scientifique*, 2e série, vol. 3, pp. 170-542, 1872-1873.
Later republished as Claude Bernard, *Leçons sur les Phénomènes de la Vie Communs aux Animaux et aux Végétaux* (Lectures on the phenomena of communal life in animals and plants, edited by Albert Dastre), J.-B. Baillière, Paris, 1879, vol. 2, Lecture 27, "Vitalisime physico-chimique" (Physico-chemical vitalism), p. 484.
In Russian: Claude Bernard, *Kurs Obshey Fisiologii* (Course in General Physiology), St. Petersburg, 1878, quoted in A.V. Nagorny, V.N. Nikitin, I.N. Bulankin, *Problema Starenia I Dolgoletia* (The Problem of Aging and Longevity), Gosudarstvennoe Izdatelstvo Medizinskoy Literatury, Moscow, 1963, Ch. XXI, V.N. Nikitin, "Problema Udlinenia Zhizni" (The problem of life-extension), p. 730.
And in the English translation used here: Claude Bernard, *Lectures on the phenomena common to animals and plants*, Translated by Hoff HE, Guillemin R, Guillemin L, published by Charles C Thomas, Springfield, IL, 1974, p. 253.)
Cf. Paracelsus' statement:
"Destruction perfects that which is good; for the good cannot appear on account of that which conceals it."
(Quoted in M. M. Pattison Muir, *The Story of Alchemy and the Beginnings of Chemistry*, London, 1913.)
As Pattison Muir summarized the alchemical theory of revival:
"Destruction must precede revivification, death must come before resurrection."
(M. M. Pattison Muir, *The Story of Alchemy and the Beginnings of Chemistry*, Hodder and Stoughton, London, 1913, Ch. I, pp. 9-20, Ch. X, pp. 122-139, reprinted in *Project Gutenberg*, http://www.gutenberg.org/files/14218/14218-h/14218-h.htm.)

[622] The application of "tissue therapy" and the theory of rejuvenative "auto-catalysis" were indeed not endemic to Soviet Russia. In 1921, the Austrian researcher Gottlieb Haberlandt (1854-1945) emphasized the importance of tissue degradation products ("wound hormones") for wound healing. In 1923, the Japanese researcher Yoneji Miyagawa (1885-1959) showed that by injections of different doses of tissue suspensions, animal organs can be either stimulated or inhibited. He therefore came to the conclusion that organs are partly regulated by products of their own degradation, and introduced the concept of "auto-regulation." The French-American biologist Alexis Carrel also extensively investigated the growth-promoting activity of protein degradation products – "trephones" and "desmones."
Yet, in the USSR, following Tushnov, "tissue therapy" ("protein/lysotherapy") was promoted and applied perhaps in the widest degree, not only in medicine as a non-specific, "general resistance" therapy, but also in veterinary care and animal breeding to increase milk, meat and wool productivity.
(*Bolshaya Sovetskaya Enziklopedia* (The Great Soviet Encyclopedia in 30 volumes), Sovetskaya Enziklopedia, Moscow, 1969-1978. "Proteinoterapia" (protein therapy) http://bse.sci-lib.com/article093541.html, "Tkanevaya Terapia" (tissue therapy) http://bse.sci-lib.com/article110899.html.)
The emphasis on the necessity of "products of degradation" or "metabolic waste products" for a continuous normal biological function seems to have been later discarded by various "damage" theories of aging and by certain proposals for rejuvenation, such as the recent Aubrey de Grey's "Strategies for Engineered Negligible Senescence," that would like to dispose of all the "garbage" and "junk" that "cells generate." (Aubrey de Grey and Michael Rae, *Ending Aging. The Rejuvenation Breakthroughs That Could Reverse Human Aging in Our Lifetime*, St. Martin's Press, NY, 2007, Ch. 7, "Upgrading the Biological Incinerators," pp. 101-133.)

[623] B. P. Tokin, *Regenerazia I Somaticheskiy Embriogenez* (Regeneration and somatic embryogenesis), Izdatelstvo Leningradskogo Gosudarstvennogo Universiteta (Leningrad State University Press), Leningrad, 1959.

[624] In his comment on Sergey Bryukhonenko's "experiments in the revival of organisms," the British biologist J.B.S. Haldane emphasized that, beside the revival techniques, "Bryukhonenko shares the credit for the methods of human blood transfusion which were first developed in the Soviet Union and are now practiced in this country, which have saved so many lives during the war."
(J.B.S. Haldane, foreword to *Experiments in the Revival of Organisms* (documentary), Techfilm, Moscow, 1940, http://www.archive.org/details/Experime1940.)

[625] Demikhov V.P., *Peresadka zhiznenno vazhnikh organov v experimente* (Experimental transplantation of vital organs), Medgiz, Moscow, 1960, pp. 139-149, 157-170.

[626] Valery Kadzhaya, "Telo I delo" (The body and the deed), *Ogonek*, No. 4 (5031), January 21, 2008, http://www.ogoniok.com/5031/31/; "Lenin, Mao i drugie mumii" (Lenin, Mao, and other mummies, photo gallery), *Topnews.ru*, 07.02.2007, http://www.topnews.ru/photo_id_891_2.html; Vladimir Mayakovsky, "Vladimir Ilyich Lenin," 1924, http://az.lib.ru/m/majakowskij_w_w/text_0480.shtml.

[627] As Bekhterev asserted:
"If the nervous-psychic activity can be reduced to energy, then we should accept that the law of the conservation of energy – which was posited by [Julius] Mayer and then supported by [Hermann] Helmholtz, and which is now commonly accepted – should fully apply also in relation to the nervous-psychic activity or related activity. …
This is why the so called afterlife, that is, life beyond the boundary of the bodily form of human personality, undoubtedly exists, either in the form of individual immortality, understood as a certain synthesis of nervous-psychic processes manifested in a given personality, or in the form of immortality of a more general character, since the *content of human personality*, spreading as a special stimulus to the breadth and depth of human society, so to say, inflowing into other living beings and being transmitted in descent to future humanity, does not have an end as long as at least a single living human creature exists on this earth."
(V.M. Bekhterev, *Bessmertie Chelovecheskoy Lichnosti kak Nauchnaya Problema* (The immortality of human personality as a scientific problem), Moscow, 1918, reprinted in Altea, St. Petersburg, 1999, emphasis added, http://www.igp.ru/2000_noo/Behterev_1918/bexter1.html.)

[628] Sherbakov Pavel Vasilievich, "Anabioz kak put k bessmertiu" (Anabiosis as a way to immortality), *Moskovsky Komsomolets*, September 04, 2001 (the article includes a personal testimony about Termen's freezing experiments, http://www.ordodeus.ru/Ordo_Deus9Anabioz_PkB_Shcherbakov.html); Mikhail Batin, Anna Kirik (Eds.), Bessmertie (Immortality), Komandor, Yaroslavl, 2007, p. 72.

[629] N.E. Pizyk, *Bogomolets. Seria: Zhizn Zamechatelnikh Ludey* (Bogomolets. The life of remarkable people series), Molodaya Gvardia, Moscow, 1964; N.N. Sirotinin, *A.A. Bogomolets*, Vyshaya Shkola, Moscow, 1967.

[630] Alexander Bogomolets, *The Prolongation of Life*, translated by Peter V. Karpovich and Sonia Bleeker, Essential Books, Duel, Sloan and Pearce, NY, 1946, pp. 50-51.
The book was first published in Russian as "Prodlenie Zhizni" (The prolongation of Life), Kiev, 1938.

[631] Another possible claimant to priority may have been the conference on aging, held at Woods Hole, Massachusetts, on June 25-26, 1937, led by Edmund V. Cowdry and Lawrence K. Frank, funded by the Josiah Macy Jr. Foundation, and jointly sponsored by the Union of American Biological Societies and the National Research Council.
Yet, unlike the Kiev conference whose materials were widely disseminated and highly influential in the USSR, the Woods Hole conference appears to have been restricted to a narrow circle (about 20 participants) and its proceedings were not published. The networking at the Woods Hole conference, however, later led to the publication of Cowdry's *Problems of Ageing* in 1939, the first American joint publication on aging. (E.V. Cowdry (Editor), *Problems of Ageing. Biological and Medical Aspects*, Williams and Wilkins, Baltimore MD, 1939, p. xi.)

Also, in the Woods Hole conference, unlike the Kiev conference, the major subject was the elucidation of basic processes of aging, and the subject of extending longevity was relegated to the background (see the following chapter on American life-extensionism). Thus, the Kiev conference may be still considered the world's first, both with regard to publicity and emphasis on life extension.

[632] Alexander Bogomolets, *The Prolongation of Life*, translated by Peter V. Karpovich and Sonia Bleeker, Essential Books, Duel, Sloan and Pearce, NY, 1946, first published in Russian as "Prodlenie Zhizni" (The prolongation of Life), Kiev, 1938; *Starost. Trudy Konferenzii po Probleme Geneza Starosti I Profilaktiki Prezhdevremennogo Starenia Organisma. Kiev 17-19 Decabria. 1938*, Izdatelstvo Akademii Nauk USSR, Kiev 1939 (Aging. Proceedings of the Conference on the Problem of the Genesis of Aging and Prophylaxis of the Organism's Untimely Aging, Kiev, December 17-19, 1938, Publication of The Ukrainian Soviet Socialist Republic Academy of Sciences, Kiev, 1939).

[633] According to Vadim Zakharovich Rogovin, *Stalinski NeoNEP* (Stalin's Revival of the New Economic Policy), Iskra, Moscow, 1994, Ch. 1. "Economic Liberalization in the USSR," Ch. 2. "World Capitalism in the 1930s" – the year 1938 was characterized by massive purges in the USSR, yet also marked a peak in Soviet pre-war industrialization and prosperity (http://trst.narod.ru/).

[634] Alexander Bogomolets, *The Prolongation of Life*, translated by Peter V. Karpovich and Sonia Bleeker, Essential Books, Duel, Sloan and Pearce, NY, 1946, first published in 1938 (hereafter referred to as "Bogomolets, 1946").

[635] Bogomolets, 1946, pp. 22-23,

[636] Bogomolets, 1946, p. 83.

[637] Quoted in Bogomolets, 1946, pp.79-80.

The reference is to Elie Metchnikoff, *The Nature of Man: Studies in Optimistic Philosophy*, translated by P.C. Mitchell, Putnam, NY, 1903, Ch. 10 "Introduction to the Scientific Study of Old Age," pp. 245-246 (first published in French in 1903).

Metchnikoff also spoke about the "cytotoxic sera" in Elie Metchnikoff, *The Prolongation of Life: Optimistic Studies*, Putnam, New York, 1908, p. 148-149 (first published in French in 1907).

[638] Bogomolets, 1946, p. 66.

[639] Bogomolets, 1946, p. 70.

[640] Bogomolets explained the workings of rejuvenative colloidoclasia:

"during the colloidoclasia that occurs in a transfusion, complex electric processes take place, a sort of electric storm, between the protein particles of the donor and those of the recipient. In consequence the electric tension of the surface of many particles is lowered. They become less stable, clump together and form a precipitate. …

this precipitate which consists of the older, less active particles, disintegrates. Thus damage to the cellular colloids, colloidoclasia, causes precipitation, within the cells, of the more used particles and hastens their consequent dissolution. This enables the cells to free themselves of old elements in the cytoplasm.

When these elements disintegrate, substances apparently are formed that resemble the already mentioned autocatalyzers which intensify metabolism and other functions of the cells. As we can see, it is not only stimulation but rejuvenation of the cell that follows the colloidoclastic shock caused by blood transfusion." (Bogomolets, 1946, pp. 76-77.)

[641] Bogomolets, 1946, p. 89.

[642] Bogomolets continued:

"This is far from being the solution of the problem of cancer therapy. As heretofore, cancer has to be removed surgically. But experiments have proved that, by stimulating the reticulo-endothelial system with our anti-reticular cytotoxic serum, it is possible to increase the resistance of the organism to cancer.

The only way to cure cancer is to diagnose it as early as possible and then to remove it surgically. ACS does not cure cancer that has reached a clinically recognizable, serious stage of development."

(Bogomolets, 1946, pp. 82, 88.)

[643] Oleg Bogomolets, *Vlianie ACS na zazhivlenie ran* (The Influence of ACS on Wound Healing), The Ukrainian Academy of Sciences, Kiev, 1944.

[644] A. Bogomolets, radio broadcast made in January 1942, quoted in Nina Emelianovna Pizyk, *Bogomolets. Seria: Zhizn Zamechatelnikh Ludey* (Bogomolets. The life of remarkable people series), Molodaya Gvardia, Moscow, 1964, p. 194.

See also, Vasily Kalita, "Milliony solnechnikh dney, otvoevannikh u bolezni" (Millions of sunny days won from disease), *Zdorovia Ukraini. Medichna Gazeta* (Health Ukraine. Medical Newspaper), No. 6, March 2006, http://health-ua.com/articles/1269.html.

[645] Bogomolets, 1946, p. 86.

[646] In Bogomolets' 'future research' program, the methods of cytotoxic stimulation were planned to be applied not only to connective tissue, but also to "parenchymal" (functional organ) tissues, thus realizing Metchnikoff's original idea.

The question of dosages was far from being resolved:

"Strong and specific stimulating effects of the cytotoxic serum on the reticulo-endothelial system should be considered as well established. But it is also possible that, through a prolonged cytotoxic stimulation, the connective tissue will excessively proliferate and then become sclerotic. The question of proper dosage will be of decisive importance."

(Bogomolets, 1946, p. 85.)

[647] Bogomolets, 1946, pp. 84-85.

[648] Bogomolets, 1946, pp. ii, vii, ix

[649] Greg Fahy, "An Interview with Dr. Denham Harman," *Life Extension Magazine*, January 1998, http://www.lef.org/magazine/mag98/jan-interview98.html.

[650] Thus, Straus's group (1946) found that "stimulation of the healing of experimentally produced fractures in rabbits was induced with small ("stimulating") doses of antireticular cytotoxic serum and depression of healing was produced with large ("depressing") doses as claimed by the Soviet investigators."

And according to Joseph Skapier's experiments on the use of the ACS in the treatment of Hodgkin's lymphoma cancer (1947), although the ACS "did not prove to be a curative agent in this disease," yet it produced "improvement in the general condition in a majority of the cases treated." Therefore, Skapier concluded, "A.C.S. may be considered a valuable adjunct to other accepted methods of therapy in the treatment of Hodgkin's disease."

(Reuben Straus, Moris Horwitz, Daniel H. Levinthal, Arthur L. Cohen and Mildred Runjavac, "Studies on Antireticular Cytotoxic Serum. III. Effect of ACS on the Healing of Experimentally Produced Fractures in Rabbits," *The Journal of Immunology*, 54, 163-177, 1946; Joseph Skapier, "Therapeutic Use of Anti-Reticular Cytotoxic Serum (A.C.S) in Hodgkin's Disease," *Cancer Research*, 7, 369-371, 1947.)

[651] Patrick M. McGrady, *The Youth Doctors*, Coward-McCann, NY, 1968, p. 289.

[652] As Bürger summarized his findings regarding the use of Bogomolets' anti-reticular cytotoxic serum:

"The hopes to use this serum to combat premature aging and to be able to prolong human life have not been fulfilled... We also tested the Bogomolets serum in several cases. It has no measurable effect on the metabolism of the aged persons" (Max Bürger, *Altern und Krankheit* (Aging and Disease), Zweite Auflage (2nd Edition), Veb Georg Thieme, Leipzig, 1954 (1947), p. 280).

Leake similarly concluded that "there is no clear evidence that ACS is particularly useful in the treatment of premature aging or in extending the life span or delaying the onset of degenerative conditions in old age" (Leake C.D., "Russian and Iron Curtain Proposals for Geriatric Therapy," *Geriatrics* 14(10), 670-673, 1959, reprinted in Geraldine M. Emerson (Ed.), *Benchmark Papers in Human Physiology, Vol. 11, Aging*, Dowden, Hutchinson and Ross, Stroudsburg PA, 1977, pp. 110-113).

And Gardner and Speaker's verdict, based on 160 studies, was that "The action of ACS, both cytotoxic for large doses and stimulating for small doses, have been verified. [sic.] ... The stimulatory effects in wound and fracture healing have been verified. The relief of pain in human cancer ...[has] been demonstrated. ... On the other hand, ACS has been shown to have little value in treating arthritis, and there is no verification of its antiaging claims." (Thomas S. Gardner and David M. Speaker, "A critical evaluation of anti-reticular cytotoxic serum (ACS)," *Texas Reports on Biology and Medicine*, 9, 448-490, 1951, quoted in Leake, 1959, p. 111.)

[653] As defined by the American National Research Council:

"Monoclonal antibodies (mAb) are used extensively in basic biomedical research, in diagnosis of disease, and in treatment of illnesses, such as infections and cancer. Antibodies are important tools used by many investigators in their research and have led to many medical advances.

Producing mAb requires immunizing an animal, usually a mouse; obtaining immune cells from its spleen; and fusing the cells with a cancer cell (such as cells from a myeloma) to make them immortal, which means that they will grow and divide indefinitely.

A tumor of the fused cells is called a hybridoma, and these cells secrete mAb. The development of the immortal hybridoma requires the use of animals; no commonly accepted nonanimal alternatives are available."

(*Monoclonal Antibody Production. A Report of the Committee on Methods of Producing Monoclonal Antibodies. Institute for Laboratory Animal Research. National Research Council*, National Academy Press, Washington DC, 1999, p. vii.)

[654] Quoted in Albert Rosenfeld, *Prolongevity. A report on the scientific discoveries now being made about aging and dying, and their promise of an extended human life span – without old age*, Alfred A. Knopf, New York, 1976, pp. 3-4.

[655] "Stalin. Istoricheskie Anekdoty" (Stalin. Historical anecdotes), *Velikie Vlastiteli Proshlogo*, Moscow, 2004, http://vlastitel.com.ru/znaete/stalin.html; http://www.rodon.org/ra2/aaabinsnpj.htm.

[656] Yury Vilensky, "Neizvestny Bogomolets" (The unknown Bogomolets), *Zerkalo Nedeli*, 14(235), 10-16 April, 1999 (the article includes an interview with Alexander Bogomolets' son, Oleg) http://www.zn.ua/3000/3150/21213/.

[657] Nikita Sergeevich Khrushchev, *Vremya, Lyudi, Vlast - Vospominania* (Time, People, Power - Memoirs), Moskovskie Novosti, Moscow, 1999 (1969), Book 1. Part 1. Ch. "Snova Na Ukraine" (In Ukraine again), pp. 170-172, Ch. "Ukraina – Moskva. Perekrestki 30kh godov" (Ukraine – Moscow. Crossroads of the 1930s), p. 206, reprinted in the Archive Project Voennaya Literatura (Military literature, http://militera.lib.ru/memo/russian/khruschev1/index.html).

[658] Yury Vilensky, "Neizvestny Bogomolets" (The Unknown Bogomolets), *Zerkalo Nedeli*, 14(235), 10-16 April, 1999, http://www.zn.ua/3000/3150/21213/.

[659] *Starost. Trudy Konferenzii po Probleme Geneza Starosti I Profilaktiki Prezhdevremennogo Starenia Organisma. Kiev 17-19 Decabria. 1938*, Izdatelstvo Akademii Nauk USSR, Kiev 1939 (Aging. Proceedings of the Conference on the Problem of the Genesis of Aging and Prophylaxis of the Organism's Untimely Aging, Kiev, December 17-19, 1938, Publication of The Ukrainian Soviet Socialist Republic Academy of Sciences, Kiev, 1939), Alexander Bogomolets "Vstupitelnoe Slovo" (Opening statement), pp. 5-6.

The conference proceedings will be hereafter referred to as "*Starost*, 1938."

[660] Dmitry Fedorovich Chebotarev, "Hochetsia verit, chto nash ogromny trud prigoditsia potomkam" (I want to believe that our immense work will be useful to our descendants), *Nauchno-Practicheskiy Zhurnal 'Medichni Vsesvit' – Mir Mediziny* (The Scientific-practical Journal "Medical World"), 1(1), 2001 (an interview with Dmitry Chebotarev by Elena Ludwig, http://www.socion.net.ua/med_journal/articles/interview/chebotarev.htm).

[661] A.V. Nagorny, "K voprosu o faktorakh, obuslovlivayushikh dlitelnost zhizni" (On the question of factors determining the duration of life), in *Starost*, 1938, p. 157.

[662] "Resoluzia" (Resolution of the Conference on the Problem of Aging and Prophylaxis of Untimely Aging), in *Starost*, 1938, p. 454.

[663] A.V. Nagorny, "K voprosu o faktorakh, obuslovlivayushikh dlitelnost zhizni" (On the question of factors determining the duration of life), in Starost, 1938, pp. 158, 170.

[664] M. Mühlmann, *Über die Ursache des Alters. Grundzüge der Physiologie des Wachstums mit besonderer Berücksichtigung des Menschen* (On the cause of aging. Fundamentals of growth physiology with a special consideration of people), Bergmann, Wiesbaden, 1900.

[665] Moisey Samuilovich Milman (Mühlmann), "Genez Starosti" (Genesis of Aging), in *Starost*, 1938, p. 52.

[666] Justus von Liebig, *Familiar Letters on Chemistry, In Its Relations to Physiology, Agriculture, Commerce, and Political Economy*, Fourth Edition, Revised and Enlarged, Edited by John Blyth, MD, Walton and Maberly, London, 1859 (first published in German in 1843), p. 342.

[667] Max Bürger, *Altern und Krankheit* (Aging and Disease), Zweite Auflage (2nd Edition), Veb Georg Thieme, Leipzig, 1954 (1947), p. 34.

[668] Denham Harman, "Aging: a theory based on free radical and radiation chemistry," *Journal of Gerontology*, 11(3), 298-300, 1956.

[669] Konstantin Pavlovich Buteyko, *Metod Buteyko. Opyt Vnedrenia v Medizinskuyu Praktiku* (The Buteyko Method. The Experience of Application in Medical Practice), Patriot, Moscow, 1990; Vladimir K. Buteyko and Marina M. Buteyko, *The Buteyko Theory about a Key Role of Breathing for Human Health, Scientific Introduction to the Buteyko Therapy for Experts* (in Russian and English), Buteyko Co LTD, Voronezh, 2005.

[670] *Rukovodstvo po Hyperbaricheskoy Oxygenazii. Teoria I Praktika Klinicheskogo Primenenia* (Hyperbaric Oxygenation: a manual. The theory and practice of clinical application), S.N. Efouni (Ed.), Akademia Medizinskikh Nauk SSSR (Academy of Medical Sciences of the USSR), Medizina, Moscow, 1986, S.N. Efouni, et al., "Hyperoxia. Patofiziologicheskie aspekty lechebnogo i toxicheskogo vozdeystvia hyperbaricheskogo kisloroda (Hyperoxia. Pathophysiological aspects of therapeutic and toxic effects of hyperbaric oxygenation), pp. 29-56; Richard D. Vann, *Oxygen Toxicity Risk Assessment*, Office of Naval Research, Arlington, VA, 1988; N. Bitterman, "CNS oxygen toxicity," *Undersea and Hyperbaric Medicine*, 31 (1), 63-72, 2004.

[671] A.V. Nagorny, "K voprosu o faktorakh, obuslovlivayushikh dlitelnost zhizni" (On the question of factors determining the duration of life), in *Starost*, 1938, p. 165; Slonaker, J. R., "The normal activity of the albino rat from birth to natural death, its rate of growth, and duration of life," *Journal of Animal Behavior*, 2, 20-42, 1912.

The confusion regarding the role of physical activity for longevity, introduced by Slonaker, has continued in American life-extensionist literature. Generally, physical exercise has been regarded as beneficial for longevity. Yet, there have been conflicting findings.

Thus, it was found that athletes live longer than normal insured men, but shorter than "physically underdeveloped" people. (Louis I. Dublin, "Longevity of college athletes," *Harper's Monthly Magazine*, 157, 229-238, 1928.)

It was also found that athletes live longer in general. (Martti J. Karvonen, "Endurance sports, longevity and health," *Annals of the New York Academy of Sciences*, 301, 653-655, 1977.)

And it was also found that athletes live shorter in general. (Peter V. Karpovich, "Longevity and athletics," *Research Quarterly*, 12, 451-455, 1941.)

And there were also found no significant differences. (Henry J. Montoye, et al., *The Longevity and Morbidity of College Athletes*, Indianapolis, 1957.)

The results also varied widely depending on the type of sports, level of athleticism, period of practice, and many other factors. (Anthony P. Polednak (Ed.), *The Longevity of Athletes*, Charles C. Thomas, Springfield IL, 1979.)

It was also shown that "blue-collar," physically active workers live shorter than sedentary "white-collar" workers. But this was explained by the assumption that the "white-collar" workers were able to exercise regularly, in a protected environment, and with sufficient rest. (Charles L. Rose and Michel L. Cohen, "Relative importance of physical activity for longevity," *Annals of the New York Academy of Sciences*, 301, 671-702, 1977.)

(These works are reviewed in William G. Bailey, *Human Longevity from Antiquity to the Modern Lab*, Greenwood Press, Westport CN, 1987, "Athleticism and Exercise," pp. 98-104.)

One of the latest in confusing findings is Howard Friedman and Leslie Martin's *The Longevity Project. Surprising Discoveries for Health and Long Life from the Landmark Eight-Decade Study*, Hudson Streen Press, Penguin Group, NY, March 2011.

Based on the analysis of a group of 1500 subjects, Friedman and Martin basically suggest that many factors popularly associated with decreased longevity – such as lack of exercise, demanding careers, anxiety, risk-taking, lack of religion, being unmarried, unsociability, pessimism, stress-producing Type A behavior, various dietary taboos – should be taken with a large grain of salt. For example, the more cheerful and relaxed people tended to live shorter than "prudent and persistent" individuals (p. 9). The book also suggests that strenuous exercise does not necessarily lead to greater longevity (pp. 105-106).

[672] G.V. Folbort, A.V. Semernina, "Izmenenie rabotosposobnosti centralnikh elementov, obespechivayushikh vyshuyu nervnuyu deatelnost, pri starenii u sobak" (Changes in the working capacity of the central elements responsible for the high nervous activity, in aging dogs), *Starost*, 1938, pp. 199-200; N.K. Zolnikova, "Izmenenia dinamiki processov istoshenia i vosstanovlenia u starykh zhivotnikh" (Changes in the dynamics of the processes of exhaustion and recovery in aged animals), *Starost*, 1938, p. 244.

[673] N.B. Medvedeva, "Ob izmeneniakh vodno-belkovogo sostava tkaney v starosti" (On changes in the water-protein content of tissues in aging); G.V. Derviz, "Biochimicheskie izmenenia v organizme i v obmene ego veshestv vo vremia starenia" (Biochemical changes in the organism and in its material metabolism during aging), *Starost*, 1938, pp. 207-212, 213-220.

[674] V.N. Nikitin, "Vozrastnie izmenenia v sinteze i raspade belkov v zhivotnom organisme" (Age-related changes in the synthesis and degradation of proteins in an animal organism), *Starost*, 1938, p. 235.

[675] A.V. Nagorny, "K voprosu o borbe so stareniem organisma" (On the question of the struggle with the aging of the organism), *Starost*, 1938, p. 168.

[676] A.V. Nagorny, "K voprosu o borbe so stareniem organisma" (On the question of the struggle with the aging of the organism), *Starost*, 1938, p. 169.

[677] S.S. Khalatov, "Starost i Xantomatoz" (Aging and Xanthomatosis), *Starost*, 1938, p. 70.

[678] V.I. Solntzev, "Vozrastnie izmenenia lipoidnogo obmena" (Age-related changes of lipid metabolism), *Starost*, 1938, p. 400.

[679] M.D. Gatzanyuk, "Ob izmeneniakh lipociticheskogo koeffizienta s vozrastom" (On changes of the lipocytic coefficient with age), *Starost*, 1938, pp. 228-232.

[680] I.M. Turovets, L.I. Pravdina, "O nekotorykh osobennostiakh obmena veshestv u starikov" (On some characteristics of metabolism in the aged), *Starost*, 1938, pp. 328-329.

[681] Low density lipoproteins were first isolated by the American biochemist John Lawrence Oncley in 1950. (Robert E. Olson, "Discovery of the Lipoproteins, Their Role in Fat Transport and Their Significance as Risk Factors," *Journal of Nutrition*, 128 (2), 439S-443S, 1998.)

[682] V.I. Solntzev, "Vozrastnie izmenenia lipoidnogo obmena" (Age-related changes of lipid metabolism), *Starost*, 1938, pp. 400-402.

[683] V.A. Elberg, "Pitanie i Starost" (Nutrition and Aging), *Starost*, 1938, pp. 391-393.

[684] N. Anitchkow, S. S. Chalatow, "Ueber experimentelle Cholesterinsteatose und ihre Bedeutung für die Entstehung einiger pathologischen Prozesse" (On experimental cholesterol steatosis and its significance for the origin of some pathological processes), *Zentralblatt für allgemeine Pathologie*, 24, 1-9, 1913.

[685] R.E. Kavetsky, "Starost i Rak" (Aging and Cancer), *Starost*, 1938, p. 192.

[686] R.E. Kavetsky, "Starost i Rak" (Aging and Cancer), *Starost*, 1938, p. 195.

[687] N.A. Shereshevsky, "Starost i endokrinnaya systema" (Aging and the endocrine system), *Starost*, 1938, p. 37; V.H. Vasilenko, R.M. Mayzlish, "Kolebania vydelenia prolana v sviazi s vozrastom" (Oscillations in prolan excretion in relation to age), *Starost*, 1938, pp. 420-421, 426-427.

[688] The Canadian/French Félix d'Herelle (1873-1949) first suggested the anti-bacterial effect of bacteriophage viruses in 1917, and described their therapeutic applications in 1921, while working at the Pasteur Institute (1911-1921).
Yet, in 1934-1936, he developed the bacteriophage therapy in Tbilisi, Soviet Georgia, in cooperation with George Grigorievich Eliava (1892-1937, executed by shooting as an "enemy of the people").
The George Eliava Institute of Bacteriophage, Microbiology and Virology, Tbilisi, exists to the present and the bacteriophage therapy is still available in pharmacies in the Republic of Georgia.
(The George Eliava Institute of Bacteriophage, Microbiology and Virology, http://www.eliava-institute.org/; Félix d'Herelle, *Le Bactériophage: Son Rôle dans l'Immunité* (The bacteriophage: its role in immunity), Masson, Paris, 1921; Thomas Häusler, *Viruses vs. Superbugs. A Solution to the Antibiotics Crisis?* Macmillan, NY, 2006.)

[689] Alexander Gavrilovich Gurwitsch (1874-1954) discovered "mitogenic" (a.k.a. mitogenetic or morphogenetic) radiation (*mitogeneticheskoe izluchenie*) in 1923. This radiation in the ultraviolet spectrum area (194-250 nm), emitted by living organisms or induced externally, was shown to stimulate cell division, hence growth and regeneration.
Subsequently, by the 1940s, extensive studies were conducted, measuring the levels of mitogenic radiation in different physiological and pathological conditions and at different ages, and attempting to enhance it in order to diminish work fatigue, produce regeneration and rejuvenation. Such research was done by S.N. Braines, E.A. Goldenberg, A.A. Kisel, L.Ya. Blern, L.Ya. Bliakher, I.K. Zamaraev, and others.
(S.Y. Salkind, "Mitogeneticheskie luchi i ich znachenie v medizine" (Mitogenic rays and their significance for medicine), in *Problemy teoreticheskoy i prakticheskoy mediziny* (Problems of Theoretical and Practical Medicine), Biomedgiz, Moscow-Leningrad, 1936, reprinted in *Medznanie*, 2007, http://www.medznanie.ru/lib/problema.html; http://www.medznanie.ru/lib/problema114.html.)

[690] V.G. Shipachev, "K voprosu o borbe so stareniem organisma" (On the question of the struggle with the aging of the organism), *Starost*, 1938, p. 411.

[691] V.G. Shipachev, "K voprosu o borbe so stareniem organisma" (On the question of the struggle with the aging of the organism), Starost, 1938, pp. 412-415.

[692] Steven Horne, "Colon cleansing: a popular, but misunderstood natural therapy," *Journal of Herbal Pharmacotherapy*, 6(2), 93-100, 2006; Anton Emmanuel, "Current management strategies and therapeutic targets in chronic constipation," *Therapeutic Advances in Gastroenterology*, 4(1), 37-48, 2011; Laura E. Matarese, Douglas L. Seidner, Ezra Steiger, "The Role of Probiotics in Gastrointestinal Disease," *Nutrition in Clinical Practice*, 18, 507-516, 2003.

[693] A.V. Leontovich, "Nekotorie materialy o vese mozga cheloveka raznykh vozrastov v biometricheskoy obrabotke" (Some materials on the weight of the human brain at different ages in biometric processing), *Starost*, 1938, p. 64

[694] N.D. Yudina, "Characteristika morfologicheskogo sostava krovi pri fiziologicheskom starenii" (Characteristics of the morphological blood content in physiological aging), *Starost*, 1938, p.180

[695] "Resoluzia" (Resolution of the Conference on the Problem of Aging and Prophylaxis of Untimely Aging), in *Starost*, 1938, pp. 454-455.

[696] N.D. Strazhesko, "Zakluchitelnoe Slovo" (Concluding statement), *Starost*, 1938, pp. 455-456.

[697] "Prenia" (Conference Debates), *Starost*, 1938, pp. 437-438.

[698] "Prenia" (Conference Debates), *Starost*, 1938, pp. 451-452.

[699] A.D. Speransky, *Elementy Postroenia Teorii Mediziny* (Elements in the Construction of a Medical Theory), Izdatelstvo Vsesoyuznogo Instituta Experimentalnoy Mediziny - VIEM (Publication of the All-Union Institute of Experimental Medicine), Moscow, 1937, p. 261.

[700] These words have been commonly attributed to the German physiologist and hygienist Max Rubner (1854-1932): "Die ganze Kunst, das menschliche Leben zu verlängern, besteht zuerst darin, es nicht zu verkürzen" (The whole art of prolonging human life consists first of all in not shortening it). Rubner is thus cited, for example, in Ernst Lautenbach, *Medizin Zitate Lexikon* (Medical citations lexicon), Iudicium, München, 2010, p. 25.

Yet, this sentiment was also expressed earlier by the Austrian physician and philosopher Ernst von Feuchtersleben (1806-1849), in his *Zur Diätetik der Seele* (Dietetics of the Soul, Verlag von Carl Gerold, Wien, 1851, p. 146, first published in 1838): "Das ganze Geheimnis, sein Leben zu verlängern, besteht darin, es nicht zu verkürzen" (the whole secret of prolonging his life consists in not shortening it).

[701] *Izvestia* (News Moscow), March 25, 1943.

[702] V.Ya. Alexandrov, *Trudnye Gody Sovetskoy Biologii. Zapiski Sovremennika* (The difficult years of Soviet Biology. Notes of a Contemporary), Nauka, St. Petersburg, 1993, http://vivovoco.rsl.ru/VV/BOOKS/ALEXANDROV/CHAPTER_1.HTM.

[703] The basic concepts in the teachings of Peter Kuzmich Anokhin (1898-1974) were "Afferent synthesis" (the adaptive reaction of the organism as a whole), "phylogenetic memory" stored in "silent genes" or "silent neurons" (*molchashie geny/neurony*), and "anticipatory reflection of reality" (*operezhayshee otrazhenie deystvitelnosti*) or the ability of the organism to anticipate future evens and act purposefully, based on accumulated genetic memory.

According to Anokhin, natural cycles and influences, acting on the organisms during millions of years of evolution, become engrained (fixed) in the structure of the organism, thanks to which the organism becomes capable of "anticipatory reflection." This anticipatory ability makes the organism stable and robust. (P.K. Anokhin, "Operezhayushee otrazhenie deystvitelnosti" (Anticipatory reflection of reality), *Voprosy Filosofii* (Questions of philosophy), 7, 97-111, 1962, reprinted in P.K. Anokhin, *Izbrannie Trudy. Filosofskie aspecty teorii funkzionalnoy systemy* (P.K. Anokhin. Selected Works. Phylosophical aspects of the theory of a functional system), Nauka, Moscow, 1978, pp. 7-26.)

Anokhin also made some more conventional life extensionist statements, such as "I see in the massive development of physical culture and sport one of the best options in the battle for human health, his creative activity and longevity" (quoted in A.Yu. Kozhevnikov, G.B. Lindberg, *Mudrost Vekov. Rossia* (The wisdom of centuries. Russia), Neva, St. Petersburg-Moscow, 2006, p. 131).

Anokhin's ideas were developed by his students, such as R.I. Kruglikov and V.B. Shvyrkov.

Thus according to Kruglikov, "All the current activities of the organism" are determined by its "accumulated history" including "genetic memory" (R.I. Kruglikov, *Prinzip determinizma i deiatelnost mozga* (The principle of determinism and the activity of the brain), Nauka, Moscow, 1988, pp. 146-147).

And according to Shvyrkov, "Different specializations of neurons of various brain structures reflect ... the history of successful behavioral acts, accumulated by the organism in the process of trial and error during phylogenesis and individual leraning" (V.B. Shvyrkov, *Vvedenie v obyektivnuyu psychologiu. Neuronalnie osnovy psichiki* (Introduction to objective psychology. The neuronal foundations of the psyche), Institut Psichologii. Rossiyskaya Academia Nauk (Institute of Psychology of the Russian Academy of Sciences), Moscow, 1995, http://www.keldysh.ru/pages/mrbur-web/misc/efs_book/).

Peter Anokhin's daughter, Irina Petrovna Anokhina (b. 1932) has also been a prominent brain researcher, acting as the director of the Institute of Medico-Biological Problems of Narcology in Moscow. Her research interests included an extensive study of genetic predispositions to psychological characteristics (such as mental illnesses and addictions).

Presently, Peter Anokhin's grandson, and Irina's son, Konstantin Vladimirovich Anokhin (b. 1957), head of the Laboratory of Neurobiology of Memory at the Institute of Normal Physiology of the Russian Academy of Medical Sciences, Moscow, explores the possibility of decoding, storing and transmitting the information contained in the human brain. (Konstantin Anokhin, "Kody Mozga" (Codes of the Brain), presentation at the Moscow Polytechnical Institute, February 11, 2010, http://woodash.ru/?p=7017.)

[704] Ivan Ivanovich Schmalhausen, *Problema Smerti I Bessmertia* (Проблема Смерти и Бессмертия - The Problem of Death and Immortality), Gosudarstvennoe Izdatelstvo, Moscow-Leningrad, 1926, http://scilib.narod.ru/Biology/Schmalgausen/.

[705] Boris Mikhailovich Zavadovsky, *Problema Starosti I Omolozhenia v Svete Noveyshikh Rabot Steinacha, Voronova I drugikh Avtorov* (Проблема Старости и Омоложения - The Problem of Aging and Rejuvenation in the light of recent works of Steinach, Voronoff and other authors), Moscow, 1921, reprinted by SovLit, 2007, http://www.ruthenia.ru/sovlit/j/102.html.

[706] N.K. Koltsov, "Smert, starost, omolozhenie" (Смерть, Старость, Омоложение - Death, aging, rejuvenation), in N.K. Koltsov (Ed.), *Omolozhenie. Sovremennie Problemy Estestvoznania* (Rejuvenation, a special volume in Modern Problems in Natural Sciences), vol. 2, 1923, pp. 5-27; N.K. Koltsov, "Predislovie Redaktora" (Editor's Introduction) and N.K. Koltsov, "Noveyshaya Amerikanskaya literatura v oblasti operativnogo omolozhenia cheloveka" (The newest American literature on operative human rejuvenation), in N.K. Koltsov (Ed.), *Omolozhenie. Sovremennie Problemy Estestvoznania* (Rejuvenation, the second special volume on Rejuvenation in Modern Problems in Natural Sciences), vol. 16, 1924, pp. 5-10, 124-147; N.K. Koltsov, *Omolozhenie Organizma po Metodu Steinacha* (The organism's rejuvenation by Steinach's method), Vremia, Peterburg, 1922.

[707] *Sessia VASHNIL - 1948. O polozhenii v biologicheskoy nauke. Stenograficheskiy otchet. Vsesoyuznaya Akademia Selskokhoziaystvennykh Nauk Imeni V.I. Lenina* (The 1948 session of V.I. Lenin's All-Union Academy of Agricultural Sciences. On the state of biological science, stenographic protocols), OGIZ – Selkhozgiz. Gosudarstvennoe izdatelstvo selskokhozayastvennoy literatury (State Publication of Agricultural Literature), Moscow, 1948, http://lib.ru/DIALEKTIKA/washniil.txt.

[708] Alexander Vasilievich Nagorny, *Starenie I Prodlenie Zhizni* (Старение и продление жизни - Aging and Life-extension), Sovetskaya Nauka (Soviet Science), Moscow, 1950, pp. 191, 218-219.

[709] I.I. Metchnikoff, *Academicheskoe Sobranie Sochineniy* (Elie Metchnikoff, Academic Collected Works, Ed. G.S. Vasezky), Academia Medizinskikh Nauk SSSR (The USSR Academy of Medical Sciences), Moscow, 1954, vol. 13, pp. 10, 40: "Predislovie ko vtoromu izdaniu 'Sorok Let Iskania Razionalnogo Mirovozzrenia'" (Introduction to the Second Edition of "Forty Years in Search of a Rational Worldview," 1914, p. 10), "Vstuplenie" (Introduction to the first edition, 1912, p. 40).
Interestingly, while the socialist sympathies are expressed in 1912, the anti-socialist stance is pronounced in 1914.

[710] V.N. Nikitin, "I.I. Metchnikoff i problema dolgoletia" (I.I. Metchnikoff and the problem of longevity), in *I.I. Metchnikoff. Sobranie Sochineniy* (Collected Works of I.I. Metchnikoff, Eds. N.N. Anichkov and R.I. Belkin), Academia Medizinskikh Nauk SSSR (The USSR Academy of Medical Sciences), Moscow, 1962, vol. 15, p. 388-389.

[711] V.N. Nikitin, "I.I. Metchnikoff i problema dolgoletia" (I.I. Metchnikoff and the problem of longevity), in *I.I. Metchnikoff. Sobranie Sochineniy* (Collected Works of I.I. Metchnikoff, Eds. N.N. Anichkov and R.I. Belkin), Academia Medizinskikh Nauk SSSR (The USSR Academy of Medical Sciences), Moscow, 1962, vol. 15, pp. 389, 399-400.

[712] R.I. Belkin "Primechania k XV tomu *I.I. Metchnikoff. Sobranie Sochineniy*" (Notes on Vol. 15 of *Collected Works of I.I. Metchnikoff*, Eds. N.N. Anichkov and R.I. Belkin, Academia Medizinskikh Nauk SSSR - The USSR Academy of Medical Sciences), Moscow, 1962, vol. 15, pp. 454, 459-460.

[713] N.S. Khrushchev, *Doklad na XXI chrezvychaynom syezde KPSS*. (Presentation at the XXI special congress of the Communist Party of the Soviet Union, January 27 – February 5, 1959), Gospolitizdat (State Political Publications), Moscow, 1959, p. 56, reprinted in the archives of the Communist Party of the Soviet Union (1898-1991), http://publ.lib.ru/ARCHIVES/K/KPSS/_KPSS.html#025.

In 1963, in *The Problem of Aging and Longevity*, Vladimir Nikitin, confirming the factual accuracy of this statement by Khrushchev, cites it as an official commendation of Soviet life-extensionism. Thanks to the "successes of the Soviet health care and social measures," Nikitin adds, "the USSR has one of the best indices of the ratio between the aged in need of material support to work-capable people" - 1:6.6, far below the USA and West-European countries.

Accordingly, "over-aging" was never considered to be an ominous social problem by Soviet gerontologists, unlike great many of their Western, and particularly American counterparts. Indeed, as will be exemplified in the next chapter on the US, the problem of overtaxing resources by the aged was foregrounded in the American literature related to life-extension.

Generally, the life-expectancy in Russia increased from 32 years in 1897, to 44 years in 1927 to 69 years in 1959.

(V.N. Nikitin, "Problema Udlinenia Zhizni" (The problem of life-extension), in A.V. Nagorny, V.N. Nikitin, I.N. Bulankin, *Problema Starenia I Dolgoletia* (The Problem of Aging and Longevity), Gosudarstvennoe Izdatelstvo Medizinskoy Literatury, Moscow, 1963, pp. 690-692.)

[714] *Programma KPSS, prinyataya XXII syezdom KPSS* (The Program of the Communist Party of the Soviet Union, approved by the XXII Party Congress), Chast 2. Zadachi Kommunisticheskoy Partii Sovetskogo Soyuza po Stroitelstvu Kommunisticheskogo Obshestva. Kommunism – Svetloe budushee vsego chelovechestva (Part 2. The tasks of the Communist Party of the Soviet Union in the building of the communist society. Communism is the bright future for all humanity), II. Zadachi Partii v oblasti podyema materialnogo blagosostoyania naroda. G' "Zabota o zdorovie i uvelichenii prodolzhitelnosti zhizni" (Section 2, The tasks of the Party in the area of raising the material well being of the people, Subsection 4(G), Care for health and the prolongation of life, http://leftinmsu.narod.ru/polit_files/books/III_program_KPSS_files/090.htm).

[715] *Pravda*, 5/IV, 1939, quoted in Alexander Nagorny, *Starenie I Prodlenie Zhizni* (Aging and Life-extension), Sovetskaya Nauka (Soviet Science), Moscow, 1950, p. 178.

[716] Chebotarev tells the following story of the Institute's establishment:

In 1957, a journalist from "Vecherniy Kiev" (The Evening Kiev), Sviatoslav Ivanov, published an article about the world's first institute of gerontology and geriatrics in Romania led by Ana Aslan and their rejuvenating experiments with Novocain (Gerovital). The article was republished in "Literaturnaya Gazeta" (Literary Newspaper) and inspired Voroshilov's interest. Voroshilov then suggested to the Health minister Maria Kovrigina (1910-1995) to send a committee to Bucharest to study the experience of the Romanian comrades (the committee included Chebotarev). The committee report was enthusiastic. After a special meeting at the USSR Ministry of Health, an order was issued by the presidium of the USSR Academy of Medical Sciences to establish the Institute of Gerontology and Experimental Pathology in Kiev, in May 1958.

(Dmitry Fedorovich Chebotarev, "Hochetsia verit, chto nash ogromny trud prigoditsia potomkam" (I want to believe that our immense work will be useful to our descendants), *Nauchno-Practicheskiy Zhurnal 'Medichni Vsesvit' – Mir Mediziny* (The Scientific-practical Journal "Medical World"), 1(1), 2001 (an interview with Dmitry Chebotarev by Elena Ludwig, http://www.socion.net.ua/med_journal/articles/interview/chebotarev.htm).

[717] There may be several speculative historical-geographical explanations for the relative prevalence of life-extensionist ideas in Ukraine. The sociologist Peter Burke generally described the 16th-17th century Ukraine as "a region favouring freedom and equality, a refuge for rebels and heretics" due to Ukraine's geographical

position at the periphery of the Russian empire, and at the intersection between Poland, Russia and Turkey. (Peter Burke, *History and Social Theory* (Second Edition), Cornell University Press, Ithaca, New York, 2005, p. 87.)

One might also add the great ethnic diversity of Ukraine, with over 100 national minorities, including the large communities of Jews, Romanians, Greeks, Germans, Serbs, and many other groups – that may have fostered intellectual fermentation and cross-fertilization.

Alternatively, Ukraine may be viewed as the historical center and birthplace of the Russian state (the Kiev Principality of the 10th century and earlier), and hence a major origin of "Russian ideas."

Another possible speculation may relate to Ukraine's geography, namely its relatively moderate climate (inspiring thoughts of stability and continuity?). Mild and constant climate was actually the prevalent rationale for the emergence of life-extensionist aspirations in California. (See Kenneth Thompson, "The idea of longevity in early California," *Bulletin of the New York Academy of Medicine*, 51(7), 805-816, 1975.) Though a little father North, the situation of Ukraine may not be entirely dissimilar.

Interestingly, beside Ukraine (Kiev, Kharkov and Odessa), several other major historical centers of European life-extensionism (such as Paris, Stuttgart, Vienna, Kishinev, Bucharest, and Zurich) were at about the same moderate latitudes (all at ~47-50°N, except for Bucharest at ~44.4°N). Of course, there were also many exceptions.

[718] According to Alexey Nikolayevich Severzov (1917), under diminished struggle for existence, long-lived organisms will be favored and the population will shift toward a greater life-span. (Severzov A.N., "O faktorakh, opredeliayushikh prodolzhitelnost zhizni mnogokletochnikh zhivotnikh" (The factors determining the longevity of multicellular animals), *Russkiy Zoologicheskiy Zhurnal* (The Russian Zoological Journal), 2(3-4), 65-77, 1917.)

Interestingly, Alexey Severzov's son, Sergey Alexeevich Severzov, a Moscow biology professor, continued to research his father's topic and in 1941 suggested that "the less is the mortality of individuals in the struggle for existence, the greater is the longevity of the species, the less varied are its population numbers, and the less prolific the species becomes." (Severzov S.A., *Dinamika naselenia i prisposobitelnaya evoluzia zhivotnikh* (Population dynamics and adaptive evolution of animals), Izdatelstvo Academii Nauk SSSR (The USSR Academy of Sciences Press), Moscow, 1941, p. 7.)

And Alexey Severzov's father and Sergey's grandfather, the zoologist Nikolay Alexeevich Severzov (1827-1885), already in 1855 noted the "polarity between nutrition and maintenance of the organism, on the one hand, and molting and child birth, on the other." He further suggested that "depending on the prevalence of the one or the other manifestation, the breed is maintained either by a rapid reproduction or by an individual longevity, but never by both simultaneously." (Severzov, N.A. *Periodicheskie yavlenia v zhizni zverey, ptiz i gad Voronezhskoy gubernii* (Periodical phenomena in the life of animals, birds and reptiles of the Voronezh district), Izdatelstvo Academii Nauk SSSR (The USSR Academy of Sciences Press), Moscow, 1950 (first published in 1855), p. 216.

Do not confuse the three Severzovs! (Quoted in Yuri Konstantinovich Duplenko, *Starenie - Ocherki Razvitia Problemy* (Aging - The Development of the Problem), Nauka, Leningrad, 1985, pp. 39, 50-51.)

[719] Ivan Ivanovich Schmalhausen, *Problema Smerti I Bessmertia* (The Problem of Death and Immortality), Gosudarstvennoe Izdatelstvo, Moscow-Leningrad, 1926, http://scilib.narod.ru/Biology/Schmalgausen/.

[720] Institut Gerontologii Academii Medichnich Nauk Ukraini (The Institute of Gerontology of the Ukrainian Academy of Medical Sciences, http://geront.kiev.ua/history.htm); *Uspekhi Gerontologii* (Advances in Gerontology, Journal of the Gerontological Society of the Russian Academy of Sciences, published since 1997, http://www.gersociety.ru/information/uspexi/); Istoria i Deiatelnost Sekzii Gerontologii pri Moskovskom Obshestve Ispytateley Prirody (History and Activity of the Gerontology Section of the Moscow Society of Natural Scientists, http://gerontology-explorer.narod.ru/efba3ff8-6f89-4cc5-8700-060321f7f681.html;

V.N. Anisimov, M.V. Soloviev, *Evoluzia Concepcy v Gerontologii* (The Evolution of Concepts in Gerontology), Aesculap, Saint Petersburg, 1999, 3. "Perspectivy razvitia gerontologii v Rossii" (Perspectives for the development of gerontology in Russia), reprinted in the *Russian Biomedical Journal*, www.Medline.ru, vol.3, http://www.medline.ru/public/art/tom3/geront/pt3-3.phtml#pt3.

[721] Dmitry Fedorovich Chebotarev, "Hochetsia verit, chto nash ogromny trud prigoditsia potomkam" (I want to believe that our immense work will be useful to our descendants), *Nauchno-Practicheskiy Zhurnal 'Medichni Vsesvit' – Mir Mediziny* (The Scientific-practical Journal "Medical World"), 1(1), 2001 (an interview with Dmitry Chebotarev by Elena Ludwig, http://www.socion.net.ua/med_journal/articles/interview/chebotarev.htm).

[722] V.F. Kuprevich, "Priglashenie k Bessmertiu" (An Invitation to Immortality), *Technika – Molodezhi* (Technology for the Youth), Issue 1, pp. 8-9, 1966, http://technica-molodezhi.ru/docs/Archive/TM_01_1966.

[723] Another idea proposed by the Soviet immortalists at that time was that potential immortality can be achieved by a cyclic mode of gene expression (Eike Libbert, GDR, 1976; Victor Gudoshnikov, 1983; Nikolay Isaev, 1985). In the 1960s-1970s, Prof. Peter Anokhin suggested the possibility of transmitting the information acquired during one's life to the progeny as a way to preserve or immortalize one's personality (P.K. Anokhin, 1962, 1973). (Reviewed in Sergey Ryazanzev, *Tanatologia* (Thanatology), Vostochno-Evropeyskiy Institut Psychoanaliza (The East-European Institute of Psychoanalysis), Saint-Petersburg, 1994, pp. 336, 339-340.)

Other possibilities for achieving human immortality were raised. The possibility of a potentially indefinite "cybernetic" regulation of the human organism via man-machine synergy was discussed as early as the 1970s. All elements of the human organism, including the brain, were deemed replaceable, if only knowing how to maintain their harmonious interactions.

Thus, Academician Nikolay Amosov (1913-2002, since 1960 director of the Biocybernetics Department of the Kiev Institute of Cybernetics), perceived human life in terms of biological and social programs that can be adjusted by an external "intellect system." The system was intended to rectify errors in the program operation. For example, it could remove disturbing/excessive elements and supplement deficient elements, thus enforcing a stable homeostasis in the organism. Indeed, Amosov recognized that even within the regulating "intellect system" errors are inevitable, hence "true" immortality is impossible, but radical life extension would be entirely feasible.

There was even raised the possibility of connecting the human brain to an artificial brain, uploading human personality traits for further backup storage and transferring it into a new artificial or biological body.

Increasing longevity and man-machine synergy were topics of central interest for the Soviet space program. Consequently, Academician Vadim Alexandrovich Trapeznikov (1905-1994, since 1959 Chair of the USSR National Committee on Automatic Regulation) thought it were possible to create learning and reasoning space probes, based on human experiences and reasoning patterns.

(Viktor Pekelis, *Kiberneticheskaya Smes* (Cybernetic mixture), Znanie (Knowledge), Moscow, 1973, pp. 163-164, 170; Nikolay Amosov, "Zapiski iz Budushego" (Notes from the Future), *Nauka I Zhizn* (Science and Life), 9-12, 1965.)

[724] T.A. Belova, C.E. Borisov, "Lev Vladimirovich Komarov – Ucheny-Gerontolog I Obshestvenny Deyatel" (Lev Vladimirovich Komarov – Scientist-Gerontologist and Social Activist), *Vestnik Gerontologicheskogo Obshestva Rossiyskoy Academii Nauk*, 11-12 (33-34), December 2000 (*Bulletin of the Gerontological Society of the Russian Academy of Sciences*, http://gerontology.euro.ru/herald/33-34.html); L.V. Komarov, H. Le Compte, "People can and must live not by decades but by centuries," *Acta Gerontologica et Geriatrica Belgica*, 10(2-3), 87-97, 1972.

[725] The full journal title was: *Rejuvenation, the official journal of the International Association on the Artificial Prolongation of the Human Specific Lifespan.*

[726] L.V. Komarov, H. Le Compte, "People can and must live not by decades but by centuries," *Acta Gerontologica et Geriatrica Belgica*, 10(2-3), 87-97, 1972.

[727] Nikolay Amosov, "Zapiski iz Budushego," *Nauka I Zhizn*, 9-12, 1965 (Notes from the Future, first published in the journal *Nauka I Zhisn* - Science and Life, 1965).

The book was translated into English in 1970: N. Amosoff, *Notes from the Future* (translated by Geroge St. George), Simon and Schuster, New York, 1970 (used here).

[728] Techniques for anabiosis by freezing were investigated in the Soviet Union in the late 1920s by Lev Sergeevich Termen, and in the 1960s by Amosov, and later by Academician Vladimir Vasilievich Kovanov, and others.

(Pavel Vasilievich Sherbakov, "Anabioz kak put k bessmertiu" (Anabiosis as a way to immortality), *Moskovsky Komsomolets*, September 04, 2001, http://www.ordodeus.ru/Ordo_Deus9Anabioz_PkB_Shcherbakov.html.)

[729] N. Amosoff, *Notes from the Future* (translated by Geroge St. George), Simon and Schuster, New York, 1970, pp. 149, 301.

[730] N.M. Amosov, *Experiment po Preodoleniu Starosti*, Stalker, Donezk, 2003 (An Experiment in the Overcoming of Aging); Amosov N.M., *Preodolenie Starosti* (The Overcoming of Aging), Bud Zdorov! (Be Healthy!), Moscow, 1996; *Mysly I Serdze* (Thoughts and Heart), Stalker, Donezk, 1998 (1964) is an autobiographical account of the physician's struggle with death.

Beside Amosov, another popular contemporary author on healthy heart and longevity was Alexander Alexandrovich Mikulin (1895-1985), the foremost Soviet aircraft engine designer (the engines of the "Mikulin AM" series served in the aircrafts MIG, ANT, TU, IL, etc.). Mikulin emphasized "vibration gymnastics" and the electrostatic and electrodynamic equilibrium for the prolonged maintenance of the human engine. (Alexander Alexandrovich Mikulin, *Aktivnoe dolgoletie. Moya Sistema Borby So Starostyu* (Active longevity. My system of fighting aging), Fizkultura I Sport, Moscow, 1977.)

[731] N.M. Amosov, *Experiment po Preodoleniu Starosti*, Stalker, Donezk, 2003 (An Experiment in the Overcoming of Aging), pp. 6-7.

[732] N.M. Amosov, *Experiment po Preodoleniu Starosti*, Stalker, Donezk, 2003 (An Experiment in the Overcoming of Aging), pp. 6-7.

[733] N.M. Amosov, *Experiment po Preodoleniu Starosti*, Stalker, Donezk, 2003 (An Experiment in the Overcoming of Aging), "Oslozhnenie" (Complications), pp. 119-124.

[734] On October 19, 1913, after a series of heart troubles, Metchnikoff wrote:

"Preparing for the end, I am happy that I foresee it courageously, calmly. ... When considering my life, I find that I spent it as "orthobiotically" as possible. It may seem that to die at 69 years and 5 months is premature, but it should not be forgotten that I began to live very early ... all my life I worried, even burned. ... Generally, I am happy that my life was not spent in vain, and I am comforted by my belief that my entire worldview is correct."

And just before his death, at the age of 71, on June 18, 1916 (Metchnikoff died on July 15), he continued to search for life prolonging factors:

"the duration of life is at least to a certain extent related to heredity ... as it is known, one's profession influences the duration of life ... All this indicates that my scientific life has ended and confirms that my orthobiosis indeed reached a desirable limit."

(Elie Metchnikoff, "Vyderzhki iz dnevnika s zapisiami samonabludeniy" (Notes from the self-observation journal), in *I.I. Metchnikoff. Sobranie Sochineniy* (*Collected Works of I.I. Metchnikoff*, Eds. N.N. Anichkov and R.I. Belkin), Academia Medizinskikh Nauk SSSR (The USSR Academy of Medical Sciences), Moscow, 1962, vol. 16, pp. 303, 308-309.)

[735] Yuri Konstantinovich Duplenko, *Starenie - Ocherki Razvitia Problemy* (Aging - The Development of the Problem), Nauka, Leningrad, 1985, p. 58.

[736] V.N. Anisimov, M.V. Soloviev, *Evoluzia Concepcy v Gerontologii* (The Evolution of Concepts in Gerontology), Aesculap, Saint Petersburg, 1999 (reprinted in *The Russian Biomedical Journal*, vol. 3, http://www.medline.ru/public/art/tom3/geront/), "Perspectivy razvitia gerontologii v Rossii" (Perspectives for the development of gerontology in Russia); V.V. Bezrukov, U.K. Duplenko, "Vydayushiysia Ukrainski Fisiolog i Gerontolog V.V. Frolkis" (The outstanding Ukrainian physiologist and gerontologist V.V. Frolkis), *Uspekhi Gerontologii* (Advances in Gerontology), vol. 4, 2000, reprinted in *Rossiyski Biomedizinski Zhurnal – The Russian Biomedical Journal*, vol. 3, 2002, http://www.medline.ru/public/art/tom3/art39.phtml.

[737] The following mechanisms of life-extension (vitauct) were outlined by Frolkis:

"The main mechanisms of vitauct are: DNA reparation, the perfection of gene regulation in the course of evolution, gene redundancy, the growth of the potentialities of the protein-synthesizing system, the great mitotic potential of the cells, an increase in their number, the establishment of a relationship between protein biosynthesis and the state of the cell membrane, the origination of the antioxidant systems, microsomal oxidation, the trophic influences of the nervous system and many connective-tissue elements, and the development of many current adaptive regulatory shifts during aging at all the levels of biological organization (polyploidy, compensatory hyperfunction, the activation of several metabolic cycles and local humoral systems, the growth of sensitivity to several mediators, etc.)"

Frolkis concluded:

"A reasonable mode of life, or Metchnikoff's orthobiosis, naturally helps to prolong life. However, gerontological achievements show that not only the struggle against life-reducing factors, but also the search for effective means of retarding aging and activating vitauct will help to prolong human life."

(Vladimir Frolkis, *Aging and Life-Prolonging Processes*, Springer-Verlag, Wien, 1982, pp. 341-342, 344.)

[738] The ways toward life-extension through organic "regulation," suggested by the works of Alexey Olovnikov and Vladimir Dilman, played a central role in the history of Soviet life-extensionism.

Alexey Matveyevich Olovnikov (b. 1936) of the Moscow Institute of Biochemical Physics of the Russian Academy of Sciences, was the first to propose (c. 1970) the mechanism of cell death as a result of DNA end sequence (telomere) shortening and hypothesized the enzymatic restoration of these end points as a possible way of cell immortalization (the telomerase hypothesis).

Olovnikov did not share the Nobel Prize in medicine of 2009 "for the discovery of how chromosomes are protected by telomeres and the enzyme telomerase," even though he was nominated. The Nobel Prize was awarded to the Americans Elizabeth H. Blackburn, Carol W. Greider, and Jack W. Szostak.

(Regarding Olovnikov's priority, see Michael D. West, *The Immortal Cell. One Scientist's Quest to Solve the Mystery of Human Aging*, Doubleday, NY, 2003, pp. 77-78, and Stephen Hall, *Merchants of Immortality, Chasing the Dream of Human Life Extension*, Houghton Mifflin, Boston, 2003, pp. 42-48.)

In a recent theory, Olovnikov suggested that aging is related not so much to the shortening of telomeres, but to the shortening of what he termed "redumeres" (or "redusomes") – hypothetical short loops in the chromosomal DNA, containing copies of regulatory genes. The redumeres become shortened under the influence of hormone release from the pineal gland. The shortening of the redumeres (also referred to as "chronomeres") in the hypothalamus is said to be particularly crucial for aging, as the hypothalamus is considered by Olovnikov to be the main biological clock of aging.

According to this theory, the peak of the hormone release, shortening the redumeres, occurs at new moon when the gravitational pull is the strongest. Thus the moon acts as a pacemaker of human development and eventually kills us.

Olovnikov envisioned a new type of anti-aging pharmacology, that would "regulate the luno-sensory physiological system of the organism," "use products of the redumeres" and "interrupt signals leading the brain to aging."

(Alexey Olovnikov, interview by Galina Kostina, "Luna, kotoraya nas ubivaet" (The moon that kills us),

Expert Ukraina, Nauka I Technologia (Expert Ukraine, Science and Technology), No. 22(118), June 4, 2007, http://www.expert.ua/articles/12/0/3924/;

A. M. Olovnikov, "The Redusome Hypothesis of Aging and the Control of Biological Time during Individual Development," *Biochemistry* (Moscow), 68(1), 2-33, 2003, Translated from *Biokhimiya*, 68 (1), 7-41, 2003, http://www.chronos.msu.ru/nameindex/olovnikov.html.)

Olovnikov's recent theory of the "redumeres' shortening" in the hypothalamus as the driving "spring in the biological clock of aging" develops the work of yet another prominent Soviet life-extensionist, Prof. Vladimir Mikhailovich Dilman (1925-1994) of the Leningrad Institute of Oncology. According to Dilman's "elevation theory of aging," hormone levels of the hypothalamus play a crucial role in the onset of aging (1958).

In Dilman's theory, aging is said to be due to "the elevation of the sensitivity threshold of the hypothalamus-pituitary complex to regulating signals." Based on this model, Dilman considered aging as a mere disease of hormonal imbalance, and hence a subject to artificial balancing, for example by pineal polypeptides or insulin-cholesterol-lowering drugs.

The significance of the hypothalamus in senescence was later investigated by the Americans W. Donner Denckla and Caleb Finch (c. 1975), who referred to the hypothalamus as a hormonal "aging clock."

(Vladimir Mikhailovich Dilman, "O vozrastnom povyshenii deiatelnosti nekotorikh hypotalamicheskikh centrov," *Trudy Instituta Physiologii imeni I.P. Pavlova*, 7, 326-336, 1958 (On senescent elevation of the activity of some hypothalamic centers, in Works of I.P. Pavlov's Institute of Physiology); V.M. Dilman, *Bolshie biologicheskie chasy (vvedenie v integralnuyu mediziny)*, Znanie, Moscow, 1982 (The great biological clock – introduction into integral medicine, http://lib.ru/NTL/MED/STARENIE/startenie.txt_with-big-pictures.html);

V.M. Dilman, "Aging, Rate of Aging and Cancer: A Search for Preventive Treatment," in *The Aging Clock. The Pineal Gland and Other Pacemakers in the Progression of Aging and Carcinogenesis. Third Stromboli Conference on Aging and Cancer*, Annals of the New York Academy of Sciences, vol. 719, pp. 454-455, 1994;

W. Donner Denckla, "Time to die," *Life Sciences*, 16(1), 31-44, 1975; Caleb Finch, "Neuroendocrinology of aging: A view of an emerging area," *Bioscience*, 25(10), 645-650, 1975.)

[739] Vladimir Frolkis, *Starenie i Uvelichenie Prodolzhitelnosti Zhisni* (Aging and Life-prolongation), Nauka, Leningrad, 1988, pp. 232-233, reprinted at *Starenie.ru*, http://starenie.ru/biblioteka/index.php.

[740] Dmitry Fedorovich Chebotarev, "Hochetsia verit, chto nash ogromny trud prigoditsia potomkam" (I want to believe that our immense work will be useful to our descendants), *Nauchno-Practicheskiy Zhurnal 'Medichni Vsesvit' – Mir Mediziny* (The Scientific-practical Journal "Medical World"), 1(1), 2001 (an interview with Dmitry Chebotarev by Elena Ludwig, http://www.socion.net.ua/med_journal/articles/interview/chebotarev.htm).

[741] N. Amosoff, *Notes from the Future* (translated by Geroge St. George), Simon and Schuster, New York, 1970, pp. 206, 211, 213, 245.

[742] In contrast to American studies, Soviet scientific developments were seen by Amosov and his compatriots as generally benign. Thus, for example, Amosov envisioned an artificial "intellect system" capable of enforcing material homeostasis for life-extension. There were also suggestions (by Academician Vadim Trapeznikov and others) to create learning and reasoning space probes. Such projects were considered feasible and desirable for the Socialist society.

At the same time, the concept of "cyborgization" – or inserting the human brain into an artificial body, suggested by the Americans Manfred Clynes and Nathan Kline in 1960 – was rejected as anti-humanistic. Even though intimately related to Amosov's "intellect systems," the American concept of "cyborgization" was attacked, it seems for the simple reason of being proposed by the Americans. In the words of the physiologist Academician Vasily Vasilievich Parin, "this anti-humane project has nothing to do with the noble purposes of exploring and conquering space. We need the space not for machines but for people."

Furthermore, the American idea of human nuclear transfer (i.e. human cloning) was also vilified as a possible future means of selection based on class and wealth. At the same time, Kuprevich's "viruses of immortality" were seen as compatible with communist progress.

Thus, even closely related (and markedly far-reaching) scientific visions could be discriminated on the sheer basis of political ideology.

(Viktor Pekelis, *Kiberneticheskaya Smes* (Cybernetic mixture), Znanie, Moscow, 1973, pp. 163-164, 170; Nikolay Amosov, *Zapiski iz Budushego*, in *Nauka I Zhizn*, 9-12, 1965 (Notes from the Future, first published in the journal *Nauka I Zhisn* - Science and Life, 1965.)

[743] N. Amosoff, *Notes from the Future* (translated by Geroge St. George), Simon and Schuster, New York, 1970, pp. 262, 273, 274.

[744] Yuri Konstantinovich Duplenko, *Starenie - Ocherki Razvitia Problemy* (Aging - The Development of the Problem), Nauka, Leningrad, 1985, pp. 20-21.

[745] Vladimir Frolkis, *Starenie i Uvelichenie Prodolzhitelnosti Zhisni* (Aging and Life-prolongation), Nauka, Leningrad, 1988, p. 6.

[746] Nathan W. Shock, *The International Association of Gerontology. A Chronicle. 1950 to 1986*, Springer Publishing Company, NY, 1988, pp. 180-185.

[747] "Lysenkovzy podnimayut golovy," *Chronica Tekushih Sobyty. Novosti Samizdata*, Vypusk 6, 28 Fevralia, 1969, Istoriko Prosvetitelskiy Centr Memorial, Archiv ("The Lysenkoists raise their head," in *The Chronicle of Current Events. Samizdat News*, Issue 6, February, 1969, reprinted in the archive of the Historical-Educational Center "Memorial," Moscow, 2005, http://www.memo.ru/HISTORY/).

[748] Medvedev, Z.A., "Aging and longevity: new approaches and new perspectives," *Gerontologist*, 15(3),196-201, 1975; Medvedev, Z.A., "On the immortality of the germ line: genetic and biochemical mechanism," *Mechanisms of Ageing and Development*, 17(4), 331-59, 1981; Medvedev, Z.A., "The past and the future of experimental gerontology," *Archives of Gerontology and Geriatrics*, 9(3), 201-13, 1989.

[749] The high standing members of the party apparatus and of the administrative, intellectual and cultural elite received by far the best medical treatment in the country (perhaps in the world) at the "Fourth Department of the Ministry of Health of the Soviet Union" (Chetvertoe Glavnoe Upravlenie Minzdrava SSSR, a.k.a. "Kremlevka" or the "Kremlin clinic").

The patients of the "Fourth Department" (about 68,000 persons) were the longest living group in the country, with an average life span of about 80 years, while the average for the rest of the country was about 70.

As the Fourth Department's director, Academician Evgeny Ivanovich Chazov (b. 1929) explained the department's "special" success:

"The most important factors are prevention, early diagnosis and regular clinical examinations. These are the three factors of success."

(Evgeny Ivanovich Chazov, "I v 80 let zhizn chudesna" ("Life is wonderful also at 80," interview by Vladimir Kozhemiakin), *Argumenty I Facty* (Arguments and Facts), No. 24, June 10, 2009, http://www.aif.ru/health/article/27402; Vladimir Khatskelevich Khavinson (interview of 20.10.2005), *Nauka Protiv Starenia* (Science against aging), http://www.starenie.ru/news/detail.php?ID=913.)

[750] D.F. Chebotarev, *Perspectivy razvitia isslevodaniy po gerontologii - 1986-1990*, Institut Gerontologii Academii Medizinskikh Nauk SSSR, Kiev, 1986 (Perspectives for developing the research in gerontology – 1986-1990, The Institute of Gerontology of the USSR Academy of Medical Sciences, Kiev, 1986);

D.F. Chebotarev, *Programma "Prodlenie Zhizni" – Vazhny Etap v Razvitii Sovetskoy Gerontologii*, Institut Gerontologii Academii Medizinskikh Nauk SSSR, Kiev, 1986, pp. 47-57 (The "Life Extension" Program – An important stage in the development of Soviet Gerontology);

Yuri Konstantinovich Duplenko, *Starenie - Ocherki Razvitia Problemy* (Aging - The Development of the Problem), Nauka, Leningrad, 1985, pp. 133-134, 159-161.

[751] The Party "Main directives for the development of health care" for the period 1987-2000, also planned to increase the production of pharmaceuticals by a factor of 2 by 1995 until "full satisfaction of all therapeutic needs," and increase the production of medical technology 3.5 times by 1997.
(A.G. Natradze, "Medizinskaya Promyshlennost" (Medical Industry), *Kratkaya Medizinskaya Enziklopedia* (Brief Medical Encyclopedia), Sovetskaya Encyclopedia, Moscow, 1989, vol. 2, p. 142.)

[752] Politichesky doklad CK KPSS. Doklad Generalnogo Sekretaria CK KPSS M.S. Gorbacheva. II. Uskorenie sozialno-economicheskogo razvitia strany – strategicheskiy kurs (M.S. Gorbachev. Political address on behalf of the Central Committee of the Communist Party of the Soviet Union. II. Acceleration of the socio-economic development of the country – the strategic course), in *Materialy XXVII syezda kommunisticheskoy partii Sovetskogo Soyuza* (Materials of the 27th congress of the Communist Party of the Soviet Union, [10 million copies in circulation]), Izdatelstvo Politicheskoy Literatury (Political Literature Press), Moscow, 1986, pp. 24-25.

[753] "Programma Kommunisticheskoy Partii Sovetskogo Soyuza. Novaya Redakzia" (The Program of the Communist Party of the Soviet Union. The New Edition), "Zadachi KPSS po sovershenstvovaniu sozializma i postepennomu perehodu k kommunizmu" (The Tasks of the Communist Party for the perfection of socialism and gradual transition to communism), in *Materialy XXVII syezda kommunisticheskoy partii Sovetskogo Soyuza* (Materials of the 27th congress of the Communist Party of the Soviet Union), Izdatelstvo Politicheskoy Literatury, Moscow, 1986, p. 153.

[754] Michael S. Bernstam and Alvin Rabushka, *From Predation to Prosperity: How to Move from Socialism to Markets*, Ch. 1 "Free and Not So Free to Charge: Income Redistribution and Russia's GDP Contraction, 1992-98, and Recovery, 1999-2007," Hoover Institution, Stanford University, 2008, http://www.hoover.org/publications/books/8261;
Sergey Kapitsa "Nastoyashee i budushee nauki v Rossii" (The past and present of Russian science), *Svobodnaya Mysl* (The Free Thought), 4, 1994; Elena Vodopianova, "Sudby Rossiyskoy Nauki" (The Fate of Russian Science), *Svobodnaya Mysl*, 21, 2005 (reprinted in the archives of the Russian Academy of Sciences, www.ras.ru).
Specifically regarding longevity research and advocacy, the USSR All-Union "Life-extension program," that had included institutes from several republics of the Soviet Union, disintegrated after the disintegration of the Union.

[755] In 1986, in the USSR/Soviet Russia, the average life-expectancy at birth peaked at 69.4-70.13 years, according to different estimates.
(World Bank Country Data, www.worldbank.org/data/; Nation Master Database, http://www.nationmaster.com/graph/hea_lif_exp_at_bir_tot_yea-life-expectancy-birth-total-years&date=1986; The State Statistics Committee of the Russian Federation (Gosudarstvenny Komitet Rossiyskoy Federazii po Statistike) http://www.gks.ru/bgd/regl/B08_16/IssWWW.exe/Stg/html1/02-09.htm; Human Mortality Database, http://www.mortality.org.)
This increase in life-expectancy (compared to 68 years in 1984) may be at least partially attributed to Gorbachev's "Measures to overcome drinking and alcoholism" of 1985, that restricted the production and sales of alcohol, or perhaps some other invigorating effects of Gorbachev's "Acceleration."
The value of 69-70 years was preserved until 1990 (i.e. up until the end of the Soviet Union) and then, by 1994, it declined to 63.85 years (57.42 years for men, and 71.08 years for women, according to the Russian Federation Statistics Committee). From 1995 to 2005 the average life-expectancy fluctuated at about 65 years, and thereafter it increased slighly.
As of 2011, the average life-expectancy in Russia was still about 66.29 years, according to the "CIA World Factbook," thus having fallen from one of the highest to one of the lowest ranks among the countries of the world in terms of life-expectancy. (https://www.cia.gov/library/publications/the-world-factbook/geos/rs.html.)

The estimate for 2011 of the State Statistics Committee of the Russian Federation was more optimistic: ~68.5 years (http://www.gks.ru/free_doc/new_site/population/demo/progn7.htm).

In a slightly different estimate, in April 2013 it was reported that in 2012 the average life expectancy was 69.70 years, which was interpreted as a maximum, yet portending the beginning of a new phase of decline (http://izvestia.ru/news/547880).

In May 2014, it was reported that the life-expectancy in Russia was at a plateau of 69 years. The estimates of the Russian government were more optimistic (above 70 years).
(http://medportal.ru/mednovosti/news/2014/05/20/283life/; http://medportal.ru/mednovosti/news/2014/04/08/053longlive/; http://kommersant.ru/doc/2474265; http://ria.ru/society/20140418/1004478479.html.)

Yet, in any case, the general trend of decline and stagnation in life-expectancy, compared to the Soviet period, is clear.

[756] Among the prominent emigrants were Leonid and Natalia Gavrilov, authors of *The Biology of Life Span* (Biologia Prodolzhitelnosti Zhizni, first published in Russian in 1986, http://imquest.alfaspace.net/index_html.htm). They presently work at the Center on the Demography and Economics of Aging, University of Chicago, http://longevity-science.org/; http://www.spc.uchicago.edu/coa/.)

[757] In 2005, the Russian President Vladimir Putin mandated increased funding for "National Priority Projects" including such programs as Health, Education, Housing and Agriculture. In 2006, the program "Health" was to receive 62.6 billion roubles (~US$2 billion) (*Soviet Pri Presidente Rossiyskoy Federazii po realizazii prioritetnykh nazionalnikh proektov* - The Russian Federation Presidential Council for the Realization of National Priority Projects, http://national.invur.ru/index.php?id=101).

In 2007, the general funding for science was set to increase from 100 billion roubles in 2007 to 170 billion in 2010, including the state program to develop Nanotechnology – with about 130 billion roubles initial planned budget, ~30 billion roubles (~US$1 billion) yearly.
(*Ministerstvo Finansov Rossiyskoy Federazii* - The Russian Federation Ministry of Finance, http://www1.minfin.ru/; The Russian Corporation of Nanotechnologies – Rusnano, http://www.rusnano.com/; *Kommersant*, 109 (3685), 26.06.2007, http://www.kommersant.ru/doc.aspx?DocsID=777978.)

A plan was also set in motion to increase support for the Russian Academy of Sciences - "Plan Reformirovania Rossiyskoy Akademii Nauk" (Governmental Decree for Reforming the Russian Academy of Sciences, December 27, 2007, reported in the online Archive of the Russian Academy of Sciences, http://www.ras.ru/).

Several large governmental programs in support for science were initiated in 2010-2012. These included:

The 2010 plan to invest "90 billion roubles (US$3 billion) into higher education and market-oriented university research over the next decade, on top of an annual university research budget of about 20 billion roubles," programs to set up "national research universities" and a new science city at Skolkovo near Moscow; the 12 billion rouble program to attract world-class expatriate and foreign scientists to Russian universities. (http://www.nature.com/news/2010/100427/full/4641257a.html; http://www.nature.com/news/2010/100616/full/465858a.html.)

In spring 2012, there were announced the 2012-2015 five-billion-rouble (US$165-million) "Global Education" program aimed to train Russian students abroad, and the "mega-grants" program aimed "to increase the amount of grant money distributed by funding agencies from around 15 billion roubles (US$500 million) a year to 25 billion roubles by 2018" (http://www.nature.com/news/putin-promises-science-boost-1.10223; http://www.nature.com/news/go-west-young-russian-1.10633).

On April 24, 2012, there was signed the 1.18 trillion roubles (US$40 billion) "Complex Program of Biotechnology Development in Russia (2012-2020)" including 150 billion roubles (US$5 billion) for biomedicine (http://www.bsbanet.org/en/news/files/Biotechnology-development-programme-2020-Russia-en.php#unique-entry-id-41).

In March 30, 2013, there was ratified the ambitious program entitled "Strategy for the Development of Medical Sciences in the Russian Federation for the Period up to 2025" by the Russian Ministry of Health, that apparently purported to become the Russian equivalent of the American National Institutes of Health programs. (Notably, almost no mention of aging and longevity was made in that program.) (http://www.imbb.soramn.ru/?current=248; https://www.rosminzdrav.ru/docs/; https://www.rosminzdrav.ru/docs/mzsr/orders/1509; http://rosminzdrav.ru/health/62/Strategiya_razvitiya_meditcinskoj_nauki.pdf.)

Yet, notably, in 2013 a reform of the Russian Academy of Sciences was enacted that has been considered by many as restrictive and controversial for the development of the Russian science (http://www.nature.com/news/russian-roulette-1.13315).

[758] According to the poll of 2007 of the Foundation of Public Opinion (Fond Obshestvennogo Mnenia), about 90% of Russian citizens believed in the possibility of some degree of life extension and the necessity of a state program directed toward that purpose (http://bd.fom.ru/report/map/dominant/dominant2007/dom0711/domt0711_4/d071122).

According to polls collected by the Russian Transhumanist Movement (Rossiyskoe Transhumanisticheskoe Dvizhenie – a major Russian grass roots group for radical life-extension), by January 2010, between 15 to 30% of Russian citizens believed in the theoretical possibility of physical immortality. (https://groups.google.com/group/rtd-main/browse_thread/thread/3d5911e3a489140f?hl=ru.)

In January 2012, the Russian aspirations for immortality were further estimated by the poll conducted by the Russian analytical "Levada Center" for the "Russia2045 Strategic Social Initiative," before the international congress "Global Future 2045" held in Moscow in February 2012. The poll included 1600 responders over 18 years old, from 45 regions of Russia, with a standard deviation of 0.035.

According to this poll, 32% of Russian citizens wanted radical life extension for themselves (23% wanted to live "several times longer than people live now," 9% wanted to live "unlimitedly long, as I wish, up to immortality," 64% wanted to live "as long as allotted by nature now). Interestingly, 44% wanted radical life extension for their close ones (29% wished their loved ones to live "several times longer" and 15% wished them "immortality").

Moreover, according to this poll, 45% of Russian citizens "would support a public association for radical life extension."

(http://www.levada.ru/06-02-2012/kak-dolgo-khotyat-zhit-rossiyane.)

In August 2012, a poll was conducted by the Canadian CBC News network, in relation to Russia2045's project of pursuing cybernetic immortality via creating artificial avatars for human consciousness. The question was "If you had the opportunity to live forever - albeit cybernetically - would you do it?" As of August 15, 2012, about 42,000 people responded, about 35.5% said "Yes, at any and all cost" and 17% more said "Yes, but only if I could afford it comfortably." Thus, about 52.5% of responders were in favor of immortality. Note the Russian inspiration for this poll.

(http://www.cbc.ca/news/yourcommunity/2012/07/human-immortality-could-be-possible-by-2045-say-russian-scientists.html.)

[759] Mikhail Batin, Anna Kirik (Eds.), *Bessmertie* (Immortality), Komandor, Yaroslavl, 2007, pp. 18-33.

[760] Since its establishment, the "Russia2045 Strategic Social Initiative" (sometimes referred to in English just as the "2045 Initiative," http://2045.com/) has generated numerous media and public relations appearances. On February 17-20, 2012, it held the international congress "Global Future 2045" in Moscow (http://gf2045.com/). On August 22, 2012, this group proposed the formation of a political party, entitled "Evolution 2045," aimed at achieving "cybernetic immortality" (http://evolution.2045.com/).

[761] In May 2011, "Russia2045" was endorsed by the American actor Steven Seagal (b. 1952), who acted as the movement's representative in the US. In an open letter to the Russian Prime Minister Vladimir Vladimirovich Putin (b. 1952), Seagal wrote the following:

"Dear Vladimir Putin, … Not so long ago, I learned from my Russian friends about "Russia 2045" a social movement, and I decided to join it and represent it in the US. The main technology of this movement, as you probably know, is dedicated to integration of the existing artificial human body technology and the ability to produce a new one. For these purposes apparently there are plans to create a network of Research and Development facilities, with HQ based in Russia. The success of this project will significantly increase people's life span and quality of life. …

For me, as a Buddhist, this is one of the greatest things the world has ever seen. I am proud to be part of this movement. …

Vladimir Vladimirovich, I know you as a prominent world leader and as a person who has already done great things for Russia. … It seems as though now you are placing more emphasis on life expectancy and life extension issues. I can see that you are actively working hard on coming up with solutions that lead Russia into the future confidently and setting an example in the world with the technology that will most certainly make for a better world" (http://www.2045.com/articles/28724.html).

[762] Mikhail Batin, *Lekarstva ot Starosti. Argumenty Nauchnogo Immortalisma* (Cures for Aging. Arguments of Scientific Immortalism), Za Uvelichenie Prodolzhitelnosti Zhizni (Public Movement for the Prolongation of Life), Izdatelstvo I.V.Balabanov, Yaroslval, 2007, p. 51.

[763] M. Batin's Blog *Chelovechestvo+* (Humanity+), "Partia Prodlenia Zhizni" (Life Extensoin Party), July 20, 2012, http://m-batin.livejournal.com/148089.html; Maria Konovalenko, "Russians Create the "Longevity Party,"" Institute for Ethics and Emerging Technologies, July 26, 2012, http://ieet.org/index.php/IEET/more/konovalenko201207261.

In December 2012, the Longevity Party initiative inspired the creation of the International Longevity Alliance (http://longevityalliance.org/the-brussels-summit-of-longevity-activists/).

[764] Rossiyskoe Transhumanisticheskoe Dvizhenie (Russian Transhumanist Movement, http://www.transhumanism-russia.ru/); Fond Vechnaya Molodost (Eternal Youth Foundation, http://www.vechnayamolodost.ru/); Fond Nauka za Prodlenie Zhizni (Science for the Prolongation of Life Foundation, a.k.a. Science Against Aging Foundation, http://www.scienceagainstaging.org/fund.html); Obshestvennoe Dvizhenie Za Uvelichenie Prodolzhitelnosti Zhizni (Public Movement for the Prolongation of Life, led by Mikhail Batin: http://m-batin.livejournal.com/; http://moikompas.ru/compas/zupg; http://web.archive.org/web/20080608232143/http://www.zazhizn.ru/).

Life extension has also been promoted by The Russian Academy of Scientific and Longevity Medicine, affiliated with the World Anti-Aging Academy of Medicine (http://www.waaam.org/member_organizations.php).

[765] Interesting work has been done by Dr. Valery Zuganov. He has discovered an anti-aging means in the gills of Salmon. The life span of the salmon is strictly determined genetically (they die immediately after spawning). However, when their gills are infested by larva of the *Margaritiferidae* mollusk, their life-span is reported to increase by another year. Dr. Zuganov produced extracts from the mollusk larva and tested them on himself, externally and internally, with reported benefits. A commercial rejuvenating/anti-cancer preparation "Leivrus Elixir Arctica+" was subsequently developed by Zuganov, based on an extract from the three-spined stickleback fish and salmon infected by the mollusk larva.

(http://www.arctic-plus.ru/opisanie-preparata.)

Even more unorthodox treatments have been tried. In the 1940s-1950s, Alexey Dorogov developed the ASD (Anti-septic Stimulator of Dorgov) which was produced by heat sublimation of tissues of frogs and cattle. Dorogov's work has been continued to the present by his daughter Olga Dorogova. ASD is now commercially available.

(*Istoria Preparata ASD* (The History of the ASD preparation), Armavirskaya Biofabrika, 2007, http://www.vetlek.ru/articles/?id=57; http://netler.ru/articles/asd.htm; Olga Dorogova, "ASD: Proshloe, Nastoyashee, Budushee" (ASD: Past, Present, Future, interview by E. Pecherskaya), *Pomogi Sebe Sam* (Help Yourself), March 2006, http://www.fpss.ru/gazeta/world/4124/.)

In the 1970s, Victor Saraev developed the FAM ("Physiologically Active Metabolites," a preparation based on tissue extracts from young rats, roosters and rams undergoing severe stress, for example suffucation (the amount of life-saving substances was supposed to increase when the animals were struggling for life). Such preparations continued to be advertised by Saraev until very recently.

(Victor Saraev, interview by Natalia Leskova, "Hotite Zhit 100 Let?" (Do you want to live to 100?), *Vechernaya Moskva*, 103 (24148), 10.06.2005, http://www.vmdaily.ru/article/11975.html; Victor Saraev, interview by Vladimir Gendlin, "Farsh Molodosti" (The mincemeat of youth), *Kommersant*, 4 (208), 03.02.1999, http://www.kommersant.ru/doc.aspx?DocsID=22423.)

Alexey Mesentzev and Victor Kostukov of Novosibirsk worked in the 2000s on yet other forms of organotherapeutic/opotherapeutic preparations, or as they are now commonly termed "biogenic stimulators," "regulatory peptides" or "biologically active substances."

(*Kupit Bessmertie* (Buying immortality), 1TV Documentary, 2008, http://www.1tv.ru/documentary/fi5671/fd200910241350.)

Further widely marketed organotherapeutic nostrums include "cat's claw" and "shark's cartilage."

The therapeutic use of human embryonic stem cells for "anti-aging" and "regenerative medicine" has been proceeding in Ukraine (not so much in Russia) way ahead of the West, apparently forgoing the lengthy Western processes of safety regulation and consensus forming. One of the more famous clinics has been the EmCell in Kiev led by Prof. Alexander Smikodub, who began the therapeutic application of human embryonic stem cells in 1987, and has since treated thousands of patients.

(http://www.emcell.com/; http://news.bbc.co.uk/2/hi/programmes/this world/4438820.stm.)

A very curious rejuvenating measure is that of Vladimir Volkov, "Doctor of the International University of Fundamental Sciences" and founder of the "International Institute for Immortality Ltd" from St. Petersburg. The title of his work is "Photon and Proton Technology of Immortality." His rejuvenation method, as stated by the author, involves "promoting photosynthesis in the human body," and his "elixir of life" is some kind of "a water acid with a clever secret which allows to accelerate the process of integration of a proton into the cell" (apparently there have been no peer-reviewed publications or clinical trials). The actual content of Volkov's remedy is very obscure; apparently it is some form of diluted sulphuric acid. The most curious thing is that Volkov's remedy has been widely commercially available, from Russia to the US. In the US, it has been marketed as "H3O" by Alpha-Omega Labs.

("H3O" http://www.herbhealers.com/store/h3ointro.htm; "Vladimir Volkov: The Immortality is Possible" http://www.doctorvolkov.ru/01eng.html; http://www.doctorvolkov.ru/01.html; "The Immortals speak Russian," Moscow University Alumni Club, 2005, http://www.moscowuniversityclub.ru/home.asp?artId=5994; *Introduction of Dr. Volkov's Key to Longevity and Wellness*, 2007, http://web.archive.org/web/20070515204201/http://www.the7thfire.com/vibrational_healing/anti-aging_longevity/intro_to_Dr_Volkov.html.)

Drinking acids is also highly recommended by Boris Vasilievich Bolotov, especially drinking diluted solutions of hydrochloric acid and fruit vinegars. Bolotov believes that "rejecting the knowledge of magic

[the white, the black and the red] … is illogical and irrational. … From the point of view of healing the society and the individual, red [astral body] magic is preferable." Bolotov is a member of the newly set up "Russian-folk Academy of Sciences" (Russkaya Academia Nauk) and is referred to as an "Academician."
Quite a few such new "Public Academies of Sciences" have sprung up in New Russia (over 90 as of 2008, including the Academy of Juvenology, Academy of Energo-Informational Sciences, and Academy of Medico-Technical Sciences) – to be distinguished from the Russian Academy of Sciences (Rossiyskaya Academia Nauk) or the Russian Academy of Medical Sciences (Rossiyskaya Academia Medizinskikh Nauk).
(http://www.boris-bolotov.org.ua/; Boris Bolotov, *Shagi K Dolgoletiu* (The Steps Toward Longevity), Piter, Moscow, 2006, pp. 23-33, 77-79, 105; Mikhail Akhmanov, "Iskusitelniy Titul" (the tempting title), Open Economy, Moscow, September 9, 2008, http://www.opec.ru/docs.aspx?id=385&ob_no=87636.)
The "acid-drinking" school of rejuvenation stands in a peculiar opposition to the "alkali-drinking" school, going back to the original suggestions of Olga Lepeshinskaya in the 1940s-1950s, recommending solutions of sodium bicarbonate (baking soda) and alkalinized mineral waters. Alkalinized and acidic waters have even been termed in Russia the "water of life" (*zhivaya voda*) and the "water of death" (*mertvaya voda*) respectively.
(*Zhivaya Voda* (The water of life), 2010, article collection, http://paralife.narod.ru/health/voda/voda_contents.htm.)
In the West, notable proponents of the "alkalinization" school have been Theodore Baroody and Ray Kurzweil.
(Ray Kurzweil and Terry Grossman, *Fantastic Voyage. Live Long Enough to Live Forever*, Plume, NY, 2005, "The importance of being alkaline," pp. 45-48; Theodore Baroody, *Alkalize or Die: Superior Health Through Proper Alkaline-Acid Balance*, Eclectic Press, Waynesville NC, 1999.)
Additional types of the "water of life" are the frozen and thawed or otherwise "structured" waters:
E.g. Sergey Nikitin's "Cryodynamics – Rejuvenation by Ice" (http://kriodinamika.ru/); "thawed water," alongside some dozen other types of "healing waters" with different mineral components, ranging from gold and silver to silicon (http://www.vodoobmen.ru/talayawater.html); and water from various "water structuring filters" (http://rezonator.net/water/; http://www.filtry.com.ua/index.php/main/index/0/306). The filters' manufacturers often refer to the work of Stanislav Zenin, the world's first doctorate on "water memory" (http://bio.fizteh.ru/student/diff_articles/memory_water.html).
A series of life-extensionist propositions emerged in Russia based on Peter Garyaev's "wave genetics," attempting to directionally/specifically affect genetic traits by radiation at a specific frequency (to be distinguished from non-directional induction of mutations by radiation) or using irradiation at "healing frequencies" assumed to be characteristic of healthy organs.
(Peter Petrovich Garyaev, *Volnovoy Genom* (The wave genome), Obshestvennaya Polza, Moscow, 1993; Peter Petrovich Garyaev, *Volnovoy Geneticheskiy Kod* (The wave genetic code), Izdatcenter, Moscow, 1997, http://rusnauka.narod.ru/lib/author/garyaev_p_p/1/.)
A multitude of "electromagnetic" ("bioresonance," "phototherapeutic," "electrotherapeutic," "magnetizing" or "ionizing") devices for rejuvenation and general health improvement have been developed and marketed in Russia, e.g.:
"Air-ionizing lamp of Chizhevsky" http://www.ion.moris.ru/; "Water ionizer" http://www.medicaltech.ru/water.php; Trans-cranial brain electro-stimulator "Medapton" http://medprom.ru/medprom/mpp_0002780; Photo-therapeutic "Bioptron" http://bioptron.mk.ua/; "Biorythm harmonizer" http://zovz.ru/angel_z/index.html; "Self-Controlled Energy-Neuro-Adaptive Regulator" http://www.scenar-revenko.ru/; "Bio-Electro-Magnetic-Energy-Regulator" http://bemer3k.ru/index.php/kak-rabotaet-bemer.html; "Electromagnetic field therapy device" http://www.detaplus.ru/ru/fordoctors/emt_t.php; "Electromagnetic resonator" http://galactic.org.ua/pr-nep/Fiz-b3.htm; "Electromagnetic therapy by fixed frequency" http://imedis.ru/pages/99; Djuna

Davitashvili's "Bio-Corrector Djuna 1" for "effective treatment of all diseases" http://www.djuna.ru/, and many more.

Several groups and authors focused on "mental affirmation" as a path toward radical longevity.

This is the way of the Institute of Human Self-Repair (Institut Samovosstanovlenia Cheloveka) established by Mirzakarim Norbekov and Yuri Hvan, as well as Norbekov's Academy of Longevity (Academia Dolgoletia).

(http://www.anoisch.com/; http://akademiyadolgoletiya.com/.)

Another method is the "Divine, Healing, Rejuvenating Affirmations" developed by Georgi Nikolaevich Sytin.

(G.N. Sytin, *Bozhestvennie Iszelyayushie Omolazhivayushie Nastroi* (Divine, Healing, Rejuvenating Affirmations), 2007, http://www.sytin-gn.ru/; http://www.bibliotekar.ru/418/; http://mugs.by.ru/.)

Yet another is Sergey Konovalov's "Information-Energetic Teaching."

(http://spiral-ssk.ru/about_doctor/; S.S. Konovalov, *Kniga, kotoraya lechit. Preodolenie starenia. Informazionno-energeticheskoe uchenie* (The book that heals. Overcoming Aging. The Information-Energetic Teaching), Olma-Press, Moscow, 2002.)

More rejuvenative "affirmative"/"psycho-energetic" techniques have been taught at Andrey Levshinov's Academy (Academia Andreya Levshinova).

(http://www.levshinov.de/index-2.html; Andrey Levshinov, *Sistemy Ozdorovlenia Zemli Russkoy* (Health Improvement Systems of the Russian Land), Olma Press, Moscow, 2002.)

Perhaps the most extreme case of evangelizing the power of "affirmation" is that of Grigori Grabovoy, founder of the "Teaching of Salvation and Universal Harmony," who claimed to be able to resurrect the dead.

(http://www.grigori-grabovoi.ru/index2.htm; http://www.drugg.kiev.ua/; http://www.drugg.ru/.)

Of course, some researchers of aging and longevity would not wish to be mentioned anywhere near some of the proponents of some of the methods of "mental affirmation" or "electromagnetic" or "pharmaceutical/organotherapeutic" rejuvenation. Still these projects do promote the idea of radical life extension on the massive popular level, and therefore their inclusion is warranted.

[766] As an example of university-based anti-aging research, Alexey Moskalev of Syktyvkar State University has been working on pharmacological methods of activating genes responsible for protection against stress and correlated with longevity (http://aging-genes.livejournal.com/).

An additional private initiative for anti-aging regenerative medicine (mainly focusing on cell therapy) has been the Institute of Biology of Aging in Moscow (http://bioaging.ru/). One of the leading figures at the institute has been Alexander Zhavoronkov, who has also led several other anti-aging research projects, such as Aging Research Portfolio (http://agingportfolio.org/).

Yet more private companies and research centers are involved in a search for anti-aging drugs, such as Quantum Pharmaceuticals (http://q-pharm.com/) and Tartis (http://www.rusventure.ru/ru/innovative_projects/detail.php?ID=11790).

[767] Some research institutions have included: National Gerontological Center http://www.ngcrussia.org/ ; http://www.ngcrussia.org/ngc.htm ; The Russian Institute of Gerontology http://www.niigeront.org/ ; Evolutionary Cytogerontology Sector of Moscow State University http://gerontology.bio.msu.ru/about_eng.htm and others.

A major research direction in those institutions has been to determine markers of biological and physiological aging, from the molecular and cellular to the organ and organism level, and to test potential geroprotective (anti-aging) treatments (drugs and regimens) according to their influence on those markers.

(Vyacheslav N. Krutko, Vitaly I. Dontsov, Tatiana M. Smirnova, "Teoria, Metody i Algoritmy Diagnostiki Starenia" (Theories, methods and algorithms for diagnosing aging), *Trudy ISA RAN* (Proceedings of the Institute of Systems Analysis of the Russian Academy of Sciences), 13, 105-143, 2005 (Russian); Alexander

N. Khokhlov, "Does aging need its own program, or is the program of development quite sufficient for it? Stationary cell cultures as a tool to search for anti-aging factors," *Current Aging Science*, 6, 14-20, 2013.)

[768] Vladimir Skulachev, "Starenie – atavism, kotory sleduet preodolet" (Aging is an atavism which must be overcome), Moscow University Seminar, 21.02.2005, http://azfor.narod.ru/geront/Skulachev.htm; Vladimir Skulachev (interview by Sergey Leskov), "Chelovek budet Zhit do 800 Let i Umirat ot Neschastnich Sluchaev" (Man will live to 800 years and die of accidents), *Izvestia*, 28.11.2003, http://nauka.izvestia.ru/analysis/article37639.html.

[769] Skulachev, V.P., et al, "An attempt to prevent senescence: a mitochondrial approach," *Biochimica et Biophysica Acta*, 1787(5), 437-61, 2009.

[770] E A Liberman, V P Topaly, L M Tsofina, A A Jasaitis, V P Skulachev, "Mechanism of Coupling of Oxidative Phosphorylation and the Membrane Potential of Mitochondria," *Nature*, 222, 1076-1078, 14 June 1969;

Biomedical Project "Skulachev Ions" http://en.skq-project.ru/doc/index.php?ID=81.

Skulachev's collaborator in that study, Prof. Efim Arsentievich Liberman (1925-2011) of the Moscow Institute for Information Transmission Problems, would later propose "Cytomolecular Computing" (1972) and later still the science of "Haimatics" (2003) from the Hebrew "Haim" (Life) "unifying biology, physics and mathematics" (http://efim.liberman.ru/index.html).

Efim Liberman died in September 2011 in Jerusalem.

[771] According to Skulachev's recent report in the Russian journal *Biochimia* (where Skulachev is the editor in chief):

"Very low (nano- and subnanomolar) concentrations of 10-(6'-plastoquinonyl) decyltriphenylphosphonium (SkQ1) were found to prolong lifespan of a fungus (Podospora anserina), a crustacean (Ceriodaphnia affinis), an insect (Drosophila melanogaster), and a mammal (mouse). In the latter case, median lifespan is doubled if animals live in a non-sterile vivarium."

(Anisimov VN, ..., Skulachev VP, "Mitochondria-targeted plastoquinone derivatives as tools to interrupt execution of the aging program. 5. SkQ1 prolongs lifespan and prevents development of traits of senescence," *Biochemistry* (Moscow), 73(12), 1329-42, 2008.)

In another report, the claims are more moderate, but still very promising:

"In the fungus Podospora anserina, the crustacean Ceriodaphnia affinis, Drosophila, and mice, SkQ1 prolonged lifespan, being especially effective at early and middle stages of aging. In mammals, the effect of SkQs on aging was accompanied by inhibition of development of such age-related diseases and traits as cataract, retinopathy, glaucoma, balding, canities, osteoporosis, involution of the thymus, hypothermia, torpor, peroxidation of lipids and proteins, etc. SkQ1 manifested a strong therapeutic action on some already pronounced retinopathies, in particular, congenital retinal dysplasia."

(Skulachev, V.P., et al, "An attempt to prevent senescence: a mitochondrial approach," *Biochimica et Biophysica Acta*, 1787(5), 437-61, 2009.)

As of 2014, the results of wide clinical trials in humans have been still prospective, except for some ophthalmologic applications.

[772] As of September 2008, the billionaire Oleg Deripaska (b. 1968, the Director General of "Russian Aluminium") could no longer fund Skulachev's project due to Deripaska's financial difficulties.

In 2009, the project support passed over to the "Rostock Investment Group" led by Alexander Chikunov (b. 1963, in 2004-2008 a director of The Unified Energy System of Russia – the country's major electric power holding company). The Rostock Investment Group has "a particular interest in prolonging an individual's life span."

Moreover, in February 2010, funding for Skulachev's project was approved by the Russian Corporation of Nanotechnologies – RUSNANO (Rusnano would invest ~710 million rubles, within the general project budget of 1.83 billion roubles, or ~$60M).

(http://www.grostock.ru/en/about/; http://www.grostock.ru/i/?tag=skulachev; http://www.grostock.ru/en/i/; http://www.rusnano.com/Post.aspx/Show/25184.)

[773] http://www.rusnanonet.ru/nanoindustry/medicine/goods/74137/; http://www.visomitin.ru/; http://skq-project.ru/news/view.php?ID=76.
In April 2014, clinical trials of the SkQ-based eye-drops Visomitin started in the US http://en.skq-project.ru/news/view.php?ID=109

[774] Programma Kommunisticheskoy Partii Rossiyskoy Federazii (The Program of the Communist Party of the Russian Federation, http://kprf.ru/party/program/).

[775] I.V. Vishev, *Na Puti k Prakticheskomu Bessmertiu* (On the way to practical immortality), Moscow, 2002, pp. 151, 153, 159.

[776] Boris Georgievich Rezhabek (interview), "Nanoroboty, Nanobacterii i Bessmertie" (Nanorobots, Nanobacteria and Immortality), *Vzgliad Zdorovie* (Health View, April 19, 2010, http://health.vz.ru/columns/2010/4/19/521.html); Boris Georgievich Rezhabek, "Aritmologia Mifov. Systematika Religiy (Arythmology of Myths. Systematics of Religions), *Rossia, Tzerkov, Apocalypsis* (Russia, Church, Apocalypse, http://www.apocalyptism.ru/Rejabec-myphs.htm);
Muzey-Biblioteka Nikolaya Fedorovicha Fedorova (N.F. Fedorov's Museum-Library, Moscow, http://www.nffedorov.ru/mbnff/index.html); Centr Russkogo Kosmizma g. Borovska (The Center of Russian Cosmism in Borovsk, Kaluga District, http://www.admobl.kaluga.ru/main/news/pressa/detail.php?ID=69751).
Regular "Fedorov's readings" have been held in the town of Borovsk, the residence place of both Nikolay Fedorov (1829-1903) and his student, the founder of Russian space exploration, Konstantin Eduardovich Tsiolkovsky (1857-1935).

[777] Vladimir Khavinson, interview by Olga Sagan, "Ludi dozhdia" (Rain men), in Mikhail Batin, Anna Kirik (Eds.), *Bessmertie* (Immortality), Komandor, Yaroslavl, 2007, pp. 38-51.

[778] Recently Khavinson claimed:
"A number of small peptides have been isolated from different organs and tissues and their analogues (di-, tri-, tetrapeptides) were synthesized from the amino acids. It was shown that long-term treatment with some peptide preparations increased mean life span by 20-40%, slow down the age-related changes in the biomarkers of aging and suppressed development of spontaneous and induced by chemical or radiation carcinogens tumorigenesis in rodents."
(Anisimov, V.N., Khavinson, V.Kh., "Peptide bioregulation of aging: results and prospects," *Biogerontology*, 11(2),139-49, 2010.)
In another recent report, the claims for the animal studies were more modest and conditional:
"Injection of epithalone (synthetic Ala-Glu-Asp-Gly peptide; subcutaneously 0.1 microg/rat 5 times a week from the age of 4 months until natural death) virtually did not change the mean lifespan of male rats, but was associated with a significant (p<0.05) normalization of population aging rate and hence, time of mortality rate doubling in groups exposed to natural or constant illumination."
(Vinogradova IA, Bukalev AV, Zabezhinski MA, Semenchenko AV, Khavinson VKh, Anisimov VN, "Geroprotective effect of ala-glu-asp-gly peptide in male rats exposed to different illumination regimens," *Bulletin of Experimental Biology and Medicine*,145(4), 472-7, 2008.)
Some 15 years ago Khavinson and Anisimov made similar claims regarding the prospects for their pineal gland preparation "epithalamin":
"Thus, during the last two decades a wide spectrum of the biological activity of the low-molecular-weight pineal preparation epithalamin was observed. … Long-term exposure to epithalamin was followed by an increase in the life span of mice and rats, by a slowdown of the age-related switching-off of reproductive function and immune function decline, and by the inhibition of spontaneous carcinogenesis and that induced by chemicals or ionizing radiation."

(V.N. Anisimov, V.Kh. Khavinson, V.G. Morozov, "Twenty years of study on effects of pineal peptide preparation: Epithalamin in experimental gerontology and oncology," in *The Aging Clock. The Pineal Gland and Other Pacemakers in the Progression of Aging and Carcinogenesis. Third Stromboli Conference on Aging and Cancer*, Annals of the New York Academy of Sciences, vol. 719, pp. 483-495, 1994.)

Either now, 15 or 35 years ago, despite some highly reassuring experimental reports (e.g. reporting a notable decrease in human mortality with pineal gland and thymus peptides), wide "clinical use of peptide bioregulators for prevention of premature aging" in humans is still in the realm of "prospects."

(Khavinson VKh, Anisimov VN, "35-year experience in research of peptide regulation of aging," *Advances in Gerontology*, 22(1), 11-23, 2009; Vladimir Kh. Khavinson, Vyacheslav G. Morozov, "Peptides of pineal gland and thymus prolong human life," *Neuroendocrinology Letters*, 24, 234-240, 2003.)

[779] Benjamin Franklin, "Letter to Joseph Priestley, Feb. 8, 1780," in *Papers of Benjamin Franklin*, The Packard Humanities Institute, Yale, 2005, vol. 31, p. 455, http://franklinpapers.org/franklin/framedVolumes.jsp.

[780] Joseph T. Freeman, "The History of Geriatrics," *Annals of Medical History*, 10, 324-335, 1938.

[781] Charles Sedgwick Minot of Harvard Medical School attributed senescence to an increase in the amount of cytoplasm in relation to the nucleus, and the corresponding differentiation of the cytoplasm, whereas undifferentiated cells were seen as potentially immortal. This was apparently one of the first morphogenetic cellular theories of aging (1908, 1913).

Minot's view that aging is due to the gradual relative increase of the cytoplasm was directly opposite to the earlier view of the German researcher Richard Hertwig (1889) who believed that, during development, it is rather the relative size of the nucleus that increases.

Thorburn Brailsford Robertson (1884-1930, from 1905 to 1919 working at the University of California, Berkeley) synthesized the two views and suggested that, in embryonic development, nuclear substance increases, and in the later period of growth and differentiation it decreases. To explain this phenomenon, T.B. Robertson, Wolfgang Ostwald and F.F. Blackman (1908-1913) suggested that the processes of growth in general follow the laws of autocatalysis. As summarized by Charles Manning Child of the Department of Zoology, University of Chicago, in *Senescence and Rejuvenescence* (1915):

"An autocatalytic reaction is one in which one or more of the products of the reaction act as catalyzers and so increase the velocity of the reaction. ... This remains true until products of the reaction begin to decrease its velocity... *If growth is a process of this kind, the rate of growth must increase up to a certain maximum as growth proceeds and then, after maintaining this maximum for a longer or shorter time, must decrease*" (Child, 1915, pp. 446-448, emphasis added).

There were two kinds of autocatalytic growth posited: "one the autostatic in which the autocatalyst is decreasing in amount, the other the autokinetic in which it is increasing in amount." Accordingly, the changes in the cytoplasm to nucleus ratio were assumed to be due to the different stages of autocatalysis: "The early period of embryonic development in which the nuclear substance is increasing and the yolk decreasing is of the autostatic type, while the later period of cytoplasmic growth and differentiation is of the autokinetic type" (Child, 1915, p. 454). Child correlated these two stages with the states of rejuvenation and aging correspondingly.

Later, Aldred Scott Warthin, professor of pathology at the University of Michigan, within the framework of the "involution" theory (summarized in 1929), attributed aging to the "gradually weakening energy-charge set in action by the moment of fertilization," while immortality of the germ plasm "rests upon the renewal of this energy charge from generation to generation."

These and other pioneers of the cellular morphogenetic theory of aging and development – H.S. Jennings, T.H. Montgomery, E.G. Conklin, and others – agreed in the fundamental premise that senility derives from body differentiation and, insofar as differentiation is only a specific state of living organisms, senile death is not a necessary outcome of life processes.

(Charles Manning Child, *Senescence and Rejuvenescence*, The University of Chicago Press, Chicago, Illinois, 1915; Thorburn Brailsford Robertson, *The Chemical Basis of Growth and Senescence*, Lippincott, Philadelphia, 1923; Aldred S. Warthin, *Old Age, the Major Involution; the Physiology and Pathology of the Ageing Process*, Hoeber, New York, 1929; Alex Comfort, *The Biology of Senescence*, Butler & Tanner, London, 1956, pp. 5,7.)

[782] According to C.M. Child, in lower animals, a great extent of rejuvenation and life-extension occurs following starvation and reduction of body weight. In higher animals and man a similar rejuvenative effect occurs, but to a limited extent:

"rejuvenescence by reduction is limited in the higher animals, for reduction in this forms soon ends in death, so that there is at present no immediate prospect of our being able to rejuvenate ourselves to any great degree, or to retard senescence or delay death to any great extent by any such means."

(Charles Manning Child, *Senescence and Rejuvenescence*, Chicago University Press, Chicago, 1915, pp. 299, 310.)

[783] Jacques Loeb, "Natural death and the duration of life," *The Scientific Monthly*, 12, 578-585, 1919 (quoted); Jacques Loeb and John Howard Northrop, "On the influence of food and temperature upon the duration of life," *Journal of Biological Chemistry*, 32, 103-121, 1917; Jacques Loeb and John Howard Northrop, "What determines the duration of life in metazoan," *Proceedings of the National Academy of Sciences USA*, 3, 382-86, 1917; Jacques Loeb and John Howard Northrop, "Is there a temperature coefficient for the duration of life," *Proceedings of the National Academy of Sciences USA*, 2, 456-457, 1916.

Jacques Loeb also argued for the theoretical possibility of potential immortality in Jacques Loeb, *Artificial Parthenogenesis and Fertilization*, The University of Chicago Press, Chicago, Illinois, 1913, p. 1; Jacques Loeb, *The Dynamics of Living Matter*, The Columbia University Press, NY, 1906, pp. 222-223; Jacques Loeb, *The Mechanistic Conception of Life: Biological Essays*, The University of Chicago Press, Chicago, 1912, p. 210.

[784] Ignatz Leo Nascher, "Geriatrics," *New York Medical Journal*, 90, 358-359, 1909.

[785] Ignatz Leo Nascher, *Geriatrics, the Diseases of Old Age and their Treatment. Including Physiological Old Age, Home and Institutional Care, and Medicolegal Relations*, P. Blakiston's Son & Co, Philadelphia, 1914, p. v.

[786] Ignatz Leo Nascher, *Geriatrics, the Diseases of Old Age and their Treatment. Including Physiological Old Age, Home and Institutional Care, and Medicolegal Relations* (Second Edition), Kegan Paul, London, 1919, p. vi.

[787] Ignatz Leo Nascher, *Geriatrics, the Diseases of Old Age and their Treatment. Including Physiological Old Age, Home and Institutional Care, and Medicolegal Relations*, P. Blakiston's Son & Co, Philadelphia, 1914, p. v-vi.

[788] The chronic under funding appears to be as old as life extensionism itself, going back to the origins of Chinese Taoist immortalism. Thus, one of the earliest Chinese alchemists, Ko Hung (Ge Hong, 283-343 CE), wrote:

"I suffer from poverty and lack of resources and strength; I have met with much misfortune. There is nobody at all to whom I can turn for help. The lanes of travel being cut, the ingredients of the medicines are unobtainable. The result is that I have never been able to compound these medicines I am recommending. When I tell people today that I know how to make gold and silver, while I personally remain cold and hungry, how do I differ from the seller of medicine for lameness who is himself unable to walk? It is simply impossible to get people to believe you. Nevertheless, even though the situation may contain some unsatisfactory elements, it is not to be rejected in its entirety. Accordingly, I am carefully committing these things to writing because I wish to enable future lovers of the extraordinary and esteemers of truth, through reading my writings, to consummate their desires to investigate God [the Immortal Tao]."

(*Alchemy, Medicine, Religion in the China of A.D. 320: The Nei P'ien of Ko Hung (Pao-p'u tzu)*, translated by James R. Ware, Massachusetts Institute of Technology Press, Cambridge MA, 1966, Ch. 16 "The Yellow and the White," p. 262.)

And the very first known author of Chinese alchemy, the man reputed for the invention of gun-powder, Wei Boyang (Wei Po-Yang), said c. 142 CE:

"I have abandoned the worldly route and forsaken my home to come here. I should be ashamed to return if I could not attain the *hsien* [immortality]."

(*Lieh Hsien Ch'üan chuan, Complete Biographies of the Immortals*, Tenney L. Davis, 1932, p. 214.)
"O, the sages of old!" Wei Boyang said "They held in their bosoms the elements of profundity and truth. ... Their sympathy for those of posterity, who might have a liking for the attainment of the Tao (Way), led them to explain the writings of old with words and illustrations. They couched their ideas in the names of stones and in vague language so that only some branches, as it were, were in view and the roots were securely hidden. Those who had access to the discourses wasted their own lives over them. The same path of misery was followed by one generation after another with the same failure. If an official, his career was cut short; if a farmer, his farm was cluttered with weeds; if a merchant, his trade was abandoned; if an ambitious scholar, his family became destitute – in the vain attempt. These grieve me and have prompted the present writing. Although concise and simple, yet it embraces the essential points. The appropriate quantities [and processes] are put down for instruction together with confusing statements. However, the wise man will be able to profit by it by using his own judgement."
(Wei Po-Yang, "Ts'an T'ung Ch'i" ["The akinness of the three", i.e. of the alchemical processes, the Book of Changes, and the Taoist doctrines], Chapter LXII, pp. 257-258, in *An ancient Chinese treatise on alchemy entitled Ts'an T'ung C'hi, Written by Wei Po-Yang about 142 A.D., Now Translated from the Chinese into English by Lu-Ch'iang Wu, With an Introduction and Notes by Tenney L. Davis*, Massachusetts Institute of Technology, Cambridge, Mass., *Isis*, Vol. 18, No. 2, Oct. 1932, pp. 210-289.)

[789] Stewart H. Holbrook, *The Golden Age of Quackery*, Macmillan Co., NY, 1959.

[790] Morris Fishbein, *The Medical Follies. An Analysis of the Foibles of Some Healing Cults, Including Osteopathy, Homeopathy, Chiropractic, and the Electronic Reactions of Abrams, with Essays on the Antivivisectionists, Health Legislation, Physical Culture, Birth Control, and Rejuvenation*, Boni and Liveright, New York, 1925; Morris Fishbein, *The New Medical Follies. An Encyclopedia of Cultism and Quackery in These United States, with Essays on the Cult of Beauty, The Craze for Reduction, Rejuvenation, Eclecticism, Bread and Dietary Fads, Physical Therapy, and a Forecast as to the Physician of the Future*, Horace Liveright, New York, 1927.

[791] Morris Fishbein, *The New Medical Follies*, 1927, pp. 140-141.

[792] One can but remember the story of the King of France Louis XV (1710-1774), who was spared blood-letting in childhood and hence remained alive, and at the end of his life refused smallpox inoculation which caused his death. Thus, therapeutic nihilism both saved and killed him.
(*What Life Was Like During The Age Of Reason: France, AD 1660-1800*, Time-Life Books, Alexandria, 1999, pp. 26, 37, 49; Ian Glynn, Jenifer Glynn, *The Life and Death of Smallpox*, Cambridge University Press, NY, 2004, p. 78; *Outline of Great Books*, Edited by Sir J. A. Hammerton, Wise & Co, New York, 1937, vol. 1, "The Later Years of King Louis XV" (from Voltaire), http://www.publicbookshelf.com/public_html/Outline_of_Great_Books_Volume_I/; Voltaire, "De la Mort de Louis XV et de la fatalité" (On the death of Louis XV and on fatality, 1774), in *Oeuvres Completes de Voltaire* (Complete works of Voltaire), vol. 33, De l'Imprimerie de la Société Littéraire-Typographique, 1785, pp. 113-123.)

[793] Lukutate was originally disseminated in Germany, by Wilhelm Hiller of Hanover, and became widely promulgated in the US in the 1920s by the Lukutate Corporation of America, established in New York, later renamed The Durian Company.
Lukutate (later renamed "Dur-Inda"), was advertised as to be made of "mineral rich" and "vitamin rich fruits of India." According to the vague company allegations, it primarily contained such fruits as durian, and perhaps also salabmisri, amalaki, papaya, sawo, tamarind, arhat fruit, etc. But the exact composition was never entirely ascertained.
As the advertisements claimed, Lukutate contained a "revitalizing vitamin" and had a "rejuvenating action." The "native tribes" who partook of it "live practically disease-free lives way beyond the century mark and yet retain the functions and appearance of much younger people."

(American Medical Association, Bureau of Investigation, "Lukutate: Another Rejuvenating Nostrum, from Orient via Germany," *Journal of the American Medical Association*, 94 (4), 281-282, 1930; American Medical Association Bureau of Investigation, "Dur-Inda or Lukutate Redivivus. Physicians and Others are Solicited to Buy Stock in a Nostrum Concern," *Journal of the American Medical Association*, 98 (17), 1493, 1932.)

[794] The schools included:

Aerotherapy; Alereos system (using baths and massages and condemning drug therapy as poisoning); Astral healing (medical advice based on horoscopes); Autohemic therapy (using a solution made of the patient's own blood); Autology ("a system of stereotyped hygienic and dietetic advice, ... another preachment of Ecclesiastes' urge for moderation in all things"); Auto-Science (using sera and psychological suggestion); Autotherapy (based on the homeopathic principle *similia similibus curantur* – "like cures like," employing, for example, filtered sputum to cure tuberculosis);

Bio-Dynamo-Chromatic-Diagnosis and Therapy (initiated in 1878 by Edwin Babbitt and developed in the 1910s-1920s by George Starr White, using "color rays, magnetic forces, and oxygen and medicated vapor inhalations"); Biological Blood-Washing (a form of naturopathy);

Chiropractic (founded in 1895 by the Canadian-born American Daniel David Palmer, whose theory emphasized adjusting "spiritual energy" or life-force streams in the body via spinal adjustments); Chirothesians (a mixture of religious and chiropractic healing); Christian philosophers; Christian Science (established by Mary Baker Eddy in 1821, emphasizing mental/faith healing and condemning intrusive therapy); Chromopathy (using George White's color healing system); Chromotherapy (another form of color healing); Couéism (originated in France by Émile Coué in 1882, and later widely spread in the US, a "self mastery by conscious auto-suggestion");

Defensive diet (dietary advice, opposing salt, fruits and starches); Divine Science (another form of faith healing); Dowieism (established in 1900 by Alexander Dowie, healing was done by the laying on of hands);

Eclecticism (discarding mineral drugs and emphasizing the use of plant remedies); Eddyism (the same as Christian Science); Electric Light Diagnosis and Therapy; Electro-Homeopathy; Electro-Naprapathy; Electronic Therapy (developed by Albert Abrams since 1910, using "the dynamizer" or the "oscilloclast" to treat patients by appropriate electronic "vibrations"); Electrotherapy; Emmanuel Movement (moral indoctrination for healing); Erosionism (the system of James A.M. McLean of California, booming in the 1910s-1920s, employing a combination of physical, metaphysical and spiritual healing);

Faith Healing;

Geotherapy (using "little pads of earth"); Gland therapy (also known as opotherapy, the main stem of contemporary rejuvenation attempts, based on the use of endocrine gland extracts and sometimes gland transplantations, see the following sections of this chapter);

Homeopathy (started in 1810 by Samuel Christian Hahnemann in Germany);

Irido-Diagnosis (eye diagnosis, diagnosing the entire body according to patterns of the iris);

Jewish Science (professed by Rabbis Alfred Moses and Morris Lichtenstein since the 1900s, the Jewish counterpart of Christian Science);

Kneipp Cure (a form of Naturopathy, founded by Sebastian Kneipp, Germany, in 1886, emphasizing hydrotherapy);

Leonic Healing (mystical healing, involving horoscopic services); Limpio Comerology (clean eating);

Charles B. McFerrin's system (dietary advice, asserting that all disease is due to malnutrition); James A.M. McLean's system (a combination of physical, metaphysical and spiritual healing); Mysticism;

Naprapathy (an offshoot of chiropractic, believing that "nerve function is impaired by the contraction of the connective tissue," defying drugs and surgery, and teaching that contracted ligaments can be cured by adjusting the spine vertebra); Naturopathy (using fasting, hot and cold baths, relaxation, exercise, health resorts, magnetic healing, mud packs, milk, vegetarian and fruitarian diets, etc.); Naturology (another name for Naturopathy); New Thought (a combination of systems, including Christian Science and Jewish Science, a successor of the "Transcendental Movement" of the 1830s, expounded in 1906 by William Walker Atkinson in *Thought Vibration*, using mental affirmation, Yoga, relaxation, visualization, music, vitalic breathing, etc.);

Orificial therapy (application of pressure and relaxation on the various openings of the body); Osteopathy; Pathiatry (similar to Osteopathy, using spinal adjustment, traction and massage, administered by the patient himself); Poropathy (no medicine should be taken through the stomach and no surgery is allowed, treatment is done by lotions administered through pores); Practo-therapy (intestinal irrigation);

Quartz therapy (the use of crystals and Ernst Krohmayer's ultraviolet lamp);

Frederick L. Rawson's School (denying everything the person "does not want");

Sanatology (acidosis and toxicosis are said to be the major causes of disease); Scientific Christianity (another form of faith healing, the movement's magazine, titled *Unity*, circulated in 185,000 copies); Somapathy (adjusting nerves emerging from the spinal chord, using heat and cold); Spectrochromism (wearing clothing according to a personal color); Spiritualism (diseases, coming from evil spirits, were exorcised); Spondylotherapy (the same as Abrams' "Electronic Therapy");

Telathermy (electric and sound vibrations); Theophonism (the same as McLean's system, combining physical therapy and mystical healing); Theosophy (founded by Helena Blavatsky in 1875 in New York, healing through telepathy, spiritualism, relaxation, prayer); Tropo-therapy (dietary advice);

Zodiac Therapy (an offspring of "Aero-Therapy-Astral Healers," dispensing medical advice based on horoscopes); Zonotherapy (dividing the body into zones and applying pressure).

(Morris Fishbein, *The New Medical Follies*, 1927, pp. 16-64.)

Additional highly popular systems included gymnastic techniques, such as the "Swedish medical gymnastics" (established by the Swedish practitioner Per Henrik Ling around 1813, and popularized in the US after the publication of the English edition of Anders Wide's *Hand-book of Medical and Orthopedic Gymnastics*, 1899); and the technique developed by Frederick Matthias Alexander (the "kinesthetic" or "Alexander Technique," publicized around 1918, attempting to overcome limitations of thought and improve posture and movement).

Popular life-extending dietary systems included that of Horace Fletcher (moderation in diet and thorough mastication, publicized in 1896) and a host of "natural" and vegetarian diets, such as those of Sylvester Graham (founding *The Graham Journal of Health and Longevity* in 1837), John Harvey Kellogg (coining the term "health foods" in 1892), Otto Carque (1904), Paul Bragg (1929) and others.

[795] Morris Fishbein, *The New Medical Follies*, 1927, pp. 61-63.

Of special note in this list are the more obscure terms, such as Prana-Yama, Zoism, Biopneuma and Vitapathy (vitalistic teachings, akin to animal magnetism, attempting to draw in and manipulate healing energies), and Tripsis (trituration, a system of drug purification and dilution).

Sacrology, Sacrognomy and Nervauric Therapeutics were basically local massage systems, first proposed by Joseph Rodes Buchanan (1814-1899), professor of cerebral physiology at the Eclectic College of Medicine in Cincinnati, Ohio, in his books *Outlines on the Neurological System of Anthropology* (1854) and *Therapeutic Sacrognomy* (1884). Spondylotherapy was presumed to heal by electronic "vibrations" (Albert Abrams' "Electronic Therapy"). Chirothesia was a form of chiropractic.

The other terms seem to be still quite familiar in alternative/complementary medicine.

[796] John Harvey Kellogg, *Plain Facts for Old and Young*, Segner & Condit, Burlington, Iowa, 1881 (first published in 1879, reprinted in *Project Gutenberg*).

According to the 1910 edition (by Kellogg's own Good Health Publishing Company, Battle Creek, Michigan), 300,000 copies were "sold by subscription."

The relation between controlling sex and increasing longevity was strongly emphasized (Kellogg, 1910, p. 143).

[797] Francis Galton (1822-1911):

"A healthy and long-lived family may be defined by the patent facts of ages at death, and number and ages of living relatives, within the degrees mentioned above, all of which can be verified and attested. A knowledge of the existence of longevity in the family would testify to the stamina of the candidate, and be an important addition to the knowledge of his present health in forecasting the probability of his performing a large measure of experienced work."
(Francis Galton, *Inquiries into Human Faculty and its Development*, Second Edition, J. M. Dent & Co. (Everyman), New York, 1907, p. 212, reprinted in the online Galton archives at http://galton.org/, originally published in 1883 by Macmillan, London.)
Charles Benedict Davenport (1866-1944):
"When Dr. O.W. [Oliver Wendell] Holmes was asked for specifications for a long life he advised, in effect, first to select longlived grandparents. This advice accords with a widespread opinion that longevity is inheritable. … while longevity is not a biological unit of inheritance a person belonging to a long lived family is a better "risk" for a life insurance company than a person belonging to a short lived family."
(Charles Benedict Davenport, *Heredity in Relation to Eugenics*, Henry Hold and Company, New York, 1911, pp. 47-48.)

[798] Michael Vincent O'Shea and John Harvey Kellogg, *Making the Most of Life*, The Macmillan Company, New York, 1915, pp. 40-41, 47.

[799] J.H. Kellogg was an active member of the American Temperance Movement, having been appointed in 1878 president of the American Health and Temperance Association.
The movement existed since the late 18th century (the American Temperance Society was established in 1826), and triumphed in the all-state alcohol prohibition of 1920-1933.

[800] John Harvey Kellogg, *The Living Temple*, Good Health Publishing Company, Battle Creek, Michigan, 1903, p. 374.

[801] Michael Vincent O'Shea and John Harvey Kellogg, *Making the Most of Life*, The Macmillan Company, New York, 1915, Ch. "Living long and well," p. 258.

[802] John Harvey Kellogg, *Autointoxication or Intestinal Toxemia*, The Modern Medicine Publishing Co., Battle Creek, Michigan, 1918, p. 44.

[803] John Harvey Kellogg, *Autointoxication or Intestinal Toxemia*, The Modern Medicine Publishing Co., Battle Creek, Michigan, 1918, p. 307.

[804] *The New York Times*, December 16, 1943, Obituary, "J. H. Kellogg Dies; Health Expert, 91." As the obituary stated:
"Battle Creek, Mich., Dec. 15 – Dr. John Harvey Kellogg, surgeon, health authority, developer of the Battle Creek Sanitarium and founder of the food business which later became the W. K. Kellogg Company, died here last night at the age of 91, nine years short of the century goal which he had set for himself."

[805] John Harvey Kellogg, *Autointoxication or Intestinal Toxemia*, The Modern Medicine Publishing Co., Battle Creek, Michigan, 1918, p. 307.

[806] *The Tym Club, Otherwise known as the Two Hundred Year Club, founded upon the Plan of Universal Life as Evolved in the Two Hundred Year Method*, Ralston Company, Washington, 1905. A longer version was first published in 1889.
Interestingly, the 1905 edition title bears no author's name, and only on page 122 it is revealed that the author is Edmund Shaftesbury, alias for Webster Edgerly.

[807] *The Tym Club, Otherwise known as the Two Hundred Year Club, founded upon the Plan of Universal Life as Evolved in the Two Hundred Year Method*, Ralston Company, Washington, 1905, pp. 12, 17, 100, 108-109, 118.

[808] *The Tym Club, Otherwise known as the Two Hundred Year Club, founded upon the Plan of Universal Life as Evolved in the Two Hundred Year Method*, Ralston Company, Washington, 1905, pp. 123-127.

[809] Quoted in Alfred Armstrong, "The Writings of Webster Edgerly" 03/02/2008, *Odd Books*, http://oddbooks.co.uk/edgerly/review.html.

[810] Kenneth Thompson, "The idea of longevity in early California," *Bulletin of the New York Academy of Medicine*, 51(7), 805-816, 1975.
According to Thompson, the life-prolonging properties commonly mentioned were:
"The mildness of the winters, the abundance of sunshine, the lack of thunderstorms, the low incidence of cloudiness, the moderate and highly seasonal precipitation, the fogginess of the coasts, and the dryness and 'purity' of the atmosphere" and/or "mild, equable marine air – of a cool medium, with a constancy of meteorological conditions, and a dry, warm soil, with the greatest possible facility for ventilation, and outdoor exercise."

[811] Marion Thrasher, *Long Life in California*, M.A. Donohue & Company, Chicago, 1915, pp. 9, 10, 110, 30.

[812] Marion Thrasher, *Long Life in California*, M.A. Donohue & Company, Chicago, 1915, p. 112.

[813] Peter Charles Remondino, "Climate in its Relation to Longevity," *Occident Medical Times*, 4, 278, 1890; Peter Charles Remondino, *The Mediterranean shores of America. Southern California: its climatic, physical, and meteorological conditions*, F.A. Davis, Philadelphia, 1892, pp. 1-6.

[814] Marion Thrasher, *Long Life in California*, M.A. Donohue & Company, Chicago, 1915, pp. 19, 34, 68.

[815] Brown-Séquard, C.E. "Des effets produits chez l'homme par des injections sous-cutanées d'un liquide retiré des testicules frais de cobaye et de chien" (Effects in man of subcutaneous injections of freshly prepared liquid from guinea pig and dog testes), *Comptes Rendus des Séances de la Société de Biologie*, Série 9, 1, 415-419, 1889, reprinted and translated in Geraldine M. Emerson (Ed.), *Benchmark Papers in Human Physiology, Vol. 11, Aging*, Dowden, Hutchinson and Ross, Stroudsburg PA, 1977, pp. 68-76.

[816] C.E. Brown-Séquard, "Du Role Physiologique et Thérapeutique d'un suc extrait de testicules d'animaux d'après nombre de faits observes chez l'homme" (The physiological and therapeutic role of animal testicular extract based on several experiments in man), *Archives de Physiologie Normale et Pathologique*, 1 (5), 739-746, 1889, reprinted in *Benchmark Papers in Human Physiology, Vol. 11*, 1977, p. 100.

[817] This account was given by the New York physician James Joseph Walsh (1865-1942) in *Old-Time Makers of Medicine. The Story of The Students And Teachers of the Sciences Related to Medicine During the Middle Ages*, Fordham University Press, New York, 1911, pp. 67-68, footnote 1.
Walsh refers here to the late 1890s-early 1900s. By the time of the writing (1911), medical science, according to him, had outgrown that phase:
"We ourselves, however, within a little more than a decade, had a phase of opotherapy – how much less absurd it seems under that high-sounding Greek term – that was apparently very learned in its scientific aspects yet quite as absurd as many phases of old-time therapy, as we look at it."

[818] Stewart H. Holbrook, *The Golden Age of Quackery*, Macmillan Co., NY, 1959, pp. 69-75.

[819] Benno Romeis, *Altern und Verjüngung. Eine Kritische Darstellung der Endokrinen "Verjüngungsmethoden", Ihrer Theoretischen Grundlagen und der Bisher Erzielten Erfolge*, Verlag von Curt Kabitzsch, Leipzig, 1931 (Aging and Rejuvenation. A Critical Presentation of Endocrine "Rejuvenation Methods," Their Theoretical Foundations and Up-to-Date Successes), pp. 1809-1873.

[820] Benno Romeis, *Altern und Verjüngung*, 1931, p. 1839.

[821] Harry C. Sharp, "Vasectomy as a means of preventing procreation in defectives," *Journal of the American Medical Association*, 53, 1897-1902, December 4, 1909.

[822] "America was First in Gland Grafting: Dr. Lydston's Account of His Own and Other Experiments Antedating Voronoff's. Operated Upon Himself And Was First In the World to Record the Results of Such a Procedure," *New York Times*, 15 August, 1920; David Hamilton, *The Monkey Gland Affair*, Chatto and Windus, London, 1986, pp. 34-35.
Victor Darwin Lespinasse (1878-1946), a Chicago urologist, was also credited for the performance of the first neuroendoscopic procedure – the intracranial intraventricular endoscopy in 1913. He was apparently

named to fit the world's highest elite; his son was named Victor King Lespinasse. (Grant JA, "Victor Darwin Lespinasse: a biographical sketch," *Neurosurgery*, 39, 1232-1233, 1996; Lespinasse VD, "Transplantation of the testicle," *Journal of the American Medical Association*, 61, 1869-1870, 1913.)

Max Thorek (1880-1960), though born in Hungary, also worked in Chicago. (Max Thorek, *A Surgeon's World*, Lippincott, Philadelphia, 1943; Max Thorek, "The present position of testicle transplantation in surgical practice: A preliminary report," *Endocrinology*, 6, 771-775, 1922.)

George Frank Lydston (1858-1923) was another Chicago urologist. (Dirk Schultheiss and Rainer M. Engel, "G. Frank Lydston (1858–1923) revisited: androgen therapy by testicular implantation in the early twentieth century," *World Journal of Urology*, 21(5), 356-363, 2003; Lydston GF, "Sex gland implantation," *Journal of the American Medical Association*, 66, 1540-1543, 1916.

Levi Jay Hammond and Howard Anderson Sutton were Philadelphia surgeons. Hammond apparently was also one of the pioneers of kidney transplantation in humans. (Susan E. Lederer, *Flesh and Blood: Organ Transplantation and Blood Transfusion in Twentieth-Century America*, Oxford University Press, Oxford, 2008, p. 30; Levi J. Hammond and Howard A. Sutton, "An Abstract Report of a Case of Transplantation of a Testicle," *International Clinics*, 22, 150-154, 1912; "Dr. Hammond Gives Patient New Kidney," *New York Times*, 14 November, 1911.)

[823] L. L. Stanley, "An Analysis of one thousand testicular substance implantations," *Endocrinology*, 6 (6), 787–794, 1922.

[824] Serge Voronoff, *Rejuvenation by Grafting*, translated by Fred F. Imianitoff, George Allen and Unwin Ltd, London, 1925, 1931, p. 34.

[825] John K. Hutchens, "Notes on the Late Dr. John R. Brinkley, Whom Radio Raised to a Certain Fame," *New York Times*, June 07, 1942; R. Alton Lee, *The Bizarre Careers of John R. Brinkley*, University Press of Kentucky, Lexington, 2002.

[826] See David Hamilton, *The Monkey Gland Affair*, Chatto and Windus, London, 1986.

[827] American Eugenics Society, *Eugenics Quarterly*, Volumes 3-4, 1956, p. 112.
Further on American Eugenics (yet with little or no reference to the subject of longevity), see:
Edwin Black, *War Against the Weak: Eugenics and America's Campaign to Create a Master Race*, Thunder's Mouth Press, New York, 2004 (2003); Daniel J. Kevles, *In the Name of Eugenics: Genetics and the Uses of Human Heredity*, Harvard University Press, Cambridge MA, 1997 (first published in 1986); Robert Peel (Ed.), *Essays in the History of Eugenics*, The Galton Institute, London, 1998.

It is of necessary note that surgical sterilization that originated in these experiments is now practiced by millions of people worldwide. (http://www.nichd.nih.gov/publications/pubs/vasectomy_safety.cfm; http://apps.who.int/rhl/fertility/contraception/lxhcom/en/index.html.)

Of additional note are the findings that, contrary to the original expectations of the rejuvenators, vasectomy does not appear to have any great effect on longevity. According to Giovannucci E, Tosteson TD, Speizer FE, Vessey MP, Colditz GA, "A long-term study of mortality in men who have undergone vasectomy," *New England Journal of Medicine*, 326(21),1392-8, 1992, after vasectomy, mortality somewhat decreases in the short term, but in the longer term remains approximately the same.

[828] Samuel Hopkins Adams, "The Great American Fraud," *Collier's Weekly*, October 7, 1905, Reprinted by Robert W. McCoy (Ed., 1927-2010), *Museum Of Quackery*, http://www.museumofquackery.com/ephemera/oct7-01.htm.

[829] US Food and Drug Administration. US Department of Health and Human Services. "Legislation," http://www.fda.gov/RegulatoryInformation/Legislation/default.htm.

[830] Abraham Flexner, *Medical Education in the United States and Canada: A Report to the Carnegie Foundation for the Advancement of Teaching*, Bulletin No. 4, The Carnegie Foundation for the Advancement of Teaching, New York City, 1910, pp. 158, 163, 166.

[831] Mark D. Hiatt, Christopher G. Stockton, "The Impact of the Flexner Report on the Fate of Medical Schools in North America After 1909," *Journal of American Physicians and Surgeons*, 8(2), 37-40, 2003.

[832] American Medical Association, Bureau of Investigation, "Lukutate: Another Rejuvenating Nostrum, from Orient via Germany," *Journal of the American Medical Association*, 94 (4), 281-282, 1930; American Medical Association, Bureau of Investigation, "Dur-Inda or Lukutate Redivivus. Physicians and Others are Solicited to Buy Stock in a Nostrum Concern," *Journal of the American Medical Association*, 98 (17), 1493, 1932; American Medical Association, Bureau of Investigation, "Misbranded 'Patent Medicines,'" and "The Clayton E. Wheeler Fraud. Another Piece of Mail-Order Quackery Debarred from the Mails," *Journal of the American Medical Association*, 103 (14), 1084-1085, 1934; American Medical Association, Bureau of Investigation, "Another "Rejuvenator" Fraud. Ekater-Enger Treatment Banned from the Mails," *Journal of the American Medical Association*, 114 (3), 271-272, 1940.

[833] Ignatz Leo Nascher, *Geriatrics, the Diseases of Old Age and their Treatment. Including Physiological Old Age, Home and Institutional Care, and Medicolegal Relations*, P. Blakiston's Son & Co, Philadelphia, 1914, p. v-vi.

[834] S.E. Chaillé, "Longevity," *New Orleans Medical and Surgical Journal*, 16, 417-424, 1859, p. 419; S.P. Cutler, M.D., Holly Springs, Mississippi, "Physiology and Chemistry of Old Age," *New Orleans Journal of Medicine*, 23(1), 96-104, 1870; M.F. Legrand, "Peut-On Reculer le Bornes de la Vie Humaine?" (Can we reverse the limitations of human life?), *L'Union Médicale*, 13 (5), 65-71, 1859.
The theory was presented in 1858 by Édouard Robin to the French Academy of Sciences in the address "Sur le causes de la vieillesse et de la mort sénile" (On causes of aging and senile death). Reprinted in Geraldine M. Emerson (Ed.), *Benchmark Papers in Human Physiology, Vol. 11, Aging*, Dowden, Hutchinson and Ross, Stroudsburg PA, 1977, pp. 28-47.

[835] Gerald J. Gruman, "C.A. Stephens (1844-1931) - Popular Author and Prophet of Gerontology," *The New England Journal of Medicine*, 254, 658-660, 1956.

[836] Charles Asbury Stephens, *Natural Salvation (Salvation by Science): Immortal Life on the Earth From the Growth of Knowledge and the Development of the Human Brain*, The Laboratory, Norway Lake, Maine, 1910 (first published in 1903), Part I. Ch. IV. "Immortal Life. Practical Methods," pp. 33, 109-140.
The following research programs for life-extension were outlined by Stephens:
(1) "Experiments in combining, or concentrating the nervous energy of several persons in one current, for transmission to another person, with a view to producing curative effects." This basically implied electro-magnetic therapy or manipulation of a physicalized "vital force."
(2) "Care and control of the organism during sleep." The program entailed sleep enhancement, for example, by specific electric stimulation or oxygenation, to strengthen the regenerative ability of rest.
(3) "Extended observations of unicellular life, in respect to nutrition." The goal was to develop "chemically perfect" nutrition "doing no inherent damage to the protoplasmic structure or constitution" whereby "a cell may gain potential" without a "steady draught on cellular energy." According to Stephens, "it is along the line of improved food, as well as regeneration of the somatic cell, that we must look for happier and longer life."
(4) Adequate blood supply to the tissues, invigoration of the blood flow.
(5) Thorough immunization.
(6) Stimulation and maintenance of neuro-humoral and energy homeostasis, chiefly by supplementation, "to rectify, fortify and give tonicity, or stimulus, to the tissue cells, themselves, by inoculations with the products of certain glands of the animal organism, or by reagents artificially produced of the nature of those gland products, or by 'serums' from the blood of animals which have been thus inoculated, fortified and rectified."
Cell therapy (cell implants) are suggested as a form of supplementation: "even a few living cells from one of these glands can be transplanted from the body of a healthy animal to the human organism, and made to live there, the necessary rectification of the organic harmony and tonicity will be produced."

(7) "The control of brain cell energy, meaning by the word control the art of generating it, or of deriving it from the lower animals, and applying it at will, to the human organism, for purposes of vital reinforcement and the cure of disease."

That would involve material and energy supplementation to enhance the brain function, its regulatory power and vitality, which may in turn improve the organism's overall resistance to illness:

"Disease from invasions of microbic life may be promptly suppressed by re-inforcement of the sufferer's vital powers by adjuvant cell energy from without, thereby heightening the aura of life, that restorative *vis medicatrix naturae* which has ever to be summoned by the physician."

(8) Investigation of multi-cellular reproduction, to ascertain "whether germ cell elements are reduplicated or generated *ab initio*, … or whether, as many believe, the cells of the parent organism are depleted by an irreplaceable out-go of such elements, and that old age is the result of the drain set upon the organic cells by the cells of the reproductive apparatus of the body."

Knowing the economy of reproduction is, accordingly, highly important for planning "the husbandry and renewal of the cell life of the body" and may have practical implications for individual longevity and for "marriage and the large communal life of the nation in the future."

(9) Understanding "how deeply and how radically, the life of the brain and nervous system controls and maintains the life of the other tissues of the organism" may help to regulate and stabilize the organism.

(10) Regeneration and protection of the skin can help to defend the organism against external damage.

Furthermore, a number of Stephens' research strategies refer to scientific organization and networking: the division of labor; bibliographic research, accumulating "useful data, scattered about the world"; field studies of longevity; dedicated publications; sustained scientific discussions; sharing individual life-style experiments; distant communications.

[837] E. Maupas, "Recherches expérimentales sur la multiplication des infusoires ciliés" (Experimental studies on the multiplication of ciliated infusoria), *Archives de zoologie expérimentale et générale*, 6 (2), 165-277, 1888.

[838] Charles Asbury Stephens, *Natural Salvation (Salvation by Science): Immortal Life on the Earth From the Growth of Knowledge and the Development of the Human Brain*, The Laboratory, Norway Lake, Maine, 1910 (first published in 1903), hereafter referred to as "Stephens, 1910."

[839] Stephens, 1910, pp. 102-106.

[840] Stephens, 1910, pp. 99-105.

[841] Stephens, 1910, p. 106.

[842] Stephens, 1910, p. 47.

[843] Stephens, 1910, p. 66.

[844] *Maimonides' [1135-1204] Commentary on the Mishna, Tractate Sanhedrin*, Chapter 10, translated by Fred Rosner, Sepher-Hermon Press, New York, 1981.

[845] Stephens, 1910, p. 101.

[846] Stephens, 1910, pp. 109-110.

[847] Stephens, 1910, p. 117.

[848] Stephens, 1910, p. 117.

[849] George W. Corner, *A History of the Rockefeller Institute. 1901-1953. Origins and Growth*, The Rockefeller Institute Press, NY, 1964.

[850] The concept of potential immortality, of death not being an indispensable part of life, was widely accepted, and furnished a strong ideological basis for the early 20th century life-extensionists.

Thus, a crucial dichotomy in Metchnikoff's writings was between potential immortality and natural death. Metchnikoff believed that the so called natural death (death that is unavoidable, inherent, programmed into a living being) rarely occurs in nature. Most deaths are "violent," caused by pernicious yet combatable agents.

Moreover, the living matter has a potential for immortality. Thus, most unicellular organisms are immortal, each dividing into two offspring identical to the parent. Even though after several divisions the cells often show signs of degradation, after conjugation they become completely rejuvenated and ready for a new series of divisions.

In nutrient-rich media, cells do not even show signs of degeneration and seem to be able to divide indefinitely. In addition to conjugation, their rejuvenation could be achieved by other kinds of stimulation (mechanical, chemical or electrical).

Thus, Metchnikoff suggested that many types of cells and unicellular organisms appear to be immortal.

Likewise, in higher organisms (including man), there are "immortal elements." Certain tissues and cells (such as the liver) are capable of regeneration and differentiation throughout the entire life span.

But most importantly, "sex elements," spermatozoa and eggs, are potentially immortal. These immortal elements were also termed "germ cells," "germinal plasma" and "hereditary substance" (now we would call them "genes").

Similarly to protozoa, these immortal elements in humans appeared to exhibit responsive behavior and bear "psychological" traits, therefore Metchnikoff saw them as carriers of a material "immortal soul."

(Elie Metchnikoff, *Etudy o Prirode Cheloveka* (Etudes on the Nature of Man), The USSR Academy of Sciences Press, Moscow, 1961 (1903), XI "Vvedenie v nauchnoe izuchenie smerti" (Introduction to the Scientific Study of Death), pp. 214-232.)

Freud, in *Beyond the Pleasure Principle*, also used these findings as a biological basis for the "Life Drive" concept (part VI).

Both Metchnikoff and Freud drew heavily on August Weismann's seminal works *On the Duration of Life* (*Über die Dauer des Lebens*, Gustav Fischer, Jena, 1882) and *On Life and Death* (*Über Leben und Tod*, Gustav Fischer, Jena, 1892).

Weismann suggested the immortality of germinal plasma as opposed to the mortal "soma" (body). Weismann posited that death is not inherent to life, but is rather a later biological acquisition needed for evolutionary development (in order to dispose of unfit, inferior organisms).

Agreeing that death is not a necessary component of life, Metchnikoff refuted the latter tenet of Weismann about death's evolutionary necessity and benefits (especially for human beings). He also disagreed with the fundamental distinction between germinal and somatic living substance. The entire plant or animal body may in principle have a potential for immortality or drastically increased life-span.

The concept of potential immortality gained evidence from several lines of research.

First of all, immortality was almost universally acknowledged for unicellular organisms. Several researchers, indeed, asserted that the division of protozoa is a form of death, as the mother organism disappears in the division (the view professed by Jürgen Harms in "Verjüngung und Verlängerung des Lebens" (Rejuvenation and Prolongation of Life), in *Leben, Altern, Tod* (Life, Aging, Death), Senckenberg-Bücher II. Hugo Bermühler Verlag, Berlin, 1926, pp. 40-56).

Some researchers observed senescent deterioration following a prolonged series of cell divisions. Among the earliest supporters of the "cell division" limit was Émile Maupas, the French "prince of protozoologists" who in 1888 described the senescent changes due to cell propagation: the decrease in cell size, the degeneration of the mouth opening and, interestingly enough, the degeneration of the cell nucleus.

(Émile Maupas, "Recherches experimentales sur la multiplication des infusoires cilies" (Experimental studies of the multiplication of ciliated infusoria), *Archives de zoologie expérimentale et générale*, 6(2), 165-277, 1888.)

The latter observation was among the first to suggest a crucial role of the cell nucleus in senescence. The Belgian mathematician, philosopher and biologist Joseph Delboeuf (1891) and the Russian zoologist Vladimir Mikhailovich Shimkevich (1893) believed that senescence results from the damage accumulating in the nucleus due to "the imperfection of cell division." On the other hand, continuing the idea of the German zoologist Otto Bütschli (1882) that senescence is the result of exhaustion of some mysterious "life enzyme," the Russian physiologist Ivan Romanovich Tarkhanov suggested in 1891 that aging is due to the depletion of the nuclear substance.

(Quoted in Yuri Konstantinovich Duplenko, *Starenie - Ocherki Razvitia Problemy* (Aging - The Development of the Problem), Nauka, Leningrad, 1985, "Kletochnie mechanismy starenia – pervonachalnie idei i ich sovremennoe razvitie" (Cellular mechanisms of aging: initial ideas and their modern development), pp. 59-70.)

Many researchers were, however, unimpressed by the concept of "disappearance in division" and strongly affirmed the inherent immortality of unicellular organisms (protozoa), that is, their unlimited continuity in essentially the same form.

Moreover, the deterioration of cell lines with the passage of time, if it at all occurred, could be countered by fairly simple physical and chemical stimuli, and their potential for immortality would seem to be restored.

The unlimited division of protozoa (without conjugation) was demonstrated in the unicellular ciliate *Paramecium* by Lorande Loss Woodruff of Yale (1911). (Woodruff L.L., "Two Thousand generations of Paramecium," *Archiv für Protistenkunde*, 21(5), 263-266, 1911; Woodruff L.L., "The Life Cycle of Paramecium when Subjected to a Varied Environment," *American Naturalist*, 42 (500), 520-526, 1908.)

Max Hartmann of Berlin (1926), by periodically reducing the size of an amoeba (just amputating a part of it from time to time), was able to prevent the propagation of the animal and thus to maintain it practically indefinitely, evidencing the potential immortality of a protozoan *individual*.

By inhibiting the division of the ciliate *Ionium pectorale* using concentrated medium salts, Hartmann was able to grow organisms several times their normal size and several times their normal life-span. Such large, long-lived organisms were, however, ultimately doomed to death (presumably due to a decreased cell surface to volume ratio, hindering their nourishment and the removal of waste products). In contrast, the organisms whose size was periodically reduced were made immortal.

(Max Hartmann, "Tod, Fortpflanzung und Verjüngung" (Death, Reproduction and Rejuvenation), in *Leben, Altern, Tod* (Life, Aging, Death), Senckenberg-Bücher II. Hugo Bermühler Verlag, Berlin, 1926, pp. 57-68.)

Furthermore, the American zoologist Charles Manning Child posited in *Senescence and Rejuvenescence* (1915) the essential analogy between the division of protozoa and the asexual reproduction by fission in more complex, multi-cellular organisms, such as rotifers, hydra, planaria or asexually reproducing plants. These multi-cellular organisms thus were also viewed as potentially immortal.

(Charles Manning Child, *Senescence and Rejuvenescence*, Chicago University Press, Chicago, 1915, pp. 293-314.)

August Weismann reserved potential immortality in higher vertebrates to germ cells only, while somatic cells were believed to be rendered mortal by the virtue of their differentiation. Accordingly, the immortality of the species could only be maintained through the union of sex cells.

The American Aldred Scott Warthin formulated this condition for immortality (1929):

"age, the major involution, is due primarily to the gradually weakening *energy-charge* set in action by the moment of fertilization… The immorality of the germ plasm rests upon the renewal of this energy charge from generation to generation."

(Warthin A.S., *Old Age, The Major Involution; The Physiology and Pathology of the Ageing Process*, Hoeber, NY, 1929, quoted in Alexander Comfort, *The Biology of Senescence*, Butler & Tanner, London, 1956, p. 5.)

A similar condition for immortality, requiring the periodical replenishment of materials and energy for continuous cell maintenance, was set earlier by the American Herbert Spencer Jennings (Jennings H.S. "Age, death and conjugation in the light of work on lower organisms," *Popular Science Monthly*, 80, 563-577, June 1912).

According to Jennings' condition, immortality is ensured by "nuclear reorganization" – either by amphimixis (fusion of male and female gametes), endomixis (periodic nuclear reorganization as in ciliate protozoa), or conjugation (fusion or exchange of nuclear material, without cell fusion, as in bacteria).

In Jennings' view, the transfer of nuclear materials between the cells enables their repair and immortalization through a process of supplementation. Jennings wrote:

"Nature has employed the method of keeping on hand a reserve stock of a material essential to life; by replacing at intervals the worn out material with this reserve, the animals are kept in a state of perpetual vigor. ... It is not mating with another individual that avoids [death]; but replacement of the worn [nuclear] material by a reserve."

(Jennings H.S., *Life and Death, Heredity and Evolution in Unicellular Organisms*, Gorham Press, Boston, 1920, p. 233.)

A number of researchers went further, asserting that potential immortality is a fundamental property of all living matter, not restricted either to the nucleus or to germ cells, but also present in non-nucleated cells and even in somatic cells.

Thus, Woodruff established the unlimited ability to divide for several strains of ciliates, *Urostyla grandis* and *Paramecium caudatum*, without a micronucleus.

Even more strikingly, several lines of evidence pointed to the potential immortality of nucleated, somatic cells, such as comprise the body of higher animals and man. The latter phenomena were mainly studied at the Rockefeller Institute for Medical Research, where the "somatic immortalist" school was perhaps the strongest in the world (see below).

[851] "Died: John Davison Rockefeller, 97, of old age," *Newsweek*, 9, 1937; "John D. Rockefeller Dies at 97 in His Florida Home," *The New York Times*, May 24, 1937.

[852] Naturally, the Institute researchers (such as Pearl and Carrel) extolled the Rockefeller Foundation, on which their livelihood depended (e.g. Raymnd Pearl, *The Biology of Death*, J.B. Lippincott Company, Philadelphia, 1922, p. 238; Alexis Carrel, *Man, The Unknown*, Burns & Oates, London, 1961 (1935), p. 226).

[853] "Jacques Loeb" in Isaac Landman (Ed.), *The Universal Jewish Encyclopedia*, vol. 7, Ktav Publishing House, NY, 1969; Fred Skolnik (Ed.), *Encyclopedia Judaica*, Second Edition, vol. 13, Thomson Gale, NY, 2007.

[854] Jacques Loeb, *The Dynamics of Living Matter*, The Columbia University Press, NY, 1906, pp. 222-223; Raymond Pearl, *The Biology of Death*, J.B. Lippincott Company, Philadelphia, 1922, "Conditions of Cellular Immortality," pp. 51-78.

[855] Jacques Loeb, *The Mechanistic Conception of Life: Biological Essays*, The University of Chicago Press, Chicago, 1912, p. 210.

[856] Alexis Carrel, "On the permanent life of tissues outside the organism," *Journal of Experimental Medicine*, 15, 516-528, 1912; Alexis Carrel and Charles Lindbergh, *The Culture of Organs*, Paul B. Hoeber, NY, 1938.

[857] In 1961, Leonard Hayflick and Paul Moorhead demonstrated that human cells derived from embryonic tissues can only divide a limited number of times in culture (Hayflick L., Moorhead P.S., "The serial cultivation of human diploid cell strains," *Experimental Cell Research*, 25, 585-621, 1961).

Later on, Hayflick, in "The limited *in vitro* lifetime of human diploid cell strains," *Experimental Cell Research*, 37, 614-636, 1965, confirmed the "limited *in vitro* multiplication of many kinds of cultured cells ... causally unrelated to conditions of cell culture" and advanced "the hypothesis that the finite lifetime of diploid cell strains *in vitro* may be an expression of aging or senescence at the cellular level."

In that article, he battered Carrel's findings:

"One possible exception to this generalization was the highly popularized development from Carrel's laboratory wherein it was claimed that a population of cells derived from embryonic chick heart tissue was kept in serial cultivation for 34 years. ...

There is serious doubt that the common interpretation of Carrel's experiment is valid. An alternative explanation of Carrel's experiment is that the method of preparation of the chick embryo extract, used as a source of nutrient for his culture and prepared daily under conditions permitting cell survival, contributed new, viable, embryonic cells to the chick heart strain at each subcultivation or feeding. ...

In any event, Carrel's experiment has never been confirmed."

Nonetheless, Hayflick too accepted the notion of potential immortality. He distinguished several kinds of immortal cells (such as embryonic and germ cells and cell lines that are "tumor-like" or derived from tumor) that divide limitlessly:

"It remained for [George and Margaret] Gey in 1936 and [Wilton R.] Earle in 1943 to demonstrate that cell populations derived from a number of mammalian tissues, including human tissue, could unequivocally be kept in a state of rapid multiplication for apparently indefinite periods of time….

The acquisition of potential for unlimited cellular division or the escape from senescent-like changes by mammalian somatic cells, even *in vivo*, can only be achieved by cells which have altered and assumed properties of cancer cells. This applies equally well to normal mammalian somatic cells growing *in vivo* or *in vitro*."

[858] Alexis Carrel, *Man, The Unknown*, Burns & Oates, London, 1961 (1935), p. 140.

[859] Piere L. du Noüy, *Biological Time*, Methuen, London, 1936.

[860] Carrel A. and Ebeling A.H., "Age and multiplication of fibroblasts," *Journal of Experimental Medicine*, 34, 599-623, 1921.

[861] Leo Loeb, "Transplantation and Individuality," *Biological Bulletin*, 50, 143-180, 1921.

[862] Several contemporary European longevity researchers concluded in favor of the possibility of life-extension thanks to potential cell immortality. Thus, the German biologist Robert Rössle (1923) hypothesized that "Death is due not to body cells, but to their organization. … From this standpoint it follows that the individual is capable of life-extension, beyond his growth and development and long after the end of the period of sexual maturity" (Robert Rössle, *Wachstum und Alter* (Growth and Aging), 1923, quoted in Dimu Kotsovsky, "Das Alter in der Geschichte der Wissenschaft" (Aging in the history of science), *Isis*, 20(1), 220-245, Nov. 1933).

The ultimate conclusion was reached by the French/Russian Sergey Metalnikov, in *Problema Bessmertia i Omolozhenia v Sovremennoy Biologii* (*The Problem of Immortality and Rejuvenation in Modern Biology*, 1917), published in French as *La Lutte Contre La Mort* (*The Struggle against Death*). If individual cells are potentially immortal, Metalnikov argued, the organism composed of them can be made potentially immortal:

"If our claims are true that the organism is composed of immortal cells that retain their capacity for multiplication until senescence, that senescence itself is not the result of wear and tear of the living organism, but only an adaptation developed for the sake of the species, then all attempts of biologists and physicians to find means by which it may be possible to rejuvenate the organism and combat senescence, can be considered practically feasible and scientifically grounded."

(Sergey Metalnikov, *Problema Bessmertia I Omolozhenia v Sovremennoy Biologii* (The Problem of Immortality and Rejuvenation in Modern Biology), Slovo, Berlin, 1924 (1917), p. 144; *La Lutte Contre La Mort* (The Struggle against Death), Gallimard, Paris, 1937, p. 203.)

To the present time, the existence of immortal cells gives perhaps the greatest hope for an eventual subduing of aging and death. (See, for example, Michael D. West, *The Immortal Cell. One Scientist's Quest to Solve the Mystery of Human Aging*, Doubleday, NY, 2003.)

[863] Raymond Pearl, *The Biology of Death*, J.B. Lippincott Company, Philadelphia, 1922, pp. 48-49.

[864] Considering the various disproportions, the American researcher Charles Sedgwick Minot (1908) suggested that during growth and differentiation, the volume of the cell nucleus diminishes in proportion to the rest of the cell body (the cytomorphosis theory of aging). Hence he hypothesized that growth inevitably leads to aging.

(Charles S. Minot, *The Problem of Age, Growth and Death; a Study of Cytomorphosis, Based on Lectures at the Lowell Institute*, March 1907, G.P. Putnam's Sons, New York and London, 1908.)

The Russian biologist Ivan Ivanovich Schmalhausen (1926) proposed that aging results from a limited growth of different parts of the body, i.e. the "limitation of the growth and form of an animal by certain norms."

(I. I. Schmalhausen, *Problema Smerti I Bessmertia* (The Problem of Death and Immortality), Gosizdat, Moscow, 1926.)

The British marine biologist George Parker Bidder (1925, 1932) similarly perceived senescence as due to "some mechanism to stop natural growth so soon as specific size is reached."

(Bidder, G.P. "The mortality of plaice," *Nature*, 115, 495, 1925; Bidder, G.P. "Senescence," *British Medical Journal*, 2, 5831, 1932, quoted in Alexander Comfort, *The Biology of Senescence*, Butler & Tanner, London, 1956, pp. 11-12.)

The British biologist Julian Sorrell Huxley (1924, 1932) related death to differential growth of organs, whose disproportions make survival impossible.

(Quoted in Medawar, P.B, *An Unsolved Problem of Biology*, H.K. Lewis, London, 1952, pp. 11, 20-21.)

The German anatomist Hans Friedenthal (1910) believed that the "cephalization factor" or the proportion of the brain weight to the total mass of the body protoplasm determines the longevity of the organism. Animals with a smaller "cephalization factor" (relatively smaller brain) appeared to live shorter.

(Friedenthal, H, "Über die Gültigkeit der Massenwirkung für den Energieumsatz der lebendigen Substanz" (On the relevance of the mass effect on the energy turnover of living matter), *Zentralblatt für Physiologie*, 24, 321-327, 1910.)

The Americans Thorburn B. Robertson and L.A. Ray (1919), slightly modifying Metchnikoff's theory, maintained that the proportion of "cellular elements" and "connective tissue elements" limits the life-span (the "cellular elements" were said to be more vital for survival).

(Robertson T.B. and Ray L.A., "Experimental Studies on Growth. XV. On the growth of relatively long lived compared with that of relatively short lived animals," *Journal of Biological Chemistry*, 52, 71-107, 1920.)

The Russian biologist Moisey Samuilovich Mühlmann assumed the major cause of senescence to be nerve cell degeneration, which was said to result from a disproportion between nutrient supply through the cell surface and nutrient demand by the cell volume (1900, 1911, 1924). According to Mühlmann, cell compartments nearer the cell surface are better nourished than those near the center, hence the "underprivileged" compartments degenerate.

(Mühlmann M.S., *Über die Ursache des Alterns: Grundzüge der Physiologie des Wachstums mit besonderer Berücksichtigung des Menschen* (On the Cause of Aging: Fundamentals of the Physiology of Growth with a special consideration of Man), Bergmann, Wiesbaden, 1900; Mühlmann M.S., "Das Altern und der physiologische Tod" (Aging and physiological Death), *Sammlungen für Anatomie und Physiologie*, 1, 455, 1911; Milman M.S., *Uchenie o Roste, Starosti I Smerti* (The Theory of Growth, Aging and Death), Baku, 1926.)

[865] Albert I. Lansing, "General Physiology," in *Cowdry's Problems of Ageing, Biological and Medical Aspects*, edited by Albert I. Lansing, The Williams and Wilkins Company, Baltimore, 1952 (first published in 1939), pp. 3-23; Dimu Kotsovsky, "Das Alter in der Geschichte der Wissenschaft" (Aging in the history of science), *Isis*, 20(1), 220-245, Nov. 1933; Max Bürger, *Altern und Krankheit* (Aging and Disease), Zweite Auflage, Georg Thieme, Leipzig, 1954 (first published in 1947); Alexander Comfort, *The Biology of Senescence*, Butler & Tanner, London, 1956.

The idea that aging is due to an accumulation of metabolic waste products goes back to the Austrian/Jewish physician Max Kassowitz (1889). He asserted that aging is caused by the diminishing of the body's assimilatory capacity. The latter results from an accumulation of metabolic products, or "metaplasms," building up during the degradation of the protoplasm, and yielding inert substances and tissues. According to Kassowitz, the metaplasms include "products of regressive metamorphosis," "fiber cartilage" and "homogenous intercellular substance." Hence the major disharmony, leading to aging and death, is the "protoplasm" being clogged by the "metaplasm."

(Max Kassowitz, *Allgemeine Biologie. Aufbau und Zerfall des Protoplasmas* (General Biology. Synthesis and Degradation of the Protoplasm), Perles, Wien, 1899, first published in 1889.)

[866] T.H. Montgomery, "On Reproduction, Animal Life Cycles and the Biological Unit," *Transactions of the Texas Academy of Sciences*, 9, 75, 1906.

Similar ideas were proposed earlier by the German zoologist Carl Friedrich Jickeli (1902) who postulated that metabolism is an incomplete process and, as a result of the incomplete utilization of materials, injurious wastes accumulate.

(Carl F. Jickeli, *Die Unvollkommenheit des Stoffwechsels als Veranlassung für Vermehrung, Wachstum, Differenzierung, Rückbildung und Tod der Lebewesen im Kampfe ums Dasein* (The incompleteness of metabolism as a cause of reproduction, growth, differentiation, involution and death in the struggle for existence), R. Friedländer & Sohn, Berlin, 1902.)

[867] Charles Manning Child, *Senescence and Rejuvenescence*, Chicago University Press, Chicago, 1915, pp. 293-314.

Among those who searched for toxins that cause senescence, was the Czech/Austrian botanist Hans Molisch (1913) who attributed a chief role in deterioration to the accumulation of calcium in the cell.

According to the American botanist Harris M. Benedict (1915), the permeability of the aging cell is reduced, increasing the retention of metabolic waste products.

[868] There was a flurry of publications on the colloidal theory of aging in the 1920s in Europe.

A series of articles on the subject was published by the Czech biologist Vladislav Růžička (1870-1934), the founder of the Institute of Biology at Prague University and president of the Czech Eugenics Society. His articles appeared under the general heading "Beitrage zum Studium der Protoplasmahysteresis und der hysterischen Vorgänge - Zur Kausalität des Alterns" (Contributions to the study of protoplasm hysteresis and hysteretic processes as a cause of aging).

Prominent advocates of this theory, in its various modifications, were the Romanian neurologist Georges Marinesco (1863-1938, the chief opponent of Metchnikoff's theory of neural damage by phagocytosis), the Spanish biochemist Antonio de Gregorio Rocasolano (1873-1941), the French inventor and biologist Auguste Lumière (1862-1954), as also the German physiologist Rudolf Ehrenberg (1884-1969), the Indian researchers N.R. Dhar and Satya Prakash, and the Russian/Soviet researchers E.S. Bauer, A.A. Kisel and A.V. Blagoveshensky.

Among the American proponents of the colloidal theory were C.M. Child, T.B. Robertson and H. Wasteneys.

Sources:

Robertson, T.B., Wasteneys, H., "On the changes in lecithin-content which accompany the development of sea-urchin eggs," *Archiv für Entwicklungsmechanik*, 37, 485-496, 1913; Robertson T.B., "On the nature of the autocatalyst of growth," *Archiv für Entwicklungsmechanik*, 37, 497-508, 1913; Child, C.M., *Senescence and Rejuvenescence*, Chicago University Press, Chicago, 1915;

Georges Marinesco, "Méchanisme chimico-colloidal de la sénilité et le problème de la mort naturelle" (Chemical-colloidal mechanisms of aging and the problem of natural death), *Revue Scientifique de la France et de l'étranger*, Paris, 1, 673-679, 1914;

Vladislav Růžička, "Beitrage zum Studium der Protoplasmahysteresis und der hysterischen Vorgänge (Zur Kausalität des Alterns) I. Die Protoplasmahysteresis als Entropieerscheinung" (Contributions to the study of protoplasm hysteresis and hysteretic processes as a cause of aging. I. Protoplasm hysteresis as an entropic phenomenon), *Archiv für Mikroskopische Anatomie und Entwicklungsmechanik*, Bonn, 101, 459-482, 1924; Vladislav Růžička, "Beitrage zum Studium der Protoplasmahysteresis und der hysterischen Vorgänge (Zur Kausalität des Alterns) XII. Das Lezithin als Schutzkolloid" (Contributions to the study of protoplasm hysteresis and hysteretic processes as a cause of aging. XII. Lecithin as a protective colloid), *Wilhelm Rouxs Archiv für Entwicklungsmechanik der Organismen*, Berlin, 112, 262-270, 1927;

Auguste Lumière, *La Vie, La Maladie et La Mort, Phénomènes Colloïdaux* (Life, Disease and Death as Colloidal Phenomena), Masson & Cie, Libraires de L'Académie de Médecine, Paris, 1928; Auguste Lumière, *Sénilité et Rajeunissement* (Aging and Rejuvenation), Librairie J.-B. Baillière et Fils, Paris, 1932.

[869] Max Bürger, in *Altern und Krankheit* (Aging and Disease), Zweite Auflage, Veb Georg Thieme, Leipzig, 1954, pp. 17-18 (1947), provides examples of the theories of exogenous damage leading to aging:
Thus, the German nuclear physicist Paul Kunze (1933) thought that organic tissues are being constantly destroyed by cosmic radiation.
The American chemists I.W.D. Hakh and E.H. Westling (1934) believed the accumulation of heavy water did most damage.
The Dutch physiologist Hendrik Zwaardemaker (1927) presumed the inhalation of radioactive substances (such as radium) destroys the living substance.
The Hungarian bacteriologist Gyula Daranyi (1930) proposed that gravity played a part in the formation of congestive sediments.
[870] Jacques Loeb, "Natural death and the duration of life," *The Scientific Monthly*, 12, 578-585, December 1919; Raymond Pearl, *The Biology of Death*, J.B. Lippincott Company, Philadelphia, 1922, pp. 47-48.
[871] Vilfredo Pareto, *The Mind and Society*, Harcourt, Brace and Company, New York, 1935, vol. 4, Ch. XIII, "The Social Equilibrium in History," section 2068, pp. 1435-1436 (translated from the original Italian *Trattato di Sociologia Generale*, 1916);
Frederick Winslow Taylor, *The Principles of Scientific Management*, Harper, NY, 1911;
Peter Burke, *History and Social Theory* (Second Edition), Cornell University Press, Ithaca, New York, 2005, Ch. 4. "Central Problems. Functionalism," pp. 128-131;
Tim Armstrong, *Modernism, Technology and the Body. A Cultural Study*, Cambridge University Press, Cambridge, 1998, Part I. "The Regulation of Energies" Ch. 2. " Waste Products," pp. 43-44.
[872] Alexis Carrel, *Man, The Unknown*, Burns & Oates, London, 1961 (1935), pp.143-148.
[873] The uncertainties regarding the "limits" to the human life-span have remained to a later date, reaching a peak in the late 1970s - early 1980s.
Thus, Nathan Keyfitz (1913-2010) of Harvard upheld the so-called "Taeuber Paradox" (named after Conrad Taeuber, 1906-1999, Chief of the Population Division of the US Census Bureau) which suggested that aging organisms become generally frail, and therefore, if an aging person will not die of one age-related disease, he will die of another. Consequently, even the entire elimination of a major age-related disease, such as cancer, will not significantly increase the general life-expectancy, as some new disease will come in its place. As Keyfitz wrote:
"A cure for cancer would only have the effect of giving people the opportunity to die of heart disease."
This was a major motivation damper for attempting to find cures against age-related diseases.
(Nathan Keyfitz, "What difference would it make if cancer were eradicated? An examination of the Taeuber Paradox," *Demography*, 14 (4), 411-418, 1977.)
However, in the same article, Keyfitz stated:
"If cancer, heart disease, etc., are merely alternative ways in which the aging of body cells makes itself manifest, then eradicating any one of them may make little difference. The proper entity to attack is the process of aging itself."
Elsewhere, Keyfitz further advocated for increasing research into cellular senescence, as an underlying general cause of many age-related diseases, rather than into any specific single disease. Such research would allow us "to break through the barrier that now seems to be set at about 80 years."
(Nathan Keyfitz, "Improving life expectancy: An uphill road ahead," *American Journal of Public Health*, 68, 954-956, 1978.)
Keyfitz's view on the intrinsic limit to life-expectancy was disputed by Arthur Schatzkin (1948-2011) of Mount Sinai Hospital, New York, and the National Institute of Cancer. Schatzkin suggested that a single health care measure can improve the outcome for several age-related diseases and the elimination of several risk factors can have a cumulative effect, leading to a significant life-extension.
(Arthur Schatzkin "How long can we live? A more optimistic view of potential gains in life expectancy,"

American Journal of Public Health, 70, 1199-1200, 1980.)

Further, the Stanford gerontologists James F. Fries and Lawrence M. Crapo, in *Vitality and Aging. Implications of the Rectangular Curve* (W.H. Freeman and Co., NY, 1981) posited that "the maximum life span is fixed at about 100 years, and the median life span is fixed at about 85 years."

This "natural" life span is determined by "a steady decline in homeostasis and organ reserve in many vital systems." This decline in organ reserves results in an exponential increase in the mortality rate with age.

The authors predicted that the society will increasingly progress toward a "rectangular survival curve" when health care measures will produce a "compression of morbidity."

This basically implied that people will remain healthy until the age of 85 and then collapse and die rapidly, saving on national health care expenditures. The authors called such a scenario "a celebration of life."

Still, the authors believed that "if we could understand the aging process at the cellular level and its underlying molecular mechanisms, then we might be able to alter the information units in the cells and thereby alter the life span."

(Fries and Crapo, *Vitality and Aging*, 1981, pp. 135-142; James F. Fries, "Aging, Natural Death, and the Compression of Morbidity," *The New England Journal of Medicine*, 303, 130-135, 1980.)

Several authors disputed Fries and Crapo's conclusions and argued that an approximation of a perfect "rectangular survival curve" will not be possible, and health care measures can produce an extension of life expectancy far beyond the 85 year "limit."

(Edward L. Schneider, Jacob A. Brody, "Aging, natural death and the compression of morbidity: Another view," *The New England Journal of Medicine*, 309, 854-856, 1983.)

Further limits on the human life span – biological, technological, ethical and economic – were proposed by J. Michael McGinnis, of the US Department of Health and Human Services, in 1985.

Yet, McGinnis concluded with "an essentially optimistic perspective." For example, regarding the limits imposed on life-extension by the deficiencies of available technology, the pace of technological change is said to be rapid and "the limits of today will become the opportunities of tomorrow."

(J. Michael McGinnis, "The limits of prevention," *Public Health Reports*, 100, 255-260, 1985.)

The question of the "limits" to the human life span and life expectancy has remained open to the present.

Thus, quite recently, the gerontologists Stuart Jay Olshansky of the University of Illinois at Chicago and Bruce Alfred Carnes of the University of Oklahoma, in *The Quest for Immortality. Science at the Frontiers of Aging* (2001), generally discarded pretty much all the means for life-extension (anti-oxidants, hormones, etc.) that were currently on commercial offer. They suggested that "given the current state of medical technology, life expectancy at birth will not rise above 85 years."

But the qualification regarding "the current state of medical technology" seems to be crucial. "The legitimate science of aging," the authors maintained, "has already led to remarkable extensions of life for many people. We can expect the hard work of researchers and medical practitioners to add to that success in the future." For example, "compounds like [the anti-oxidant] WR-2721 [amifostine] that enhance the ancient mechanisms of cellular maintenance and repair could prove to be the *elixir vitae* mankind has sought for thousands of years."

At any rate, "The ongoing debate over limits to life expectancy is not likely to be resolved any time soon."

(S. Jay Olshansky and Bruce A. Carnes, *The Quest for Immortality. Science at the Frontiers of Aging*, W.W. Norton and Co., NY, 2001, pp. 142, 149, 211, 219.)

[874] Robertson, T.B., "On the nature of the autocatalyst of growth," *Archiv für Entwicklungsmechanik*, 37, 497-508, 1913; Robertson T.B., Wasteneys, H. "On the changes in lecithin-content which accompany the development of sea-urchin eggs," *Archiv für Entwicklungsmechanik*, 37, 485-496, 1913; Vladislav Růžička, "Beitrage zum Studium der Protoplasmahysteresis und der hysterischen Vorgänge (Zur Kausalität des Alterns) XII. Das Lezithin als Schutzkolloid" (Contributions to the study of protoplasm hysteresis and

hysteretic processes as a cause of aging. XII. Lecithin as a protective colloid), *Wilhelm Rouxs Archiv für Entwicklungsmechanik der Organismen*, Berlin, 112, 262-270, 1927.

[875] Thorburn Brailsford Robertson's summative work on aging and longevity, *The Chemical Basis of Growth and Senescence* (Lippincott, Philadelphia, 1923), appeared after his move from the US in 1919 to become chair of physiology at Adelaide University, Australia.

Born in Edinburgh, Scotland, and educated in Australia, Robertson, after the formative period of his career in the US (from 1905 to 1919) and return to Australia, became apparently one of the first Australian experimental gerontologists and longevity researchers.

[876] Wilder D. Bancroft and John R. Rutzler Jr, "Reversible coagulation in living tissue. XII," *Proceedings of the National Academy of Sciences USA*, 20, 501-509, 1934.

[877] Albert I. Lansing, "General Physiology," in *Cowdry's Problems of Ageing, Biological and Medical Aspects*, edited by Albert I. Lansing, The Williams and Wilkins Company, Baltimore, 1952 (1939), pp. 3-23.

[878] Wilhelm Roux, in *Der Kampf der Teile im Organismus* (1881), postulated that aging and disease are due to "The Struggle of the Parts of the Organism," and that the path toward health and longevity consists in harmonizing these parts and diminishing the struggle.

Similar views were expounded by the Russian scientist Nikolay Alexandrovich Kholodkovsky in 1882.

(Wilhelm Roux, *Der Kampf der Teile im Organismus* (The Struggle of the Parts in the Organism), Wilhelm Engelmann, Leipzig, 1881; Kholodkovsky N.A. "Tod und Unsterblichkeit in der Thierwelt" (sic. Death and immortality in the animal world), *Zoologischer Anzeiger*, 5(11), 264-265, 1882.)

[879] Herbert Spencer, *The Principles of Ethics*, Vol. 1, Ch. 9, "Parenthood," 1897 (reprinted in the Online Library of Liberty, http://oll.libertyfund.org/).

[880] In 1882, the German zoologist Otto Bütschli suggested that death results from a depletion of a "vital enzyme" (*lebensferment* - life ferment) allocated to the individual at birth.

The German physiologist Max Rubner (1908) found that most mammals, during their life course, consume about the same amount of energy per kilogram body weight (approximately 200,000 calories) and die when the "life-energy" quota is spent.

According to Rubner, for man the value is outstandingly high (~725,000 calories), but is a preordained quantity nonetheless. Intense metabolic activity was, in this view, fatal.

(Otto Bütschli, "Gedanken über Leben und Tod" (Thoughts on Life and Death), *Zoologischer Anzeiger*, 5(103), 64-67, 1882; Max Rubner, *Das Problem der Lebensdauer und seine Beziehungen zu Wachstum und Ernährung* (The problem of Life-duration and its relations to Growth and Nutrition), München, 1908.)

[881] Jacques Loeb, "Über den Temperaturkoeffizienten für die Lebensdauer kaltblütiger Thiere" (On the temperature coefficients for the duration of life of cold-blooded animals), *Pflugers Archiv für Gesammte Physiologie*, 124, 411-426, 1908; Jacques Loeb and John Howard Northrop, "On the influence of food and temperature on the duration of life," *The Journal of Biological Chemistry*, 32, 103-121, 1917; Jacques Loeb and J. H. Northrop, "What Determines the Duration of Life in Metazoa?" *Proceedings of the National Academy of Sciences*, 3, 382-386, 1917; Jacques Loeb and J. H. Northrop, "Is there a Temperature Coefficient for the Duration of Life?" *Proceedings of the National Academy of Sciences*, 2, 456-457, 1916.

[882] Raymond Pearl and Sylvia L. Parker, "Experimental Studies on the Duration of Life. V. On the Influence of Certain Environmental Factors on Duration of Life in Drosophila," *American Naturalist*, 56, 385-405, 1922; Raymond Pearl and Sylvia L. Parker, "Experimental Studies on the Duration of Life. X. The Duration of Life of Drosophila Melanogaster in the Complete Absence of Food," *American Naturalist*, 58, 193-218, 1924.

[883] Thomas B. Osborne and Lafayette B. Mendel, "The Resumption of growth after long-continued failure to grow," *The Journal of Biological Chemistry*, 23, 439-454, 1915; T.B. Osborne, L. B. Mendel, E.R. Ferry, "The effect of retardation of growth upon the breeding period and duration of life of rats," *Science*, 45, 294-295, 1917.

[884] F. Rous, "The Influence of diet on transplant and spontaneous tumors," *Journal of Experimental Medicine*, 20, 433-451, 1914.

[885] John A. Garraty, *The Great Depression: An Inquiry into the Causes, Course, and Consequences of the Worldwide Depression of the Nineteen-Thirties, as Seen by Contemporaries and in Light of History*, Harcourt Brace Jovanovich, NY, 1986.

[886] C. M. McCay, W.E. Dilly, M.F. Crowell, "Growth rates of brook trout reared upon purified rations, upon skim milk diets, and upon combinations of cereal grains," *Journal of Nutrition*, 1, 233-246, 1929; Clive McCay, "The Effect of Retarded Growth Upon the Length of Life Span and upon the Ultimate Body Size," *Journal of Nutrition*, 10, 63-79, 1935.

[887] Richard Weindruch, Roy L. Walford, *The Retardation of Aging and Disease by Dietary Restriction*, Charles C. Thomas, Springfield, Illinois, 1988; Luigi Fontana, Linda Partridge, Valter D. Longo, "Extending Healthy Life Span – From Yeast to Humans," *Science*, 328 (5976), 321-326, April 16, 2010.

Notably, despite the almost universal life-extension by calorie restriction in model organisms – from yeast to mice; in August 2012, it was reported that calorie restriction failed to prolong life in rhesus monkeys. Yet, even this study reported health benefits from the calorie restriction.

(Julie A. Mattison, …., Rafael de Cabo, "Impact of caloric restriction on health and survival in rhesus monkeys from the NIA study," *Nature*, Published online 29 August 2012.)

That study contradicted an earlier one that did show both health and longevity benefits.

(Ricki J. Colman, …, Richard Weindruch, "Caloric restriction delays disease onset and mortality in rhesus monkeys," *Science*, 325, 201-204, 2009.)

Moreover, doubts have been expressed whether calorie restriction would produce a significant prolongation of life in humans. No such long-term empirical study was ever performed, due to its necessary length and complexity. Yet, a failure was predicted.

(John P. Phelan, Michael R. Rose, "Why dietary restriction substantially increases longevity in animal models but won't in humans," *Aging Research Reviews*, 4(3), 339-350, 2005.)

Nonetheless, even now, some health benefits were reported for human subjects practicing calorie restriction.

(Edward P. Weiss, Luigi Fontana, "Caloric restriction: powerful protection for the aging heart and vasculature," *American Journal of Physiology – Heart and Circulatory Physiology*, 301(4), H1205-1219, 2011.)

Of additional note is that moderation in energy consumption (calorie restriction) also implies a limitation of the organism's growth. Yet, it was reported that animals exhibiting continuous growth often show no signs of aging. Still, such non-aging, continually growing animals can die due to lack of resources needed to sustain such a growth.

(Dr. João Pedro de Magalhães, Institute of Integrative Biology, University of Liverpool, UK, "Some Animals Age, Others May Not," *Senescence.info*, Accessed September 2012, http://senescence.info/aging_animals.html.)

[888] Raymond Pearl, *The Biology of Death*, 1922, p. 226.

[889] Walter Cannon spoke of "extreme activities, which are wasteful of energy" specifically regarding the correlation of the body and of the society (p. 299), and more concerning the various "margins of economy" (p. 120), and "mechanisms of preserving homeostasis" by "storage" and "reserves" (pp. 47-48, 171, 272), in Walter Cannon, *The Wisdom of the Body*, Norton, NY, 1932.

In *The Stress of Life* (1956), Hans Selye developed and summarized the concept of the "vital capital":

"True age depends largely on the rate of wear and tear, on the speed of self-consumption; for life is essentially a process which gradually spends the given amount of adaptation energy that we inherited from our parents.

Vitality is like a special kind of bank account which you can use up by withdrawals but cannot increase by deposits. Your only control over this most precious fortune is the rate at which you make your withdrawals.

The solution is evidently not to stop withdrawing, for this would be death. Nor is it to withdraw just enough for survival. For this would permit only a vegetative life, worse than death. The intelligent thing to do is to withdraw generously, but never expend wastefully."
(Hans Selye, *The Stress of Life*, McGraw-Hill, NY, 1956, p. 274.)
Selye's "limit" on the "vital capital," however, is modifiable:
"As far as we know, our reserve of adaptation energy is an inherited finite amount, which cannot be regenerated. On the other hand, I am sure we could still enormously lengthen the average human life-span by living in better harmony with natural laws. ...
If its amount is unchangeable, we may learn more about how to conserve it. If it can be transmitted, we may explore means of extracting the carrier of this vital energy – for instance, from the tissues of young animals – and trying to transmit it to the old and aging."
(Hans Selye, *The Stress of Life*, 1956, pp. 276, 303-304.)

[890] Raymond Pearl, *The Biology of Death*, 1922, pp. 17, 227.

[891] The 92 distinguished members of the Institute's Hygiene Reference Board, included:
Rupert Blue, Surgeon-General, US Public Health Service (The Institute's Public Health Administration division); George Blumer, Dean of Yale Medical School; John H. Kellogg, superintendent of the Battle Creek Sanatorium; Colonel William J. Mayo, ex-president of the American Medical Association; George H. Simmons, Secretary of the American Medical Association (Medicine and Surgery division); Colonel Walter B. Cannon, Professor of Physiology, Harvard University, a leader in the research of homeostasis; Richard Pearce, Secretary of the Medical Advisory Committee, American Red Cross; Colonel William Welch, Dean of the School of Hygiene and Public Health, Johns Hopkins University (Division of Chemistry, Bacteriology, Pathology, Physiology, Biology); Henry W. Farnam, Professor of Economics, Yale University (statistics); Alexander Graham Bell, the inventor of the telephone and chairman of the US Eugenics Record Office; Charles B. Davenport, Director of the Eugenics Record Office; Winfield Scott Hall, professor of physiology, Northwestern University, Chicago (the Eugenics division).
The Institute's Foreign Advisory Board included the following distinguished Professors:
the pathologist John George Adami of Canada; Carlos Fernandez Pena of Chile (President of the Chilean National Educational Association and the National League Against Alcoholism); Sir. Thomas Oliver of England (a world leading expert in occupational medicine); the chemical biologist and specialist in nutrition Armand Gautier of France; the psychiatrist, hygienist and researcher of aging Leonardo Bianchi of Italy; Shibasaburo Kitasato of Japan (the co-discoverer of the bubonic plague bacterium and co-developer of serum therapy against diphtheria and anthrax); and the physiologist Ivan Pavlov of Russia.

[892] The board of directors included heads of companies and foundations, such as:
Lee K. Frankel, Head of the Welfare Department, Metropolitan Life Insurance Company; Robert W. Forest, trustee of the Russell Sage Foundation; Charles H. Sabin, vice president of the Guaranty Trust Company;
and bankers such as Frank A. Vanderlip, president of the National City Bank; Francis R. Cooley of Hartford, Connecticut, and Henry A. Bowman of Springfield, Massachusetts.
("National Society to Conserve Life. Life Extension Institute Formed to Teach Hygiene and Prevention of Disease. Large Capital Behind It," *New York Times*, Dec. 30, 1913.)

[893] The Institute in fact developed from the "Human Life Extension Committee" established by the Association of Life Insurance Presidents, after their meeting in February 1909, hearing a key note by Prof. Irving Fisher (1867-1947), titled "The Economic Aspect of Lengthening Human Life."

[894] Given the professional scope and public impact, the Life Extension Institute may be well considered the world's first large-scale life-extensionist *public health institution*.
Yet, the world priority as the first large-scale life-extensionist *scientific research institute* (or gerontological institute in the modern sense) may be still attributed to Dimu Kotsovsky's Institute for the Study and

Combat of Aging, founded in Kishinev, Moldova-Romania, in 1933, which emphasized basic experimental studies of aging and longevity and acted as a hub of international cooperation in this field of research.

[895] Irving Fisher and Eugene Lyman Fisk, *How to Live. Rules for Healthful Living Based on Modern Science. Authorized by and Prepared in Collaboration with the Hygiene Reference Board of the Life Extension Institute Inc, Fifteenth Edition, Completely Revised, Enlarged and Reset*, Funk and Wagnalls Company, New York, 1919 (hereafter referred to as "Irving Fisher and Eugene Lyman Fisk, *How to Live*, 1919").

[896] In the *Report on National Vitality. U. S. Mortality Statistics* of 1908, Fisher wrote:
"With better Hygiene and superior Eugenics, and the proper methods of living, the extreme limit of life might be more frequently attained, and that after many generations, the average of humanity might perhaps approximate to the limit of 110 to 150 years."
(Irving Fisher, *Report on National Vitality. U. S. Mortality Statistics*, 1908, quoted in Marion Thrasher, *Long Life in California*, M.A. Donohue & Company, Chicago, 1915, p. 36.)

[897] Irving Fisher and Eugene Lyman Fisk, *How to Live*, 1919, p. vi.

[898] Irving Fisher and Eugene Lyman Fisk, *How to Live*, 1919, pp. 138-139.

[899] "National Society to Conserve Life. Life Extension Institute Formed to Teach Hygiene and Prevention of Disease. Large Capital Behind It," *New York Times*, Dec. 30, 1913.

[900] Irving Fisher and Eugene Lyman Fisk, *How to Live*, 1919, p. x.

[901] Jean Finot, *The Philosophy of Long Life* (translated by Harry Roberts), John Lane Company, London and New York, 1909, pp. 77-78, first published in French as *La Philosophie de la Longévité*, Schleicher Freres, Paris, 1900.

[902] American Eugenics Society, *Eugenics Quarterly*, vol. 3-4, 1956, p. 112; Edwin Black, *War Against the Weak: Eugenics and America's Campaign to Create a Master Race*, Thunder's Mouth Press, New York, 2004 (2003); Daniel J. Kevles, *In the Name of Eugenics: Genetics and the Uses of Human Heredity*, Harvard University Press, Cambridge MA, 1997 (first published in 1986); Robert Peel (Ed.), *Essays in the History of Eugenics*, The Galton Institute, London, 1998.

[903] Raymond Pearl, *The Biology of Death*, 1922, pp. 227-228.

[904] Irving Fisher and Eugene Lyman Fisk, *How to Live*, 1919, pp. 416-418, 441-443.

[905] Ludwig Roemheld provided the following figures:
"The Life Extension Institute was founded in New York in 1913, with the only aim to care for the prolongation of life. The Metropolitan Life Insurance Company, from February 1914 until July 1921, gave $225,000 for examinations in this Institute, with the result that the number of deaths of the insured and accordingly the money expenses of the company were significantly reduced."
(Ludwig Roemheld, *Wie verlängere ich mein Leben?* (How do I prolong my life?) Ferdinand Enke Verlag, Stuttgart, 1941, first presented in 1933, p 10.)
And according to Gerhard Venzmer,
"One of the greatest life insurance companies [the Metropolitan], ... from 1907 to 1927 gave the sum of 32 million dollars so that its insured would be enlightened about a rational, life-prolonging life-style. The success was extraordinary: the mortality of the insured decreased so greatly that not only were the expenses for this education fully returned, but another 43 million dollars were saved!"
(Gerhard Venzmer, *Lang leben und jung bleiben!* (Live long and stay young), Franckhsche Verlag, Stuttgart, 1937 (republished in 1941), p. 87.)

[906] "National Society to Conserve Life. Life Extension Institute Formed to Teach Hygiene and Prevention of Disease. Large Capital Behind It," *New York Times*, Dec. 30, 1913;
Carol A.S. Thompson, *The History of the Conspiracy Against Tobacco*, "Life Extension Institute," Madison, Wisconsin, http://www.smokershistory.com/LEI.htm.

[907] "Life Extension Institute acquires executive health group; Nation's largest provider of physical examination and preventive health care services," *Business Wire*, May 25, 1995, http://www.encyclopedia.com/doc/1G1-16904649.html.

[908] EHE International, http://www.eheintl.com/, http://www.eheintl.com/about_history.jsp.

[909] Hoffman B, "Health Care Reform and Social Movements in the United States," *American Journal of Public Health*, 93(1), 75-85, 2003; Robert Whaples, "Hours of Work in U.S. History," *EH.Net Encyclopedia*, edited by Robert Whaples, August 15, 2001, http://eh.net/encyclopedia/article/whaples.work.hours.us; "The 1930s-1940s: Medicine and Health: Overview," *American Decades*, Encyclopedia.com, 2001, http://www.encyclopedia.com/doc/1G2-3468301278.html; http://www.encyclopedia.com/doc/1G2-3468301630.html.

[910] Louis Israel Dublin [New York, 1882-1969], "Longevity in retrospect and in prospect," pp. 100-119, in E.V. Cowdry (Editor), *Problems of Ageing. Biological and Medical Aspects*, Williams and Wilkins, Baltimore MD, 1939.

[911] *The Townsend Crusade: An Impartial Review of the Townsend Movement and the Probable Effects of the Townsend Plan*, The Committee on Old Age Security of the Twentieth Century Fund, Inc., NY, 1936; Landis, J. T., "Old-age movements in the United States," in R. J. Havinghurst, *Social Adjustment in Old Age. Social Science Research Council*, NY, 1946, pp. 64-66; "'Youth' and 'Age' may be future political parties," *Science News Letter*, Washington, 35, 360, 1939.

[912] Arthur M. Schlesinger Jr., *The Coming of the New Deal, 1933–1935*, Houghton Mifflin, Boston, 1958; Seymour Martin Lipset and Gary Marks, "How FDR Saved Capitalism," in *It Didn't Happen Here: Why Socialism Failed in the United States*, W.W. Norton & Co., NY, 2001; Eric Rauchway, *The Great Depression and the New Deal*, Oxford University Press, Oxford, 2007; Robert S. McElvaine, *The Great Depression: America, 1929–1941*, Three Rivers Press, NY, 1993; Alan Brinkley, *Liberalism and Its Discontents*, Harvard University Press, Harvard, 1998; Ronald L. Heinemann, *Depression and New Deal in Virginia*, The University Press of Virginia, Charlottesville, VA, 1983.

[913] Milton Wainwright, *Miracle Cure: The Study of Penicillin and the Golden Age of Antibiotics*, Blackwell, Cambridge MA, 1990; Roy Porter, *The Greatest Benefit to Mankind: A Medical History of Humanity*, W.W. Norton & Company, NY, 1998 (1997), pp. 455-457; "Discovery of Penicillin," American Chemical Society, 2010, http://portal.acs.org/portal/acs/corg/content.

[914] Nathan W. Shock (Ed.), *A Classified Bibliography of Gerontology and Geriatrics*, Stanford University Press, Stanford CA, 1951, "Rejuvenation," pp. 74-80, "Theories," pp. 80-86; *Supplement One 1949-1955*, 1957; *Supplement Two 1956-1961*, 1963 (reprinted in 1980 by Stanford University Press).

[915] One of the most persistent skeptics of rejuvenation attempts was Morris Fishbein (1889-1976), from 1924 to 1950 editor of the *Journal of the American Medical Association*.

[916] Edwin J. Dingle, *Breathing Your Way to Youth*, Mental Physics School of Wisdom, NY, 1931; Teofilo De la Torre, *Psycho-Physical Regeneration, Rejuvenation and Longevity*, Lemurian Press, Milwaukee WI, 1939; Rose J. Ballinger, *Brain Waves: Healing and Rejuvenation by Magnetic Currents of Brain Waves*, Meador, Boston, 1946.

[917] Herbert M. Shelton, *Living Life to Live it Longer: A Study in Orthobionomics, Orthopathy and Healthful Living*, Kessinger Publishing, Oklahoma City, 1926; Herbert M. Shelton, *The Hygienic System*, Library of New Atlantis, San Antonio, Texas, 1935; Otto Carqué, *Vital Facts About Foods: A Guide to Health and Longevity*, Los Angeles, 1933 (reprinted in 1974 by Health Research, Pomeroy WA); Adelle Davis, *You Can Stay Well*, Stationers' Hall, London, 1939; Robert B. Pearson, *The Prolongation of Life Through Diet*, Health Inc., Denver Col., 1941; Gayelord Hauser, *Look Younger, Live Longer*, Farrar, Straus and Co., NY, 1950.

[918] Raymond Pearl and Ruth DeWitt Pearl [Raymond Pearl's daughter], *The Ancestry of the Long-Lived*, Johns Hopkins Press, Baltimore, 1934; Jean Brierley, "An Exploratory Investigation of the Selective Value of Certain Genes and Their Combinations in Drosophila," *Biological Bulletin*, 75, 475-493, 1938; Brenda Stoessiger, "On the inheritance of duration of life and cause of death," *Annals of Eugenics*, 5, 105-178, 1933.

[919] Raymond Pearl, *The Rate of Living, Being an Account of Some Experimental Studies in the Biology of Life Duration*, A.A. Knopf, NY, 1928; Clive McCay, "The Effect of Retarded Growth Upon the Length of Life Span and upon the Ultimate Body Size," *Journal of Nutrition*, 10, 63-79, 1935.

[920] The notion of aging and deterioration as due to disharmony is ancient. It was prevalent in the writings of Hippocrates, Galen and their followers. In the humoralist/Galenic tradition, it was the balance of the Four Humors – the black bile, yellow bile, phlegm and blood – that needed to be maintained. In alchemy, senescence was perceived as due to an "imbalance" of the primary elements; and the Philosopher's stone, comprising all the vital essences and elements, was intended to restore this balance.
(Gerald J. Gruman, *A History of Ideas about the Prolongation of Life. The Evolution of Prolongevity Hypotheses to 1800*, Transactions of the American Philosophical Society, Volume 56 (9), Philadelphia, 1966, "The Alchemists," pp. 49-67.)
The search for harmony, equilibrium and constancy, has continued. These concepts were often used interchangeably, since a harmonious, balanced state of equilibrium implied maintaining the constancy of some measurable property or structure.
In 1878, Claude Bernard famously posited that "the fixity of the internal environment is the condition for free, independent life."
Since the publication of Wilhelm Roux's *Der Kampf der Teile im Organismus* (The Struggle of the Parts in the Organism, 1881), the "disequilibrium" was mainly seen in terms of a "struggle" or "competition" of organs and tissues.
Various processes of regulation of physiological equilibrium (maintaining the constancy of body metabolism and function) were established by the neurophysiologists Ivan Sechenov (1878), Michael Foster (1885), Walter Holbrook Gaskell (1886), Patrick Geddes and John Arthur Thomson (1889), James Baldwin (1901), Charles Sherrington (1906), Thomas Graham Brown (1911) and Ivan Pavlov (1928).
(Roy Porter, *The Greatest Benefit to Mankind: A Medical History of Humanity*, W.W. Norton & Company, NY, 1998 (1997), Ch. XVII, "Medical Research," Subsection "Neurology," pp. 534-545; Georges Canguilhem, "The Development of the Concept of Biological Regulation in the Eighteenth and Nineteenth Centuries," in *Ideology and Rationality in the History of the Life Sciences*, translated from French by Arthur Goldhammer, Cambridge MA, The Massachusetts Institute of Technology Press, 1988, pp. 81-102; Chen Bu, "Homeostasis and Chinese Traditional Medicine: Commenting on Cannon's The Wisdom of The Body," *Chinese Studies in the History and Philosophy of Science and Technology*, Fan Dainian and Robert S. Cohen (Eds.), translated by Kathleen Dugan and Jiang Minshan, Kluwer Academic Publishers, Dordrecht, The Netherlands, 1996, pp. 115-124.)
By the time of Cannon's writing, body equilibrium appeared to many to be a necessary condition of life, and it was believed that shifts in it could be counteracted, and thus its prolonged maintenance was conceivable. The body equilibrium was perceived in terms of substance and energy expenditure and replenishment. Accordingly, death was seen as due to bio-economic imbalance or inadequate supply of nutrients and energy, and longevity derived from the ability to restore or balance the deficits.
The dynamic opposition was posited between processes of destruction and regeneration, both measurable and subject to intervention. Various metabolic processes were considered as favorable or unfavorable for life preservation. The constructive anabolism (synthesis) was opposed to catabolism (degradation). Anabolism was associated with youth and vigor, with life itself, yet it was driven by catabolism.
The acid-base and oxidation-reduction mechanisms posited in the 1920s (both the acid-base theory of Johannes Bronsted/Thomas Lowry and that of Gilbert Lewis, were presented in 1923), as well as the discovery of ATP in 1929 by Karl Lohmann, showed how energy expended in life processes is replenished to sustain life.
Freud's *Beyond the Pleasure Principle* (1920) also essentially considered body equilibrium and energy balance. The "life drive" (or instincts "exercising pressure towards a prolongation of life" or the power of "Eros"

integrating the living substance) and the "death instinct" ("pressure towards death" or the internal drive "to return to the state of quiescence") basically reflected the concepts of stimulation as opposed to relaxation. These were seen by Freud as dialectically opposed processes in maintaining life, and there were shown no built-in controls that would make the victory of the death drive inevitable.
(Sigmund Freud, *Beyond the Pleasure Principle*, Translated by James Strachey, Liveright, New York, 1976, first published in 1920.)
Yet, the formal concept of homeostasis, building on the earlier notions of stable equilibrium, and its implications for aging and longevity, were introduced by Walter Cannon.
(Walter B. Cannon, "Reasons for Optimism in the Care of the Sick," *The New England Journal of Medicine*, 199 (13) 593-597, 1928; Walter B. Cannon, "Organization for Physiological Homeostasis," *Physiological Reviews*, 9, 399-431, 1929; Walter B. Cannon, *The Wisdom of the Body*, Norton, NY, 1932; Walter B. Cannon, "Aging of homeostatic mechanisms," in E.V. Cowdry (Editor), *Problems of Ageing. Biological and Medical Aspects*, Williams and Wilkins, Baltimore MD, 1939, pp. 623-641.)
The relation between homeostasis and longevity was further developed by Hans Selye.
(Hans Selye, *The Stress of Life*, McGraw-Hill, NY, 1956; Hans Selye, "Stress and Aging," *Journal of the American Geriatrics Society*, 18, 669-680, 1970.)
Biological equilibrium was further discussed in terms of order and organization, as a low entropy (high order) system maintaining its orderliness by increasing the disorder of the environment (as explained in Erwin Schrödinger's *What is Life?* Cambridge University Press, Cambridge, 1996, first published in 1944).
Implicitly, such "entropy parasitism" (in Schrödinger's words, "feeding on negative entropy") seemed to present no inherent physical constraints to its maintenance, and thus theoretically could be sustained indefinitely.

[921] W.B. Cannon, "Physiological regulation of normal states: some tentative postulates concerning biological homeostatics," in Auguste Pettit (Ed.), *Charles Richet: Ses amis, ses collègues, ses élèves* (Charles Richet, his friends, colleagues and students), Editions Médicales, Paris, 1926, pp. 91-93.

[922] Walter B. Cannon, *The Wisdom of the Body*, Norton, NY, 1932, p. 24.

[923] Walter B. Cannon, *The Wisdom of the Body*, Norton, NY, 1932, p. 302.

[924] Hans Selye, "A syndrome produced by diverse nocuous agents," *Nature*, 138, 32, 1936.

[925] When discussing the "evolution of intercellular altruism," Selye maintained that "the strength of the whole depended upon all its parts."
Further, he argued that "We invariably die because one vital part has worn out too early in proportion to the rest of the body. Life, the biologic chain that holds our parts together, is only as strong as its weakest vital link. When this breaks – no matter which vital link it be – our parts can no longer be held together as a single living being."
Therefore, according to Selye, one possible way toward life-extension is "*deviation*, the frequent shifting-over of work from one part to the other. The human body – like the tires on a car, or the rug on a floor – wears longest when it wears evenly."
Selye was "certain that the natural human life-span is far in excess of the actual one" since he "*[did] not think anyone has ever died of old age yet*" (emphasis in the original). Reaching that "natural" life-span, far exceeding our present one, "would be the ideal accomplishment of medical research." But even this "natural" life-span can be overcome, if "discovering how to regenerate adaptation energy."
(Hans Selye, *The Stress of Life*, McGraw-Hill, NY, 1956, "To die of old age," pp. 276-277, "The evolution of intercellular altruism," pp. 282-283, "The road ahead," pp. 303-305.
Selye argued along similar lines in *Stress Without Distress*, Hodder and Stoughton, London, 1974, "Stress and Aging," pp. 93-99, "Reflections on egotism," pp. 61-70.)

[926] Walter B. Cannon, *The Wisdom of the Body*, Norton, NY, 1932, p. 295.

[927] Walter B. Cannon, *The Wisdom of the Body*, Norton, NY, 1932, pp. 299-300.

[928] Walter B. Cannon, *The Wisdom of the Body*, Norton, NY, 1932, pp. 302-303.
[929] Cannon specifically referred to the following means of economic stabilization:
"Among these suggestions are the establishment of a national economic council or a business congress or a board of industries or of trade associations, representing key industries or the more highly concentrated industries, and endowed (in some of the schemes) with mandatory power to coordinate production and consumption for the benefit of wage earners;
provision for regularity and continuity of employment, with national employment bureaus as an aid, with unemployment insurance as a safety device, and with planned public works as a means of absorbing idle workmen;
incentives for the preservation of individual initiative and originality in spite of the dangers of fixed organization;
shortening of the working time and prohibition of child labor; the raising of the average industrial wage…"
(Walter B. Cannon, *The Wisdom of the Body*, Norton, NY, 1932, p. 303.)
[930] Walter B. Cannon, *The Wisdom of the Body*, Norton, NY, 1932, pp. 303-304.
[931] Walter B. Cannon, *The Wisdom of the Body*, Norton, NY, 1932, p. 305.
[932] Walter B. Cannon, *The Wisdom of the Body*, Norton, NY, 1932, p. 303.
[933] C.C. Furnas, *The Next Hundred Years. The Unfinished Business of Science*, Reynal & Hitchcock, New York, 1936, hereafter referred to as "C.C. Furnas, 1936".
[934] C.C. Furnas, 1936, Ch. 10, "The Price of Progress," p. 136.
[935] C.C. Furnas, 1936, Ch. 31, "The Life of Assurance," pp. 390-391.
[936] C.C. Furnas, 1936, Ch. 1, "The Battle of Eugenics," pp. 7-22.
[937] C.C. Furnas, 1936, Ch. 1, "The Battle of Eugenics," p. 93.
[938] C.C. Furnas, 1936, Ch. 7, "What of Death?" p. 98.
[939] C.C. Furnas, 1936, Ch. 7, "What of Death?" p. 100.
[940] The institutionalizing events in aging and longevity research in France, Germany, Romania, Switzerland and Russia, have been discussed in the previous chapters.
[941] Edmund V. Cowdry (Editor), *Problems of Ageing. Biological and Medical Aspects*, Williams and Wilkins, Baltimore MD, 1939.
The second edition appeared in 1942 (edited by E.V. Cowdry), and the third, edited by Albert I. Lansing, was published in 1952 by Williams and Wilkins, Baltimore MD.
[942] Recently, the Gerontological Society's largest section has been "Social and Behavioral Sciences" (comprising over a half of the society's 5,400+ members, as of 2012), in addition to the "Social Research, Policy and Practice" section.
Concurrently, the goals of life-extension have hardly ever been mentioned, either in these or the relatively smaller "Biological Sciences" and "Health Sciences" sections.
(The Gerontological Society of America, http://www.geron.org/, http://www.geron.org/Membership.)
[943] Nathan W. Shock, *The International Association of Gerontology. A Chronicle. 1950 to 1986*, Springer Publishing Company, NY, 1988, Ch. 1 "Testing the waters," pp. 1-31.
[944] American Geriatrics Society (AGS) www.americangeriatrics.org/; Gerontological Society of America (GSA) www.geron.org; American Association of Retired Persons (AARP) www.aarp.org; Duke University Center for the Study of Aging and Human Development, http://www.geri.duke.edu/; National Institute on Aging (NIA) http://www.nia.nih.gov/.
[945] Additionally, Richard Nixon signed the National Cancer Act of 1971, declaring "war on cancer" and providing a major impetus for anti-cancer research worldwide.
(http://www.csicop.org/si/show/war_on_cancer_a_progress_report_for_skeptics/; http://legislative.cancer.gov/history/phsa/1971/; http://legislative.cancer.gov/history/1937.)

946 Public Law 93-296, 93rd Congress, 2nd Session, 1974, "Research on Aging Act of 1974," in *United States Code Congressional And Administrative News*, *93rd Congress - Second Session*, 1974, http://www.govtrack.us/congress/bills/93/hr6175.
The NIA founding director, Dr. Robert Butler (1927-2010), later grounded the American Association for Aging Research (1979), The Alliance for Aging Research (1986) and The International Longevity Center (1990). Further see Robert Butler, *The Longevity Revolution: The Benefits and Challenges of Living a Long Life*, Perseus Books, NY, 2008.

947 Born and educated in Canada, after his graduation from the University of Toronto in 1909, Cowdry moved to the US and worked successively at the University of Chicago, Johns Hopkins University, Rockefeller Institute, and the Washington University School of Medicine in St. Louis, Missouri. In 1939 he became the director of research at Barnard Free Skin and Cancer Hospital in St Louis. In 1941 he was appointed head of the Department of Anatomy and in 1950 became the director of the Wernse Cancer Research Laboratory at the Washington University School of Medicine, St Louis.
(J. T. Freeman, "Edmund Vincent Cowdry, creative gerontologist: memoir and autobiographical notes," *The Gerontologist*, December 24(6), 641-645, 1984.)

948 Edmund V. Cowdry (Editor), *Problems of Ageing. Biological and Medical Aspects*, Williams and Wilkins, Baltimore MD, 1939, hereafter referred to as "Cowdry, *Problems of Ageing*, 1939."

949 Cowdry, *Problems of Ageing*, 1939, Lawrence K. Frank [General Education Board and Josiah Macy Foundation, New York, 1890-1968], "Foreword", p. xviii.

950 Cowdry, *Problems of Ageing*, 1939, Clive M. McCay [Cornell University, Ithaca, New York, 1898-1967], "Chemical Aspects of Ageing," pp. 572, 585, 617.
Langstroth L., "Relation of the American dietary to degenerative diseases," *Journal of the American Medical Association*, 93, 1607-1613, 1929.

951 Cowdry, *Problems of Ageing*, 1939, Walter B. Cannon [Harvard Medical School, Boston, MA, 1871-1945], "Aging of homeostatic mechanisms," pp. 636, 539.

952 Cowdry, *Problems of Ageing*, 1939, Edmund V. Cowdry [Washington University, St. Louis, 1888-1975], "Ageing of tissue fluids," pp. 642, 686-690.

953 Cowdry, *Problems of Ageing*, 1939, William de Berniere MacNider [University of North Carolina, Chapel Hill, 1881-1951], "Ageing processes considered in relation to tissue susceptibility," pp. 695-716.

954 *Cowdry's Problems of Ageing. Biological and Medical Aspects*, 3rd Edition, edited by Albert Ingram Lansing [1915-2002, Washington University School of Medicine, St. Louis], The Williams and Wilkins Company, Baltimore, 1952, Edmund V. Cowdry, "Foreword" p. vi.

955 Cowdry, *Problems of Ageing*, 1939, Louis Israel Dublin [Metropolitan Life Insurance Company, New York, 1882-1969], "Longevity in retrospect and in prospect," pp. 100-119 (in the 1st edition); and in the 3rd edition of *Cowdry's Problems of Ageing*, 1952, Louis I. Dublin, "Longevity in retrospect and in prospect," pp. 203-220.

956 Robert H. Binstock, "The War on 'Anti-Aging Medicine'," *The Gerontologist*, 43(1), 4-14, 2003; Imre Zs-Nagy, "Is consensus in anti-aging medical intervention an elusive expectation or a realistic goal?" *Archives of Gerontology and Geriatrics*, 48(3), 271-5, 2009.

957 Nathan W. Shock, *The International Association of Gerontology. A Chronicle. 1950 to 1986*, Springer Publishing Company, NY, 1988.

958 In the First World Congress of Gerontology on July 10-12, 1950, in Liege, Belgium, 14 countries were represented by 113 attendees. The Second Congress in 1951 in St. Lois, USA, hosted 655 attendees from 51 countries. The 18th World Congress in 2005 in Rio de Janeiro, Brazil, had 4000 attendees from 63 countries. The 19th Congress in 2009 in Paris, France, included 5,800 participants from 91 countries (http://www.iagg.info/history/past-world-congresses). The latest 20th IAGG World Congress took place on June 23-27, 2013, in Seoul, Korea, and included 4,289 participants from 86 countries (http://www.iagg2013.org/Eng/20th-1.php).

The Association presidents included consecutively:
Lucien Brull of Belgium (1950-1951), Edmund Vincent Cowdry of the US (1951-1954), Joseph Harold Sheldon of the UK (1954-1957), Enrico Greppi of Italy (1957-1960), Louis Kuplan of the US (1960-1963), Torben Geill of Denmark (1963-1966), Walter Doberauer of Austria (1966-1969), Nathan Shock of the US (1969-1972), Dmitry Chebotarev of the USSR (1972-1975), David Danon of Israel (1975-1978), Mototaka Murakami of Japan (1978-1981), Hans Thomae of West Germany (1981-1983), Ewald Busse of the US (1983-1989), Samuel Bravo Williams of Mexico (1989-1993), Edit Beregi of Hungary (1993-1997), Gary Andrews of Australia (1997-2001), Gloria Gutman of Canada (2001-2005), Renato Maia Guimaraes of Brazil (2005-2009), Bruno Vellas of France (2009-2012), Heung Bong Cha of South Korea (2013-2017). The president elect for the next term is John Rowe of the US, and the 21st World Congress of IAGG is set to take place in 2017 in San Francisco, US (http://www.iagg.info/organization/executive; http://www.iagg.info/news-iagg).

[959] Nathan W. Shock, *The International Association of Gerontology. A Chronicle. 1950 to 1986*, Springer Publishing Company, NY, 1988, "Appendix II. Official Rules of the International Association of Gerontology. December 3, 1950", pp. 304-311.

[960] Gordon Wolstenholme, "The CIBA Foundation," *Science*, 117 (3039), 3, March 27, 1953.

[961] *CIBA Foundation Colloquia on Ageing, Vol. 1, General Aspects*, Wolstenholme G.E.W. and Margaret P. Cameron (Eds.), J. & A. Churchill Ltd, London, 1955; *Man and His Future: A CIBA Foundation Volume*, Gordon Wolstenholme (Ed.), Little, Brown and Co., Boston, 1963.

[962] Nathan W. Shock (Ed.), *A Classified Bibliography of Gerontology and Geriatrics*, Stanford University Press, Stanford CA, 1951; *Supplement One 1949-1955*, 1957; *Supplement Two 1956-1961*, 1963, category "Theories" (reprinted in 1980 by Stanford University Press).

[963] According to the data of the Organisation for Economic Co-operation and Development (OECD, published in 2011), as of 2009, the total expenditures on health in the US were approximately $ 2.3 Trillion, or about 17.4% of the country's Gross Domestic Product (GDP).
For comparison, in 2009, the total health expenditure in Japan was about $375 billion (8.5% GDP), Germany - 370 billion (11.6% GDP), France $300 billion (11.8% GDP), UK - $205 billion (9.8% GDP), Spain - $133 billion (9.5% GDP) and Russia ~$70 billion (4.8% GDP).
According to the *OECD Health Data 2011*, as of 2009, with the estimated $7,960 per capita yearly health expenditures (17.4% GDP), the US also led the world in relative terms, followed by Norway (~$5,350, 9.6% GDP) and Switzerland ($5,150, 11.4% GDP). (*Eurostat* http://www.oecd.org/document/16/0,3746,en_2649_37407_2085200_1_1_1_37407,00.html; http://epp.eurostat.ec.europa.eu/portal/page/portal/statistics/search_database.)
For comparison, in 2009, the global health care spending was about $5.5 Trillion (~10% global GDP), and the world average health care expenditure per capita was about $900, with the lowest value of about $20 per person a year (3% GDP) in Eritrea, Africa.
Also according to the slightly diverging estimate of the *World Health Statistics 2011*, as of 2008 the health expenditure in the US was the highest: ~$7,150 per capita (15.2% GDP), followed by Monaco (~$6,000, 3.6% GDP) and Luxemburg ($5,750, 6.8% GDP). (*World Health Statistics 2011*, WHO Department of Health Statistics and Informatics, May 13, 2011, http://www.who.int/whosis/whostat/2011/en/index.html; http://www.who.int/countries/en/; http://www.who.int/nha/en/.)

[964] Compared to the large general expenditures on health care and health research, in the US, the expenditures on the basic research of aging and longevity have been relatively negligible.
After its inauguration in 1974, the National Institute on Aging (NIA) started its first budget in 1976 with $19 million, the lowest within the National Institutes of Health – NIH (~0.8% within the total $2.3 billion NIH budget).

Since then, the research of aging has remained relatively low on the priority lists. In 1993, the NIH budget appropriation was $10.3 billion, the NIA received $400 million (3.8%). In 2001, NIH - $20.5 billion, NIA – $786 million (3.8%). In 2002, NIH - $23.3 billion, NIA – $892 million (3.7%).

In 2003, NIH - $27.1 billion (after an almost $4 billion increase compared to 2002), NIA - $994 million (~3.7%). In this year, within the NIA, only about $150 million (0.55% of the NIH budget) went to the "Biology of Aging" program, dealing with basic research of aging and experimental life-extension.

In 2008, NIH received $29.6 billion, NIA - $1.05 billion (~3.55%), Biology of Aging - $176 million (0.59%). In 2010, NIH - $31.2 billion, NIA - $1.110 billion (~3.56%), Biology of Aging - $182 million (0.58%). In 2011, NIH – $30.9 billion, NIA - $1.100 billion (3.56%), Biology of Aging - $175 million (0.57%). And in 2012, NIH – $30.87 billion, NIA – $1.102 billion (3.57%), Biology of Aging - $176.15 million (0.57%).

For 2013, the "Biology of Aging" program was requested to increase by $97,000 (from $176.154 million to $176,251), and the total NIA budget was about to increase from 1,102 to 1,110 (by about 8 million). The overall NIH budget request for 2013 was $30.860 billion (as published in 2012, with 0.57% of it dedicated to the biology of aging).

For 2014, the situation for the NIA was hoped to be brighter. Thus, for 2014, the NIA budget was set to increase from $1,110 to about $1,193 million. Most of the $83 million increase was to be given to the Neuroscience division of the NIA, including an $80 million increase for Alzheimer's disease (AD) related work, in no small measure thanks to Alzheimer's disease patients advocacy.

Under the deficit of effective and active advocacy for research on biological aging as a root cause of age-related diseases, the budget for the biology of aging program for 2014 would remain almost the same, or only slightly improve, rising from $179.725M to $180.575M (an increase of about $850,000, but still about 0.57% of the overall NIH budget request of $31.331 billion).

These figures are somewhat confounded by the estimated $1.7 billion effective cut in NIH budget via "sequestration", coming into effect in March 2013, and effectively reducing the NIH Budget to about $29 billion. According to some estimates, the National Institute of Aging was actually about to lose $86 million due to the sequestration (http://publichealthfunding.org/uploads/Sequestration_Impacts.pdf). Also "enacted" sums often differ from "actual" sums.

Still, in any case, nowhere in the world are there any aging research programs of comparable magnitude.

(Budget details of the NIH, NIA and "The Biology of Aging Program" are found at *National Institutes of Health, Office of Budget, Appropriations History by Institute/Center (1938 to Present)* http://officeofbudget.od.nih.gov/approp_hist.html; *National Institute on Aging. Budget Requests*, http://www.nia.nih.gov/AboutNIA/BudgetRequests/; *Association of American Universities. Budget and Appropriations*, http://www.aau.edu/budget/article.aspx?id=4660; Randy Barrett, "Aging Research Funding Holds Steady in '04 Request," *Sage Crossroads*, March 10, 2003, http://www.sagecrossroads.net/node/346; Chris Mooney, "What's Next for Longevity Research?" *Sage Crossroads*, May 19, 2003, http://www.sagecrossroads.net/node/337; *Longecity*, March 2005, http://www.longecity.org/forum/topic/5729-cureagingorg/.)

For comparison, in the European Union, as published in 2011, within the proposed Multi-Annual Financial Framework for 2014-2020 (MFF) with the budget of €1,025bn, about €80 billion were to be allocated to the Horizon 2020 (H2020) program, which would succeed the EU Framework Programme for Research (FP7) as Europe's main instrument for funding research and innovation. Out of the originally planned €80 billion (as published in 2011), about €3.1 billion were to be allocated to the program on "Future and Emerging Technologies" (within the Excellent Science Division) and about €8.033 billion were to be allocated to "Health, demographic change and wellbeing program" (within the Societal Challenges division). Yet no specific mention of biology of aging, not to mention life-extension, could be found in these programs (http://ec.europa.eu/research/index.cfm; http://ec.europa.eu/research/horizon2020/pdf/press/horizon_2020_budget_constant_2011.pdf).

As of July 2013, the agreed budget for Horizon2020 program was set to decrease to about 70 billion, yet there was still no mention of biology of aging.

Also, within the European Innovation Partnership on Active and Healthy Ageing (EIP AHA), as of 2012, the three major components were 1) Prevention, Screening and Early Diagnosis; 2) Care and Cure; and, 3) Active Ageing and Independent Living. Yet, there was almost no explicit mention of fundamental or applied biomedical research on biology of aging and longevity either. http://ec.europa.eu/research/innovation-union/index_en.cfm?section=active-healthy-ageing&pg=about

In many other places of the world, however, research of aging generally and biology of aging in particular seem to be considered of even less importance, and not considered as a budget item at all. Thus, undoubtedly the US holds an indisputable leadership in the field, despite the current limited funding.

[965] One of the notable recent campaigns to increase funding for aging research was the "Longevity Dividend" campaign started in 2006.

(See S. Jay Olshansky, Daniel Perry, Richard A. Miller, and Robert N. Butler, "In Pursuit of the Longevity Dividend: What Should We Be Doing To Prepare for the Unprecedented Aging of Humanity?" *The Scientist*, 20(3), p. 28, March 2006, http://www.the-scientist.com/article/display/23191/; *Pursuing the Longevity Dividend. Scientific Goals for an Aging World*, September 12, 2006, including a full list of the campaign signatories: http://www.edmontonagingsymposium.com/files/eas/Longevity_Dividend_Signatories.pdf.)

Another campaign was *The A4M Twelve-Point Actionable Healthcare Plan*, proposed in July 2009 by the American Academy of Anti-Aging Medicine - A4M.

(Ronald M. Klatz, Robert M. Goldman, Joseph C. Maroon, Nicholas A. DiNubile, Michael Klentze, *The A4M Twelve-Point Actionable Healthcare Plan: A Blueprint for A Low Cost, High Yield Wellness Model of Healthcare by 2012*, A4M, 22 July, 2009, p. 4, http://www.waaam.org/twelve_points_summary.php.)

Lobbying and fund-raising efforts have also been carried out by Friends of the National Institute on Aging, http://friendsofnia.ning.com/ and other groups.

[966] Istvan Szabo (Budapest, Hungary), "Purposes and aims of a projected Old Age Institute," *Altersprobleme*, 1(2), 49-50, 1937.

[967] Examples of political dissent included the American biologist, the Nobel Laureate in physiology/medicine of 1946, Hermann Joseph Muller (1890-1967), who was infatuated with Soviet Socialism and briefly sojourned in the USSR in the 1920s and 1930s. Hermann Muller's sympathetic views on life extension are exemplified in *Man and his Future*, 1963 (Hermann J. Muller, "Genetic progress by voluntarily conducted germinal choice," in *Man and His Future: A CIBA Foundation Volume*, Gordon Wolstenholme (Ed.), Little, Brown and Co., Boston, 1963, pp. 247-262).

Other Western life-extensionists, such as the French Felix d'Herelle (1873-1949) and the Austrian Paul Kammerer (1880-1926), also sympathized with communism and for brief periods worked in Soviet Russia.

(Felix d'Herelle's proponency of Metchnikoff's theory of aging is referred to in Scott H. Podolsky, "Cultural Divergence: Elie Metchnikoff's Bacillus bulgaricus Therapy and His Underlying Concept of Health," *Bulletin of the History of Medicine*, 72 (1), 1-27, 1998.

Paul Kammerer's enthusiastic outlooks on longevity and rejuvenation are exemplified in his book, *The Inheritance of Acquired Characteristics*, Boni and Liveright, New York, 1924, XLIX "Inheritance and Old Age," pp. 306-316, L "Inheritance and Rejuvenation," pp. 317-327.)

In Russia too, examples of political dissent among life-extensionists were not numerous. These included the anti-Czarist revolutionary activity of Alexander Bogdanov (1873-1928). Notably, however, Bogdanov's theory of "Tectology" or the "Universal Science of Organization," first formulated during his revolutionary years, propagandized the ideals of social and physiological stability and planning, which, according to him, could not be realized under Aristocracy. After the establishment of the Soviet state, his ideas became normative within the mainstream state ideology.

Other examples were the escape from Soviet Russia of Vladimir Korenchevsky (1880-1959) to England in 1919 and the defection of Abram Zalmanov (1876-1964) to France in 1921. Still, after the emigration, both Korenchevsky and Zalmanov strongly emphasized the importance of a stable social and biological organization.

More recently, in 1974, there was the exile from the USSR to the UK of the dissident gerontologist Zhores Medvedev (b. 1925).

Another example of dissension was the assimilation in the US of the Iranian transhumanist philosopher Fereydun M. Esfandiary (1930-2000) in the times of unrest in Iran in the early 1950s.

In Britain, Alex Comfort (1920-2000) was well known for his anarchistic views (though he was quite well established economically).

[968] Examples of local social adaptations of life-extensionism in Russia and Germany were discussed in greater detail in the previous chapters. A few proponents and their works may be recalled.

In Russia: Alexander Alexandrovich Bogomolets, *The Prolongation of Life* (translated by Peter V. Karpovich and Sonia Bleeker, Essential Books, Duel, Sloan and Pearce, NY, 1946, first published in Russian in 1938); Ivan Mikhailovich Sarkizov-Serazini, *Put k Zdoroviu, Sile I Dolgoy Zhizni* (The Way to Health, Strength and Long Life, Fizkultura i Sport, Moscow, 1987, first published in 1956, including Sarkizov-Serazini's articles written in the 1930s-1940s).

In Germany: Ludwig Roemheld, *Wie verlängere ich mein Leben?* (How do I prolong my life?), Ferdinand Enke Verlag, Stuttgart, 1941; Johannes Steudel, "Zur Geschichte der Lehre von den Greisenkrankheiten," (The history of the study of the diseases of old age), *Sudhoffs Archiv für Geschichte der Medizin und der Naturwissenschaften*, 35, 1-27, 1942; Gerhard Venzmer, *Lang leben und jung bleiben!* (Live long and stay young), Franckh'sche Verlagshandlung, Stuttgart, 1941.

As discussed previously in this chapter, in the US, marked examples of social adaptation can be found in Walter Cannon, *The Wisdom of the Body*, Norton, NY, 1932, "Epilogue: Relations of Biological and Social Homeostasis," pp. 287-306; Clifford Cook Furnas, *The Next Hundred Years. The Unfinished Business of Science*, Williams and Wilkins, New York, 1936.

[969] Nathan W. Shock (Ed.), *A Classified Bibliography of Gerontology and Geriatrics*, Stanford University Press, Stanford CA, 1951; *Supplement One 1949-1955*, 1957; *Supplement Two 1956-1961*, 1963, categories "Rejuvenation" and "Theories" (reprinted in 1980 by Stanford University Press).

In Shock's bibliography, the "rejuvenation" category includes about 270 works, and the "theories" category includes about 350 works. Altogether, Nathan Shock's *Classified Bibliography* (including the 2 supplements) lists over 52,000 items.

A valuable bibliographic selection is "The Nathan Shock list" of 201 major publications in biogerontology (reproduced in full in George T. Baker and W. Andrew Achenbaum, "A Historical Perspective of Research on the Biology of Aging from Nathan W. Shock," *Experimental Gerontology*, 27, 261-273, 1992).

A highly valuable annotated English bibliography is William G. Bailey's *Human Longevity from Antiquity to the Modern Lab. A Selected, Annotated Bibliography*, Greenwood Press, Westport CN, 1987 (approximately 1,400 entries).

A thorough on-line bibliography is *Medlina Anti-aging, Longevity and Rejuvenation Books* (over 1500 entries, from the sixteenth century, up to and including 2003, http://replay.waybackmachine.org/20050208215831/http:/medlina.com/anti-aging_longevity_books.htm).

About 3,000 books on longevity were listed on Amazon.com as of May 2011. As of May 2012, there were about 3,500 books on the subject. In August 2013, there were about 4,400, and in July 2014 – over 5,000 books (http://www.amazon.com/).

Over 120 recent books related to radical life extension are listed in *Religion and the Implications of Radical Life Extension*, Edited by Derek Maher and Calvin Mercer, Palgrave Macmillan, NY, 2009, "Expanded Bibliography and Other Resources" (http://www.ecu.edu/religionprogram/mercer/docs/rler.pdf).

Another large bibliography is *Longevity. Webster's Timeline History. 1586-2007*, edited by Philip M. Parker, ICON Group International, San Diego, CA, 2009, including over 3,200 entries.

Generally, in any book on aging and longevity, the bibliographies are usually massive, reflecting the complexity of the problem.

[970] Benno Romeis, *Altern und Verjüngung. Eine Kritische Darstellung der Endokrinen "Verjüngungsmethoden", Ihrer Theoretischen Grundlagen und der Bisher Erzielten Erfolge*, Verlag von Curt Kabitzsch, Leipzig, 1931 (Aging and Rejuvenation. A Critical Presentation of Endocrine "Rejuvenation Methods," Their Theoretical Foundations and Up-to-Date Successes); Max Bürger, *Altern und Krankheit* (Aging and Disease), Zweite Auflage, Georg Thieme, Leipzig, 1954 (1947).

[971] Elie Metchnikoff, *Etudy o Prirode Cheloveka* (Etudes on the Nature of Man), The USSR Academy of Sciences Press, Moscow, 1961 (1903), Ch. X. "Vvedenie v nauchnoe izuchenie starosti" (An introduction to the scientific study of aging), p. 201.

[972] Denham Harman, "Aging: a theory based on free radical and radiation chemistry," *Journal of Gerontology*, 11(3), 298-300, 1956.

[973] The development of synthetic hormones followed the works of Adolf Butenandt, Lavoslav Ruzicka and Ernst Laqueur, conducted between 1931 and 1935. (Eugen Steinach and John Loebel, *Sex and Life: Forty Years of Biological and Medical Experiments*, Faber, London, 1940, pp. 212-213.)

Beside rejuvenation, hormone replacements often had a humbler aim of correcting endocrine diseases and imbalances, and were mass-produced for that purpose in the 1930s by pharmaceutical concerns, such as the American Eli Lilly and Co., the German Merck, and Bayer. Yet their intended use for rejuvenation was also widespread.

[974] Betty L. Rubin, R. I. Dorfman, G. Pincus, "17-Ketosteroid Excretion in Aging Subjects," in *CIBA Foundation Colloquia on Ageing, Vol. 1, General Aspects* (Wolstenholme G.E.W. and Margaret P. Cameron, Eds.), J. & A. Churchill Ltd, London, 1955, pp. 126-137; *Man and His Future: A CIBA Foundation Volume*, Gordon Wolstenholme (Ed.), Little, Brown and Co., Boston, 1963, Discussion "Health and disease," p. 230.

[975] Robert Wilson, *Feminine Forever*, M. Evans & Co., NY, 1966; Robert Ettinger, *The Prospect of Immortality*, Doubleday, NY, 1964, pp. 58-59; Allan L. Goldstein and Abraham White, "Thymosin and Other Thymic Hormones: Their Nature and Roles in the Thymic Dependency of Immunological Phenomena," *Contemporary Topics in Immunobiology*, 2, 339-350, 1973; Law I.W, Goldstein AL, White A, "Influence of thymosin on immunological competence of lymphoid cells from thymictomized mice," *Nature*, 219, 1391-1392, 1968.

[976] The causative role of the hypothalamus in aging was proposed earlier by the Russian gerontologist Vladimir Dilman, and then expanded by the American researchers. (Vladimir Mikhailovich Dilman, "O vozrastnom povyshenii deiatelnosti nekotorikh hypotalamicheskikh centrov," *Trudy Instituta Physiologii imeni I.P. Pavlova*, 7, 326-336, 1958 (On senescent elevation of the activity of some hypothalamic centers, in Proceedings of I.P. Pavlov's Institute of Physiology); W. Donner Denckla, "Time to die," *Life Sciences*, 16(1), 31-44, 1975; Caleb Finch, "Neuroendocrinology of aging: A view of an emerging area," *Bioscience*, 25(10), 645-650, 1975.)

[977] Durk Pearson and Sandy Shaw, *Life Extension. A Practical Scientific Approach. Adding Years to Your Life and Life to Your Years*, Warner Books, NY, 1982, Appendix E "What is the government doing about aging research?" pp. 558, 566.

[978] The idea of organ transplantation is indeed very ancient.

According to the ancient Chinese legend, the physician Bian Que (Pien Ch'iao or Qin Yueren, ca. 700-350 BCE) transplanted hearts. (*Taoist Teachings. Translated from the book of Lieh-Tzü with Introduction and Notes by Lionel Giles*, J. Murray, London, 1912, Book 5, "The Questions of Pang," pp. 81-83, http://www.sacred-texts.com/tao/tt/index.htm.)

According to the ancient Indian epic Mahabharata (commonly dated 400-500 BCE and attributed to Vyasa), the body of the Magadha king Jarasandha, was fused from two halves. (The fusion of Jarasandha's body and his killing by breaking his body in two, are described in *The Mahabharata*, Book 2: Sabha Parva, Kisari Mohan Ganguli, tr., 1883-1896, "Rajasuyarambha Parva," Section 17, "Jarasandhta-badha Parva," pp. 40-41, Section 24, pp. 53-54, http://www.sacred-texts.com/hin/m02/index.htm.

According to the Indian epic Ramayana (sometimes dated 300-500 BCE), the mutilated nose and ears of the asura princess Surpanakha, sister of Ravan and Khara, could be restored. (The mutilation story is related in *Ramayan of Valmiki, Translated Into English Verse by Ralph T. H. Griffith*, 1870-1874, Book 3, Cantos 18-19, reprinted at http://www.sacred-texts.com/hin/rama/index.htm. The possibility of restoring Surpanakha's facial features is mentioned in Kampan's version of the *Ramayana*, according to Kathleen M. Erndl, "The Mutilation of Surpanakha," in Paula Richman (Ed.), *Many Rāmāyanas: The Diversity of a Narrative Tradition in South Asia*, University of California Press, Berkeley, 1991, p. 75, http://publishing.cdlib.org/ucpressebooks/.)

Actual methods of skin transplantaiton to adhere severed earlobes and restore mutilated noses, are described in the ancient Indian medical treatise, *Sushruta Samhita* (The physician Sushruta's collection), dated approximately c. 300-800 BCE. (*An English translation of the Sushruta samhita, based on original Sanskrit text, Edited and published by Kaviraj Kunja Lal Bhishagratna*, Calcutta, 1907, Vol. 1, Ch. 16, pp. 141-154.)

According to the Christian legend, the 3rd century physicians and saints, brothers Cosmas and Damian, living in Asia Minor and martyred c. 287-303 CE, were able to transplant legs. This operation was depicted in many paintings, most famously in the 15th-16th century works by Fra Angelico, Jaume Huguet, Meister des Stettener, and others. (See Harry Hayes, *An Anthology of Plastic Surgery*, Aspen Publishers, NY, 1986, pp. 40-41.)

[979] The histories of transplantation that *do not* mention the early 20th century rejuvenative transplantation surgeries, include:

Nadey S. Hakim, Vassilios E. Papalois (Eds.) *History of Organ and Cell Transplantation*, Imperial College Press, London, 2003; Roy Porter, *The Greatest Benefit to Mankind: A Medical History of Humanity*, W.W. Norton & Company, NY, 1998 (1997), Ch. XIX "Surgery," Subsection "Transplants," pp. 618-623; G. James Cerilli, *Organ Transplantation and Replacement*, Lippincott, NY, 1988; David Petechuk, *Organ Transplantation*, Greenwood Publishing Group, NY, 2006, Ch. 1. "A Brief History of Transplantation," pp. 3-24; Peter J. Morris, *Tissue Transplantation*, Churchill Livingstone, Philadelphia, 1982, "A History of Transplantation," p. 1-13.

The two latter works just include a brief mention of Voronoff (p. 19 in Petechuk and p. 5 in Morris), but only very cursorily, creating an impression of Voronoff's work as a wild weed by the highway of medical progress, rather than as a pioneering attempt of live tissue transplantation in humans.

David Hamilton, *A History of Organ Transplantation: Ancient Legends to Modern Practice*, University of Pittsburgh Press, Pittsburgh, 2012, provides a fuller account, yet rather unsympathetic, generally referring to the rejuvenation attempts as "Anarchy in the 1920s," pp. 126-153.

[980] In the 1950s, the understanding of immune rejection and tolerance mechanisms became far advanced upon the original theories of the blood groups (1901) and of the rhesus factor (1937) posited by Karl Landsteiner (1868-1943, Germany/US, the Nobel Laureate in physiology/medicine of 1930).

The developments in the study of immuno-suppression were due to Peter Medawar (UK), Frank Macfarlane Burnet (Australia) – Nobel Laureates of 1960; Baruj Benacerraf (Venezuela/US), Jean Dausset (France),

George D. Snell (US) – Nobel Laureates of 1980; James W. Black (UK), Gertrude B. Elion (US), George H. Hitchings (US) – Nobel Laureates in physiology/medicine of 1988.

[981] The live human kidney was apparently first successfully transplanted in 1954 by Joseph E. Murray, Hartwell Harrison, David Hume, and John Merril (US).

(The Russian Yuri Voronoy transplanted kidney from a dead donor in 1933. And the Americans Alexis Carrel and Levi Jay Hammond attempted a kidney transplant from a dead donor in 1911.)

The first heart valve was introduced in 1955 by Gordon Murray (Canada). In 1956, the bone marrow transplants (carrying adult stem-cells) were first effectively applied against leukemia by E. Donnall Thomas (US). Cardiac bypass grafts were introduced by Robert Goetz (US) in 1960.

The first transplantations of the liver were done by Thomas E. Starzl (1963, US), the lung by James D. Hardy (1963, US), hand by Roberto Gilbert Elizalde (1964, Ecuador), pancreas by Richard Carlton Lillehei and William D. Kelly (1966, US).

The human heart was first transplanted by the South African Christiaan Barnard in 1967. Shortly afterwards, on May 25, 1968, heart transplantation was performed in Richmond, Virginia, by Richard Lower.

In 1963, Robert J. White of Cleveland, Ohio, performed a full head transplant in a monkey.

(Further see Charles W. Hewitt, W. P. Andrew Lee, Chad R. Gordon (Eds.), *Transplantation of composite tissue allografts*, Springer, NY, 2008.)

[982] Attaran S.E., et al., "Homotransplantation of the testis," *Journal of Urology*, 95, 387-389, 1966; Sturgis S.H., Castellanos H., "Ovarian homografts in organic filter chambers," *Annals of Surgery*, 156, 367-76, 1962.

Only in the early 2000s, there were widely announced the "first" successful human transplantations of the patients' own frozen ovaries and testes (autotransplants) with subsequent childbearing.

(Gaia Vince, "Man fathers child after testicular transplant," *New Scientist*, 28 February 2001, http://www.newscientist.com/article/dn1851-man-fathers-child-after-testicular-transplant.htm; Shaoni Bhattacharya, "First baby born after ovarian transplant," *New Scientist*, 24 September 2004, http://www.newscientist.com/article/dn6444-first-baby-born-after-ovarian-transplant.html.)

[983] Philip Siekevitz, "Man of the Future," *The Nation*, vol. 187, Sept. 13, 1958, pp. 126-127; John Paul, "Culturing Animal Cells," in *Penguin Science Survey B*, S.A. Barnett and Anne McLaren (Eds.), Penguin Books, London, 1963 – quoted in Robert Ettinger, *The Prospect of Immortality*, Doubleday, NY, 1964, "Organ culture and regeneration," p. 56.

[984] The techniques of facial reconstructive surgery were perfected during WWII (notably by Archibald McIndoe, New Zealand, and others), and after the war applied to the civilian population.

Laser eye surgery was first performed by Francis L'Esperance in 1963 (US), and laser surgery/phototherapy also became a common method for skin rejuvenation, following the work of Endre Mester (Hungary, 1967). In the 1960s, synthetic fillers started to be used for skin renovation and augmentation (famously by Thomas Cronin, 1962, US, and others). (Sperber PA, "Chemexfoliation and silicone infiltration in the treatment of aging skin and dermal defects," *Journal of the American Geriatrics Society*, 12, 594-601, 1964; Baker TJ, Gordon HL, "Current approaches to facial rejuvenation," *Southern Medical Journal*, 58, 1077-1082, 1965.)

Though, arguably, cosmetic surgery is no cure for aging and does not extend the human life-span, its advanced application formed an adjunct to life-extensionism, as it involved a radical reconstruction of the human form and restoration of youthful appearance.

According to the US National Library of Medicine PubMed database, as of 2014, scientific articles relating to "rejuvenation" referred predominantly to cosmetic medicine, particularly skin rejuvenation. The sense in which the term "rejuvenation" was used in the 1920s, namely, the reversing of the aging process, or restoring the physiological and functional abilities of the aged, is now very seldom present in biomedical literature. (PubMed "Rejuvenation," http://www.ncbi.nlm.nih.gov/pubmed?term=rejuvenation.)

Only with the inauguration of the journal *Rejuvenation Research* in 2004 (reaching an impact factor of 8.571 in 2005) by the Cambridge biogerontologist Aubrey de Grey, did the term seem to begin resuming its earlier meaning in scientific literature. Articles published in *Rejuvenation Research* now account for a small portion of papers pertaining to "rejuvenation" on PubMed.

Still, most articles on "rejuvenation" on PubMed (rising from about 200 in 2005 to about 300 in 2009, 2010, and 2011, to about 400 in 2012 and 2013, with over 4000 papers published altogether through July 2014) predominantly deal with cosmetic medicine.

Now the term "anti-aging" has been commonly substituted for "rejuvenation" in its original sense. Cosmetic treatments too play a prominent role in "anti-aging," but general revitalization is also emphasized. Thus, for example, the journal *Rejuvenation Research* was formerly named *Journal of Anti-Aging Medicine* (with Michael Fossel as the editor in chief, 1998-2003).

According to PubMed, an expansion of articles on anti-aging occurred in the 1980s. Prior to 1980, just about 5 articles are listed (out of over 1500 articles on "anti-aging" up to and including July 2014).

According to PubMed, in the 1980s, articles referring to anti-aging were published in the following journals: *Journal of Gerontology; Mechanisms of Ageing and Development; Experimental Gerontology; Experimental Aging Research; Gerontology; Archives of Gerontology and Geriatrics; Journal of Nutrition; Journal of Traditional Chinese Medicine; Chinese Journal of Modern Developments in Traditional Medicine (Zhong Xi Yi Jie He Za Zhi);* and later in *International Journal of Neuroscience; Aging (Milano); Biochemistry (Moscow); Indian Journal of Experimental Biology; Acta Medica Okayama; Folia Pharmacologica Japonica (Nippon Yakurigaku Zasshi).*

(PubMed "Anti-aging" http://www.ncbi.nlm.nih.gov/pubmed?term=anti-aging.)

[985] The term "bionics" was coined by the American physician and military bio-engineer Jack E. Steele in 1958.

Some of the key figures in the artificial organ development in the 1950s included:

Charles Hufnagel developed the artificial heart valve (1951, US). Paul Zoll, Earl Bakken and Clarence Walton Lillehei (older brother of the pancreas transplantation pioneer Richard Carlton Lillehei) perfected the cardiac pace-makers (1952, US). John Gibbon (1953, US) and Willem Kolff (1957, US) were chiefly responsible for the development of the heart and lung machine. The artificial kidney was perfected by Willem Kolff in 1955, after his immigration to the US in 1950, though he developed the first crude prototype of the dialysis machine in 1945 in his native Netherlands. In 1956, Ite Boerema (Netherlands) successfully used hyperbaric oxygen chambers for resuscitation.

Some of the key developments of the 1960s:

The first balloon angioplasty (arterial stent, using a balloon-tipped catheter) for the treatment of atherosclerotic vascular disease, was conducted by Charles Dotter and Melvin Judkins of the University of Oregon, Portland, in 1964. (Later, in 1977, the German surgeon Andreas Gruentzig first performed coronary stent angioplasty.) Artificial hip replacements were designed by John Charnley (1962, UK). Soft contact lenses were developed by Otto Wichterle and Drahoslav Lim (1960, Czechoslovakia). The first prototypes of biosensors and artificial blood were made by Leland Clark (1962, US). The possibilities of creating synthetic skin were investigated by John F. Burke (1965, US).

In the 1960s, a series of "robotic arms" were constructed for the first time:

The "Unimate" (developed by George Devol, 1961, US), the "Rancho Arm" (developed in 1963 at Rancho Los Amigos Hospital in Downey, California, as an aid for the handicapped), the "Stanford Arm" (Victor Scheinman, 1969), and others, marked the advent of industrial robotics and indicated the possibilities of robotic prosthetics in humans. The actual robotic/bionic arm transplants were to emerge only in the late 1990s-early 2000s, e.g. the Edinburgh Modular Arm System (UK, 1998), and the Rehabilitation Institute of Chicago's "Bionic Arm" (US, 2001).

[986] Already in 1885, German scientists, Max von Frey and Max Gruber, experimented with the first artificial heart-lung apparatus for organ perfusion in animal studies. In 1913, experimental hemodialysis was attempted by John Abel, Leonard Rowntree and Benjamin Turner at Johns Hopkins Hospital, Baltimore,

US. (Rüdiger Kramme (Ed.), *Medizintechnik: Verfahren, Systeme, Informationsverarbeitung* (Medical technology: methods, systems, information processing), Springer, Berlin, 2002, p. 345.)

The first experimental resuscitation devices were developed in the 1900s-1930s by Russian scientists: Alexey Kuliabko (1902), Fyodor Andreev (1913), Sergey Bryukhonenko and Sergey Chechulin (1928), and Vladimir Demikhov (1937).

And perhaps the highest homage for the development of artificial organs and life-support systems is due to the French/American Alexis Carrel – the world pioneer and long-time leader in tissue and organ culture and transplantation, including out-of-body organ maintenance and treatment, using novel surgical techniques, nutritive media and the perfusion pump that he developed in cooperation with Charles Lindbergh (1935).

[987] Katherine Ott, David Harley Serlin, David Serlin, Stephen Mihm (Eds.), *Artificial Parts, Practical Lives: Modern Histories of Prosthetics*, New York University Press, NY, 2001.

[988] Not only transplantable organs were preserved through hypothermia (down to zero temperature), but entire organisms were preserved at extremely low temperatures with the aid of simple anti-freezes, such as glycerol and dimethyl sulfoxide.

In the late 1940s-early 1950s, mammals were revived from freezing temperatures (down to 0°C), and lower animals (insects, amphibians and fish) and viable human sperm and eggs were preserved in liquid nitrogen (-196°C) and liquid helium (-269°C).

The field was led by the French biologist Jean Rostand and a group of British researchers with the National Institute for Medical Research in London – Alan S. Parkes, Christopher Polge, Audrey Smith and James Lovelock. The research of the British group started in 1941 as a military assignment. In the 1950s-1960s, the techniques of cryopreservation were rapidly and enthusiastically adopted in the US.

(J. Rostand, "Glycérine et résistance du sperme aux basses températures" (Glycerol and resistance of sperm at low temperatures), *Comptes rendus des séances de l'Academie des Sciences*, 222, 1524-1525, 1946; James Lovelock, "Diathermy apparatus for the rapid rewarming of whole animals from 0°C and below," *Proceedings of the Royal Society B*, 147, 545, 1957; Audrey Smith, *Biological Effects of Freezing and Supercooling*, Williams & Wilkins, Baltimore, 1961; Polge C, Smith AU, Parkes AS, "Revival of spermatozoa after vitrification and dehydration at low temperatures," *Nature* 164(4172), 666, 1949; Lovelock JE, Bishop MW, "Prevention of freezing damage to living cells by dimethyl sulphoxide," *Nature* 183(4672), 1394-1395, 1959; Robert Ettinger, *The Prospect of Immortality*, Doubleday, NY, 1964.)

[989] Among those who doubted far-reaching advances of transplantation medicine, the British gerontologist Alexander Comfort feared that any single organ replacement could disturb the homeostasis of the body as a whole, "rather as replacement of a faulty component in an old radio may restore voltages to their correct original levels and blow out several other components which can no longer stand them" *(Man and His Future*, 1963).

According to Comfort, only "a radical interference with the whole process of ageing… will give us 150 or 200-year lifespans."

Comfort's skepticism regarding the potential uses of replacement therapy, was however countered as vague and defeatist. Thus, Comfort's teacher, the Nobel Prize-winning immunologist Peter Medawar, responded to Comfort's "[doubt] about using replacement – transplantation methods, for example – as a cure or palliative for sterile deterioration, because ageing is essentially due to a multiple failure of homoeostasis" and argued that "that is altogether too vague a description of ageing, because all human afflictions and infections in fact could be described as failures of homoeostasis. I don't think it will be possible to repudiate the idea of using replacement therapy until we have a theory of ageing, which we haven't got at present."

More generally and aggressively, Ettinger referred to cryobiologists who are principally opposed to whole body preservation and replacement as "poor devils [that] will have to hope for their own failure!"

(Alexander Comfort, "Longevity of Man and his Tissues," in *Man and His Future: A CIBA Foundation Volume*, Gordon Wolstenholme (Ed.), Little, Brown and Co., Boston, 1963, pp. 218, 230; Robert Ettinger, *The Prospect of Immortality*, Doubleday, NY, 1964, p. 167.)

[990] Quoted in Robert Ettinger, *The Prospect of Immortality*, Doubleday, NY, 1964, p. 54.

The term "biological engineering" was apparently coined in 1958 by Edward Tatum, the American geneticist and Nobel Laureate for the "discovery that genes act by regulating definite chemical events." (Edward Tatum, Nobel Lecture, December 11, 1958, http://nobelprize.org/nobel_prizes/medicine/laureates/1958/tatum-lecture.html.)

The term "genetic engineering" was introduced by the American biochemist and molecular geneticist Rollin D. Hotchkiss in 1965. (Rollin D. Hotchkiss, "Portents for a Genetic Engineering," *Journal of Heredity*, 56 (5), 197-202, 1965.)

The term "tissue engineering" was apparently first proposed by the American mechanical engineer Yuan-Cheng Fung in 1985, and later adopted at a meeting of the American National Science Foundation (NSF) in Washington, District of Columbia, in 1987, and at the subsequent NSF workshop in Granlibakken Resort, California, in 1988 (http://www.nsf.gov/pubs/2004/nsf0450/emergence.htm).

The MIT biologist Eugene Bell (1918-2007) has been commonly regarded as the "father of tissue engineering." (E. Bell, et al., "Living Tissue Formed in vitro and Accepted as Skin-Equivalent Tissue of Full Thickness," *Science*, 211 (4486), 1052-1054, 6 March, 1981; "Eugene Bell, 'father of tissue engineering,' dies at 88," *MIT News*, July 12, 2007, http://web.mit.edu/newsoffice/2007/obit-bell-0712.html.)

[991] Robert Ettinger, *The Prospect of Immortality*, Doubleday, NY, 1964, p. 50.

[992] Robert Ettinger, *The Prospect of Immortality*, Doubleday, NY, 1964, p. 135.

[993] See, for example, Michael D. West, *The Immortal Cell. One Scientist's Quest to Solve the Mystery of Human Aging*, Doubleday, NY, 2003 (e.g. Ch. 6 "The Philosopher's Stone" and Ch. 8 "The Abolition of Death," pp.126-158, 209-232); Ray Kurzweil and Terry Grossman, *Fantastic Voyage. Live Long Enough to Live Forever*, Plume, NY, 2005, especially the descriptions of prospective stages or "bridges" of biomedical technology: "Bridge Two" (replacement therapy) and "Bridge Three" (man-machine synergy) throughout the book; Aubrey de Grey and Michael Rae, *Ending Aging. The Rejuvenation Breakthroughs That Could Reverse Human Aging in Our Lifetime*, St. Martin's Press, NY, 2007 (e.g. Ch. 11, "New Cells for Old," pp. 238-273).

[994] Thus, for example, the German physiologist Benno Romeis criticized hormone replacements for rejuvenation, yet maintained that these therapies have some value in treating hormonal imbalances. The German internist Max Bürger critically reassessed both hormonal replacements and Bogomolets' "Anti-Reticular Cytotoxic Serum" and found their delivery to be far short of their promise. Still, according to him, they were not entirely ineffectual.

(Benno Romeis, *Altern und Verjüngung. Eine Kritische Darstellung der Endokrinen "Verjüngungsmethoden", Ihrer Theoretischen Grundlagen und der Bisher Erzielten Erfolge* (Aging and Rejuvenation. A Critical Presentation of Endocrine "Rejuvenation Methods," Their Theoretical Foundations and Up-to-Date Successes), Verlag von Curt Kabitzsch, Leipzig, 1931; Max Bürger, *Altern und Krankheit* (Aging and Disease), Zweite Auflage, Georg Thieme, Leipzig, 1954 (First published in 1947, Part A, Ch. 5. "Aging Theories," pp. 8-41, and Ch. 25 "Macrobiotics," pp. 279-282).

[995] Nathan W. Shock (Ed.), *A Classified Bibliography of Gerontology and Geriatrics*, Stanford University Press, Stanford CA, 1951; *Supplement One 1949-1955*, 1957; *Supplement Two 1956-1961*, 1963, categories "Rejuvenation" and "Theories" (reprinted in 1980 by Stanford University Press).

[996] Albert I. Lansing, "General Physiology," in *Cowdry's Problems of Ageing. Biological and Medical Aspects*, 3rd Edition, Albert I. Lansing (Ed.), The Williams and Wilkins Company, Baltimore, 1952, p. 4 (the book's first edition was in 1939, and the second was in 1942); Alexander A. Bogomolets, *The Prolongation of Life*, translated by Peter V. Karpovich and Sonia Bleeker, Essential Books, Duel, Sloan and Pearce, NY, 1946, pp. 18-23, 83; Max Bürger, *Altern und Krankheit*, 1954, pp.16-17; Vladimir Nikitin, "I.I. Metchnikoff i

problema dolgoletia" (I.I. Metchnikoff and the problem of longevity), in *Collected Works of I.I. Metchnikoff* (Ed. N.N. Anichkov and R.I. Belkin), The USSR Academy of Medical Sciences, Moscow, 1962, vol. 15, pp. 375-400.

[997] Auguste Lumière, *La Vie, La Maladie et La Mort, Phénomènes Colloïdaux* (Life, Disease and Death as Colloidal Phenomena), Masson & Cie, Libraires de L'Académie de Médecine, Paris, 1928, pp. 63-71; Auguste Lumière, *Sénilité et Rajeunissement* (Aging and Rejuvenation), Librairie J.-B. Baillière et Fils, Paris, 1932; Auguste Lumière, *La Renaissance de la Médecine Humorale*, Deuxie edition (The Renaissance of Humoral Medicine), Imprimerie Léon Sézanne, Lyon, 1937; Auguste Lumière, "La Médecine Humorale et ses Résultats" (Humoral Medicine and its Results), in *Médecine Officielle et Médecines Hérétiques* (Official Medicine and Heretical Medicines), "Présences" Plon, Paris, 1945, pp. 53-62.

[998] Raymond Pearl, *The Biology of Death*, J.B. Lippincott Company, Philadelphia, 1922, Ch. VI "The Inheritance of Duration of Life in Man," pp. 150-185.

In the late 1930s through the late 1940s, the inheritance of longevity was studied by the Swiss biologists Alfred Vogt (1938) and Ernst Hanhart (1939) with a particular focus on age-related eye diseases in twins, and by the Finnish Eeva Jalavisto (1948, 1951), cited in Max Bürger's *Altern und Krankheit*, 1954, pp. 278-279.

[999] Max Bürger, *Altern und Krankheit* (Aging and Disease), Zweite Auflage, Georg Thieme, Leipzig, 1954 (1947), pp. 7, 561.

[1000] The studies on polypeptide chains by William Lawrence Bragg, John Kendrew, Max Perutz (1950, UK), Linus Pauling and Robert Corey (1950, 1951, US) and others, and, most momentously, the positing of the Double Strand DNA model by James Watson and Francis Crick (1953, UK), elucidated the molecular structure of proteins and genes, and marked the beginning of the era of molecular biology.

Pivotal works on protein and DNA molecular structure included:

Bragg WL, Kendrew JC, and Perutz MF, "Polypeptide chain configurations in crystalline proteins," *Proceedings of the Royal Society of London A*, 203, 321-357, 1950; Pauling L and Corey RB, "Two hydrogen-bonded spiral configurations of the polypeptide chain," *Journal of the American Chemical Society*, 72, 5349, 1950; Pauling L and Corey RB, "The structure of synthetic polypeptides," *Proceedings of the National Academy of Sciences USA*, 37, 241-250, 1951; Watson JD, Crick FH, "Molecular structure of nucleic acids: a structure for deoxyribose nucleic acid," *Nature*, 171 (4356), 737-738, April 1953.

Histories of Molecular Biology include:

Soraya de Chadarevian, *Designs for Life: Molecular Biology After World War II*, Cambridge University Press, Cambridge, 2002; Jack D. Dunitz, "Linus Carl Pauling. February 28, 1901 – August 19, 1994," *Biographical Memoirs of Fellows of the Royal Society*, 42, 316-338, 1996; Michel Morange, *A History of Molecular Biology* (translated from French by Matthew Cobb), Harvard University Press, Cambridge MA, 2000.

The constructs of molecular biology were largely derived from the application of quantum mechanics in life-sciences and the development of such analytical techniques as X-ray crystallography, ultracentrifugation, scattering spectroscopy, electron microscopy, electrophoresis, and partition chromatography.

The application of quantum mechanics in life-sciences was chiefly associated with the works of Max Delbrück (1942) and Erwin Schrödinger (1944). (Erwin Schrödinger, *What is Life?* Cambridge University Press, Cambridge, 1996, first published in 1944.)

The development of the analytical techniques of X-ray crystallography is commonly attributed to Max von Laue, William Henry Bragg and his son William Lawrence Bragg (1912), Linus Pauling (1931), and Dorothy Hodgkin (1937).

Ultracentrifugation was designed by Theodor Svedberg (1925).

Scattering spectroscopy was perfected by Chandrasekhara V. Raman, Grigory Landsberg and Leonid Mandelstam (1928).

Electron microscopy was introduced by Ernst Ruska and Max Knoll (1931).

Electrophoresis was advanced by Arne Tiselius (1937).

Partition chromatography was developed by Richard Synge and Archer Martin (1941).

These developments profoundly affected theories of aging and anti-aging strategies.

[1001] Norman G. Anderson, "Approaches to aging on the molecular level," in Bernard L. Strehler (Ed.), *The Biology of Aging. A Symposium Held at Gatlinburg, Tennessee, May 1-3, 1957, Under the Sponsorship of the American Institute of Biological Sciences and with support of the National Science Foundation*, Waverly Press, Baltimore, 1960, pp.105-112 (hereafter referred to as "Strehler, *The Biology of Aging,* 1960"); Carol Kahn, *Beyond the Helix. DNA and the Quest for Longevity*, Times Books, NY, 1985.

[1002] Denham Harman "Aging: A Theory based on Free Radical and Radiation Chemistry," *Journal of Gerontology*, 11, 298-300, 1956; Hamilton LA and Olcott HS, "Antioxidants and the autoxidation of fats. V. Mode of action of anti- and pro-oxidants," *Journal of the American Oil Chemists' Society*, 13, 127-129, 1936.

[1003] Vitamins were applied for anti-aging much earlier than the positing of the "Free Radical" theory of aging. (Gardner TS, "The design of experiments for the cumulative effects of vitamins as anti-aging factors," *Journal of the Tennessee Academy of Sciences*, 23(4), 291-306, 1948.)

The theory and application of anti-oxidants may have owed to the established massive pharmaceutical industry, particularly vitamin production, in the US, Europe, Japan and the USSR. As many vitamins (such as A, C, D, E, several B vitamins) in time proved to possess anti-oxidant properties, their application for anti-aging, derived from the Free Radical theory, did not appear to involve any great technological or logistic change, but rather a conceptual renovation. Antioxidant-rich (vitamin-rich) foodstuffs or supplements were relatively inexpensive and could be mass-produced and widely consumed. Rich industrial/agricultural countries, generally capable of mass production and distribution, had a clear advantage in developing this field.

By 1941, all 13 major vitamins were discovered and synthesized, thanks to the works of Nikolai Lunin (1881), Takaki Kanehiro (1884), Christiaan Eijkman (1897), William Fletcher (1905), Frederick Gowland Hopkins (1906), and Umetaro Suzuki (1910) – introducing the notion of some "vital" or "accessory factors" in foods, other than just proteins, carbohydrates and fats. Casimir Funk coined the term "vitamines" in 1912.

Elmer V. McCollum identified "fat-soluble A" and "water-soluble B" vitamins in 1913-1915. Edward Mellanby isolated vitamin D in 1922. Herbert Evans and Katherine Bishop discovered vitamin E in 1922. Albert Szent-Györgyi identified ascorbic acid (vitamin C) in 1932. Edward Doisy and Henrik Dam isolated vitamin K in 1935. George Minot, William Murphy and George Whipple discovered cyanocobalamin (vitamin B12) in 1926. Otto Warburg, Paul Karrer and Richard Kuhn introduced riboflavin (vitamin B2) in 1932-1933. Lucy Wills introduced folic acid (B9) in 1933. Paul Gyorgy discovered pyridoxine (B6) in 1934. Conrad Elvehjem isolated niacin (B3) in 1937 – to name just a few researchers among many.

Pharmaceutical industry rapidly put these discoveries into practice:

In 1934, the Swiss pharmaceutical concern Hoffmann – La Roche first began the mass production of vitamin C (under the trade name Redoxon), using Tadeus Reichstein's synthesis method.

The German corporation E.Merck began mass-producing vitamin D (trade name Vigantol) in 1927, and vitamin C (trade name Cebion) in 1934. In 1936-1937, the American Merck and Company went into mass production of thiamin – vitamin B1, using the synthesis technique developed by Robert R. Williams, and riboflavin - vitamin B2, using Max Tishler's method.

(The American "Merck and Co." was then unrelated to the German "E. Merck - Merck KGaA." During WWI, in 1917, Merck's American branch was expropriated by the US government, and since then the two companies have existed independently.)

In 1935, the USSR Vitamins Institute was established to coordinate vitamins production in the Soviet Union.

(On the history of vitamins development, see Linus Pauling, *How To Live Longer And Feel Better*, W.H.

Freeman and Company, NY, 1986, Ch. 7. "How vitamins were discovered," pp. 47-54; Fred Aftalion, *A History of the International Chemical Industry*, Translated by Theodor Benfey, University of Pennsylvania Press, Philadelphia, 1991, "Nutritional disorders and the emergence of vitamins," pp. 162-164; Stuart Anderson, *Making Medicines: A Brief History of Pharmacy and Pharmaceuticals*, Pharmaceutical Press, London, 2005, p. 166.)

[1004] The cross-linkage theory of aging originated in Bjorksten's work on cross-linking in thin protein gels, as Chief Chemist for Ditto Inc. (a subsidiary of Eastman Kodak) in Chicago, Illinois (Johan Bjorksten, "Recent developments in protein chemistry," *Chemical Industries*, 48, 746-751, 1941; Johan Bjorksten, "Chemistry of duplication," *Chemical Industries*, 49, 2, 1942; http://www.bjorksten.com/history.php).

Yet, the theory rose to prominence in gerontology after the publication of Bjorksten's "A common molecular basis for the aging syndrome," *Journal of the American Geriatrics Society*, 6(10), 740-748, 1958, and later works, such as Johan Bjorksten, "Aging: a Positive Approach," *The Chemist*, 36(12), 1-12, 1959; Johan Bjorksten, "The Crosslinkage Theory of Aging," *Journal of the American Geriatrics Society*, 16(4), 408-427, 1968. A conjoined contribution to the cross-linkage theory was made by the Swiss gerontologist Fritz Verzár (c. 1955-1960, see the previous section on Verzár) and others. (Fritz Verzár, "Veränderungen der thermoelastischen Kontraktion von Sehnenfasern im Alter" (Changes in thermoelastic contraction of tendon fibers in aging), *Helvetica Physiologica Acta*, 13, 64-67, 1955; Fritz Verzár, "Das Altern des Kollagens in der Haut und in Narben" (Aging of collagens in the skin and scars), *Schweizerische medizinische Wochenschrift*, 91, 1234-1236, 1961.)

[1005] Nandy K, Bourne GH, "Effect of centrophenoxine on the lipofuscin pigments in the neurones of senile guinea-pigs," *Nature*, 210, 313-314, 1966; Zeman W, "The neuronal ceroid-lipofuscinoses – Batten-Vogt syndrome: a model for human aging?" *Advances in Gerontological Research*, 3, 147-170, 1971.

[1006] Johan Bjorksten, *Longevity, 2: Past, Present, Future*, Jab Publishing, Charleston SC, 1987, p. 104; Johan Bjorksten, *Longevity, a Quest: An Odyssey*, Bjorksten Research Foundation, Madison Wisconsin, 1981.

[1007] Aubrey de Grey and Michael Rae, *Ending Aging. The Rejuvenation Breakthroughs That Could Reverse Human Aging in Our Lifetime*, St. Martin's Press, NY, 2007, Ch. 9, "Breaking the Shackles of AGE," pp. 165-199.

[1008] Gioacchino Failla, "The aging process and cancerogenesis," *Annals of the New York Academy of Sciences*, 71, 1124-35, 1958; Gioacchino Failla, "The aging process and somatic mutations," in Strehler, *The Biology of Aging*, 1960; Leo Szilard, "On the Nature of the Aging Process," *Proceedings of the National Academy of Sciences USA*, 45(1), 30-45, January 1959.

[1009] Bernard Strehler, "Origin and comparison of the effects of time and high-energy radiation on living systems," *Quarterly Review of Biology*, 34(2), 117-142, 1959.

[1010] Vincent J. Cristofalo, "An Overview of the Theories of Biological Aging," in *Emergent Theories of Aging*, James E. Birren, Vern L. Bengtson (Eds.), Springer, New York, 1988, pp. 118-127.

[1011] Leslie E. Orgel, "The maintenance of the accuracy of protein synthesis and its relevance to aging," *Proceedings of the National Academy of Sciences USA*, 49, 517-521, 1963.

[1012] Holliday R, Tarrant GM, "Altered enzymes in aging human fibroblasts," *Nature*, 238, 26-30, 1972.

[1013] Carol Kahn, *Beyond the Helix. DNA and the Quest for Longevity*, Times Books, NY, 1985; Albert Rosenfeld, *Prolongevity. A report on the scientific discoveries now being made about aging and dying, and their promise of an extended human life span – without old age*, Alfred A. Knopf, New York, 1976, "The Cell as a Machine that Wears Out," pp. 30-42; Hart RW and Setlow RB, "Correlation between deoxyribonucleic acid excision-repair and life-span in a number of mammalian species," *Proceedings of the National Academy of Sciences USA*, 71, 2169-2173, 1974.

[1014] Albert Rosenfeld, *Prolongevity. A report on the scientific discoveries now being made about aging and dying, and their promise of an extended human life span – without old age*, Alfred A. Knopf, New York, 1976, "Heredity and Immunity," pp. 75-87, "Decelerated Aging in the Lab," pp. 106-118; Roy L. Walford, *Leukocyte Antigens and Antibodies*, Grune and Stratton Inc., New York, 1960; Roy L. Walford, *The Immunological Theory of Aging*, Munksgaard, Copenhagen, 1969; Richard Weindruch, Roy L. Walford, *The Retardation of Aging and Disease by*

Dietary Restriction, Charles C. Thomas, Springfield, Illinois, 1988; Roy L. Walford, *Beyond the 120-Year Diet: How to Double Your Vital Years*, Four Walls Eight Windows, New York, 2000.

[1015] Hayflick L, Moorhead PS, "The serial cultivation of human diploid cell strains," *Experimental Cell Research*, 25, 585-621, 1961; Leonard Hayflick, "The limited *in vitro* lifetime of human diploid cell strains," *Experimental Cell Research*, 37, 614-636, 1965.

[1016] The theories of mortality developed in the 1950s built on much earlier concepts, such as Benjamin Gompertz's law of an exponential increase in mortality rate with age, posited in 1825, or Karl Pearson's population mortality statistics of 1900. Yet, they were more elaborate, quantitatively relating the rate of (molecular) damage to the rate of mortality.

Several theories of mortality were formulated between 1952 and 1960:

According to the [Elaine] Brody-[Gioacchino] Failla theory, the "mortality rate is inversely proportional to the vitality."

In the [Henry] Simms-[Hardin] Jones theory, death was attributed to "autocatalytic accumulation of damage and disease," where "the lessening of vitality [is regarded] as the accumulation of damage" and "the rate at which damage is incurred is proportional to the damage that has already been acquired in the past."

George Sacher's theory stated that "death occurs when a displacement of the physiologic state extends below a certain limiting value."

And the [Bernard] Strehler-[Albert] Mildvan theory claimed that "the rate of death is assumed to be proportional to the frequency of stresses which surpass the ability of a subsystem to restore initial conditions."

Though these theories did not seem to refer to any specific goals of life-extension, they were designed for a thorough quantitative elucidation of the aging process, deemed a necessary precondition for any actual intervention.

(Albert S. Mildvan and Bernard L. Strehler, "A critique of theories of mortality," in Strehler, *The Biology of Aging*, 1960, pp. 216-235; Bernard Strehler, "Fluctuating energy demands as determinants of the death process (A parsimonious theory of the Gompertz function)," in Strehler, *The Biology of Aging*, 1960, pp. 309-314.)

Later on, the "Reliability theory" of aging, longevity and mortality, was proposed by Leonid Gavrilov and Natalia Gavrilov (1987, 2001). This theory "predicts that even those systems that are entirely composed of non-aging elements (with a constant failure rate) will nevertheless deteriorate (fail more often) with age, if these systems are *redundant* in irreplaceable elements."

The implications of the "reliability theory" for practical life-extension were more explicit:

"It also follows from this model that even small progress in optimizing the processes of ontogenesis and increasing the numbers of initially functional elements (j) can potentially result in a remarkable fall in mortality and a significant improvement in lifespan. This optimistic prediction is supported by experimental evidence of increased offspring lifespan in response to protection of parental germ cells against oxidative damage just by feeding the future parents with antioxidants. …

Increased lifespan is also observed among the progeny of parents with a low respiration rate (proxy for the rate of oxidative damage to DNA of germ cells…). …

The model also predicts that early life events may affect survival in later adult life through the level of initial damage. This prediction proved to be correct for such early life indicators as parental age at a person's conception …and the month of person's birth."

(Leonid A. Gavrilov, Natalia S. Gavrilova, "The Reliability Theory of Aging and Longevity," *Journal of Theoretical Biology*, 213, 527-545, 2001.)

[1017] The evolutionary theories of aging that appeared in the 1950s had direct implications for identifying specific causes of aging and raised questions regarding potential interventions:
Is aging due to a definite number of evolutionarily inherited "death genes" that could be potentially switched off? Could the self-preservation mechanisms that emerged in the course of evolution be harnessed to combat aging? What would be the trade-offs between harmful and beneficial effects of life-prolonging genetic modifications, in terms of reproduction, variability and survival? Or is aging the result of an incalculable number of random mutations, defying any scrutiny and manipulation? These questions, arising from the evolutionary theories, seem to have continued to concern researchers to the present.
(On evolutionary theories of aging, see:
Peter B. Medawar, *An Unsolved Problem of Biology*, H.K. Lewis, London, 1952; Peter B. Medawar, "The definition and measurement of senescence," in *CIBA Foundation Colloquia on Ageing, Vol. 1, General Aspects* (Wolstenholme G.E.W. and Margaret P. Cameron, Eds.), J. & A. Churchill Ltd, London, 1955, pp. 4-15; George C. Williams, "Pleiotropy, natural selection, and the evolution of senescence," *Evolution*, 11, 398-411, 1957 (an abridged version of this paper appears in Strehler, *The Biology of Aging*, 1960, pp. 332-337); George A. Sacher, "Relationship of lifespan to brain weight and body weight in mammals," in *Ciba Foundation Colloquia on Aging*, Churchill, London, 1959, vol. 5, pp. 115-133; George A. Sacher "Longevity and Aging in Vertebrate Evolution," *BioScience*, 28 (8), 497-501, 1978; Richard Cutler, "Evolution of human longevity and the genetic complexity governing aging rate," *Proceedings of the National Academy of Sciences USA*, 72 (11), 4664-4668, 1975.
The earliest works on the evolutionary basis of aging and longevity were:
August Weismann, *Über die Dauer des Lebens* (On the duration of life), G. Fischer, Jena, 1882, and August Weismann, *Über Leben und Tod* (On life and death), G. Fischer, Jena, 1884 (these works appear in English translation in August Weismann, *On Heredity*, Clarendon Press, Oxford, 1889); Elie Metchnikoff, *Etudy o Prirode Cheloveka* (Etudes On the Nature of Man), The USSR Academy of Sciences Press, Moscow, 1961 (First published in 1903), XI "Vvedenie v nauchnoe izuchenie smerti" (An introduction to the scientific study of death), pp. 214-245.
More recent works are:
Vladimir Frolkis, *Aging and Life-Prolonging Processes*, Springer-Verlag, Wien, 1982; Vladimir Frolkis, *Starenie i Uvelichenie Prodolzhitelnosti Zhisni* (Aging and Life-prolongation), Nauka, Leningrad, 1988; Michael R. Rose, *Evolutionary Biology of Aging*, Oxford University Press, Oxford, 1991; Tom Kirkwood, *Time of Our Lives. The Science of Human Aging*, Oxford University Press, Oxford, 1999; Steven N. Austad, *Why We Age: What Science Is Discovering about the Body's Journey Through Life*, John Wiley, NY, 1997; George M. Martin, "How is the evolutionary biological theory of aging holding up against mounting attacks?" *American Aging Association Newsletter*, March 2005, http://www.americanaging.org/news/mar05.html.)
Evolutionary theories of aging will be further discussed in the section on British longevity research.

[1018] The foundational works were: Norbert Wiener, *Cybernetics, or Control and Communication in the Animal and the Machine*, Wiley, NY, 1948; Shannon C.E. and Weaver W., *Mathematical Theory of Communication*, University of Illinois Press, Urbana, 1949.

[1019] Hirsch S., "Senescence, entropy, and cybernetics: a clarification of basic concepts in gerontological research," in *The 3rd Congress of the International Association of Gerontology. London. 1954. Old Age in the Modern World*, E&S Livingstone Ltd., London, 1955, pp. 622-627; Bellman R., Harris T., "On the theory of age-dependent stochastic branching processes," *Proceedings of the National Academy of Sciences USA*, 34, 601-604, 1948; Still J.W., "Are organismal aging and aging death necessarily the result of death of vital cells in the organism? A cybernetic theory of aging," *Medical Annals of the District of Columbia*, 25, 199-204, 1956; Hubert P. Yockey, "On the role of information theory in mathematical biology," Ch. 11 in *Radiation Biology and Medicine*, Addison Wesley, Reading MA, 1958, pp. 250-282; Hubert P. Yockey, "The use of information theory in aging and radiation damage," in Strehler, *The Biology of Aging*, 1960, pp. 338-347; Symposium on

Information Theory in Biology, Gatlinburg, Tennessee, October 29-31, 1956, Edited by Hubert P. Yockey, Pergamon Press, NY, 1958, Part V, "Aging and Radiation Damage," pp. 293-356.

[1020] Robert S. Ledley and Lee B. Lusted, "Reasoning foundations of medical diagnosis: Symbolic logic, probability, and value theory aid our understanding of how physicians reason," *Science*, 130 (3366), 9-21, 3 July 1959; Warner H.R., Toronto A.K., Vezsy L.G., Stephenson R.A., "Mathematical approach to medical diagnosis: application to congenital heart disease," *Journal of the American Medical Association*, 177, 177-83, 1964.

Later on, information theory (entropy measurement) has been specifically suggested for diagnosing age-related diseases and conditions and even for diagnosing aging itself.

(Lewis A. Lipsitz, Ary L. Goldberger, "Loss of 'complexity' and aging: potential applications of fractals and chaos theory to senescence," *Journal of the American Medical Association*, 267, 1806-1809, 1992; Ary L. Goldberger, C.-K. Penga, Lewis A. Lipsitz, "What is physiologic complexity and how does it change with aging and disease?" *Neurobiology of Aging*, 23, 23-26, 2002; David Blokh and Ilia Stambler, "Estimation of heterogeneity in diagnostic parameters of age-related diseases," *Aging and Disease*, 5, 2014; David Blokh and Ilia Stambler, "Information theoretical analysis of aging as a risk factor for heart disease," *Aging and Disease*, 6, 2015, http://www.aginganddisease.org/.)

Information theory was also employed for a variety of cell parameters for age-related cancer detection:

David Blokh, Ilia Stambler, Elena Afrimzon, Yana Shafran, Eden Korech, Judith Sandbank, Ruben Orda, Naomi Zurgil, Mordechai Deutsch, "The information-theory analysis of Michaelis-Menten constants for detection of breast cancer," *Cancer Detection and Prevention*, 31, 489-498, 2007; David Blokh, Ilia Stambler, Elena Afrimzon, et al., "Comparative analysis of cell parameter groups for breast cancer detection," *Computer Methods and Programs in Biomedicine*, 94(3), 239-249, 2009; David Blokh, Naomi Zurgil, Ilia Stambler, et al., "An information-theoretical model for breast cancer detection," *Methods of Information in Medicine*, 47, 322-327, 2008; David Blokh, Elena Afrimzon, Ilia Stambler, et al., "Breast cancer detection by Michaelis-Menten constants via linear programming," *Computer Methods and Programs in Biomedicine*, 85, 210-213, 2007.

[1021] "Discussion Session IX. Methodology, Information Theory, Design and Approach," in Strehler, *The Biology of Aging*, 1960, pp. 93-94.

[1022] Murray Eden, "An analogy between probabilistic automata and living organisms," in Strehler, *The Biology of Aging*, 1960, pp.167-169.

[1023] Donald M. MacKay, "Machines and societies," in *Man and His Future: A CIBA Foundation Volume*, Gordon Wolstenholme (Ed.), Little, Brown and Co., Boston, 1963, pp. 153-168 and p. 232.

[1024] Manfred E. Clynes and Nathan S. Kline, "Cyborgs and space," *Astronautics*, 13, 26-27, 74-76, September 1960, http://www.scribd.com/doc/2962194/Cyborgs-and-Space-Clynes-Kline; Jerome B. Wiesner, "Electronics and Evolution," *Proceedings of the Institute of Radio Engineers*, 50, 5, May 1962; Robert Ettinger, *The Prospect of Immortality*, Doubleday, NY, 1964, "The solid gold computer," pp. 105-110.

The possibilities of cybernetic immortality were advanced, among others, by Marvin Lee Minsky (b. 1927), a pioneer of Artificial Intelligence research, working in the 1960s at the Massachusetts Institute of Technology. As of 2014, among other positions, Marvin Minsky was listed as a scientific advisor for the Alcor Life Extension Foundation.

The current honorary scientific director of the American Federation for Aging Research (as of 2014), George M. Martin (b. 1927), spoke in 1971 about an "interim solution for immortality" via "suitable techniques of 'read-out' of the stored information from cryobiologically preserved brains into nth generation computers" (Martin G.M., "Brief proposal on immortality: an interim solution," *Perspectives in Biology and Medicine*, 14 (2), 339, 1971).

Cyborgization was also discussed in Alvin Toffler [b. 1928], *Future Shock*, Bantam Books, NY, 1971 (1970). Curiously, in Toffler's considerations of "transience and novelty," radical life extension seems to be a very minor concern. Yet, Toffler does say that "Through application of the modular principle – preservation of the whole through systematic replacement of transient components – we may add two or three decades to the average life span of the population" ("The transient organ," p. 178). Furthermore, "The brain can be isolated from its body and kept alive after the 'death' of the rest of the organism" ("The cyborgs among us," pp. 180-183).

The motifs of the indefinitely prolonged replacement of body parts and man-machine symbiosis recur in F.M. Esfandiary, *Up-Wingers. A Futurist Manifesto*, Popular Library, Toronto, 1977 (1973), pp. 173, 180; Ray Kurzweil, *The Age of Spiritual Machines. When Computers Exceed Human Intelligence*, Viking, NY, 1999; Ray Kurzweil, *The Singularity Is Near. When Humans Transcend Biology*, Penguin Books, 2005; James Hughes, *Citizen Cyborg: Why Democratic Societies Must Respond to the Redesigned Human of the Future*, Westview Press, London, 2004, and other works.

[1025] Durk Pearson and Sandy Shaw, *Life Extension. A Practical Scientific Approach. Adding Years to Your Life and Life to Your Years*, Warner Books, NY, 1982, "Aging research at the Atomic Energy Commission" p. 560, "You can protect yourself from pollution" pp. 264-265; Saul Kent, *The Life-Extension Revolution. The Definitive Guide to Better Health, Longer Life, and Physical Immortality*, William Morrow and Company, NY, 1980, pp. 120-121, 131.

Protection against radiation seems to have remained a major concern in life-extensionist regimens. Thus, according to Roy Walford, "Evidence strongly suggests that CRON [Calorie Restriction with Optimal Nutrition], works to extend life span and prevent disease by harmoniously upregulating the multiple systems that protect the body from the various forms of stress: the stress of high temperature, of infection, of food shortage, of irradiation, of exposure to toxic compounds" (Roy L. Walford, *Beyond the 120-Year Diet: How to Double Your Vital Years*, Four Walls Eight Windows, New York, 2000, p. 163).

[1026] John Burdon Sanderson Haldane, "Biological Possibilities for the Human Species in the Next Ten Thousand Years," in *Man and His Future: A CIBA Foundation Volume*, Gordon Wolstenholme (Ed.), Little, Brown and Co., Boston, 1963, p. 339.

[1027] Robert Ettinger, *The Prospect of Immortality*, Doubleday, NY, 1964, "The view from the Olympus: How rich can we get?" pp. 110-112.

[1028] John Burdon Sanderson Haldane, "Biological Possibilities for the Human Species in the Next Ten Thousand Years," in *Man and His Future*, 1963, p. 354; Robert Heinlein, *Methuselah's Children*, Signet, NY, 1958, p. 45.

[1029] The wish to increase the scope of the research of aging and longevity to the level of nuclear research was expressed by the British gerontologist Vladimir Korenchevsky in 1946 and 1952 (Vladimir Korenchevsky, "Conditions desirable for the rapid progress of gerontological research," *British Medical Journal*, 2, 468, Sept. 28, 1946; Vladimir Korenchevsky, "The International Association of Gerontology and Rapid Progress of Gerontology," *British Medical Journal*, 1, 375-376, Feb.16, 1952).

Recently, the desire to establish a longevity research project on the scale of the "Manhattan Project" or "Apollo Project" was expressed by Aubrey De Grey ("Aubrey de Grey wants to wish you a happy 200th birthday," interview by Andrey Kobilnyk, *FirstScience.com*, 4 Feb, 2008, http://www.firstscience.com/home/articles/humans/aubrey-de-grey-wants-to-wish-you-a-happy-200th-birthday_42775.html, and in numerous other interviews and public presentations by Aubrey de Grey).

[1030] Linus Pauling, *How To Live Longer And Feel Better*, W.H. Freeman and Company, NY, 1986, Ch. 29 "A happy life and a better world," pp. 271-274; Linus Pauling, *No More War*, Dodd, Mead and Co., NY, 1983 (first published in 1958), Ch.10 "A proposal: Research for Peace," pp. 223-237; Linus Pauling, "Observations on Aging and Death," *Engineering and Science Magazine*, California Institute of Technology, Pasadena, May 1960; Alex Comfort, *The Biology of Senescence*, Butler & Tanner, London, 1956, pp.191-194; Alex Comfort, *A Wreath for the Living*, Routledge, London, 1942, Alex Comfort, *The Signal to Engage*, Routledge, London, 1946 (collections of Comfort's anti-war, "pro-life" poems).

[1031] The adaptation of life-extensionist views to the recent scientific advancements is further exemplified by the Yugoslavian historian of medicine Mirko Dražen Grmek (*On Ageing and Old Age. Monographiae Biologicae*, 5, 2, Den Haag, 1958):

"The application of certain ideas of modern physics to gerontologic problems opens new, quite unexpected aspects. Ageing of organisms may be understood as a special example of general laws determining the direction and irreversibility of all the processes in nature (for instance [Karl] Braun-[Lois] Le Chatelier's principle and the thermodynamic law of entropy).

[Hermann] Minkowski's and [Albert] Einstein's concept of time as an equal coordinate in the pluridimensional universe offers fascinating theoretical possibilities to change the direction of irreversible processes, and consequently, at long last, – the transmutation of elements, that ancient goal of alchemists, having already been attained – to make their next dream, rejuvenation, come true. "The philosophers' stone" may well be contained in radioactivity. Does the future hold the discovery of the "elixir of life" also in store for mankind?" (p. 7).

[1032] Roger Bacon, in the *Letter Concerning the Marvelous Power of Art and of Nature*, asserted that "the possibility of the prolongation of life is confirmed by the consideration that the soul naturally is immortal and capable of not dying. … Therefore it follows that this shortening [of life] is accidental and may be remedied wholly or in part" (Chapter 7, "Prolongation of Life").

Interestingly, in that letter, Roger Bacon also spoke about the possibility of constructing mechanically powered cars, boats, flying, submerging and weight-lifting machines (Chapter 4, "Extraordinary Mechanical Inventions").

In *Opus Majus*, in evidence for the possibility of life-prolongation, Roger Bacon cited the longevity of antediluvian patriarchs, the life-prolonging properties of regulated conduct and adherence to hygiene, as well as the reported cases of super-longevity among men and animals. The conservation of the "vital heat and moisture" was, according to Roger Bacon, essential for such durability.

The possibility of a major prolongation of life was further affirmed in Roger Bacon's *The Cure of Old Age and Preservation of Youth – De Prolongatione Vitae*.

These works are quoted in Gerald J. Gruman, *A History of Ideas about the Prolongation of Life. The Evolution of Prolongevity Hypotheses to 1800*, Transactions of the American Philosophical Society, Volume 56 (9), Philadelphia, 1966, "Latin Alchemy," pp. 62-67; and in Johann Heinrich Cohausen, *Hermippus Redivivus or the Sage's Triumph over Old Age and the Grave*, J. Nourse, London, 1748, pp.16-18, 42-45, first published in Latin in 1742.

The *Letter Concerning the Marvelous Power of Art and of Nature* is published in full in English: *Letter Concerning the Marvelous Power of Art and of Nature and Concerning the Nullity of Magic*, translated from the Latin by Tenney L. Davis, The Chemical Publishing Co., Easton, Pa., 1923, and in Latin: "Epistola Fratris Rogerii Baconis de Secretis Operibus Artis et Naturae, et de Nullitate Magiae," in *Fr. Rogeri Bacon Opera quaedam hactenus inedita*, Edited by J.S. Brewer, Longman, London, 1859, vol. 1, Appendix I., pp. 523-551.

Further, cf. *The cure of old age and preservation of youth, by Roger Bacon, a Franciscan frier. Translated out of Latin, with annotations and an account of his life and writings by Richard Browne*, London, 1683; *The Opus Majus of Roger Bacon*, translated by Robert Belle Burke, Russell & Russell, NY, 1962, first published in 1928, vol. 2, Part 6 "Experimental Science," "Chapter on the Second Prerogative of Experimental Science. Example 2. Prolongation of Human Life," pp. 617-626.

[1033] Francis Bacon, in *De Vijs Mortis - An Inquiry Concerning the Ways of Death, the Postponing of Old Age, and the Restoration of the Vital Powers*, elaborated on the preservation of "vital heat" and "vital moisture" and expressed the conviction that, by this method, radical life-extension can be achieved, as "anything that can be constantly fed, and by feeding be wholly restored is, like the vestal flame, potentially everlasting" (Francis Bacon, "De Vijs Mortis – An Inquiry Concerning the Ways of Death, the Postponing of Old Age, and the

Restoring of Vital Powers," in *The Oxford Francis Bacon. VI. Philosophical Studies c.1611-c.1619*, Clarendon Press, Oxford, 1996, p. 271).

The aspiration to extreme longevity recurs in Francis Bacon's *Novum Organum* (1620), *The History of Life and Death* (1623) and *The New Atlantis* (1627).

In *The New Atlantis*, the devising of the means for extending longevity is one of the central research programs of the residents: "And amongst them we have a water which we call Water of Paradise, being, by that we do to it made very sovereign for health, and prolongation of life" (http://www.gutenberg.org/files/2434/2434-h/2434-h.htm).

In *Novum Organum*, Bacon spoke of herbal remedies and opiates which condense and quiet the spirits, "whereby they contribute no little to the cure of diseases and prolongation of life" (Aphorisms, Book 2, L, http://www.constitution.org/bacon/nov_org.htm).

Bacon's *The History of Life and Death* (the third part of the *Instauratio Magna – The Great Renewal*) abounds with recipes for life extension, including gold, pearls, crystals, bezoar stone, ambergris, saffron and other herbal remedies, and comprises a large compendium of hygienic and dietary theories and recommendations for nourishing the spirits and the body and thus extending longevity (http://www.sirbacon.org/historylifedeath.htm).

[1034] William Godwin, in *An Enquiry Concerning Political Justice and its Influence on Modern Morals and Manners* (1793), argued for the possibility and desirability of physical immortality. He stated:

"Let us here return to the sublime conjecture of [Benjamin] Franklin, that 'mind will one day become omnipotent over matter.' If over all other matter, why not over the matter of our own bodies? ... In a word, why may not man be one day immortal?"

(William Godwin, *An Enquiry Concerning Political Justice and its Influence on Modern Morals and Manners*, G.G.J. and J. Robinson, London, 1793, vol. 2, Book 8, Ch. 7 "Of Population," p. 862, and in the third corrected edition of 1798 (Robinson, London), vol. 2, Book 8, Ch. 9, Appendix. "Of Health, and the Prolongation of Human Life," p. 520.)

These beliefs of Godwin's goaded Thomas Malthus to write his famous *Essay on the Principle of Population. On the Speculations of Mr. Godwin, M. Condorcet, and Other Writers* (1798), battering against the indefinite prolongation of life.

Curiously, William Godwin's daughter, Mary Shelley, authored *Frankenstein; or, The Modern Prometheus* (1818), a resounding warning against the pursuit of immortality (a child's revolt?).

[1035] William Winwood Reade, in *The Martyrdom of Man* (1872), avowed the belief that, in the future, "disease will be extirpated; the causes of decay will be removed; immortality will be invented. And then, the earth being small, mankind will migrate into space, and will cross the airless Saharas which separate planet from planet, and sun from sun" (William Winwood Reade, *The Martyrdom of Man*, Watts & Co, London, 1924, p. 423, first published in 1872).

Interestingly, this belief in unlimited scientific progress coincided with Reade's equally firm belief in unchangeable human nature and perpetual inequality:

"Human nature cannot be transformed by a *coup d'état*, as the Comtists and Communists imagine. ... as long as men continue unequal in patience, industry, talent, and sobriety, so long there will be rich men and poor men – men who roll in their carriages, and men who die in the streets" (p. 417).

[1036] Samuel Butler, *Essays on Life, Art and Science*, Chelsea House, NY, 1983, "The Deadlock in Darwinism," p. 336 (initially published in 1889, republished in 1908).

[1037] Spencer generally did not seem to express any hopes for a drastic extension of the human life span. Moreover, according to him, the achievement through public health policy of "bodily life [that] is lower in quality than it was, though greater in quantity" could become a serious social ill (Herbert Spencer, The Study of Sociology, Ch. 14, "Preparation in biology," Henry S. King, London, 1873).

Nevertheless, he did claim that an "increase of longevity ... even alone yields conclusive proof of general amelioration" (Herbert Spencer, *From Freedom to Bondage*, 1891).

According to Spencer, "to have a large surplus of vital energy implies a good organization – an organization likely to last long" (Herbert Spencer, *The Study of Sociology*, Ch. 5. "Objective difficulties," Henry S. King, London, 1873).

The citations are from Spencer's works reprinted in the Online Library of Liberty, http://oll.libertyfund.org.

[1038] Herbert Spencer, *The Principles of Ethics*, Vol. 2, Ch. 1. "Animal ethics," 1897 (reprinted in The Online Library of Liberty, http://oll.libertyfund.org); Albert Dastre, *La Vie et la Mort* (Life and Death), Flammarion, Paris, 1907 (first published in 1903), p. 348.

[1039] Herbert Spencer, *The Data of Ethics*, Ch. 3. "Good and Bad Conduct," § 9, Williams and Norgate, London, 1879, reprinted in the Online Library of Liberty, http://oll.libertyfund.org.

[1040] Among the works promoting life-extension, John Harris (b. 1945, director of the Institute for Science, Ethics and Innovation at the University of Manchester, UK, and a founding director of the International Association of Bioethics), wrote: *Clones, Genes and Immortality. Ethics and the Genetic Revolution*, Oxford University Press, Oxford, 1998; John Harris, "Immortal Ethics," *Annals of the New York Academy of Sciences*, 1019, 527-534, 2004.

Max More (b. Max T. O'Connor, 1964, educated in Oxford, currently working in the US, founder of the Extropy Institute) authored "Transhumanism: Toward a Futurist Philosophy" (1990, 1996, http://www.maxmore.com/transhum.htm); and "Principles of Extropy" (2003, http://www.extropy.org/principles.htm), containing definitive and strong life-extensionist sentiments.

Nick Bostrom (born in Sweden in 1973, as of 2014 director of the Future of Humanity Institute at Oxford University, http://www.fhi.ox.ac.uk/, co-founder of the World Transhumanist Association), called to work and fight for life extension in "The Fable of the Dragon-Tyrant," *Journal of Medical Ethics*, 31 (5), 273-277, 2005 (reprinted at http://www.nickbostrom.com/fable/dragon.html); and Nick Bostrom, "Transhumanist FAQ" (1998, 2003, http://humanityplus.org/learn/transhumanist-faq/).

The second co-founder of the World Transhumanist Association in 1998 is the British philosopher David Pearce, director of Better Living Through Chemistry (BLTC) Research, Brighton, UK (http://www.bltc.info/). He too has been a strong supporter of radical life-extension.

[1041] Thomas Malthus, *An Essay on the Principle of Population, as it Affects the Future Improvement of Society with Remarks on the Speculations of Mr. Godwin, M. Condorcet, and Other Writers*, J. Johnson, London, 1798 (reprinted in Project Gutenberg, http://www.gutenberg.org/files/4239/4239-h/4239-h.htm).

[1042] It has been often suggested that Malthus' social theory was the forerunner of Darwin's theory of evolution by natural selection. (See, for example, Ernst Mayr, *The Growth of Biological Thought: Diversity, Evolution, and Inheritance*, Harvard University Press, Cambridge, Massachusetts, 1982, p. 479.)

I would still contend that Malthus' views were profoundly and explicitly anti-evolutionary, insofar as, according to Malthus, current phenotypic traits (such as the current life-span, length of limbs, ability to run, resistance to diseases, etc) were seen as more or less fixed in stone (rigidly predetermined) and any attempts to radically modify these traits were anticipated to fail.

[1043] With the contemporary population of Britain of about 7 million (in 1798 as of Malthus' writing), and assuming the doubling of the population every 25 years (as it then took place in the United States), and assuming food supplies to grow in an arithmetic progression – in 50 years, Malthus expected, the maximal population that the Island could support would be 21 millions, everyone above this number would be starved.

For a contemporary comparison, as of 2011, the UK population was about 62 million. In 1961-2000, the UK population grew from 52 million to 60 million (15% increase). In the same period, the yield of wheat increased in the UK from ~3.5 ton to ~8 ton per hectare (128% increase) and was the largest in the world. Concurrently, the gross domestic product per capita increased from $1,451 to $24,150 (over 1500%

increase). The life-expectancy in the UK increased in that period from 70.85 to 78.04 years (over 10% increase). (http://www.nationmaster.com/; http://faostat.fao.org/site/567/DesktopDefault.aspx#ancor www.mortality.org.)

Thus, contrary to Malthus' expectations, increasing longevity was by no means accompanied by insupportable over-population and starvation.

Extensive empirical data disproving Malthus' theory are given in Colin Clark, "Agricultural productivity in relation to population," in *Man and His Future: A CIBA Foundation Volume*, Gordon Wolstenholme (Ed.), Little, Brown and Co., Boston, 1963, pp. 23-35; and in Erik Stokstad, "Will Malthus Continue to be Wrong?" *Science*, 309 (5731), 102, 2005.

[1044] To give a few examples of the anti-life-extensionist stance:

In 1905, the renowned Canadian physician William Osler (1849-1919) spoke of the "uselessness of men above sixty years of age."

(William Osler, "Farewell address on leaving the Johns Hopkins University" (1905), *Scientific American*, March 25, 1905, reproduced in full in Stanley Hall, *Senescence, the Last Half of Life*, D. Appleton & Company, New York, 1922, pp. 3-5.)

In 1925, the American physician, Editor in Chief of the *Journal of the American Medical Association*, Morris Fishbein (1889-1976) wrote that "there has been, however, but little average prolongation of life beyond the age of seventy, and there is not the slightest scientific reason to believe that there ever will be."

(Morris Fishbein, *The Medical Follies*, Boni and Liveright, New York, 1925, p. 234.)

Later on, Norbert Wiener (1894-1964), the author of the theory of cybernetic regulation in living organisms and machines, reported a discussion of "the possibility of a radical attack upon the degenerative disease known as old age," the possibility of reaching "the day – perhaps not too far in the future – when the time of inevitable death should be rolled back, perhaps into the indefinite future, and death would be accidental." Wiener admitted that "the weight of the names supporting it …was too great to allow me to reject the suggestion out of hand."

Yet, he was deeply concerned by potential consequences of life extension:

"Consoling as the suggestion may seem at first sight, it is in reality very terrifying, and above all for the doctors. For if one thing is clear, it is that humanity as such could not long survive the indefinite prolongation of all lives which come into being. Not only would the nonselfsupporting part of the humanity come to outweigh the part on which its continued existence depends, but we should be under such a perpetual debt to the men of the past that we should be totally unprepared to face the new problems of the future."

(Norbert Wiener, *God and Golem, Inc. A Comment on Certain Points where Cybernetics Impinges on Religion*, The MIT Press, Cambridge, Massachusetts, 1964, pp. 66-67.)

The Australian immunologist, the Nobel Laureate in medicine of 1960 and the author of the "intrinsic mutagenesis" theory of aging (1974), Frank Macfarlane Burnet (1899-1985) "doubt[ed] very much whether anything worthwhile would be gained by extending the human life span beyond its present bracket of 70 to 100 years – and that if we wanted this extension of life, I am deeply sceptical about our chance of ever achieving it." And furthermore, "death in the old should be accepted as something always inevitable and sometimes as positively desirable."

(Frank Macfarlane Burnet, *The Biology of Aging*, Auckland University Press, Auckland NZ, 1974, pp. 63, 66.)

Leonard Hayflick (b. 1928), the world-renowned American gerontologist, the author of the "cell division limit" theory of aging, admitted that in the 1960s he "would have been unwise to identify [himself] as a gerontologist doing basic research." Until very recently, Hayflick has claimed that "No intervention will slow, stop, or reverse the aging process in humans." And even if it were somehow possible, "the problems created by having the power to arrest or even slow the aging process could be enormous and damaging to both the individual and society in general." Hayflick candidly admitted that he studies the aging process for sheer scientific curiosity, without any intention to ever interfere in it.

(Leonard Hayflick, "Address to the Select Committee on Aging, Washington, Feb, 1978," quoted in William G. Bailey, *Human Longevity from Antiquity to the Modern Lab*, Greenwood Press, Westport CN, 1987, p. ix; Leonard Hayflick, "'Anti-aging' is an oxymoron," *Journal of Gerontology*, 59(6), B573-578, 2004; Leonard Hayflick, *How and Why we Age*, Ballantine Books, NY, 1994, "No More Aging: Blessing or Nightmare?" pp. 336-338.)

More recent examples of the anti-life-extension attitude can be found in the views of Leon Kass (former President G.W. Bush's advisor on bioethics), Daniel Callahan (researcher of medical ethics and public policy, president of the Hastings Center for Bioethics), Koichiro Matsuura (Director General of UNESCO, 1999-2009), Reverend Jerry Falwell (televangelist, founder of the Moral Majority movement), William Hurlbut (a member of the US President's Council on Bioethics, 2002-2009), Sherwin Nuland (professor of surgery at Yale), Francis Fukuyama (member of the US President's Council on Bioethics, from 2001 to 2005) or Michael Shermer (psychologist and philosopher, founder of the Skeptics Society and *Skeptic* magazine).

These authors tended to rationalize mortality and warned against radical life-extension, using a standard available panoply of ideations, highly reminiscent of Malthus' arguments: death as a provider of meaning and excitement, injustice of radical life-extension, overpopulation, limitations to human potential, unsafety of any intervention into human nature, etc.

(See Leon Kass, "L'Chaim and Its Limits: Why Not Immortality?" *First Things*, 113, 17-24, May 2001; Daniel Callahan, *What Price Better Health? Hazards of the Research Imperative*, University of California Press, Berkeley, 2003, Ch.3. "Is research a moral obligation? The war against death," pp. 64-66; Koïchiro Matsuura, "Of sheep and men," *The Daily Star*, 4 (113), September 16, 2003; "Jerry Falwell Attacks Life Extension Foundation," *Life Extension Magazine*, October 2003; "Bioethicist William Hurlbut on the dangers of radical lifespan extension," *US News and World Report*, 5/28/2004; Sherwin Nuland, *How We Die: Reflections on Life's Final Chapter*, Knopf, NY, 1994, Sherwin Nuland, "Do You Want to Live Forever?" and Jason Pontin's editorial "Against Transcendence," *Technology Review*, February 2005; Francis Fukuyama, *Our Posthuman Future. Consequences of the Biotechnological Revolution*, 2002, Ch. 4. "The prolongation of life," pp. 57-71, Francis Fukuyama, "The World's Most Dangerous Ideas: Transhumanism," *Foreign Policy*, 144, 42-43, 2004; Michael Shermer, "Hope Springs Eternal," *Scientific American Magazine*, June 27, 2005, Michael Shermer, "The Immoralist," *Science*, 332 (6025), 40, 2011.)

[1045] The standard set of ethical arguments for and against radical life-extension is reviewed by the American bioethicist Robert Veatch.

He presents two "Cases for Life."

The first "rationalist case for life" suggests that radical life extension is compatible with the valuation of an individual life, with the "prudent, personal self-interest of the rational person." For example, life-prolongation is desired for the accomplishment of one's worldly projects and enjoyment of "future states."

In the "Social-eschatological case for life" radical life extension is desired as a part of the millennial expectations for a perfect society, "conquering the evil of death."

On the other hand, five common objections are presented and countered. "The Cases for Death and Against Immortality" include arguments from "Death as relief from suffering," "Death as relief from boredom," "Death as a source of meaning," "Death as a force for progress," and "Natural death as a comforting fiction."

First of all, death is not a solution against suffering. Suffering is not perceived as inevitable, and the ability of human beings to actively influence their fate and relieve suffering is emphasized. Essentially, it is not prolonged suffering that is desired, but rather prolonged health.

Regarding boredom and loss of meaning, it is argued that life may carry a meaning of its own, independent of death, that it is hard to place a temporal limit on the love and enjoyment of life, and that human beings are entitled to choose a prolonged existence.

The benefits of short life-spans for human progress are disputed: the potential for learning will be increased by longer life-spans, and such a prolonged "cultural adaptation" may be sufficient and necessary for the survival of the society. Moreover, rationally controlled development and care for the survival of the weak may be more advantageous for progress than blind Darwinian selection.

Concerning the inexorable "natural" limit to the human life, it is argued that however comforting a reconciliation with death may be, it should not replace an active quest for life preservation.

(Robert Veatch, *Death, Dying, and the Biological Revolution. Our Last Quest for Responsibility*, Yale University Press, New Haven CT, 1977, Ch. 8. "Natural death and public policy," pp. 293-305.)

The British bio-ethicist John Harris adds to these the questions of "justice" (whether prolonging life for select groups is justifiable) and "identity" (whether the incessant transformations of the body and mind permit us to speak of a long-term preservation of identity).

Regarding justice, Harris asserts that the inability to provide a good to all should not prevent providing it to some, and he expects a wide and equitable sharing of future medical technologies.

Regarding the loss of identity during a prolonged life history, he argues for the continuity of human existence.

(John Harris, "Immortal Ethics," presented at the International Association of Biogerontologists (IABG) 10th Annual Conference "Strategies for Engineered Negligible Senescence," Queens College, Cambridge, UK, September 17-24, 2003, reprinted in *Strategies for Engineered Negligible Senescence: Why Genuine Control of Aging May Be Foreseeable* (Ed. Aubrey de Grey), *Annals of the New York Academy of Sciences*, 1019, 527-534, June 2004.)

[1046] To give a few examples of the negative representation of radical life extension in English fiction:

Jonathan Swift's *Gulliver's Travels* (1726) features the immortal Struldbrugs eternally suffering from mental and physical debility. (Jonathan Swift, *Gulliver's Travels into Several Remote Nations of the World*, "Part III: A Voyage to Laputa, Balnibarbi, Luggnagg, Glubbdubdrib, and Japan," Chapter X, reprinted in Project Gutenberg, http://www.gutenberg.org/files/829/829-h/829-h.htm.)

Mary W. Shelley's *Frankenstein; or, The Modern Prometheus* (1818), urged the reader to learn "how dangerous is the acquirement of knowledge." The book promulgated the notions that "destruction and infallible misery" ensue "the capacity of bestowing animation," and that there is no "rest but in death" (Mary W. Shelley, *Frankenstein; or, The Modern Prometheus* (1818), Penguin Popular Classics, 1994, pp. 51, 214).

In William Henry Hudson's *A Crystal Age* (1887), there is a drink that promises "*a new life*" and a "*cure.*" The protagonist understands that this is "a panacea for all diseases, even for the disease of old age, so that a man may live two hundred years, and still find some pleasure in existence." But first he has "no wish to last so long." When he eventually decides to drink the potion, it proves to be a deadly poison. (http://www.gutenberg.org/dirs/etext05/cryst10h.htm.)

In Herbert George Wells' *The Time Machine* (1895), the ageless Eloi are inferior to present day humans, while in his *Men Like Gods* (1923), the long-lived Utopians are disturbingly superior.

In Bram Stoker's *Dracula* (1897), the immortal vampire is downright monstrous. In Oscar Wilde's *The Picture of Dorian Grey* (1890), the physically rejuvenated Dorian is morally corrupt.

Though the dystopian visions of life-extension were very prominent in England, they were not endemic. In *The Makropulos Secret* (*Věc Makropulos*, 1922) by the Czech novelist Karel Čapek, the long-lived Helen is plagued by eternal ennui. Under the conviction of the complete worthlessness of prolonged boredom, Helen's recipe for rejuvenation is burned.

The fiction of the American author Howard Lovecraft too is largely dystopian: *The Call of Cthulhu* (1928) features an immortal monster and its minions, while his *Herbert West – Reanimator* (1922) continues the *Frankenstein* tradition of Mary Shelley, featuring the creation of monsters in the pursuit of immortality.

The major source of dystopian visions, however, seems to have been in England, influencing writers in other countries. Thus, H.G. Wells' writings clearly informed the science fiction novels of the Russian author Mikhail Bulgakov that were explicitly opposed to attempts at life enhancement and life extension. Wells' *The Island of Doctor Moreau* (1896) bears a clear stamp on Bulgakov's *The Heart of a Dog* (1925) and Wells' *The Food of the Gods* (1904) thematically underscores Bulgakov's *Fatal Eggs* (1925).

Notably, there appears a surge of literary interest in the subject, in England and elsewhere, in the aftermath of WWI. The writings are profoundly apprehensive of the potentially dangerous outcomes of increasingly powerful science and technology, of their ability to dramatically disturb the status quo. The futuristic treatments of life-extension reflected present anxieties.

[1047] Bernard Shaw, *Man and Superman. A Comedy and a Philosophy*, Penguin Books, 1994 (first published in 1903), available online at the Internet Archive, http://www.archive.org/details/mansupermancomed00shawrich, and at Project Gutenberg, http://www.gutenberg.org/files/3328/3328-h/3328-h.htm.

Man and Superman praised the sustaining "Life Force." Yet, Shaw emphasized that the "life force" is neither static nor sterile. The mere perpetuation of an "illusory form" or "appearance" of life – with all the enjoyment and "glamor," yet without intellectual and spiritual growth, work, moral responsibility, duty, hope and prayer – is the very essence of hell:

"But in hell old age is not tolerated. It is too real. … Well, here [in hell] we have no bodies: we see each other as bodies only because we learnt to think about one another under that aspect when we were alive; and we still think in that way, knowing no other. But we can appear to one another at what age we choose. You have but to will any of your old looks back, and back they will come. … Hell, in short, is a place where you have nothing to do but amuse yourself. … [In hell] the humbug of death and age and change is dropped because here WE are all dead and all eternal."

In contrast, contemplation, mastery of reality, and work for the preservation of life, belong in the realm of heaven:

"heaven is the home of the masters of reality… In the Heaven I seek, no other joy [than contemplation]. But there is the work of helping Life in its struggle upward. Think of how it wastes and scatters itself, how it raises up obstacles to itself and destroys itself in its ignorance and blindness. It needs a brain, this irresistible force, lest in its ignorance it should resist itself. … And there you have our difference: to be in hell is to drift: to be in heaven is to steer."

(Bernard Shaw, *Man and Superman. A Comedy and a Philosophy*, Penguin Books, 1994 (1903), Act Three, pp. 124-177.)

[1048] Bernard Shaw, *Back to Methuselah*, in *Bernard Shaw Selected Plays*, Dodd, Mead and Company, Vail-Ballow Press Inc., New York, 1949 (first published in 1921, first performed in 1922), available online at Project Gutenberg, http://www.gutenberg.org/cache/epub/13084/pg13084.html.

Back to Methuselah explored and encouraged the efforts for the strengthening of the "Life Force" and "Prolongation of Life" on earth.

[1049] James Hilton [1900-1954], *Lost Horizon* (1933), written and first published in London, reprinted in Project Gutenberg of Australia, http://gutenberg.net.au/ebooks05/0500141h.html#e01.

[1050] Aldous Huxley's *Brave New World* (1932) and *After Many a Summer Dies the Swan* (1939) are commonly considered as paragons of dystopian fiction, warning against biotechnological interventions into human capabilities.

(See, for example, Tim Armstrong, *Modernism, Technology and the Body. A Cultural Study*, Cambridge University Press, Cambridge, 1998, p. 63; S.L. Rosen, "Alienation as the Price of Immortality: The Tithonus Syndrome in Science Fiction and Fantasy," p. 130, Howard V. Hendrix, "Dual Immortality, No Kids: The Dink Link between Birthlessness and Deathlessness in Science Fiction," p. 184-186, in George Slusser, Gary Westfahl, Eric Rabkin (Eds.), *Immortal Engines. Life Extension and Immortality in Science Fiction and Fantasy*, The University of Georgia Press, Athens GA and London, 1996.)

Yet, *Brave New World*, I would argue, also contains a strong life-extensionist sentiment.

In *Brave New World*, biotechnology is used by the society first and foremost for pleasure, and not for the prolongation of life or intellectual growth. The residents of the "New World" undergo "death conditioning" to inculcate the idea that death is a natural, good and pleasant event:

"Death conditioning begins at eighteen months. Every tot spends two mornings a week in a Hospital for the Dying. All the best toys are kept there, and they get chocolate cream on death days. They learn to take dying as a matter of course."

Those who prolong their days and become old are the savages who live at the Reservation. The "civilized" ones are simply not allowed to age and live long, but die young and healthy at the age of sixty:

"That's because we don't allow them to be like that [growing old]. We preserve them from diseases. We keep their internal secretions artificially balanced at a youthful equilibrium. We don't permit their magnesium-calcium ratio to fall below what it was at thirty. We give them transfusion of young blood. We keep their metabolism permanently stimulated. So, of course, they don't look like that. Partly," he added, "because most of them die long before they reach this old creature's age. Youth almost unimpaired till sixty, and then, crack! the end."

It is unclear why the balancing of the metabolism would lead to a pre-determined early demise, but the passage seems to imply that this is due to an excessive and permanent stimulation. Thus, *Brave New World* appears to be an adverse reaction to contemporary reductionist rejuvenation attempts (that were indeed often followed by a rapid deterioration) and not to life-extension *per se*.

The work is also opposed to the eugenics-type selectivity that was wide-spread in American and British life-extensionist literature until the 1930s, but not to improvements of human capabilities *per se*. Thus, the question why the World Controllers do not breed everyone to become members of the highest, most capable "Alpha" caste is answered by Mustapha Mond, Resident World Controller for Western Europe:

"Because we have no wish to have our throats cut," he answered. "We believe in happiness and stability. A society of Alphas couldn't fail to be unstable and miserable. Imagine a factory staffed by Alphas — that is to say by separate and unrelated individuals of good heredity and conditioned so as to be capable (within limits) of making a free choice and assuming responsibilities. Imagine it!'"

Thus, *Brave New World* does not seem to be in an outright opposition to life-extension as such, life-enhancement as such, or even bio-technology as such, but rather against particular biotechnological methods (e.g. reductionist rejuvenation) and their particular social purposes (e.g. mass pacification and selectivity) that were emerging by the time of Huxley's writing.

(Aldous Huxley, *Brave New World and Brave New World Revisited*, Harper Perennial, New York, 1965, first published in 1932, pp. 125, 84, 170-171.)

In Aldous Huxley's *After Many a Summer Dies the Swan* (1939), on the other hand, the pursuit of longevity is generally opposed.

In *After Many a Summer*, Dr. Obispo succinctly explains the nature of the life-extensionist pursuit, as compared to general medical practice:

"Longevity, the doctor explained, as they left the room. That was his subject. Had been ever since he left medical school. But of course, so long as he was in practice, he hadn't been able to do any serious work on it. Practice was fatal to serious work, he added parenthetically. How could you do anything sensible when you had to spend all your time looking after patients? Patients belonged to three classes: those that imagined they were sick, but weren't; those that were sick, but would get well anyhow; those that were sick and would be much better dead. For anybody capable of serious work to waste his time with patients was simply idiotic."

Funded by the millionaire Jo Stoyte, who is willing to pay to forestall his end, Dr. Obispo is able to continue his longevity research and bring it to fruition.

Coming from a family of biologists and having undergone some training in biology, Aldous Huxley's speculations about the theoretical possibilities of life-extension are perhaps on a par with any contemporary popular work on the subject. Thus, relying on Metchnikoff's theory, Dr. Obispo observes that carps live long and hence their intestines must contain some anti-toxic, pro-biotic micro-flora. An extract from the carp intestines becomes the elixir of longevity.

Yet, its attainment comes at the expense of morality. Dr. Obispo is a shameless profiteer, and Jo Stoyte is a lecherous, egotistic tyrant, and eventually a murderer ('accidentally' killing Peter Boone, Obispo's idealistic research assistant, for which act Obispo agrees to cover up for a fee).

The administration of the elixir and the attainment of super-longevity lead to the gradual devolving into a mindless ape-like creature. Or, in Huxley's terms, it leads to "evolving" from a neotenic (juvenile) human form into the advanced (mature) form of an ape. And Stoyte willingly chooses to go on this path.

(The idea of neoteny (juvenile human appearance) apparently derived from the "fetalization theory" of the Dutch anatomist Lodewijk 'Louis' Bolk (1866-1930). Cf. the speculations on neoteny by the American biologist Stephen Jay Gould (1941-2002) in his book *Ontogeny and Phylogeny* (Harvard University Press, Cambridge MA, 1977), Ch. 10 "Retardation and Neoteny in Human Evolution" pp. 352-404.)

Thus, the pursuit of life-extension, associated not only with a loss of human values but even with a loss of human appearance, could not be represented as less attractive. *After Many a Summer* became a strong link in the chain of the British literary tradition resisting life-extension.

(Aldous Huxley, *After Many a Summer Dies the Swan*, Avon Publishing Co. Inc., NY, 1952, pp. 65-66, 266, first published in 1939.)

[1051] In William Olaf Stapledon's (1886-1950) *Last and First Men: A Story of the Near and Far Future* (1930, UK), extended longevity is a desirable trait of many of the future stages of human evolution. Yet, several concerns are raised.

Thus, in one of the future evolutionary forms ("the Fifth Men"), "when they had actually come into possession of the means to make themselves immortal, they refrained, choosing rather merely to increase the life-span of succeeding generations to fifty thousand years. Such a period seemed to be demanded for the full exercise of human capacity; but immortality, they held, would lead to spiritual disaster."

(Ch. 12. "The last terrestrials," 1. "The cult of evanescence," reprinted in Project Gutenberg of Australia, http://gutenberg.net.au/ebooks06/0601101h.html.)

[1052] For example, in 1936, the Welsh poet Dylan Thomas (1914-1953) came up with the passionate slogan "And death shall have no dominion" and in a later date (1951) urged "Do not go gentle into that good night. Old age should burn and rave at close of day. Rage, rage against the dying of the light."

Yet, for Dylan Thomas, a confirmed drinker, the command to "rage against the dying of the light" did not seem to entail any practical suggestions for life-extension. The above lines by Dylan Thomas have been often cited by life-extensionists as a motto. Yet, one may doubt whether the poet himself would describe himself as a "life-extensionist."

Later on, in a somewhat analogous vein, the American actor and writer Woody Allen (b. Allan Stewart Konigsberg, 1935) made some strikingly "immortalist" statements, such as "The enemy is not the Chinese or the guy next door to you, the enemy is out there. That's what *Death* is all about" or "I don't want to achieve immortality through my work, I want to achieve it through not dying" (Eric Lax, *On Being Funny: Woody Allen and Comedy*, Charterhouse, NY, 1975, pp. 227, 232), or "The National Rifle Association declared death a good thing" (*Bananas*, 1971).

Yet, apparently Allen has not been involved in any public promotion of longevity research or practice (whatever his personal life-style might be). As Bob Brakeman from the Cryonics Institute wondered about

Allen's "immortalism," "is he going to remain forever one of those people who talk a good game but do nothing about it, or will he finally act upon his supposed principles?... Has he given a nickel to any immortalist cause?"
(Bob Brakeman, "Should Woody Allen be Taken Seriously?" *Cryonics Institute*, January 16, 2000, http://www.cryonics.org/woody.html.)
There are more examples, among English-speaking literati, of the apparent mental disconnection between words and deeds with reference to life-extension, as well as cases of dissociation between resistance to death and support of longevity. Thus, the German-American political theorist Hannah Arendt (1906-1975), stated in *The Human Condition* (1958):
"the wish to escape the human condition, I suspect, also underlies the hope to extend man's life-span far beyond the hundred-year limit. ... There is no reason to doubt our abilities to accomplish such an exchange, just as there is no reason to doubt our present ability to destroy all organic life on earth." And furthermore, "the striving for immortality ... originally had been the spring and center of the *vita activa*" (pp. 2-3, 21).
However, these statements did not seem to bear any reference to life-extension research, advocacy or practice. In fact, the book only briefly mentioned the word "medicine" twice (pp. 91, 128) in the lists of various human activities.
(Hannah Arendt, *The Human Condition*, Second Edition, The University of Chicago Press, Chicago, 1998, first published in 1958.)
[1053] The social, ethical and psychological problems contingent on radical life-extension continued to haunt English fiction writers to a later date.
For example, *Trouble with Lichen* by John Wyndham (1960, UK) portrays an interplay of greed and magnanimity, secrecy and altruism, surrounding the discovery of a life-extending ("metabolism slowing") elixir made of lichen. The discovery was made by a lucky chance, modeled on the "chance" discovery of penicillin by Alexander Fleming (1928), and biological mechanisms are given only a minor consideration. The social implications of the discovery, however, were thought through carefully. Some of the potential problems associated with the wide application of the elixir would include "the contriving, the intriguing, the bribery, perhaps fighting even – that would come of people trying to get in first to grab even a few extra years, and the chaos that would follow in a world which is already overpopulated, with a birth-rate far too high. The whole prospect was and is quite appalling."
(John Wyndham, *Trouble with Lichen*, Ballantine Books, New York, 1960, p. 38.)
Yet the book ending sounds an optimistic note. The remedy, the inventors hope, when it becomes possible to synthesize it in large quantities, will not be used only to make profit for its creators, will not be reserved for financial tycoons or state leaders, but is promised to all humanity. With the support of a large governmental grant, there was "*no* doubt that British brains, British purpose, and British know-how would succeed – and succeed in the very near future – in producing a supply of the Antigerone for every man and woman in the country who wished to use it." International competition and massive public demand will only increase the chances for success: "Both the Americans and the Russians have now made bigger allocations for research... the children have been promised their sweets, and they'll raise hell if they're not forthcoming. But they will forthcome."
(John Wyndham, *Trouble with Lichen*, Ballantine Books, New York, 1960, pp. 155-158.)
Very similar concerns are expressed in *Bug Jack Barron* by the American writer Norman Spinrad (1969). Interestingly, the book was first serialized in England, in the magazine *New Worlds*. The "immortality treatment," developed by scientists in employ of the billionaire Benedict Howards, involves no less than killing children to transplant their glands, and Howards intends to use his own invulnerability for no less than taking over the world, using the "immortality treatment" as the most undeniable bribe. The conspiracy is foiled by the TV host Jack Barron, who will afterwards strive to make life-extending treatments universally available "for everyone without murder."

(Norman Spinrad, *Bug Jack Barron*, The Overlook Press, Woodstock and New York, 2004, first published in 1969, p. 245.)

It is in fact very difficult to find a contemporary work of literary fiction on radical life-extension that would not contain a strongly negative, critical element. English science fiction seems to have been a major, if not *the* major source of discourse on the political, social and ethical implications of life-extension, pervasively emphasizing the dystopian element.

The very few works that are predominantly positive, or 'explicitly' life-extensionist, include *The First Immortal* (1998) by the American entrepreneur and author James Halperin. The novel essentially portrays the triumphant march of cryonics, nanotechnology, artificial intelligence and cloning, making physical immortality standard. The work envisions that life-extending technologies will be eventually made available to all: "Let us make victory against death a prospect for all, lest it become a prospect for none." As the book asserts, the "argument [for cryonics and life-extension] does not wrangle between provision for the poor versus the sanctity of individual achievement. And while consequential, neither is morality today's primary topic." The primary topic is rather the developing of survival technologies, as fast as possible. The book makes the reader aware that life is enjoyable now, and must continue to be enjoyable in the future, for as long as possible.

(James L. Halperin, *The First Immortal*, Random House Publishing, New York, 1998, http://coins.ha.com/tfi/.)

Poul Anderson's *The Boat of a Million Years* (1989) also emphasizes a positive outlook, asserting that exploratory human endeavor may have no temporal boundaries. In the book, there are prolonged debates on whether knowing of the possibility of physical immortality will be beneficial or disastrous for the world. It is assumed that radical life extension for "everybody" will be overall beneficial, but "let the change come gradually, with forewarning." Eventually, not only everybody on earth is made immortal, but it also becomes possible to resurrect (model) the dead.

(Poul Anderson, *The Boat of a Million Years*, Tor Books, NY, 1991, quote p. 334, first published in 1989.)

Halperin's and Anderson's books are rare examples of pro-longevity science fiction. On the contrary, the arguments against life-extension have continued to dominate the genre.

There have been many recent examples of opposition:

In Phylis Dorothy James' *The Children of Men* (1992, UK), the prevalence of longevity research is a sign of a dying civilization. In Kim Stanley Robinson's *The Mars Trilogy* (1992, 1992, 1996, US), the "longevity treatment" is available only to the rich on "Terra," contributing to a social divide, and eventually to a world war. "The treatment" also entailed overpopulation, memory loss and prolonged boredom. Jorge Luis Borges' *The Immortal* (1962, Argentina), Natalie Babbitt's *Tuck Everlasting* (1975, US), Arkady and Boris Strugatsky's *Five Spoons of Elixir* (1985, Russia) again and again raise the specter of prolonged boredom, following the "Struldbrugs" tradition of Swift's *Gulliver's Travels*.

In many other works of science fiction and fantasy, those who pursue immortality are either depraved, evil or outright monstrous, and their doom is sealed. The motif is prevalent in movies, from *Death Becomes Her* (1992, US) where the immortals are ridiculed, to *Interview with the Vampire* where they are feared and pitied (1994, US), and novels from John Ronald Reuel (J.R.R.) Tolkien's *The Lord of the Rings* (1937-1949) where the immortal monster needs to be contained, to Joanne Kathleen (J.K.) Rowling's *Harry Potter and the Philosopher's Stone* (1997, UK) where immortality is the main villain's desire, and not only is the monster thwarted, but the immortality-conferring Philosopher's Stone is destroyed as well, for safety.

Gerald J. Gruman refers to the latter tradition as "apologist," an attempt to depreciate what we cannot have, to rationalize away or even apologize for our mortality.

(Gerald J. Gruman, *A History of Ideas about the Prolongation of Life. The Evolution of Prolongevity Hypotheses to 1800*, Transactions of the American Philosophical Society, Volume 56 (9), Philadelphia, 1966, II "Apologism. Conclusion," pp. 19-20.)

The works under consideration, however, rather present themselves as earnest attempts to foresee the social consequences of radical life-extension.

[1054] Sir Thomas Malory (1405-1471), *Morte Darthur*, Book VI: "The Noble Tale of the Sangreal" (Caxton XIII–XVII); Book VIII: "The Death of Arthur" (Caxton XX–XXI).
Consider particularly Book XII. Ch. 2, Book XIII. Ch. 7, on healing and replenishment by the Sangreal, and Book XXI, Ch. 5-7 on King Arthur's entering Avilion, interpretable as a state of suspended animation or joining the immortals (cf. the Chinese legend about the Island of the immortals – Peng-Lai). Malory's *Morte Darthur* was completed in 1469-1470, and first published in 1485 by William Caxton (reprinted in Project Gutenberg, http://www.gutenberg.org/files/1252/1252-h/1252-h.htm).

[1055] Shaw's *Back to Methuselah*, including the play and the prefacing manifesto on the sociology of science, presented life-extension as a positive phylogenetic quest of humanity.
According to Shaw, far from being boring or socially disturbing, the quest for "voluntary longevity" is a noble and humane cause. Shaw considered *Back to Methuselah* to be his highest literary achievement and a contribution to the cause. He saw it as a "Metabiological Pentateuch" relating ethics to biology and survival. According to Shaw, the major rationale for life extension is that, at the present, we die in our infancy, not having enough time to mature and learn. Consequently, a society composed of infants is childish and prone to self-harm. Accordingly, the striving to reach a 300 years life-span (a distant yet imaginable goal) was suggested as the program for human development.

Shaw's argument was in direct opposition to the frequent contention that early mortality is necessary for the progress of humanity as a species, freeing the way for new people and ideas, and that younger, less established intellectuals are more likely to be receptive to new ideas than the old.

Such "ageist" convictions were expressed, among others, by Antoine Lavoisier, Charles Darwin and Thomas Huxley, and were later summarized by "Planck's principle" according to which "A new scientific truth does not triumph by convincing its opponents and making them see the light, but rather because its opponents eventually die and a new generation grows up that is familiar with it" (Max Planck [1858-1947], *Scientific Autobiography and Other Papers*, Philosophical Library, NY, 1950, "A Scientific Autobiography," pp. 33-34).

Shaw's work provided a counterargument. According to Shaw, the life of a long-lived individual is of no less value than that of a short-lived one, perhaps even more. Shaw contended that a person is capable of activity, learning, intellectual adjustment and contribution throughout life. (Shaw's 94 year long, productive life is a good example.) Thus he rejected the notion that the elderly need to die to free space and/or resources for the young. In Shaw's line of argument, the knowledge and experience of the long-living can mean survival for the entire human community:

"If on opportunist grounds Man now fixes the term of his life at three score and ten years, he can equally fix it at three hundred, or three thousand, or even at the genuine Circumstantial Selection limit, which would be until a sooner-or-later-inevitable fatal accident makes an end of the individual. All that is necessary to make him extend his present span is that tremendous catastrophes such as the late war shall convince him of the necessity of at least outliving his taste for golf and cigars if the race is to be saved."

(Bernard Shaw, *Back to Methuselah*, in *Bernard Shaw Selected Plays*, Dodd, Mead and Company, Vail-Ballow Press Inc., New York, 1949 (first published in 1921, first performed in 1922), "Voluntary Longevity," p. xix.)

"Planck's principle" that younger scholars are as a whole more receptive, innovative and risk-taking has also been refuted by recent empirical studies in sociology of science.
(See David L. Hull, Peter D. Tessner and Arthur M. Diamond, "Planck's Principle," *Science*, 202, 717-723, 1978; John T. Blackmore "Is Planck's 'Principle' True?" *The British Journal of the Philosophy of Science*, 29(4), 347-349, 1978; David Rier, "Gender, Lifecourse and Publication Decisions in Toxic-Exposure Epidemiology: 'Now!' Versus 'Wait a Minute!'" *Social Studies of Science*, 33, 269-300, 2003.)

[1056] Shaw began his crusade to overcome current biological limitations in *Man and Superman* (1903), where he promoted humane, positive eugenics or deliberate selection for health and longevity via "intelligently controlled, conscious fertility," while defending and caring for the less able. (Bernard Shaw, *Man and Superman. A Comedy and a Philosophy*, Penguin Books, 1994, first published in 1903, p. 226.)

Shaw's eugenic inclinations corresponded to the wide contemporary popularity of eugenics in the UK, US, in fact all across the world (see the previous discussion on the eugenic beliefs of the early twentieth century American life-extensionists).

In *Back to Methuselah* (1921), eugenic rationales are also present. The very first "long-lived" protagonists choose their mates "in cold blood because their children will live three hundred years" (Shaw, *Back to Methuselah*, 1949 (1921), Part III, p. 127).

Yet, Shaw's emphasis in *Back to Methuselah* seems to have shifted in the direction of behaviorism, implying that a person, through education, conditioning, power of the will and perseverance, can achieve almost any objective, including extended longevity. (The formal notion of "behaviorism" was proposed later, in John B. Watson, *Behaviorism*, Norton, New York, 1925.)

[1057] The form of life-extensionism suggested in *Back to Methuselah* was mainly predicated on the power of the will and wish fulfillment by perseverance, and was strongly affiliated with William Godwin's belief in the "omnipotence" of mind over matter, and with Henri Bergson's concept of *Élan vital*, a vital impetus strengthened by the will.

To recall, Metchnikoff argued that the Bergsonian "religion of the will" is antagonistic to scientific, materialistic and reductionist life-extensionism. In contrast, the system for life prolongation proposed in *Back to Methuselah* may be associated with "idealistic" or "holistic" life-extensionism. It implied that the mind can directly exercise a regulating influence on the body, strengthen its Life Force and prolong its existence. "Discouragement" was named as a major cause of death.

This approach goes back to *Summa Theologica* (1265-1274) by Thomas Aquinas (1225-1274) who saw the impurity of the soul as the major cause of mortality. Thomas Aquinas argued that "in the state of innocence man would have been immortal" (Summa Theologica, 1, Question 97, Article 1) and "Death and other bodily defects are the result of sin" (2:1, Question 85, Articles 5-6). The soul's weakness caused the Fall and brought about physical death, hence the soul's purification and strengthening can reverse the effect and bring forth immortality of the body. In such an ideal state of the soul, a level of a prophet can be achieved, allowing an insight into the spiritual world while in the physical body ("The manner in which prophetic knowledge is conveyed," 2:2, Question 173).

One of Thomas Aquinas' spiritual suggestions for life-extension is honoring one's parents: "Now we owe the favor of bodily life to our parents after God: wherefore he that honors his parents deserves the prolongation of his life, because he is grateful for that favor" (2:2 Question 122, Article 5). (Thomas Aquinas, *Summa Theologica*, reprinted by The New Advent, http://www.newadvent.org/summa/1097.htm.)

Similar views were held by Roger Bacon (*Opus Majus*, 1266, see above) and by Albertus Magnus (1193-1280) in his treatises *On Youth and Old Age* and *On Life and Death*.

In *Back to Methuselah*, Shaw directly alludes to the willed loss of immortality in the Garden of Eden in Part I, Act 1, "In the Beginning."

The existence of psychosomatic, psycho-regulatory and placebo effects on health status is now generally undoubted, yet these effects have been often considered to be limited.

(Anne Harrington (Ed.), *The Placebo Effect: An Interdisciplinary Exploration*, Harvard University Press, Cambridge MA, 1997.)

Already Thomas Malthus, in his polemics with William Godwin and Nicolas Condorcet, argued that, unlike inoculation, a mere will to live is insufficient to save a person infected with smallpox or the plague (Malthus, *Essay on the Principle of Population*, 1798).

According to Metchnikoff, psychosomatic healing is limited to "a few neurological disorders" (Elie Metchnikoff, *Etudy o Prirode Cheloveka* (Etudes on the Nature of Man), The USSR Academy of Sciences Press, Moscow, 1961 (1903), p. 178).

Yet, in Shaw's optimistic view, the direct power of the mind over the body is unlimited and can produce miracles of complete healing and extreme longevity.

Shaw was not alone in this belief, as a great variety of health systems of his time, from "Christian Science" of Mary Baker Eddy (since 1821) to "Jewish Science" of Rabbi Morris Lichtenstein (since the early 1900s), hailed the miracle of psycho-regulation. This concept has remained popular to the present time and has been manifested in diverse 'meditative' schools of life extension, believing that the right state of the mind, meditation and self-suggestion, desire and tranquility, can cure all ailment and prolong life. (See for example, Deepak Chopra, *Ageless Body, Timeless Mind. The Quantum Alternative to Growing Old*, Harmony Books, New York, 1993.)

[1058] In Shaw, the effect of the mind on the body was believed to be mediated by a biological field, electromagnetic or of some yet unknown kind.

The Vital Force has been known by many names: the Chinese *chi*, the Hindu *prana*, the Latin *spiritum vitae*, the Greek *aether*, *entelecheia* or *physis* (*vis medicatrix naturae*, the restoring and balancing power of nature), and more recently Franz Mesmer's *Animal Magnetism*, Carl Reichenbach's *Odic Force*, Henri Bergson's *Élan vital*, and many others.

In Shaw and other "vitalist" life-extensionists at the beginning of the twentieth century, the Life Force assumes a scientific-like appearance of a field of energy, a "physical force" measurable and expressed by "algebraic equations" (Bernard Shaw, *Back To Methuselah*, Part IV, Act 2, pp. 176-177).

In the fifth part of *Back to Methuselah*, Shaw portrays the future of the human species, "as far as thought can reach," when physical immortality is virtually achieved. Based on the principle "ontogeny recapitulates phylogeny," the development of individuals in that remote utopian era reflects the overall evolutionary path of humankind.

After birth and until two years of age, people of that epoch play and make love (according to Shaw, this period corresponds to our entire present life span, before we die in our infancy). After that, until 5 years of age, people are interested in arts and sciences (here again Shaw mocks what he believes to be the present infantile state of art and science). Then until about 50 years of age, human beings begin to experiment with their life force and learn to consciously modify their body, for example, grow several heads or arms (here Shaw valorizes direct mind control over artifice). The ultimate stage beyond that is the pure energy state, the state of vital force freed from matter – at this stage, our offspring become immortal energy "vortices" and as such can, for example, travel to other planets. This state seems to be close to the "thin," "ethereal," "astral" body of spiritualism (as in Helena Blavatsky's *The Secret Doctrine*, 1888), and is one step short of the traditional idealistic disembodied mind or soul.

The bio-field theory was then not altogether unscientific, though not particularly mainstream. In its supposition, Shaw was close to a very influential contemporary branch of Soviet science which, even though negating the existence of the soul, seriously investigated bio-field effects and attempted to scientifically test para-psychological phenomena, as for example did Vladimir Bekhterev and Leonid Vasiliev in the 1920s. (Bekhterev V.M., *Hypnos, vnushenie, telepatia* (Hypnosis, Suggestion, Telepathy – collected works), Mysl, Moscow, 1994; Vasiliev L.L., *Tainstvennie Yavlenia Chelovecheskoy Psichiki* (Mysterious phenomena of the human psyche), Gospolitizdat, Moscow, 1963.)

Attempts to induce regeneration by radiation and electromagnetic stimulation were made in the 1920s by Alexander Gurwitsch and Alexander Lubishev. (A.A. Lubishev, A.G. Gurwitsch, *Dialog o Biopole – Shornik* (A dialogue on the biofield – collected works), Ulyanovsk, 1998.)

Photographing "biological fields" was introduced by Semen Kirlian in 1939 and patented in 1949. (Kirlian V.H., Kirlian S.D., *V mire chudesnikh razriadov* (The world of miraculous charges), Znanie, Moscow, 1964.)

In the 1920s-early 1930s, the Russian scientists Alexander Chizhevsky (Tchijevsky) and Alexey Voynar (Voynard) hoped to retard senescence by breathing negatively ionized air. (Tchijevsky and Voynard, "The ageing of the organism retarded by the inspiration of negatively ionised air," quoted in "Societies and Academies," *Nature*, 134, 546-548, October 6, 1934.)

The idea of a "life-ray" (vitality enhanced by irradiation) was present in Michael Bulgakov's *The Fatal Eggs* (1924).

In the West, in the 1920s, there were famous biofield-related studies by the Italian psychologist Ferdinando Cazzamalli and the German Hans Bender.

(F. Cazzamalli, *Les Ondes electro-magnetiques en correlation avec certains Phenomenes psycho-sensoriels. Comptes Rendues de III-e Congres International de Recherches Psychiques* (Electromagnetic waves and their correlation with certain psycho-sensory phenomena. Reports of the Third International Congress of Psychic Research), Paris, 1928; Hans Bender, *Zum Problem der außersinnlichen Wahrnehmung. Ein Beitrag zur Untersuchung des räumlichen Hellsehens mit Laboratoriumsmethoden* (On the problem of extrasensory perception. A contribution to the study of spatial clairvoyance by laboratory methods), J.A. Barth, Leipzig, 1936.)

The Italian physicist Giacondo Protti determined the loss of "vital radiation in senescence" and its renewal upon transfusion of young blood. (Protti G., *L'emoinnesto intramuscolare e le radiazioni vitali nella vecchiaia e nell'esaurimento* (Intramuscular grafting and vital radiation in senescence and exhaustion), Ulrico Hoepli, Milano, 1931.)

Moreover, various electromagnetic devices were used to stimulate nerve function, activate tissues, increase blood flow, improve sleep, etc.

Diathermy, or electric heating and stimulation of the blood flow, formed an indispensable part of many rejuvenation techniques.

The rays of radium were employed for healing, laying down the grounds for radio-therapy. (Benjamin M. Duggar (Ed.), *Biological Effects of Radiation*, McGraw-Hill, NY, 1936.)

At the time of Shaw's writing, a host of other health improvement and rejuvenation systems, most notably, Homeopathy, Osteopathy, Chiropractic, Magnetic and Electronic Healing, were predicated on "adjusting," "augmenting" and "transmitting" of the "vital energy."

The interest in the "biological energy fields" and their applications to aging and longevity, have continued until now (see the section "The present time" of the current investigation).

[1059] The power of the "Will" was considered by Shaw to be so great as to positively influence not only the longevity of the willful individuals, but even the longevity of their offspring.

In *Back to Methuselah*, Shaw emphasizes the "where there is a wish, there is a way" approach, yet he does not go so far as to suggest a complete absence of mediation between the will and its material execution. Rather, the mediation is described in terms of neo-Lamarckian inheritance, i.e. the inheritance of acquired characteristics.

Shaw was much influenced by the neo-Lamarckian convictions, promulgated at the end of the 19th century by Professor of Physiology Ewald Hering of Prague, and the English intellectuals – professor of Natural History Marcus Manuel Hartog, the naturalist John Jenner Weir and mostly by the novelist Samuel Butler – who argued for the inheritance of acquired characteristics.

Thus, Butler maintained that "heredity is only a mode of memory," which explains the phenomena of aging and longevity: After the age of reproduction, the organism falls into disarray because it can no longer "fall back" on reliable "memorable" activity patters. In other words, after the age of reproduction the organism becomes incompetent in performing its vital functions, since "we cannot suppose offspring to remember anything that happens to the parent subsequently to the parent's ceasing to contain the offspring within itself."

Accordingly, "those organisms that are the longest in reaching maturity should on the average be the longest-lived, for they will have received the most momentous impulse from the weight of memory behind them."
(Samuel Butler, *Essays on Life, Art and Science,* Chelsea House, NY, 1983, first published in 1889, republished in 1908, "The Deadlock in Darwinism," pp. 322-323, 333-336.)
Shaw as well believed that physical traits acquired by learning are inherited, but went even further and suggested that when longevity is desired strongly enough, the desire will embody in the actual life extension in future generations.
Shaw dismissed the scientific evidence that acquired characteristics are not inherited, for example, that cutting off mice tails does not produce short-tailed mice. According to Shaw, short tails are not produced due to the lack of conscious will on the part of the mice.
About the time of appearance of Shaw's *Back to Methuselah*, the Austrian/Jewish biologist Paul Kammerer (1880-1926) offered a neo-Lamarckian theory for improving longevity.
Paul Kammerer's positive views on the possibility of rejuvenation and inheritance of longevity were expressed in his books: *The Inheritance of Acquired Characteristics* (1924), *Rejuvenation and the Prolongation of Human Efficiency* (1923, first published in German in 1921), and *Death and Immortality* (1923).
Kammerer, in his terms, sought "reconciliation" between "Lamarckism" and "Mendelism" and saw himself as a true follower of "Darwinism," which for him implied not so much the weeding out of the unfit by negative selection, but mainly the positive role of the environment in shaping new life forms.
(Kammerer, *The Inheritance of Acquired Characteristics*, NY, 1924, "Darwinism and Socialism," p. 261.)
In this view of evolution, according to Kammerer, the offspring's ability to regenerate, their immunity and longevity, could be improved by manipulating the internal and external environment of the parents, particularly, by manipulating their sex glands.
Consistently with this view, Kammerer was a firm supporter and popularizer of Steinach's method of rejuvenation by sex gland stimulation.
He was an integral member of the international rejuvenation movement of the 1920s.
Kammerer's *Rejuvenation and the Prolongation of Human Efficiency* was dedicated in "sincere friendship" to the chief of American rejuvenators, Dr. Harry Benjamin, and was prefaced by a letter from Steinach who praised the book as a "splendid treatise... the best that has been written along this line of research." Kammerer's work, Steinach said, "greatly advanced our research work" and, thanks to it, Steinach himself was "now able to better observe and apprehend the intricate biological connections."
(Kammerer, *Rejuvenation and the Prolongation of Human Efficiency*, NY, 1923, p. i .)
Kammerer made a strong connection between the "inheritance of acquired characteristics" and "rejuvenation." According to him, animal experiments indicated that descendants of the parents subjected to sex gland rejuvenation, not only were not inferior to those of younger parents, but might have even become more physically and sexually vigorous (studies of Ottmar Wilhelm, Paul Lengemann, Gustav Fritsch, Eugen Steinach and Paul Kammerer's own are cited in evidence).
Moreover, the "acquisitions of the psyche" were believed to be inheritable. According to Kammerer, at the present, reproduction at a very advanced age may be detrimental to the vitality of the offspring. Successful rejuvenation, however, can change the situation dramatically: Not only the acquired vigor will be transmitted to the progeny, but the descendants will also benefit from the hereditary transmission of the wisdom of the long-living parents, they will inherit their will to live and longevity.
(Kammerer, *The Inheritance of Acquired Characteristics*, NY, 1924, "Inheritance and rejuvenation," pp. 317-327.)
(The publication of Paul Kammerer's *The Inheritance of Acquired Characteristics*, translated by A. Paul Maerker-Branden, Boni and Liveright, New York, 1924, appears to predate the German original: Paul Kammerer, *Neuvererbung oder Vererbung erworbener Eigenschaften*, Seifert Verlag, Stuttgart-Heilbronn, 1925. Paul Kammerer's *Rejuvenation and the Prolongation of Human Efficiency*, translated by A. Paul Maerker-Branden, Boni and

Liveright, New York, 1923, was first published in German as *Über Verjüngung und Verlängerung des persönlichen Lebens*, Deutsche Verlagsanstalt, Stuttgart-Berlin, 1921. Paul Kammerer's *Tod und Unsterblichkeit* (Death and Immortality), Ernst Heinrich Moritz, Stuttgart, 1923, apparently was not translated into English.)

In his belief in the inheritance of acquired characteristics, Shaw again exhibits a strong affinity with the Soviet life sciences of the time, many of whose representatives were essentially neo-Lamarckist.

(On neo-Lamarckism in Soviet biology, yet disregarding its implications for longevity research, see David Joravsky, *The Lysenko Affair*, Harvard University Press, Cambridge MA, 1970.)

Both Shaw and Kammerer were ardent socialists.

Kammerer's works were widely disseminated in post-Revolutionary Russia, including his *Death and Immortality* (Smert i Bessmertie, Moscow, 1925) and *Rejuvenation and the Prolongation of Life* (Omolazhivanie i Dolgovechnost, Peterburg, 1922). He was a close friend of Anatoly Lunacharsky (1875-1933), Soviet Russia's Chief Commissar of Education, who made a movie about Kammerer's work. In 1926, Kammerer set out to establish a laboratory in Moscow and was about to move to the USSR permanently.

(Arthur Koestler, *The Case of the Midwife Toad*, Hutchinson, London, 1971, pp. 15-17.)

Shaw, in turn, apart from the tribute to Soviet science in the Preface to *Back to Methuselah*, dedicated a copy of it to "Nicolas Lenin the only European ruler who is displaying the ability character and knowledge proper to his responsible position."

(Quoted in Michael Holroyd, *Bernard Shaw. Vol. 3. 1918-1950. The Lure of Fantasy*, Penguin, Harmondsworth, 1988, p. 54.)

Many works of Soviet biology from the 1920s through the 1940s were essentially neo-Lamarckist, including those of Clement Timiriazev (1843-1920), Ivan Michurin (1855-1935), and most notoriously Trofim Lysenko (1898-1976). Lysenko's chief supporter in the field of gerontology was Alexander Nagorny who upheld Lysenko's tenet about the "assimilation of external conditions."

(Alexander Vasilievich Nagorny, *Starenie I Prodlenie Zhizni* (Aging and Life-extension), Sovetskaya Nauka (Soviet Science), Moscow, 1950, pp. 191, 218-219.)

In the neo-Lamarckist school, it was staunchly believed that plants and animals can be taught survival skills, for example, crops can be conditioned to withstand cold or draught and become more long-lived, and that those skills will be transmitted to the progeny. This possibility was indicated in Timiryazev's and Michurin's studies on plants.

Moreover, Ivan Pavlov's and his pupil Nikolay Studentzov's experiments suggested the possibility of inheriting conditioned reflexes in animals, i.e. the transmission of newly-learned behavior and experience via presumed changes in the hereditary substance. These findings of Pavlov were enthusiastically enlisted into the armory by the European life-extensionists, Kammerer and Kotsovsky.

(Pavlov, I.P., "New Researches on Conditioned Reflexes," *The Bulletin of the Battle Creek Sanitarium and Hospital Clinic*, 19 (1), 1-4, December 1923; Dimu Kotsovsky, "Allgemeine vergleichende biologie des Alters. Genese des Alters" (General comparative biology of aging, the genesis of aging), *Reviews of Physiology, Biochemistry and Experimental Pharmacology*, 1(31), 132-164, 1931; Paul Kammerer, *The Inheritance of Acquired Characteristics*, Boni and Liveright, New York, 1924, "Acquisitions by the Psyche and their inheritance," p. 191.)

Of course, these studies mainly concerned modifications of animals and plants, apparently with no direct attempts at human experimentation. Yet, they had plain implications for the Soviet ideal of creating a New Man, resistant to diseases and long-lived, through indoctrination and amelioration of the environment.

In Kammerer, the projections of animal experiments on man were explicit. In Kammerer, a healthy life-style of the parents was suggested to positively affect the health and longevity of the offspring.

Shaw too appeared to be very certain that the acquired drive to live long will be embodied in the progeny.

The scientific evidence that the neo-Lamarckists produced is now generally disregarded as invalid. Yet, neo-Lamarckism has not died.

Neo-Lamarckism resurfaced in the 1960s with the emergence of the theory of "non-neuronal memory transfer."
(James McConnell, "Memory Transfer through Cannibalism in Planaria," *Journal of Neuropsychiatry*, 3, 42-48, 1962.)
To the present time, and particularly in Russia, neo-Lamarckism represents an enticing, though hardly scientifically supported, branch of life-extensionism, suggesting the possibility of transmitting acquired longevity and even personal immortalization by passing on the information acquired during one's life to the progeny.
(Sergey Ryazanzev, *Tanatologia – Nauka o Smerti* (Thanatology – The Study of Death), Vostochno-Evropeyskiy Institut Psychoanaliza (The East-European Institute of Psychoanalysis), Saint-Petersburg, 1994, p. 336.)
The spirit (though not the name) of neo-Lamarckism was recently raised by the studies investigating the effects of high parental age on the health and longevity of the offspring. The effects were found to be either positive or negative.
(See Eisenberg DTA, Hayes MG, Kuzawa CW, "Delayed paternal age of reproduction in humans is associated with longer telomeres across two generations of descendants," *Proceedings of the National Academy of Sciences USA*, June 11, 2012; Gavrilov LA, Gavrilova NS, "Parental age at conception and offspring longevity," *Reviews in Clinical Gerontology*, 7, 5-12, 1997.)

[1060] Bernard Shaw, "Preface" to *The Doctor's Dilemma* (1906), in *Prefaces by Bernard Shaw*, Constable and Company Ltd., London, 1934, pp. 237-281.
(The Preface to *The Doctor's Dilemma* is available online at http://www.online-literature.com/george_bernard_shaw/doctors-dilemma/0/.)

[1061] The reluctance to become involved in practical rejuvenation and laboratory research of aging may have been a part of the contempt of many British "patrician" holistic physicians toward laboratory medicine generally, as described by the British medical historian Christopher Lawrence.
(Christopher Lawrence, "Still incommunicable: Clinical holists and medical knowledge in interwar Britain," in Christopher Lawrence, George Weisz (Eds.), *Greater Than the Parts: Holism in Biomedicine, 1920-1950*, Oxford University Press, Oxford, 1998, pp. 94-111.)
In addition, the great influence of the anti-vivisectionist movement in Britain in the early 20th century is well known.
(Anita Guerrini, *Experimenting with Humans and Animals: From Galen to Animal Rights*, The Johns Hopkins University Press, Baltimore, 2003.)
Furthermore, the attitude toward practical rejuvenation may have been not unlike the attitude toward practical eugenics. When comparing the eugenic movements in Britain and the US, the American historian Daniel J. Kevles notes that even though the theory of eugenics originated on the British soil, its practice was much wider spread in the US. He explains this by the much greater authority given to scientific experts in practical matters in the US.
(Daniel J. Kevles, *In the Name of Eugenics: Genetics and the Uses of Human Heredity*, Harvard University Press, Cambridge MA, 1997 (1986), p. 101.)

[1062] A noticeable increase in the dissemination of science-fictional works related to life extension appears to have occurred in the 1960s, not only in English science fiction, but also in other countries.
In several works, the noble quest for physical immortality, and the Herculean feats performed after its attainment, motivated the plot. Thus, for example, in *Perry Rhodan* (the world's most prolific German science fiction novel series, created in 1961 by Karl-Herbert Scheer and Clark Darlton), the physical invulnerability granted to the protagonists enable the continuation of the cosmic saga for millennia.
The perennial search for the immortalizing "phoenix blood" underscores Tezuka Osamu's *Phoenix* (a Japanese manga comics series started in 1966).

In Robert Zelazny's *This Immortal* (1966, US), the protagonist's "high survival potential," and immortalization through a "mutation" or through the unspecified life-extending "Sprung-Samser" treatment, would make him able to indefinitely defend the Earth.

In Dan Simmons's *Hyperion Cantos* series (1989, 1990, 1996, 1997, US), the regenerating and resurrecting "cruciform" symbiotes make the protagonists virtually indestructible in continuous space wars.

In these works of "soft science fiction"/"space operas," specifying methods for life-extension appears to be of minor importance, and the "immortalization" is just an instrument for plot development and a means of impunity in cosmic battles, in the eternal conflict between the forces of destruction and preservation.

(See http://en.wikipedia.org/wiki/Immortality_in_fiction.)

Other works of science fiction appear to have placed a greater emphasis on discussions of methodology, and their visions concerning potential ways to life-extension often diverged radically. The dispute between reductionist and holistic approaches to life-extension was transferred from scientific and popular-scientific works onto the pages of science fiction.

On the side of "holism," after Shaw, the concept of a biological energy field provided the terminology to describe a direct control of the mind over the body, that was believed to be necessary for extreme longevity. In this "holistic" line of thought, Colin Wilson's novel *The Philosopher's Stone* (1971, UK) explicitly propagandizes the power of meditation and intellectual activity for fostering individual vitality. In this novel, the life-extension device was initially conceived as a brain-stimulating electrode implant. But such a contrivance was then considered unnecessary, as the same effects could be produced directly by the power of the mind. By employing enhanced conscious "control of the prefrontal cortex," Wilson believes, "It is man's destiny to be immortal." Furthermore, this ability "may be transmitted directly to the child." This is Shaw's vision reincarnated.

(Collin Wilson, *The Philosopher's Stone*, Crown Publishers, NY, 1971, pp. 267-268.)

Tom Robbins's *Jitterbug Perfume* (Bantam Books, NY, 1984) too seems to side with the "holistic approach," emphasizing hygiene, alternative medicine (particularly aromatherapy), and the revitalizing power of the mind. The main protagonist, Alobar, achieves immortality by following a simple hygienic regimen: continuous and balanced breathing, hot baths, moderate nutrition with a great emphasis on beets, and much sex.

But the most important element in Alobar's life-extending regimen is the believing mind:

"The body is the servant o' the mind... May be the main reason your Alobar lived on was because he believed he could" (p. 296).

Robbins further expounds on the revitalizing power of the mind and proposes a theory where the human consciousness (and the brain) evolve from the reptilian through the mammalian and ultimately toward the "floral consciousness" that is compatible with "a data-based soft technology":

"A floral consciousness and an immortalist society are ideally suited for one another. (Flowers have superior powers of renewal, and the longevity of trees is celebrated. The floral brain is the organ of eternity)" (p. 368).

Apparently, Robbins wished his concurrent "Flower Children" revolution to go on forever.

The crucial role of expanded consciousness for vitality was later professed by the American health guru, Deepak Chopra, in *The Way of the Wizard* (1995). The book contains literary dialogues between Merlin the Wizard and King Arthur, and conveys the ideas about the immediate and unmediated conscious control over life processes and about the illusory nature of death.

(Deepak Chopra, *The Way of the Wizard. Twenty Spiritual Lessons in Creating the Life You Want*, Harmony Books, New York, 1995.)

This work complements Chopra's other self-help books on attaining longevity through the power of "awareness."

(Deepak Chopra, *Grow Younger, Live Longer. Ten Steps to Reverse Aging*, Harmony Books, New York, 2001;

Deepak Chopra, *Ageless Body, Timeless Mind. The Quantum Alternative to Growing Old*, Harmony Books, New York, 1993.)

Thus, the "holistic" approach to life extension, emphasizing the "mind-over-matter" control, has persisted in literary (English) fiction throughout the century: from Shaw to Wilson, to Robbins, and later to Chopra.

On the other hand, the "reductionist" materialistic approach to life-extension appears to play a greater part in works tending toward "hard science fiction."

The science fictional visions, it appears, followed the scientific ones, and not the other way around.

Thus, in Robert Heinlein's *Methuselah's Children* (Signet, NY, 1958), extreme longevity is achieved by direct biomedical interventions: eugenic selection for longevity and rejuvenative biotechnology. Rejuvenation mainly involved the transfusion of young blood "grown outside the body" and a variety of "subsidiary techniques" such as tooth budding, growth inhibiting, hormone therapy, cosmetic treatments, etc. (p. 154).

These techniques bear a striking resemblance to Bogdanov's and Protti's earlier experiments on rejuvenative blood transfusion, Steinach's hormone therapy and Alexis Carrel's experiments on genetic selection, tissue culture and transplantation.

The work also features prolonged body preservation by "cold rest" and "cold sleep," and a refreshing chamber (the "fresher") providing personal hygiene, bathing and massage, as well as a "sleep surrogate" pill (again an echo of Carrel's and Lindbergh's experiments on hibernation). The prospects Heinlein poses are quite optimistic.

A series of reductionist rejuvenation methods is presented in Larry Niven's *Ringworld* (1970), including the anti-aging drug "boosterspice" made of genetically engineered ragweed (*Ambrosia*), replacement organs, electromagnetic sleep induction ("Sleep Headsets"), the "stasis field," and the "Autodoc" – a device for the automatic diagnosing and balancing of one's metabolism and automatic surgical repairs.

There are more suggestions in Niven's *A World Out of Time* (1976), including cryo-preservation ("corpsicles"), mind transfer into an "empty" body, "RNA shots" to transfer personal memories, the "Young Forever" treatment permanently arresting development at a juvenile stage (neoteny). The inspirations from contemporary science can be perceived.

(See *Technovelgy*, 2003-2007, for a list of life-extending devices in science fiction, http://web.archive.org/web/20071024140911/www.technovelgy.com/.

Cf. the sections "The 1950s-1960s. The evolution of rejuvenation methods" and "Theories of Aging" of the present study.)

Both Heinlein and Niven seem to emphasize reductionist biological rejuvenation, using stimulants, biological tissue replacements and bio-preservers, developing ideas from contemporary biology.

Other authors valorized durable "cybernetic organisms," indefinitely maintainable through prosthetic replacements, man-computer synergy or personality uploading into a computer, that were inspired by recent advances in bionics and cybernetics. (See the latter sections of this study on the relation of life-extensionism to the research on bionic transplantation and cybernetics of the 1950s-1960s.)

Thus, for example, revival becomes possible thanks to bionic replacements in Martin Caidin's novel *Cyborg* (1972, the basis for *The Six Million Dollar Man* TV series, US, broadcasted in 1974-1978).

Life extension is achieved by continuous artificial organ ("artiforg") replacements in Philip Dick's *Ubik* (1969, US), whereby the age is "impossible any more to tell by looks, especially after ninety."

The suggestion of the unlimited replacement of all biological constituents by artificial organs also emerges in Stanislaw Lem's *Do You Exist, Mr. Johns?* (1955, Poland).

The idea of "cyborg" immortality extended further, suggesting that a human personality can be indefinitely preserved in the form of uploaded or recorded "constructs" (William Gibson, *Neuormancer*, 1984, US).

Isaac Asimov's "Robot Novels" – *The Caves of Steel* (1954), *The Naked Sun* (1957), and *The Bicentennial Man* (1976, US) – also explore cybernetic immortality, whereby personality can be indefinitely maintained in a "positronic brain" of platinum and iridium, supported by a mechanical body.

In Arthur Clarke's *2001: A Space Odyssey* (1968, UK), immortality is achieved either through personality uploading or ascendance to a disembodied (ethereal) state.

The biotechnological and cybernetic paths to immortality envisioned by English science fiction appear to be equally reductionist – viewing the human body as a machine subject to indefinite maintenance through the replacement of its parts.

Yet, there appears to be an opposition between the biotechnological and cybernetic methods.

Even though equally opposed to the "holistic" approach with its somewhat old fashioned views on hygiene and the somewhat mystical perceptions of "wholeness" and the revitalizing power of the "mind" – the "cybernetic" and "biological" approaches essentially diverged on the attitude to the materials that would make up the replacement parts.

The "biological" techniques may have appeared more traditional and hence may have invoked less resistance in the reader than the "bionic"/"cybernetic" ones.

The opposition is exemplified in Frank Herbert's *Dune* (1965, US), where computer technology is banned and humanity evolves solely through biological enhancement.

Bruce Sterling's cyberpunk novel *Schismatrix* (1985, US) envisions a direct confrontation between "shapers" or proponents of genetic engineering and "mechanists" or supporters of technological/cybernetic enhancement.

In David Brin and Gregory Benford's *Heart of the Comet* (1986, US), the "programmer," the "doctor" and the "spacer" represent three lines of scientific development.

As the above examples indicate, the current scientific and popular-scientific disputes on potential life-extending methods – particularly as regards the future promise of holistic hygiene and psychological regulation on the one hand, and reductionist bio-transplantation and prosthetics on the other – were projected in science-fictional works.

Thus the science-fictional treatments of longevity techniques seem to have been not unlike retrofitting – extending new features on the present designs. As much as being a means to attenuate "future shock," to accustom the mind to possible technological novelties, they appear in an equal, perhaps a greater measure, to advertise current technological developments.

In a similar manner, extending longevity seems to have acted as an effective hyperbolic plot device, serving to project the current social, political and ideological patterns and trends far into the future.

The explicitly life-extensionist works (such as Halperin's *The First Immortal*, 1998, US) viewed the current basic social pattern as generally beneficent and worthy of indefinite preservation. As evidenced earlier, this has been the general life-extensionist stance in any genre, throughout the century and throughout diverse national settings.

Not only was the prospect of radical longevity seen optimistically as an opportunity to learn and develop, but paradoxically also as a means to conserve one's loves and habits.

Other science-fictional works are rather emotionally "neutral." But great many others are predominantly pessimistic, viewing extending life-spans as a means to expand current social evils – egotism, inequality, boredom, scarcity, and so forth – into monstrous proportions.

The latter attitude seems to have been most pronounced in British science fiction. American authors generally appear somewhat more optimistic, and giving a greater supportive weight to far-reaching technologies.

Thus, the works of science fiction related to life extension, as well as life-extensionist works in the scientific and popular scientific genres, broadcasted the present social, ideological and technological patterns and controversies.

[1063] Bernard Shaw, "Preface" to *English Prisons under Local Government by Sidney and Beatrice Webb, with Preface by Bernard Shaw*, Longmans, Green and Co., London, 1922, pp. vii-lxxiii; in a later revised reprint, Bernard Shaw, "Imprisonment (English Local Government by Sidney and Beatrice Webb)," in *Prefaces by Bernard*

Shaw, Constable and Company Ltd., London, 1934, pp. 282-320 (hereafter referred to as "Shaw, 1922" or "Shaw, 1934"), quotes: Shaw, 1922, pp. xxxii, xxxiv, lxxii, Shaw, 1934, p. 319.

[1064] Bernard Shaw, "Preface to *On the Rocks*," in *Prefaces by Bernard Shaw*, Constable and Company Ltd., London, 1934, pp. 351-375 (hereafter referred to as "Shaw, 1933"), quote: p. 352.

[1065] Bertrand Russell, "The Menace of Old Age" (1931), pp. 18-20, "On Euthanasia" (1934), pp. 267-268, in Bertrand Russell, *Mortals and Others, American Essays 1931-1935, Volumes I and II*, Routledge Classics, London and New York, 2009 (first published in 1975); Bertrand Russell, "How to Grow Old" (written in 1944), in Bertrand Russell, *Portraits from Memory: And Other Essays*, Simon and Schuster, NY, 1956, pp. 50-53.

[1066] Lisa Yount, *Physician-Assisted Suicide and Euthanasia*, Facts on File Inc, NY, 2000, p. 9.

[1067] Bernard Shaw, *Back to Methuselah*, in *Bernard Shaw Selected Plays*, Dodd, Mead and Company, Vail-Ballow Press Inc., New York, 1949 (first published in 1921, first performed in 1922), Part II, p. 80, available online at Project Gutenberg, http://www.gutenberg.org/cache/epub/13084/pg13084.html.

[1068] Herbert George Wells, *Men Like Gods* (1923), reprinted at http://gutenberg.net.au/ebooks02/0200221.txt.

[1069] Herbert George Wells, *Anticipations of the Reaction of Mechanical and Scientific Progress Upon Human Life and Thought*, Second Edition, Chapman and Hall, London, 1902 (first published in 1901), p. 308, http://www.gutenberg.org/files/19229/19229-h/19229-h.htm.

[1070] Recently, the practice of euthanasia in Britain appears to have become overwhelming, according to Steve Doughty, "Top doctor's chilling claim: The NHS kills off 130,000 elderly patients every year," *London Daily Mail*, 21 June 2012, http://www.dailymail.co.uk/news/article-2161869/Top-doctors-chilling-claim-The-NHS-kills-130-000-elderly-patients-year.html.

I am not aware of estimates of the numbers of openly or covertly euthanized people in other countries.

[1071] Elie Metchnikoff, *Etudy o Prirode Cheloveka* (Etudes on the Nature of Man), The USSR Academy of Sciences Press, Moscow, 1961 (1903), Ch. X. "Vvedenie v nauchnoe izuchenie starosti" (An introduction to the scientific study of aging), p. 201.

[1072] Shaw, 1922, pp. xxx, xxxi, lxxi, lvii.

[1073] Shaw, 1933, p. 358.

[1074] Shaw, 1922, pp. xxvi, xxxiv, xxxvii, xliii, xliv.

[1075] Shaw, 1933, pp. 355, 359, 361.

[1076] Bernard Shaw, *Man and Superman. A Comedy and a Philosophy*, Penguin Books, 1994 (first published in 1903), "The Revolutionist's Handbook. VIII. The Conceit of Civilization," p. 240, available online at the Internet Archive, http://www.archive.org/details/mansupermancomed00shawrich.

[1077] Shaw, 1933, p. 362.

[1078] Shaw, 1922, pp. xxvii-xxviii, xxxiii, lvi, lxxi.

[1079] Ignatz Leo Nascher, *Geriatrics, the Diseases of Old Age and their Treatment. Including Physiological Old Age, Home and Institutional Care, and Medicolegal Relations*, P. Blakiston's Son & Co, Philadelphia, 1914, p. v.

[1080] Shaw, Back to Methuselah, 1949 (1921), Part IV, Act 1, p. 171.

[1081] H.G. Wells, *Men Like Gods* (1923), http://gutenberg.net.au/ebooks02/0200221.txt.

[1082] On the "lukewarm" reception of *Back to Methuselah*, see Michael Holroyd, *Bernard Shaw. Vol. 3. 1918-1950. The Lure of Fantasy*, Penguin, Harmondsworth, 1988, pp. 53-59.

[1083] *The Metamorphoses* (8 AD) by Ovid (43 BC - 17 AD) provides perhaps the most inclusive and eclectic collection of arguments and warnings against human attempts to attain immortality. The impact of these warnings can be still felt today.

The story of Medea, the sorceress who possessed the skill of rejuvenation, is perhaps one of the most memorable (*The Metamorphoses*, Book VII). Medea uses her skill both for good (when she rejuvenates Jason's father, Aeson) and for evil (when she, by a show of medical skill, tricks Pelias' daughters into killing their father). The evil outcome of her power is however emphasized.

Most ancient sources mention only the murder of Pelias, with no reference to the rejuvenation of Aeson, as for example, in Euripides' *Medea* (431 BC), Apollodorus' *Library* (c. 140 BC, the most comprehensive surviving collection of Greek mythology), and Pausanias' *Description of Greece* (c. 143-176 AD). Sir James George Frazer names only the lost *Nostoi* (Νόστοι - *Returns*, attributed to Stesichorus, 632-552 BC) and Ovid's *Metamorphoses* as sources explicitly mentioning the rejuvenation of Aeson.
(J.G. Frazer, Note 4 to Apollodorus, *Library and Epitome*, Translated, edited and notes by Sir James George Frazer, William Heinemann, London, 1921, 1.9.27, reprinted at the Perseus Digital Library, Tufts University, Boston, MA, http://www.perseus.tufts.edu/.)
Further, Ovid reiterates the archetype of Tithonus, the forever suffering immortal human (the story of Tithonus, who gained eternal life, but not eternal youth, first appeared in the *Homeric Hymn to Aphrodite*, c. 7th or 8th century BC). In Ovid's version (Book XIV), it is the Sibyl who is doomed to an eternity of senility and misery. Like Tithonus, in her bargain with the gods, she forgot to ask for eternal youth.
Asclepius, who could heal the sick and even revive the dead, is largely sympathized with (Book II). And yet, he is struck by Jove's lightening, apparently for violating the divine order.
Also the immortal centaur Chiron "shall feel such pain ... as will make you pray for the knack of death" (Book II).
There is, in addition, the story of Glaucus, who eats a magic herb, turns into a green merman and lives (rather unhappily) ever after in the sea (Book XIII).
Perhaps a single positive victory over death, in *The Metamorphoses*, is that of Hercules, gloriously ascending to physical immortality and joining the company of the gods (Book IX). But most human attempts to emulate the immortality of the gods end in disaster.
The collection of arguments against physical immortality includes Ovid's claim that a poet can only achieve immortality through his immortal work (a thought earlier expressed by Theognis, 5th century BC, Theognis, "Immortality in Poetry," *Ancient Literary Criticism. The Principal Texts in New Translations*, edited by D.A. Russel and M. Winterbottom, Oxford University Press, Oxford, 1972, p. 3).
Ovid's collection also includes the notion of an afterlife of spirits or shadows (Book XIV), reducing the fear of death.
Yet, it seems, the strongest argument employed against physical immortality lies in the idea of the "metamorphosis" itself.
There is just no sense in preserving life for an extremely prolonged period, because there is, in Ovid, no such thing as a stable life form. Life undergoes incessant transformation: now one is a girl, the next moment a spider (Book VI). The opening part of the *Metamorphoses* – "Bodies, I have in mind, and how they can change to assume new shapes" – suggests the inevitable change and destructibility of the body. Also, there can be no preservation of personality, for there is no stable personality (or no personality at all): the protagonists of the *Metamorphoses* change their perceptions and attitudes in a flash. There is just nothing but a constant recombination of elements. The eternal dynamic transformations and being a part of the eternal whole constitute, accordingly, the only real immortality.
Thus, as can be seen, the representation of the pursuit of radical life-extension (in effect synonymous with the striving for physical immortality) in classical Greco-Roman literature was every bit as dystopian and antagonistic as in modern English literature.
(The translation used here is Ovid, *The Metamorphoses*, Translated by David Slavitt, The Johns Hopkins University Press, Baltimore and London, 1994.)

[1084] Shaw, 1922, pp. lvi, lxvi.
[1085] Shaw, 1933, pp. 351-352.
[1086] Shaw, 1922, xxxiv-xxxvii, xl.
[1087] John Harris, "Immortal Ethics," presented at the International Association of Biogerontologists (IABG) 10th Annual Conference "Strategies for Engineered Negligible Senescence," Queens College, Cambridge,

UK, September 17-24, 2003, reprinted in *Strategies for Engineered Negligible Senescence: Why Genuine Control of Aging May Be Foreseeable* (Ed. Aubrey de Grey), *Annals of the New York Academy of Sciences*, 1019, 527-534, June 2004.

[1088] Norman Spinrad, *Bug Jack Barron*, The Overlook Press, Woodstock and New York, 2004, first published in 1969, p. 245.

[1089] Poul Anderson, *The Boat of a Million Years*, Tor Books, NY, 1991, p. 334, first published in 1989. See also:
Ilia Stambler, "The pursuit of longevity – The bringer of peace to the Middle East," *Current Aging Science*, 6, 25-31, 2014,
http://benthamscience.com/journal/abstracts.php?journalID=cas&articleID=122295.

[1090] James L. Halperin, *The First Immortal*, Random House Publishing, New York, 1998, reprinted at http://coins.ha.com/tfi/.

[1091] Edmund Gayton, *Art of longevity, or, A diaeteticall institution*, Oxford, 1659.

[1092] John Smith, MD, *The Pourtract of Old Age: Wherein is Contained a Sacred Anatomy Both of Soul, and Body, and a Perfect Account of the Infirmities of Age Incident to Them Both. Being a Paraphrase upon the Six former Verses of the 12. Chapter of Ecclesiastes*, J. Macock, London, 1666. Quoted in Joseph T. Freeman, "The History of Geriatrics," *Annals of Medical History*, 10, 324-335, 1938, p. 326.

[1093] William Harvey, "Anatomical Examination of the Body of Thomas Parr," in *The Works of William Harvey*, The Sydenham Society, London, 1847, pp. 589-592.

[1094] Thomas Sydenham, "On Gout," in *The Works of Thomas Sydenham*, The Sydenham Society, London, 1848, vol. 2, p. 129.

[1095] Joseph T. Freeman, "The History of Geriatrics," *Annals of Medical History*, 10, 324-335, 1938.

[1096] Sir John Sinclair [1754-1835], *The Code of health and longevity; or, A concise view, of the principles calculated for the preservation of health, and the attainment of long life*, J. Murray, London, 1807 (with subsequent editions in 1816 and 1818); *The Abernethian Code of Health and Longevity, ... Founded on the principles and practice of John Abernethy, Esq. F.R.S.* [M.R.C.S., 1764-1831], J. Williams, London, 1829 (with a later edition in 1836).
An additional popular work was Joel Pinney, Esq., *A Key to the Attainment of the Full Term of Life*, Longman, London, 1860 (an earlier version was published in 1856, entitled *The Duration of Human Life, and its three eras: showing the possible causes that have shortened the lives of the human race, and the barriers that prevent a return to the longevity of the early Patriarchs*, 1856).
A famous 19th century work doubting the existence of supercentenarians and hence the possibility of the prolongation of life beyond a century, was William John Thoms [1803-1885], *Human longevity, its facts and its fictions*, Murray, London, 1873.

[1097] James Crichton-Browne, *The Prevention of Senility and a Sanitary Outlook*, MacMillan, NY and London, 1905, p. 54.

[1098] Robert Boyle was famous for suggesting the rejuvenative "Boyle's Gold Tincture" ("Tinctura Auri Boylaei").
George Thomson developed a whole series of rejuvenative preparations, such as the "essential oil of myrrh, aloe and saffron," the "serene amber oil," etc.
(Johann Heinrich Cohausen, *Tentaminum physico-medicorum curiosa decas de vita humana theoretice et practice per pharmaciam prolonganda* (Curioous physico-medical theoretical and practical attempts to prolong the decades of human life by pharmacological means), M.A. Fuhrmannum, Osnabrück, 1714 (1699), pp. 23, 25, 109, 116, 121.)

[1099] David Hamilton, *A History of Organ Transplantation: Ancient Legends to Modern Practice*, University of Pittsburgh Press, Pittsburgh, 2012, pp. 31-48. Notably, in his studies of transplantation, Hunter was careful to note age-related changes in vitality.

[1100] As Hunter described the purpose of his experiments on freezing and thawing of animals, conducted in

1766:

"I had imagined that it might be possible to prolong life to any period by freezing a person in the frigid zone, as I thought all action and waste would cease until the body was thawed. I thought that if a man would give up the last ten years of his life to this kind of alternate oblivion and action, it might be prolonged to a thousand years; and by getting himself frozen every hundred years, he might learn what happened during his frozen condition."

Yet, as his freezing and thawing experiments were not successful, he became "undeceived."

(John Hunter, *Lectures on the Principles of Surgery* [delivered in 1786-1787], Ch. VIII. "On the Heat of Animals," in *The Works of John Hunter, F.R.S*, with notes, edited by James F. Palmer, Longman, London, 1835, vol. 1, p. 284.)

[1101] Elie Metchnikoff, "Mirosozerzanie i Medizina" (The worldview and medicine), first published in *Vestnik Evropy* in 1910, in *I.I. Metchnikoff, Akademicheskoe Sobranie Sochineniy* (The Academic Collection of Works), Edited by G.S. Vasezky, Gosudarstvennoe Izdatelstvo Medizinskoy Literatury (State Medical Literature Press), Moscow, 1954, vol. 13, p. 204-205.

Elsewhere Metchnikoff further valorized minimally invasive methods, such as "replacement of harmful intestinal bacteria by beneficial bacteria" over Lane's intestinal surgery.

(Elie Metchnikoff, "Dolgovechnost ili prodolzhitelnost zhizni" (Longevity or the Span of Life), in *Collected Works of I.I. Metchnikoff* (Ed. N.N. Anichkov and R.I. Belkin), The USSR Academy of Medical Sciences, Moscow, vol. 15, 1962, p. 361.)

[1102] As stated by the UK Home Secretary, William Joynson-Hicks (1865-1932) in 1928, regarding "Dr. Voronoff's Treatment" – "No such experiments are conducted in this country at all" (*Hansard, the Official Report of debates in Parliament*, HC Debate, 14 June, 1928, vol. 218, cc1148-9, http://hansard.millbanksystems.com/commons/1928/jun/14/dr-voronoffs-treatment).

[1103] Norman Haire, *Rejuvenation. The Work of Steinach, Voronoff and Others*, George Allen & Unwin, London, 1924.

[1104] Richard Ellmann, *W.B. Yeats's Second Puberty. A lecture delivered at the Library of Congress on April 2, 1984*, Library of Congress, Washington, 1985; David Boyd Haycock, *Mortal Coil. A Short History of Living Longer*, Yale University Press, New Haven and London, 2008, pp. 179, 184-185; Chandak Sengoopta, *The Most Secret Quintessence of Life. Sex, Glands, and Hormones, 1850-1950*, The University of Chicago Press, Chicago and London, 2006, p. 87.

[1105] Bogdanov A.A., "O fiziologicheskom kollektivizme" (On physiological collectivism), in *Statyi A. A. Bogdanova, opublikovannie v Vestnike Mezhdunarodnogo Instituta A. Bogdanova* (Articles by A. A. Bogdanov, reprinted in the Bulletin of the International Alexander Bogdanov Institute), 13(1), 14(2), 2003, Bogdanov Institute, Yekaterinburg, http://www.bogdinst.ru/vestnik/v14_05.htm.

[1106] Roy Porter, *The Greatest Benefit to Mankind: A Medical History of Humanity*, W.W. Norton & Company, NY, 1998 (1997), p. 619.

[1107] *Man and His Future: A CIBA Foundation Volume*, Gordon Wolstenholme (Ed.), Little, Brown and Co., Boston, 1963, p. 230.

[1108] James Lovelock, "Diathermy apparatus for the rapid rewarming of whole animals from 0°C and below," *Proceedings of the Royal Society B*, 147 (929), 545, 1957; Audrey Smith, *Biological Effects of Freezing and Supercooling*, Williams & Wilkins, Baltimore, 1961; Polge C, Smith AU, Parkes AS, "Revival of spermatozoa after vitrification and dehydration at low temperatures," *Nature* 164 (4172), 666, 1949; Lovelock JE, Bishop MW, "Prevention of freezing damage to living cells by dimethyl sulphoxide," *Nature*, 183 (4672), 1394-1395, 1959; Robert Ettinger, *The Prospect of Immortality*, Doubleday, NY, 1964.

[1109] Gerhard Venzmer, *Lang leben und jung bleiben!* (Live long and stay young), Franckhsche Verlag, Stuttgart, 1937 (republished in 1941), pp. 43, 47.

[1110] Charles Robert Darwin [1809-1882], *Notebook E, Transmutation of Species, 1838-1839*, p. 184, transcribed at *Darwin Online*, http://darwin-online.org.uk/.

Thomas Andrew Knight's (FRS, 1759-1838) original statement of 1816 was:

"The general law of nature appears to be that no living organized being shall exist beyond a limited term of years; and that law must be obeyed. It is nevertheless in the power of man to extend the lives of individual vegetable beings far beyond the period apparently assigned by nature; and parts of the same annual plant may be preserved through many years, perhaps through ages, though it cannot be rendered immortal."

(Thomas Andrew Knight, "Upon the Advantages of Propagating from the Roots of old ungrafted Fruit Trees," *Transactions of the Horticultural Society of London*, Volume 2, 1822, p. 252, first read on December 3, 1816.)

[1111] Darwin C. R., *The Variation of Animals and Plants under Domestication*, 2nd Edition, John Murray, London, 1875, vol. 1, pp. 455, 476, vol. 2, pp. 56, 376, reprinted in *Darwin Online*, http://darwin-online.org.uk/.

[1112] E. Ray Lankester, *On Comparative Longevity in Man and the Lower Animals*, Macmillan and Co., London, 1870, "5. Inherent Death," pp. 30-45.

[1113] Charles Darwin's grandfather, the physician and early proponent of the idea of evolution, Erasmus Darwin (1731-1802) was indeed keenly interested in life extension.

In *The Temple of Nature; or, the Origin of Society. A Poem, with Philosophical Notes* (T. Bensley, London, 1803), Erasmus Darwin asserted:

"The great and principal means to prevent the approach of old age and death, must consist in the due management of the quantity of every kind of stimulus."

That is to say, irritating stimuli should be neither excessive nor deficient for organs' function. This idea was similar to Hufeland's "consumption" and "renovation" of the life force.

(E. Darwin, *The Temple of Nature*, 1803, Canto II. 1.4. Note 7. II "Means of preventing old age.")

The same idea recurs in Erasmus Darwin's *The Botanic Garden* (1789, Part II, l. 387). It reappears again in his *Zoonomia* (1796, Vol. 1. Sect. XXXVII. IV. "Old age and death from inirritability," V. "Art of producing long life").

(Reprinted in *Project Gutenberg*, http://www.gutenberg.org/ebooks/26861.)

[1114] Francis Galton [1822-1911], *Inquiries into Human Faculty and its Development*, Second Edition, J. M. Dent & Co. (Everyman), New York, 1907, p. 212, reprinted in the online Galton archives at http://galton.org/, originally published in 1883 by Macmillan, London.

[1115] Mary Beeton, Karl Pearson, "Data for the problem of evolution in man. II. A first study of the inheritance of longevity and the selective death rate in man," *Proceedings of the Royal Society of London*, 65, 290-305, 1899; M. Beeton, G.U. Yule, Karl Pearson, "Data for the problem of evolution in man. II. On the correlation between duration of life and the number of offspring," *Proceedings of the Royal Society of London*, 67, 159-179, 1900; Natalia S Gavrilova and Leonid A Gavrilov, "Human longevity and reproduction. An evolutionary perspective," in *Grandmotherhood: The Evolutionary Significance of the Second Half of Female Life*, Rutgers University Press, New Brunswick, NJ, USA, 2005, pp. 59-80.

[1116] It has been often assumed by gerontologists that the evolutionary origin of aging and death was first explored by the German biologist August Weismann (1834-1914).

In 1882, Weismann posited that, in contrast to undifferentiated immortal cells (protozoa and germ cells), highly differentiated tissues have a diminished ability to replicate and restore themselves, thus aging and death are the evolutionary price paid for the organisms' differentiation and complexity.

According to Weismann, "death is not an essential attribute of living matter." However, aging and natural death are necessary conditions for multicellular evolution: without them aged organisms would fill the earth, exhausting all the resources and stifling the emergence of new generations and new life forms. Hence, aging is an evolutionary, selected feature.

(August Weismann, *Über die Dauer des Lebens* (On the duration of life), G. Fischer, Jena, 1882; *Über Leben und Tod* (On life and death), G. Fischer, Jena, 1884.

These works appear in English translation in August Weismann, *Essays Upon Heredity and Kindred Biological Problems*, Authorized Translation Edited by Edward B. Poulton, Selmar Schönland, and Arthur E. Shipley, Clarendon Press, Oxford, 1889, reprinted by Dabor Science Publications, New York, 1977, hereafter referred to as "Weismann, 1889.")

In Weismann's own words:

"Death is to be looked upon as an occurrence which is advantageous to the species as a concession to the other conditions of life, and not as an absolute necessity, essentially inherent in life itself. Death, that is the end of life, is by no means, as is usually assumed, an attribute of all organisms."

(Weismann, 1889, "The Duration of Life," p. 25.)

Weismann further speculated on the evolutionary factors that would contribute to the prolongation of life in higher animals, such as an extended period of reproduction and extended period of rearing the young.

Longevity, according to Weisman, is determined by the "replacement of cells by multiplication" and can be extended by increasing the number of somatic cell multiplications.

Life extension, according to Weismann, is "obviously of use to man, for it enables the old to care for their children, and is also advantageous in enabling the older individuals to participate in human affairs and to exercise and influence upon the advancement of intellectual powers, and thus to influence indirectly the maintenance of the race."

Weismann's general conclusion was:

"Death itself, and the longer or shorter duration of life, both depend entirely on adaptation. Death is not an essential attribute of living matter; it is neither necessarily associated with reproduction, nor a necessary consequence of it."

(Weismann, 1889, "Life and Death," pp. 145-146, 154-159.)

Yet, as attested by Edward Bagnall Poulton, the editor of the first and only English translation of Weismann's "The Duration of Life" and "Life and Death" of 1889, very similar conclusions were reached earlier by the British biologist Alfred Russel Wallace (1823-1913), the independent co-discoverer of evolution by natural selection.

In an unpublished note written "sometime between 1865 and 1870" that Wallace sent to Poulton after reading Weismann's translation proofs, Wallace claimed:

"if individuals did not die they would soon multiply inordinately and would interfere with each other's healthy existence. ... [Natural death] would be in any case for the advantage of the race and would therefore, by natural selection, soon become established as the regular course of things, and thus we have the origin of *old age, decay*, and *death*."

(Emphasis in the original, A.R. Wallace, "The Action of Natural Selection in Producing Old Age, Decay, and Death," c. 1865-1870, reproduced in full by Poulton in August Weismann, *Essays Upon Heredity and Kindred Biological Problems*, Clarendon Press, Oxford, 1889, p. 23.)

As Poulton relates, Wallace's note had been forgotten until recalled by Weismann's argument.

Elie Metchnikoff was apparently the first to systematically argue against Weismann's theory.

(Elie Metchnikoff, *Etudy o Prirode Cheloveka* (Etudes On the Nature of Man), Izdatelstvo Academii Nauk SSSR (The USSR Academy of Sciences Press), Moscow, 1961 (first published in 1903), "11. Vvedenie v nauchnoe izuchenie smerti" (An introduction to the scientific study of death), pp. 214-232.)

Metchnikoff agreed with Weismann that death is not a necessary prerequisite of all life, as unicellular organisms and germ cells are potentially immortal. However, Metchnikoff did not believe that natural death can be an evolutionary advantage. According to him, "normal aging" and "natural death" almost never occur in nature. Weakened organisms are eliminated by external causes – predation, diseases, accidents, competition – with a very slight chance for them to "age naturally" or die a "natural death."

And if aging and natural death almost never occur in nature, then natural selection cannot operate on them, let alone select for them.
(Metchnikoff, *Etudes On the Nature of Man*, pp. 216-218.)
Regarding humans, Metchnikoff strongly asserted that we all die a "violent" and not a "natural death," and that if we are ever able to combat the pathogens that cause our premature demise, human life can be greatly prolonged.
Interestingly, Metchnikoff traced the beginning of publicity of his theory of aging to England, to his presentation on April 22, 1901, at the Manchester Literary and Philosophical Society, where he "laid out a program of investigations aimed to unravel the problem of aging, the problem that had seemed almost intractable."
(Elie Metchnikoff, "Borba so Starcheskim Pererozhdeniem" (The struggle against the degeneration of senescence), in *I.I. Metchnikoff. Sobranie Sochineniy* (Collected Works), Eds. N.N. Anichkov and R.I. Belkin, The USSR Academy of Medical Sciences, Moscow, 1962, vol. XV, pp. 346-350.)
Clearly, the British soil was very receptive to evolutionary notions, and, as the further examples will show, the evolutionary theory of aging has since remained a central topic in British longevity science.
[1117] Since the 1920s, president of the UK Marine Biological Association, George Parker Bidder (1863-1953), further advanced the evolutionary concept of aging. Following Wallace's original proposition of "a limit of growth" as a cause of aging (as quoted in Weismann, 1889, p. 23), Bidder perceived senescence as due to "some mechanism to stop natural growth so soon as specific size is reached."
(George Parker Bidder, "The mortality of plaice," *Nature*, 115, 495, 1925; G.P. Bidder, "Senescence," *British Medical Journal*, 2, 5831, 1932; quoted in Alexander Comfort, *The Biology of Senescence*, Butler & Tanner, London, 1956, pp. 11-13.)
The inherent limitation of growth, being necessary for the survival of the species, yet eventually leading to senile degeneration, was also suggested by the British biologists Julian Sorrell Huxley in 1932 and later by Mary Whitear in 1952. These authors observed that in consequence of a prolonged differential growth of organs, their proportions may become structurally and functionally grotesque, leading to death (for example, the claw of a crab may grow to 10 times its body weight). Therefore the evolutionary emergence of extremely long lived animals appeared unlikely.
(Huxley J.S., *Problems of Relative Growth*, Methuen, London, 1932, Mary Whitear, "Internode Length in the Skin Plexuses of Fish and the Frog," *Quarterly Journal of Microscopical Science*, 93 (3), 307-313, 1952.
J.S. Huxley's and Whitear's findings are referred to in Peter Brian Medawar, *An Unsolved Problem of Biology*, H.K. Lewis, London, 1952, pp. 11, 20-21.)
In 1952, Peter Medawar of the University College, London, developed the evolutionary "Mutation Accumulation" theory of aging. (Self-admittedly, the theory was influenced by J.B.S. Haldane's ideas expressed in 1941 on the genetic conservation of George Huntington's chorea – a neurodegenerative genetic disease, manifesting late in adult life.) Medawar's theory was predicated on "an argument based upon combining an innate potential immortality with a contingent real mortality."
According to the theory, the accumulation of mutations acting late in life (and thus making little difference for individual reproductive success early in life) is the main cause of senescence.
(Peter B. Medawar, *An Unsolved Problem of Biology*, H.K. Lewis, London, 1952, quote from p. 23; Peter B. Medawar, "The definition and measurement of senescence," in *CIBA Foundation Colloquia on Ageing, Vol. 1, General Aspects*, Wolstenholme G.E.W. and Margaret P. Cameron (Eds.), J. & A. Churchill Ltd, London, 1955, pp. 4-15; J.B.S. Haldane, *New Paths in Genetics*, George Allen and Unwin, London, 1941.)
A refinement of the evolutionary theory was offered by the American biologist George Christopher Williams.
According to Williams' "Antagonistic Pleiotropy" concept (1957), it is not just the mere accumulation of late-acting mutations that causes senescence, but the very same genes that aid survival and reproduction in

an early period of life, can be damaging and cause senescence in a later period.
(George C. Williams, "Pleiotropy, natural selection, and the evolution of senescence," *Evolution*, 11, 398-411, 1957. An abridged version of this paper appears in Strehler, *The Biology of Aging*, 1960, pp. 332-337.)
John Maynard Smith generally sided with Williams, but added that natural selection may "synchronise" different causes of ageing.
(John Maynard Smith, "Review Lectures on Senescence. I. The Causes of Ageing," *Proceedings of the Royal Society of London B. Biological Sciences*, 157, 115-127, 1962.)
Medawar, Williams and Maynard Smith essentially accepted both Weismann's (Wallace's) fundamental premise that natural selection must play a role in developing senescence and fully agreed with Metchnikoff's qualification that senescence cannot be directly selected for (as it seldom occurs in nature). According to Medawar and Williams, senescence is a result of evolutionary "neglect," that is, the accumulation of hereditary deficiencies ("inherited disharmonies," in Metchnikoff's terms) that become pronounced late in life.
Later on, in 1977, Thomas Burton Loram Kirkwood of Newcastle University, UK, suggested the "Disposable Soma" theory.
This was an extension of the Antagonistic Pleiotropy theory, yet Kirkwood emphasized the terms of energy expenditure. According to this theory, the expenditure of energy on reproduction (early in life) is more important for evolutionary success than the energy expended on prolonged body/soma maintenance (throughout the life history). In this view, the body is "disposable": most energy resources are spent on reproduction at the expense of individual longevity.
(Kirkwood T.B.L., "Evolution of aging," *Nature*, 270, 301-304, 1977; Tom Kirkwood, *Time of Our Lives. The Science of Human Aging*, Oxford University Press, Oxford, 1999.)
Later still, this view was expanded and experimentally verified by yet another English biologist, Michael Robertson Rose, who showed that delayed reproduction in Drosophila flies drastically increases their life-span (more than twice).
Having received his PhD in 1979 at the University of Sussex, England, Rose later moved to the US (presently he works at the University of California, Irvine). He is now one of the foremost proponents of the evolutionary (bioeconomic) approach to radical life-extension.
Recently, in 2008, Rose coined the term SENSE – Strategies for Engineered Negligible Senescence Evolutionarily – a conceptual transformation of Cambridge's Aubrey de Grey's "SENS" anti-aging research program (see below on SENS). Rose's SENSE approach aims to validate evolutionary pathways leading to extreme longevity (as in the case of drosophila files in Rose's experiments) and then proceed to establishing the physiological mechanisms that may be produced in such evolutionary pathways.
(Michael R. Rose, *Evolutionary Biology of Aging*, Oxford University Press, Oxford, 1991; Michael R. Rose, "Making SENSE: strategies for engineering negligible senescence evolutionarily," *Rejuvenation Research*, 11(2), 527-534, 2008; National Institute on Aging, *Aging Under the Microscope: A Biological Quest*, 2008, http://www.nia.nih.gov/HealthInformation/Publications/AgingUndertheMicroscope/chapter02.htm.)
Notably, there has also been recently a resurgence of the 'Neo-Weismannist' view on the evolutionary theory of aging.
Its proponents continue to argue for the evolutionary advantage and hence specific selection (programming) for aging – instead of random disarray due to evolutionary "neglect" late in life.
On the optimistic side, the 'Neo-Weismannist' view has often implied that the genetic 'program' of aging that emerged in the past course of evolution, can in principle be 'cancelled.'
Yet, apparently, this school of thought has been more popular in the US and Russia, than in the UK.
(See, for example, Theodore C. Goldsmith [US], "Aging as an Evolved Characteristic – Weismann's Theory Reconsidered," *Medical Hypotheses*, 62(2), 304-308, 2004; Joshua Mitteldorf [US], "Ageing selected for its own sake," *Evolutionary Ecology Research*, 6, 937-953, 2004; Vladimir Petrovich Skulachev [Russia], "Aging is a

Specific Biological Function Rather than the Result of a Disorder in Complex Living Systems: Biochemical Evidence in Support of Weismann's Hypothesis," *Biochemistry (Moscow)*, 62(11), 1191-1195, Nov 1997.)

[1118] Alex Comfort, *The Biology of Senescence,* Butler & Tanner, London, 1956, pp. 189-191 (hereafter referred to as "Comfort, 1956").

[1119] Comfort, 1956, pp. 164, 168.

[1120] Comfort, 1956, pp. 6-8.

[1121] Comfort, 1956, p. 198.

[1122] Comfort, 1956, p. 191.

[1123] Alexander Comfort, "Longevity of Man and his Tissues," in *Man and His Future: A CIBA Foundation Volume*, Gordon Wolstenholme (Ed.), Little, Brown and Co., Boston, 1963, p. 218.

[1124] "John Maynard Smith on ageing research," Interview by Robin Holliday, *Biogerontology*, 1(2), 185-189, 2000.

[1125] John Maynard Smith, "Prolongation of the life of *Drosophila subobscura* by a brief exposure of adults to a high temperature," *Nature*, 181, 496-497, 1958; John Maynard Smith, "The effects of temperature and of egg-laying on the longevity of *Drosophila subobscura,*" *The Journal of Experimental Biology*, 35, 832-842, 1958; Jean M. Clarke and John Maynard Smith, "The genetics and cytology of *Drosophila subobscura*. XI. Hybrid vigour and longevity," *Journal of Genetics*, 53, 172-180, 1955; John Maynard Smith, "Review Lectures on Senescence. I. The Causes of Aging," *Proceedings of the Royal Society of London B. Biological Sciences*, 157, 115-127, 1962; "John Maynard Smith on ageing research," Interview by Robin Holliday, *Biogerontology*, 1(2), 185-189, 2000.

[1126] Sona Rosa Burstein, "Gerontology: a Modern Science with a Long History," *Post Graduate Medical Journal* (London), 22, 185-190, 1946; Nathan W. Shock, *The International Association of Gerontology. A Chronicle. 1950 to 1986*, Springer Publishing Company, NY, 1988, pp. 5, 9.

[1127] Institut Istorii Estestvoznania i Techniki imeni S.I. Vavilova Rossiyskoy Academii Nauk (S.I. Vavilov's Institute for the History of Natural Sciences and Technology of the Russian Academy of Sciences), V.P. Borisov, et al., *Rossiyskie uchenie i inzhenery emmigranty: 1920-50-e gody* (Russian emigrant scientists and engineers of the 1920s-1950s), Moscow, 2004, "Korenchevsky, Vladimir Georgievich," http://www.ihst.ru/projects/emigrants/korenchevskii.htm.

[1128] Nathan W. Shock, *The International Association of Gerontology. A Chronicle. 1950 to 1986*, Springer Publishing Company, NY, 1988, pp. 5-8.

Moreover, Edmund Cowdry considered Korenchevsky as a "father of the field" of gerontology (E.V. Cowdry, "V. Korenchevsky, father of gerontology," *Science*, 130, 1391-1392, Nov 20, 1959).

[1129] Sona Rosa Burstein, "Gerontology: a Modern Science with a Long History," *Post Graduate Medical Journal* (London), 22, 185-190, 1946; Nathan W. Shock, *The International Association of Gerontology. A Chronicle. 1950 to 1986*, Springer Publishing Company, NY, 1988, pp. 5-9; Vladimir Korenchevsky, "Gerontology in the United Kingdom," *Journal of Gerontology*, 9, 79-83, 1954.

[1130] Nathan W. Shock, *The International Association of Gerontology. A Chronicle. 1950 to 1986*, Springer Publishing Company, NY, 1988, pp. 9-11.

[1131] See Nathan W. Shock, *The International Association of Gerontology. A Chronicle. 1950 to 1986*, Springer Publishing Company, NY, 1988; W. Andrew Achenbaum, *Crossing Frontiers: Gerontology Emerges as a Science*, Cambridge University Press, Cambridge, 1995.

Since the World Congress of Gerontology in Liège, Belgium has retained its central place in the organization of European gerontology. Thus, in 2011, the European Innovation Partnership on Active and Healthy Ageing was initiated in Brussels. (ec.europa.eu/active-healthy-ageing.)

[1132] Vladimir Korenchevsky, "Conditions desirable for the rapid progress of gerontological research," *British Medical Journal*, 2 (4473, Sept. 28), 468, 1946.

By statistical analysis elsewhere, Korenchevsky arrived at a figure of 113 years as an approximate "natural" human life-span, which is seldom reached (quoted in Mirko D. Grmek, *On Ageing and Old Age, Basic Problems and Historic Aspects of Gerontology and Geriatrics*, Monographiae Biologicae, 5, 2, Den Haag, 1958, pp. 35-36).

[1133] Vladimir Korenchevsky, "Rejuvenative, or Preventive and Eliminative Treatment of Senility," *Geriatrics*, 5(6), 297-302, 1950 (first read at the seventh annual meeting of the American Geriatrics Society, NY, June 1, 1950).

[1134] Vladimir Korenchevsky, "The International Association of Gerontology and Rapid Progress of Gerontology," *British Medical Journal*, 1 (4754, Feb.16), 375-376, 1952 (this paper was first read at the opening session of the Second International Gerontological Congress, held on September 9, 1951, in St. Louis, US), p. 297.

[1135] Vladimir Korenchevsky, "Rejuvenative, or Preventive and Eliminative Treatment of Senility," *Geriatrics*, 5(6), 297-302, 1950.

[1136] Vladimir Korenchevsky, "Rejuvenative, or Preventive and Eliminative Treatment of Senility," *Geriatrics*, 5(6), 297-302, 1950, pp. 299, 300.

[1137] Vladimir Korenchevsky, "Conditions desirable for the rapid progress of gerontological research," *British Medical Journal*, 2 (4473, Sept. 28), 468, 1946.

Essentially the same conditions were proposed in Vladimir Korenchevsky, "The International Association of Gerontology and Rapid Progress of Gerontology," *British Medical Journal*, 1 (4754, Feb.16), 375-376, 1952.

[1138] In 2014, the chair of the British Society for Research on Aging was assumed by Prof. Helen Griffith of Aston University. (http://www.theguardian.com/science/2014/jul/09/ageing-revolution-must-benefit-all.)

[1139] British Council for Ageing, http://www.bcageing.org.uk/; British Society for Research on Ageing, www.bsra.org.uk; British Society of Gerontology, http://www.britishgerontology.org/; British Geriatrics Society, www.bgs.org.uk; International Association of Gerontology – Member Organizations, http://www.iagg.info/organization/council.

[1140] The International Longevity Centre – United Kingdom, with Baroness Sally Greengross as the Chief Executive (as of 2014), works "primarily with central government" and is supported by some 30 foundations and corporations, including the Nuffield foundation, Age Concern and Help the Aged (Age UK), Pfizer, Merck, British Petroleum, and more.

(International Longevity Centre – UK, http://www.ilcuk.org.uk/; The Nuffield Foundation, http://www.nuffieldfoundation.org/; Merck Institute of Ageing and Health, http://www.merck.com/; Age Concern, http://www.ageuk.org.uk/; Help the Aged, www.helptheaged.org/uk.

On July 1, 2011, Age Concern and Help the Aged joined to form Age UK, http://www.ageuk.org.uk/.)

[1141] Speakers at the 2010 Royal Society Conference "The New Science of Ageing" included Professors Andrzej Bartke, Nir Barzilai, Cynthia Kenyon (US), Gillian Bates, Tom Kirkwood, Richard Faragher, Janet Lord (UK), and others.

Yet another conference that may exemplify the increasing scope and prestige of longevity research in the UK was held by the British Society for Research on Ageing on July 15-16, 2010 in Newcastle upon Tyne. The conference was entitled "Systems Biology of Ageing" and the topics included the role of mitochondria and reactive oxygen species in aging, dietary restriction for life-extension, repair of DNA damage, application of stem cells for anti-aging, regulation and recycling in aging – all showing a fairly optimistic and pro-active attitude.

Further conferences related to life-extension, have been promoted by the British Society for Research on Aging: http://www.bsra.org.uk/taxonomy/term/16.

(The Royal Society, May 2010: The New Science of Ageing, http://royalsociety.org/May-2010-The-new-science-of-ageing/; The Oxford Institute for Ageing, http://www.ageing.ox.ac.uk/aggregator/sources/7; British Society for Research on Ageing, 60th Scientific Meeting, *Systems Biology of Ageing*, 15th & 16th July 2010, Newcastle upon Tyne, http://bsra.org.uk/.)

[1142] British Longevity Society, http://www.thebls.org/; Marios Kyriazis, "What is Anti-aging Medicine?" *International Antiaging Magazine*, March 2006, pp. 33-36, http://www.antiaging-magazine.com/articles/article_7_2.html; Marios Kyriazis MD, *Extreme Lifespans though Perpetual-Equalising Interventions (Human Biological Immortality)*, 2010, http://www.elpistheory.info/page13.htm.

[1143] Other central British anti-aging organizations have been the British Society of Anti-Ageing Medicine (BSAAM, http://www.bsaam.com/) and British Anti-Ageing Association (BAAA, http://www.britishantiageing.com/).

[1144] Aubrey de Grey and Michael Rae, *Ending Aging. The Rejuvenation Breakthroughs That Could Reverse Human Aging in Our Lifetime*, St. Martin's Press, NY, 2007; Methuselah Foundation: http://www.mprize.org/; https://www.mfoundation.org/; http://www.methuselahfoundation.org/; SENS Foundation - Strategies for Engineered Negligible Senescence: http://www.sens.org; *Rejuvenation Research*: http://www.liebertonline.com/rej.

[1145] Aubrey de Grey and Michael Rae, *Ending Aging. The Rejuvenation Breakthroughs That Could Reverse Human Aging in Our Lifetime*, St. Martin's Press, NY, 2007, p. 7.

[1146] Personal communication to the author (December 29, 2011). Indeed, currently, a large portion of PubMed search results on "Rejuvenation" would refer to the journal *Rejuvenation Research* edited by de Grey.

[1147] *Man and His Future: A CIBA Foundation Volume*, Gordon Wolstenholme (Ed.), Little, Brown and Co., Boston, 1963 (hereafter referred to as "*Man and His Future*, 1963").

[1148] Some of the credentials of the conference participants included the following.
From the US:
Joshua Lederberg – the Nobel prize winning molecular biologist, the discoverer of genetic recombination of bacteria; Hermann Joseph Muller – Nobel laureate and discoverer of X-ray mutagenesis; Albert Szent-Györgyi – Nobel Laureate and synthesizer of vitamin C; Hilary Koprowski – inventor of the first polio vaccine; Gregory Pincus – inventor of the oral contraceptive pill; Carleton Stevens Coon – president of the American Association of Physical Anthropology; Hudson Hoagland – President of the American Academy of Arts and Sciences.
From the United Kingdom:
Francis Harry Compton Crick – the Nobel laureate and co-discoverer of the DNA double helix structure; Peter Brian Medawar – the Nobel Prize winning immunologist, co-discoverer of acquired immunological tolerance; John Burdon Sanderson Haldane, FRS – one of the founders of population genetics; Julian Sorell Huxley, FRS – a foremost evolutionary biologists; Alan Sterling Parkes – one of the founders of cryobiology; Donald MacCrimmon MacKay – a pioneer of artificial intelligence research; Colin Grant Clark – director of the Agricultural Economics Research Institute, Oxford; Jacob Bronowski – mathematician and biologist, Director-General of the Process Development Department, British National Coal Board; Alex Comfort – one of England's foremost researchers of the biology of senescence, from University College, London, and others.
The few representatives from other countries included:
John Fleming Brock of South Africa – a nutrition scientist, the World Health Organization's consultant on nutrition (born and educated in England, until 1938 he worked as an assistant director of research in medicine at Cambridge University); Artur Glikson of Israel – an architect and ecologist, vice-president of the Landscape Planning Committee of the International Union for Conservation of Nature and Natural Resources; Brock Chisholm of Canada – Director-General of the World Health Organization (1948-1953) and a few others.

[1149] J.B.S. Haldane, self-admittedly, was less concerned about postponing death than about improving living conditions:
"I am more worried by the prison than the sentence."

Yet, he nonetheless believed that "Shaw, in *Back to Methuselah*, was correct as to the social value of longevity."
(J.B.S. Haldane, "Biological possibilities for the human species in the next ten thousand years," *Man and His Future*, 1963, pp. 339, 341, 353, 360.)
Generally, Haldane was quite fond of unconventional truths.
Thus regarding the little appreciated finding that most life tables "consistently overestimate future death rates" (such pessimistic estimates "find favor" with insurance companies mainly dealing with "payment of fixed sums at death, rather than payment of life annuities to the aged"), Haldane stated:
"I suppose the process of acceptance will pass through the usual four stages: (i) this is worthless nonsense; (ii) this is an interesting, but perverse, point of view; (iii) this is true, but quite unimportant; (iv) I always said so."
(J.B.S. Haldane, "The Truth about Death," Review, *Journal of Genetics*, 58, 463-464, 1963.)
Cf. the words of the American psychologist William James (1842-1910):
"I fully expect to see the pragmatist view of truth run through the classic stages of a theory's career. First, you know, a new theory is attacked as absurd; then it is admitted to be true, but obvious and insignificant; finally it is seen to be so important that its adversaries claim that they themselves discovered it."
(William James, *Pragmatism: A New Name for Some Old Ways of Thinking*, 1907, Lecture VI, Pragmatism's Conception of Truth, reprinted in Project Gutenberg, http://www.gutenberg.org/dirs/etext04/prgmt10.txt; also quoted in L.V. Komarov, H. Le Compte, "People can and must live not by decades but by centuries," *Acta Gerontologica et Geriatrica Belgica*, 10(2-3), 87-97, 1972, p. 93.)
The latter saying by William James has often been quoted (misquoted?) as "A new idea is first condemned as ridiculous and then dismissed as trivial, until finally, it becomes what everybody knows" (e.g. *Quotes by James, William, quotationsbook.com*, 2012, p. 8, http://quotationsbook.com/quotes/author/3769/).
[1150] The cited papers from *Man and His Future*, 1963, are:
Julian Huxley, "The future of man – evolutionary aspects," p. 8; Colin Clark, "Agricultural productivity in relation to population," pp. 34-35; John F. Brock, "Sophisticated diets and man's health," pp. 40, 43; Gregory Pincus, "Control of reproduction in mammals," p. 90; Alan S. Parkes, "The sex-ratio in human populations," pp. 96-98; Artur Glikson, "Man's relationship to his environment," pp. 134, 138, 141, 146; Carleton S. Coon, "Growth and development of social groups," pp. 124-126; Donald M. MacKay, "Machines and societies," pp. 155-157, 160, 162; Albert Szent-Györgyi, "The promise of medical science," p. 190; Hilary Koprowski, "Future of infectious and malignant diseases," p. 197; Alex Comfort, "Longevity of man and his tissues," pp. 218, 227; *Discussion* "Health and disease," p. 230 (Medawar), 231 (Pincus), 232 (MacKay), 233 (Haldane), 240 (Szent-Gyorgyi), 241 (Crick); Hermann J. Muller, "Genetic progress by voluntarily conducted germinal choice," pp. 255, 260; Joshua Lederberg, "Biological future of man," pp. 265-269; *Discussion* "Eugenics and genetics," pp. 274-277 (Crick), 285 (Bronowski); Hudson Hoagland, "Potentialities in the control of behavior," p. 314; Brock Chisholm, "Future of the mind," pp. 320-321; J.B.S. Haldane, "Biological possibilities for the human species in the next ten thousand years," pp. 339, 341, 353, 360; *Discussion* "Ethical considerations," p. 362 (Medawar, Lederberg).
[1151] "Man's effective control of his own reproductive capacity," Pincus asserted, "is inescapable."
Parkes went even further in foreseeing the capabilities of reproduction control:
"A sex ratio 1:1, from the point of view of maximum reproductivity, is simply a waste of male biomass. …
Possibly, therefore, in man a ratio of one male to ten females would ensure maximum reproductivity in relation to biomass and a ratio of this kind might therefore expect to have a substantial survival value."
The cryopreservation procedures that Parkes developed for the long term storage and subsequent manipulation of reproductive cells, would, according to him, play a crucial role in controlling reproduction, both with regard to the quantities and the qualities of the offspring.

It may be safe to say that Pincus' and Parkes' vision has now been for the most part realized (except perhaps for radical changes in gender ratios):

In 1963, Pincus reported the test results for his contraceptive preparation "Enovid" in a few hundred volunteers. By 2000, more than 100 million women worldwide were using oral contraceptives.

By 1963, Parkes wrote, "there have been reports in America of one or two children being produced as a result of artificial insemination with preserved spermatozoa." By 2006, more than 3 million babies were born using in vitro fertilization worldwide.

Cold storage, according to Parkes, would also be useful for "long-term preservation of mammalian cells" generally, as needed for transplantation. This anticipation also came true.

(Gregory Pincus, "Control of reproduction in mammals," p. 90; Alan S. Parkes, "The sex-ratio in human populations," pp. 96-98, in *Man and His Future*, 1963; Blackburn RD, Cunkelman A, Zlidar VM, et al., "Oral contraceptives – an update," *Population Reports A*, 9, 1-39, 2000, http://info.k4health.org/pr/a9/a9print.shtml; Caroline Ryan, "More than 3m babies born from IVF," *BBC News*, June 21, 2006, http://news.bbc.co.uk/2/hi/health/5101684.stm; Arne Sunde, "Europe's declining population and the contribution of ART" [assisted reproduction technology], *Pharmaceuticals Policy and Law*, 9, 79-89, 2007.)

[1152] In 1961, about the time *Man and His Future* was published, the yield of wheat in the UK was ~3.5 ton per hectare (3,500 kilograms dry weight grain per 10,000 square meters).

Allowing for the 500 kg of food per person per year (1.370 kg per person per day) to come exclusively from nutritious crops, that yield would very roughly suffice for 7 people per hectare to be fed from a single harvest, equivalent to 1400 square meters per person.

(Notably, the yield of wheat in the UK increased to 8000 kg per hectare in 2000. In 2011, the world's greatest yield of cereal grains generally was almost 19,000 kg per hectare, and was achieved in Oman. The data are from Food and Agriculture Organization Statistics, http://faostat.fao.org/site/567/DesktopDefault.aspx#ancor; and World Bank, http://data.worldbank.org/indicator/AG.YLD.CREL.KG?order=wbapi_data_value_2010+wbapi_data_value+wbapi_data_value-last&sort=desc.)

Yet, Clark also made some allowances for meat and milk consumption, on the contemporary North American scale, as follows:

Cereals, sugar, etc. – 500 square meters per person (smpp); 45 kg. pig and poultry meat – 500 smpp; 45 kg. beef and mutton – 400 smpp; 250 kg. milk – 400 smpp; altogether – 1800 square meters per person, or 5.5 persons per hectare.

Furthermore, Clark admitted that with optimal rates of photosynthesis (using specially grown algae or crops) and utilizing the entire crop biomass, the land requirements can be diminished much further:

"algae ..... rarely have a yield of 20 grams dry matter per square metre per day. ... the familiar radish and broccoli can grow at 40 to 45 grams of dry matter per square metre per day. ... the corn cockle has been found to grow at 57. ...

If we settle for 50 grams per square metre per day [182,000 kg per hectare per year], then the growth of each person's requirements of food and fibre requires only 27 square meters" [370 people per hectare].

Also "There have been cases of fertilized tropical grasslands producing at the rate of 80 tons [80,000 kg] dry weight per hectare per year."

In Clark's estimate, the area of usable agricultural land in the world is "8,200 million hectares" (82 million square kilometers) or approximately half of the Earth's dry land area (~148.94 million square kilometers), out of ~510.072 million square kilometers of the entire Earth surface area, including the water surface.

(Colin Clark, "Agricultural productivity in relation to population," in *Man and His Future*, 1963, pp. 30-35.)

[1153] Colin Clark, "Agricultural productivity in relation to population," pp. 34-35, in *Man and His Future*, 1963.

[1154] Artur Glikson, "Man's relationship to his environment," pp. 134, 138, 141, 146; Donald M. MacKay, "Machines and societies," pp. 155-157, 160, 162, in *Man and His Future*, 1963.

[1155] Carleton S. Coon, "Growth and development of social groups," pp. 124-126, in *Man and His Future*, 1963.

[1156] Hudson Hoagland, "Potentialities in the control of behavior," p. 314; Brock Chisholm, "Future of the mind," pp. 320-321, in *Man and His Future*, 1963.

[1157] *Man and His Future* was evidently not an isolated, uniquely emerging work, but a link in the chain of futuristic literary tradition.

Thus, Julian Huxley's and J.B.S. Haldane's papers in *Man and His Future* clearly continued the authors' earlier interest in life enhancement and life extension.

In 1957, Julian Huxley introduced the concept of "transhumanism" – the quest to surmount the "present limitations" of our "nasty, brutish and short" existence.

See Julian Huxley, "Transhumanism," in *New Bottles for New Wine*, Chatto & Windus, London, 1957, pp. 13-17, reprinted by the World Transhumanist Association: http://www.transhumanism.org/index.php/WTA/more/huxley/.

(The notion that life in the state of nature is "solitary, poor, nasty, brutish and short" was originated by Thomas Hobbes (1588-1679) in *Leviathan*, 1651, Ch. 13, "On the Natural Condition of Mankind," http://www.gutenberg.org/files/3207/3207-h/3207-h.htm#2H_4_0115.)

Haldane too was a long-time believer in the perfectibility of the human species, as for example expressed in his *Daedalus* (1923), where he spoke about prolonging muscular and mental work, and extending "a woman's youth."

(J.B.S. Haldane, "Daedalus or Science and the Future, A Paper read to the Heretics, Cambridge, on February 4th, 1923," http://cscs.umich.edu/~crshalizi/Daedalus.html.)

Another critical figure in the tradition of British scientific futurism was John Desmond Bernal, the pioneer of X-ray crystallography and Fellow of the Royal Society.

In *The World, the Flesh & the Devil* (1929), Bernal spoke about the "capacity for indefinite extension" of human capabilities:

"there would be the sending parts of the television apparatus, tele-acoustic and tele-chemical organs, and tele-sensory organs of the nature of touch for determining all forms of textures. Besides these there would be various tele-motor organs for manipulating materials at great distances from the controlling mind."

These technological extensions will lead to a radical extension of life, and ultimately to immortality:

"Death would still exist for the mentally-directed mechanism we have just described; it would merely be postponed for three hundred or perhaps a thousand years, as long as the brain cells could be persuaded to live in the most favorable environment, but not forever.

But the multiple individual would be, barring cataclysmic accidents, immortal, the older component [sic] as they died being replaced by newer ones without losing the continuity of the self, the memories and feelings of the older member transferring themselves almost completely to the common stock before its death."

(John Desmond Bernal, *The World, the Flesh & the Devil. An Enquiry into the Future of the Three Enemies of the Rational Soul*, 1929, III. "The Flesh," Reprinted at the Marxists Internet Archive Library, UK, http://www.marxists.org/archive/bernal/works/1920s/soul/.)

On the American side, among others, Hermann Joseph Muller was a long-time, well-known proponent of eugenic human betterment, famous for his attempt to advocate for eugenic selection in the Soviet Union.

(See John Glad, "Hermann J. Muller's 1936 Letter to Stalin," *The Mankind Quarterly* 43 (3), Spring 2003, pp. 305-319.)

[1158] Examples of recent Transhumanist and Life-extensionist literature include:

Ronald Bailey "Forever Young. The New Scientific Search for Immortality," *Reason*, August 2002; Nick Bostrom, "The Fable of the Dragon Tyrant," *Journal of Medical Ethics*, 31(5), 273-277, 2005; John Harris,

Clones, Genes and Immortality. Ethics and the Genetic Revolution, Oxford University Press, Oxford, 1998; Michael D. West, *The Immortal Cell, One Scientist's Quest to solve the Mystery of Human Aging*, Doubleday, NY, 2003; Aubrey de Grey and Michael Rae, *Ending Aging. The Rejuvenation Breakthroughs That Could Reverse Human Aging in Our Lifetime*, St. Martin's Press, NY, 2007; Stanley Shostak, *Becoming Immortal, Combining Cloning and Stem-Cell therapy*, State University of New York Press, NY, 2002; Ben Bova, *Immortality: How Science is Extending Your Life Span and Changing the World*, Harper Collins, NY, 1998; Bruce Klein (Ed.), *The Scientific Conquest of Death*, LibrosEnRed, Buenos Aires, 2004; Ramez Naam, *More Than Human: Embracing the Promise of Biological Enhancement*, Broadway Press, NY, 2005; James Hughes, *Citizen Cyborg: Why Democratic Societies Must Respond to the Redesigned Human of the Future*, Westview Press, Cambridge MA, 2004; Kevin Warwick, *Cybernetic organisms: our future?*" Proceedings IEEE, 87(2), 387-389, 1999; Eliezer Yudkowsky, *Creating Friendly AI*, Singularity Institute, 2001; Ray Kurzweil, *The Age of Spiritual Machines*, Viking, NY, 1999; Ray Kurzweil and Terry Grossman, *Fantastic Voyage. Live Long Enough to Live Forever*, Plume, NY, 2005; Ray Kurzweil, *The Singularity Is Near: When Humans Transcend Biology*, Penguin Books, New York, 2005.

[1159] Until his death, Dr. Robert Butler remained a leader of aging and longevity research and public advocacy, fighting against what he termed "ageism" – the discrimination based on age and the neglect of the health and well being of the aged considered by some to be unworthy of scientific and societal efforts.

In addition to the founding of the International Longevity Center in 1990, among the existing organizations that Butler founded or helped to found, there are:

The Alzheimer's Association (initially named Alzheimer's Disease and Related Disorders Association, 1980), which introduced the problem of Alzheimer's disease to the wide public, http://www.alz.org/; the American Association for Geriatric Psychiatry (1978) http://www.aagpgpa.org/; The American Federation for Aging Research (1981) http://www.afar.org/; and The Alliance for Aging Research (1986) http://www.agingresearch.org/.

Presently The Global Alliance of the International Longevity Centers, initiated by Butler, includes organizations in 12 countries: USA, Japan, France, UK, Dominican Republic, India, South Africa, Argentina, the Netherlands, Israel, Singapore and Czechoslovakia. (http://www.ilcusa.org/pages/about-us/global-partners.php; http://ilcindia.org/global_alliance%20.html.)

Butler was also a director of Biotime Inc., a biotech company (with Dr. Michael David West as the CEO) engaged in developing blood plasma volume expanders, and regenerative medicine using cultured human embryonic stem cell therapy "to rebuild cell and tissue function lost to degenerative disease or injury" (http://www.biotimeinc.com/AboutBT.htm).

Following the publication of Butler's Pulitzer Prize-winning book *Why Survive? Being Old in America* (1975), he continued to popularize longevity research and care in his book *The Longevity Revolution: The Benefits and Challenges of Living a Long Life*, Perseus Books, NY, 2008.

[1160] International Longevity Center – USA, http://www.ilcusa.org/pages/about-us.php.

[1161] The American Aging Association, http://www.americanaging.org/history.html; http://www.articledoctor.com/anti-aging/american-anti-aging-association---the-american-anti-aging-association-2157.

[1162] Gerontology Research Group, http://www.grg.org/.

[1163] According to the Geroscience Interest Group mission statement:

"Basic research in animal models has demonstrated the plasticity of lifespan. Most importantly, it has shown that often, extension of lifespan is accompanied by a delay in the appearance and progression of morbidity, as well as a slowing in age-related functional decline.

That is, slowing the aging processes leads to an increase in healthspan, the portion of life spent in good health. Yet many fundamental issues remain that need to be addressed and understood, not the least of which is translation research and application of these findings to the human population."

(Geroscience Interest Group, http://sigs.nih.gov/geroscience/Pages/default.aspx.)

In 2012, The Ellison Medical Foundation, the Gerontological Society of America, the American Aging Association and the Alliance for Aging Research commissioned a "white paper" from distinguished American gerontologists, health economists and demographers (Dana Goldman, David Cutler, Jay Olshansky, Eileen Crimmins, George Martin, and others), further encouraging research into aging as the main cause of chronic diseases, disability and mortality. The paper was published in October 2013.
(Goldman DP, Cutler D, Rowe JW, Michaud PC, Sullivan J, Peneva D, Olshansky SJ, "Substantial health and economic returns from delayed aging may warrant a new focus for medical research," *Health Affairs*, 10, 1698-1705, 2013; http://www.healthspancampaign.org/about-us/partners/.)
Generally, in the US, the assertion that degenerative aging is the main root cause and risk factor of age-related diseases (such as cancer, heart disease, type 2 diabetes and neurodegenerative diseases) has represented the main argumentative strategy to advocate for increased emphasis and investment into biological research of aging for the purpose of healthy life extension.
(Rae MJ, Butler RN, Campisi J, de Grey ADNJ, Finch CE, Gough M, Martin GM, Vijg J, Perrott KM, Logan BJ, "The demographic and biomedical case for late-life interventions in aging," *Science Translational Medicine, 2 (40)*, 40cm21, 2010, http://stm.sciencemag.org/content/2/40/40cm21.full.)

[1164] Since about 2001 (possibly related to the September 11, 2001 terrorist attack on the US), under the direction of Anthony J. Tether (2001-2009), Regina E. Dugan (2009-2012) and Kaigham J. Gabriel (2012), DARPA has been increasingly interested in biological and medical research. By 2012, DARPA's research projects related to life-extension, included the following:

The "Biochronicity" project has aimed to understand and manipulate biological time, including the manipulation of the life-span, "in pursuit of breakthroughs in managing the effects of time on human physiology." Potential applications might include the understanding and "managing" of such processes as "cell-cycle progression, growth, metabolism, aging and cell death," rejuvenation and suspended animation.
(http://www.darpa.mil/Our_Work/DSO/Programs/Biochronicity.aspx;
http://nextbigfuture.com/2011/07/darpa-project-seeks-immortality.html;
http://www.wired.com/dangerroom/2011/07/darpa-life-master-clock/.)

The "BioDesign" project has sought to create synthetic organisms with a controllable life-span, implying that such organisms could be made potentially immortal, but could be killed when necessary with a built-in "self-destruct option."
(http://www.wired.com/dangerroom/2010/02/pentagon-looks-to-breed-immortal-synthetic-organisms-molecular-kill-switch-included/.)

The Blood Pharming program has intended to manufacture blood for transfusion. The program's stated objective has been "to develop an automated culture and packaging system that yields transfusable levels of universal donor red blood cells (RBCs) from progenitor cell sources."
(http://www.darpa.mil/Our_Work/DSO/Programs/Blood_Pharming.aspx;
http://www.popsci.com/technology/article/2010-07/darpas-synthetic-blood-flows-lab-fda-could-be-battlefields-soon.)

Living Foundries has been a Synthetic Biology program aimed to manufacture tailor-made biological materials with "novel capabilities," to specifically design "fuels and medicines," to "program and engineer biology."
(http://www.darpa.mil/Our_Work/MTO/Programs/Living_Foundries.aspx;
http://www.wired.com/dangerroom/2012/05/living-foundries/.)

Dialysis-Like Therapeutics has been intended to filter out all toxins and pathogens from blood using a portable device.
(http://www.darpa.mil/Our_Work/MTO/Programs/Dialysis-Like_Therapeutics_(DLT).aspx;
http://mariakonovalenko.wordpress.com/2011/03/21/darpa-wants-machine-to-suck-all-your-blood-out-other-fun-stuff/.)

Additional developments have included a body-cooling glove to reverse fatigue, regeneration induced by electromagnetic fields and "bone morphogenetic proteins," improved enzymatic digestion, fortifying hormone therapies, trans-cranial magnetic and electric stimulation to reduce sleep requirements and improve cognitive performance, additional physiotherapeutic ways to induce the restorative and potentially life-extending "slow-wave" deep sleep, and more. All these could serve the aged.
(http://www.wired.com/wired/archive/15.03/bemore.html;
http://www.scribd.com/doc/124630444/New-Scientist-Sleep-and-Dreaming-Slumber-at-the-Flick-of-a-Switch.)

Programs such as Unconventional Therapeutics, Pathogen Defeat, Autonomous Diagnostics to Enable Prevention and Therapeutics (ADEPT), Predicting Health and Disease (PHD), Maintaining Combat Performance, Preventing Violent Explosive Neurologic Trauma (PREVENT), Tactical Biomedical Technologies, and others – have been intended to produce breakthroughs in diagnosis and therapy, all of which could in principle be used in the fight against aging (in case they are feasible and safe).

Furthermore, human replacement parts have been developed vigorously by the department.

Thus, the Reliable Neural-Interface Technology program has been dedicated to wire the brain for a better control of prosthetic limbs. Revolutionizing Prosthetics has intended to develop "fully integrated and functional limb replacements." There was also a program entitled "On Demand Custom Body Implant and Prosthesis." And the "Avatar Program" has developed "telepresence and remote operation of ground systems" in the "synergistic partnership between machine and operator."
(Unclassified. Department of Defense Fiscal Year (FY) 2013 President's Budget Submission, February 2012, Defense Advanced Research Projects Agency, Justification Book Volume 1, Research, Development, Test & Evaluation, Defense-Wide, available at http://www.darpa.mil/.)

Perhaps the most explicitly anti-aging program of the US Department of Defense has been the "Optimized Human Performance: Mitochondrial Energetics."

Its purpose has been to "develop metabolic supplements to optimize adenosine triphosphate production in eukaryotes" (such as quercetin, green tee and B vitamins).

The following justifications have been provided for the program:

"The modern Army is constrained by biology. Highly qualified and very experienced soldiers routinely leave the Army because they are old; their physical and/or cognitive performance capabilities are significantly less than that of a 20 year old. The biological basis of this reduction in performance capability may be an injury, but in most cases is simply due to the reduced efficiency of old mitochondria, resulting in reduced levels of energy (adenosine triphosphate) provided to the body to power cognitive and physical tasks. The ability to stimulate mitochondrial energy production would extend the time that soldiers remain fit for duty, boost soldier physical and performance capabilities, and expand the age range of suitable recruits. It would also eliminate the current dichotomy of the ideal soldier being optimized both for youth (high performance capabilities) and experience. …

The world contains approximately 4.2 billion people over the age of twenty. Even a small enhancement of cognitive capacity in these individuals would probably have an impact on the world economy rivaling that of the internet. The commercial market for a compound that could reverse the effects of aging on human energetics would be more than significant. The cost of Social Security in the U.S. is expected to approach 7% of the gross domestic product (GDP); reducing this cost by any significant degree would also have substantial impact on federal obligations and expenditures."
(http://www.dodtechmatch.com/DOD/Opportunities/SBIRView.aspx?id=A08-T006.)

For 2013, the entire requested budget of the US Department of Defense for DARPA's Research, Development, Test and Evaluation was ~$2.817 Billion (compared to ~$2.816 in 2012). As of 2012, the budget included ~$38M for Basic Operational Medical Research Science, ~$95M for Biomedical Technology, ~$30M for Biological Warfare Defense, and ~$220M for Materials and Biological Technology.

[1165] The National Institute on Aging of the US National Institutes of Health, at Bethesda, Maryland, is perhaps the world's largest institutional body supporting research on aging and longevity.
It includes the Biology of Aging program which studies the fundamental mechanisms of aging, alongside other programs, such as Behavioral and Social Research; Neuroscience and Neuropsychology of Aging; and Geriatrics and Clinical Gerontology Program, http://www.nia.nih.gov/research/.
The National Institute on Aging has its own public advocacy group: Friends of the National Institute on Aging, http://friendsofnia.ning.com/.
The largest public support group is the American Association of Retired Persons (AARP), with about 37 million members as of April 2014, http://www.aarp.org/health/.
Additional governmental organizations are:
The US Administration on Aging (AoA) http://www.aoa.gov/; United States Department of Health and Human Services, Disability, Aging and Long-term Care, http://aspe.hhs.gov/_/office_specific/daltcp.cfm; US Senate Special Committee on Aging http://aging.senate.gov/; US Environmental Protection Agency, Aging Initiative http://www.epa.gov/aging/.
Some international programs include: World Health Organization Aging and Life Course Projects, http://www.who.int/ageing/projects/en/; United Nations Action Plan for a World Aging Rapidly, www.un.org/ageing/.
Lists of dozens of additional American gerontological societies can be found at: http://www.agingresearch.org/section/resources/; http://www.all.org/charities/; http://www.apa.org/pi/aging/resources/organizations/index.aspx; http://sociologyindex.com/aging_and_gerontology.htm.

[1166] The Life Extension Foundation (LEF) was established in 1980 by Saul Kent and William Faloon, the latter has been the LEF current president (as of 2014). The Life Extension Foundation has been mainly involved in research and sales of anti-aging supplements. As of 2005, it comprised over 100,000 members, and its *Life Extension Magazine* was circulated in over 300,000 copies. Thus it has been the largest life-extensionist organization in the world (www.lef.org).
The LEF appears to have been waging an on-going legal and public relations war with the US Food and Drug Administration, which the LEF presents as a major bureaucratic obstacle on the path of developing and applying life-saving therapies.
(Saul Kent, "The Life Extension Foundation's Victory over the FDA," *Life Extension Magazine*, February 1996, http://www.lef.org/fda/victory.htm; http://www.lef.org/research.)

[1167] The American Academy of Anti-Aging Medicine (A4M), founded in 1992 by Ronald Klatz and Robert Goldman, as of 2012, included more than 22,000 members from 105 countries, among which 85% were physicians (MD, DO, MBBS), 12% scientists, researchers and health practitioners, and 3% government officials, members of the press and general public.
The A4M has been holding many world congresses on anti-aging, preventive, integrative and regenerative medicine, which are perceived to be closely related. (http://www.worldhealth.net/about-a4m/; http://www.worldhealth.net/pages/events/.)

[1168] The list of organizations promoting anti-aging medicine is very extensive. It includes:
the American Board of Longevity Medicine and Sciences; Life Extension Core of Information; American Academy of Rejuvenative Medicine; American Academy of Regenerative Medicine; American Academy of Longevity Medicine, http://www.americanacademyoflongevitymedicine.com/; International Academy of Longevity, Medicine; International Academy of Anti-Aging Medicine, http://www.antiagingforme.com/; The World Anti-Aging Academy of Medicine, http://www.waaam.org/ and many more.

[1169] As of 2008, the global anti-aging market was estimated at $162.2 billion, with the "disease segment" generating $66 billion, and the "appearance segment" of $64.4 billion. The global market was projected to grow to $274.5 billion in 2013, and $291.9 Billion by 2015.
According to a Global Industry Analysis:
"Key players dominating the global anti-aging products market include Allergan Inc, Alberto Culver Company, Avon Products Inc, Beiersdorf, Bio Pharma US Corp, Bayer Schering Pharma AG, Chanel SA, Christian Dior, Clarins, Elizabeth Arden Inc, Ella Bache, Estee Lauder Inc, F. Hoffmann-La Roche Ltd, GlaxoSmithKline Plc, General Nutrition Centers Inc, Henkel KgaA, Jan Marini Skin Research Inc, Johnson & Johnson, Janssen Pharmaceutica Products LP, Neutrogena Corporation, L'Oréal SA, Merck & Company Incorporated, NeoStrata Company Inc, Novartis International AG, Orlane SA, Procter & Gamble, Pfizer Incorporated, Revlon Inc, Robanda International, Shiseido Co. Ltd, SkinMedica Inc, Unilever PLC, Valeant Pharmaceuticals International, Woodridge Labs Inc, Wyeth and Zosano Pharma Inc., among others."
(*Global Anti-Aging Products Market to Reach $291.9 Billion by 2015, According to New Report by Global Industry Analysts*, WorldHealth.net, February 19, 2009, http://www.worldhealth.net/news/global_anti-aging_products_market_to_rea/;
BBC Research. *Anti-aging Products and Services: The Global Market*, December 24, 2009, http://www.bccresearch.com/report/HLC060A.html.)
In a more recent report, of August 2013, it was estimated that in 2012, the total market for antiaging products and services was $249.3 billion in 2012, and was expected to increase to nearly $261.9 billion by the end of 2013 and $345.8 billion in 2018, http://www.bccresearch.com/market-research/healthcare/antiaging-products-services-hlc060b.html.
Some estimates are more modest, yet still range in hundreds of billions of dollars. According to an estimate made in May 2014, the anti-aging market was valued at $122.3 billion in 2013, and was expected to rise to $191.7 billion by 2019, https://www.linkedin.com/today/post/article/20140508134947-339157087-global-anti-aging-market-to-be-worth-usd-191-7-billion-by-2019.
Scores of American suppliers of dietary supplements, relating to attempts at life-extension, are listed at the National Sanitation Foundation (NSF) International, http://www.nsf.org/business/search_listings/; Longecity "Supplier Listlinks" http://www.imminst.org/forum/index.php?showtopic=6791; Inforret, http://www.inforret.com/Health_Care/Nutritional_Supplements_Vitamins.html; Icis, http://www.icis.com/Search/ProductNumber/63482/WorldWide/Nutritional+Food+Supplements.htm and many other sites.
Some resources and listings on supplements can be found, among others, at the US National Library of Medicine. Dietary Supplements Labels Database, http://dietarysupplements.nlm.nih.gov/dietary/; International Bibliographic Information on Dietary Supplements of the National Institutes of Health http://ods.od.nih.gov/Health_Information/IBIDS.aspx; Food and Drug Administration http://www.fda.gov/Food/DietarySupplements/default.htm; Snake Oil, a database rating evidence-based efficacy of supplements or most often lack thereof, http://www.informationisbeautiful.net/play/snake-oil-supplements/, and many other sources.
Life extension has been a big business in America and around the world. According to a survey of the US National Wellness Association (2001) "Exploring Consumer Attitudes About Dietary Supplement Barometer Survey" – "Today, six in ten Americans (59 percent) report taking dietary supplements on a regular basis."
Among them, 50% said that their reason for taking the supplements is "to live longer" (http://nationalwellnessassociation.com/News/Survey_Results.htm).
According to a Nielsen Global Online Survey (October, 2008), 56% of Americans used vitamins and dietary supplements, exceeded only by the Philippines and Thailand (66%). 44% of Americans claimed to use dietary supplements daily, compared to the lowest daily use in Spain and Italy (8%). Globally, ~40% of the

people were found to use vitamins and dietary supplements. The average use in North America was the highest (54%), followed by Asia Pacific (43%), The Middle East (33%), Europe (30%), and Latin America (28%). The proportions slightly varied in 2009. (http://th.nielsen.com/news/20090317.shtml; http://en-us.nielsen.com/content/nielsen/en_us/insights/consumer_insight/issue_16/half_of_americans.html.)

[1170] The coining of the term "regenerative medicine" has been commonly attributed to Dr. Leland R. Kaiser's paper of 1992, "The Future of Multihospital Systems," *Top Healthcare Finance*, 18, 32-45, 1992.

[1171] Michael David West (b. 1953) has played a leading part in developing the longevity industry. He was the founder and director of Geron Corporation (1990-1998), working on therapeutic cloning (nuclear transfer) aimed to create the types of cells needed for transplantation. The company was a pioneer of telomerase diagnostics. Insofar as the shortening of DNA end-segments, the telomeres, leads to a termination of cell divisions, and the enzyme telomerase restores the telomeres' length – the level of telomerase would indicate cells' replicative potential. Other lines of work included telomerase inhibition as a potential anti-tumor therapy; and telomerase stimulation as a potential way to immortalize human cells and postpone aging.

In 1999, Geron Corporation acquired Roslin Biomed, the Scottish company established by Ian Wilmut, the creator of the first cloned mammal, Dolly the sheep (1996). According to Wilmut, the chief element in cloning is to "reset" the cell's biological clock to a youthful stage.

(Stephen Hall, *Merchants of Immortality, Chasing the Dream of Human Life Extension*, Houghton Mifflin, Boston, 2003, p. 167; Rick Weiss, "Scientists Achieve Cloning Success," *Washington Post*, A01, February 24, 1997.)

In 1998, West became President and Chief Scientific Officer at Advanced Cell Technology (1998-2007) specializing in developing human stem cell technology for regenerative and anti-aging medicine, including the development of methods for stem-cell production that would not destroy the human embryo.

Since 2007, West has been the CEO of BioTime, a company engaged in developing blood plasma volume expanders, and regenerative medicine, developing therapies based on human embryonic stem cell technology "to rebuild cell and tissue function" lost to degenerative disease, aging or injury.

(The companies' websites: http://www.geron.com/; http://www.advancedcell.com/; http://biotimeinc.com/; http://www.michaelwest.org/about.htm; Michael D. West, *The Immortal Cell. One Scientist's Quest to Solve the Mystery of Human Aging*, Doubleday, NY, 2003.)

[1172] Among the foremost companies dedicated to slowing down aging, there has been *Sirtris Pharmaceuticals*, founded by Leonard (Lenny) P. Guarente (b. 1952) in 2004, and, as of 2012, co-chaired by Guarente, Guarente's pupil David A. Sinclair, and Crhistoph Westphal.

Sirtris, a GlaxoSmithKline (GSK) company, "is developing small molecule drugs that target the sirtuins, a family of seven enzymes associated with diseases of aging. Modulation of these enzymes offers the promise of drug discovery in multiple therapeutic areas."

Guarente was the first to show that the expression of the Sirtuin genes is correlated with high longevity in animal models. The Sirtuin expression has been suggested by Guarente to be a cell defense mechanism, "a universal regulator of survival," stabilizing DNA structure. Hence, the activation of the sirtuins in humans (SIRT1 specifically) by pharmacological means (such as Resveratrol and its analogs, including SRT-1720, SRT-2104, MR-3, R-4 and other drugs) appeared promising to Guarente and his school.

"A family feud" was reported in 2004 between Guarente and Sinclair. As *Science Magazine* reported, "Lenny Guarente and his former postdoc David Sinclair can dramatically extend the life span of yeast. They're battling over how this works, and competing head-to-head to grant extra years to humans." Now they join forces in the Sirtris' Scientific Advisory Board.

On February 22, 2012, one of David Sinclair's former collaborators at Harvard Medical School, Dr. Haim Y. Cohen of Bar-Ilan University, Israel, reported an about 16% increase in the life-span of male mice due to inducing over-expression of the Sirtuin 6 gene. As a *Nature* commentator noted, "At last, a member of the celebrated sirtuin family of proteins has been shown to extend lifespan in mammals — although it's not the one that has received the most attention and financial investment."

(Lenny Guarente, *Ageless Quest, One Scientist's Search for Genes That Prolong Youth*, Cold Spring Harbor Laboratory Press, Cold Spring Harbor, New York, 2003, pp. 59, 115; Jennifer Couzin, "Aging Research's Family Feud," *Science*, 303, 1276-1278, February 27, 2004; Sirtris Pharmaceuticals, http://www.sirtrispharma.com/;

Yariv Kanfi, ..., Haim Y. Cohen, "The sirtuin SIRT6 regulates lifespan in male mice," *Nature*, February 22, 483(7388), 218-221, 2012; Heidi Ledford, "Sirtuin protein linked to longevity in mammals," *Nature News*, February 22, 2012, http://www.nature.com/news/sirtuin-protein-linked-to-longevity-in-mammals-1.10074.)

[1173] Another leader in life-extension research industry has been Elixir Pharmaceuticals, founded in 1999, by Lenny Guarente and Cynthia Jane Kenyon (b. 1955). (As of 2010, Guarente and Kenyon were not listed in the company board of directors.)

According to the company history, "The Company's scientific founders identified that interactions between specific genes and enzymes can slow the aging process, and we are developing compounds that stimulate these interactions and will be used to treat a range of diseases of aging, including metabolic disease."

The products of Elixir Pharmaceuticals include:

agonists of Ghrelin (a stimulant of hunger) to treat cachexia; and antagonists of Ghrelin (e.g. Leptin) to treat obesity and type 2 diabetes; Glinides that increase the fusion of insulin granules with the cell membrane to treat type 2 diabetes; inhibitors of the Sirtuin 1 (Sirt1) gene to treat the neurodegenerative Huntington's disease and cancer; and activators of the Sirtuin 1 gene (such as Resveratrol and its successors) to treat obesity, type 2 diabetes, and generally to postpone aging.

In her experiments on nematode worms *Caenorhabditis Elegans*, suppressing the expression of the DAF2 gene (Decay Accelerating Factor 2, which normally codes for an insulin-like receptor), Cynthia Kenyon was able to increase the life span of the animals 2 to 5 times, while preserving their youthful activity. The hope has been to reproduce this success in humans.

(Cynthia Kenyon Lab, University of California, San Francisco, http://kenyonlab.ucsf.edu/; Elixir Pharmaceuticals, http://www.crunchbase.com/company/elixir-pharmaceuticals, http://web.archive.org/web/20090426070528/http://www.elixirpharm.com/.)

In 2013, Cynthia Kenyon joined Calico (California Life Company) – a subsidiary of Google Inc. dedicated to research and development for human life extension (established in September 2013) – as a part of its leading team, alongside Arthur Levinson, Robert Cohen, Hal Barrron, and David Botstein (http://nextbigfuture.com/2013/11/google-anti-aging-company-calico.html).

[1174] A leader in developing regenerative medicine has been Anthony Atala (b. 1958) of the Wake Forest Institute for Regenerative Medicine at Winston-Salem, North Carolina. His research has focused on "tissue engineering" to grow organs for transplantation. Atala developed apparently the first such laboratory-grown organ (a bladder) that was transplanted into a human in 1999.

Some of the methods investigated at the Institute for Regenerative Medicine include growing tissues on degradable "scaffolds," "printable organs" and the use of stem cells harvested from the amniotic fluid of pregnant women and then directed to differentiate into the organs of choice. The explicit purpose of this research, according to Atala, is human life extension.

Atala has also been the Scientific Founder and Chairman of Research & Development at Tengion Inc., a company specializing in regenerative medicine.

(Anthony Atala, "Life extension by tissue and organ replacement," in *The Future of Aging: Pathways to Human Life Extension* (Gregory M. Fahy, Michael D. West, L. Stephen Coles and Steven B. Harris, Eds.), Springer, NY, 2010, Ch. 17, pp. 543-572; The Wake Forest Institute for Regenerative Medicine, http://www.wakehealth.edu/wfirm/; Tengion Inc, http://www.tengion.com/.)

Generally, the field of "growing organs/tissues" for transplantation has been rapidly developing. Some of the "grown" or "tissue-engineered" organs and tissues have included: muscles, nerves, artificial liver and pancreas, bladder, cartilage (ears, noses), airways (trachea), blood vessels, skin, bone marrow, artificial bone, penile tissue, mucosa, and more. (http://en.wikipedia.org/wiki/Tissue_engineering.)

[1175] John Craig Venter (b.1946) has been a pioneering figure of genomics and synthetic biology. Venter's company *Celera* was responsible for the first complete human genome sequencing in 2000. On May 20, 2010, J. Craig Venter Institute announced the creation of the first self-replicating synthetic bacterial cell, with synthetic DNA. Back in 2003, Craig Venter and his team from the Institute of Biological Energy Alternatives in Rockville, Maryland, also produced one of the first synthetic viruses, the Phi-X174 bacteriophage, with a genome consisting of about 5,000 base pairs.

Eventually, in March 2014, Craig Venter founded "Human Longevity Inc." whose "goal is to extend and enhance the healthy, high-performance lifespan and change the face of aging" by harnessing "for the first time, the power of human genomics, informatics, next generation DNA sequencing technologies, and stem cell advances".

(http://www.jcvi.org/cms/press/press-releases/full-text/article/first-self-replicating-synthetic-bacterial-cell-constructed-by-j-craig-venter-institute-researcher/; http://news.bbc.co.uk/2/hi/health/3268259.stm; http://news.bbc.co.uk/2/hi/2122619.stm;
https://www.celera.com/celera/pr_1056578007;
http://www.humanlongevity.com/; http://www.humanlongevity.com/human-longevity-inc-hli-launched-to-promote-healthy-aging-using-advances-in-genomics-and-stem-cell-therapies/)

[1176] Prof. George Church, of the Department of Genetics of Harvard Medical School, a pioneer of genomic sequencing and synthetic biology, has co-founded several companies related to health and longevity, including Warp Drive Bio, Alacris, Gen9 Bio and others.

Church's life-extensionist views have been explicit:

Ed Regis, George Church, "The Recipe For Immortality. An expert in synthetic biology explains how people could soon live for centuries," *Discover Magazine*, October 17, 2012 http://discovermagazine.com/2012/oct/20-the-recipe-for-immortality; George Church, "Brain healthspan extension", March 4, 2013 https://www.youtube.com/watch?v=--TfH82bTi8.

[1177] Beside "regenerative medicine," another common key term in the pursuit of life-extension has been "slowing down/reversing the aging process." Some companies working in the first decade of the 21st century to find the cure for aging included:

Alteon, focusing on breaking down crosslinked advanced glycation end-products (AGE), with Kenneth Moch as the CEO, http://www.alteon.com/; BioMarker Pharmaceuticals, founded by Saul Kent and Xi Zhao-Wilson, with Dr. Andrzej Bartke as the scientific advisor, focusing on substances that may mimic caloric restriction and developing DNA micro-arrays to test for molecular markers of aging, http://www.biomarkerinc.com/; GeriGene Medical Corporation, working on anti-aging drugs and tissue engineering, with Don Kleinsek as the president, http://www.gerigene.com/; Legendary Pharmaceuticals, attempting to repair and reverse progressive damage to mitochondria and lysosomes, with John Furber as the president, http://www.legendarypharma.com/; LifeGen Technologies, aiming to discover the genetic basis of the aging process, founded by Richard Weindruch, a long-time leader of calorie restriction research, http://www.lifegentech.com/; Longenity, founded by Preston (Pete) Estep, using genomics and bioinformatics to study aging, http://www.longenity.com/; Phoenix Biomolecular, focusing on telomeres and telomerase research to develop applications for regenerative medicine, founded by Kenneth Weiss, http://www.phoenixbiomolecular.com/; Prana Biotechnology, focusing on drugs against degenerative diseases of the eye and the brain, presided over by Geoffrey Kempler, http://www.pranabio.com/; Sierra Sciences, focusing on telomerase, with Pierluigi Zappacosta as the CEO, http://www.sierrasci.com/; Genescient, led by Gregory Benford, searching for genes correlating with longevity (mainly in drosophila flies) and attempting to stimulate those genes by pharmaceuticals, http://www.genescient.com/; InSilico Medicine, led by Alex Zhavoronkov, developing geroprotective (anti-aging) drugs via bioinformatic analysis of gene expression, http://insilicomedicine.com/.

(Mainly based on http://whoswho.senescence.info/corp.php. Also, a large list of life-extensionist companies is available at http://www.agingportfolio.org/index/cac.)

[1178] In September 2013, the IT concern Google (headed by Larry Page and Sergey Brin) announced the establishment of Calico (California Life Company) that would be dedicated to human life extension. As of 2014, the company has been led by Arthur Levinson, the former Chief Executive Officer of the biotechnology company Genentech and board member of the information technology company Apple, accompanied by such captains of high tech and biotech industry as Robert Cohen (an oncology expert), Hal Barrron (an expert in pharmaceuticals development), David Botstein (an expert in genomics), and Cynthia Kenyon (and expert in genetics of aging). As of 2014, the research program of Calico was yet unclear. (http://www.calicolabs.com/.)

[1179] In August 2012, great financial difficulties were reported for Tengion, apparently due to the reluctance of capital-holders to invest in regenerative medicine without the assurance of immediate and substantial profits. Other companies developing regenerative and anti-aging medicine have recently met with the same scarcity of investments. Thus, funding problems were reported for Geron Corporation in November 2011 (which was the reason to discontinue its clinical trials of human embryonic stem cell therapy for spinal cord injury). Biotime ran into financial difficulties in April 2012. After the takeover of Sirtris by GlaxoSmithKline in 2008, and Elixir Pharmaceuticals by Novartis in 2009, the development of Sirtris and Elixir seems to have slowed down and news from those companies have become relatively scarce.

Difficulties in the development of other (probably all) stem-cell, regenerative and anti-aging start-up companies have been reported as well.

Generally, investments into life sciences decreased from 2008 to 2012, by over $5 billion, from ~$7.8 billion a year in 2007 and 2008, to $2.5 billion in 2012.

(http://www.biospace.com/News/tengion-inc-were-almost-out-of-cash-could-shut/270174;
http://www.biospace.com/News/geron-corporation-halts-stem-cell-work-cuts-38-of/240698;
http://seekingalpha.com/article/490841-biotime-is-running-out-of-time-short-sell-idea;
http://www.biospace.com/News/glaxosmithkline-halts-development-of-sirtris/204203;
http://www.physicventures.com/news/elixir-inks-500m-novartis-option-deal-and-closes-12m-financing;
http://seekingalpha.com/article/607751-stem-cell-hype-has-yet-to-breed-profits;
http://news.yahoo.com/hunt-anti-aging-pill-134718294.html;
http://venturebeat.com/2013/04/15/fenwickwest-study-finds-funding-for-life-sciences-continues-to-slow/.)

[1180] Many hopes for increasing longevity have been recently pinned on the research of the telomerase enzyme, that normally lengthens the end segments of chromosomes (the telomeres) and thereby increases the number of cell divisions, consequently raising tissue regenerative potential (as well as the potential for cancer development).

Thus, in November 2010, it was announced that a Harvard team led by Prof. Ronald A. DePinho, succeeded in reversing the aging process by telomerase activation in telomerase deficient mice.

(Mariela Jaskelioff, ..., Ronald A. DePinho, "Telomerase reactivation reverses tissue degeneration in aged telomerase-deficient mice," *Nature*, 469, 102-106, January 6, 2011, first published online on November 28, 2010; Ian Sample, "Harvard scientists reverse the ageing process in mice – now for humans," *Guardian*, November 28, 2010, http://www.guardian.co.uk/science/2010/nov/28/scientists-reverse-ageing-mice-humans.)

On May 15, 2012, researchers at the Spanish National Cancer Research Centre (CNIO) in Madrid, directed by Maria Blasco, claimed to have produced a 24% increase in the life-span of adult mice, accompanied by a delay in the onset of age-related diseases and "without increasing the incidence of cancer," using "telomerase gene therapy," i.e. inserting a DNA-modified, telomerase-coding virus into animal cells to increase the cells' telomerase enzyme production.

In their earlier study of 2008, the authors claimed, "Remarkably, over-expression of TERT [Telomerase Reverse Transcriptase] in the context of mice engineered to be cancer resistant ([expressing the genes] Sp53/Sp16/SARF/TgTERT mice) is sufficient to decrease telomere damage with age, delay aging and increase median longevity by 40% (Tomas-Loba et al, 2008)."

Apparently, neither of these studies has yet been reproduced by other groups, and it is too early to speak of any human applicability.

Maria Blasco is also the Chief Scientific Advisor at the company called Life Length (http://www.lifelength.com/), attempting to test for genetic predispositions to longevity in humans by measuring telomere length, among other biological markers of aging and longevity. Several American scientists have participated in the company, including Jerry Shay, Sandy Chang, Kathleen Collins, William Andrews, Mark Rosenberg, and others.

Blasco's group additionally suggested the use of various pharmaceutical and nutraceutical "Telomerase Activators" ("TA's" such as TA-65 made from the root of *Astragalus membranaceus* herb) to postpone aging. Other suggested "Telomerase Activators" have included fish oil, resveratrol, statins, milk thistle, horny goat weed, grape-seed extract, turmeric root extract, ashwagandha root extract, hacopa extract, N-acetyl cysteine, pomegranate fruit extract, DL-alpha lipoic acid, ginseng root extract, blue-berry fruit extract, berberine rhizome extract, bilberry fruit extract, green tea extract, white tea extract, black tea leaf, acacia bark extract, placental extract, etc. Even tobacco has been reported to produce telomerase activation.

Yet, the efficacy of such supplements (especially for telomerase regeneration, and especially for humans) seems to be very uncertain.

As a recent popular review in *The Scientist* noted, "Whether telomere shortening mediates human aging – and conversely, whether telomere elongation may reverse aging or prevent age-related diseases – are still controversial issues."

(Bruno Bernardes de Jesus, ..., Maria A. Blasco, "Telomerase gene therapy in adult and old mice delays aging and increases longevity without increasing cancer," *EMBO [The European Molecular Biology Organization] Molecular Medicine*, 4(8), 691-704, 2012; "First Gene Therapy Successful Against Aging-Associated Decline: Mouse Lifespan Extended Up to 24% With a Single Treatment," *ScienceDaily*, May 14, 2012, http://www.sciencedaily.com/releases/2012/05/120514204050.htm; Tomas-Loba A, ..., Blasco MA, "Telomerase reverse transcriptase delays aging in cancer-resistant mice," *Cell*, 135, 609- 622, 2008; Bruno Bernardes de Jesus, ..., Maria A. Blasco, "The telomerase activator TA-65 elongates short telomeres and increases health span of adult/old mice without increasing cancer incidence," *Aging Cell*, 10(4), 604-621, 2011; Rodrigo Calado and Neal Young, "Telomeres in Disease," *The Scientist: Magazine of the Life Sciences*, May 1, 2012, http://the-scientist.com/2012/05/01/telomeres-in-disease/; Hyeon Woo Yim, ..., Jack A. Taylor, "Smoking is associated with increased telomerase activity in short-term cultures of human bronchial epithelial cells," *Cancer Letters*, 246(1), 24-33, 2007.)

[1181] As of 2014, several companies have made initial attempts to provide commercial testing for chances of longevity, based on determining "longevity markers" or "longevity genes."

Such companies included:

23andMe, https://www.23andme.com/health/Longevity/; Spectracell Laboratories, http://www.spectracell.com/; Telome Health, http://www.telomehealth.com/; Life Length, http://www.lifelength.com/, and others.

As of May - July 2010, Boston University also planned to establish a company to provide genetic testing for chances of exceptional longevity (*Genomeweb*, July 7, 2010, http://www.genomeweb.com/blog/waiting-uproar-over-long-life-genetic-testing). Yet, as of July 2011, the Boston University "Scientists retract report on predicting longevity" (*New York Times*, July 22, 2011, http://www.nytimes.com/2011/07/23/science/23retract.html).

Despite the setbacks, hopes have been mounting to achieve a more precise and "personalized" assessment of one's chances for health and increased life-span.

These hopes have been inspired by the rapidly developing and increasingly cheaper gene sequencing technology.

As of 2012-2013, companies, such as Ion Torrent, Oxford Nanopore Technologies and others, strove to establish the entire DNA sequence of the entire personal human genome for less than $1,000 and within hours, as compared to 13 years and $3 billion it took to accomplish the Human Genome Project – the successful complete human genome sequencing announced in 2003. (http://www.technologyreview.com/biomedicine/39458/?p1=MstRcnt; http://www.sciencedaily.com/releases/2012/05/120522152655.htm.)

This milestone of sequencing an entire human genome for $1000 was said to be reached by the American company Illumina in January 2014.
(http://www.technologyreview.com/news/523601/does-illumina-have-the-first-1000-genome/)

However, there have also been doubts whether the great amassment of disparate genetic data would soon produce much useful clinical information.
(http://www.scientificamerican.com/article.cfm?id=taking-genomes-peronally; http://www.medpagetoday.com/MeetingCoverage/AACR/31985.)

Rather, an ever greater emphasis has been given to "epigenetic" (environmental) regulation of gene expression. (Gravina S, Vijg J., "Epigenetic factors in aging and longevity," *Pfluegers Archiv: European Journal of Physiology*, 459(2), 247-258, 2010; Greer EL, Maures TJ, Ucar D, Hauswirth AG, Mancini E, Lim JP, Benayoun BA, Shi Y, and Brunet A, "Transgenerational epigenetic inheritance of longevity in Caenorhabditis Elegans," *Nature*, 479 (7373), 365-371, October 2011, http://www.alzforum.org/new/detail.asp?id=2939.)

Still, there have been strong hopes to find genes correlating with high longevity, and perhaps more importantly, to somehow "activate" those genes via adjustments of the environment and life style, or even by pharmacological means.

Several such potential "longevity gene tests" and "longevity genes activators" have been publicized. Thus, for example, *Telegraph.co.uk* reported in May 2010:

"Prof [Nir] Barzilai [of Albert Einstein College of Medicine, New York] and his team have already identified a number of [longevity] genes among the centenarians. Laboratories are now working on creating a drug which mimics the effects of three of them - two that increase the production of so-called 'good' cholesterol in the body, which reduces the risk of heart disease and stroke, and a third that helps prevent diabetes. Testing could begin by 2012, he said, with it appearing on the market "within five or 10 years". He predicted: "People will take a pill, starting at 40, and their lives will be longer."
(Stephen Adams, "Don't worry about your health if you want to see 100," *Telegraph.co.uk*, 11 May, 2010, http://www.telegraph.co.uk/news/7706710/Dont-worry-about-your-health-if-you-want-to-see-100.html; http://www.fightaging.org/archives/2010/05/an-interview-with-nir-barzilai.php.)

It is unclear from the press releases which "drugs" were meant, presumably they would involve the "targeting" of such genes as sirtuins, insulin growth factor (IGF), cholesteryl ester transfer protein (CETP), tumor suppressor gene - p53, "regeneration inhibitor" - p21, or telomerase.

[1182] An additional gene-therapeutic approach explored in relation to life-extension, has been the use of RNA interference to block the production of undesirable proteins, without a direct intervention into genes.
(Raoul C, Barker S, Aebischer P, "Viral-based modelling and correction of neurodegenerative diseases by RNA interference," *Gene Therapy*, 13 (6), 487-495, 2006; Copeland JM, et al., "Extension of Drosophila life span by RNAi of the mitochondrial respiratory chain," *Current Biology*, 19(19), 1591-1598, 2009; Fire A, Xu S, Montgomery M, Kostas S, Driver S, Mello C, "Potent and specific genetic interference by double-stranded RNA in Caenorhabditis Elegans," *Nature*, 391(6669), 806-811, 1998.)

[1183] An example of the work on the removal of senescent cells: Darren J. Baker, ..., Jan M. van Deursen, "Clearance of p16^{Ink4a}-positive senescent cells delays ageing-associated disorders," *Nature*, 479, 232-236, 10 November 2011.

An example of the work on stimulation of cell division/regeneration/immortalization: Atsuhiko T. Naito, ..., Issei Komuro, "Complement C1q Activates Canonical Wnt Signaling and Promotes Aging-Related Phenotypes," *Cell*, 149 (6), 1298-1313, 8 June 2012. Notice the potential impairment of immunity in the process of "immortalization."

Also notice that the increased cell multiplication (regeneration) carries the risk of uncontrolled cell multiplication (cancer). Conversely, suppression of cancer can carry the risk of increased cell death (cell senescence). Thresholds, balances and tradeoffs between processes of revival and dying seem to be vital.

See, for example, Steven S. Foster, Saurav Dea, Linda K. Johnson, John H. J. Petrini, Travis H. Stracker, "Cell cycle- and DNA repair pathway-specific effects of apoptosis on tumor suppression," *Proceedings of the National Academy of Sciences USA*, 109 (25), 9953-9958, June 19, 2012.

Interestingly, there has been some evidence that the incidence of cancer (a proliferative disease) is indeed inversely related to the incidence of Alzheimer's (a degenerative disease), i.e. older people with cancer were indicated to have a reduced risk of Alzheimer's and vice versa. However, both diseases increase exponentially with age, having the deteriorative process of aging as their common root. (Musicco M, Adorni F, Di Santo S, Prinelli F, Pettenati C, Caltagirone C, Palmer K, Russo A., "Inverse occurrence of cancer and Alzheimer disease: A population-based incidence study," *Neurology*, 81, 322-328, 2013.)

Still, the balance between cell death and regeneration has been successfully manipulated to produce tissue repair.

(C. Fuchs, ..., M. Hengstschlager, "Tuberin and PRAS40 are anti-apoptotic gatekeepers during early human amniotic fluid stem cell differentiation," *Human Molecular Genetics*, 21(5), 1049-1061, March 2012; E. R. Porrello, ..., H. A. Sadek, "Regulation of neonatal and adult mammalian heart regeneration by the miR-15 family," *Proceedings of the National Academy of Sciences*, 110(1), 187-192, January 2013.)

Some highly promising developments have been reported, for both regeneration of organs within the body and growing transplantable organs outside of the body.

For example, regeneration of the thymus in mice was reported – important also due to the role of the thymus in producing native immune T cells which are diminished in the aged persons, causing their increased susceptibility to infection and low response to vaccination.

(Bredenkamp N, Nowell CS, Blackburn CC, "Regeneration of the aged thymus by a single transcription factor," *Development*, 141 (8), 1627-1637, 2014;
http://www.theguardian.com/science/2014/apr/10/scientists-create-living-organ-in-mice-in-world-first-breakthrough.)

Large functional organs, such as the entire human lung, were also grown.

(Nichols JE, Niles J, Riddle M, Vargas G, Schilagard T, Cortiella J, et al. "Production and assessment of decellularized pig and human lung scaffolds," *Tissue Engineering Part A*, 19, 2045-2062, 2013; http://www.worldhealth.net/news/first-lab-grown-human-lung/.)

Yet effective and widely available "regenerative medicine" remains to be desired.

[1184] Recently, human stem cells, in principle capable to differentiate into any desired tissue, have been widely perceived as the main potential building blocks for "regenerative medicine" aimed to rebuild age-related loss of body structure and function.

Generally, as of 2009, the global market for regenerative medicine, including cell therapy (particularly stem cell therapy) and tissue engineering and repair, was estimated at about US$2-5 billion. Over 500 companies were involved worldwide, among them over 100 companies investigated stem cell therapies. The overall field of regenerative medicine was estimated to grow by about 30% yearly to reach more than $11 billion by 2020.

(*Regenerative Medicine. Industry Briefing*, MaRS Advisory Services, Toronto, 2009.)
At the beginning of 2012, it was estimated that there were about "400 regenerative products on the market" and "600 more in development" (http://www.inc.com/eric-markowitz/immortality-the-next-great-investment-boom.html).
In 2014, the expectations further increased, yet varied. Some estimates claimed that the global regenerative medicine market would grow from $16.4 billion in 2013 to $67.6 billion by 2020, and some estimated that it would grow just from $2.6 billion in 2012 to $6.5 billion in 2019.
(http://www.prnewswire.co.uk/news-releases/regenerative-medicine-market-is-expected-to-reach-676-billion-global-by-2020-265200791.html;
http://www.bloomberg.com/article/2014-07-17/a3DpgePCra_I.html;
Mason C, Brindley DA, Culme-Seymour EJ, Davie NL, "Cell therapy industry: billion dollar global business with unlimited potential," *Regenerative Medicine*, 6(3), 265–272, 2011.)
Yet, it seems that, apart from the use of adult stem cells in bone marrow transplants to treat leukemia (started in the 1960s), there have been few practical clinical applications of stem cells, either adult or embryonic, in either "regenerative" or other fields.
Nonetheless, research in this area has been wide and several therapies deriving from stem cells and their products have entered clinical trials, including therapies against auto-immune diseases, diabetes, neurological disorders, cancer, and other degenerative and age-related conditions.
(Daniele Lodi, Tommaso Iannitti, Beniamino Palmieri, "Stem cells in clinical practice: applications and warnings," *Journal of Experimental & Clinical Cancer Research*, 30, 9, 2011, http://www.jeccr.com/content/30/1/9 ; Alan Trounson, Rahul G. Thakar, Geoff Lomax, Don Gibbons, "Clinical trials for stem cell therapies," *BMC Medicine*, 9, 52, 2011, http://www.biomedcentral.com/1741-7015/9/52; Megan Scudellari, "The Little Cell That Could," *The Scientist*, July 1, 2012, http://the-scientist.com/2012/07/01/the-little-cell-that-could/; NIH Clinical Trials with Stem Cells, http://clinicaltrials.gov/ct/search?term=stem+cell&submit=Search.)
Apparently, the world's first officially approved stem-cell therapy was Prochymal (remestemcel-L) from Osiris Therapeutics Inc., for the treatment of acute graft-vs-host disease (GvHD) in children, approved in Canada in May 2012.
("World's First Approved Stem Cell Drug: Osiris Receives Marketing Clearance from Health Canada for Prochymal. Historic decision offers hope to children suffering from life-threatening GvHD, Osiris Therapeutics Inc. Press Release, May 17, 2012, at Regenerative Medicine Forum, http://us1.campaign-archive1.com/?u=513b92eb0bf6c17f06cca149d&id=7d9bdc735c.)
However, some bad news for the therapy came in July 2012, when it was revealed that Osiris falsified some of the results of Prochymal's clinical trials in heart attack patients.
(Adam Feuerstein, "Osiris Therapeutics, Inc. (OSIR) Fibs About Clinical Trial Results," *BioSpace*, 7/3/2012, http://www.biospace.com/news_story.aspx?NewsEntityId=265698.)
A further ramification of regenerative stem cell therapy has been the use of induced pluripotent stem cells (iPSC).
In July 2013, it was announced that the first clinical trial was initiated in Kyoto, Japan, with iPSC for the treatment of age-related macular degeneration.
(http://www.japantimes.co.jp/news/2013/07/12/national/japan-health-ministry-subpanel-oks-ips-cell-clinical-trial/#.UgSzUNKTTbM.)
The research of iPSC has been pioneered by Shinya Yamanaka of Kyoto University who received the Nobel Prize in Physiology or Medicine together with John Gurdon "for the discovery that mature cells can be reprogrammed to become pluripotent." http://www.nobelprize.org/nobel_prizes/medicine/laureates/2012/press.html.
The common rationale for preferring the use of iPSC in regenerative and anti-aging medicine is that the

technique allows the "reprogramming" of any cell of the body to become any tissue in need of regeneration (thus, among other benefits, avoiding the danger of immune rejection by the host, and bypassing the need to use embryonic tissues). However, with this technique as well, the practicality and actual benefits of application have been uncertain.

[1185] There have been many methods investigated for improving mitochondrial function:

Anti-oxidants have been inserted into mitochondria to eliminate oxidative damage at its origin.

(Skulachev, V.P., et al, "An attempt to prevent senescence: a mitochondrial approach," *Biochimica et Biophysica Acta*, 1787(5), 437-61, 2009.)

Some experiments aimed to create a "backup" copy of mitochondrial DNA in the cell nucleus (the so-called allotopic expression of mitochondrial DNA in the nucleus) to produce mitochondrial proteins guarded against mitochondrial DNA mutations.

(Aubrey de Grey and Michael Rae, *Ending Aging. The Rejuvenation Breakthroughs That Could Reverse Human Aging in Our Lifetime*, St. Martin's Press, NY, 2007, "Getting off the grid," pp. 77-100; Crystel Bonnet, ..., Marisol Corral-Debrinski, "Allotopic mRNA localization to the mitochondrial surface rescues respiratory chain defects in fibroblasts harboring mitochondrial DNA mutations affecting complex I or V subunits," *Rejuvenation Research*, 10(2), 127-144, 2007.)

Also, healthy donor mitochondrial DNA has been injected in place of native diseased mitochondrial DNA. The process is known as mitochondrial DNA protofection or cytoplasmic transfer or mitochondrial gene replacement therapy.

(Rafal Smigrodzki and Francisco R. Portell, "Mitochondrial Manipulation as Treatment for Aging," Ch. 16, in *The Future of Aging: Pathways to Human Life Extension* (Gregory M. Fahy, Michael D. West, L. Stephen Coles and Steven B. Harris, Eds.), Springer, NY, 2010, pp. 521-541; Shaharyar M. Khan and James P. Bennett, "Development of mitochondrial gene replacement therapy," *Journal of Bioenergetics and Biomembranes*, 36(4), 387-393, 2004.)

Notably, donor mitochondrial DNA has been inserted not only into adult human cells, but also into embryonic human cells, in fact producing genetically modified human babies. In this procedure, the mitochondrial DNA donor serves as the "third parent" in addition to the maternal and paternal nuclear DNA donors.

This reproductive/eugenic technology has been considered as a method for treating mitochondrial diseases generally, and as a preventive measure against an early onset of aging in particular. Note that the mitochondrial DNA passes on to the offspring only via the mother (the egg producer), through the cytoplasm of the egg, and the native mitochondrial DNA will likely be destroyed.

(The seminal work in the field is Jason A. Barritt, Steen Willadsen, Carol Brenner and Jacques Cohen, "Cytoplasmic transfer in assisted reproduction," *Human Reproduction Update*, 7 (4), 428-435, 2001.)

Conversely, instead of injecting healthy donor mitochondrial DNA into an egg of a person having a mitochondrial disease, there were also attempts to inject the diseased donor's nuclear DNA into a healthy egg, leaving the donor's diseased mitochondrial DNA behind.

In this nuclear transfer/cloning procedure, it is the native nuclear DNA of the egg that will be destroyed, but the native mitochondrial DNA will remain in the offspring.

Among other implications, this would mean that the produced babies will not be the identical clones of either the mitochondrial or nuclear DNA donors (parents). (The only exception would be of course women self-fertilizing themselves (donating both the egg mitochondrial DNA and the somatic nuclear DNA), but that would not solve their mitochondrial disease problems or aging.)

(Lyndsey Craven, ..., Douglass M. Turnbull DM, "Pronuclear transfer in human embryos to prevent transmission of mitochondrial DNA disease," *Nature*, 465 (7294), 82-85, May 2010.)

Furthermore, mutations in mitochondrial DNA were also repaired by targeting corrective RNA into the mitochondria.

(Geng Wang, ..., Carla M. Koehler, "Correcting human mitochondrial mutations with targeted RNA import," *Proceedings of the National Academy of Sciences USA*, March 12, 2012.)

An additional wide assortment of genetic and pharmacological approaches to improve mitochondrial function has been proposed.

The pharmacological approaches have included:

Supplementation with components of the respiratory oxidative phoshorylation system – such as CoQ10, pyruvate, succinate, idebenone, menadione, vitamins C and K, carnitine, Copper, folate, riboflavin, thiamine, etc.; as well as various anti-acidic, anti-toxic, and anti-oxidant substances – such as bicarbonate, dichloroacetate, dialysis reagents, diuretics, lipoic acid, vitamins E and B, CoQ10; Manganese tetrabutylammonium perchlorate (Mn-TBAP), quercetin, etc.

(Eric A. Schon and Salvatore DiMauro, "Medicinal and Genetic Approaches to the Treatment of Mitochondrial Disease," *Current Medicinal Chemistry*, 10, 2523-2533, 2003.)

Moreover, the sufficient supply of oxygen to living tissues has been a vital concern for improving mitochondrial energy production and sustaining life generally.

(See, for example, John N. Kheir, ..., Francis X. McGowan, "Oxygen Gas-Filled Microparticles Provide Intravenous Oxygen Delivery," *Science Translational Medicine*, 4, 140ra88, 2012.)

It can be added, however, that enhanced mitochondrial energetics should not necessarily lead to a longer life-span, in agreement with Hufeland's old principle that "strengthening, carried too far, may tend to accelerate life, and consequently, to shorten its duration." Consider for example, the acceleration of apoptotic cell death by cell mitogenic stimulation.

(See Ara B. Hwang, Seung-Jae Lee, "Regulation of life span by mitochondrial respiration: the HIF-1 and ROS connection," *Aging* (Albany NY), 3(3), 304-310, 2011; Jean H. Overmeyer and William A. Maltese, "Death Pathways Triggered by Activated Ras [rat sarcoma proteins] in Cancer Cells," *Frontiers in Bioscience*, 16, 1693-1713, 2011; Christina K. Speirs,..., Bo Lu, "Harnessing the cell death pathway for targeted cancer treatment," *American Journal of Cancer Research*, 1(1), 43-61, 2011; *Hufeland's Art of Prolonging Life*, Edited by Erasmus Wilson, Lindsay & Blakiston, Philadelphia, 1867 (1796), pp. IX-X.)

In fact, it was directly suggested that actually inhibiting mitochondrial function, for example by reducing *mrps-5* protein expression, increases the life span (as shown in *Caenorhabditis Elegans* worms). The mitochondrial inhibition was suggested to trigger mitochondrial unfolded protein response (UPR) which was considered adaptive against stress. The mitochondria inhibition for life extension was also attempted by pharmacological means, such as antibiotics, like doxycyline and rapamycin (Riekelt H. Houtkooper, ..., Johan Auwerx, "Mitonuclear protein imbalance as a conserved longevity mechanism," *Nature*, 497, 451-457, 2013; http://www.the-scientist.com/?articles.view/articleNo/35673/title/Inhibit-Mitochondria-to-Live-Longer-/).

[1186] As noted earlier, various commercial anti-aging companies and their preparations now count by the hundreds. And the search for the "perfect," multi-componental and "personalized" life-extending nutritional supplements is far from over.

The supplements can be quite sophisticated. Just one example is a drug promoted in April 2011 by a newsletter from the American Academy of Anti-Aging Medicine - A4M (sent from the A4M-affiliated Medical Conferences International Inc., at Boca Raton, Florida). The drug was advertised as belonging to "The Next Generation of Anti-Ageing Treatments" and was called "Human Citoplacell 2G" from Biocell Ultravital, Switzerland.

It was said to be an "oral complex which, when taken regularly, will reactivate hypophysis function, thus achieving improved production of HGH [human growth hormone], also known as the "Youth Hormone," thanks to the action of a select set of amino acids and catalyst peptides."

It also promised "anti-oxidant protection, improved cognitive faculties and increased energy and general well-being."

The formula contained a "mix of amino acids composed of: alanine, arginine, aspartic acid, phenylalanine, glutamic acid, glycine, histidine, lysine, methionine, serine, threonine, tryptophan, tyrosine, valine, standardised ginkgo biloba extract, tecoma curialis extract, coenzyme Q-10, Royal Jelly extract, freeze-dried soy protein, DHEA, KH3, riboflavin, prenenolone, and thiamine."

According to the company's description, the compound "Cellorgane Multi-Complex 2G" (listed under the heading of "Human Citoplacell 2G") contained:

"Extract of Adult Stem Cells Freeze-dried during Embryonic Phase/Testicles, Pituitary Gland, Adrenal Gland, Whole Brain, Embryoblasts, Thymus Gland, Hypothalamus, Liver, Spleen, Bone Marrow, Pancreas, Kidney, Lung, Arteries, Veins, Blood Vessels, Heart, Spinal Cord, Muscle, Cartilage, Skin, Eyes, Adenosine Desaminase, Superoxide Dismutase, Glutathione Reductase, Glutathione S-transferase, Glutathione Peroxidase, Distilled Water, Phenol."

According to yet another description by the company, "Cellorgane is made of young sheep or bovine organs (not older than 6 months) worked into an ultrafiltrate compound with extraordinary healing and preventive potential."

According to an earlier product description of 2004 (which however could not be found on accessing the company website in 2011-2012), Human Citoplacell also contained "Human placenta" and "thymus extract" ["from young calves"].

(http://www.biocellultravital.com/en/prods_bio/products_citoplacell_clinical_info.php; http://www.biocellultravital.com/en/prods_bio/products_cellorgane_clinical_info.php; http://www.casewatch.org/fdawarning/prod/2004/belleza.shtml.)

The efficacy of any such supplements seems to be very uncertain. The interactions of the different components and their working dosages, hence their safety, seem to be equally undetermined.

As said by the patriarch of Swiss rejuvenators, Paracelsus (1493-1541) – "The dose makes the poison." Or more precisely: "Alle Dinge sind Gift, und nichts ist ohne Gift; allein die dosis machts, daß ein Ding kein Gift sei" – "All things are poison and there is nothing without poison; only the dose makes that a thing is not a poison."

("Die dritte Defension wegen des Schreibens der neuen Rezepte," in *Septem Defensiones 1538*, Theophrast Paracelsus: Werke, Band 2, Medizinische Schriften, Besorgt von Will-Erich Peuckert, Wissenschaftliche Buchgesellschaft, Darmstadt, 1965, p. 510 – The third defense on the writing of new recipes, in The Seven Defenses, 1538, Theophrast Parcelsus Works, Medical writings, edited by Will-Erich Peuckert, Academic Press, Darmstadt, 1965, vol. 2, reprinted at http://www.zeno.org/Philosophie/M/Paracelsus/Septem+Defensiones.)

[1187] Recently, some of the highest profile potential anti-aging substances (or "geroprotectors") have been the anti-biotic Rapamycin, the anti-diabetic Metformin, and the red-wine derivative Resveratrol.

(David E. Harrison, …, Richard A. Miller, "Rapamycin fed late in life extends lifespan in genetically heterogeneous mice," *Nature*, 460, 392-395, July 16, 2009; Frauke Neff, …, Dan Ehninger, "Rapamycin extends murine lifespan but has limited effects on aging," *Journal of Clinical Investigations*, 123(8), 3272-91, July 2013; James M. Flynn, …, Simon Melov, "Late life rapamycin treatment reverses age-related heart dysfunction," *Aging Cell*, July 2013;

Alejandro Martin-Montalvo, …., Rafael de Cabo, "Metformin improves healthspan and lifespan in mice," *Nature Communications*, 4, July 2013; Georges Mairet-Coello, … , Franck Polleux, "The CAMKK2-AMPK Kinase Pathway Mediates the Synaptotoxic Effects of Aβ Oligomers through Tau Phosphorylation," *Neuron*, 78 (1), 94-108, 10 April 2013;

B. P. Hubbard, …, D. A. Sinclair, "Evidence for a Common Mechanism of SIRT1 Regulation by Allosteric Activators," *Science*, 339 (6124), 1216-1219, March 2013; Lasse Gliemann, …, Ylva Hellsten, "Resveratrol Blunts the Positive Effects of Exercise Training on Cardiovascular Health in Aged Men," *Journal of Physiology*, July 2013.)

In 2013, considerable attention was given to several additional substances with suggested potential anti-aging and life-extending properties. The number of publications on life extension has also been growing. Though, there seem to have been few fundamentally novel, paradigm-shifting conceptual breakthroughs.

Thus, one substance that drew attention was high molecular weight hyaluronan polysaccharide (HMW-HA). It is a natural lubricant in the extracellular matrix of the body, and it has been commonly used for the treatment for arthritis (usually in injections) and in anti-wrinkle skin care products. Now it has been implicated as a protective substance against cancer in naked mole rats, and may also explain their remarkable longevity (up to 30 years compared to other mammals of comparable size, such as mice and rats that only live 2-3 years).

(Xiao Tian, …, Vera Gorbunova and Andrei Seluanov, "High-molecular-mass hyaluronan mediates the cancer resistance of the naked mole rat," *Nature*, 499, 346-349, 2013.)

Other potential pharmacological ways to life extension could be by blocking the NF-ϰB (nuclear factor kappa-light-chain-enhancer of activated B cells) in the hypothalamus to reduce inflammation, as well as supplementing with gonadotropin-releasing hormone (GnRH) to promote neurogenesis.

(Guo Zhang, …, Dongsheng Cai, "Hypothalamic programming of systemic ageing involving IKK-β, NF-ϰB and GnRH," *Nature*, 497, 211–216, 2013.)

Several additional hormonal interventions against aging have been investigated, such as the Growth Differentiation Factor 11 (GDF11, also known as Bone Morphogenetic Protein 11) found in high concentrations in the blood of young mice.

(Francesco S. Loffredo, …, Richard T. Lee, "Growth Differentiation Factor 11 Is a Circulating Factor that Reverses Age-Related Cardiac Hypertrophy," *Cell*, 153(4), 828-839, 2013; Manisha Sinha, … , Amy J. Wagers, "Restoring systemic GDF11 levels reverses age-related dysfunction in mouse skeletal muscle," *Science*, 344 (6184), 649-652, 2014.)

A wide variety of other potential anti-aging and life-extending pharmaceuticals has been tested and discussed across the world. A special attention has been given to sugar lowering (anti-glycemic) and lipid lowering (statin), anti-coagulant, and anti-inflammatory medications.

(E.g. Stephen R. Spindler, Patricia L. Mote, James M. Flegal, and Bruce Teter, "Influence on Longevity of Blueberry, Cinnamon, Green and Black Tea, Pomegranate, Sesame, Curcumin, Morin, Pycnogenol, Quercetin, and Taxifolin Fed Iso-Calorically to Long-Lived, F1 Hybrid Mice," *Rejuvenation Research*, 16(2), 143-151, April 2013; http://www.eha2012.org/.)

Yet the substances and the dosages effective for human life extension, as well as their potential side-effects, have been still uncertain. It may be a long way before creating a "youth pill" and ascertaining its long-term efficacy.

Furthermore, there have been some advances toward pharmacological treatment of progeria or "accelerated aging" – increasing the hope also to find interventions against "normal aging".

The anti-progeria treatments often employed pharmacological inhibitors of enzymes involved in progeria, such as inhibitors of farnesyltransferase or inhibitors of isoprenylcysteine carboxyl methyltransferase (ICMT).

(Leslie B. Gordon, …, Mark W. Kieran, "Clinical trial of a farnesyltransferase inhibitor in children with Hutchinson–Gilford progeria syndrome," *Proceedings of the National Academy of Sciences USA*, 109(41), 16666-16671, October 2012; Mohamed X. Ibrahim, …, Martin O. Bergo, "Targeting Isoprenylcysteine Methylation Ameliorates Disease in a Mouse Model of Progeria," *Science*, 340 (6138), 1330-1333, 16 May 2013.)

There have also been advances toward pharmacological treatment of Alzheimer's disease, a major, perhaps even "generic" age-related condition.

Some promising substances included the anti-inflammatory and neuroprotective Cannabinoid type 2 (CB2) agonists, such as 1-((3-benzyl-3-methyl-2,3-dihydro-1-benzofuran-6-yl) carbonyl) piperidine (MDA7), as well as vaccines and adjuvants against amyloid beta and tau aggregates in the brain that presumably cause Alzheimer's disease.

(Jiang Wu, ..., Mohamed Naguib, "Activation of the CB2 receptor system reverses amyloid-induced memory deficiency," *Neurobiology of Aging*, 34(3), 791-804, March 2013; Jean-Philippe Michaud, ..., Serge Rivest, "Toll-like receptor 4 stimulation with the detoxified ligand monophosphoryl lipid A improves Alzheimer's disease-related pathology," *Proceedings of the National Academy of Sciences*, 110(5), 1941-1946, January 2013.)

Still, the advances in the treatment of major age-related diseases have been limited, perhaps even minimal. According to many life-extensionist experts, to achieve any significant progress in the treatment of aging-related diseases, the very process of aging will need to be addressed and treated at the root.

(Michael J. Rae, Robert N. Butler, Judith Campisi, Aubrey D. N. J. de Grey, Caleb E. Finch, Michael Gough, George M. Martin, Jan Vijg, Kevin M. Perrott, and Barbara J. Logan, "The demographic and biomedical case for late-life interventions in aging," *Science Translational Medicine*, 2 (40), 40cm21, 2010.)

[1188] Clearly, the currently known healthy lifestyle approaches are of only limited value for human lifespan extension. Still, they are demonstrably associated with extended longevity, giving the benefit of additional several years of life, that may also increase the chances to live to the emergence of groundbreaking life-extending technologies, if they were ever to emerge.

The literature on the subject of healthy life prolonging lifestyles has been enormous. Just for example: Rizzuto, D., Orsini, N., Qiu, C., Wang, H-X, and Fratiglioni, L., "Lifestyle, social factors, and survival after age 75: population based study," *British Medical Journal*, 345, e5568, 2012.

[1189] American research centers for healthy aging (not necessarily "radical" life-extensionist, but rather with varying degrees of hopefulness) have included:

Buck Institute for "Extending Healthy Years of Life" (Novato, California) http://www.buckinstitute.org/; Kronos Longevity Research Institute (KLRI, Phoenix, Arizona) http://www.kronosinstitute.org; Linus Pauling Institute (affiliated with Oregon State University) http://lpi.oregonstate.edu/; Vitae Institute, http://www.vitaeinstitute.org/; Supercentenarian Research Foundation, for research into the biology of aging, http://www.supercentenarian-research-foundation.org/; Healthy Aging Research Networks, University of Washington, http://depts.washington.edu/harn/; Brookdale Center for Healthy Aging and Longevity, http://www.brookdale.org/; Stanford School of Medicine, Division of Endocrinology, Gerontology and Metabolism, http://endocrinology.stanford.edu/research/; Center on Social Gerontology and the Aging Revolution at Trinity University, Texas, www.trinity.edu/~mkearl/geron.html; Gerontology Research Group at the University of California, Los Angeles, http://www.grg.org/ and many more.

In other countries, there have been comparatively fewer such centers, such as, for example, Aging Research Center (Waterloo, Ontario, Canada) http://www.arclab.org/ or National Aging Research Institute Incorporated (NARI, Parkville, Australia) http://www.mednwh.unimelb.edu.au/.

Organizations for the advancement of medical research, including research of aging, regenerative and rejuvenative medicine, have included:

The American Association for the Advancement of Science, http://www.aaas.org/; The Coalition for the Advancement of Medical Research, http://www.camradvocacy.org/; Biotechnology Industry Organization (BIO) http://www.bio.org/; Research! America (An Alliance for Discoveries in Health) http://www.researchamerica.org/; Do It Yourself Biology (An Institution for the Amateur Biologist) http://diybio.org/, and others.

As of 2014, a list of about 300 prominent scientists working on postponing aging in the US and around the world was given in *Senescence.info* by João Pedro de Magalhães of the Institute of Integrative Biology, University of Liverpool, UK (http://whoswho.senescence.info/people.php; www.senescence.info; http://pcwww.liv.ac.uk/~aging/).

A list of 370 scientists working on different strategies for life extension across the world (mainly in the US) can be found in *Nauchnie Trendy Prodlenia Zhizni* (Scientific trends for life extension), Mikhail Batin, Olga Martynyuk, Leonid Gavrilov, et al. (Eds.), Nauka Za Prodlenie Zhizni (Science Against Aging) and Rostock Group, Moscow, 2010 (the names and references in English are on pp. 263-397). A shorter English version is *25 Scientific Ideas of Life Extension*, 2010. Both works are available at http://www.scienceagainstaging.com/; http://grostock.ru/.

[1190] "Biologist Vera Gorbunova to Lead $9.5 Million Multi-Institution Longevity Research Project," University of Rochester, April, 23, 2014, http://www.newswise.com/articles/biologist-vera-gorbunova-to-lead-9-5-million-multi-institution-longevity-research-project; University of California at Los Angeles (UCLA) Longevity Center, http://www.semel.ucla.edu/longevity; Immortality Project, University of California at Riverside, http://www.sptimmortalityproject.com/.

A general impression of the scope of various experimental life extension programs can be obtained by the amount of relevant studies. As of July 2014, the site ClinicalTrials.gov listed over 170,000 studies with locations in all 50 states of the US and in 187 countries. Among them, ~1732 related to "aging", and just about 117 related to "longevity", 40 to "rejuvenation", 16 to "anti-aging", and 13 to "life-extension" (https://clinicaltrials.gov/).

[1191] Aubrey de Grey's SENS program (Strategies for Engineered Negligible Senescence) comprises "seven major types of therapy addressing seven major categories of aging damage":

Damage from cell loss and tissue atrophy are to be countered by adding stem cells and tissue engineering (RepleniSENS); Nuclear (epi-)mutations leading to cancer will be neutralized by the removal of telomere-lengthening machinery (OncoSENS); Mutant mitochondria will be backed up by allotopic expression of 13 proteins in the nucleus (MitoSENS); Death-resistant cells will be removed by targeted ablation (ApoptoSENS); Tissue stiffening will be prevented by compounds breaking Advanced Glycation End-products – AGE-breakers (GlycoSENS) and by tissue engineering; Extracellular aggregates will be cleaned up by immunotherapeutic clearance (AmyloSENS); and Intracellular aggregates will be dissolved by novel lysosomal hydrolases (LysoSENS).

(http://www.sens.org/sens-research/research-themes.)

In addition to the SENS Foundation (founded in 2009 and renamed in 2013 SENS Research Foundation), Aubrey de Grey also founded The Methuselah Foundation (founded in 2000), "seeking, supporting and rewarding science that extends healthy lifespan." The Methuselah Mouse Prize (MPrize) of the Methuselah Foundation is specifically dedicated to finding radical life-extending interventions in mice, on the assumption that the effective postponement of aging in mammalian models will drastically increase the scientific and societal efforts to find similarly effective means for humans.

(The Methuselah Foundation websites: https://www.mfoundation.org/; http://www.methuselahfoundation.org/; http://www.mprize.org/;
the SENS Research Foundation current website: http://www.sens.org;
the former SENS Foundation website (available at the Internet Archive, *www.archive.org*): http://web.archive.org/web/20060106031237/http://www.gen.cam.ac.uk/sens/time.htm#mememe.)

In 2004, de Grey established and became the editor in chief of the journal *Rejuvenation Research*, the official journal of the European Society of Preventive, Regenerative and Anti-Aging Medicine (ESAAM) and of the World Federation of Preventive and Regenerative Medicine. The journal reached an impact factor of 8.571 in 2005, which however declined to 4.225 in 2010 and 2.919 in 2012. *Rejuvenation Research* was formerly called *Journal of Anti-Aging Medicine* (1998-2003), with Michael Fossel of Stanford University (presently with the Michigan State University) as the editor in chief.

(*Rejuvenation Research*: http://www.liebertonline.com/rej.)

De Grey has vigorously popularized his ideas in public media. Interviews included CBS 60 Minutes, BBC, New York Times, Fortune Magazine, Popular Science, CNN Global Health Show, and more. Largely thanks to these popularization efforts, Aubrey de Grey has become one of the most influential leaders of the contemporary life-extensionist movement. Promoting SENS and the Methuselah Foundation has become a wide-spread life-extensionist enterprise.

De Grey's central public appeal is for increasing funding, since without funding there will be no research, and without research there will be no healthy life extension. In contrast to many academic institutions, the SENS project seeks to enlist the "common people" – students and lay volunteers – in research, fund raising and public relations. From this communal effort, de Grey and his supporters believe, the scientific effort will receive a boost. From the "grass roots" democratic support, it is hoped, governmental support will eventually follow.

Yet, presently, a major portion of SENS funding came from the American billionaire, the co-founder of PayPal, Peter Andreas Thiel, who pledged to SENS some $3.5M in 2006 (for a period through 2009): $500K direct funds, and $3M matching funds (at 50% to additional donations). As of December 2009, despite thousands of declared SENS supporters, public donations appeared to have come short of matching Thiel's pledge. (*Longecity Newsletter*, December 2009, http://www.imminst.org/newsletter/ImmInstNewsletter_2009_12.htm.)

As of July 2010, all the received donations were about $5.196M and, including pledge commitments, the donations totaled $14.373M. As of July 2011, the amount of funding did not seem to significantly increase: the received donations were ~$6.170M and, together with the pledge commitments, the total donations were $14.578M. As of July 2012, the amount of donations (including a newly established "New Organ MPrize") remained almost exactly the same: the received donations were ~$6.370M, with the total of ~$14.770M. As of July 2013, the funding increased only slightly: the received donations were ~$6.052M, and the total was $15.082M. In July 2014, the received donations were ~$6.607M, and the total was $15.2142M
(http://www.mprize.org/index.php?pagename=mj_donations_funding.)

However, in July 2013, it was also widely announced that Aubrey de Grey donated $13M of his own inheritance to SENS, essentially doubling the donations. (http://nextbigfuture.com/2013/07/aubrey-de-grey-donates-13-million-of.html.)

Yet, regardless the amount of funding, the public interest in radical life-extension that Aubrey de Grey's has raised, in the UK, the US and the world, is unprecedented.

(De Grey ADNJ, *The Mitochondrial Free Radical Theory of Aging*, Landes Bioscience, Austin TX, 1999; Aubrey de Grey, "Gerontologists and the media: the dangers of over-pessimism," *Biogerontology* 1(4), 369-370, 2000; Aubrey de Grey (Ed.), *Strategies for Engineered Negligible Senescence: Why Genuine Control of Aging May Be Foreseeable*, Annals of the New York Academy of Sciences, 1019, June 2004; Jonathan Weiner, *Long for This World: The Strange Science of Immortality*, Ecco, New York, 2010; Aubrey de Grey and Michael Rae, *Ending Aging. The Rejuvenation Breakthroughs That Could Reverse Human Aging in Our Lifetime*, St. Martin's Press, NY, 2007; http://www.sens.org/users/aubrey-de-grey.)

Beside SENS, the projects supported by the Methuselah foundation include:

Organovo Inc.; *Silverstone Solutions*; *Halcyon Molecular*; and *My Bridge 4 Life* (http://mfoundation.org/?pn=mj_mlife_sciences):

Organovo Inc. is a "regenerative medicine company that applies proprietary technology to 'print' new organs." Their work is intended "to quickly and cheaply produce organs and body tissue for transplantation and testing, thereby reducing the need for animal test subjects and greatly speeding up the process by which treatments can be used for humans." (http://www.organovo.com/.)

(The concept of organ printing was originated by Vladimir Mironov of the Department of Cell Biology and Anatomy, Medical University of South Carolina, see Vladimir Mironov, Thomas Boland, Thomas Trusk, Gabor Forgacs, Roger R. Markwald, "Organ printing: computer-aided jet-based 3D tissue engineering," *Trends in Biotechnology*, 21 (4), 157-161, April 2003; http://cba.musc.edu/Faculty/MironovV.htm.)

Silverstone Solutions "delivers efficient, cost-effective technology solutions to the medical and healthcare industries. The flagship product, Matchmaker, is a clinical application that allows hospitals and transplant organizations to more quickly and accurately match patients in need of kidney transplants who have a qualified, incompatible donor with an alternate compatible donor. The software utilizes a series of proprietary algorithms to generate all potential paired donations and also optimizes the greatest number of possible transplants for an entire pool of paired-exchange candidates." (http://www.silverstonesolutions.com/.)

Halcyon Molecular develops ultra-high-throughput, low-cost DNA sequencing technology (genomics). "By making DNA sequencing significantly faster, cheaper, more accurate, and more comprehensive, Halcyon intends to significantly increase the scope and depth of genetic research, accelerating the development of biotechnologies that improve health and longevity." (http://www.halcyonmolecular.com/.)

My Bridge 4 Life is a "wellness network designed to help patients, caregivers, supporters and individuals create a personalized wellness plan." (http://www.mybridge4life.com/.)

[1192] Immortality Institute (ImmInst, founded in 2002 by Bruce Klein) has been a leading longevity community network. It has been a discussion forum on life-extension, with over 25,000 participants (as of July, 2014). More than 50,000 topics have been discussed, including the scientific aspects of life-extension, such as bioscience (theories of aging, genetics, stem cells, medicine and diseases), nutritional supplements, life-style, computer science (brain-computer interfaces, artificial intelligence and technological singularity), cryonics, nanotechnology, physics and space; as well as social and ideological implications of life-extension, discussing immortalist philosophy, politics and law, spirituality and religion, society and economics, environment and global risks.

The forum has also been a platform for communal projects (books, conferences, fund-raising, regional organizations). Continuous efforts were made to change the organization's name to reflect its major preoccupation with life-extension, rather than "immortality." The latter term, though attractive, was perceived as fairly meaningless and exceedingly provocative. Eventually, in December 2010, ImmInst was renamed "Longecity. Advocacy and Research for Unlimited Lifespans."

(www.imminst.org; http://www.longecity.org/forum/; http://www.imminst.org/cureaging/; http://www.longecity.com/communities.php.)

Various public (but not very numerous) advocacy groups for life-extension have been active, including:
The supporters of the Methuselah Foundation, led by Aubrey de Grey, "seeking, supporting and rewarding science that extends healthy lifespan," http://www.methuselahfoundation.org/, and fans of the Strategies for Engineered Negligible Senescence – SENS, http://www.sens.org/forum; The Coalition to Extend Life – an American political lobbying effort to increase funding for longevity research, led by Thomas Mooney, http://www.coalitiontoextendlife.org/; Maximum Life Foundation, led by David Kekich, http://www.maxlife.org/; The Campaign for Aging Research, http://www.healthyyears.org/; various Transhumanist networks, e.g. http://humanityplus.org/; http://www.transhumanism.org/campus/; and the Longevity Party, http://www.facebook.com/groups/longevity.party/, stemming into the International Longevity Alliance, http://longevityalliance.org/, https://www.facebook.com/groups/longevity.alliance/, and additional groups http://www.longevityforall.org/groups/.

Some additional relatively small groups have included:
Immortality International, founded in 2001 by Mitch Ronco, http://replay.web.archive.org/20090510021812/http://immortalityonline.org/dnn/default.aspx; Society for Universal Immortalism, http://www.universalimmortalism.org/; Society for Venturism, http://www.venturist.info/; National Rejuvenation Foundation (2001-2007) http://web.archive.org/web/20010419142005/http://www.nrfnews.com/; Timeship, http://www.timeship.org/; The Elysian Enterprise, http://www.elysianenterprise.com/; the extremely provocative Fuck Death Foundation, http://www.fuckdeath.org/; and a few others.

The support groups have also included:
The Calorie Restriction (CR) Society, which "supports people who practice calorie restriction for future longevity and current health," http://www.crsociety.org/; Foresight Nanotech Institute, with a dominant emphasis on life-extension through developing nanomedicine, http://www.foresight.org/resources/, http://www.nanomedicine.com/; The Lifeboat Foundation, including the "LifePreserver" program, specifically exploring new life-extension methods, http://lifeboat.com/ex/life.preserver; KurzweilAI.net, including "Biomed/Longevity" as one of the central categories (until 2008, this category was termed "Living Forever") http://www.kurzweilai.net/articles?t=51.

Projects publicizing life-extension have included:
The Longevity Meme/Fight Aging (related outreach projects, established in 2000 and 2002, respectively, by Reason, http://www.longevitymeme.org/, http://fightaging.org/); Extreme Longevity (founded in 2011 by Lyle Dennis, http://extremelongevity.net/); Immortal Life (founded in 2013 by Hank Pellissier, http://immortallife.info/).

Betterhumans (2001-2008), led by James Clement, was an online magazine on transhumanism, human enhancement and life extension, http://web.archive.org/web/20080531081344/http://www.betterhumans.com/. Since 2008 Betterhumans LLC has been publishing *Humanity Plus Magazine*, also with a strong emphasis on life-extension, http://www.hplusmagazine.com/.

Several scientific blogs have explored the topic of life-extension, such as Longevity Science by Leonid Gavrilov (University of Chicago) http://longevity-science.blogspot.com/; Senescence.info by Joao Pedro de Magalhaes (University of Liverpool) www.senescence.info; Biosingularity by Derya Unutmaz (NYU School of Medicine) http://biosingularity.wordpress.com/; and Ouroboros, "a community weblog for biologists of aging" http://ouroboros.wordpress.com/.

The establishment of Ouroboros in 2006 was a reaction to the closure of the research portal on aging and longevity – SAGE Knowledge Environment (SAGE KE, 2001-2006) which was set up by the American Association for the Advancement of Science (AAAS, the publishers of *Science* magazine) in collaboration with Stanford University Libraries. SAGE KE was established to provide "resources pertaining to aging-related research." Yet, in 2006 the site was closed due to the lack of funding and it now exists only as an archive, http://sageke.sciencemag.org/.

Another on-line repository of research information on aging and longevity has been SAGE Crossroads, started in 2003 at the initiative of the US Alliance for Aging Research (AAR) together with the American Association for the Advancement of Science. As of 2014, it has been still operational, though not particularly active – http://www.sagecrossroads.com/.

A more active and informative resource has been SAGEWEB developed by the University of Washington, http://sageweb.org/.

These online resources supplement some of the veteran journals concerned with aging and longevity, such as: *Journal of the American Geriatrics Society, Journal of Gerontology, The Gerontologist, Mechanisms of Aging and Development* and a couple of dozen others.

(Ouroboros, "Journals" http://ouroboros.wordpress.com/; ISI Web of Knowledge, http://science.thomsonreuters.com/mjl/.)

[1193] Prominent American intellectuals who have expressed sympathy with the striving toward radical life-extension include:
Ben Bova (science fiction writer, former president of the Science Fiction and Fantasy Writers of America), William Haseltine (CEO of Human Genome Sciences), Arthur Caplan (Chair of the UN Advisory Committee on Human Cloning), Philip Emeagwali (one of the fathers of the Internet), Christine Peterson (president of the Foresight Nanotech Institute and the coiner of the term "open source software"), Ray Kurzweil (the developer of the optical character recognition technology), Marvin Minsky (the cofounder of

the MIT Artificial Intelligence laboratory), Michio Kaku (the cofounder of the string field theory) and others.

(Ben Bova, *Immortality: How Science is Extending Your Life Span and Changing the World*, Harper Collins, NY, 1998; "Dr. William Haseltine on Regenerative Medicine, Aging and Human Immortality, Report Interview," *Life Extension Magazine*, July 2002, http://www.lef.org/magazine/mag2002/jul2002_report_haseltine_01.html; Arthur Caplan, "It's not immoral to want to be immortal," 4/25/2008, *MSNBC.com*, http://www.msnbc.msn.com/id/23562623/ns/health-aging/t/its-not-immoral-want-be-immortal/; Philip Emeagwali, "My Search for the Holy Grail of Immortality," delivered at the conference of the Black Data Processing Association, Augusta, Georgia, April 26, 2003 http://emeagwali.com/speeches/immortality/my-search-for-the-holy-grail-of-immortality.html; Christine Peterson, "Life Extension: Good News, Bad News, Surprising News," presented at the 7th Alcor Conference, April 13, 2007, http://www.acceleratingfuture.com/people-blog/category/people/christine-peterson/; Ray Kurzweil, *The Singularity Is Near: When Humans Transcend Biology*, Penguin Books, New York, 2005, Ch. 5, "GNR: Three Overlapping Revolutions," pp. 205-298; Marvin L. Minsky, "Will Robots Inherit the Earth?" *Scientific American*, October, 1994, revised version, http://web.media.mit.edu/~minsky/papers/sciam.inherit.html; Michio Kaku, *Visions: How Science will Revolutionize the Twenty-first Century and Beyond*, Oxford University Press, Oxford, 1999, Ch. 10 "To Live Forever?" pp. 200-219.)

[1194] Life-extensionism is a defining and unifying idea of the otherwise loosely organized and little coordinated transhumanist intellectual movement. Transhumanists have had varying (usually rather low) levels of public and scientific impact. Yet altogether they increase the public pressure in favor of technological enhancement and life-extension. Some transhumanist groups are listed below.

The Extropy Institute (1988-2005) was one of the first transhumanist groups. It advocated the use of technology to fight disorder (entropy), i.e. increase order ("extropy"), including such tasks as augmenting intelligence, improving social systems, optimizing physiology and extending healthy life.

The Extropy Institute was established by Dr. Max More (b. Max T. O'Connor) of the University of Southern California, together with Prof. Tom Bell. They were later joined by Prof. Hans Moravec, Dr. Ralph Merkle, Dr. Roy Walford, Fereydun M. Esfandiary (pseudonym FM-2030) and Prof. Bart Kosko. More joined in later: Dr. Robert A. Freitas, Jr., Dr. Aubrey de Grey, Ray Kurzweil, Dr. Christine Peterson, Dr. Michael Shapiro, Dr. Gregory Stock, Prof. Marvin Minsky, Dr. Michael D. West, Dr. Lee Silver and others. Natasha Vita-More served as the institute's last acting president.

In 2006, the Extropy Institute ceased operation, as it considered its mission "to bring great minds together to incubate ideas about emerging technologies, life extension and the future" to have been "essentially completed." Nonetheless, most of the Institute's major actors have retained the leading positions in the transhumanist/life-extensionist movement, and "extropianism" – the struggle against entropic forces of destruction and for increasing complexity and durability – remains a powerful concept within transhumanist/life-extensionist philosophy.

(http://www.extropy.org/history.htm; http://www.extropy.org/future.htm.)

The World Transhumanist Association (WTA), established in 1998 by Nick Bostrom and David Pearce, has advocated the ethical use of technology to expand human capabilities. In 2008, the organization changed its name to Humanity Plus (H+), to improve its public branding. According to their website, as of 2012, the organization included about 6,000 members, from more than 100 countries, but mostly from the US and EU (http://humanityplus.org/about/mission/).

In 2006, the "Campaign for Longer, Better Lives" became one of the three central campaigns of the WTA/H+ community, aiming to "promote a multinational research program to develop therapies to slow aging." The other 2 campaigns were the "Campaign for a Future Friendly Culture" to "encourage balanced and constructive portrayals of longevity, human enhancement and emerging technologies in popular culture" and the "Campaign for the Rights of the Person," including universal access to life-extending technologies.

(The current H+ website: http://humanityplus.org; the former (archived) website: www.transhumanism.org.)

The life-extensionist views of the founders of the World Transhumanist Association have been explicit. Nick Bostrom's life-extensionist manifesto is "The Fable of the Dragon-Tyrant," *Journal of Medical Ethics*, 31 (5), 273-277, 2005, reprinted at http://www.nickbostrom.com/fable/dragon.html. David Pearce's "Hedonistic Imperative" involves the development of "radical antiaging technologies" (http://www.abolitionist.com/; http://www.hedweb.com/; http://www.bltc.com/).

One of the leading democratic transhumanist/techno-progressive think tanks has been the Institute for Ethics and Emerging Technologies. It has been led by Dr. James J. Hughes of Trinity College, Connecticut, formerly the secretary/executive director of the World Transhumanist Association (2002-2009). The Institute's credo is that "technological progress can be a catalyst for positive human development so long as we ensure that technologies are safe and equitably distributed." One of the central programs is to promote "Longer, Better Lives," geared to "addressing objections to life extension", "coordinating and developing consultation with senior citizens groups and organizations of the disabled to help them challenge ageist and ableist attitudes that discourage the full utilization of health technology." (http://ieet.org/index.php/IEET/life; http://jetpress.org/; http://changesurfer.com/.)

Another prominent transhumanist organization has been the Lifeboat Foundation (started by the financier Eric Klien in 2002), dedicated to researching potential ways for safeguarding humanity from existential threats, both technological and ecological. Its advisory boards (as of July 2014) included over 2,000 scientists (mostly PhD's). One of the programs – the "LifePreserver" – specifically explores new life-extension methods, such as the work of Dr. Aubrey de Grey on strategies for engineered negligible senescence and the work of Dr. Robert A. Freitas Jr. on nanomedicine. As of July 2014, the foundation's Biotech/Medical Board listed over 350 scientists, many of whom were explicit life-extensionists, and its Life Extension Board included over 150 scientists, many of whom were also members of the Biotech/Medical Board (http://lifeboat.com/ex/main.)

In addition to transhumanist political and scholarly organizations, several transhumanist art projects have been created, depicting the beauty of technologically-enhanced, long-lived humans. (See, for example, Anders Sandberg's "Transhumanist Resources" http://www.aleph.se/Trans/Cultural/Art/index.html; Transhumanist Arts and Culture http://www.transhumanist.biz/.) The Transhumanist Arts movement has been led by Natasha Vita-More (born Nancie Clark, http://www.natasha.cc/).

Further lists of transhumanist organizations can be found at http://www.aleph.se/Trans/Org/org_page.html; http://www.extropy.org/resources.htm; http://humanityplus.org/get-involved/chapters-of-humanity/; http://www.transhumanism-russia.ru/content/view/347/121/. The vast majority of transhumanist organizations involve the explicit advocacy of radical life extension.

Science fiction fan communities represent yet another broad public group interested in radical life extension. Though, unlike Transhumanist organizations, specific advocacy for research is rarely directly implied in these groups, the idea that science fiction very rapidly becomes science fact is prevalent, including biomedical advances. (See for example, Technovelgy, 2003-2007, http://web.archive.org/web/20071024140911/www.technovelgy.com/.)

In general, futurist societies have expressed marked interest in life extension, for example, The World Future Society (http://www.wfs.org/; http://beta.wfs.org/search/node/longevity; http://www.wfs.org/futurist.htm).

[1195] Foresight Nanotech Institute, established in 1986 by the founder of nanotechnology, Kim Eric Drexler, and currently directed by Christine Peterson, has been a leading think tank dedicated to research on nanotechnology – the manipulation of matter on an atomic or molecular (nanometer) scale (moving ever further toward "homeopathic" scales).

The institute has promoted potential positive applications of nanotechnology, such as molecular manufacturing and nanomedicine, and investigated social implications and methods of protection against possible ecological and health risks posed by nanotechnology.

Life-extension is one of the institute's chief areas of study, insofar as nanotechnology is hoped to provide profound biomedical interventions: precise drug delivery and imaging; nano-robots or "nanobots" for DNA repair or other kinds of "molecular repair"; "artificial phagocytes" or "artificial immune/microbivore/scavenging cells" against poisons, tissue waste products, infectious agents and cancer cells; "artificial respirocytes" for improved oxygen delivery that could dramatically reduce mortality insofar as great many cases of death are caused by a lack of oxygen supply, etc.

(Foresight Nanotech Institute, http://www.foresight.org/resources/.)

Robert A. Freitas Jr. is one of the leading visionaries of nano-medicine, and the author of a series of books on the subject. (Robert A. Freitas, Jr, *Nanomedicine*, Landes Bioscience, Austin TX, 1999, 2003, http://www.nanomedicine.com/.)

The anticipated beauty of nano-medicine and the various types of healing "artificial cells" ("nanobots") can be envisioned at the Nanomedicine Art Gallery of the Foresight Institute. (http://www.foresight.org/Nanomedicine/Gallery/.)

The Center for Responsible Nanotechnology, directed by Chris Phoenix and Jamais Cascio, is another prominent think tank that aims to "raise awareness, expedite examination and creation of nanotechnology." Beside promoting the positive aspects of nanotechnology, it seeks ways to prevent potential existential threats that may be caused by it, such as self-replicating nano-robots potentially capable of reducing living matter into "gray goo."

The goal is to develop "global immunity" against potential nano-technological pathogens, starting at a political/regulatory level to prevent dangerous developments and, if this fails, proceeding to technical solutions, such as "anti-bodies" against the spread of nano-poisons or uncontrollably self-replicating nano-bots. (http://www.crnano.org/.)

Recent and prospective developments in nanotechnology have been reported, among many other sources, at Nanotechnology News Network (http://nanonewsnet.com/category/health/) or Nanowerk (http://www.nanowerk.com/phpscripts/n_news.php).

Scores of nanotechnology organizations currently working across the world are listed at http://www.dmoz.org/Science/Technology/Nanotechnology/Research_Groups_and_Centers/; http://en.wikipedia.org/wiki/List_of_nanotechnology_organizations.

From its inception, nanotechnology has been linked to cell and tissue repair. Thus, the original vision of the founder of nanotechnology, Eric Drexler (b. 1955), was truly far-reaching and was directly related to life extension.

In *Engines of Creation* (1986), the formative work on nanotechnology, Drexler wrote:

"Because molecular machines will be able to build molecules and cells from scratch, they will be able to repair even cells damaged to the point of complete inactivity."

This capacity for cell repair will eventually lead to the defeat of aging:

"Imagine someone who is now thirty years old. In another thirty years, biotechnology will have advanced greatly, yet that thirty year old will only be sixty. Statistical tables which assume no advances in medicine say that a thirty year old U.S. citizen can now expect to live almost fifty more years - that is, well into the 2030s. Fairly routine advances (of sorts demonstrated in animals) seem likely to add years, perhaps decades, to life by 2030. The mere beginnings of cell repair technology might extend life by several decades. In short, the medicine of 2010, 2020, and 2030 seems likely to extend our thirty-year-old's life into the 2040s and 2050s. By then, if not before, medical advances may permit rejuvenation. Thus, those under thirty (and perhaps substantially older) can look forward - at least tentatively - to medicine's overtaking their aging process and delivering them safely to an era of cell repair, vigor, and indefinite lifespan."

(K. Eric Drexler, *Engines of Creation: The Coming Era of Nanotechnology*, Anchor Books, New York, 1986, Ch. 7. Agents of Healing. Sct. A Disease Called "Aging"; Ch. 8. Long Life in an Open World. Sct. Progress in Life Extension, reprinted at http://e-drexler.com/d/06/00/EOC/EOC_Table_of_Contents.html.)

The possibility of extreme miniaturization of technology, "maneuvering things atom by atom," was raised even earlier by the Nobel Laureate in Physics of 1965, Richard Phillips Feynman (1918-1988), in his famous presentation "There's Plenty of Room at the Bottom" (presented at the annual meeting of the American Physical Society, December 29, 1959, http://www.zyvex.com/nanotech/feynman.html).

It is not surprising that also for Feynman, the strong believer in our ability to profoundly manipulate matter, overcoming death did not appear to be something entirely out of reach. As he wrote:

"there is nothing in biology yet found that indicates the inevitability of death. This suggests to me that it is not at all inevitable, and that it is only a matter of time before the biologists discover what it is that is causing us the trouble and that that terrible universal disease or temporariness of the human's body will be cured."

(Richard P. Feynman, "What Is and What Should be the Role of Scientific Culture in Modern Society," presented at the Galileo Symposium in Italy, in 1964, in Richard P. Feynman, *The Pleasure of Finding Things Out: The Best Short Works of Richard P. Feynman*, Perseus Books, NY, 1999, p. 100.)

[1196] Drexler's anticipation of 1986 that "the medicine of 2010, 2020, and 2030 seems likely to extend our thirty-year-old's life into the 2040s and 2050s" may have been overly optimistic. Yet, studies on the use of nanotechnology in biology and medicine, and in life-extension research in particular, have been progressing: There have been numerous reports on the experimental use of "nano-particles" – particularly gold, silver and carbon nano-particles – to destroy cancer cells.

See, for example, Geoffrey von Maltzahn, ..., Sangeeta N. Bhatia, "Computationally Guided Photothermal Tumor Therapy Using Long-Circulating Gold Nanorod Antennas," *Cancer Research*, 69(9), 3892-3900, 2009; O.V. Salata, "Applications of nanoparticles in biology and medicine," *Journal of Nanobiotechnology*, 2, 3, 2004; L. Zhang, ..., O.C. Farokhzad, "Nanoparticles in Medicine: Therapeutic Applications and Developments," *Clinical Pharmacology & Therapeutics*, 83 (5), 761-769, 2008; S. Karve, ..., A.Z. Wang, "Revival of the abandoned therapeutic wortmannin by nanoparticle drug delivery," *Proceedings of the National Academy of Sciences USA*, 109 (21), 8230-8235, 2012.

Further information on nanomedicine is available at the National Cancer Institute's Alliance for Nanotechnology in Cancer, http://nano.cancer.gov/; European Technology Platform on Nanomedicine, http://www.etp-nanomedicine.eu/public; American Society for Nanomedicine, http://www.amsocnanomed.org/; European Society for Nanomedicine, http://www.esnam.org/, and many other sources.

Nano-technology has also been directly implicated in life-extension research proper.

Thus in April 2012, in a study from the University of Paris, France, in collaboration with the University of Carthage, Tunisia, it was announced:

"Here we show that oral administration of C60 [a spherical nano-particle composed of 60 carbon atoms, a.k.a. "buckyball" or "buckminsterfullerene"] dissolved in olive oil (0.8 mg/ml) at reiterated doses (1.7 mg/kg of body weight) to rats not only does not entail chronic toxicity but it almost doubles their lifespan." (There have been, however, several concerns about the study's accuracy.)

The study mainly attributed the life-prolonging effect to the nano-particles' anti-oxidant activity:

"The effects of C60-olive oil solutions in an experimental model of CCl4 intoxication in rat strongly suggest that the effect on lifespan is mainly due to the attenuation of age-associated increases in oxidative stress."

Yet other therapeutic mechanisms were also suggested, strongly reminiscent of Auguste Lumière' "Anthrotherapy" or immunization, desensitization, microstimulation, colloidoclasia and detoxification by small carbon particles. According to the authors:

"Since 1993 countless studies showed that [60]fullerene (C60) and derivatives exhibit paramount potentialities in several fields of biology and medicine [1] mainly including specific DNA cleavage, imaging [2], UV and radioprotection [3], antiviral, antioxidant, and anti-amyloid activities [1,4e7], allergic response [8] and angiogenesis [9] inhibitions, immune stimulating and antitumour effects [10,11], enhancing effect on neurite outgrowth [12], gene delivery [13], and even hair-growing activity [14]."

(Tarek Baati, …, Fathi Moussa, "The prolongation of the lifespan of rats by repeated oral administration of [60] fullerene," *Biomaterials*, 33(19), 4936-4946, 2012.)

Moreover, there have even been announced the first operating medical nanorobots, mainly intended to assist in precise drug delivery and mainly against cancer, acting as prototypes of artificial immune cells, such as those described in the study from Harvard University, US, and Bar Ilan University, Israel.

(Shawn M. Douglas, Ido Bachelet, George M. Church, "A Logic-Gated Nanorobot for Targeted Transport of Molecular Payloads," *Science*, 335 (6070), 831-834, 17 February 2012.)

[1197] The cryonics movement, endeavoring to freeze people after death for future revival, is inherently related, but not synonymous with either immortalism or life-extensionism and represents one of the various branches of the wider movement.

Data on cryonics is available from the major American cryonics institutions: Alcor Life Extension Foundation (www.alcor.org) and Cryonics Institute (www.cryonics.org).

Despite the wide publicity, the actual number of people involved is rather miniscule. As of January 2012, about 260 people were cryopreserved (compared to about 230 as of January 2011) and about 2,000 signed up for cryonic suspension (http://www.cryonics.org/comparisons.html#Size).

Also contrary to a wide public belief, cryonics is not profitable.

The Alcor Life Extension Foundation, located in Scottsdale, Arizona, has been a world leader in cryonics services, research and development. According to one of the Alcor directors, the nano-technologist Ralph Merkle, future nanotechnological devices for cellular repair will be able to repair the damage caused by the cryogenic suspension. Thus, the feasibility of revival after the freezing is hoped to increase in time.

(Ralph C. Merkle, "The molecular repair of the brain," *Cryonics*, 15 (1&2), January and April, 1994, reprinted at http://www.merkle.com/cryo/techFeas.html; http://www.alcor.org/sciencefaq.htm.)

The Cryonics Institute, in Clinton Township, Michigan, was founded in 1967 by Robert Chester Wilson Ettinger (1918-2011). Ettinger is considered to be the "father" of cryonics, having authored *The Prospect of Immortality* (1964), the book that first introduced the idea of human cryopreservation to the broad public (reprinted at Cryonics Institute, http://www.cryonics.org/book1.html).

The Cryonics Institute offers information on cryonics and cryopreservation services at prices lower than Alcor, respectively $28K and $200K for whole body preservation, as of January 2012 (http://www.cryonics.org/comparisons.html#Prices).

A prominent cryonics-related organization has been *21st Century Medicine*, a company that set as its goal "Expanding the Boundaries of Preservation Science" with Gregory M. Fahy as the chief science officer and Brian G. Wowk as the senior physicist. The company specializes in bio-preservation technology required for developing biopharmaceuticals, assisted reproductive technology and transplantation. Both Gregory M. Fahy and Brian G. Wowk believe that techniques for whole body preservation will be eventually perfected as well.

As the company describes its products:

"21CM develops custom preservation solutions for a variety of applications across a wide range of temperatures. Our base solutions currently include hypothermic organ preservation solutions, ice blocking agents, ice crystal suppression polymers, heart resuscitation solutions and multiple cryoprotectant mixtures.

Our platform vitrification technology enables successful cryopreservation in systems as simple as individual cells and as complex as whole mammalian organs. It is likely that our advanced formulas can be used to produce better results after freezing and thawing as well."

(21CM, http://www.21cm.com/products.html; Gregory M. Fahy, "Precedents for the biological control of aging: Experimental postponement, prevention, and reversal of aging processes," in *The Future of Aging: Pathways to Human Life Extension* (Gregory M. Fahy, Michael D. West, L. Stephen Coles and Steven B. Harris, Eds.), Springer, NY, 2010, pp. 127-226; Brian Wowk, *Cryonics: Reaching for Tomorrow*, Alcor Life Extension Foundation, Scottsdale, Arizona, 1991.)

There have been cryonics groups and companies in countries other than the US:

Kriorus in Russia, http://www.kriorus.ru/en, http://old.kriorus.ru/english.html; EUCrio in Portugal, http://www.eucrio.eu/; and Cryonics UK in England, http://www.cryonics-uk.com/.

Yet the scope of their services has been incomparably smaller than even the limited scope of such services in the US.

Apart from Alcor and Cryonics Institute, cryonics services have been provided in the US by the American Cryonics Society, http://www.americancryonics.org/; Suspended Animation Inc., http://www.suspendedinc.com/; and Trans Time Inc., http://www.transtime.com/.

As of January 2012, about 110 subjects were frozen by Alcor and Cryonics Institute each, and about 40 subjects frozen by all the other companies altogether, mainly by ACS and Kriorus. (http://www.cryonics.org/comparisons.html#Size.)

One of the primary American advocacy groups for cryonics has been the Immortalist Society, established by Robert Ettinger in 1967. Since then, it has published the magazine *The Immortalist* dedicated to various aspects of life extension. The magazine was renamed *Long Life* in 1970 and *Long Life: Longevity Through Technology* in 2006 and is now published on line: http://www.cryonics.org/immortalist/.

Communication and information platforms for the cryonics community include:

CryoNet, http://www.cryonet.org/; the Cryonics Society, http://www.cryonicssociety.org/; Depressed Metabolism, http://www.depressedmetabolism.com/; the Society for Universal Immortalism, www.universalimmortalism.org/, and a few others.

Another American pro-cryonics group is the Life Extension Society. The society was formed in 1964 by Ev Cooper (1926-1983), the author of *Immortality: Physically, Scientifically, Now* (1962, under the pseudonym Nathan Duhring). Cooper's book predated and expressed ideas very similar to Ettinger's *Prospect of Immortality* (1964) regarding cryopreservation as an insurance policy for future life-extension. Yet Ettinger's book was much wider publicized.

(Ev Cooper, *Immortality: Physically, Scientifically, Now*, 1962, reprinted at http://www.depressedmetabolism.com/2008/05/27/ev-coopers-cryonics-classic-published-online/.)

The Life Extension Society has existed until the present.

(http://keithlynch.net/les/index.html; http://www.benbest.com/cryonics/history.html.)

Generally, the field of cryopreservation has been burgeoning:

Over 9000 articles on "Tissue Cryopreservation" were listed on PubMed as of July 2014.

Advances in the field have included successful revivals after freezing at cryogenic temperatures, ranging from complete mammalian organs (such as kidneys and livers) to some human tissues and complete human embryos.

In February 2012, it was announced that a complete fertile plant was revived from fruit tissue frozen for 30,000 years at permafrost temperatures.

Sources:

"Tissue Cryopreservation" on PubMed database, as of June 2012, http://www.ncbi.nlm.nih.gov/pubmed?term=tissue%20cryopreservation; "Scientists' Open Letter on

Cryonics," endorsed by over 60 scientists, and referencing ground-breaking papers on cryopreservation, published by Longecity (Immortality Institute), 2004-2006, http://www.imminst.org/cryonics_letter/.
Further see:
Gregory M. Fahy, ..., Eric Zendejas, "Cryopreservation of organs by vitrification: perspectives and recent advances," *Cryobiology*, 48(2), 157-178, 2004; Zohar Gavish, Menachem Ben-Haim, Amir Arav, "Cryopreservation of whole murine and porcine livers," *Rejuvenation Research*, 11(4), 765-772, 2008; Bianca Polchow, ..., Cora Lueders, "Cryopreservation of human vascular umbilical cord cells under good manufacturing practice conditions for future cell banks," *Journal of Translational Medicine*, 10(1), 98, 16 May, 2012; Leyre Herrero, Monica Martinez, Juan A Garcia-Velasco, "Current status of human oocyte and embryo cryopreservation," *Current Opinion in Obstetrics and Gynecology*, 23(4), 245-50, 2011; Svetlana Yashina, ..., David Gilichinsky, "Regeneration of whole fertile plants from 30,000-y-old fruit tissue buried in Siberian permafrost," *Proceedings of the National Academy of Sciences USA*, 109(10), 4008-4013, 2012.
Alternatives to cryopreservation have been sought, such as chemical preservation or "chemopreservation" (http://www.alcor.org/Library/html/chemopreservation.html; http://www.brainpreservation.org/).

[1198] The term "cyborg" or "cybernetic organism" – a symbiosis between a human and a computer or machine – was introduced in Manfred E. Clynes and Nathan S. Kline, "Cyborgs and space," *Astronautics*, 13, 26-27, 74-76, September 1960, http://www.scribd.com/doc/2962194/Cyborgs-and-Space-Clynes-Kline.
Since then, the topic of "cyborgs" has been widely discussed and the possibility of immortality of such organisms has been often termed "cybernetic immortality."
Recently, the idea of cyborgization has been given a great boost by the research of Kevin Warwick, professor of cybernetics at the University of Reading, UK. Warwick has been acclaimed as "the first cyborg," having computer sensors connected to his nervous system in 2002.
Warwick also expressed the belief that "humans will be able to evolve by harnessing the superintelligence and extra abilities offered by the machines of the future, by joining with them." And he "had liked the ideas about uploading the human brain into silicon."
(Kevin Warwick, *I, Cyborg*, University of Illinois Press, Urbana IL, 2004, pp. 4, 62; http://www.kevinwarwick.com/.)
Many companies and research centers now actively explore brain scanning, brain modeling and brain-computer interfaces which are considered to be important tools in the project of cyborgization.
(Some resources on brain scanning/imaging include: http://www.brainmapping.org/; http://www.med.harvard.edu/AANLIB/home.html; http://www.asnweb.org/i4a/pages/index.cfm?pageid=1.
Some resources on brain modeling/reverse-engineering: http://domino.watson.ibm.com/comm/research_projects.nsf/pages/bmc_modeling.index.html; http://bluebrain.epfl.ch/.
And some resources on brain-computer interfaces/mind control: http://openeeg.sourceforge.net/doc/; http://www.phypa.org/; http://www.emotiv.com/; http://www.gtec.at/.)
Many books have been written referring to cybernetic organisms and future cybernetic immortality, including:
Marvin Minsky, *The Society of Mind*, Simon & Schuster, NY, 1988; Hans Moravec, *Mind Children: The Future of Robot and Human Intelligence*, Harvard University Press, Cambridge MA, 1990; Eliezer Yudkowsky, *Creating Friendly AI*, Singularity Institute, 2001; James Hughes, *Citizen Cyborg: Why Democratic Societies Must Respond to the Redesigned Human of the Future*, Westview Press, Cambridge MA, 2004; Kevin Warwick, *I, Cyborg*, University of Illinois Press, Urbana IL, 2004; Ray Kurzweil, *The Age of Spiritual Machines*, Viking, NY, 1999; Ray Kurzweil, *The Singularity Is Near: When Humans Transcend Biology*, Penguin Books, New York, 2005, and others.

[1199] Notably, apart from straightforward "cyborgization" (hoping to achieve robust human-machine "hybrids"), the very increased interaction with technology was surmised to have a more indirect effect for extending longevity, namely, via enhancing brain resilience and plasticity and increasing intelligence (the "Flynn effect"). (Marios Kyriazis, "Information-sharing, adaptive epigenetics and human longevity," Quantitative Biology, Arxiv, July 20, 2014, http://arxiv.org/abs/1407.6030; Richard D. Fuerle, "A possible explanation for the Flynn effect," Majority Rights, January 11, 2008.)

[1200] The idea of future human-computer symbiosis into an extremely long-lived life form (a.k.a. "cybernetic immortality") is very prominent among the proponents of the Singularitarian intellectual movement, even though it is also prevalent in the Transhumanist/Humanity Plus and Extropian movements. These movements are often indistinguishable from each other in this regard and for the most part involve the same supporters, advocating for the radical enhancement of human nature by technology. The leader of the Singularitarian movement has been the American computer scientist, industrialist and inventor Ray Kurzweil (b. 1948), the winner of the US National Medal of Technology of 1999 and the recipient of 19 honorary doctorates.

In the Singularitarian view, thanks to the exponentially accelerating technological development, particularly the acceleration of information technology and artificial intelligence (following Gordon E. Moore's law for the accelerated development of computing technology, posited in 1965), there will be (soon) reached a point in time, the point of "Singularity," where artificial intelligence will surpass human intelligence and technological capabilities will expand beyond anything we can presently imagine.

At that point, there will be a fundamental change of the nature of the human body and mind, and of the nature of the society. As at the point of physical singularity the normal laws of physics may radically change, so at the point of technological singularity human capabilities may explode beyond anything that is currently held normative.

The accelerated development of information technology is believed to be the driving force propelling the acceleration in other converging technological areas, including biology and medicine.

In the singularitarian credo, the human mind will eventually merge with strong Artificial Intelligence, and the human body will merge with versatile and powerful artificial extensions. This merging will, in effect, enable the creation of virtually immortal beings having next to unlimited knowledge and control over nature.

Kurzweil thus defined the Singularity:

"The Singularity will represent the culmination of the merger of our biological thinking and existence with our technology, resulting in a world that is still human but transcends our biological roots."

"The Singularity" Kurzweil maintained "will allow us to transcend these limitations of our biological bodies and brains. We will gain power over our fates. Our mortality will be in our own hands."

(Ray Kurzweil, *The Singularity Is Near: When Humans Transcend Biology*, Penguin Books, New York, 2005, p. 9, a new edition appeared in 2007.)

Kurzweil has led several public programs advertising "singularity" and "artificial intelligence." For example, Kurzweil Artificial Intelligence Net is a massive repository of information on all aspects of singularity, including news, analysis and essays on emerging trends in many technologies: artificial intelligence, nanotechnology, genetic engineering and life extension (www.kurzweilAI.net).

In 2008, Kurzweil set up the Singularity University at NASA Ames Research Park, Silicon Valley, California, with the help of generous donations from the aerospace entrepreneur Peter Diamandis, Google's co-founder Larry Page, and others (http://singularityu.org/).

Several web-sites have been dedicated to promoting Kurzweil's vision (http://www.singularity.com/; http://www.fantastic-voyage.net/; http://www.rayandterry.com/index.asp).

A list of about a dozen companies founded by Kurzweil can be found at the site of Kurzweil Technologies Inc. (http://www.kurzweiltech.com/aboutray.html.)

The media coverage of Kurzweil's ideas has been extensive, e.g. Lev Grossman, "2045: The Year Man Becomes Immortal," *Time*, February 10, 2011, http://www.time.com/time/health/article/0,8599,2048138,00.html; Ashlee Vance, "Merely Human? That's So Yesterday," *New York Times*, June 12, 2010, http://www.nytimes.com/2010/06/13/business/13sing.html, among many other appearances.

Several popular web-logs have been dedicated to Technological Singularity and related subjects: Singularity Hub, http://singularityhub.com/; Blogging the Singularity, http://www.bloggingthesingularity.com/; Singularity.org, http://singularity.org; Singularity Symposium, http://www.singularitysymposium.com/; Singularity Weblog, http://www.singularityweblog.com and others.

A list of prominent supporters of the notion of Singularity can be found at "The Accelerating Future People Database" http://www.acceleratingfuture.com/people/ or "Top 10 singularitarians of all times" http://singularityblog.singularitysymposium.com/top-10-singularitarians/.

It is important to note that the area of discussions of the singularitarians is not restricted to "cyborgism" and "artificial intelligence," but also involves ethics, sociology and a wide spectrum of biological interventions for life-extension that would allow us to survive until the advent of cybernetic immortality.

(E.g. Ray Kurzweil and Terry Grossman, *Fantastic Voyage. Live Long Enough to Live Forever*, Plume, NY, 2005; Ray Kurzweil and Terry Grossman, *Transcend. Nine Steps to Living Well Forever*, Rodale Books, NY, 2009.)

The topic of biological life-extension features prominently in Kurzweil's works and in virtually all singularitarian/transhumanist organizations, the vast majority of whose members are resolute life-extensionists. Yet, insofar as the conservation of the "biological body" is only seen as an intermediate stage or "bridge" on the path toward cybernetic immortality, singularitarianism represents a distinct branch of life-extensionism, quite novel and unorthodox by "traditional" life-extensionist standards.

[1201] The idea of "technological singularity" was introduced in 1993 by Vernor Steffen Vinge, Professor at the Department of Mathematical Sciences, San Diego State University, California (b. 1944). Yet, the idea is clearly a part of a more general meliorist tradition positing an inexorable and rapid progressive development of humanity as a whole.

In Vinge, the implications of the concept of "singularity" are quite far-reaching:

"Within thirty years, we will have the technological means to create superhuman intelligence. Shortly after, the human era will be ended. ... I think it's fair to call this event a singularity ("the Singularity" for the purposes of this paper). It is a point where our old models must be discarded and a new reality rules."

Radical life extension is an integral part of this vision:

"It could be a golden age that also involved progress (overleaping Stent's barrier). Immortality (or at least a lifetime as long as we can make the universe survive) would be achievable."

(Vernor Vinge, "Technological Singularity," *Whole Earth Review*, 81, 89-95, 1993, reprinted at http://mindstalk.net/vinge/vinge-sing.html. The original version of this article was presented at the VISION-21 Symposium sponsored by NASA Lewis Research Center and the Ohio Aerospace Institute, March 30-31, 1993.)

According to Vinge's own admission, cognate ideas appeared even earlier in John von Neumann: "Von Neumann even uses the term singularity, though it appears he is thinking of normal progress, not the creation of superhuman intellect" and in Gunther S. Stent, *The Coming of the Golden Age: A View of the End of Progress*, The Natural History Press, NY, 1969.

The idea of the Singularity can be traced further back to the work of the French Jesuit priest, paleontologist and geologist, Pierre Teilhard de Chardin (1881-1955).

De Chardin introduced the concept of the "Omega Point," the culmination of the complexity of the "Noosphere" – the sphere of the reigning mind.

Based on Christian eschatology, the ultimate resistance to death is a central feature of the "Omega point":

"We have finally to banish the specter of Death from our horizon. And this we are enabled to do by the idea (a corollary, as we have seen, of the mechanism of planetization) that ahead of, or rather in the heart of, a universe prolonged along its axis of complexity, there exists a divine center of convergence. That nothing may be prejudged, and in order to stress its synthesizing and personalizing function, let us call it the *point Omega*. Let us suppose that from this universal center, this Omega point, there constantly emanate radiations hitherto only perceptible to those persons whom we call 'mystics'"

(Pierre Teilhard de Chardin, *The Future of Mankind*, translated by Norman Denny, Harper & Row, New York and Evanston, 1959, Chapter 6: Life and the Planet, first presented in March 1945, reprinted at Religion on Line, http://www.religion-online.org/showchapter.asp?title=2287&C=2167.)

The "synthesizing and personalizing" Omega point encourages human development and constitutes its pinnacle.

De Chardin rebelled against ultimate destruction of humanity, but rather envisioned "not an ending of the ultra-human but its accession to some sort of *trans-humanity* at the ultimate heart of things."

(Emphasis added, de Chardin, *The Future of Mankind*, 1959, Chapter 21: From the Pre-Human to the Ultra-Human: The Phases of a Living Planet, originally "Du Pre-Humain a l'Ultra Humain," written in Paris in 1950 and first published in *Almanach des Sciences* in 1951.)

This was apparently one of the earliest uses of the term "transhumanity" – the transcendence of human capabilities. An even earlier use was made by de Chardin himself, in 1949, when he spoke about "man ... 'trans-humanizing' himself by developing his potentialities to the fullest extent (de Chardin, *The Future of Mankind*, 1959, Chapter 16: The Essence of the Democratic Idea: A Biological Approach, written in 1949).

Thus de Chardin's writings predated the use of the term "Transhumanism" by Julian Huxley (Julian Huxley, "Transhumanism," in *New Bottles for New Wine*, Chatto & Windus, London, 1957, pp. 13-17, http://www.transhumanism.org/index.php/WTA/more/huxley); Fereydun M. Esfandiary (FM-2030, *Are You a Transhuman?* Warner Books, New York, 1989); and Max More (b. Max T. O'Connor, "Transhumanism: Toward a Futurist Philosophy," 1990, 1996, http://www.maxmore.com/transhum.htm).

De Chardin was apparently also one of the first to mention the term "singularity" in a context somewhat compatible with the present use. Thus, in *Les singularités de l'Espèce Humaine* (The Singularities of the Human Species, Masson, Paris, 1955, first published in 1954), he spoke about the "original singularity" (la singularité originelle), the "present singularity" (la singularité présente) and the "terminal singularity" of the human species (la singularité terminale de l'espèce humaine).

De Chardin considered the possibility of humanity's indefinite spatial and temporal expansion:

"what duration may we not look for in the case of Man, that favored race which, by its intelligence, has succeeded in removing all danger of serious competition and *even in attacking the causes of senescence at the root*" (emphasis added).

Yet, he ultimately favored not the outward temporal or spatial expansion, but rather the inward concentration of thought, that would accumulate and augment our vitality:

"For if by its structure Mankind does not dissipate itself but concentrates upon itself; in other words, if, alone among all the living forms known to us, our zoological phylum is laboriously moving towards a *critical point of speciation*, then are not all hopes permitted to us in the matter of survival and irreversibility [immortality]?"

(Chapter 22: The End of the Species, emphasis in the original, in Pierre Teilhard de Chardin, *The Future of Mankind*, translated by Norman Denny, Harper & Row, New York and Evanston, 1959, reprinted at Religion on Line, http://www.religion-online.org/showbook.asp?title=2287.)

The influence of Teilhard de Chardin' idealistic philosophy on Transhumanist and Singularitarian thought and on the development of the concept of the "Noosphere" has been generally acknowledged.
(See Nick Bostrom, "The Transhumanist FAQ. What is a transhumanism and the transhuman?" 1999-2003, http://www.extropy.org/faq.htm; Eric Steinhart, "Teilhard de Chardin and Transhumanism," *Journal of Evolution and Technology*, 20 (1), 1-22, 2008, http://jetpress.org/v20/steinhart.htm.)
Notably, about the same time as de Chardin's work, the idea of the "Noosphere" constantly progressing toward an ever greater complexity and power was also developed by the Russian thinker Vladimir Ivanovich Vernadsky (1863-1945).
Yet the author of the concept of the "Noosphere" is uncertain. Its origin is sometimes traced back to around 1922, and attributed either to Teilhard de Chardin or sometimes to his friend Édouard Le Roy, both of whom were at the time attending Vernadsky's lectures at the Sorbonne University in Paris.
(See Georgy S. Levit, Lennart Olsson, ""Evolution on Rails": Mechanisms and Levels of Orthogenesis," *Annals for the History and Philosophy of Biology*, 11, 97-136, 2006, p. 125.)
Vernadsky himself gave credit to the notions advanced around 1856-1859 about the ever increasing "cephalization" (proposed by the American geologist James Dwight Dana, 1813-1895) and "psychozoic era" (the term suggested by the American geologist Joseph LeConte, 1823-1901) as precursors of the idea of the "Noosphere."
(V.I. Vernadsky, "Neskolko Slov o Noosphere" (A few words about the Noosphere), *Uspekhi Sovremennoy Biologii* (Successes of Modern Biology), 18(2), 113-120, 1944, reprinted at http://www.trypillya.kiev.ua/vernadskiy/noosf.htm.)
In Vernadsky too, radical life extension was a central element of the progress. Thus, in *The Scientific Thought as a Planetary Phenomenon* (1938), Vernadsky maintained that one of the humanity's crucial "tasks" is the "prolongation of life, and diminishment of diseases for all humanity."
(Vladimir Ivanovich Vernadsky, *Nauchnaya Mysl kak Planetnoe Yavlenie* (Scientific thought as a planetary phenomenon), Nauka, Moscow, 1991, Ch. 2.31 (written c. 1937-1938), reprinted in *Biblioteka Maxima Moshkova Lib.Ru*, http://lib.ru/FILOSOF/WERNADSKIJ/mysl.txt.)
Thus, in all these proponents of progressive human development – Vernadsky, Teilhard de Chardin, Huxley, Esfandiary, More, Bostrom, Vinge, and Kurzweil – the next "singular" leap of human evolution will entail radical life extension.
[1202] The early precursors of the idea of "Singularity," Vernadsky and Teilhard de Chardin, did not seem to specifically refer to computers or information technology.
(Even though de Chardin did speak in general terms of "a higher state of consciousness diffused through the ultra-technified, ultra-socialized, ultra-cerebralized layers of the human mass." And Vernadsky did generally relate "the complete settlement of the biosphere by humans" with "speed of communications, successes in transportation technology, the possibility of immediate transmission of thought and its simultaneous discussion everywhere on the planet."
See Pierre Teilhard de Chardin, *The Future of Mankind*, 1959 (1950), Chapter 19: On The Probable Coming of an 'Ultra-Humanity' http://www.religion-online.org/showchapter.asp?title=2287&C=2180; Vladimir Ivanovich Vernadsky, *Nauchnaya Mysl kak Planetnoe Yavlenie* (Scientific thought as a planetary phenomenon), Nauka, Moscow, 1991, Ch. 2.14 (written c. 1937-1938), reprinted in *Biblioteka Maxima Moshkova Lib.Ru*, http://lib.ru/FILOSOF/WERNADSKIJ/mysl.txt.)
In contrast, in Vinge, Kurzweil and other singularitarians, the accelerated development of information technology and of artificial intelligence is absolutely central for human progress generally and for life-extension in particular.
The singularitarians believe that various types of technologies do progress at an accelerating rate: nanotechnology, biotechnology, cognitive technology and information technology. Moreover, the technologies converge, with each kind of technology directly affecting and accelerating the development of

all the other technologies. Yet, information technology and artificial intelligence are valorized in the Singularitarian vision, as these are believed to be vital for the development of all areas of human knowledge. The singularitarians (such as Kurzweil) will have us believe that human-level or above-human-level artificial intelligence and man-machine synergy are inevitable, given the current trend of accelerating development of information technology. It is this technology, the singularitarians anticipate, that will help us find effective life-extending means through biological data mining, will provide a comprehensive register of our personal biological traits for personalized therapy, for a precise "fine-tuning" or balancing of the deficiencies of our metabolism to prolong our existence, and will direct the repair of our body.

In more radical visions, human personality will indefinitely survive as a "computer-uploaded" information pattern (the "mind file").

Kurzweil asserts:

"If we are diligent enough in maintaining our mind file, making frequent backups, and porting to current formats and mediums, a form of immortality can be attained, at least for software-based humans."

(Kurzweil, *The Singularity Is Near*, 2005, p. 325.)

If desired, the information patterns constituting human personality can be transferred into a forever renewable or replaceable cloned or robotic body, or they may stay forever in the hard disk – the nature of the material substrate is of less significance than the "information patterns" it contains.

Kurzweil writes:

"We will continue to have human bodies, but they will become morphable projections of our intelligence. In other words, once we have incorporated MNT [molecular nano-technology] fabrication into ourselves, we will be able to create and re-create different bodies at will" (p. 324).

Accelerating technologies, infused by information technology, will also enable the indefinite survival of the society at large. They will provide for unlimited energy and material resources, enable earth exploration and space expansion, give a better education by computerized teaching systems and expand service by robotic labor. In short, they will carry humanity to the next stage of individual and social evolution, with unlimited capabilities and wealth.

In some stage, at the point of "Singularity," Kurzweil believes the machines will succeed us entirely: in the best scenario our memories will be a component of machine intelligence (probably a very small component), and in the worst the machines will entirely supersede the human race.

Such a level of change may be too great to accept for a "traditional life-extensionist" wanting to be around in the same (or very similar) body and environment (that is to say, it may induce an incapacitating "future shock").

Still, even the far-reaching singularitarian visions may be seen as straightforward continuations of the existing social and ideological backgrounds.

If de Chardin's path to immortality proceeded from what he knew well, namely Christian mysticism; the path to immortality envisioned by the singularitarians is based on what *they* know well, namely, computer technology, particularly the research and development, business and advertising aspects of private high tech companies, expecting a perpetual market "boom" without an eventual "bust."

In other words, despite the professed belief that at the *future* point of singularity, the social and technological development of the human species will become "unpredictable," the Singularitarians, as a rule, seem to overemphasize the predictability, desirability and even inexorable, almost deterministic continuity of *current* technological and social trends.

[1203] One of the leaders in the research of potential implications of emerging strong artificial intelligence has been the Singularity Institute for Artificial Intelligence. The institute's declared mission is "to ensure the development of friendly Artificial Intelligence, for the benefit of all mankind," "to prevent unsafe Artificial Intelligence from causing harm" and "to encourage rational thought about our future as a species." The institute conducts a variety of research programs for the development of artificial intelligence and holds special summits on singularity. It was founded in 2000 by Eliezer Yudkowsky, who remains one of its leading researchers. (www.singinst.org; http://www.singularitysummit.com/; http://yudkowsky.net/; http://wiki.lesswrong.com/wiki/Rationality_materials.)

A number of programs for developing artificial intelligence (AI) has been initiated by Dr. Ben Goertzel: Artificial General Intelligence (AGI) Research Institute, attempting to "foster the creation of powerful and ethically positive Artificial Intelligence" http://www.agiri.org/wiki/Main_Page; OpenCog – a software project aiming to build an open source thinking machine, http://opencog.org/; and companies, such as Novamente LLC, dedicated to developing "general" ["human level"/"friendly"] artificial intelligence, as well as more practical AI applications, http://novamente.net/; Biomind LLC, dedicated to using AI as a research instrument for life extension, that is, as a method for extracting clinical and biological knowledge from high-volume genetic data and applying advanced AI for systems biology modeling, http://www.biomind.com/. As of 2011, Goertzel also served as the chairman of Humanity Plus, formerly the World Transhumanist Association (http://humanityplus.org/about/board/; http://www.goertzel.org/). Other projects on artificial intelligence have included Adaptive Artificial Intelligence, led by Peter Voss, http://www.adaptiveai.com/; Acceleration Studies Foundation Metaverse Roadmap, among other topics exploring the behavior of "virtual humans" http://metaverseroadmap.org/, http://accelerating.org/; Association for the Advancement of Artificial Intelligence, http://www.aaai.org/home.html, and others.

Several breakthroughs in developing artificial intelligence have been recently announced:

In 2011, the supercomputer "Watson" developed by IBM defeated the human world champion in the trivia game "Jeopardy!" (http://www.research.ibm.com/deepqa/faq.shtml), following the victory of the IBM supercomputer "Deep Blue" over the human world champion in chess in 1997 (http://www-03.ibm.com/ibm/history/exhibits/vintage/vintage_4506VV1001.html).

Incidentally, as of 2013-2014, the main use and the main line of IBM Watson's development have been in medical diagnosis, which has been hoped to far surpass the capabilities of a human physician. So far it has been mainly employed for the diagnosis of cancer, yet its further application for an extensive study of aging and longevity appears likely. (http://www.wired.co.uk/news/archive/2013-02/11/ibm-watson-medical-doctor; http://www.businessinsider.com/ibms-watson-may-soon-be-the-best-doctor-in-the-world-2014-4.)

Through 2014, ever more powerful types of artificial intelligence have been developed by other large Information Technology companies like Microsoft and Google.
(http://www.zdnet.com/advancing-artificial-intelligence-microsoft-deploys-corgis-to-beat-google-on-imaging-7000031601/;
http://www.cnet.com/news/google-scientists-find-evidence-of-machine-learning/.)

In July 2013, it was estimated that some AI systems had the level of intelligence comparable to that of a 4 year old child (http://phys.org/news/2013-07-smart-year-old.html#nwlt).

Also as of 2011-2014, there were several announcements that the Turing test (proposed in 1950 by the British computer scientist Alan Turing to test whether computers/chat-robots can fool humans to think that they are human too during a conversation) was either partly passed or on the verge of being passed. (http://www.newscientist.com/article/dn20865-software-tricks-people-into-thinking-it-is-human.html; http://www.wired.com/wiredscience/2012/04/turing-test-revisited/;
http://www.smh.com.au/national/education/has-the-turing-test-really-been-passed-20140710-zt2h3.html.)

And it was also announced that some preliminary forms of Artificial General Intelligence have been created (http://www.hutter1.net/; http://wp.goertzel.org/?page_id=325; http://agi-conf.org/2014/).

[1204] Some developments in robotics may have direct implications for life-extension, such as robotic surgery or computerized diagnosis, including micro-machines for diagnosis and repair that are capable of traveling within the human body (in the gastro-intestinal tract or the blood stream).

Moreover, vital artificial organs, such as the heart, created with computer assistance, could provide spare parts for aging and failing human organs.

Other developments, such as robotic limbs, brain-computer interfaces and exogenous skeletons, may be perceived as usable not so much for "life-extension" as for "life-enhancement" – the improvement of mental and physical capabilities and the quality of life, especially for disabled and disadvantaged people.

Yet, such applications too may aid human survival, hence it is extremely difficult to distinguish between means for "life extension," "life enhancement" and "disease treatment."

Eventually, it has been sometimes envisioned that *all* human organs or the *entire* human body could be maintained, replaced or backed up by durable robotic organs.

The borders between "science fiction" and "science fact" and between "long term" and "short term" developments become blurred, as can be appreciated from several recent headlines in the development of medical robotics and artificial organs:

The robotic arm transplants appeared in the late 1990s-early 2000s, e.g. the Edinburgh Modular Arm System (UK, 1998) and the Rehabilitation Institute of Chicago's "Bionic Arm" (US, 2001). In 2007, robotic arms became commercially available (I-Limb, UK, $18,000).

In 2009, robotic arms were already reported to provide a sense of touch to the patient (SmartHand 2009, developed by scientists from Ireland, Italy, Iceland, Denmark and Israel). On May 16, 2012, it was reported in *Nature* that a stroke-paralyzed woman was able to operate a robotic arm solely by an exercise of thought using a brain-computer interface (the BrainGate).

As of May 2014 the prices for commercial robotic prosthetic arms were said to start from $6000, yet could be produced by individuals ("do it yourself") for as cheap as $200 using 3-dimensional printing.

(http://news.bbc.co.uk/2/hi/health/154545.stm; http://robotnews.wordpress.com/2005/10/09/bionicle-man-1998-2/; http://www.canada.com/theprovince/news/story.html?id=ddd3c484-b2d1-42ba-9aad-6be933d964cd; http://www.ric.org/research/accomplishments/Bionic/; http://www.elmat.lth.se/~smarthand/; http://www.ynetnews.com/articles/0,7340,L-3803524,00.html; http://www.nature.com/news/mind-controlled-robot-arms-show-promise-1.10652; http://www.gizmag.com/go/4282/; http://www.iflscience.com/technology/students-use-3-d-printer-produce-prosthetic-arm-200.)

In August 2012, the South African double-amputee sprinter Oscar Pistorius (b. 1986), using high-performance carbon-fiber prosthetic legs, competed in the World Olympics, in London. Earlier, he was disqualified from the World Olympics 2008 in Beijing for "unfair advantage.".

(http://www.engadget.com/2008/01/17/prosthetic-limbed-runner-disqualified-from-olympics/; http://www.huffingtonpost.com/rami-hashish/oscar-pistorius-prosthetic-legs-advantage_b_1766417.html; http://www.oscarpistorius.com/; http://www.telegraph.co.uk/sport/olympics/athletics/9454624/Oscar-Pistorius-knocked-out-of-London-2012-Olympics-but-his-achievements-will-resound-for-years-to-come.html.)

Robotic surgery was introduced around the late 1980s (PUMA-560, Probot, Robodoc, etc.). In 2000, the Da Vinci Surgical System became the world's first robot approved by the US Food and Drug Administration for general laparoscopic operations, and has since been used in tens of thousands of operations. As of 2009, the price of the Da Vinci Surgical System (Da Vinci Si) was about $1.75 Million.

In 2010, a fully robotic procedure (prostatectomy) took place at McGill University, Montreal, Canada, using Da Vinci and the robotic anesthetizing system called "McSleepy."

In 2014, the price for an upgraded Da Vinci system (Da Vinci Xi) ranged from $1.85 million to $2.3 million.

(http://www.roboticoncology.com/history/; http://www.engadget.com/2010/10/21/first-all-robot-surgery-performed-at-mcgill-university/; http://muhc.ca/newsroom/news/mcsleepy-meets-davinci; http://www.advisory.com/daily-briefing/2014/04/02/fda-oks-first-major-da-vinci-upgrades-in-five-years).

Micro-robots capable of moving inside the patient' gastrointestinal tract (such as PillCam) and even in the bloodstream have been developed. The current applications have been mainly diagnostic, yet possibilities for surgical and other therapeutic interventions using such devices have also been raised. (http://www.givenimaging.com/en-us/Pages/GivenWelcomePage.aspx; http://www.haaretz.com/news/israeli-scientists-unveil-mini-robot-that-can-travel-through-bloodstream-1.224111; http://www.kurzweilai.net/tiny-robots-allow-for-minimally-invasive-heart-surgery; http://www.geek.com/science/cell-sized-micro-robot-will-carry-drugs-in-your-bloodstream-1570919/.)

Mechanical power suits (a.k.a. "powered armor" or "powered exoskeletons" such as those developed by Argo/Rewalk, Cyberdyne, Ekso, RB3D, and others) have allowed paralyzed persons to walk and normal persons (soldiers) to carry up to and above 100 kilograms load, without fatigue. As of 2011, the exoskeleton HAL-5 from Cyberdyne, Japan, was said to cost between $14,000 and $19,000. In June 2014, the FDA approved the exoskeleton "ReWalk" (from Argo Medical Technologies, Israel) for use by the paralyzed (priced about $70,000).

(http://www.exoskeleton-suit.com/; http://www.upgradeyourbody.com/biotech-directory/bionics/exoskeletons/; http://news.cnet.com/8301-27083_3-20043544-247.html; http://robohub.org/3-exoskeleton-companies-go-public-2/.)

After the first transplantation of the entire artificial heart in 1982 (Jarvik-7, designed by Robert Koffler Jarvik of the University of Utah), new improved models have been developed. For example, by May 2014, there were made 1300 implants of the Total Artificial Heart from SynCardia, Tucson, Arizona. As of 2010, the cost of The Syncardia Total Artificial Heart was about $125,000 and about $18,000 a year to maintain. As of 2014, the prices seem not to have fallen.

(http://www.syncardia.com/Overview.html; http://www.cbsnews.com/2100-18563_162-6507572.html; http://www.syncardia.com/2014-press-releases/1300th-implant-of-the-syncardia-total-artificial-heart-was-performed-in-april-2014/itemid-1658.html.)

As of 2014, some of the artificial/bionic/robotic organs under development – with varying degrees of success – have included:

brain pace-makers/stimulators, artificial neurons, electronic hippocampus brain segment, cardia and pylorus valves, corpora cavernosa (penile implants), artificial ovaries, uterus, bones, joints, teeth (using high-performance materials), skin (including the addition of epidermal sensory patches), blood (mainly experimenting with perfluorocarbon-based oxygen carriers), heart (including continuous flow/"no-pulse" models), lung, kidney, liver, pancreas, bladder, intestine, trachea, robotic arms, hands, fingers, legs, feet and knees, eye (retinal implants), eardrums (cochlear implants), inner ear (vestibular prosthesis), tongue (taste sensor), nose (olfactory sensors), and more.

(http://www.popsci.com/taxonomy/term/43414/all; http://www.asaio.com/; http://en.wikipedia.org/wiki/Artificial_organ.)

The Russian consortium "Russia2045" has explicitly pronounced the desire to create an entire artificial human body (or "avatar") with the stated purpose to achieve immortality.

(http://2045.com/; http://2045.com/tech/; http://2045.com/experts/.)

[1205] A number of organizations have pursued the goal of cybernetic immortality, even though not necessarily identifying themselves as "singularitarians." Of course all these organizations are strongly allied.

The Terasem Movement (since 2002 presided over by Martine Rothblatt) has concentrated on the "uploading" of human personalities ("mind files") for indefinite storage and transfer to a different medium. It was established "for the purpose of educating the public on the practicality and necessity of greatly extending human life, consistent with diversity and unity, via geoethical nanotechnology and personal cyberconsciousness, concentrating in particular on facilitating revivals from biostasis." The Terasem movement aspires to the indefinite preservation of human personality by such means as entire brain scans and conservation of personal information and memorabilia (http://www.terasemcentral.org/about.html).

Cybernetic Immortality has also explored "the art of transferring one's consciousness to a computer like substrate, thus potentially defeating death" (http://cyberneticimmortality.com/).

And so has done the Principia Cybernetica Web established by Cliff Joslyn, Valentin Turchin and Francis Heylighen, who "raise the banner of cybernetic immortality" and argue that "the human being is, in the last analysis, a certain form of organization of matter. ... This organization can survive a partial – perhaps, even a complete – change of the material from which it is built" (http://pespmc1.vub.ac.be/cybimm.html).

The same purpose of reincarnation of a digitized human mind into a different material substrate has been pursued by Settleretics, established by the Russian neuro-cyberneticist Yan Korchmaryuk (http://settleretics.ru/) and others.

[1206] Several scientific projects have sought to perpetuate human experiences and memories in a computer substrate, such as the "MyLifeBits" project by C. Gordon Bell of Microsoft Research.
(Gordon Bell, *Your Life, Uploaded: The Digital Way to Better Memory, Health, and Productivity*, Plume, NY, 2010, previously published as *Total Recall: How the e-Memory Revolution Will Change Everything*, Dutton, NY, 2009, http://research.microsoft.com/en-us/um/people/gbell/.)

InnerSpace Foundation, led by Preston (Pete) Estep, has been dedicated to neuroengineering approaches for the enhancement of memory. Its "memory prize will be awarded for a device that allows storage and later retrieval of memory information" (http://www.innerspacefoundation.org/prizes.htm).

Kurzweil has had his own project for the indefinite conservation of memories, called DAISI – Document and Image Storage Invention. (Ray Kurzweil, *The Singularity Is Near: When Humans Transcend Biology*, Penguin Books, New York, 2005, p. 328.)

A similar objective of memory-conservation has been pursued by the Long Now Foundation (http://longnow.org/) and Brain Preservation Foundation (http://www.brainpreservation.org/).

Another memory recording project was run in 2003 by The US Defense Advanced Research Projects Agency (DARPA), called "LifeLog." Yet apparently the project was terminated in 2004.
(*DARPA. Unclassified Fiscal Year (FY) 2004/FY 2005 Biennial Budget Estimates. February 2003*, volume 1, page 93, http://www.darpa.mil/; http://www.wired.com/politics/security/news/2004/02/62158.)

Ever improved interfaces between humans and machines, or more specifically Brain-Computer Interfaces (BCI), including both external (caps) and internal devices (brain implants), often with the explicit purpose of memory restoration, storage, enhancement and manipulation, have been developed on the road to cyborgization.
(http://www.ncbi.nlm.nih.gov/pmc/articles/PMC3497935/;
http://www.popsci.com/category/tags/brain-implants;
http://www.newscientist.com/article/dn3488-worlds-first-brain-prosthesis-revealed.html#.U9EpTfn0DW8; http://emotiv.com/;
http://www.kurzweilai.net/artificial-hippocampal-system-restores-long-term-memory;
http://science.dodlive.mil/2014/07/15/darpas-memory-recovery-program/.)

Of course, the computerized conservation of discretized memory is not tantamount to the conservation of embodied and integrated human personality, which is pursued by "traditional life-extensionism."

[1207] In 2013, there was announced the start of additional massive brain mapping and modeling projects: The E.U. government-funded Human Brain Project (with the budget of $1.3 billion over 10 years, launched in January 2013, http://www.humanbrainproject.eu/) and the U.S. government-funded Brain Research through Advancing Innovative Neurotechnologies (BRAIN) Initiative (launched in April 2013, starting with $100 million in 2014, further estimated to receive ~$300 million per year for over ten years (http://www.whitehouse.gov/the-press-office/2013/04/02/fact-sheet-brain-initiative).
http://www.nature.com/news/brain-simulation-and-graphene-projects-win-billion-euro-competition-1.12291; http://blogs.nature.com/news/2013/04/obama-launches-ambitious-brain-map-project-with-100-million.html.)

One of these projects' central, explicit and immediate purposes is to map and model the brain in order to develop techniques for the diagnosis and treatment of neurodegenerative (age-related) diseases.

Yet, in case the projects succeed toward their explicit purpose – "the complete modeling of the human brain" – it can be quite imaginable that a "computer brain" emulating or surpassing the "human brain" can be created, with untold (potentially either fateful or fatal) implications for the "human" counterpart.

[1208] In the "quantified self" approach, all imaginable parameters of the human body are hoped to be measured, uploaded into a computer and then analyzed to find the best correlates for health and longevity. This information can then be used to balance out various deficiencies in the organism.

This is seen by many as a feasible path to healthy life extension. Yet, also clearly, the approach is somewhat reductionist and restrictive, as there can be no measuring of "all the possible parameters," hence the models will always be incomplete. Will they be sufficient to provide useful information to extend longevity? Also, even though the "quantified self" approach may have immediate practical implications, for the long term it can also be seen as a step toward "cyborgization" and "personality uploading". As the human body will be constantly measured, balanced by the machines and networked, the "quantified persons" will be clearly at risk of losing their autonomy. Will there be workable voluntary and non-detrimental possibilities to "opt out"? Could the same health and longevity benefits be attained without losing autonomy?

Some current trends in the use of the "quantified self" approach for life extension include:

The use of supercomputers (such as IBM Watson) for health data analysis (e.g. http://www-03.ibm.com/innovation/us/watson/watson_in_healthcare.shtml); massive collection of biometric data from the population (e.g. http://data.euro.who.int/hfadb/; http://www.euro.who.int/en/what-we-do/data-and-evidence/databases/european-health-for-all-database-hfa-db2); the development of ever smaller, cheaper and more personalized, diverse and widely accessible health measurement devices (e.g. http://www.qualcommtricorderxprize.org/; http://nokiasensingxchallenge.org/competition-details/overview); and the proliferation of Quantified Self groups and initiatives (e.g. http://www.thehumanmemomeproject.com/; http://quantifiedself.com/2011/04/quantify-your-way-to-health-and-longevity/).

Apprehensions put aside, the Quantified Health (or Quantified Self) industry has been rapidly expanding. As of April 2014, over 70% of Americans regularly tracked their health status, using portable health data devices. Trends in Europe and other industrialized areas were similar to the US, indicating a large and ongoing increase in the Quantified Health market. In February 2012 it was estimated that about 21 million wearable health and fitness devices were sold in 2011, and it was anticipated that the number could reach 170 million devices by 2017. In January 2014, it was expected that over 90 million wearable health devices would be shipped in a year. The trend toward expansion appears to continue. The use of the Quantified Self applications to study aging and longevity has been intensifying.

(http://finance.yahoo.com/news/doctors-note-70-percent-people-130000743.html;
https://www.abiresearch.com/press/ninety-million-wearable-computing-devices-will-be-;
https://www.abiresearch.com/press/wearable-sports-and-fitness-devices-will-hit-90-mi;
http://www.iol.co.za/scitech/technology/software/computing-your-longevity-1.1722143#.U9E9hvn0DW9.)

[1209] Safety has been a vital concern for scientists involved in life-extension research. Yet, it has been an overwhelming concern for those who would rather not be involved in such research, for those who believe that the risks of any potential interventions into the environment or into human nature would outweigh the benefits almost *a priori*. The latter position has been often termed "bio-conservatism" – a popular polemic term in transhumanist literature.

Several organizations have been noted for their cautious or "bio-conservative" approach to technology generally, and to biological and life-extending technology in particular, including:

The Center for Genetics in Society, http://www.geneticsandsociety.org/; Council for Responsible Genetics, http://www.councilforresponsiblegenetics.org/; Erosion, Technology and Concentration (ETC) group, http://www.etcgroup.org/; Center for Bioethics and Culture, http://www.cbc-network.org/; Center for

Bioethics and Human Dignity, http://cbhd.org/; Institute on Biotechnology and the Human Future, http://www.thehumanfuture.com/, and several others.
(According to "Overview of Biopolitics" published by the Institute for Ethics and Emerging Technologies, http://ieet.org/index.php/IEET/biopolitics.
See also James J. Hughes, "The Politics of Transhumanism," 2002, http://www.changesurfer.com/Acad/TranshumPolitics.htmp; and Joel Garreau, *Radical Evolution. The Promise and Peril of Enhancing Our Minds, Our Bodies – and What It Means to Be Human*, Doubleday, NY, 2005.)
The "bio-conservative" attitude has been well expressed already since the time of the *Panchatantra* ("The Five Principles") – an ancient Indian collection of fables, sometimes dated c. 300 BCE and attributed to Vishnu Sharman. Of special interest in this collection is the story about the three scholars who endeavored to revive a dead lion – "to assemble the skeleton," to "supply skin, flesh, and blood," to give it "the breath of life" – but were opposed by their unlearned yet sensible friend. As the story goes:
"one of them said: "A good opportunity to test the ripeness of our scholarship. Here lies some kind of creature, dead. Let us bring it to life by means of the scholarship we have honestly won." ... the man of sense advised against it, remarking: "This is a lion. If you bring him to life, he will kill every one of us." "You simpleton!" said the other, "it is not I who will reduce scholarship to a nullity." "In that case," came the reply, "wait a moment, while I climb this convenient tree. "When this had been done, the lion was brought to life, rose up, and killed all three. But the man of sense, after the lion had gone elsewhere, climbed down and went home."
(*The Panchatantra*, Translated from the Sanskrit by Arthur W. Ryder, The University of Chicago Press, Chicago, Illinois, 1955 (1925), Book 5 – Ill-Considered Action, "The Lion-Makers," pp. 442-444, http://oaks.nvg.org/pt74.html#lion-makers.)
A common transhumanist and life-extensionist argument against the cautious and "bio-conservative" approach has been that often inactivity can be more deadly than activity.
[1210] There have been several projects to select the most promising directions for life extension. Some examples include: "*The Common Denominator. A Research Agenda to Slow Aging and Slow Disease*" by the Alliance for Aging Research (AAR), http://www.agingresearch.org/content/article/detail/8029/; http://www.healthspancampaign.org/research/research-agenda/; "*Strategies for Engineered Negligible Senescence*" of the SENS Research Foundation, http://www.sens.org/research/introduction-to-sens-research; "*Human Life Extension Program*" of the Science for Life Extension Foundation, http://www.scienceagainstaging.com/ENG/index_ENG.html;
https://www.fightaging.org/archives/2011/07/biomarkers-of-aging-and-age-related-conditions.php;
"*Anti-aging Strategies*" of Aging Research Portfolio, http://agingportfolio.org/; http://agingportfolio.org/index/category/id/134; "*The Biology of Aging Platform*" of the Russian Transhumanist Movement, http://sciencevsaging.org/; http://transhumanism-russia.ru/content/view/654/121/; "*The Lifespan Application*" of the Denigma project, http://www.denigma.de/lifespan/ and some others.
Also quite a few books about longevity contain breakdowns of promising directions for longevity research, for example: Gregory M. Fahy, Michael D. West, L. Stephen Coles and Steven B. Harris (Eds.), *The future of aging: Pathways to human life extension*, Springer, NY, 2010 or even earlier Vladimir V. Frolkis and Khachik K. Muradian, *Experimental Life Prolongation*, CRC Press, Boca Raton, 1991, and many other texts.
Yet, there has been an observable deficit of integrative capacity and quantitative evidential basis for selecting particular promising research directions and their combinations. Also, as a rule, the existing projects have mostly considered scientific, but not social determinants of life-extension research and its prospective results.
Furthermore, as of 2014, several legislative proposals have been suggested in various countries (i.e. Korea, Israel, Ukraine, and the US), arguing for the necessity to "make a plan" for the society to cope with the

problem of aging, while taking into consideration the development of biomedical research and technology (listed at the International Longevity Alliance Linkedin group and site: http://denigma.de/url/3H; Law proposals in support of aging research and regenerative medicine in different countries; http://longevityalliance.org/law-proposals-in-support-of-aging-research-and-regenerative-medicine-in-different-countries/). Yet, so far, these proposals have not produced any apparent results in terms of legislation or implementation.

[1211] There have been many spiritually-oriented groups, emphasizing the power of meditation for longevity, including:

"People Unlimited," previously "People Forever," a.k.a. "Physical Immortality," founded in 1971 in Scottsdale, Arizona, and since then headed by Bernadeane Brown, Charles Paul Brown and James Strole (www.peopleunlimited.biz); "Rebirthing-Breathwork" inspired by Leonard Orr and Sondra Ray (http://www.rebirthingbreathwork.com/node/39); "The International Physical Immortality Project" led by Robert Hedges (http://web.archive.org/web/20041019230107/http://www.anycities.com/immortality/physical921immortality.html); "Jhershierra's Board" presided over by Jhershierra Jelsma of Houston, Texas (http://replay.web.archive.org/20071011014254/http://www.jhershierra.com/), and others.

Longevity has also been a topic of interest at the Institute of Noetic Sciences, emphasizing consciousness-based healing (http://www.noetic.org/search/?q=longevity). A fellow of the Institute of Noetic Sciences, Dr. Rupert Sheldrake had a strong interest in the research of cell aging and cell immortality (Rupert Sheldrake, "The Ageing, Growth and Death of Cells," *Nature*, 250 (5465), 381-385, August 2, 1974).

The "spiritual" approach has often involved discussions of "life energies," hence the fascination with various "bio-energetic/bio-electromagnetic" means for life-extension. One of the recent fads among some "spiritually-oriented" life-extensionists was Alex Chiu's magnetic "Immortality Rings" (www.magneticdiscovery.com). The massive promotion of the "immortality devices" became a sub-movement in and of itself.

As of January 22, 2012, over 700 American alternative medicine academies, schools, associations, boards, centers, councils, foundations, institutes, information services, and other organizations, were listed by QuackWatch. Many of these organizations emphasize energetic healing and mind healing, and many relate to healthy longevity (http://www.quackwatch.com/04ConsumerEducation/nonrecorg.html).

The "spiritual" or "electromagnetic" means for life-extension have been usually either ignored or criticized by members of "materialistic" groups, though some attempt to synthesize reductionist-biological and holistic-meditative approaches.

For example, Ray Kurzweil and Terry Grossman's book *Live Long Enough to Live Forever*, though overall entirely materialistic and mechanistic, recommending a wide range of chemical and biological supplements, concludes with a section on meditation (Ray Kurzweil and Terry Grossman, *Fantastic Voyage. Live Long Enough to Live Forever – The Science Behind Radical Life Extension*, Plume, NY, 2005, pp. 272-276).

On the other hand, in the less materialistic circles, even though biological mechanisms are considered, drawing in the life-prolonging vital energy through the power of meditation is valorized (e.g. Deepak Chopra, *Ageless Body, Timeless Mind. The Quantum Alternative to Growing Old*, Harmony Books, New York, 1993).

Longevity (even immortality) charms and spells are widely distributed, as they have been distributed since the ancient times. But presently, rather than drawing in the divine power or propitiating the gods of longevity, they are largely assumed to draw in the revitalizing "cosmic energy." Some examples of longevity charms available: http://primaltrek.com/.

Sir James George Frazer, in *The Golden Bough*, gives a striking example of employing sympathetic/homeopathic magic (based on the principle "like produces like") for increasing longevity. As Frazer reports:

"To ensure a long life the Chinese have recourse to certain complicated charms, which concentrate in themselves the magical essence emanating, on homoeopathic principles, from times and seasons, from persons and from things. ... Amongst the clothes there is one robe in particular on which special pains have been lavished to imbue it with this priceless quality. It is a long silken gown of the deepest blue colour, with the word "longevity" embroidered all over it in thread of gold. As the garment purports to prolong the life of its owner, he often wears it, especially on festive occasions, in order to allow the influence of longevity, created by the many golden letters with which it is bespangled, to work their full effect upon his person."
(Sir James George Frazer, *The Golden Bough. A Study of Magic and Religion*, Third Edition (in 2 volumes), Part 1, "The Magic Art and the Evolution of Kings," vol. 1, Macmillan and Co., London, 1920 (first published in 1890), pp. 168-169.)

It is as if by writing the word "Longevity" so many times, the actual span of life will be increased.

Yet, even among the most materialistic life-extensionists, spreading the word about longevity is often considered to be of high potential influence. There is even a term for this: "spreading the longevity meme."

(A "meme" is a transmissible 'gene of culture' defined by the Merriam-Webster dictionary as "an idea, behavior, style, or usage that spreads from person to person." The term originated in Richard Dawkins' *The Selfish Gene*, Oxford University Press, Oxford, 1989, first published in 1976, pp. 192-194.)

Spreading the "longevity meme," it is believed, will eventually lead to real longevity. But of course, in the materialistic view, the power of the symbol does not consist in drawing upon divine or magic influences, but rather it is assumed to motivate people to search for life-extension.

Thus properly motivated, someone among the carriers of the "longevity meme" will be able to figure out the correct recipe for the elixir.

12.12 Materialists, atheists and opponents of established religions have indeed represented a considerable proportion of life extensionists.

Thus, for example, the American novelist Alan Harrington (1918-1997), in *The Immortalist*, began the argument with these words:

"Death is an imposition on the human race, and no longer acceptable." In the chapter "Satan, Our Standard-Bearer," Harrington continued: "We created the Devil to express our most radical and dangerous intent. Through history he has been the host, the standard-bearer of man's aspiration to become immortal and divine" (Alan Harrington, *The Immortalist*, Celestial Arts, Millbrae, California, 1977, first published in 1969, pp. 3, 67).

Some "rationalist" life-extensionist religions have been proposed. For example, the "Society for Universal Immortalism" defines itself as "a progressive religion that holds rationality, reason, and the scientific method as central tenets of our faith. We reject supernatural and mystical forces as solutions to the problems that face us. It is upon the shoulders of humanity that our destiny rests. We welcome all who wish to take part in engineering our eternal future!" (http://www.universalimmortalism.org/). Another group, the "Society for Venturism," was initiated in 1986 as "Church of Venturism" with the aim "to advocate and promote the worldwide conquest of death and the continuation and enhancement of life through technological means, including cryonic suspension" (http://www.venturist.info/).

Yet it would be a great mistake to think that life-extensionism is an exclusively atheist enterprise (particularly in the US). Rather, as the following examples will show, various forms of life-extensionism have been adjusted to a variety of religions, according to the backgrounds of the proponents. Thus, life-extensionism transcends religious boundaries, and may be seen as a unifying pursuit, yet adapted to particular existing ideological environments.

12.13 Many Christian believers have been opposed to physical immortality, to therapeutic cloning or other kinds of far-reaching "meddling" with human nature, however not to life-extension *per se*. A (moderately) prolongevist stance is apparent in the activities of the "Priests for Life" movement (www.priestsforlife.org). Some Christians have even aspired to physical immortality. Thus, members of Brown's spiritual group

"People Unlimited" have exhibited strong Christian affiliations. Other Christian groups have embraced radical life-extending technologies with all their heart, such as Christian Transhumanism, championed by James McLean Ledford. (http://www.hyper-evolution.com/; http://christian-transhumanism.blogspot.co.il/; http://xtransoc.blogspot.com/2008/08/bishop-kuhn-in-sl.html.)

The Mormon Transhumanist Association, as of 2013 presided over by Lincoln Cannon, has been devoted to life-extension and life-enhancement, promoting "active faith in human exaltation through charitable use of science and technology" (http://transfigurism.org/). A prominent life-extension activist, Shannon Vyff, belongs to the Unitarian Universalist Church (http://lifeboat.com/ex/bios.shannon.vyff). The co-founder and CEO of the Methuselah Foundation, David Gobel, belongs to the Jehovah's Witnesses Church (http://diyhpl.us/~bryan/irc/extropians/www.lucifer.com/exi-lists/extropians.1Q99/1008.html; http://www.mfoundation.org/?pn=mj_about_who).

Incomparably more influential, encyclicals of Pope John Paul II (Karol Józef Wojtyła, 1920-2005) hailed bio-medical research as "a field which promises great benefits for humanity," capable of discovering "ever more effective remedies: treatments which were once inconceivable." Furthermore, he acknowledged and encouraged the ability of medical science to prolong life, "not only to attend to cases formerly considered untreatable and to reduce or eliminate pain, but also to sustain and prolong life even in situations of extreme frailty" (Ioannes Paulus PP. II, *Evangelium Vitae, On the Value and Inviolability of Human Life*, March 25, 1995, http://www.vatican.va/holy_father/john_paul_ii/encyclicals/documents/hf_jp-ii_enc_25031995_evangelium-vitae_en.html).

The Pope Benedict XVI (Joseph Aloisius Ratzinger, b. 1927) addressed radical life extension, the possibility that "Sooner or later it should be possible to find the remedy not only for this or that illness, but for our ultimate destiny – for death itself," with unprecedented earnestness, with mixed sympathy and concern (*Easter Vigil, Homily of His Holiness Benedict XVI, Saint Peter's Basilica, Holy Saturday, 3 April 2010* http://www.vatican.va/holy_father/benedict_xvi/homilies/2010/documents/hf_ben-xvi_hom_20100403_veglia-pasquale_en.html).

[1214] Neither are Islamic thinkers inherently opposed to radical life-extension.

Thus, the book *Al-Imam al-Mahdi, The Just Leader of Humanity* by Ayatollah Ibrahim Amini (b. 1925, a foremost Islamic scholar, since 1999 Vice President of the Assembly of Experts of the Leadership of the Islamic Republic of Iran), includes the chapter "The Research About Longevity."

The necessity to pursue longevity research is derived from the desire to explain and emulate the remarkable longevity of Al-Mahdi – مهدي – the messianic "Last Imam" who, in the belief of the Twelver Shi'a Muslims (the largest branch of Shi'a Islam) will come to protect mankind and, together with Jesus, will bring peace and justice to the world.

According to this tradition, the Last Twelfth Imam, Muhammad al-Mahdi, was born c. 869 CE (255 AH - anno hegirae), and has not died but lives in "occultation." Biological science is required to explain this fact and make such great longevity a gift to humanity. As the book states:

"There is no such age fixed for human life the transgression of which would be impossible. …. All the above observations in the medical and biological sciences make it possible for human beings to expect to discover the secret of longevity and overcome old age one day. Moreover, it has prompted them to continue their research until the goal is reached. There is hope that scientific research into understanding the mystery of longevity will also lead to uncovering the secret of the long life of the Qa'im [al-Mahdi] from the Family of the Prophet (peace be upon him and his progeny). Let us hope that day will come soon."

These were the words of Dr. Abu Turab Nafisi, Professor and Chair of the School of Medicine, University of Isfahan, and they were cited approvingly.

(Ayatollah Ibrahim Amini, *Al-Imam al-Mahdi, The Just Leader of Humanity*, Ch. 9 "The Research About Longevity," translated by Dr. Abdulaziz Sachedina, professor of Religious Studies at the University of Virginia, 1996, reprinted at "Al-Islam" – The Ahlul Bayt Digital Library Project, Spring Lake Park, MN – a repository of Islamic cultural resources, http://www.al-islam.org/mahdi/nontl/Chap-9.htm; http://ibrahimamini.ir/english/.)

Other Islamic scholars agree. Thus according to the Imam Reza website (affiliated to the Ahlul Bayt – 'People of the House' – Global Center for Information), the Islamic tradition acknowledges the possibility of extended life spans, such as those of Noah, Jesus, Khidhr, or Dajjal. Hence, "There is no dispute amongst theists and followers of Divine Religions about the possibility of extended longevity and that there is no limitation on the human life span."

The views of great Islamic thinkers on the subject are unambiguous:

"Khwajah Nasir al Deen Tusi has said: 'Extended life spans have occurred for other than al-Mahdi (p.b.u.h.) and been recorded, and for this very reason it is pure ignorance to consider his longevity as improbable.'

Allamah Tabataba'i stated: 'There are no intellectual reasons or rules to denote the impossibility of an extended life span; therefore, we cannot deny it.'

The great Islamic philosopher, Avicenna has said: 'Consider as possible whatever you hear about the strange things until you have no reason to reject it.'"

The article continues:

"As we have seen, the Holy Qur'an, the noble traditions, intellect, and history, provide proof of the possibility and the existence of extended longevity. …

From a biological, medical or scientific point of view, the human life span does not have a specific time frame where exceeding it would be considered impossible. No scientist up to now has stated that a specified amount of years is the maximum limit of the human life span after which death would be certain. Indeed some scientists, from the east and west, old and new, have stipulated that the human life span is not limited and in fact humans can have power over their deaths by delaying it and thus extending their life spans. This scientific hypothesis encourages scientists to research and administer tests day and night in hope of success. Through these tests they have proved that death, is similar to other illnesses because it is an effect of natural causes which, if they could be discovered and altered, death can be delayed. Just as scientists have been able to discover remedies for different illnesses through research, they can do the same for death."

("The Long Life Span of Imam Mahdi (A.S.)" *Imam Reza*, 2012, http://www.imamreza.net/eng/imamreza.php?id=7127;

See also YaNabi.com – Reviving the spirit of Islam, http://www.yanabi.com/index.php?/topic/424211-cryonics-life-extension-cloning-immortality/; Aisha Y. Musa (Florida International University), "A Thousand Years, Less Fifty: Toward a Quranic View of Extreme Longevity," in Calvin Mercer and Derek F. Maher (Eds.), *Religion and the Implications of Radical Life Extension*, Macmillan Palgrave, New York, 2009, pp. 123-131.)

[1215] Life-extensionism has strong roots in the Jewish religious tradition, insofar as in Judaism, human life has been an absolute and supreme value.

Thus the principle "ve-chai bahem" – וחי בהם – viz. the obligation to live by the commandments and not to die by them, is strongly emphasized.

(Leviticus 18:5; Talmud – Masechet (Tractate) Sanhedrin 74a; Talmud – Masechet Yoma 85b; the translation of the Talmud used here is *English Babylonian Talmud*, Rabbi Dr. J. H. Hertz, Rabbi Dr. I Epstein, et al. (Eds.), Talmudic Books, 2012, at http://halakhah.com/.)

The value of human life is illustrated by the saying that "whosoever preserves a single soul [*any* soul, according to most manuscript versions of the Talmud], scripture ascribes merit to him as though he had preserved a complete world" (Talmud – Masechet Sanhedrin 37a).

The obligation to preserve life ("pikuach nefesh") is so important that it overrides all other obligations and observances (such as Shabbat, Fast, etc.), in fact it overrides all commandments of the Torah. As the Talmud states, "there is nothing that can stand before the duty of saving life."

(Talmud – Masechet Yoma 82a; also Talmud – Masechet Yoma 84b-85b; Talmud – Masechet Sanhedrin –

74a.)

The only exceptional cases, in which a person is said to be obliged to sacrifice one's life, but not transgress, are: idolatry, forbidden sexual practices, and murder. Yet, in some attenuating circumstances and according to some Rabbis, even the former two prohibitions can be excused to preserve life. In contrast, murder of innocent people (for example to use their body parts to sustain one's life) is prohibited under any circumstances, as it contradicts the very principle of the preservation of life (to be distinguished from the killing of an aggressor in self defense which is permitted).

A related principle is "ein dokhin nefesh mipney nefesh" – "do not reject a soul for another soul" (Mishnah – Ohalot 7:6). That is, one cannot curtail some person's life to preserve another person's life. It can be added that an implication of this is that one cannot reject the preservation of life for the aged in favor of the preservation of life in other diseases. All causes of death are equal, and one cannot reject one for another.

("Pikuach Nefesh" (Saving a life), in *Encyclopedia of Jewish Medical Ethics* (Hebrew), compiled and edited by Abraham Steinberg, The Shlezinger Institute, Jerusalem, 1996, vol. 5, pp. 390-392, 404-406.)

In the Jewish religious rules of conduct – the Halacha (Halakhah), "tumah" (the unholiness, evil or impurity) means simply "the negation of life," hence the prohibition of murder and of bloodshed, and the laws of "tumah ve'taharah" (or ritual purity).

("Tameh met" (unholiness of death), "Tumah" (unholiness), in *Talmudic Encyclopedia. A Digest of Halachic Literature and Jewish Law from the Tannaitic Period to the Present Time* (Hebrew), edited by Rabbi Meyer Berlin, Talmudic Encyclopedia Institute, Jerusalem, 1997, vol. 19, pp. 450-507.)

Moreover, the Talmud equates between evil, Satan and death: "Satan, the evil prompter, and the Angel of Death are all one" (Talmud - Baba Bathra 16a).

All these concepts are directly supportive of life-extension, insofar as life-preservation, life-saving and life-extension are logical equivalents.

Reaching farther, super-longevity, rejuvenation, and even immortality and revival, are prominent concepts in the Jewish tradition:

Mortality, the main tragedy of the Fall, was not the original and inexorable destiny of humankind (Genesis 3:17-24).

The extreme longevity of antediluvian patriarchs is admired, ranging from 365 years for Enoch to 969 years for Methuselah (Genesis 5:1-32).

According to the Talmud, "Until Abraham there was no [signs of] old age" (Talmud – Masechet Sanhedrin 107b).

In the Torah, longevity is the main prize for observing the commandments (without a direct mentioning of an afterlife – Exodus 20:12, Leviticus 26:3, Deuteronomy 5:33).

In other books of the Tanakh (Torah, Neviim, Ketuvim – Torah, Prophets and Writings – the corpus of what has been sometimes called "The Old Testament"), the prophet Elijah attained physical immortality (the ascension in the chariot of fire – 2 Kings 2:11).

Ezekiel could revive the dead (the vision of the resurrection of dry bones – Ezekiel 37:1-14; also in the Talmud – Masechet Sanhedrin 92b). The prophecy continues: "And David, my servant, will be their prince forever" (Ezekiel 37:25, *The Bible: New International Version*).

King David (conventionally dated c. 1040-970 BCE) practiced rejuvenation (by proximity to young maidens – 1 Kings 1:1-4).

"Tchiat Hametim" (resurrection in the flesh) is among the Thirteen Articles of Faith of Maimonides (1135-1204) – one of the greatest Jewish intellectual authorities, a theologian as well as a physician.

(Rabbi Moshe ben Maimon (the Rambam), *Perush Hamishna*, Masechet Sanhedrin 10 - Maimonides' Commentary on the Mishna, Tractate [Masechet] Sanhedrin, Chapter 10; Rabbi Israel Meir Lau, *Judaism Halacha Lemaaseh [Practical Halakhah]. The Oral Tradition* (Hebrew), Dfus Pele, Givataim, Israel, 1988, pp. 370-371.)

Furthermore, resurrection is a subject of the daily prayer (Amida): "Blessed are you, O Eternal, Who Resurrects the Dead." And it is given the same weight in the prayer as "Blessed are you, O Eternal, Who Heals the Sick."

According to many great Rabbis, such as Rabbi Saadia Gaon (882-942), Rabbi Moshe ben Nachman/Nachmanides (1194-1270) and Rabbi Abraham Bibago (1446-1489), the resurrection is to be followed by physical immortality.

(Dov Schwartz, *Messianism in Medieval Jewish Thought* (Hebrew), Bar-Ilan University Press, Ramat-Gan, Israel, 1997, pp. 36, 105, 142-143, 218-219.)

These examples may appear far-reaching, mystical and mythical, yet they demonstrate that in the Jewish intellectual tradition (as in many others), the pursuit of life does not seem to have any limits.

Essentially, the preservation of life is not something just to pray for, but to work for.

There is a work by Maimonides – "The Responsum on Longevity" – which is definitive of the pro-active principle for the prolongation of life. Maimonides believed that there is no predetermined limit to human life, and therefore efforts toward the prolongation of life are justified. In the "Responsum on Longevity," Maimonides stated directly:

"For us Jews, there is no predetermined end point of life. The living being exists as long as replenishment is provided [for that amount of] its substantive moisture [i.e. bodily humors] that dissolves."

In agreement with the theoretical perception that if something can be broken, it can also be fixed, Maimonides appeared to be quite pro-active:

"It is written: 'When you build a new house, you should make a parapet for your roof so that you bring not bloodshed upon your house should any man fall therefrom" [Deut. 22:8]. This phrase proves that preparing oneself, and adopting precautionary measures – in that one is careful before undertaking dangerous enterprises – can prevent their occurrence. ...

This demonstrates, however, that there is no firmly determined time for death. Moreover, the elimination of harmful things is efficacious in prolonging life, whereas the undertaking of dangerous things is the basis for shortening life" (pp. 255, 258).

(Fred Rosner, "Moses Maimonides' Responsum on Longevity," *Geriatrics*, 23, 170-178, October 1968, reprinted in Fred Rosner, *The Medical Legacy of Moses Maimonides*, Ktav, Hoboken NJ, 1998, pp. 246-258.)

Indeed this passage does not explicitly speak of immortality, but only implies the possibility of indefinite life extension.

Elsewhere in the Jewish oral tradition, the concept of potential physical immortality is explicit. There is even foreshadowing of regenerative biotechnology. Thus, for example, there is an extensive Jewish oral tradition about the "Etzem Luz" – עצם לוז – the bone of resurrection, the indestructible part of the human body from which the resurrection will proceed.

"Luz" (almond) is a very fraught mystical concept, denoting the source of resurrection and regeneration, as well as an endocrine gland and a sprout. Jacob used "Luz" (almond) rods for "bioengineering," to change the color of his sheep (Genesis 30:37-39). "Luz" is also the name of the blessed land of the immortals.

It may be sufficient to quote a remarkable article on "Luz" from *Jewish Encyclopedia* to illustrate how deeply rooted is the concept of potential immortality (and even its laboratory testing) in the Talmud and Midrash (orally transmitted legends):

"LUZ - Name of a city in the land of the Hittites [a territory restricted to the hills of Canaan-Israel or broadly referring to Anatolia-Asia Minor], built by an emigrant from Beth-el, who was spared and sent abroad by the Israelitish invaders because he showed them the entrance to the city (Judges i. 26). "Luz" being the Hebrew word for an almond-tree, it has been suggested that the city derived its name from such a tree or grove of trees. Winckler compares the Arabic "laudh" ("asylum"). Robinson ("Researches," iii. 389) identifies the city either with Luwaizah, near the city of Dan, or (ib.iii. 425) with Kamid al-Lauz, north of

Heshbon (now Hasbiyyah); Talmudic references seem to point to its location as somewhere near the Phenician coast (Sotah 46b; Sanh. 12a; Gen. R. lxix. 7).

Legend invested the place with miraculous qualities. "Luz, the city known for its blue dye, is the city which Sennacherib entered but could not harm; Nebuchadnezzar, but could not destroy; *the city over which the angel of death has no power; outside the walls of which the aged who are tired of life are placed, where they meet death*" (Sotah 46b); wherefore it is said of Luz, "the name thereof is unto this day" (Judges i. 26, Hebr.). It is furthermore stated that an almond-tree with a hole in it stood before the entrance to a cave that was near Luz; through that hole persons entered the cave and found the way to the city, which was altogether hidden (Gen. R.l.c.)."

Luz is also "Aramaic name for the *os coccyx*, the "nut" of the spinal column. *The belief was that, being indestructible, it will form the nucleus for the resurrection of the body*. The Talmud narrates that the emperor Hadrian, when told by R. Joshua that the revival of the body at the resurrection will take its start with the "almond," or the "nut," of the spinal column, *had investigations made and found that water could not soften, nor fire burn, nor the pestle and mortar crush it* (Lev. R. xviii.; Eccl. R. xii.).

The legend of the "resurrection bone," connected with Ps. xxxiv. 21 (A. V. 20: "unum ex illis [ossibus] non confringetur" - [one of those bones is unbreakable]) and identified with the cauda equina [horse tailbone] (see Eisenmenger, "Entdecktes Judenthum" [Judaism discovered], ii. 931-933), *was accepted as an axiomatic truth by the Christian and Mohammedan theologians and anatomists*, and in the Middle Ages the bone received the name "Juden Knöchlein" (Jew-bone; see Hyrtl, "Das Arabische und Hebräische in der Anatomie" [The Hebrew and Arabic elements in Anatomy] 1879, pp. 165-168; comp. p. 24). Averroes accepted the legend as true (see his "Religion und Philosophie," transl. by Müller, 1875, p. 117; see also Steinschneider, "Polemische Literatur," 1877, pp. 315, 421; *idem*, "Hebr. Bibl." xxi. 98; *idem*, "Hebr. Uebers." p. 319; Löw, "Aramäische Pflanzennamen" [Aramaic plant names] 1881, p. 320).

Possibly the legend owes its origin to the Egyptian rite of burying "the spinal column of Osiris" in the holy city of Busiris, at the close of the days of mourning for Osiris, after which his resurrection was celebrated (Brugsch, "Religion und Mythologie," 1888, pp. 618, 634). Bibliography: Jastrow, *Dict.*; Levy, *Neuhebr. Wörterb.* K."

(Emphasis added. Kaufmann Kohler, "Luz," *Jewish Encyclopedia*, in 12 volumes, 1901-1906, online reprint, http://www.jewishencyclopedia.com/view.jsp?artid=635&letter=L.)

The latter statement about potential immortality being "accepted as an axiomatic truth by the Christian and Mohammedan theologians and anatomists" is of particular interest, showing the compatibility of the religions with the concept of radical life-extension.

(See also, Fred Rosner, *Medicine in the Bible and the Talmud*, Ktav, Hoboken NJ, 1995 (1977), particularly the articles "The Balm of Gilead" and "Therapeutic Efficacy of Chicken Soup," pp. 132-139; James Joseph Walsh, *Old-Time Makers of Medicine. The Story of The Students And Teachers of the Sciences Related to Medicine During the Middle Ages*, Fordham University Press, NY, 1911, Ch. III "Great Jewish Physicians," Ch. IV "Maimonides," pp. 61-108, http://www.gutenberg.org/files/20216/20216-h/20216-h.htm.)

In more recent times, Jewish thinkers have expressed an agreement with life-extensionist goals and with biotechnological interventions generally.

Thus, in March 2000, the International Symposium "Extended Life – Eternal Life" took place in Philadelphia (www.extended-eternallife.org). The Russian journalist Michael Ettinghoff thus summarized the symposium discussion: "Christians are against immortality. Jews are for it." The Conservative American Rabbi Neil Gilman is quoted as saying at the conference that he would be ready to break Shabbat and Yom Kippur, even if they occur on the same day, for the preservation of life (*Argumeny I Fakty*, 41/322, 2000, http://gazeta.aif.ru/online/health/322/z41_13).

According to the Conservative American Rabbi Elliot N. Dorff, radical life-extension ties with Jewish expectations of the Messianic Era. At the same time Dorff did express some concerns that radical life extension will make us "even more blind to the importance of other values, such as family, enjoying life, fixing the world, and connecting with God" and it will "likely bring a variety of yet unseen problems to thwart the arrival of the Messianic era" as it will exacerbate the "overpopulation" problem. Yet, ultimately, he asserted that imaginative thinking will "prompt us to exert yet more effort in achieving the ideal world, and may we succeed!"

(Rabbi Elliot N. Dorff, "Becoming Yet More Like God: A Jewish Perspective on Radical Life Extension," in *Religion and the Implications of Radical Life Extension*, Edited by Calvin Mercer and Derek F. Maher, Macmillan Palgrave, New York, 2009, pp. 63-74.)

The Society for Jewish Science (a part of Reform Judaism), established in 1916-1921 by the American Rabbis Alfred Moses and Morris Lichtenstern, believing in the power of "affirmative" prayer for healing and longevity, exists to the present time. (http://www.appliedjudaism.org/; http://www.irenedanon.com/Rabbi.htm.)

There has also been pronounced interest in physical immortality in the literature of Chabad.

(A branch of Orthodox Hasidic Judaism, deriving the name from Chochmah, Binah, Daat – Wisdom, Understanding, and Knowledge. See, for example, Prof. Yirmiyahu Branover, "The Immortality Enzyme," *Chabad World Magazine*, 10/22/2009, http://www.chabadworld.net/; Rabbi Nissan Dovid Dubov, *To Live And Live Again. An Overview of Techiyas Hameisim Based On The Classical Sources And On The Teachings Of Chabad Chassidism*," 1995 [5756], Ch. 10, "Life after the Resurrection," http://www.sichosinenglish.org/books/to-live-and-live-again/.)

[1216] In Hinduism (or rather in the variety of religions of India designated by this term), radical life extension has been a persistent theme since a very early time.

The entire Book 9 of *The Rigveda* (c. 1700-1100 BCE) is dedicated to praises of the immortality-giving "Soma" plant.

(The plant is called "Haoma" in ancient Iranian (Aryan) religious sources, such as *Avesta*, c. 1200-200 BCE, http://www.avesta.org/ka/yt9sbe.htm.)

(*The Hymns of the Rigveda*, translated by Ralph T.H. Griffith, E.J. Lazarus and Co., Benares, 1891, Book 9, pp. 361-412, the 1896 edition is reprinted at http://www.sacred-texts.com/hin/rigveda/.)

In India, the immortal Rishis, Arhats, and the Ciranjivas (the "extremely long-lived persons") are revered to the present (http://www.veda.harekrsna.cz/encyclopedia/general.htm#6).

In Ayurveda, "the science of long life," including the special field of Rasayana (rejuvenation), the religious foundations are inalienable.

According to one of the earliest Ayurvedic texts, *The Sushruta Samhita* (Sushruta's Compilation of Knowledge, c. 800-300 BCE):

"Bramha was the first to inculcate the principles of the holy Ayurveda. Prajapati learned the science from him. The Ashvins learned it from Prajapati and imparted the knowledge to Indra, who has favoured me [Dhanvantari, an incarnation of Lord Vishnu, the protector of life and the giver of Ayurveda on earth] with an entire knowledge thereof." This knowledge was in turn "disclosed by the holy Dhanvantari to his disciple Sushruta."

According to the *Sushruta Samhita*, human life can be normally prolonged to 100 years. Yet, with the use of certain Rasayana remedies (such as Brahmi Rasayana and Vidanga-Kalpa), life can be prolonged to 500 or 800 years.

And the use of the "Soma plant, the lord of all medicinal herbs [24 candidate plants are named], is followed by rejuvenation of the system of its user and enables him to witness ten thousand summers on earth in the full enjoyment of a new (youthful) body."

(*An English translation of the Sushruta samhita, based on original Sanskrit text, Edited and published by Kaviraj Kunja Lal Bhishagratna*, Calcutta, 1907, 1911, 1916, Vol. 1, Sutrasthanam (Fundamental principles), Ch. 1, p. 8, Vol. 2, Chikitsasthanam (Therapeutics), Ch. 27, p. 518, Ch. 28, p. 525, Ch. 29, pp. 530, 536.)

The religious devotion and the pursuit of rejuvenation and radical life-extension are also present in another foundational text of Ayurveda, *The Charaka Samhita* (Charaka's Compilation of Knowledge, c. 300-100 BCE). Like Sushruta, Charaka attributes the origins of Ayurveda to the gods.

According to the *Charaka Samhita*, the normal human life-span is 100 years. Yet, the users of an Amalaka Rasayana could live many hundreds of years and the users of the Amalakayasa Brahma Rasayana could reach the life-span of 1000 years.

The great sages, who grasped perfectly the knowledge of Ayurveda, "attained the highest well-being and nonperishable life-span."

(*Charaka Samhita. Handbook on Ayurveda*, edited by Gabriel Van Loon, Durham NC, 2003, vol. 1, Cikitsasthana 1.1.75, p. 446, Cikitsasthana 1.3.3-6, p. 455, Sutrasthana 1. 27-29, p. 107.)

This tradition has now been continued by a vast number of Ayurveda practitioners, often emphasizing the spiritual and religious elements and the pursuit of life-extension. (http://www.dmoz.org/Health/Alternative/Ayurveda/.)

To name just a single example among many, Maharishi Ayurveda, founded in 1980 by Maharishi Mahesh Yogi (1914-2008), the founder of Transcendental Meditation, has produced many rejuvenating substances, including Maharishi Amrit Kalash. ("Amrit" – अमृत – means the "nectar of immortality" and "Kalash" is its vessel, http://www.mapi.com/maharishi_ayurveda/products/amrit/index.html.)

Extending longevity has also been a topic of interest at the International Society for Krishna Consciousness, founded in 1966 in New York by Swami Prabhupada (1896-1977). (http://www.veda.harekrsna.cz/.)

[1217] Buddhism too has a strong connection to life-extensionism.

The Great Buddha who grants Longevity is Amitābha, the Buddha of Infinite Light, also known as Amitāyus, the Buddha of Infinite Life. Those who invoke him will reach longevity in this realm, and will be reborn in Amitabha's Pure Land (Sukhāvatī or Dewachen in Tibetan Buddhism) where they will enjoy virtually unlimited longevity.

One of the mantras in Amitabha's praise is "Om amrita teje hara hum" (Om save us in the glory of the Deathless One hum).

Many Buddhist mantras for longevity are recited, dedicated to the great healers of old, so that a portal to their wisdom may be opened and, through their compassion, suffering will be abolished and health and longevity reached in this world.

Material means for rejuvenation and life-extension have also been developed by Buddhist physicians. For example, the Tibetan Medical and Astrology Institute of H.H. The Dalai Lama, has produced such preparations as Gaay-Pa-Sowae Chulen – the "Elixir of Rejuvenation" and Tsephel Dhutse – the "Elixir of Life" which "assists to boost body energies and prolongs the life span."

(http://www.men-tsee-khang.org/hprd/suppliment/index.htm; http://www.dharma-haven.org/tibetan/healing.htm; http://www.sutrasmantras.info/sutra04.html; Derek F. Maher, "Two Wings of a Bird: Radical Life Extension from a Buddhist Perspective," in Calvin Mercer and Derek F. Maher (Eds.), *Religion and the Implications of Radical Life Extension*, Macmillan Palgrave, New York, 2009, pp. 111-121; Luis O. Gomez, *The Land of the Bliss: The Paradise of the Buddha of Measureless Light*, Motilal Banarsidass Publishers, Delhi, 1996.)

[1218] And in Taoism, radical life extension has always been a defining, all pervasive pursuit. Taoism has been now undergoing a true revival throughout the world and particularly in the US, with many practitioners and many books emphasizing the attainment of healthy life-extension on Taoist principles.

To name just a few resources among many:

Livia Kohn, *Daoism and Chinese Culture*, Three Pines Press, Cambridge, MA, 2001, Ch. 11 "Daoism Today," pp. 187-203; Daniel P. Reid, *The Tao of Health, Sex, and Longevity: A Modern Practical Guide to the Ancient Way*,

Simon and Schuster, NY, 1989; http://www.goldenelixir.com/; http://www.levity.com/alchemy/ge_hong.html.

Generally, in the traditional Chinese household pantheon, "Longevity" (Shou, 寿) is one of the three most venerated deities, alongside "Happiness" (Fu, 福) and "Prosperity" (Lu, 禄) – altogether referred to as the "three lucky star gods - Fu, Lu, Shou."

The iconography of the Chinese god of longevity (Shou) is almost identical to that of the large-headed, scroll and elixir-carrying Japanese Shinto "lucky god" Fukurokuju (福禄寿) combining *fuku* - "happiness"; *roku* - "wealth"; and *ju* - "longevity" in a single person.

[1219] A variety of less established American religious groups have championed radical life-extension, even physical immortality.

Thus, achieving radical life-extension through biotechnology has been an expressed goal of the Raelian Movement, founded in 1974 by Claude Maurice Marcel Vorilhon (a.k.a. Raël, born in 1946 in France). Members of the Raelian movement have been awaiting technological and medical aid from extraterrestrial intelligence (alongside other recently growing groups of believers in extraterrestrials). The movement has had its headquarters in Geneva, Switzerland, and branches all over the world, including the American branches in the Hawaii, Las Vegas and Virgin Islands. According to Rael's revelations, the extraterrestrials, with whom he contacted, normally live 750-1200 years, and when they die, they "take a cell from [a] preserved sample and re-create the body in full… with all its scientific knowledge and, of course, its personality" (Rael, *Intelligent Design: Message from the Designers*, Book 1. The Book Which Tells the Truth, "7. The Elohim – The Secret of Eternity," Raelian Foundation, 2005 [first published in French in 1974], p. 109). The extraterrestrials promised to share their secrets, but the Raelians have been intensely attempting to reproduce such an immortalizing technology by their own designs.
(http://rael.org/; http://raelianews.org/.)

(Of course, the ideas about regeneration and resurrection from a piece of flesh are not new, but have been pondered for centuries. Consider, for example, the archetypal legends about the resurrection of Phoenix from its ashes or the resurrection of Osiris from severed pieces – stories recorded at least as early as Herodotus, c. 5th century BCE.)

In the religious groups associated with "The Ascended Masters Teachings" – whose roots can be traced to the doctrines of the American mystics of the 1930s-1950s, such as Guy Warren Ballard and Mark L. Prophet and further back to the theosophists of the late 19th century, such as Helena Petrovna Blavatsky and Charles Webster Leadbeater – the cult figures are humans who are believed to have ascended to physical immortality (such as Count Saint Germain, 1712-1784). The Ascended Masters are looked up to provide guidance for such an ascension. As claimed in the "Ascension Research Center" website, "Saint Germain reminded us that we can even take our physical body with us in the Ascension, if we so choose!"
(http://www.ascension-research.org/top-6-99.html; http://en.wikipedia.org/wiki/Entering_heaven_alive.)

Adherents of the Rastafari religion – emerging in the 1930s in Jamaica and revering the Ethiopian Emperor Haile Selassie ("Power of the Trinity," lived 1892-1975, ruled 1930-1974, born Duke (Ras) Tafari Makonnen) – believe in the possibility of physical immortality or "Life Everliving" in the current physical body. The Rastafari often view death as a result of lack of self-preservation or insufficient competence in self-preservation, thus not as a necessity. (George D. Chryssides, *Exploring New Religions*, Continuum, London and New York, 2001 (1999), p. 275.) Though I was unable to find a formal statement of the doctrine, the immortalist beliefs have been expressed by Rastafari artists, as for example in the case of the refusal of the singer Robert Nesta (Bob) Marley (1945-1981) to write his will, even in extreme danger (http://www.important.ca/rastafari_religion_overview.html).

In the religions identifiable as "Wicca" or "Paganism" throughout the world, from the US and UK to Russia, brews for rejuvenation and longevity and – in the exalted pristine, ancient form – even for physical immortality (e.g. the Celtic "Life Everlasting" or the Russian "Suritza" - сурица), have been widely discussed. (On the use of herbs in these religions, see for example, http://wicca.com/celtic/herbal/herbindex.htm; http://radosvet.net/11384-surica.html.)

There have been further examples of narratives on radical life extension and immortality in the religious traditions of Sumer, Egypt, the Hebrew Bible, Persia, China, Japan, Korea and the Roman Empire, in Christianity, Islam, Hinduism, Falun Gong, and Theosophy and New Age. (Such narratives have been reported in the Wikipedia article "Longevity Myths," edited by Robert Douglas Young, http://en.wikipedia.org/wiki/Longevity_myths; and some works of "mythical history" e.g. Craig Paardekooper, *Records of Human Longevity from Other Nations*, 2001, http://s8int.com/phile/page44.html; http://saturniancosmology.org/files/kings/turin5.txt.)

Potential implications of radical life extension for Protestantism, Catholicism, Islam, Judaism, Jainism, Hinduism, Buddhism and Taoism, are discussed in Calvin Mercer and Derek F. Maher (Eds.), *Religion and the Implications of Radical Life Extension*, Macmillan Palgrave, New York, 2009. (Further resources can be found at Calvin Mercer's website at the Religious Studies Program, at East Carolina State University, http://www.ecu.edu/religionprogram/mercer/.)

Rejuvenation, resurrection and extreme longevity are also recurrent themes in the native myths of North, Central and South America, and in Oceania and Africa. In all these native mythological traditions, humans were originally immortal and succumbed to death only because of negligence, ill-will or accident. (*Larousse World Mythology*, edited by Pierre Grimal, Gallery Books, NY, 1989, pp. 452, 489, 493, 502, 522, 525, 545.)

Thus, Life-extensionism can be adjusted to any ideological or religious outlook, depending on particular backgrounds. It may be that adherents of any faith can have an element of faith in life-extension.

[1220] John Glen Sperling (b. 1921) was a leader of the American "for-profit education" movement (much of his wealth coming from establishing the for-profit University of Phoenix, Arizona), and was a prominent advocate of marijuana legalization. Among other enterprises, he founded Genetic Savings & Clone, that offered pet animal gene banking and cloning services (closed in September 2009).

In 2004, Sperling promised to bequeath $3 billion of his wealth to longevity research. Yet, even now, he has endowed the Kronos Longevity Research Institute, in Phoenix, Arizona, that "conducts state-of-the-art clinical translational research on the prevention of age-related diseases and the extension of healthier human life." As of 2013, S. Mitchell Harman acted as Kronos' director and president.

The Kronos Institute has studied nutrition, exercise physiology (e.g. the effects of physical fitness on stress resilience in the aged), cardiovascular disease, endocrinology and metabolism (e.g. the effects of estrogen on atherosclerosis in women and of testosterone on atherosclerosis in men, the effects of cholesterol-reducing statins on the energy-producing coenzyme Q10, the influence of saturated fatty acids on insulin resistance, and the influence of tart cherry juice on oxidative stress and inflammation), mechanisms of aging and oxidative stress, bioinformatics and statistics of aging.

Moreover, the Institute has been developing assays for biological markers of aging that are commercially available (and naturally highly proprietary). Yet, the institute has provided community and professional education (seminars, newsletters), and has sought volunteers for its studies.

(Kronos Institute, http://www.kronosinstitute.org/; Brian Alexander, "John Sperling Wants You to Live Forever And he's Promising $3 billion to make it so," *Wired Magazine*, Issue 12.02, February 2004, http://www.wired.com/wired/archive/12.02/immortal.html.)

[1221] Lawrence Joseph (Larry) Ellison (b. 1944), the founder of the Oracle software corporation, has been another great patron of life-extension. Within the Ellison Medical Foundation (a non-profit corporation), one of the central programs was the Aging Program, with Dr. Richard L. Sprott as the founding executive director (through 2012) and Dr. Kevin J. Lee serving as the executive director in 2013.

The foundation funded a wide range of anti-aging research, involving: structural biology, molecular genetics, candidate longevity genes, immune host defense molecules in aging systems, mechanisms of free-radical-induced cell aging, gene/environment and gene/gene interactions, integrative physiology, new approaches to age-modulated disease mechanisms, and more.
(The Ellison Medical Foundation, http://www.ellisonfoundation.org/.)
However, in September 2013, it was announced that the Ellison Medical Foundation discontinued its funding programs.
(http://announcements.ovpr.uga.edu/announcements/ellison-medical-foundation-discontinues-funding-programs/.)

[1222] Peter Andreas Thiel (b. 1967), the co-founder of PayPal, as well as a major investor in Facebook and Palantir, among other ventures, has been a major sponsor of the Methuselah Foundation. (http://www.fightaging.org/archives/2006/09/help-meet-peter.php/.)
Generally, "The Thiel Foundation has supported the exploration of new frontiers in science and technology, including artificial intelligence, longevity science, cyberspace development, outer space exploration, and seasteading." Among others, it supports the longevity science projects of Cynthia Kenyon and Aubrey de Grey, and the Artificial Intelligence research projects of the Singularity Institute. (The Thiel Foundation, Science and Technology Projects, http://thielfoundation.org/index.html.)

[1223] "Maximum Life Foundation. Where Biotech, Infotech and Nanotech Meet to Reverse Human Aging by 2029" (http://www.maxlife.org/);
The Manhattan Beach Project (http://www.maxlife.org/m_beach_project.asp; http://www.manhattanbeachproject.com/Longevity-Summit/manhattan-beach-project.htm; http://www.manhattanbeachproject.com/);
Ronald Bailey, "The Methuselah Manifesto. Witnessing the launch of Immortality, Inc.?" *Reason*, November 17, 2009 (http://reason.com/archives/2009/11/17/the-methuselah-manifesto/singlepage);
"Maximum Life CEO David Kekich: the investment strategy of life extension," at Partial Immortalization, November 23, 2006
(http://pimm.wordpress.com/2006/11/23/maximum-life-ceo-david-kekich-the-investment-strategy-of-life-extension/);
Maximum Life Foundation Management: David A Kekich, Michael Riskin, L. Stephen Coles (http://www.maxlife.org/management.asp).

[1224] David A. Kekich, "Good News/Bad News," *MaxLife Newsletter, Longevity News Digest*, December 28, 2011 (http://www.maxlife.org/longevity-news.asp; http://www.agereversalinc.com/; http://marketbrief.com/age-reversal-inc; http://archive.aweber.com/longevitynews/KMrdc/h/Longevity_News_Digest_Good.htm).
Notably, quite consistently with the cautious attitude to investments that do not promise immediate profits, investments into life sciences by venture capitalists were estimated to fall steeply from 2008 to 2012, by over $5 billion, from ~$7.8 billion a year in 2007 and 2008, to $2.5 billion in 2012, http://venturebeat.com/2013/04/15/fenwickwest-study-finds-funding-for-life-sciences-continues-to-slow/

[1225] The World Transhumanist Association (WTA) may provide a representative demographic sample. As of 2006, it included about ~3,762 active members. According to the WTA statistics (*WTA News*, November 3, 2006), there were overall about 1975 members from North America and 1079 from Europe (altogether about 81% of the world-wide WTA membership which at the time included members from over 100 countries).
The US had the largest membership (1704, ~45%), followed by the UK (267, ~7.1%), Canada (241, ~6.4%), Finland (157, ~4.1%) and Australia (128, ~3.4%).
There were much fewer members in India (83), Germany (67), Sweden (67), Nigeria (55), and Italy (49).

Rather low in the list were France (33), Israel (28), Venezuela (27), Russia (18) and China (14). (*WTA News*, November 3, 2006, http://www.transhumanism.org/index.php/WTA/more/1222/.)

As of 2007, with the active membership remaining about the same (about 3,737 working email addresses out of 4642 nominally registered members), the proportions of members from different countries remained about the same as well (44% from the US, 6% from Canada, 29% from Europe).

As of 2007, about 90% were male, with a median age of 30-33, with about 75% younger than 45. ("Report on the 2007 Interests and Beliefs Survey of the Members of the World Transhumanist Association," Prepared by James J. Hughes, Ph.D., Secretary, World Transhumanist Association, Willington CT 06279 USA, January 2008, http://www.transhumanism.org/index.php/WTA/more/2007survey/.)

Later on, the relative weight of the US appears to have increased even more (even though, as of 2012, specific numbers on country memberships were no longer included).

Thus, in 2007, there were 50 existing or organizing local WTA chapters, among them 13 in the US (http://web.archive.org/web/20070207101455/transhumanism.org/index.php/WTA/global).

As of 2012, there were 45 chapters, among them 17 in the US (http://humanityplus.org/get-involved/chapters-of-humanity/).

Unlike 2007, in 2012 there were no longer any chapters in Africa (existing or organizing), in fact no chapters in any "developing country" at all.

[1226] A recent example of increasing internationalization and wider public involvement in longevity advocacy has been the International Longevity Alliance (formed in December 2012). As of August 2013, groups of the International Longevity Alliance were emerging in over 50 countries (mainly operating via social networks, but also in live meetings), with several thousand members listed in the groups (at least nominally).

(http://longevityalliance.org/; http://www.denigma.de/; http://www.longevityday.org/; https://www.facebook.com/groups/longevity.alliance/doc/429937660426393/; http://longevityalliance.org/longevity-is-the-common-language/; http://ieet.org/index.php/IEET/more/stambler20140110; http://www.longevityforall.org/groups/.)

[1227] Ronald Bailey, "The Methuselah Manifesto. Witnessing the launch of Immortality, Inc.?" *Reason*, November 17, 2009, the article mentions Wayne Allyn Root as one of the launchers, http://reason.com/archives/2009/11/17/the-methuselah-manifesto; http://www.rootforamerica.com/;

"Who Is Mary J. Ruwart?" *Life Extension Magazine*, July 2001, http://www.lef.org/magazine/mag2001/july2001_cover_ruwart.html;

http://www.ourcampaigns.com/CandidateDetail.html?CandidateID=8235;

Glenn Reynolds on life extension, http://pajamasmedia.com/instapundit/?s=life+extension;

Comprehensive Listing of Notable Libertarians, http://chelm.freeyellow.com/famous_index.html.

[1228] Of special note is President William Jefferson ("Bill") Clinton's statement "We want to live forever, and we are getting there" (presented at The Eighth Millennium Evening at the White House "Informatics meets Genomics," October 12, 1999, http://www.rand.org/scitech/stpi/ourfuture/Rosetta/millennium.html).

Of additional note is President George Walker Bush's statement: "It is wise to always err on the side of life" (presented at the Social Security Conversation in Arizona, March 21, 2005, http://www.whitehouse.gov/news/releases/2005/03/20050321-7.html).

And yet another statement by President Barack Hussein Obama: "There's no finish line in the work of science. The race is always with us – the urgent work of giving substance to hope and answering those many bedside prayers, of seeking a day when words like "terminal" and "incurable" are potentially retired from our vocabulary" (presented on March 9, 2009, at the White House, upon signing the "Executive Order Removing Barriers to Responsible Scientific Research Involving Human Stem Cells" – in fact lifting the

Bush administration's ban on such research, http://www.nytimes.com/2009/03/09/us/politics/09text-obama.html?_r=1).

[1229] The social-democratic views are represented by Dr. James Hughes of the Trinity College, Connecticut, president of the Institute for Ethics and Emerging Technologies, formerly a long-time secretary/executive director of the World Transhumanist Association (2002-2009), and author of *Citizen Cyborg: Why Democratic Societies Must Respond to the Redesigned Human of the Future* (2004).

In *Citizen Cyborg*, Hughes points out:

"Not surprisingly the countries with the longest life expectancies are those with the strongest public health systems – social democratic northern Europe and Canada, and relatively egalitarian Japan."

(Israel could be added to the list, having a strong social-democratic tradition and welfare system, and one of the highest values of life-expectancy: about 82 years as of 2009, the 5th place in the world, lagging only about a year after the first places: San Marino, Japan, Hong Kong and Switzerland, according to the World Bank data.

http://data.worldbank.org/indicator/SP.DYN.LE00.IN?order=wbapi_data_value_2009+wbapi_data_value+wbapi_data_value-last&sort=desc.

In 2010, however, Israel moved to the 9th place – could it be related to the concurrent strengthening of economic neoliberalism, after the elections of 2009?)

Longevity, Hughes argues, is positively correlated with equality, shown by the "Gini coefficient" of equality, where 0 indicates perfect equality, and 1 denotes a society "where one person has everything and everybody else have nothing."

The Human Development Index – comprised of average life-expectancy, income per person, and the level of education – is also the highest in social democracies: "the top five countries in the index are social democratic Norway, Sweden, Canada, Belgium and Australia."

According to the social-democratic perspective, government programs designed to ensure a rapid development and universal distribution of life-extending technologies, will be paramount:

"In the coming decades, however, the affluent will be able to buy extra decades, and soon may buy extra centuries. Once we cross that threshold there will be an enormous pressure to make life extension medicine universally accessible."

Furthermore:

"Getting serious about risks to the public and proposing nonmarket solutions for those risks is the only way to fight the bioLuddite agenda politically. Only believable and effective policies that guarantee technologies are safe and equitably distributed can reassure skittish publics. Panglossian assurances that all will work itself out in the market or after the Singularity won't cut it. We will face much more opposition to enhancement and radical life extension if they are only available to the rich."

(James Hughes, *Citizen Cyborg*, Westview Press, Cambridge MA, 2004, pp. 23-24, 31, 191-192, 205.)

[1230] In September 2011 and January 2012, small-scale demonstrations for radical life-extension took place, respectively, in Russia and Israel, emphasizing "social justice" in accord with the rising popularity of that topic (http://www.transhumanism-russia.ru/content/view/799/11/; http://www.singulariut.com/).

[1231] The "Longevity Dividend" campaign was originated in 2006 by some of America's foremost institutionally-affiliated gerontologists, who went on to petition the Senate to increase funding for aging research.

(S. Jay Olshansky, Daniel Perry, Richard A. Miller, and Robert N. Butler, "In Pursuit of the Longevity Dividend: What Should We Be Doing To Prepare for the Unprecedented Aging of Humanity?" *The Scientist*, 20(3), p. 28, March 2006, http://www.the-scientist.com/article/display/23191/; *Pursuing the Longevity Dividend. Scientific Goals for an Aging World*, September 12, 2006, including a full list of signatories: http://www.edmontonagingsymposium.com/files/eas/Longevity_Dividend_Signatories.pdf; *Gray is the*

New Gold: State of the Science 2009. Optimism in Aging Research, Kronos Longevity Research Institute, Phoenix, Arizona, 2009, pp. 4-6, http://www.kronosinstitute.org/publications/reports/sos_2009.cfm.)

As the list of the Longevity Dividend signatories demonstrates, most of the supporters have been in some way affiliated with government-sponsored academic institutions.

The Longevity Dividend campaign, endorsed by the academics worldwide, has attempted to dispel the fears of a dystopian society burdened by unproductive senescent populations due to life extension. The campaign has argued that healthy life extension is a social and economic good and hence has lobbied the US government to invest in longevity research.

Since then, the budget of the US National Institute on Aging (NIA), the world's main funding agency for aging research, did not seem to improve much:

From 2005 through 2008, the budget of the National Institute on Aging remained almost exactly the same: about $1.050 billion, while the general National Institutes of Health (NIH) funding increased from $28.6 billion in 2005 to $29.6 billion in 2008.

In 2010, the NIH received $31.2 billion, NIA - $1.110 billion (~3.56% of the general NIH funding); within the NIA, the Biology of Aging Program, studying the basic mechanisms of aging and life extension, received $182 million (~0.58% of the NIH budget).

In 2011, the NIH received approximately $30.9 billion, NIA - $1.100 billion (3.56%), Biology of Aging - $175 million (0.57%).

In 2012, the NIH obtained $30.87 billion, NIA – $1.102 billion (3.57%), Biology of Aging - $176.15 million (0.57%).

For 2013, the requested NIA budget would remain almost exactly the same. To be precise, the requested increase would be from $1.102128 billion to $1.102650 billion for the entire NIA ($522,000 raise), and from $176.154 million to $176.251 million for the Biology of Aging Program ($97,000 raise). In actual sums, in 2013, the NIA budget apparently increased to about $1.110 billion.

For 2014, the NIA budget was set to increase from $1.110 billion (the actual figure for 2013, which turned out to be slightly higher than the requested figure) to about $1.193 billion (almost the entire increase was dedicated to Alzheimer's disease research).

(National Institutes of Health, History of Congressional Appropriations, Fiscal Years 2000-2012, http://officeofbudget.od.nih.gov/pdfs/FY12/Approp.%20History%20by%20IC)2012.pdf; Department of Health and Human Services, National Institute of Health. National Institute on Aging (NIA), FY 2013 Budget, p. 10, FY 2014 Budget, p. 10
http://www.nia.nih.gov/sites/default/files/vol_4_tab_3_-_nia_printed_version_0.pdf;
http://www.nia.nih.gov/sites/default/files/fy2014_budget_request.pdf.)

Yet, the "Longevity Dividend" lobbying effort has continued to the present 2014.

In July 2009, The American Academy of Anti-Aging Medicine (A4M) issued its own version of the Longevity Dividend program that advocated "a healthcare model promoting innovative science and research to prolong the healthy lifespan in humans."

(Ronald M. Klatz, Robert M. Goldman, Joseph C. Maroon, Nicholas A. DiNubile, Michael Klentze, *The A4M Twelve-Point Actionable Healthcare Plan: A Blueprint for A Low Cost, High Yield Wellness Model of Healthcare by 2012*, A4M, 22 July 2009, p. 4, http://www.waaam.org/twelve_points_summary.php; "The A4M version of the longevity dividend," *Fight Aging. Longevity Meme*, August 19, 2009, http://www.fightaging.org/archives/2009/08/a4ms-version-of-the-longevity-dividend.php.)

These efforts were encouraged by the president Barak Obama's health care reform plans and some of his statements, such as "we need ... to figure out what works, and encourage rapid implementation of what works into [doctor's] practices. That's why we're making a major investment in research to identify the best treatments for a variety of ailments and conditions."

(Jesse Lee, "Why Reform, Why Now," *The White House Blog*, June 15, 2009,

http://www.whitehouse.gov/blog/Why-Reform-Why-Now/.)
As the authors of the A4M plan claimed, "At this writing, legislators on Capitol Hill are actively debating a $1.65 trillion, 10-year plan to overhaul the nation's healthcare system. The majority of the plan focuses on how to pay for health insurance, rather than formulating a comprehensive plan of action to reform healthcare itself."
(*The A4M Twelve-Point Actionable Healthcare Plan: A Blueprint for A Low Cost, High Yield Wellness Model of Healthcare by 2012*, A4M, 22 July 2009, p. 4.)
Yet, apparently, the A4M appeals were not heeded by the government.
Various organizations have now been acting in the same objective, lobbying for increased funding for aging and longevity research, through a variety of programs, including:
The American Federation for Aging Research, The Alliance for Aging Research, in partnerships with American Aging Association, Friends of the National Institute on Aging, Coalition for the Advancement of Medical Research (CAMR), Partnership to Fight Chronic Disease, and more radical groups such as the American supporters of the Methuselah Foundation, Humanity Plus, Coalition to Extend Life, Longecity, American Longevity Alliance and many others.
As all these instances demonstrate, the adaptation of life-extensionism to the existing economic and state structures is indispensable.

[1232] A telling example of confrontation between commercial and institutional longevity researchers occurred in 2002. A position statement on anti-aging medicine was then formulated by NIA-funded gerontologists Jay Olshansky, Leonard Hayflick and Bruce Carnes.
(Olshansky SJ, Hayflick L, Carnes BA, "Position statement on human aging," *Journal of Gerontology: Biological Sciences*, 57 (8), B292–B297, 2002, reprinted in *Scientific American*, May 13, 2002, http://www.scientificamerican.com/article.cfm?id=the-truth-about-human-agi.)
It was signed by 51 leading gerontologists, including Robert Butler, Aubrey de Grey, James Fries, Tom Kirkwood, David Gershon, Carol Greider, Jerry Shay, Raj Sohal, Christofer Heward, Richard Sprott, et al.
This position statement accused anti-aging medicine as being fraudulent by and large, peddling ineffectual remedies.
On the other hand, it strongly advocated massive investments in basic aging research as our only hope to find effective anti-aging interventions.
The American Academy of Anti-Aging Medicine retaliated, asserting the value of supplements and accusing the NIA of not having come up with any real therapies for decades.
(Elizabeth Pope, "51 Top Scientists Blast Anti-Aging Idea. New Position Paper Refutes Claims of Miracle Treatments," *The American Association of Retired Persons (AARP) Bulletin*, June 2002, http://web.archive.org/web/20021001170034/http://www.aarp.org/bulletin/departments/2002/health/0605_health_1.html.)
The 'petition battles' for the public support of longevity research continue.

[1233] Ray Kurzweil, *The Singularity Is Near. When Humans Transcend Biology*, Penguin Books, 2005, p. 406.

[1234] A powerful subscriber to the capitalist techno-libertarian view has been the investor Dr. Peter Diamandis (b. 1961). Among other enterprises, in the period 2009-2012, he has launched several projects aimed to develop futuristic technologies, including life-extension technologies.
One such project is the Singularity University, with the mission "to assemble, educate and inspire a cadre of leaders who strive to understand and facilitate the development of exponentially advancing technologies to address humanity's challenges" (http://singularityu.org/).
The Singularity University has included the FutureMed Program, "exploring and driving the future of medicine through exponential, game changing technologies" (http://futuremed2020.com/).
Diamandis also founded The X PRIZE Foundation, whose mission is "to bring about radical breakthroughs for the benefit of humanity, thereby inspiring the formation of new industries and the revitalization of

markets that are currently stuck due to existing failures or a commonly held belief that a solution is not possible" (http://www.xprize.org/).

The X Prize projects related to life-extension include The Archon Genomics X PRIZE "that will award $10 million to the first team to rapidly, accurately and economically sequence 100 whole human genomes to a level of accuracy never before achieved." The genomes to be sequenced will be donated by 100 centenarians (ages 100 or older) with the aim to "inspire breakthrough genome sequencing innovations and technologies that will usher in a new era of personalized medicine" (http://genomics.xprize.org/).

Another life-extension-related prize, within the X Prize framework, has been the $2.25M Sensing X Challenge, aimed to develop "affordable, personalized healthcare through sophisticated sensing technologies that put you in charge of your own health" (http://www.nokiasensingxchallenge.org/).

And yet another has been the 10$M Qualcomm Tricorder X PRIZE (named after the diagnostic device in the science fiction television series "Star Trek"), intended to develop "a portable, wireless device in the palm of your hand that monitors and diagnoses your health conditions" (http://www.qualcommtricorderxprize.org/)

The ambition has been far-reaching. In April 2012, Diamandis announced the creation of an asteroid-mining company "Planetary Resources," continuing his three decades of work in developing space-related businesses (http://www.planetaryresources.com/).

In February 2012, Peter Diamandis and the science journalist Steven Kotler, published a book detailing the enterprising technological-optimistic vision: Peter Diamandis, Steven Kotler, *Abundance – the Future is Better Than You Think*, Free Press, February 2012, http://www.diamandis.com/abundance/.

Other libertarian-inclined investors into "personalized medicine" – involving an ever increasing level of computer pattern recognition and communication – have included Google's co-founders Sergey Brin (b. 1973) and Larry Page (b. 1973).

(Cf. The Brin Wojcicki Foundation, focusing on "personalized" (genomic) medicine, http://dynamodata.fdncenter.org/990s/990search/ffindershow.cgi?id=RINF001;

And the Google Foundation, focusing on global epidemiological disease trends, http://www.google.org/foundation.html; http://www.google.org/projects.html.)

In September 2013, Google announced the creation of its subsidiary "Calico" (California Life Company) dedicated to combating aging and to life extension. As of 2014, the exact research to be conducted at the company was yet unclear to the wide public. Yet, the impact of the company establishment for legitimizing the pursuit of life extension was high. (Jay Yarow, "Google Is Launching A Company That Hopes To Cure Death, Business Insider, September 18, 2013, http://www.businessinsider.com/google-is-launching-a-company-that-hopes-to-cure-death-2013-9; Harry McCracken, Lev Grossman, "Google Vs. Death", *Time*, September 30, 2013, http://time.com/574/google-vs-death/; www.calicolabs.com.)

[1235] Historically, there have been life-extensionist utopias, such as Tommaso Campanella's *The City of the Sun* (1602), where "The length of their lives is generally 100 years, but often they reach 200" (http://www.gutenberg.org/files/2816/2816-h/2816-h.htm).

And there have also been anti-life-extensionist utopias, such as Thomas More's *Utopia* (1516), where "when any is taken with a torturing and lingering pain," they "choose rather to die since they cannot live but in much misery" (http://www.gutenberg.org/files/2130/2130-h/2130-h.htm).

But in both cases, the utopian society is preserved in a perpetual equilibrium, and the authors' values are hoped to triumph for all posterity.

[1236] Georg Wilhelm Friedrich Hegel [1770-1831], *The Philosophy of History*, Translated by J. Sibree, Colonial Press, New York, 1900 (first published in 1837), pp. 26-27, 77.

[1237] G.W.F. Hegel, *Encyclopaedia of the Philosophical Sciences*, Third and Final Edition, 1830, translated by William Wallace, first published 1873, Part II. Preliminary, "Universals apprehended in Reflection" §21n, reprinted at http://www.marxists.org/reference/archive/hegel/index.htm.

[1238] Karl Marx and Frederick Engels, *Manifesto of the Communist Party*, London, 1848, http://www.anu.edu.au/polsci/marx/classics/manifesto.html.

[1239] Livia Kohn, "Told you so: Extreme Longevity and Daoist Realization," in *Religion and the Implications of Radical Life Extension*, Edited by Calvin Mercer and Derek F. Maher, Macmillan Palgrave, New York, 2009, pp. 85-96; *The Book of Changes - The I Ching*, translated by James Legge, *Sacred Books of the East, vol. 16*, 1899, http://www.sacred-texts.com/ich/index.htm.

[1240] Thomas S. Kuhn, *The Structure of Scientific Revolutions*, Third Edition, The University of Chicago Press, Chicago, 1996 (first published in 1962), p. 25.

[1241] Bruno Latour, *Pandora's Hope. Essays on the Reality of Science Studies*, Harvard University Press, Cambridge MA, 1999, p. 58.

[1242] Steven Shapin, *The Scientific Revolution*, The University of Chicago Press, Chicago, 1996, p. 123.

[1243] Michel Foucault, *The Order of Things*, Pantheon Books, New York, 1971, reprinted by Random House / Vintage Books, New York, 1994, p. 150 (first published in 1966).

[1244] M. Norton Wise, "Mediations: Enlightenment Balancing Acts, or the Technologies of Rationalism," in Paul Horwich (Ed.), *World Changes. Thomas Kuhn and the Nature of Science*, The MIT Press, Cambridge MA, 1993, pp. 207-256.

[1245] Peter Burke, *History and Social Theory* (Second Edition), Cornell University Press, Ithaca, New York, 2005, pp. 166-167, 173.

[1246] Jean Finot, *The Philosophy of Long Life* (translated by Harry Roberts), John Lane Company, London and New York, 1909, first published in French as *La Philosophie De La Longévité*, Schleicher Freres, Paris, 1900, "The Immortal Body," pp. 122-145.

[1247] Reinhard Burger, Michael Lynch, "Evolution and Extinction in a Changing Environment: A Quantitative-Genetic Analysis," *Evolution*, 49 (1), 151-163, 1995.

[1248] In a recent evolutionary model, André C. R. Martins argued almost precisely to that effect: "when the system is completely stable, no mutation going on and no changing conditions for worse, ... it is to be expected that a population that shows senescence will be driven to extinction." However, "When conditions change, a senescent species can drive immortal competitors to extinction." The author concludes: "We age because the world changes."
(André C. R. Martins, "Change and Aging: Senescence as an adaptation," Cornell University Library, Populations and Evolution, March 23, 2011, http://arxiv.org/abs/1103.4649v1.)
Several other contemporary researchers, such as Mitteldorf, Goldsmith and Skulachev, have pondered a return to Weismann's theory, proposing a direct evolutionary selection for senescence. A major suggested reason for such a selection is that an absence of senescence would diminish the species' variability and diversity, hence impair its adaptability and evolvability.
(Theodore C. Goldsmith, "Aging as an Evolved Characteristic – Weismann's Theory Reconsidered," *Medical Hypotheses*, 62(2), 304-308, 2004; Joshua Mitteldorf, "Ageing selected for its own sake," *Evolutionary Ecology Research*, 6, 937-953, 2004; Vladimir Petrovich Skulachev, "Aging is a Specific Biological Function Rather than the Result of a Disorder in Complex Living Systems: Biochemical Evidence in Support of Weismann's Hypothesis," *Biochemistry (Moscow)*, 62(11), 1191-1195, Nov 1997.)
Furthermore, a link has been suggested between increasing longevity and the level of inbreeding (keeping the genome stable), even for humans.
(A. Montesanto, G. Passarino, A. Senatore, L. Carotenuto, G. De Benedictus, "Spatial Analysis and Surname Analysis: Complementary Tools for Shedding Light on Human Longevity Patterns," *Annals of Human Genetics*, 72(2), 253-260, 2008.)
Notably, all these authors seem to be in favor of finding effective anti-aging means for humans, suggesting that through a better understanding of the evolutionary mechanism, factors affecting longevity can be identified and manipulated.

[1249] Giuseppe Tomasi di Lampedusa, *The Leopard*, translated by Archibald Colquhoun, Pantheon Books, NY, 1960, p. 40.

[1250] Lewis Carroll, *Through the Looking-Glass*, 1871, reprinted at Project Gutenberg, http://www.gutenberg.org/files/12/12-h/12-h.htm.

[1251] Ernst Hans Josef Gombrich, *In Search of Cultural History*, Clarendon Press, Oxford, 1969, p. 30.

[1252] To cite just a few recent reports about the preoccupation with the Apocalypse in America:
Benjamin Anastas, "The Final Days," *New York Times*, July 1, 2007, http://www.nytimes.com/2007/07/01/magazine/01world-t.html; Mark Matousek, "'The Apocalypse Is Coming!' A skeptic ponders the prediction that May 21 [2011] is Judgment Day. He's not worried. Are you?" *The American Association of Retired Persons (AARP) Bulletin*, May 19, 2011, http://www.aarp.org/personal-growth/spirituality-faith/info-05-2011/apocalypse-judgment-day.html.
A large register of millennial and apocalyptic organizations, both religious and secular, active around 2000, is provided by the Boston University Center for Millennial Studies, http://www.bu.edu/mille/links.html.
See also Daniel Wojcik, *End of the World as We Know It: Faith, Fatalism, and Apocalypse in America*, New York University Press, NY, 1997.

[1253] John Updike, "When Everyone was pregnant" (1972), *Museums & Women And Other Stories*, Alfred A. Knopf, NY, 1972, p. 93, quoted in Donald J. Greiner, "Updike, Rabbit, and the myth of American exceptionalism," p. 153, in Stacey Michele Olster, Stacey Olster (Eds.), *The Cambridge Companion to John Updike*, Cambridge University Press, Cambridge, 2006.

[1254] *Talmud [Gemara] – Masechet Berachoth [Tractate on Blessings]*, 10a, in *English Babylonian Talmud*, Rabbi Dr. J. H. Hertz, Rabbi Dr. I Epstein, et al. (Eds.), Talmudic Books, 2012, at http://halakhah.com/.

[1255] It should be strongly emphasized that many transhumanists and life-extensionists are well aware of a wide variety of global existential risks – either natural threats, such as increasing solar activity, or technogenic threats, such as uncontrolled industrial pollution, or combined natural and man-made threats, such as global warming – yet they seek ways to mitigate and endure potential disasters.
(See, for example, the stated missions of the Lifeboat Foundation, http://lifeboat.com/ex/boards.)

[1256] William G. Bailey, *Human Longevity from Antiquity to the Modern Lab*, Greenwood Press, Westport CN, 1987, p. ix.

[1257] *Lao-Tse. The Tao Teh King. The Tao and Its Characteristics*, Translated by James Legge, 1880, reprinted at Project Gutenberg, Ch 1, 25. 1-3, http://www.gutenberg.org/files/216/216-h/216-h.htm.

[1258] Walter Cannon, *The Wisdom of the Body*, Norton, NY, 1932, pp. 27-28.

[1259] Richard W. Jones, *Principles of Biological Regulation. An Introduction to Feedback Systems*, Academic Press, NY, 1973, Ch. 2, "Flow Processes in the Steady State," p. 7.

[1260] Schopenhauer's exact words were:
"Die Leiche ist ein blosses Exkrement der stets bestehenden menschlichen Form."
(*Arthur Schopenhauers Sämtliche werke*, hrsg. von Dr. Paul Deussen, R. Piper & Co., München, 1913, Bd. 10. Arthur Schopenhauers handschriftlicher Nachlaß. Philosophische Vorlesungen: Hälfte 2, Metaphysik der Natur, des Schönen und der Sitten, S. 374 – The Complete Works of Arthur Schopenhauer, edited by Dr. Paul Deussen, R. Piper & Co., Munich, 1913, Vol. 10, Arthur Schopenhauer's hand-written papers, Philosophical Lectures, Half 2, The metaphysics of nature, of beauty and of manners, p. 374.)
This passage is quoted in Max Bürger, *Altern und Krankheit* (Aging and Disease), Zweite Auflage, Georg Thieme, Leipzig, 1954 (1947), p. 41.

[1261] Elie Metchnikoff, *Etudy o Prirode Cheloveka* (Etudes on the Nature of Man), Izdatelstvo Academii Nauk SSSR (The USSR Academy of Sciences Press), Moscow, 1961 (1903), Ch. 8. "Popytki filosofskich system borotsia s disharmoniami chelovecheskoy prirody" (Attempts of philosophical systems to combat the disharmonies of human nature), pp. 151-153.

[1262] Max Bürger, *Altern und Krankheit* (Aging and Disease), Zweite Auflage, Georg Thieme, Leipzig, 1954

(1947), p. 41.

[1263] Plutarch, *Theseus*, c. 75 CE, translated by John Dryden (1683), reprinted in The Internet Classics Archive, http://classics.mit.edu/Plutarch/theseus.html.

[1264] In contrast, two identically constructed "Theseus' Ships" or "copies" would not be the same, following Aristotle's principle that the same body cannot occupy separate places at the same time.

According to Aristotle's *On the Soul*, the existence of "two bodies in the same place" is impossible. And one can also infer that one body in separate places simultaneously is equally impossible. Aristotle, who equates the Soul with Vitality, establishes the immateriality of the soul precisely on this principle, since "there must be two bodies in the same place, if the soul is a body." Rather, according to Aristotle, the soul (or vitality) is a particular faculty or form of matter.

(Aristotle [384 BCE - 322 BCE], *On the Soul,* translated with notes by Walter Stanley Hett, in *Aristotle in Twenty-Three Volumes*, William Heinemann Ltd., London, 1975 (1936), Vol. 8, Book 1, Part 5, 409b1-5, p. 53.)

The preservation of each human "copy" separately would of course be necessary unless someone embraces the notion of a "quantum mechanical body," spoken of in some life-extensionist circles, which would allow a person to be in several places simultaneously or alternatively.

The physicist Frank Jennings Tipler, in *The Physics of Immortality* (1994), directly states:

"The nonidentity of systems in different spatial positions is not true in quantum mechanics, and actual physical systems are quantum systems. In particular, a human being is merely a special type of quantum system. Thus the quantum criterion for system identity applies to humans, and hence two humans in the same quantum state *are* the same person: if a "replica" of a long-dead person were made which was identical to the quantum state of the long-dead person, the "replica" would *be* that person."

Generally, Tipler expects the super-powerful and caring computers of the future to emulate the precise "quantum state" of every person who has ever lived, and hence Tipler expects "every one of us to be resurrected one day and live forever."

(Frank J. Tipler, *The Physics of Immortality: Modern Cosmology, God, and the Resurrection of the Dead*, Doubleday, NY, 1994, pp. 173, 234.)

"Quantum Immortality" and "Resurrection" are discussed in a number of other works. For example, Ben Goertzel and Stephan Bugaj refer to the "Many Worlds" interpretation of quantum mechanics, implying a sort of "quantum immortality": a person may die in one world but survive in another.

(Ben Goertzel and Stephan Vladimir Bugaj, *The Path to Posthumanity: 21st Century Technology and Its Radical Implications for Mind, Society and Reality*, Academica Press, Bethesda MD, 2006, "Physics and Immortality," pp. 342-348.)

The possibilities of resurrection, yet from a more theological perspective, are discussed in Tel Peters, Robert John Russell, and Michael Welker (Eds.), *Resurrection: Theological and Scientific Assessments*, William B. Eerdmans, Grand Rapids, Michigan, 2002.

The possibility of the body existing outside of spatial and temporal confines is popularly expounded by the holistic physician Deepak Chopra in *Ageless Body, Timeless Mind* (1993):

"To all appearances, the physical body occupies a few cubic feet of space; it serves as a fragile life-support system for seven or eight decades before it must be discarded. The quantum mechanical body, on the other hand, occupies no well-defined space and never wears out."

(Deepak Chopra, *Ageless Body, Timeless Mind. The Quantum Alternative to Growing Old*, Harmony Books, New York, 1993, p. 288.)

The ideas about "quantum immortality" and transcending spatial confines, may be too unorthodox for a "traditional" life-extensionist wishing to defend the "few cubic feet of space" in a particular "quantum state" from decay and disintegration.

[1265] Stanislaw Lem, "Czy pan istnieje, Mr Johns?" *Przekrój*, Krakow, Poland, 1955, № 553 ("Do you exist, Mr. Johns?" Translated from Polish into Russian by A. Yakushev, "Sushestvuete li vy, Mister Johns?" 1960, http://samlib.ru/w/wladimir_pro/mister_djons.shtml).

[1266] Aubrey de Grey and Michael Rae, *Ending Aging. The Rejuvenation Breakthroughs That Could Reverse Human Aging in Our Lifetime*, St. Martin's Press, NY, 2007, p. 326.

[1267] As stated by one of the pioneers in the application of Information Theory to biology, Henry Quastler (1908-1963):

"The basic concepts of information theory – measures of information, of noise, of constraint, of redundancy – establish the possibility of associating precise (although relative) measures with things like form, specificity, lawfulness, structure, degree of organization. ...

Closely related is the problem of destruction of orderliness. In biology, this is the problem of aging and decay."

(Henry Quastler, "The Domain of Information Theory in Biology," *Symposium on Information Theory in Biology, Gatlinburg, Tennessee, October 29-31, 1956*, Edited by Hubert P. Yockey, with the assistance of Robert L. Platzman and Henry Quastler, Pergamon Press, NY, 1958, pp. 187-196.)

[1268] Ray Kurzweil, *The Singularity Is Near: When Humans Transcend Biology*, Penguin Books, New York, 2005, pp. 371-372.

[1269] Ray Kurzweil, *The Singularity Is Near*, 2005, p. 323.

[1270] Ray Kurzweil, *The Singularity Is Near*, 2005, p. 406.

[1271] Ray Kurzweil, *The Singularity Is Near*, 2005, p. 258.

[1272] Vernor Vinge, "Technological Singularity," *Whole Earth Review*, 81, 89-95, 1993, reprinted at http://mindstalk.net/vinge/vinge-sing.html.

[1273] Ilia Stambler, "The unexpected outcomes of anti-aging, rejuvenation and life extension studies: an origin of modern therapies," *Rejuvenation Research*, 17, 297-305, 2014 http://online.liebertpub.com/doi/abs/10.1089/rej.2013.1527.

[1274] John Desmond Bernal, *The World, the Flesh & the Devil. An Enquiry into the Future of the Three Enemies of the Rational Soul*, 1929, III "The Flesh," Reprinted at the Marxists Internet Archive Library, http://www.marxists.org/archive/bernal/works/1920s/soul/.

[1275] I.P. Pavlov, *Dvadzatiletniy opyt objektivnogo izuchenia vyshey nervnoy deyatelnosti povedenia zhivotnikh. Uslovnie reflexy* (Twenty years of objective study of the high nervous activity of animal behavior. Conditioned reflexes), Gosudarstvennoe Izdatelstvo, Moscow, 1923, p. 77.

Quoted in A.L. Chizhevsky, *Physicheskie Factory Istoricheskogo Processa* (Physical Factors of the Historical Process), Kaluga, 1924, IV. "Vlianie geofizicheskikh i kosmicheskikh faktorov na povedenie individuv i kollektivov" (The influence of geophysical and cosmic factors on the behavior of individuals and collectives, http://www.astrologic.ru/library/chizhevsky/index.htm).

[1276] Alexis Carrel, *Man, The Unknown*, Burns & Oates, London, 1961 (1935), p. 180.

[1277] Jean Finot, *The Philosophy of Long Life* (translated by Harry Roberts), John Lane Company, London and New York, 1909, p. 278, first published in French as *La Philosophie De La Longévité*, Schleicher Freres, Paris, 1900.

www.ingramcontent.com/pod-product-compliance
Lightning Source LLC
Chambersburg PA
CBHW081102170526
45165CB00008B/2295